APPLIED MYCOLOGY AND BIOTECHNOLOGY

VOLUME 4

FUNGAL GENOMICS

APPLIED MYCOLOGY AND BIOTECHNOLOGY

VOLUME 4

FUNGAL GENOMICS

Edited by

Dilip K. Arora
Department of Botany
Banaras Hindu University
India

George G. Khachatourians
Department of Applied Microbiology and Food Sciences
College of Agriculture
University of Saskatchewan
Saskatoon, SK, Canada

ELSEVIER

2004

Amsterdam – Boston – Heidelberg – London – New York – Oxford
Paris – San Diego – San Francisco – Singapore – Sydney – Tokyo

ELSEVIER B.V.
Sara Burgerhartstraat 25
P.O. Box 211, 1000 AE
Amsterdam, The Netherlands

ELSEVIER Inc.
525 B Street, Suite 1900
San Diego, CA 92101-4495
USA

ELSEVIER Ltd
The Boulevard, Langford Lane
Kidlington, Oxford OX5 1GB
UK

ELSEVIER Ltd
84 Theobalds Road
London WC1X 8RR
UK

First edition 2004

Library of Congress Cataloging in Publication Data
A catalog record is available from the Library of Congress.

British Library Cataloguing in Publication Data
A catalogue record is available from the British Library.

ISBN: 0-444-51642-5

Transferred to digital print 2007
Printed and bound by CPI Antony Rowe, Eastbourne

Editors

Dilip K. Arora
Department of Botany
Banaras Hindu University
Varanasi, India
Fax: +91 542 2368141
Tel : + 91 542 2369570
E-mail: aroradilip@yahoo.co.in

George G. Khachatourians
Department of Applied Microbiology
and Food Sciences
College of Agriculture
University of Saskatchewan
Saskatoon, Canada
Tel : +1 306 966 5032
E- mail: khachatouria@sask.usask.ca

Contents

Contributors

Keietsu Abe	The New Industry Creation Hatchery Center (NICHe), Tohoku University, Sendai 980-8579, Japan.
Susan J. Assinder	School of Biological Sciences, University of Wales, Bangor, Gwynedd LL57 2UW, Wales, UK (s.assinder@bangor.ac.uk).
Concepción Azcón- Aguilar	Estación Experimental del Zaidín, CSIC, Granada, Spain.
Bert Bago	Centro de Investigaciones sobre Desertificación, CSIC, Valencia, Spain.
Randy M. Berka	Novozymes Biotech, Inc., 1445 Drew Avenue, Davis, California 95616-4880, USA.
Deepak Bhatnagar	Food and Feed Safety Research Unit, U.S. Department of Agriculture, Agricultural Research Service, Southern Regional Research Center, New Orleans, Louisiana 70124, USA.
Edward L. Braun	Department of Zoology, University of Florida, Gainesville, Florida 32611, USA.
Daren W. Brown	Mycotoxin Research Unit, U.S. Department of Agriculture, Agricultural Research Service, National Center for Agricultural Utilization Research, Peoria, Illinois 61604 USA.
Ignazio Carbone	Center for Integrated Fungal Research, Department of Plant Pathology, North Carolina State University, Box 7244 – Partners II Building, Raleigh, NC 27695-7244, USA.
Dee A. Carter	Discipline of Microbiology, School of Molecular and Microbial Biosciences, University of Sydney, NSW 2006, Australia (d.carter@mmb.usyd.edu.au).

Thomas E. Cleveland Food and Feed Safety Research Unit, U.S. Department of Agriculture, Agricultural Research Service, Southern Regional Research Center, New Orleans, Louisiana 70124, USA.

Katherine F. Dobinson Southern Crop Protection and Food Research Centre, Agriculture and Agri-Food Canada, London, Ontario N5V-4T3, Canada, and Departments of Biology, Microbiology and Immunology, The University of Western Ontario, London, ON, Canada (dobinsonk@agr.gc.ca).

Angelika Felk Institute of General Botany, Department of Molecular Phytopathology and Genetics (AMPIII), University of Hamburg, Ohnhorststrasse 18, D-22609 Hamburg, Germany.

Nuria Ferrol Estación Experimental del Zaidín, CSIC, Granada, Spain.

Philipp Franken Institute for Vegetable and Ornamental Plants, Grossbeeren, Germany.

Martin J. Gilbert School of Biological Sciences, University of Exeter, Washington Singer Laboratories, Perry Road, Exeter, EX4 4QG, UK.

Armelle Gollotte INRA-CMSE, 17 rue Sully-BV154021034, Dijon Cedex, France.

Katsuya Gomi The New Industry Creation Hatchery Center (NICHe), Tohoku University, Sendai 980-879, Japan.

Manuel González-Guerrero Estación Experimental del Zaidin, CSIC, Granada, Spain.

Stephen B. Goodwin U. S. Department of Agriculture, Agricultural Research Service, Department of Botany and Plant Pathology, 915 West State Street, Purdue University, West Lafayette, IN 47907-2054, USA (sgoodwin@purdue.edu).

Lucy Alexandra Harrier The Scottish Agricultural College, Edinburgh, United Kingdom.

Fumihiko Hasegawa The New Industry Creation Hatchery Center (NICHe), Tohoku University, Sendai 980-8579, Japan.

Mitsuro Hyakumachi Laboratory of Plant Pathology, Faculty of Agriculture, Gifu University, Gifu 501-1193, Japan.

Seogchan Kang Department of Plant Pathology, 311 Buckhout, The Pennsylvania State University, University Park, PA 16802, USA (sxk55@psu.edu).

Gert H. J. Kema Plant Research International B.V., P.O. Box 16, 6700 AA Wageningen, The Netherlands.

Linda Kohn Department of Botany, University of Toronto, 3359 Mississauga Rd. N., Mississauga, ON L5L 1C6, Canada (kohn@utm.utoronto.ca).

George G. Khachatourians BioInsecticide Research Laboratory, Department of Applied Microbiology and Food Science, University of Saskatchewan, Saskatoon, S7N 5A8, Canada (Khachatouria@usask.ca).

Luisa Lanfranco Università degli Studi di Torino, Torino, Italy.

Antonio Llobell Instituto de Bioquimica Vegetal y Fotosintesis, University of Sevilla/CSIC, Seville, Spain.

Katherine F. LoBuglio Harvard University Herbaria, 22 Divinity Ave., Cambridge, MA 02138, U.S.A.

Matteo Lorito Dipartimento Ar. Bo.Pa.Ve., sezione di Patologia Vegetale, Laboratori di Biocontrollo, Università di Napoli Federico II, Via Università, 100, 80055 Portici (Napoli) Italy (lorito@unina.it).

Mette Lübeck Department of Plant Biology, Plant Pathology Section, The Royal Veterinary and Agricultural University, 40 Thorvaldsensvej, DK-1871 Frederiksberg C, Denmark (mette.lubeck@plbio.kvl.dk).

Masayuki Machida Research Center for Glycoscience, National Institute of Advanced Industrial Science and Technology (AIST), Tsukuba, Ibaraki 305-8566, Japan.

Kevin T. Madden Microbia, Inc., 320 Bent St., Cambridge MA 02142 U.S.A.

Robert E. Marra Box 3020, Department of Molecular Genetics and Microbiology, Duke University Medical Center, Durham, NC 27710, USA (r.marra@duke.edu).

Enrique Monte Centro Hispano Luso de Investigaciones Agrarias, University of Salamanca, Salamanca, Spain.

Donald O. Natvig — Department of Biology, University of New Mexico, Albuquerque, New Mexico 87131, USA.

Mary Anne Nelson — Department of Biology, University of New Mexico, Albuquerque, New Mexico 87131, USA (manelson@unm.edu).

Beth A. Nelson — Novozymes Biotech, Inc., 1445 Drew Avenue, Davis, California 95616- 4880 USA.

William C. Nierman — Institute for Genomic Research, Rockville, Maryland 20850 U.S.A.

Thea C. Norman — Microbia, Inc., 320 Bent St., Cambridge MA 02142 U.S.A.

Vivienne Gianinazzi-Pearson — INRA-CMSE, 17 rue Sully-BV154021034, Dijon Cedex, France.

Robert H. Proctor — Mycotoxin Research Unit, U.S. Department of Agriculture, Agricultural Research Service, National Center for Agricultural Utilization Research, Peoria, Illinois 61604 U.S.A.

Manuel Rey — Newbiotechnic S.A., Isla de la Cartuja, Seville, Spain.

Michael W. Rey — Novozymes Biotech, Inc., 1445 Drew Avenue, Davis, California 95616-4880 USA.

John C. Royer — Microbia, Inc., 320 Bent St., Cambridge MA 02142 U.S.A (jroyer@microbia.com).

Siegfried Salomon — Institute of General Botany, Department of Molecular Phytopathology and Genetics (AMPIII), University of Hamburg, Ohnhorststrasse 18, D-22609 Hamburg, Germany.

Felice Scala — Dipartimento Ar.Bo.Pa.Ve., sezione di Patologia Vegetale, Laboratori di Biocontrollo, Università di Napoli Federico II, Via Università, 100, 80055 Portici (Napoli) Italy.

Wilhelm Schäfer — Institute of General Botany, Department of Molecular Phytopathology and Genetics (AMPIII), University of Hamburg, Ohnhorststrasse 18, D-22609 Hamburg, Germany (schäfer@botanik.uni-hamburg.de).

Darren M. Soanes School of Biological Sciences, University of Exeter, Washington Singer Laboratories, Perry Road, Exeter, EX4 4QG, UK.

Haruhisa Suga Molecular Genetics Research Center, Gifu University, Gifu 501-1193, Japan (suga@cc.gifu-u.ac.jp).

Nicholas J Talbot School of Biological Sciences, University of Exeter, Washington Singer Laboratories, Perry Road, Exeter, EX4 4QG, UK (N.J.Talbot@exeter.ac.uk).

Nai Tran-Dinh Food Science Australia, Riverside Corporate Park, North Ryde NSW 2113, Australia (Nai.Tran-Dinh@csiro.au).

Diederik van Tuinen INRA-CMSE, 17 rue Sully-BV154021034, Dijon Cedex, France.

Daniel Uribe Biotechnology Institute, National University of Colombia, Bogota, Colombia.

Raul E. Vera Orbit3 Pty Ltd, 8 Coneill Place, Forest Lodge, NSW 2037, Australia (raul@orbit3.com).

Cees Waalwijk Plant Research International B.V., P.O. Box 16, 6700 AA Wageningen, The Netherlands.

Margaret Werner- Washburne Department of Biology, University of New Mexico, Albuquerque, New Mexico 87131, USA.

Wendy T. Yoder Novozymes Biotech Inc., 1445 Drew Avenue, Davis, California 95616-4880, USA.

Jiujiang Yu Food and Feed Safety Research Unit, U.S. Department of Agriculture, Agricultural Research Service, Southern Regional Research Center, New Orleans, Louisiana 70124, USA (jiuyu@srrc.ars.usda.gov).

Elizabeth J. Zaretsky Novozymes Biotech, Inc., 1445 Drew Avenue, Davis, California 95616-4880, USA.

Preface

Genetics of fungi, since 1940s, have been instrumental in the production of industrial feedstock chemicals, enzymes, pharmaceuticals and pre- and post-harvest agriculture. Interest in the general and molecular genetics of fungi has proven to be pivotal to the development of a plethora of bulk enzymes, chemicals, agri-food commodities and human health products. Research in the genomics of a handful of fungi has matured at an unprecedented rate to allow their comprehensive review. Developments in fungal genomics should be of great significance to new strategies in ancillary fields where disciplinary crossovers of fungal genomics, genes and their regulation, expression, and engineering will have a strongest impact in dealing with agriculture, foods, natural resources, life sciences, biotechnology, informatics, metabolomics, pharmaceuticals and bioactive compounds. We are confident that applied mycology will continue to be an important beneficiary of genomic technology and concepts.

The development of fungal genome initiatives have changed our understanding of taxonomically useful genetic sequences involved in analysis of population genetics and genetic variability. Obviously characterization of newly discovered isolates require expert knowledge and depends on availability of well-defined markers. Molecular characterization of fungal genomes offers new hope. This volume analyzes the application of commonly used molecular marker systems. These systems, in conjunction with computer-based genome analysis, opens up exciting opportunities in fungal ecology, biology and genetics. A critical analysis of methods of genomic analysis have led to inferential but new knowledge of the dynamic processes leading to population divergence and speciation. This is one place where divergent literature of population genetics, evolutionary statistics and, of course, phylogeography has converged.

It is possible to detect recombination in fungi with haploid genome and either substantial asexual reproduction or with significant selfed sexual reproduction, which are not widespread throughout a phylogeny. This volume also elaborates the development of biochemical genetics, which provides a model system that established the relationship between genes and enzymes. The impact that the recent publication of a high-quality draft sequence of the *N. crassa* genome that contains about 10,000 protein-coding genes, approximately twice that in yeasts but slightly fewer than the invertebrate animals. What types of different processes were responsible for differences in gene content between *N. crassa* and the yeasts? Why did the widest array of genome defense mechanisms known for any organism came to block the productive duplication of genes, alternative genes, unexpected genes, secondary metabolites, shared apparent "pathogenicity" genes with plant pathogens, and response to environmental cues such as light in novel ways? The genome sequence for *N. crassa* is the first exciting step toward a detailed understanding of the biology of filamentous fungi.

Some of the most important pathogens of humans, insects and plants are found amongst fungi. Because of their importance to production and post-harvest agriculture, genomics of the phytopathogenic and entomopathogenic fungi continue to receive special attention. In this volume, current knowledge about the genomics and genetic variability of *Candida albicans*, the polymorphic opportunistic human pathogen of increasing medical importance, especially in immunocompromised individuals, has been covered in detail. Besides this, current understanding of the genetics and functional genomics of the most important fungal pathogens of staple food crops, rice and wheat among others are covered, including the chapters dealing with the genetics and genomics of *Aspergillus*, *Fusarium*, *Magnaporthe grisea*, *Mycosphaerella graminicola*, *Penicillium*, *Rhizoctonia*, *Trichoderma* and entomopathogenic fungi.

The fourth volume of Applied Mycology and Biotechnology, the companion to volume three, is dedicated to recent developments in fungal genomics representing a meaningful comprehensive reference set with a wide coverage of its emerging range and complex knowledge. The selections of chapters in this volume reflect the input of Editors and Editorial Board members, and as a consequence, there is considerable breadth and depth of coverage offered by a group of splendid authors. With several thousand citations, we hope that volume three and four will serve as a useful reference for knowledgeable veterans and beginners as well as for those crossing disciplinary boundaries and getting into the exciting field of biotechnology, genomics and bioinformatics of fungi.

We are indebted to the contributors for their valuable assistance in compiling this volume. Our sincere thanks to Ms. Hetty Verhagen and Ana- Bela Sa Dias of Elsevier Life Sciences for their technical assistance in editing this book.

Dilip K. Arora
George G. Khachatourians

Applied Mycology & Biotechnology
An International Series. Volume 4. Fungal Genomics

1

The Development of Genetic Markers from Fungal Genome Initiatives

Dee A. Carter[1], Nai Tran-Dinh[2], Robert E. Marra[3] and Raul E. Vera[4]
[1]Discipline of Microbiology, School of Molecular and Microbial Biosciences, University of Sydney, NSW 2006, Australia (d.carter@mmb.usyd.edu.au); [2]Food Science Australia, Riverside Corporate Park, North Ryde NSW 2113, Australia (Nai.Tran-Dinh@csiro.au); [3]Box 3020, Department of Molecular Genetics and Microbiology, Duke University Medical Center, Durham, NC 27710, USA (r.marra@duke.edu); [4]Orbit3 Pty Ltd, 8 Coneill Place, Forest Lodge, NSW 2037, Australia (raul@orbit3.com).

The lack of morphological characters has made molecular markers invaluable for studying aspects of fungal ecology, biology and genetics. Generally, markers are developed specifically for each fungal species under study using laboratory-based techniques. With a growing body of information on fungal genomes, it becomes possible to use these data to develop markers from sequenced genomes. In addition, it is possible to look at trends across genomes, which may allow more universal marker sets to be developed or may guide researchers into making the best predictions of the kinds of markers that could be useful in a given fungus. In this chapter, we review four commonly used molecular markers: microsatellites, minisatellites, interspersed repetitive sequence elements and single nucleotide polymorphisms, and discuss how computer-based genome analysis can find and analyse these using data from fungal genome initiatives.

1. INTRODUCTION

Molecular markers have become enormously important in many areas of fungal biology, including strain typing, epidemiology, population genetics, fungal detection and identification, genetic mapping, gene isolation, phylogenetics and evolutionary biology (Spitzer *et al.* 1989; Meyer *et al.* 1993; Carter *et al.* 1997; Girardin, 1997; van der Lee *et al.* 1997 Geiser *et al.* 1998; Sudarshan *et al.* 1999; McDade and Cox, 2001;). These markers are based on minor differences that accumulate in the genomes of members of a species as they diverge from one another over time. In fungi, markers can be developed from chromosomal, extrachromosomal or mitochondrial DNA. They can be present in the organism in a single copy, or repeated in multiple copies throughout the genome.

There are a number of "universal" sequences that can be used as markers in a range of different fungal genomes, with little or no prior knowledge of the genome. One widely used example is the rRNA gene, which contains both highly conserved and variable regions. Universal primers exist that can amplify regions of this gene from essentially any fungal species (White *et al.*, 1990). Primers and probes can also be developed based on highly conserved repetitive motifs found in fungal genomes. Examples of these are the M13 core minisatellite region, which occurs in tandem copies at multiple sites throughout eukaryotic genomes, and microsatellite motifs, which are likewise present throughout eukaryotic genomes (Meyer and Mitchell 1995).

For many applications, however, the information provided by universal markers is not sufficient. Conserved universal gene sequences may not allow sufficient resolution to distinguish between closely related strains. Methods based on hybridizing to or amplifying from repetitive motifs can be difficult to interpret and standardize between laboratories. Also, these types of markers may not be suited to the study in question. It is therefore frequently necessary to develop specific markers for a particular fungal species. Markers can be based on functional genes or on anonymous DNA sequences. Development of the former is usually done by using homologous primers or probes to find the gene and isolate it from the genome or from a genomic library of the organism under study. Informative, variable nucleotide positions generally occur either in the third codon position or within introns, where the amino acid sequence of the resulting protein is not affected. These markers are usually minor base changes in non-repetitive DNA such as base substitutions (often referred to as single nucleotide polymorphisms - SNPs) or small insertions or deletions (indels). Anonymous DNA sequences are obtained from an unbiased sampling of genomic DNA and these may or may not contain functional genes. The non-coding regions, being free from selective pressures, are generally more likely to tolerate sequence changes than coding sequences; these can also contain repetitive motifs such as mini- and microsatellites. Developing markers from anonymous sequences requires screening genomic libraries or randomly amplified fragments of DNA for potentially useful sequences (Landry and Michelmore 1985; Carter *et al.* 1995; Bart-Delabesse *et al.* 1998).

Marker identification is an often laborious and expensive process. At the outset of a study there may be little or no knowledge of the genome in question to guide the researcher in the choice of markers that are most likely to be present and successful. The proportion of AT vs. GC bases present, the relative amounts of single copy and repetitive DNA, whether microsatellites or minisatellites occur frequently and what sequences motifs are most commonly encountered are likely to be unknown. Any of these parameters could influence the choice of marker to be targeted. Without this information marker development is largely a case of trial and error.

Until recently there has been a paucity of sequence data available for most fungi. Reduced sequencing costs and an increasing recognition of the medical and agricultural importance of many fungal species, combined with the fact that most fungal genomes are relatively small and therefore can feasibly be sequenced in their entirety, has placed an increasing number of fungal species on the list of organisms targeted by public and private sequencing initiatives worldwide. *Saccharomyces cerevisiae* was the first eukaryotic organism to be fully sequenced and was completed in 1997 (Goffeau *et al.* 1996); the genome of *Schizosaccharomyces pombe* was completed in early 2002, and the *Neurospora crassa* genome is nearing completion. Extensive data are available for *Candida albicans, Cryptococcus neoformans*, *Aspergillus nidulans* and *Pneumocystis carinii.* Partial data based on cDNA sequences are also available for *Aspergillus flavus, Aspergillus oryzae*, *Fusarium sporotrichioides* and *Phytophthora infestans* (Yoder and Turgeon, 2001). The sequence information of all of these genomes is publicly available, and the sequencing of numerous additional species is currently underway in private research organisations.

Access to a significant amount of sequence data can be very helpful when attempting to develop molecular markers. Potentially useful genes can be identified and screened for introns, and primers can be developed directly from the sequence information, eliminating the need for lengthy optimisations using homologous primers. Microsatellite sequences can likewise be rapidly identified and their potential usefulness (eg. whether they are of a length and sequence composition that is likely to be polymorphic) assessed before any expensive experiments are conducted. Clearly, the more data available the more useful this will be, particularly if there are a number of separate strains sequenced within a given species. However, sequence data for even a relatively small proportion of a single genome from one species can allow the development of enough markers for a useful study. In this review we will present some examples of some of the approaches we have taken to develop molecular markers using genomic information. These markers are based on microsatellites, minisatellites, interspersed repetitive sequence elements, SNPs and small insertions and deletions.

2. MICROSATELLITES

Microsatellites or simple sequence repeats (SSRs) are tandem arrays of short DNA sequences composed of one to six base pair (bp) motifs. These motifs are usually repeated at least five times. They are found ubiquitously in eukaryotic genomes, including fungi, and are also found in some prokaryotic genomes (Tautz and Renz, 1984; Bruford and Wayne, 1993; Rosewich and McDonald, 1994; Field and Wills, 1996; Hancock, 1996). Microsatellites have been found in coding and noncoding regions and are co-dominantly inherited (Edwards *et al.* 1992; Bowcock *et al.* 1994; Forbes *et al.* 1995). They are often characterised by a high degree of length polymorphism; consequently the number of tandem repeats at a given locus can vary from one individual to the next. Due to these characteristics of ubiquity, co-dominant inheritance, high polymorphism and their applicability to PCR, microsatellites are very powerful genetic markers, especially for the study of closely related organisms. They have been applied in many fields of research including genome mapping, identification in ancient and forensic samples, studies of population structure, mating systems, phylogeny, linkage and conservation biology (Ashley and Dow, 1994; Tautz and Schlötterer, 1994).

Microsatellites belong to a family of repetitive DNA sequences that also includes satellite and minisatellite sequences (Charlesworth *et al.* 1994; Chambers and MacAvoy, 2000). As their name suggests, microsatellites are the smallest sequences in this family, with a maximal size of several hundred base pairs, as compared to satellite (up to several megabases) and minisatellite (0.5 to 30 kilobases) sequences. Microsatellites are highly variable, with loci commonly having ten or more alleles and heterozygosities above 0.60 (Bowcock *et al.* 1994; Deka *et al.* 1995). Microsatellites can be divided into three categories: 1) perfect repeats where the same motif is uninterrupted in a tandem array; 2) imperfect repeats where interruptions occur in the run of repeats; and 3) compound repeats where one motif is immediately followed by another. Perfect repeats have been found to be the most variable and informative, while imperfect and compound repeats have shown lower levels of polymorphism (Weber, 1990; Rassmann *et al.* 1991; Richard and Dujon, 1996).

The number of repeat units at a microsatellite locus is highly variable between individuals. It has been estimated that mutation rates at microsatellite loci vary between 10^{-2} and 10^{-6} (Dallas, 1992; Weber and Wong, 1993; Nielsen and Palsboll, 1999). This inherent instability of microsatellites has been hypothesised to be caused by two different phenomena: DNA polymerase slippage and unequal recombination (Levinson and Gutman, 1987; Schlötterer and Tautz, 1992; Eisen, 2000). In the case of DNA polymerase slippage, transient dissociation of the replicating DNA strands in the polymerase complex may be followed by misaligned reassociation of the two strands. This may result in an increase or decrease in the number of repeat units at a locus depending upon whether the misalignment occurred on the newly synthesised strand or on the template strand, respectively (Levinson and Gutman, 1987;

Richards and Sutherland, 1994). The frequency of these slippage mutations is not known, but in prokaryotes they occur more frequently than spontaneous mutations, and from indirect evidence this is also the case for eukaryotes (Tautz, 1989). Variation in repeat lengths may also result from unequal crossing-over events during recombination (Levinson and Gutman, 1987; Japupciak and Wells, 1999), but DNA polymerase slippage is generally considered to be the primary mechanism of allelic differences at microsatellite loci.

2.1 The Application of Microsatellites as Molecular Markers

The first observation of the large number and almost ubiquitous distribution of microsatellite sequences in various genomes was made by Hamada *et al.* (1982), who found hundreds of copies of $(TG)_n$ repeats in yeasts and tens of thousands in vertebrates (Hamada *et al.* 1982). Microsatellites were first recognised as locus-specific polymorphic markers by three groups simultaneously (Litt and Luty, 1989; Tautz, 1989; Weber and May, 1989). Since then, microsatellites have had various applications including DNA fingerprinting (Weir, 1996), genome mapping (Dib *et al.* 1996; Roder *et al.* 1998), paternity and relatedness testing (Queller *et al.* 1993), genetic distance measurements (Goldstein *et al.* 1995; Shriver *et al.* 1995; Slatkin, 1995), forensic science (Hagelberg *et al.* 1991), for identification of individuals of unknown origin (Shriver *et al.* 1997; Davies *et al.* 1999) and for the detection of hidden population structure (Pritchard and Rosenberg, 1999). Studies using microsatellites have looked at a wide variety of organisms including humans (Dib *et al.* 1996), animals (Taylor *et al.* 1994; Dawson *et al.* 1997; Valsecchi *et al.* 1997; Brown Gladden *et al.* 1999; Feldheim *et al.* 2001), insects (Estoup *et al.* 1993; Hughes and Queller, 1993; Goldstein and Clark, 1995; Michalakis and Veuille, 1996), plants (Condit and Hubbel, 1991; Thomas *et al.* 1994; Akkaya *et al.* 1995; Broun and Tanksley, 1996; Taramino and Tingey, 1996;) and bacteria (Lupski and Weinstock, 1992).

The use of microsatellites to analyse fungi has been limited to date. This is indicative of the trend that new molecular techniques are applied first to humans, then to bacteria, then to plants and animals and finally applied to fungi. However, microsatellites have been used in genotyping of strains and in epidemiological studies for the human fungal pathogens *Histoplasma capsulatum*, *Coccidiodes immitis* and *Candida albicans* (Field *et al.* 1996; Bretagne *et al.* 1997; Carter *et al.* 1997; Metzgar *et al.* 1998a; Metzgar *et al.* 1998b; Fisher *et al.* 2000; Carter *et al.* 2001). They have also been used to study the ecology and diversity of *Epichloë*, an endophyte of temperate grasses (Groppe *et al.* 1995; Groppe and Boller, 1997; Moon *et al.* 1999). The population structure of the anther smut fungus *Microbotryum biolaceum* was analysed using variation at five microsatellite loci (Bucheli *et al.* 2000). The increasing application of microsatellites to fungi has been apparent from a number of recent publications reporting their isolation (Fisher *et al.* 1999; Tran-Dinh and Carter, 2000; Enjalbert *et al.* 2002; Zhou *et al.* 2002).

In practical terms, microsatellites are visualised through PCR amplification using primers complementary to unique sequences flanking the microsatellite locus, followed by polyacrylamide gel electrophoresis. The length of the PCR product is determined by the location of the flanking primers, which are usually designed to amplify products of between 50 and 300 nucleotides (Tautz 1989). Allelic variation in PCR product lengths caused by variation in the number of repeat units at a microsatellite locus can be used to characterise individual strains. Differences of one repeat unit between strains can be detected as the resolution of polyacrylamide gels allows detection of single nucleotide differences. The use of several microsatellite markers usually provides sufficient polymorphism to identify individual members of a population.

There are a range of methods for analysing microsatellite data, many of which are available through "Microsat", (http://hpgl.stanford.edu/projects/microsat/) an online program capable of producing distance matrices based on allele length variations. Unfortunately, some methods (e.g. Rst; Slatkin, 1995) are not applicable to haploid organisms and therefore cannot be used for most fungal species. Microsatellites suffer some limitations, principally the problem of homoplasy, in which the same allele may arise via different means – for example, CA_{10} could be generated by the addition of a CA unit to a CA_9 allele, or by deletion of a CA unit from CA_{11}, or two units from CA_{12}, etc. For this reason microsatellites need to be treated with caution when used as phylogenetic markers, however they remain extremely important in population, epidemiology, and strain typing studies.

2.2 Traditional Methods for the Isolation of Microsatellites

The major difficulty with microsatellites is the amount of work and time required in identifying, isolating and characterising them in new taxonomic groups. For microsatellite loci to function as molecular markers, primers flanking the microsatellite loci must be designed, thus sequencing data at these loci is required. Usually the major cost of a project involving microsatellites is the time spent acquiring this sequence data. Once flanking primers have been designed and optimised, microsatellite length analysis is a very efficient method of surveying large numbers of strains in population genetic studies. Various methods of isolating microsatellites have been used. Laboratory-based methods include the traditional approach of creating and screening partial genomic DNA libraries for the presence of repeat units (Rassmann *et al.* 1991), the use of enrichment techniques to increase the likelihood of inserts containing microsatellite repeats in genomic libraries (reviewed in Zane *et al.* 2002), and PCR-based methods (Carter *et al.* 1996; Ender *et al.* 1996). With the increasing amounts of sequence data becoming available, many researchers are able to take the far easier and cheaper approach of screening published sequences and genomic sequencing projects for microsatellite repeats.

2.3 Screening Published Sequences for Microsatellite Repeats

The growing body of sequence information available for fungi now allows microsatellite loci to be identified in some fungal species by searching published DNA sequences in databases such as GenBank and EMBL. The most experimentally useful microsatellite repeat units (mono-, di-, tri-, and tetranucleotides) are simply used as queries against the published genomic data.

We have employed this method to search for microsatellites in *Aspergillus flavus* and *Aspergillus parasiticus* (Tran-Dinh and Carter 2000). Database searches of published sequences were performed using software made available through WebANGIS (Australian National Genomic Information Service; http://www.angis.org.au/pbin/WebANGIS/wrapper.pl). First, all published sequences of *A. flavus* and *A. parasiticus*, and the two closely related species, *A. oryzae* and *A. sojae* were extracted from GenBank and placed in a separate database. *A. oryzae* and *A. sojae* sequences were included because cross-species amplifications for microsatellite markers have been reported (Schlötterer *et al.* 1991; Rubinsztein *et al.* 1995; Dawson *et al.*, 1997). All possible dinucleotide and trinucleotide repeat motifs were used as queries in BLASTN searches (Altschul *et al.* 1990). A search for microsatellites with one particular repeat motif will actually search for several repeat motifs because of permutations and complementary sequences. For example searching with the repeat motif (AT) also searches for loci containing (TA) repeats and searching for (ACC) repeats will also find loci with (CCA), (CAC), (TGG), (GGT) and (GTG) repeats. Sequences containing at least five tandem repeats were chosen for further analysis. Flanking primers for microsatellite loci were designed using OLIGO version 4.0 software (National Biosciences).

Using this method flanking primers for six candidate microsatellite markers (AFPM2-7) were designed and tested on 20 isolates of *A. flavus* and 15 isolates of *A. parasiticus* (Table 1; modified from Tran-Dinh and Carter, 2000). All the microsatellite loci were polymorphic in length, with 2-11 alleles found within the population tested. Greater variation was seen within *A. flavus* than in *A. parasiticus*, evident from the greater number of alleles and the higher observed heterozygosities in *A. flavus* (Table 1). These results were consistent with previous analyses using RAPD markers (Tran-Dinh *et al.* 1999). Sequence analyses of various microsatellite loci revealed that in most cases length polymorphisms were due to variation in numbers of repeat units between the individual strains (Fig. 1a). However, in some cases variation in allele size did not correspond with the number of repeat units at the microsatellite locus and was due to deletions or insertions in DNA flanking the microsatellite (Fig. 1b). This phenomenon has been reported by other researchers (Ortí *et al.* 1997; Fisher *et al.* 2000; Carter *et al.* 2001).

Table 1. Microsatellite markers developed from DNA sequences from *Aspergillus flavus*, *A. parasiticus* and *A. oryzae*

Locus	Repeat motif	Size Range	No. of alleles		H_O[3]	
			A. f[1]	*A. p*[2]	*A. f*	*A. p*
AFPM2	$(ACT)_5T(CTC)_4$	206-266	7	6	0.81	0.74
AFPM3	$(AT)_6AAGGGCG(GA)_8$	199-217	7	4	0.75	0.67
AFPM4	$(CA)_{13}$	179-206	5	2	0.70	0.24
AFPM5	$(AG)_5AC(AG)_2$	210-338	10	7	0.82	0.82
AFPM6	$(GT)_6$	341-355	4	4	0.59	0.35
AFPM7	$(AC)_{35}$	215-276	11	9	0.85	0.84

[1]*A. flavus*; [2]*A. parasiticus*; [3] Observed heterozygosity; Table modified from Tran-Dinh and Carter (2000).

Searching published sequences for microsatellites is a simple and efficient method of isolating microsatellite loci. One limitation of this method is the reliance upon available sequencing data, especially since microsatellites have been found to be more abundant in noncoding regions (Hancock 1995; Chambers and MacAvoy 2000), and these regions are underrepresented in sequencing databases. It is advantageous to choose sequences containing ten or more repeats because shorter repeats tend to have a lower level of polymorphism and therefore are of less value as genetic markers (Weber 1990; Edwards *et al.* 1991; Primmer *et al.* 1996). With limited sequencing data, researchers may have to use sequences containing a low number of repeats, in the hope that the allele is at the low end of a polymorphic locus.

2.4 Using Genome Sequencing Projects for Isolating Microsatellites

Whole genome DNA sequencing projects are a promising resource for the development of microsatellite markers. Isolation of microsatellites from genomic sequencing projects can be done in the same way as screening published sequences from databases. The advantage of genome sequencing projects is the enormous amount of genetic data available, which offers the opportunity to examine microsatellite repeats across a whole genome, including noncoding regions where they are most abundant. Genome sequencing projects also allow researchers to investigate microsatellites that have a large number of repeats and in this way increase the likelihood of finding polymorphic markers. Fully assembled genome sequencing projects also provide ample flanking sequences surrounding microsatellite loci from which to design primers, which may not be the case with the short sequences available from other databases.

```
                   10        20        30        40        50        60
A.flavFRR4472 TACTCTGTATACCCAGTCCCAGCGGTCTGTGACTGCAGAGGTAGAGCCATAACACACACA
A.flavFRR5308 TACTCTGTATACCCAGTCCCAGCGGTCTGTGACTGCAGAGGTAGAGCCATAACACACACA
A.flavFRR5311 TACTCTGTATACCCAGTCCCAGCGGTCTGTGACTGCAGAGGTAGAGCCATAACACACACA
A.flavFRR5313 TACTCTGTATACCCAGTCCCAGCGGTCTGTGACTGCAGAGGTAGAGCCATAACACACACA
A.flavFRR5315 TACTCTGTATACCCAGTCCCAGCGGTCTGTGACTGCAGAGGTAGAGCCATAACACACACA
A.paraFRR2749 TACTCTGTATACCCAGTCCCAGCGGTCTGTGACTGCAGAGGTAGAGCCATAACACACACA
A.paraFRR4468 TACTCTGTATACCCAGTCCCAGCGGTCTGTGACTGCAGAGGTAGAGCCATAACACACACA
              ************************************************************

4472 CACACACA----------------TATTCGTTCTAATGTGCAAGATCAAATTGTTAAACC  (189)
5308 CACACACACACACACACA------TATTCGCTCTAATGTGCAAGATCAAATTGTTAAACC  (199)
5311 CACACACACACACACACACACACATATTCGTTCTAATGTGCAAGATCAAATTGTTAAACC  (205)
5313 CACACACACA--------------TATTCGTTCTAATGTGCAAGATCAAATTGTTAAACC  (191)
5315 CACACACACACACACACA------TATTCGTTCTAATGTGCAAGATCAAATTGTTAAACC  (199)
2749 CACACACA----------------TATTCGTTCTAATGTGCAAGATCAAATTGTTAAACC  (189)
4468 CACACACACACACACACACACACATATTCGTTCTAATGTGCAAGATCAAATTGTTAAACC  (205)
********                ****** ******************************
```

a: AFPM4 locus

```
                   10        20        30        40        50        60
A.flavFRR4086 GCGATAGTTTTATAATGAGTTCTTTATCTGCTTTCCCTCTCTTTCTTTCCTTTC--------
A.flavFRR5305 GCGATAGTTTTATAATGAGTTCTTTATCTGCTTTCCCTCTCCTTCTTTCCTTTC--------
A.flavFRR5309 GCGATAGTTTTATAATGAGTTCTTTATCTGCTTTCCCTCTCTTTCTTTCCTTTC--------
A.flavFRR5312 GCGATAGTTTTATAATGAGTTCTTTATCTGCTTTCCCTCTCTTTCTTTCCTTTC--------
A.paraFRR2501 GCGATAGTTT-ATAATGACTTCTTTATCTGCTTTCCTTCTCTTTCTTTTTTTTTTTAAAAAA
A.paraFRR2749 GCGATAGTTTTATAATGAGTTCTTTATCTGCTTTCCCTCTCTTTCTTTCCTTTC--------
              ********** ******* ***************** **** ******  ***

4086 --------AATACCTT-CTAAGTTGACACAATAATTAGATTTGATTAAATAACAATGAATTAAATATTAAT
5305 --------AATACCTT-CTAAGTTAAAACAATAATCAGATTTAATTAAATAACAATGAATTAAATGTTAAT
5309 --------AATACCTT-CTAAGTTGACACAATAATTAGATTTGATTAAATAACAATGAATTAAATATTAAT
5312 --------AATACCTT-CTAAGTTGACACAATAATTAGATTTGATTAAATAACAATGAATTAAATATTAAT
2501 AAAATAATAATACCTTTCTAAGTTAACACGATAGTCAGATCTGATTAAATAACAATGAATCAGATGTTCTT
2749 --------AATACCTT-CTAAGTTGACACAATAATTAGATTTGATTAAATAACAATGAATTAAATATTAAT
        ******** ******* * ** *** * **** * ***************** * ** ** *

4086 TTAGAGAGAGAGAGAGAGAGAGACAGAGAAAAAGAGAAAAAGAGAAAAAGAGAAAAAGAGAAAAAGAGAAA
5305 TTAGAGAGAGAGACAGAGAGAAAAAGAGAAAAAG-------------------------------------
5309 TTAGAGAGAGAGAGAGAGAGAGACAGAGAAAAAGAGAAAAAGAGAAAAAGAGAAAAAGAGAAAAAGAGAAA
5312 TTAGAGAGAGAGAGAGAGAGAGACAGAGAAAAAGAGAAAAAGAGAAAAAGAGAAAAAGAGAAAAAGAGAAA
2501 TTT-ATAGAGAGAGAGAGAGAGAGAGAGAGAG---------------------------------------
2749 TTAGAGAGAGAGAGAGAGAGAGACAGAGAAAAAGAGAAAAAGAGAAAAAGAGAAAAAGAGAAAAAGAGAAA
**  * *************** * *****  **

4086 AAGAGAAAAAGAGAAAAAG-----------------------------------AAATAA (274)
5305 ------------------------------------------------------AAATAA (218)
5309 AAGAGAAAAAGAGAAAAAG-----------------------------------AAATAA (274)
5312 AAGAGAAAAAGAGAAAAAGAGAAAAGGAGAAAAAGAGAAAAAGAGAAAAAGAGAAAAAGAAATAA (314)
2501 ------------------------------------------------------AAAGAT (246)
2749 AAGAGAAAAAGAGAAAAAGAGAAAAGGAGAAAAAGAGAAAAAGAGAAAAAGAGAAAAAGAAATAA (314)
                                                      *** *
```

b: AFPM5 locus

Fig. 1. Multiple sequence alignment of microsatellite loci a) AFPM4 and b) AFPM5 in different strains of *A. flavus* and *A. parasiticus*. The microsatellite repeat units are underlined. Numbers in parentheses following each sequence indicate the allele size (in base pairs).Note that at locus AFPM5, strain A.paraFRR2501 has a 15 bp insertion in the flanking sequence upstream from the microsatellite, which is partially responsible for the length variation at this locus.

Three publicly available fungal genome databases that are complete or near completion (the yeast *Saccharomyces cerevisiae* (Goffeau *et al.* 1996); http://genome-www.stanford.edu/Saccharomyces/; the fission yeast *Schizosaccharomyces pombe* http://www.sanger.ac.uk/Projecs/S_pombe/; and the filamentous fungus *Neurospora crassa* http://www-genome.wi.mit.edu/annotation /fungi/neurospora) were used in searches for all possible di-, tri- and tetranucleotide repeats using BLASTN (Altschul *et al.* 1997). Only perfect microsatellites of length 20bp or more were used as queries, since these loci have a high probability of being polymorphic. Numerous di-, tri- and tetranucleotide repeat microsatellites were found in the three genomes (Tables 2-4). Any of these loci are potential microsatellite markers and can be developed by extracting sequencing data from the databases and designing flanking primers to amplify the relevant repetitive region.

Within *S. cerevisiae*, $(AT)_n$ and $(AAT)_n$ were the most frequent repeat motifs. The *S. cerevisiae* genome is 61% A+T, which may explain the predominance of A+T motifs. The *S. pombe* genome also revealed a bias towards A+T microsatellite motifs, and likewise has a relatively high A+T content (63%). The *N. crassa* genome was found to have a much higher frequency of repetitive DNA sequences than the other two genomes and a very high diversity of motif types. This may be due in part to its larger size (~40Mb, compared to ~12.1Mb and ~11.9Mb for *S. cerevisiae* and *S. pombe*, respectively), and the fact that it has a nearly equal ratio of AT:GC.

Field and Wills (1998) used the information from the *S. cerevisiae* genome initiative to develop 20 highly polymorphic microsatellite markers. These were used to study twelve yeast strains, including seven strains of *S. cerevisiae* and five strains of closely related species. The markers were found to have between three and eleven alleles and were able to reveal intra- and interspecific variation. To our knowledge, no microsatellite markers have been developed for *S. pombe* or *N. crassa.* The high frequency of repeats shown in these genomes suggests that there is great potential for their development and use in these and other fungal species.

2.5 Automation of Genome Screens

Recently, Lim (2002) reported the development of a computer program designed to systematically screen sequencing databases for the presence of microsatellites. The search algorithm, called Magellan, was written to search for all possible mono- to hexanucleotide microsatellite motifs with at least five repeats. Starting with the first base, the program was set to search for at least five repeats of that nucleotide. If no such motif were present, it would then search for five or more repeats of the first two bases. This would continue until it either found a microsatellite, or found that there was no microsatellite array in the first thirty bases (that is, a hexanucleotide motif repeated five times). If Magellan found no repeat motif, it proceeded to the second base in the sequence and repeated the search. If a microsatellite was found, the program would continue the search from the first base in the sequence after the identified short tandem repeat. The search settings were chosen so that a given base could only occur in one microsatellite at a time. An output file is produced at the end of the search that tabulates the location of the different microsatellite loci, and provides summaries of the types and lengths of each motif type.

This allows for a very rapid assessment of the number and type of microsatellites present in genomes. As well as being useful for marker development, these data are interesting for taxonomic and evolutionary comparisons of repeat motifs in genomes (Lim *et al.* in preparation). Magellan is available on request from the authors.

Table 2. Microsatellite loci ≥20bp in *Saccharomyces cerevisiae* genome.

Repeat motif	Chromosome number	No. of microsatellites	Repeat motif	Chromosome number	No. of microsatellites
$(AC)_n$	1,5,7,12,13,16	1	$(AAT)_n$	Mitochondrion	10
	2,4,6	2		1,3,8,10	1
	10	3		2,5,6,7,12	2
Total		15		4,9	3
$(AG)_n$	1,7	1		11,13,15,16	4
	12	2	Total		46
Total		4	$(ACC)_n$	15,16	1
$(AT)_n$	1	1	Total		2
	6,9	3	$(ACT)_n$	1,15,16	1
	11,14	5	Total		3
	10	6	$(AGC)_n$	6,8,10,12,14	1
	2,3,5,8,16	7		2,9,11,13	2
	13, 15, mitochondrion	9		7	3
	12	11	Total		16
	7	13	$(AGG)_n$	9,11	1
	4	17	Total		2
Total		126	$(AGT)_n$	3,5,7,8,9,11,12	1
$(AAC)_n$	1,4,8,9	1		2,13	2
	7,16	2		4	3
	2,10,15	3		15	5
	13	5	Total		19
Total		22	$(AAAC)_n$	2,4	1
$(AAG)_n$	1,2,5,6,7,8,11,13,14	1	Total		2
	3	2	$(AAAT)_n$	4	2
	15	3		11	3
	4	5		Mitochondrion	8
	16	6	Total		13
Total		25	$(ACAT)_n$	3,13,14	1
$(ACG)_n$	7,9,11,15	1	Total		3
Total		4	$(ATTA)_n$	Mitochondrion	10
$(AAAG)_n$	11	1			

The following repeat motifs were also used as search queries but no loci were found: $(CG)_n$, $(CCG)_n$, $(AACG)_n$, $(AACT)_n$, $(AAGC)_n$, $(AAGT)_n$, $(AATC)_n$, $(AATG)_n$, $(ACAG)_n$, $(ACCA)_n$, $(ACCC)_n$, $(ACCG)_n$, $(ACCT)_n$, $(ACGC)_n$, $(ACGT)_n$, $(ACTC)_n$, $(ACTG)_n$, $(AGAT)_n$, $(AGCG)_n$, $(AGCT)_n$, $(AGGA)_n$, $(AGGC)_n$, $(AGGG)_n$, $(AGGT)_n$, $(AGTC)_n$, $(ATTA)_n$, $(CCCG)_n$, $(CCGG)_n$, $(CCTG)_n$.

3. MINISATELLITES

Minisatellites have been associated with a number of interesting features of human genome biology, which can probably also be extended to other organisms. Minisatellite sequences can be part of open reading frames and can occur in the 5' upstream region of some genes, where they appear to regulate transcription (Kennedy *et al.* 1995). They have also been found within introns where they interfere with splicing (Turri *et al.* 1995), and at imprint loci, where they are thought to play a role in imprint control (Neumann *et al.* 1995). Their apparent ability to bind to DNA binding proteins appears to be important for these regulatory and imprinting functions. Minisatellites may also form chromosome fragile sites and have been found in the vicinity of a number of recurrent translocation breakpoints (Sutherland *et al.* 1998).

Table 3. Microsatellite loci ≥20bp in *Schizosaccharomyces pombe* genome.

Repeat motif	Chromosome number	No. of microsatellites
$(AC)_n$	3	1
	2	2
	1	3
$(AG)_n$	1	1
	2	6
$(AT)_n$	3	6
	2	12
	1	17
$(AAC)_n$	1,2	1
$(AAG)_n$	2,3	2
$(AAT)_n$	2,3	4
	1	7
$(AGT)_n$	3	1
$(AGG)_n$	2,3	1
$(AAAG)_n$	1	13
$(AAAT)_n$	1	2
	2	5
$(AATC)_n$	2	1
$(AGAT)_n$	3	1
$(AGGA)_n$	3	1

The following repeat motifs were used as search queries but no loci were found: $(CG)_n$, $(ACC)_n$, $(ACG)_n$, $(ACT)_n$, $(AGC)_n$, $(CCG)_n$, $(AAAC)_n$, $(AACG)_n$, $(AACT)_n$, $(AAGC)_n$, $(AAGT)_n$, $(AATG)_n$, $(ACAG)_n$, $(ACAT)_n$, $(ACCA)_n$, $(ACCC)_n$, $(ACCG)_n$, $(ACCT)_n$, $(ACGC)_n$, $(ACGT)_n$, $(ACTC)_n$, $(ACTG)_n$, $(AGCG)_n$, $(AGCT)_n$, $(AGGC)_n$, $(AGGG)_n$, $(AGGT)_n$, $(AGTC)_n$, $(ATTA)_n$, $(CAGC)_n$, $(CCCG)_n$, $(CCGG)_n$, $(CCTG)_n$.

Like microsatellites, minisatellites are unstable and give rise to variants with increased or reduced numbers of repeats. In contrast to microsatellites, however, the repeat heterogeneity in minisatellites is thought to be primarily due to unequal recombination, which reshuffles the repeat variants. In addition, the influence of local and general biological activities appears to be important in determining the level of instability of each repeat sequence (Debrauwere *et al.* 1997). This instability occurs despite the functional roles of minisatellites, perhaps because little selection pressure exists on the maintenance of the exact length of the repeats beyond a certain minimum sequence, allowing the number of repeats in the sequence to be very flexible (Singh 1995).

One of the first minisatellite markers found to be present in a wide range of organisms was a 15 base pair sequence from the M13 filamentous phage, with the consensus 5'-GAGGGTGGXGGXTCT-3'. This was discovered when an M13 vector without any insert DNA was used as a hybridisation probe against digested human and animal DNA in the absence of any competitor DNA, and revealed a surprisingly complex, multi-banded hybridisation profile (Vassart *et al.* 1987). Subsequent studies found this sequence in the genomes of fish (Georges *et al.* 1988), plants (Rogstad *et al.* 1988), protozoa (Upcroft *et al.* 1990), fungi (Meyer *et al.* 1993) and bacteria (Huey and Hall 1989). The role, if any, of this sequence in the genome of these organisms is not known.

3.1 Use of Minisatellites as Molecular Markers

Minisatellite loci have generally been considered to be too large to assess length variation resulting from the expansion or contraction of repeat units, as is done for microsatellites. Instead, the minisatellite sequences are used as probes or primers to "fingerprint" organisms. As probes, labelled minisatellites are hybridised to DNA from the organism in question that has been digested with a restriction endonuclease, separated on an agarose gel and immobilised onto a nylon membrane by Southern transfer. Following autoradiography, bands are seen wherever the probe has found its complement; as minisatellites occur frequently this generally results in a ladder of bands, usually of varying intensities. Changes to the position or frequency of minisatellites or to the restriction endonuclease sites flanking minisatellite loci mean that different individuals or clones have differences in the banding profile. The relative similarity or difference between organisms can therefore be assessed by applying simple similarity coefficients to the profile, in which the number of bands that differ are compared to the number of bands shared.

Table 4. Microsatellite loci ≥20bp in *Neurospora crassa* sequencing contigs.

Repeat motif	No. of contigs with repeats	Repeat motif	No. of contigs with repeats
$(AC)_n$	98	$(ACAG)_n$	24
$(AG)_n$	>100	$(ACAT)_n$	26
$(AT)_n$	27	$(ACCA)_n$	18
$(AAC)_n$	82	$(ACCC)_n$	4
$(AAG)_n$	35	$(ACCG)_n$	15
$(AAT)_n$	9	$(ACCT)_n/(AGGT)_n$*	96
$(ACC)_n$	40	$(ACGC)_n$	8
$(ACG)_n$	22	$(ACTC)_n$	18
$(ACT)_n/(AGT)_n$*	5	$(ACTG)_n$	31
$(AGC)_n$	52	$(AGAT)_n$	3
$(AGG)_n$	36	$(AGCG)_n$	2
$(CCG)_n$	2	$(AGCT)_n$	2
$(AAAC)_n$	17	$(AGGA)_n$	27
$(AAAG)_n$	28	$(AGGC)_n$	30
$(AAAT)_n$	1	$(AGGG)_n$	6
$(AACG)_n$	11	$(AGTC)_n$	33
$(AACT)_n$	4	$(ATTA)_n$	2
$(AAGC)_n$	22	$(CAGC)_n$	23
$(AAGT)_n$	12	$(CCCG)_n$	5
$(AATC)_n$	14	$(CCTG)_n$	31
$(AATG)_n$	5		

* The search algorithm searched for $(ACT)_n/(AGT)_n$ and $(ACCT)_n/(AGGT)_n$ repeats simultaneously.
The following repeat motifs were also used as search queries but no loci were found: $(CG)_n$, $(ACGT)_n$, $(CCGG)_n$.

Currently, fingerprinting more commonly uses minisatellites as primers rather than probes (Meyer *et al.* 1993). This permits essentially the same analysis but requires far less DNA, and the PCR products are electrophoresed on an agarose gel and visualised directly. Successful amplification using minisatellite primers requires complementary annealing sites up to a few kilobases apart. Polymorphic amplicons result from differences in the sequence or position of minisatellites in the genome, or the insertion or deletion of DNA between minisatellite loci. As with the analyses using minisatellelites as probes, differences in amplification profiles are used to assess the relative similarities or differences between the organisms under study.

PCR-based minisatellite fingerprinting is a relatively straightforward and useful technique, however it has some limitations. The production of an informative profile of bands depends on the number and location of minisatellite sequences in the organism under study, and this cannot be established without experiment. There is also some evidence that minisatellites are clustered on certain chromosomes or on specific regions of chromosomes (Amarger *et al.* 1998), and the assessment of strain relatedness may therefore be based on only a subset of the genome and may not be representative of the organism as a whole. Finally, and probably most importantly, it can be difficult to standardise fingerprints between laboratories and even within the same laboratory if different thermocyclers are used for the PCR amplifications. The quality and quantity of template DNA, and the method used to extract template DNA, can likewise affect the banding profile. These problems are probably due to variations between the consensus sequence of the minisatellite primer and the minisatellite sequences present in the genome. The latter are likely to have diverged over time, and the extent of divergence will affect whether or not they form suitable sites for primer annealing. Generally, the PCR employs a relatively low annealing temperature and an extended number of cycles to maximise the number of bands produced, but this can result in amplification from sites with low homology that may be difficult to reproduce if the stringency of primer annealing is raised slightly.

More information about the number, position, and sequence of minisatellites would assist in optimising fingerprinting studies. Additional repetitive DNA sequences that can be used for fingerprinting are also likely to be present in some genomes. Genome data can be used to identify and analyse minisatellite loci. In this section we use genome information to examine the occurrence of the M13 sequence in *Cryptococcus neoformans,* which has been extensively fingerprinted using this primer. We also examine a computer-based method for screening genomes for repetitive sequences. Some of these may be developed for use as variable molecular markers.

3.2 Testing the Presence of a Marker: BLASTN Searches for the M13 Sequence in *Cryptococcus neoformans*

The sequence 5'-GAGGGTGGXGGXTCT-3' was used as a query in a series of BLASTN searches against the 156 contigs covering more than 18Mb of *C. neoformans* sequence. This sequence was available through the *C. neoformans* Genome Project, Stanford Genome Technology Center (http://www-sequence.stanford.edu/group/C.neoformans/index.html), and was accessed on 24 December 2002. The third assembly (011005) of nuclear DNA from strains JEC21 & B-3501a, assembled 5 October 2001 and providing 13.3X coverage of the genome was used, which was the most recent assembly on the date of accession. A BLASTN search available through the *C. neoformans* Genome Project web page automatically screens the contigs in both directions, therefore searching with the reverse complement of the sequence was unnecessary. "X" in the M13 sequence indicates the presence of either C or T (Vassart *et al.* 1987), thus four different queries were submitted. The mitochondrial DNA was also subjected to the same search but no matches were found.

Table 5 summarises the results of the searches. A total of 200 sequences were found with a high level of similarity (at least 11 out of 15 bases matching) to the M13 sequence. In general, the internal bases of the sequence aligned with the query, and mismatches occurred at the bases on the 3' and 5' ends of the sequence. Aligning sequences were then assessed for their suitability as priming sites. The 3' bases of a primer are critical for extension by *Taq* polymerase, however some mismatches can be tolerated if the 3' base of the primer is a T (Huang *et al.* 1992), which is the case for the M13 sequence. Aligning sequences were therefore considered to be potentially good candidates for primer binding sites if they were at least 11 bases long and included the ultimate or penultimate 3' bases of the M13 sequence.

Table 5. Sequences aligning with M13 minisatellite in *C. neoformans* contigs.

M13 query sequence	Number sequences aligning	Number suitable primer binding sites
GAGGGTGGCGGCTCT	46	12
GAGGGTGGTGGCTCT	81	14
GAGGGTGGCGGTTCT	63	17
GAGGGTGGTGGTTCT	100	20
Total	200	63

Two hundred sequences with significant identity to the M13 consensus sequence were found in the contigs comprising 18,058,271bp of *C. neoformans* sequence. Of these, 63 appeared to be suitable as priming sites for the M13 primer. While the frequency of M13-like sequences is considerably greater than would be expected by chance for a 15-base oligonucleotide (1 in 4^{15}, which would be expected <0.002 times in 18Mb), the density of the suitable primer sites, at 1 per 286 kb on average, is too low to predict amplification of any bands when M13 is used as a single primer. A closer analysis of the location of the suitable sites found them to sometimes cluster on single contigs; for example contig cneo011005.C81 contained four separate binding sites, and contig cneo011005.C666 contained three. However, these occurred a minimum of 11.2 kb from one another, indicating amplicons would not be produced in a fingerprinting PCR. Interestingly, no evidence was found of any tandem repetition of the M13 sequence in the *C. neoformans* contigs examined, however a thorough analysis of this was not performed. Minisatellites can be difficult to sequence and it is possible that as further sequencing and annotation are performed more M13 sequences will become apparent. It is also possible that M13 does not occur in tandem sequences in *C. neoformans*.

It appears from this analysis that the M13 primer amplified from sites in the *C. neoformans* genome that are only partly homologous. These results do not invalidate previous work using M13 as a primer in fingerprinting studies, which has been well supported using other methods, but they may help explain the difficulties encountered when trying to standardise M13-based fingerprinting among laboratories.

3.3 Finding Additional Minisatellites Using Tandem Repeats Finder

Tandem Repeats Finder (TRF) Version 3.01 (http://tandem.biomath.mssm.edu/trf/trf.html) is a web-based computer program designed to systematically search DNA sequences for two or more tandem repetitions of any sequence (Benson, 1999). Any sequence file in FASTA format can be examined, and any type of tandem repeat, from microsatellites to large satellite sequences (up to 2 kb), should be detected. For every sequence analysed two output files are generated, one with a summary table of the types and locations of the repeats that have been found, and linked to this an alignment file showing the alignment of the copies against the tandem repeat consensus sequences.

We used TRF to analyse the 821 contigs of *Neurospora crassa* Release 3 sequence (release date 2/12/02, http://www-genome.wi.mit.edu/annotation/fungi/neurospora/). Each contig is at least 2 kb, with the total release containing 38,044,343 bp of DNA sequence. Of these, 819 contained at least one tandemly repeated sequence. An example of the summary file for one contig (Contig 3.1, scaffold 1) is shown in Figure 2a. The alignment file linked to index 23274-23328 is shown in Figure 2b. Note that none of the tandem repeats found are long enough to fall under the classical description of a minisatellite (minimum size of 0.5 kb). In fact, a scan of the first 20 contigs found only three minisatellites that would fit this description; interestingly these all occurred together on one contig. This is consistent with the finding that the genomes of higher fungi typically contain relatively little repetitive DNA (Wöstemeyer and Kreibich 2002).

Tandem Repeats Finder Program written by:

Gary Benson
Department of Biomathematical Sciences
Mount Sinai School of Medicine
Version 3.01

Please cite:
G. Benson,
"Tandem repeats finder: a program to analyze DNA sequences"
Nucleic Acid Research(1999)
Vol. 27, No. 2, pp. 573-580.

Sequence: Neurospora crassa contig 3.2 (scaffold 1)
Parameters: 2 7 7 80 10 50 500
Length: 60090

Tables: 1

This is table 1 of 1 (10 repeats found)

Click on indices to view alignment

See Table Explanation in Tandem Repeats Finder Help

Indices	Period Size	Copy Number	Consensus Size	Percent Matches	Percent Indels	Score	A	C	G	T	Entropy (0-2)
2076--2107	12	2.7	12	100	0	64	6	6	62	25	1.42
10818--10842	4	6.3	4	100	0	50	48	24	0	28	1.52
23274--23328	16	3.4	16	94	0	101	69	9	21	0	1.16
23922--23963	1	42.0	1	100	0	84	0	0	0	100	0.00
24526--24573	1	48.0	1	100	0	96	100	0	0	0	0.00
24657--24683	14	1.9	14	100	0	54	66	22	11	0	1.22
38969--38996	1	28.0	1	100	0	56	0	0	0	100	0.00
48645--48724	10	8.0	10	100	0	160	40	40	0	20	1.52
54217--54247	1	31.0	1	93	0	53	96	3	0	0	0.21
56112--56141	14	2.2	13	94	5	51	33	26	40	0	1.57

Tables: 1

The End!

(a)

```
Found at i:23297 original size:16 final size:16

See Alignment Explanation in Tandem Repeats Finder Help

     Indices: 23274--23328  Score: 101
     Period size: 16  Copynumber: 3.4  Consensus size: 16

        23264 TGCTCGGCAG

        23274 AGAAACAGAAAGAAGA
            1 AGAAACAGAAAGAAGA

               *
        23290 ACAAACAGAAAGAAGA
            1 AGAAACAGAAAGAAGA

        23306 AGAAACAGAAAGAAGA
            1 AGAAACAGAAAGAAGA

        23322 AGAAACA
            1 AGAAACA

        23329 AAAAAAGAGG

Statistics
Matches: 37,   Mismatches: 2, Indels: 0
        0.95              0.05          0.00

Matches are distributed among these distances:
  16    37   1.00

ACGTcount: A:0.69, C:0.09, G:0.22, T:0.00

Consensus pattern (16 bp):
AGAAACAGAAAGAAGA
```

(b)

Fig. 2: Output files from Tandem Repeats Finder showing results for *N. crassa* genome. (a) Summary table for Contig 3.2, scaffold 1. The indices on the left denote the positions in the sequence at which the tandem repeats occur. These hyperlink to the relative alignments files, an example of which is shown in (b), which presents the alignment of a 16 bp sequence found 3.4 times in Contig 3.2 at index 23274-23328. The consensus pattern is written at the bottom of the alignment.

TRF has the option of including 500bp of flanking sequence for each repeat found, which can be used to design PCR primers to amplify potentially useful micro- and minisatellites and analyse these for variation in repeat number. This is a potentially very useful program for anyone wishing to develop these kinds of variable repeat markers from genomic sequence. In addition, the type and relative location of tandem repeats can be used to examine gross chromosomal translocations and rearrangements. Benson (1999) used TRF to analyse *S. cerevisiae* chromosomes I and VIII and found a cluster of tandem repeats elements of similar period size at the ends of both chromosomes. The sequences around these clusters also revealed close homology. Recent swapping of the chromosome ends was inferred from this result.

4. INTERSPERSED REPETITIVE ELEMENTS

Like minisatellites, interspersed repetitive elements occur throughout the genome of many eukaryotes, but are distinguished by a generally longer unit size (with a minimum of just under 100 bp; Yadon and Catcheside, 1995), and the fact that they usually do not occur in tandemly repeated series but as dispersed individual units. Most interspersed repetitive sequences appear to be derived from transposable elements and for this reason most of this section will be concentrating on fungal transposons. These can accumulate in eukaryote genomes to an extraordinary extent; for example in the human genome, which contains only 5% coding sequences, 45% of the DNA is derived from transposable elements (Buzdin *et al.* 2002). This figure rises to as much as 70% in the genomes of some plant species (San Miguel and Bennetzen, 1998). It is likely to be lower in the fungi, in which repetitive DNA is less common (Wöstemeyer and Kreibich, 2002).

Eukaryote transposons are classified into two major groups: Class 1 elements, also called retrotransposons, which transpose via RNA intermediates (Boeke and Stoye, 1997), and Class 2 elements that transpose from DNA to DNA. Class 1 retrotransposons have most commonly been detected in the fungi to date and can be divided further into three types. Two of these, the LTR retrotransposons and non-LTR retrotransposons or LINE-like retrotransposons, encode their own reverse transcriptase and are typically 5-10 kb long. LTR retrotransposons possess long terminal repeats (LTRs) at each end, whereas LINE-like retrotransposons frequently terminate in poly-A tails. SINE-like elements comprise the third type of Class 1 element. These do not possess a reverse transcriptase gene but rely on complementation by intact retrotransposons for their transposition (Kempken and Kück, 1998).

The activity of transposable elements can profoundly affect genome structure and evolution (Wöstemeyer and Kreibich, 2002). Insertion of transposons into and near genes can affect gene structure, expression, and transcriptional regulation. Recombination between transposons at different sites can cause large deletions and chromosomal rearrangements. Perhaps to avoid excessive harm to the host, which could result in their own extinction, transposons appear often to have evolved damage-control mechanisms, such as limiting their integration to specific regions of the genome. The host genome also strictly regulates expression of transposable elements, but these can be activated by various stresses (Grandbastien, 1998) such as heat shock (Ratner *et al.* 1992), UV irradiation (Rolfe *et al.* 1986) or microbial infection (Grandbastien *et al.* 1997).

4.1 Use of Transposons as Molecular Markers

The abundant and repetitive nature of many transposons makes them excellent molecular markers for a range of studies. As they are often multicopy, they can be used in the same way as microsatellites and minisatellites to type strains and provide epidemiological data. They are particularly useful as hybridisation probes as their relatively long, conserved sequences allows hybridisation at high stringency, preventing hybridisation artefacts. When used to probe fungal chromosomes separated by Pulsed Field Gel Electrophoresis, hybridisation with transposons can follow large scale translocations, deletions and rearrangements. Their use in these studies may be particularly revealing as they are thought to often contribute to these processes (Daviere *et al.* 2001). Transposons may also be isolated and used in transposon-tagging experiments, in which they are used to disrupt genes and at the same time provide selectable markers for gene isolation (Kempken and Kück, 1998). In addition, transposons can act as single locus genetic markers and as such have some advantages over traditional markers such as SNPs or RFLPs. Transposons are stable, rarely undergoing complete deletions, and as their integration is a random event it is unlikely that two different transposons will integrate at exactly the same site, thus their presence at the same locus indicates identity by descent. Moreover, the ancestral state of a transposon is known, which is the absence of the transposon. These make them potentially very useful in population genetic, phylogenetic and evolutionary studies of closely related fungal species and strains and as loci for genetic mapping (Buzdin *et al.* 2002).

4.2 Traditional Methods of Identifying and Isolating Transposons

In the absence of sequence data, laboratory-based methods have been used to identify fungal transposons. Heterologous probes or primers can be designed, based on conserved motifs from within transposon sequences (Flavel *et al.* 1992; Kachroo *et al.* 1994). Probes are used to screen libraries for transposon-containing clones, and primers are used to amplify transposon sequences from clones or from genomic DNA. Both these approaches have the limitation of only being able to detect transposons of known type. Transposon trapping identifies transposons by their insertion into a gene, eg the nitrate reductase (*niaD*) gene, leading to a change of phenotype (Daboussi *et al.* 1992). The transposon can then easily be identified by PCR amplification of mutants using primers designed from the disrupted gene sequence. Clearly this method is most suitable for active transposons, and cannot detect those in which transposition activity has degenerated. Differential hybridisation is the final method, and is used to identify highly repeated sequences, with all repetitive non-rDNA sequences subjected to further analysis (Kempken *et al.* 1995). The success of this method will depend on how much repetitive DNA that does not encode transposons is present.

4.3 Identifying Transposons from Genome Databases

Databases can be screened for transposon sequences by using queries based on conserved transposon motifs. This approach was used by Goodwin and Poulter (2001) to identify transposons in the genome of *Cryptococcus neoformans*. Retrotransposon elements were first detected by conducting a series of TBLASTN searches (protein query vs. DNA database) using the protein sequence of a wide variety of retrotransposons from other species as queries. Identified retrotransposon sequences were then used as queries in subsequent BLASN and TBLASTN searches to identify additional elements. Finally, some elements were identified from within or adjacent to other retrotransposons. The resulting retrotransposons were grouped into families based on the level of sequence identity to one another (95-100% identity in overlapping regions), and a representative for each family was then sought by building a contig of overlapping sequences. At the time of analysis, only a 2X coverage of the genome was available through the Stanford Genome Technology Center (http://www-sequence.stanford.edu /group/C.neoformands/index.html), thus it is likely that some transposons were missed, nonetheless this search revealed 15 distinct families of LTR-retrotransposons and several families of non-LTR retrotransposons. An interesting aspect of this study was a comparison of the abundance of retrotransposons among the genomes of *C. neoformans, S. cerevisiae, Sz. pombe* and *Candida albicans*, which contained 15, 4, 1 and 34 LTR families respectively. It is possible that the number of transposons tolerated in a genome is inversely related to genome size, as the genomes of *S. cerevisiae* and *Sz. pombe* are both relatively compact (12 and 14 Mb respectively) and the genome of *S. cerevisiae* has been reported to be particularly streamlined, with small intragenic regions and rare introns (Dujon, 1996). However, *C. albicans* is similarly small (16 Mb) yet contains more than twice the number of LTR families as *C. neoformans* with a genome size of 23 Mb. The analysis of additional fungal genomes, as they become available, will shed more light on the process of retrotransposon accumulation and maintenance and may allow the type and number of transposons likely to be present in unsequenced genomes to be predicted.

4.4 Finding Uncharacterised Interspersed Repetitive Elements

BLAST searches, as outlined above, suffer the same limitation as the traditional, laboratory-based screening approaches using probes and primers to conserved transposon motifs in that they can only identify sequences that are similar to those already known. We are currently developing software to search for interspersed repeats of any sequence (Vera and Carter, unpublished). The first, preliminary strategy takes as its query the first 20 bases of the sequence to be analysed and looks for perfect repeats of this oligonucleotide in the remaining sequence. It then moves along one base and looks at the next 20 base motif. This very simplistic and computation-intensive strategy needs to be significantly improved to make it workable and useful for large chromosomes. To date we have analysed chromosomes 1, 3, 6, 8, and 9 and the mitochondrial genome of *Saccharomyces cerevisiae*. Preliminary analysis of the output suggests that the mitochondrial genome is rich in repetitive sequences, whereas these are relatively uncommon in the chromosomes. These findings are again consistent with the compact nature of the *S. cerevisiae* genome (Dujon, 1996), and its relatively large, repetitive mitochondrial genome which is known to contain many microsatellites (Sia *et al.* 2000).

5. SINGLE NUCLEOTIDE POLYMORPHISMS (SNPs)

When using genomic date to find microsatellites, minisatellites and transposons, as outlined above, one is searching for the motif itself, thus a sequence from only one individual is necessary in order to identify the locus. This is then followed by a screen for polymorphisms, which no longer depends directly on gene sequence. However, with rapidly growing genome sequence databases, the discovery of other classes of markers is being facilitated by the ability to directly compare genetic sequences from more than one individual. Examples include indels, SNPs, and RFLPs. These terms are not mutually exclusive: RFLPs are often the result of SNPs (though not all SNPs result in RFLPs), but can also be the product of indels at either the restriction site itself or between two restriction sites. Because of their overwhelming abundance, and because of their tractability to automation (important for complex genetic studies requiring many thousands of genotypes), SNPs have become the focus of the greatest amount of attention in the past few years.

5.1 Structure and Utility of SNPs

A SNP is most easily defined as a single base position at which at least two different bases can be found (Figure 3). SNPs are probably the most common type of mutation found in genomic DNA (Barnes, 2002). Indeed, among the wealth of information contributed by the human genome sequencing project has been the finding that SNPs occur at least every 200-300bp. Currently, human SNP databases contain over 2 million confirmed SNPs of immediate utility to various fields of human genetics (Barnes 2002), which may represent only 20-30% of the total SNPs in the human genome. Most SNPs are reasonably assumed to be neutral, and because of their abundance there is a high probability of finding close associations between markers and heritable phenotypes. However, other approaches allow for the identification of *intragenic* SNPs using BLAST searches of public databases (Aerts *et al.* 2002), with far-reaching implications for studies such as disease susceptibility and drug efficacy. The relationship between SNP variation and phenotypic expression is being investigated by a number of researchers (Lander and Schork, 1994; Martin *et al.* 1997; LeVan *et al.* 2001).

The significance and utility of a particular SNP will depend on the context in which it is being studied. For example, at the level of populations, only base variants that occur at frequencies greater than 1% are considered significant as polymorphisms (Altshuler *et al.* 2000). However, population frequencies of markers are not important in studies such as linkage mapping or Quantitative Trait Locus (QTL) analysis, where the only criteria for a marker are that it is polymorphic between the parents of the mapping population, and that the alleles segregate as Mendelian factors among the progeny.

G – A – A – C – C – T – [G] – A – T – T – C –
G – A – A – C – C – T – [A] – A – T – T – C –

Fig. 3. Alignment of two DNA sequences showing SNP (boxed).

SNP discovery requires a comparison of homologous sequence traces from two or more individuals. Therefore, the potential of SNPs as markers has only been realised since DNA sequencing became economically feasible, and DNA sequences could be accessioned into publicly available databases. Data on human genetic variation have been accumulating for the past seven or so years, with the vast majority of human SNP data acquired over the last two years. There are currently several SNP and mutation databases on the Web; in 1998, the National Center for Biotechnology Information, run by the National Institutes of Health, established a central repository for SNPs and short insertion/deletion data, *dbSNP* (http://www.ncbi.nlm.nih.gov/SNP/), which contains information for SNPs from the human genome as well as SNPs from fourteen other species.

Given their abundance, as well as advances in technologies available for high-throughput genotyping, SNPs are becoming increasingly recognised as powerful tools in genetics research. The interest in creating a high-density SNP map of the human genome is being undertaken by a consortium of both private and public concerns (Marth, *et al.* 2001; and NCBI website mentioned above), and high-density SNP maps have been created for several genome regions (Shastry, 2002). For small-scale studies, SNP scoring and genotyping can be done using more traditional methods such as electrophoresis, direct sequencing, or mass spectrometry. However, for large-scale studies – linkage mapping, QTL analysis, genetic association analysis, clinical and pharmaceutical trials – scaling up these methods is prohibitively expensive if one is to maintain significant statistical relevance. In response, several innovative technologies have become available (Jenkins and Gibson, 2002). Many depend on Fluorescence Resonance Energy Transfer (FRET) technologies to discriminate one allele from the other. Additionally, real-time PCR may be involved, as well as techniques that take advantage of melting temperature differences created by the polymorphism. All of these methods exploit automation and either plate or microarray formats in order to increase speed and minimise experimental error. Because of the projected increase in demand for large-scale SNP genotyping, the number of methodologies will continue to grow, with perhaps a concomitant drop in effort and cost.

5.2 Use of Genome Data to Find SNPs, RFLPs and Indels in *Cryptococcus neoformans*

The *Cryptococcus* genome database comprises the sequences of two *C. neoformans* serotype D strains, B3501 and JEC21 (Heitman *et al.* 1999), and is the first fungal genome project in which interstrain genome comparisons can be made. Strain JEC21 is nearly isogenic with strain B3502, which is a sibling of B3501, both being the progeny of a cross between two highly divergent strains, NIH433 and NIH12 (Heitman *et al.* 1999). Although genetic variation between B3501 and JEC21 is half that between NIH433 and NIH12, nonetheless comparisons of homologous sequences from the two genomes have yielded a substantial number of candidate polymorphisms.

Marra and colleagues are currently creating a linkage map of *C. neoformans*, using progeny from a cross between strains B3501 and B3502. In addition to being a prerequisite to identifying the quantitative genetics underlying complex traits, the linkage map will be an aid in contig assembly, and will provide markers that will serve as anchors for the identification of centromeres. The genome sequence database was used to find polymorphic RFLPs, SNPs, and indels, to be used as genetic markers for the map and to supplement earlier mapping efforts using microsatellite markers. The goal was to find at least one marker per contig, but to have as many as three or four for the very largest contigs. Multiple markers are required for larger contigs because of the high recombination rates (exceeding 30% and as high as 48%) found between distal markers on even intermediate-length (100-kb) contigs (Marra, unpublished). Under most mapping scenarios – mapping progeny populations between 50 and 100 individuals – recombination rates greater than 30% are not significantly different from 50%, which is to say that one cannot reject a null hypothesis that the two markers are unlinked.

To identify RFLPs, Dr. Eula Fung at the Stanford Genome Technology Center developed a perl program that compares the two genome databases for restriction sites that are present in one database and absent in the other. In order to minimise false positives, only sequences that had strong support based on multiple traces were used. The initial search was for cleavage sites for six fairly common (and inexpensive) six-cutter restriction-enzymes: *Bam*H1, *Eco*RI, *Hin*dIII, *Pst*I, *Xba*I, and *Xho*I. Because a significant number of contigs remained without informative markers after this search, the search was expanded to include nine additional six-cutter sites as well as nine four-cutter sites, resulting in only about 20 contigs lacking markers. For each candidate RFLP, PCR primers were designed to amplify 500-1000bp surrounding the restriction site. Amplicons obtained from DNA of the mapping parents (B3501 and B3502) were then subjected to restriction digestion, followed by electrophoresis on agarose. Approximately 200 candidate RFLPs have been chosen to date, and more than 95% of these have been validated. Subsequent to validation, each RFLP is assayed on the mapping population, and segregation data are then used to establish linkage relationships among loci.

A comparative genomics approach has also been used to identify SNPs and indels segregating between the sequences of B3501 and B3502 (Marra, unpublished). As expected, SNPs are by far the most abundant marker; at least one SNP can be found in close proximity to most of the mapped microsatellites and RFLPs. However, although none has been validated yet, the feasibility of several technologies is currently being evaluated. Once acquired, these technologies will greatly speed the process of genotyping. Therefore, a switch from microsatellites and RFLPs to SNPs is anticipated, where possible. Most of the indels that were identified are characterised by insertions and deletions of less than 10 bp, and would therefore require resolution on acrylamide rather than agarose, making them less desirable markers than RFLPs, which in any case are more abundant.

6. THE FUTURE OF MARKER DEVELOPMENT

As the field of genomics grows and develops, new and innovative ways of using this emerging body of data will emerge. Unfortunately, many of these will require expensive equipment and reagents and may remain outside the application to fungi, which cannot be expected to attract the funding provided to human genomics. However, some technologies, such as microarrays, are decreasing in cost and hence are increasing in their application to some lower profile organisms. Here we briefly examine some of the emerging opportunities that may allow further development of molecular markers from fungal genomes.

6.1 Genome Initiatives on Additional Fungal Species

As more fungal species are sequenced is will become possible to further cross-species analyses of DNA sequence motifs. From these it is expected that "universal" fungal sequences will emerge, which can be used, like the rRNA gene unit, in phylogenetic, population biology, epidemiology and ecology studies of a range of different fungal species. Such sequences may be highly conserved genes or conserved repetitive DNA elements. This class of markers will be of particular value as they can be applied to fungi for which no sequence data are available. They may also be applicable to uncultivable fungi and those present in mixed culture. A growing need exists for such markers to augment and challenge the now considerable body of information based exclusively on rDNA studies.

6.2 Genome Initiatives on Additional Strains of Some Fungal Species

The provision of genomic data from more than one strain within a species, as outlined above, will allow cross-genome comparisons that will rapidly identify polymorphic regions such as SNPs, microsatellites and transposons. Although microsatellites and transposons can be located in a single genome, this can introduce ascertainment bias into an analysis, whereby the sequenced genome from which these markers were developed appears to be more polymorphic and/or ancestral to other genomes under comparison (Carter *et al.* 2001; Buzdin *et al.* 2002). Cross-strain analysis can also be expected to reveal gene and possibly whole chromosomal regions that are highly conserved as well as those that are more likely to be variable, allowing the researcher to choose regions to suit a particular application.

6.3 Microarrays

As well as being able to buy microarrays for target organisms such as *S. cerevisiae*, it is now possible to have these synthesised for any organism at reasonable cost (Cheung *et al.* 1999). Microarrays can include any DNA sequence that is of interest to the researcher; of particular use in marker application will be microarrays based on oligonucleotides that can detect the presence of particular alleles at SNP loci (Hoheisel and Vingron, 1999.). Used in this way it will be possible to screen many different loci simultaneously and to process many different samples in a cost-effective manner. This will be of particular use in large-scale epidemiological and population genetic analyses. Other markers, for example those targeting mating type, pathogenicity loci or other physiological or biological traits, may also be included on a microarray to allow an overall analysis of genetic and biological properties of the strains under study.

7. CONCLUSION

Screening genome sequencing data can provide researchers with a relatively quick and cost-effective method of isolating and characterising highly polymorphic genetic markers. The number of fungal genomic projects is small when compared with other organisms. Of the 379 genome project websites currently available, only sixteen (~4%) include fungal projects, and these cover only twelve fungal species. The continuing interest in fungal genetics, their importance as crop, animal and human pathogens, their relatively small size compared with other eukaryote genomes, and the gradually diminishing cost of genome sequencing mean this number can be expected to grow steadily in the near future. Molecular markers are indispensable for characterising and identifying fungal species and strains, thus their development and application will be of use to all researchers concerned with fungal biology and genetics.

Acknowledgements: We acknowledge the following genome projects, which provided data for the analyses included in this paper: the *C. neoformans* Genome Project, Stanford Genome Technology Center, funded by the NIAID/NIH under cooperative agreement U01 AI47087, and The Institute for Genomic Research, funded by the NIAID/NIH under cooperative agreement U01 AI48594; the *Neurospora* Sequencing Project. Whitehead Institute/MIT Center for Genome Research; The *Sz. pombe* genome project, Welcome Trust Sanger Institute; the *Saccharomyces* Genome Database project, Department of Genetics at the School of Medicine, Stanford University, funded by the National Human Genome Research Institute at the US National Institutes of Health. R.E. Marra was supported by USPHS NIH grants AI 25783 and AI 44975.

REFERENCES

Aerts J, Wetzels Y, Cohen N and Aerssens J (2002). Data mining of public SNP databases for the selection of intragenic SNPs. Hum Mut 20:162-173.

Akkaya MS, Shoemaker RC, Specht JE, Bhagwat AA and Cregan PB (1995). Integration of simple sequence repeat DNA markers into a soybean linkage map. Crop Sci 35:1439-1445.

Altschul SF, Gish W, Miller W, Myers EW and Lipman DJ (1990). Basic local alignment search tool. J Mol Biol 215:403-410.

Altschul SF, Madden TL, Schäffer AA, Zhang Z, Zhang Z, Miller W and Lipman DJ (1997). Gapped BLAST and PSI-BLAST: a new generation of protein database search programs. Nucl Acids Res 25:3389-3402.

Altshuler D, Pollara VJ, Cowles CR, Etten WJV, Baldwin J, Linton L and Lander ES (2000). An SNP map of the human genome generated by reduced representation shotgun sequencing. Nature 407:513-516.

Amarger V, Gauguier D, Yerle M, Apiou F, Pinton P, Giraudeau F, Monfouilloux S, Lanthrop M, Dutrillaux B and Buard J (1998). Analysis of the human, pig, and rat genomes supports a universal telomeric origin of minisatellite sequences. Genomics 52:62-71.

Ashley MV and Dow BD (1994). The use of microsatellite analysis in population biology: background, methods and potential applications, In: B Schierwater, B Streit, GP Wagner and R DeSalle, eds. Molecular Ecology and Evolution Approaches and Applications. Basel: Birkhäuser Verlag, pp. 185-201.

Barnes MR (2002). SNP and mutation data on the Web -- hidden treasures for uncovering. Comp Funct Genom 3:67-74.

Bart-Delabesse E, Humbert JF, Delabesse E and Bretagne S (1998). Microsatellite markers for typing *Aspergillus fumigatus* isolates. J Clin Microbiol 36:2413-2418.

Benson G (1999). Tandem repeats finder: a program to analyze DNA sequences. Nucl Acids Res 27:573-580.

Boeke JD and Stoye JP (1997). Retrotransposons, endogenous retroviruses, and the evolution of retroelements, In: JM Coffin, SH Hughes and HE Varmus, eds. Retroviruses. New York: Cold Spring Harbor Laboratory Press, pp. 343-435.

Bowcock A, Ruiz-Linares A, Tomfohrde J, Minch E, Kidd J and Cavalli-Sforza LL (1994). High resolution of human evolutionary trees with polymorphic microsatellites. Nature 368:455-457.

Bretagne S, Costa J-M, Besmond C, Carsique R and Calderone R (1997). Microsatellite polymorphism in the promoter sequence of the elongation factor 3 gene of *Candida albicans* as the basis for a typing system. J Clin Microbiol 35:1777-1780.

Broun P and Tanksley SD (1996). Characterisation and genetic mapping of simple sequences in the tomato genome. Mol Genet 250:29-49.

Brown Gladden JG, Ferguson MM, Friesen MK and Clayton JW (1999). Population structure of North American beluga whales (*Delphinapterus leucas*) based on nuclear DNA microsatellite variation and contrasted with the population structure revealed by mitochondrial DNA variation. Mol Ecol 8:347-363.

Bruford M and Wayne R (1993). Microsatellites and their application to population genetic studies. Curr Op Genet Devel 3:939-943.

Bucheli E, Gautschi B and Shykoff J (2000). Host-specific differentiation in the anther smut fungus *Microbotryum violaceum* as revealed by microsatellites. J Evol Biol 13:188-198.

Buzdin A, Khodosevich K, Mamedov I, Vinogradova T, Lebedev Y, Hunsmann G and Sverdlov E (2002). A technique for genome-wide identification of differences in the interspersed repeats integrations between closely related genomes and its application to detection of human-specific integrations of HERV-K LTRs. Genomics 79:413-422.

Carter DA, Burt A and Taylor JW (1995). Direct analysis of specific bands from arbitrarily primed PCR reactions, In: MA Innis, DH Gelfand and JJ Sninsky, eds. PCR Strategies. San Diego: Academic Press, pp. 325-332.

Carter DA, Burt A, Taylor JW, Koenig GL, Dechairo BM and White TJ (1997). A set of electrophoretic molecular markers for strain typing and population genetic studies of *Histoplasma capsulatum*. Electrophoresis 18:1047-1953.

Carter DA, Reynolds R, Fildes N and White TJ (1996). Future applications of PCR to conservation biology, In: TB Smith and RK Wayne, eds. Molecular Genetic Approaches in Conservation. Oxford: Oxford University Press, pp. 314-326.

Carter DA, Taylor JW, Dechairo B, Burt A, Koenig GL and White TJ (2001). Amplified single-nucleotide polymorphisms and a $(GA)_n$ microsatellite marker reveal genetic differentiation between populations of *Histoplasma capsulatum* from the Americas. Fungal Genet Biol 34:37-48.

Chambers G and MacAvoy E (2000). Microsatellites: consensus and controversy. Compar Biochem Physiol B 126:455-476.

Charlesworth B, Sniegowski P and Stephan W (1994). The evolutionary dynamics of repetitive DNA in eukaryotes. Nature 371:215-220.

Cheung VG, Morley M, Aguilar F, Massimi A, Kucherlapati R and Childs G (1999). Making and reading microarrays. Nat Genet 21:15-19.

Condit R and Hubbel S (1991). Abundance and DNA sequence of two-base repeat regions in tropical tree genomes. Genome 34:87-96.

Daboussi M-J, Langin T and Brygoo Y (1992). *Fot1*, a new family of fungal transposable elements. Mol Gen Genet 232:12-16.

Dallas J (1992). Estimation of microsatellite mutation rates in recombinant inbred strains of mouse. Mammal Genom 3:452-456.

Daviere J-M, Langin T and Daboussi M-J (2001). Potential role of transposable elements in the rapid reorganization of the *Fusarium oxysporum* genome. Fungal Genet Biol 34:177-192.

Davies N, Villablanca F and Roderick G (1999). Determining the source of individuals: multilocus genotyping in nonequilibrium populations genetics. Trends Ecol Evol 14:17-21.

Dawson R, Gibbs H, Hobson K and Yezerinac S (1997). Isolation of microsatellite DNA markers from a passerine bird, *Dendroica petechia* (the yellow warbler), and their use in population studies. Heredity 79:506-514.

Debrauwere H, Gendrel CG, Lechat S and Dutreix M (1997). Differences and similarities between various tandem repeat sequences: minisatellites and microsatellites. Biochimie 79:577-586.

Deka R, Jin L, Shriver M, Yu L, Decroo S, Hundrieser J, Bunker C, Ferrell R and Chakraborty R (1995). Population genetics of dinucleotide (dC-dA)n.(dG-dT)n polymorphisms in world populations. Am J Hum Genet 56:461-474.

Dib C, Faure S, Fizames C, Samson D, Drouot N, Vignal A, Millasseau P, Marc S, Hazan J, Seboun E, Lathrop M, Gyapay G, Morissette J and Weissenbach J (1996). A comprehensive genetic map of the human genome based on 5264 microsatellites. Nature 380:152-154.

Dujon B (1996). The yeast genome project: what did we learn? Trends Genet 12:263-270.

Edwards A, Civitello A, Hammond H and Caskey C (1991). DNA typing and genetic-mapping with trimeric and tetrameric tandem repeats. Am J Hum Genet 49:746-756.

Edwards A, Hammond H, Jin L, Caskey C and Chakraborty R (1992). Genetic variation at five trimeric and tetrameric tandem repeat loci in four human population groups. Genomics 12:241-253.

Eisen J (2000). Mechanistic basis for microsatellite instability, In: D Goldstein and C Schlötterer, eds. Microsatellites: evolution and Applications. Oxford: Oxford University Press, pp. 34-48.

Ender A, Schwenk K, Stadler T, Streit B and Schierwater B (1996). RAPD identification of microsatellites in *Daphnia.* Mol Ecol 5:437-441.

Enjalbert J, Duan X, Giraud T, Vautrin D, De Vallavieille-Pope C and Solignac M (2002). Isolation of twelve microsatellite loci, using an enrichment protocol, in the phytopathogenic fungus *Puccinia striiformis* f. sp. *tritici.* Mol Ecol Notes 2:563-565.

Estoup A, Solignac M, Harry M and Cornuet J-M (1993). Characterization of (GT)n and (CT)n microsatellites in two insect species: *Apis mellifera* and *Bombus terrestris*. Nucl Acids Res 21:1427-1431.

Feldheim KA, Gruber SH and Ashley MV (2001). Population genetic structure of the lemon shark (*Negaprion brevirostris*) in the western Atlantic: DNA microsatellite variation. Mol Ecol 10:295-303.

Field D, Eggert L, Metzgar D, Rose R and Wills C (1996). Use of polymorphic short and clustered coding-region microsatellites to distinguish strains of *Candida albicans*. FEMS Immunol Med Microbiol 15:73-79.

Field D and Wills C (1996). Long, polymorphic microsatellites in simple organisms. Proc Roy Soc Lond B 263:209-215.

Field D and Wills C (1998). Abundant microsatellite polymorphism in *Saccharomyces cerevisiae*, and the different distributions of microsatellites in eight prokaryotes and *S. cerevisiae*, result from strong mutation pressures and a variety of selective forces. Proc Natl Acad Sci USA 95:1647-1652.

Fisher MC, Koenig G, White TJ and Taylor JW (2000). A test for concordance between the multilocus genealogies of genes and microsatellites in the pathogenic fungus *Coccidioides immitis*. Mol Biol Evol 17:1164-1174.

Fisher MC, White TJ and Taylor JW (1999). Primers for genotyping single nucleotide polymorphisms and microsatellites in the pathogenic fungus *Coccidioides immitis*. Mol Ecol 8:1082-1084.

Flavel AJ, Dunbar E, Anderson R, Pearce SR, Hartley R and Kumar A (1992). *Ty1-copia* group of retrotransposons are ubiquitous and heterogenous in higher plants. Nucl Acids Res 20:3639-3644.

Forbes SH, Hogg JT, Buchanan FC, Crawford AM and Allendorf FW (1995). Microsatellite evolution in congeneric mammals: domestic and Bighorn sheep. Mol Biol Evol 12:1106-1113.

Geiser DM, Pitt JI and Taylor JW (1998). Cryptic speciation and recombination in the aflatoxin-producing fungus *Aspergillus flavus*. Proc Natl Acad Sci USA 95:388-393.

Georges M, Lequarre AS, Castelli M, Hanset R and Vassart G (1988). DNA fingerprinting in domestic animals using four different minisatellite probes. Cytogenet Cell Genet 47:127-131.

Girardin H (1997). Detection of filamentous fungi in foods. Sci Alimet 17:3-19.

Goffeau A, Berrell BG, Bussey H, Davis RW, Dujon B, Felmann H, Galibert F, Hoheisel JD, Jacq C, Johnston M, Louis EJ, Mewes HW, Murakami Y, Philippsen P, Tettelin H and Oliver SG (1996). Life with 6000 genes. Science 274:563-567.

Goldstein DB and Clark AG (1995). Microsatellite variation in North American populations of *Drosophila melanogaster*. Nucl Acids Res 23:3882-3886.

Goldstein DB, Linares AR, Cavalli-Sforza LL and Feldman MW (1995). An evaluation of genetic distances for use with microsatellite loci. Genetics 139:463-471.

Goodwin TJD and Poulter RTM (2001). The diversity of retrotransposons in the yeast *Cryptococcus neoformans*. Yeast 18:865-880.

Grandbastien M-A (1998). Activation of plant retrotransposons under stress conditions. Trends Pl Sci 3:181-187.

Grandbastien M-A, Lucas H, Morel J-B, Mhiri C, Vernhettes S and Casacuberta JM (1997). The expression of the tobacco *Tnt1* retrotransposon is linked to plant defense responses. Genetica 100:241-252.

Groppe K and Boller T (1997). PCR assay based on a microsatellite-containing locus for detection and quantification of *Epichloë* endophytes in grass tissue. Appl Environ Microbiol 63:1543-1550.

Groppe K, Sanders I, Wiemken A and Boller T (1995). A microsatellite marker for studying the ecology and diversity of fungal endophytes (*Epochloë* spp.) in grasses. Appl Environ Microbiol 61:3943-3949.

Hagelberg E, Gray IC and Jeffreys AJ (1991). Identification of the skeletal remains of a murder victim by DNA analysis. Nature 352:427-429.

Hamada H, Petrino MG and Kakunaga T (1982). A novel repeated element with Z-DNA-forming potential is widely found in evolutionarily diverse eukaryotic genomes. Proc Natl Acad Sci USA 79:6465-6469.

Hancock JM (1995). The contribution of DNA slippage to eukaryotic nuclear 18S-ribosomal-RNA evolution. J Mol Evol 40:629-639.

Hancock JM (1996). Simple sequences in a "minimal" genome. Nat Genet 14:14-15.

Heitman J, Allen B, Alspaugh JA and Kwon-Chung KJ (1999). On the origins of congenic MATalpha and MATa strains of the pathogenic yeast *Cryptococcus neoformans*. Fungal Genet Biol 28:1-5.

Hoheisel JD and Vingron M (1999) DNA chip technology. Biochemica 3:31-35.

Huang M-M, Arnheim N and Goodman MF (1992). Extenstion of base mispairs by *Taq* polymerase: implications for single nucleotide discrimination in PCR. Nucl Acids Res 20:4567-4573.

Huey B and Hall J (1989). Hypervariable DNA fingerprinting in *Escherichia coli* minisatellite probe from bacteriophage M13. J Bacteriol 171:2528-2532.

Hughes CR and Queller DC (1993). Detection of highly polymorphic microsatellite loci in a species with little allozyme polymorphism. Mol Ecol 2:131-137.

Japupciak JP and Wells RD (1999). Genetic instabilities in (CTG/CAG) repeats occur by recombination. J Biol Chem 274:23468-23479.

Jenkins S and Gibson N (2002). High-throughput SNP genotyping. Comp Funct Genom 3:57-66.

Kachroo P, Leong SA and Chattoo BB (1994). *Pot2*, an inverted repeat transposon from the rice blast fungus *Magnaporthe grisea*. Mol Gen Genet 245:339-348.

Kempken F and Kück U (1998). Transposons in filamentous fungi - facts and perspectives. BioEssays 20:652-659.

Kempken F, Schreiner C, Schörgendorfer K and Kück U (1995). A unique repeated DNA sequence in the cyclosporin-producing strain of *Tolypocladium inflatum* (ACTT 34921). Exp Mycol 19:305-313.

Kennedy GC, German MS and Rutter RJ (1995). The minisatellite in the diabetes susceptibility locus IDDM2 regulates insulin transcription. Nat Genet 9:293-298.

Lander ES and Schork NJ (1994). Genetic dissection of complex traits. Science 265:2037-2048.

Landry BS and Michelmore RW (1985). Selection of probes for restriction fragment analysis from plant genomic clones. Pl Mol Biol Rep 3:174-179.

LeVan TD, Bloom JW, Bailey TJ, Karp CL, Halonen M, Martinez FD and Vercelli D (2001). A common single nucleotide polymorphism in the CD14 promoter decreases the affinity of Sp protein binding and enhances transcriptional activity. J Immunol 167:5838-5844.

Levinson G and Gutman GA (1987). Slipped-strand mispairing: a major mechanism for DNA sequence evolution. Mol Biol Evol 4:203-221.

Lim S (2002). Microevolution in *Cryptococcus neoformans* var. *gattii* in response to antifungal treatment. PhD Thesis. University of Sydney, Sydney, Australia.

Litt M and Luty JA (1989). A hypervariable microsatellite revealed by in vitro amplification of a dinucleotide repeat within the cardiac muscle actin gene. Am J Hum Genet 44:388-396.

Lupski JR and Weinstock GM (1992). Short, interspersed repetitive DNA sequences in prokaryotic genomes. J Bacteriol 174:4525-4529.

Marth G, Yeh R, Minton D, Donaldson R, Li Q, Duan S, Davenport R, Miller RD and Kwok P-Y (2001). Single nucleotide polymorphisms in the public domain: how useful are they? Nat Genet 27:371-372.

Martin N, Boomsma D and Machin G (1997). A twin-pronged attack on complex traits. Nat Genet 17:387-392.

McDade HC and Cox GM (2001). A new dominant selectable marker for use in *Cryptococcus neoformans*. Med Mycol 39:151-154.

Metzgar D, Field D, Haubrich R and Wills C (1998a). Sequence analysis of a compound coding-region microsatellite in *Candida albicans* resolves homoplasies and provides a high-resolution tool for genotyping. FEMS Immunol Med Microbiol 20:103-109.

Metzgar D, van Belkum A, Field D, Haubrich R and Wills C (1998b). Random amplification of polymorphic DNA and microsatellite genotyping of pre and posttreatment isolates of *Candida* spp from human immunodeficiency virus-infected patients on different fluconazole regimes. J Clin Microbiol 36:2308-2313.

Meyer W and Mitchell TG (1995). Polymerase chain reaction fingerprinting in fungi using single primers specific to minisatellites and simple repetitive DNA sequences: strain variation in *Cryptococcus neoformans*. Electrophoresis 16:1648-1656.

Meyer W, Mitchell TG, Freedman EZ and Vilgalys R (1993). Hybridization probes for conventional fingerprinting used as single primers in the polymerase chain reaction to distinguish strains of *Cryptococcus neoformans*. J Clin Microbiol 31:2274-2280.

Michalakis Y and Veuille M (1996). Length variation of CAG/CAA trinucleotide repeats in natural populations of *Drosophila melanogaster* and its relation to the recombination rate. Genetics 143:1713-1725.

Moon CD, Tapper BA and Scott B (1999). Identification of *Epichloë* endophytes *in planta* by a microsatellite based PCR fingerprinting assay with automated analysis. Appl Environ Microbiol 65:1268-1279.

Neumann B, Kubicka P and Barlow DP (1995). Characteristics of imprinted genes. Nat Genet 9:12-13.

Nielsen R and Palsboll PJ (1999). Single-locus tests of microsatellite evolution: multi-step mutations and constraints on allele size. Mol Phylog Evol 11:477-484.

Ortí G, Pearse DE and Avise JC (1997). Phylogenetic assessment of length variation at a microsatellite locus. Proc Natl Acad Sci USA 94:10745-10749.

Primmer CR, Moller AP and Ellegren H (1996). A wide-ranging survey of cross-species amplification in birds. Mol Ecol 5:365-378.

Pritchard JK and Rosenberg NA (1999). Use of unlinked genetic markers to detect population stratification in association studies. Am J Hum Genet 65:220-228.

Queller DC, Strassmann JE and Hughes CR (1993). Microsatellites and kinship. Trends Ecol Evol 8:285-288.

Rassmann K, Schlötterer C and Tautz D (1991). Isolation of simple-sequence loci for use in polymerase chain reaction-based DNA fingerprinting. Electrophoresis 12:113-118.

Ratner VA, Zabanov SA, Lolesnikova OV and Vasilyeva LA (1992). Induction of the mobile genetic element *Dm-412* transpositions in the *Drosophila* genome by heat shock treatment. Proc Natl Acad Sci USA 89:5650-5654.

Richard G-F and Dujon B (1996). Distribution and variability of trinucleotide repeats in the genome of the yeast *Saccharomyces cerevisiae*. Gene 174:165-174.

Richards RI and Sutherland GR (1994). Simple repeat DNA is not replicated simply. Nat Genet 6:114-116.

Roder MS, Korzun V, Wendehake K, Plaschke J, Tixier MH, Leroy P and Ganal MW (1998). A microsatellite map of wheat. Genetics 149:2007-2023.

Rogstad SH, Patton JC, II and Schaal BA (1988). M13 repeat probe detects DNA minisatellite-like sequences in gymnosperms and angiosperms. Proc Natl Acad Sci USA 85:9176-9178.

Rolfe M, Spanos A and Banks G (1986). Induction of yeast *Ty* element transcription by ultraviolet light. Nature 319:339-341.

Rosewich UL and McDonald BA (1994). DNA fingerprinting in fungi. Method Mol Cell Biol 5:41-48.

Rubinsztein DD, Amos W, Leggo J, Goodburn S, Jain S, Li S-H, Margolis RL, Ross CA and Ferguson-Smith MA (1995). Microsatellite evolution - evidence for directionality and variation in rate between species. Nat Genet 10:337-343.

San Miguel P and Bennetzen JL (1998). Evidence that a recent increase in maize genome size was caused by the massive amplification of intergene retrotransposons. Annal Bot 82:37-44.

Schlötterer C, Amos B and Tautz D (1991). Conservation of polymorphic simple sequence loci in cetacean species. Nature 354:63-65.

Schlötterer C and Tautz D (1992). Slippage synthesis of simple sequence DNA. Nucl Acids Res 20:211-215.

Shastry BS (2002). SNP alleles in human disease and evolution. J Hum Genet 47:561-566.

Shriver MD, Jin L, Boerwinkle E, Deka R, Ferrell RE and Chakraborty R (1995). A novel measure of genetic distance for highly polymorphic tandem repeat loci. Mol Biol Evol 12:914-920.

Shriver MD, Smith MW, Jin L, Marcini A, Akey JM, Deka R and Ferrell RE (1997). Ethnic-affiliation estimation by use of population-specific DNA markers. Am J Hum Genet 60:957-964.

Sia EA, Butler CA, Dominska M, Greenwell P, Fox TD and Petes TD (2000). Analysis of microsatellite mutations in the mitochondrial DNA of *Saccharomyces cerevisiae*. Proc Natl Acad Sci USA 97:250-255.

Singh L (1995). Biological significance of minisatellites. Electrophoresis 16:1586-1595.

Slatkin M (1995). A measure of population subdivision based on microsatellite allele frequencies. Genetics 139:457-462.

Spitzer ED, Lasker BA, Travis SJ, Kobarashi GS and Medoff G (1989). Use of mitochondrial and ribosomal DNA polymorphisms to classify clinical and soil isolates of *Histoplasma capsulatum*. Infect Immun 57:1409-1412.

Sudarshan S, Davidson RC, Heitman J and Alspaugh JA (1999). Molecular analysis of the *Cryptococcus neoformans ADE2* gene, a selectable marker for transformation and gene disruption. Fungal Genet Biol 27:36-48.

Sutherland GR, Baker E and Richard RI (1998). Fragile sites still breaking. Trends Genet 14:501-506.

Taramino G and Tingey S (1996). Simple sequence repeats for germplasm analysis and mapping in maize. Genome 39:277-287.

Tautz D (1989). Hypervariability of simple sequences as a general source for polymorphic DNA markers. Nucl Acids Res 17:6463-6471.

Tautz D and Renz M (1984). Simple sequences are ubiquitous repetitive components of eukaryotic genomes. Nucl Acids Res 12:4127-4138.

Tautz D and Schlötterer C (1994). Simple sequences. Curr Op Genet Devel 4:832-837.

Taylor AC, Sherwin WB and Wayne RK (1994). Genetic variation of microsatellite loci in a bottlenecked species: the northern hairy-nosed wombat *Lasiorhinus krefftii*. Mol Ecol 3:277-290.

Thomas MR, Cain P and Scott NS (1994). DNA typing of grapevines: a universal methodology and database for describing cultivars and evaluating genetic relatedness. Plant Mol Biol 25:939-949.

Tran-Dinh N and Carter DA (2000). Characterisation of microsatellite loci in the aflatoxigenic fungi *Aspergillus flavus* and *Aspergillus parasiticus*. Mol Ecol 9:2170-2172.

Tran-Dinh N, Pitt JI and Carter DA (1999). Molecular genotype analysis of natural toxigenic and nontoxigenic isolates of *Aspergillus flavus* and *A. parasiticus*. Mycol Res 103:1485-1490.

Turri MG, Cuin KA and Porter AC (1995). Characterisation of a novel minisatellite that provides multiple splice donor sites in an interferon-induced transcript. Nucl Acids Res 23:1854-1861.

Upcroft P, Mitchell R and Boreham PFL (1990). DNA fingerprinting of the intestinal parasite *Giardia duodenalis* with the M13 phage genome. Int J Parasitol 20:319-324.

Valsecchi E, Palsboll P, Hale P, Glockner-Ferrari D, Ferrari M, Clapham P, Larsen F, Mattila D, Sears R, Sigurjonsson J, M. B, Corkeron P and Amos B (1997). Microsatellite genetic distances between oceanic populations of the humpback whale (*Megaptera novaeangliae*). Mol Biol Evol 14:355-362.

van der Lee T, De Witte I, Drenth A, Alfonso C and Govers F (1997). AFLP Linkage Map of the Oomycete *Phytophthora infestans*. Fungal Genet Biol 21:278-291.

Vassart G, Georges M, Monsieur R, Brocas H, Lequarre AS and Christophe D (1987). A sequence in M-13 phage detects hypervariable minisatellites in human and animal DNA. Science 235:683-684.

Vergnaud G and Denoeud F (2000). Minisatellites: mutability and genome architecture. Genome Res 10:899-907.

Weber JL (1990). Informativeness of human (dC-dA) (dG-dT) polymorphisms. Genomics 7:524-530.

Weber JL and May PE (1989). Abundant class of human DNA polymorphisms which can be typed using the polymerase chain reaction. Am J Hum Genet 44:388-396.

Weber JL and Wong C (1993). Mutation of human short tandem repeats. Hum Mol Genet 2:1123-1128.

Weir BS (1996). Genetic Data Analysis II. Sunderland, UK: Sinauer Associates.

White TJ, Bruns T, Lee S and Taylor JW (1990). Amplification and direct sequencing of fungal ribosomal RNA genes for phylogenetics, In: MA Innes, DH Gelfand, JJ Sninsky and TJ White, eds. PCR Protocols: a Guide to Methods and Applications. New York: Academic Press, pp. 315-322.

Wöstemeyer J and Kreibich A (2002). Repetitive DNA elements in fungi (Mycota): impact on genomic architecture and evolution. Curr Genet 41:189-198.

Yadon PJ and Catcheside DEA (1995). Guest: A 98 bp inverted repeat transposable element in *Neurospora crassa*. Mol Gen Genet 247:105-109.

Yoder OC and Turgeon BG (2001). Fungal genomics and pathogenicity. Curr Op Pl Biol 4:315-321.

Zane L, Bargelloni L and Patarnello T (2002). Strategies for microsatellite isolation: a review. Mol Ecol 11:1-16.

Zhou XD, Burgess T, De Beer ZW, Wingfield BD and Wingfield MJ (2002). Development of polymorphic microsatellite markers for the tree pathogen and sapstain agent, *Ophiostoma* Ips. Mol Ecol Notes 2:309-312.

Applied Mycology & Biotechnology
An International Series. Volume 4. Fungal Genomics

Inferring Process from Pattern in Fungal Population Genetics

Ignazio Carbone[1] and Linda Kohn[2]
[1]Center for Integrated Fungal Research, Department of Plant Pathology, North Carolina State University, Box 7244 – Partners II Building, Raleigh, NC 27695-7244, USA; [2]Department of Botany, University of Toronto, 3359 Mississauga Rd. N., Mississauga, ON L5L 1C6, Canada (kohn@utm.utoronto.ca).

Our focus in this review is on powerful new methods for determining population patterning over time and space and how from this, the dynamic processes leading to population divergence and speciation can be inferred. We focus on fungal populations, but draw from the wider literature on population genetics, evolutionary statistics, and, of course, phylogeography (see Avise, 2000). We discuss the problems of gene duplication, paralogy, orthology, and deep coalescence as challenges to finding the interface between population divergence and speciation. Our main objective, however, is to guide the reader through the key phylogenetic, nested phylogenetic, coalescent and Bayesian operations with the aid of a set of figures based on a simple, hypothetical dataset of DNA haplotypes. Phylogenetic and compatibility approaches are presented with the goal of not only detecting recombination, but of detecting recombination when it is not widespread throughout a phylogeny. This is a major challenge in fungal systems with substantial asexual reproduction or with significant selfed sexual reproduction in a haploid genome. The key feature here is that recombination can be "localized" in some but not all clades in a phylogeny and that these clades can be identified. From this, contemporary versus historical patterns of recombination can be inferred from a phylogeny. Phylogenetic approaches based on conversion of the phylogeny to a nested hierarchical statistical design are presented for fuller exploration of associations between each nested level of the phylogeny and any variable, such as geographical location, host, or symptom type. The basic operations for both testing for population subdivision based on geographical associations, and for cladistic inference of population processes are presented. Our hypothetical dataset is also used to demonstrate how genealogical relationships and population parameters can be inferred using coalescent and Bayesian methods. The basic principles of these approaches are graphically presented, along with useful references and comments on key assumptions implicit in methods currently available.

Corresponding author: L.M. Kohn

1. INTRODUCTION

Population genetics is the study of the structure of populations and of the evolutionary processes that shape these structural patterns. The patterns of distinct, divergent populations are inferred from the genetic diversity of contemporary samples made from "the field", including clinical patient populations. The evolutionary processes include mutation, gene flow, recombination, selection, and drift. Population divergence resulting from such evolutionary processes, as well as from hybridization or vicariance (fragmentation of the environment that can lead to fragmentation of populations), eventually results in speciation. Through phylogenetic and coalescent statistical models, including Bayesian approaches, we can retrospectively determine the most probable chronology of events causing population divergence and identify the most probable events responsible for this divergence. Population genomics takes the genetics of natural or experimental populations steps further to study changes in genotype and gene expression during adaptation, one of the many applications of microarray technology (Cowen *et al.* 2002; Zeyl 2000).

The fundamental source of biological variation is mutation. This variation is shuffled among individuals by genetic exchange, through sex or horizontal transfer, recombination and segregation. Natural selection, *i.e.* differential reproduction, acts on the individual, but of course the results of selection are only visible in populations. Populations of a species are dynamic; in practice, the boundary between evolving, diverging populations and speciation may be difficult to define. Populations may diverge in response to changes in population size, genetic drift (random changes in allele frequencies to which small populations are especially prone), and changes in gene flow (the movement of genes, gametes, or individuals). Genetic diversity can be described and quantified in three ways (McDonald and Linde 2002a). Nucleotide diversity within genes or genomic regions (loci) is measured as the average number of nucleotide differences per site, π, between any two randomly chosen DNA sequences from a population (Nei 1987). In contrast, the two types of genetic diversity that are major components of population structure are gene diversity, the number and frequency of alleles at a single locus in a population, and genotype diversity, the number and frequency of multilocus genotypes (distinct individuals) in a population. Increasing gene diversity results not only in additional alleles but also in an equalization of allele frequencies (McDonald and Linde 2002a). For the purposes of this review, a population is defined as a group of individuals that occupy a particular geographical space in time, share a common ancestry, undergo genetic drift together, and may eventually become reproductively, ecologically and genetically well-differentiated as species (de Queiroz 1998).

Fungal population genetics has been amply reviewed (Anderson and Kohn 1998; Burdon 1993; Leung *et al.* 1993; McDonald 1997; McDonald and Linde 2002a; Milgroom 1996). A perusal of these reviews offers a history of a field that has exploded with the development of different types of molecular markers, from isozymes to RFLPs, AFLPs, microsatellites, oligonucleotides, and single nucleotide polymophisms (SNPs), as well as with the improved implementation of several types of statistical analyses and the development of important, new statistical approaches.

Once gene and genotypic diversity are determined by means of markers as allele frequencies among single or multilocus haplotypes, a range of analyses can partition this diversity as patterns of distinct populations or subpopulations. From these patterns, inferences of gene flow or genetic drift can be made. Leung *et al.* (1993), McDermott and McDonald (1993), and Milgroom (1995) reviewed the concepts, analyses (including virulence) and standard statistical approaches to determining population structure. These include measures of genetic variation and determination of partitions (patterns) of this variation by means of F statistics, notably F_{ST} (Wright 1951) and G_{ST} (Nei 1973). Leung *et al.* (1993) introduced tree-building methods for inferring similarity among individuals.

In sexual reproduction, regular genetic exchange through mating and recombination can accelerate the evolution of new genotypes by bringing together mutations arising in different individuals. In fungi, recombination in sexual reproduction and processes of recombination outside of sex, such as parasexuality or transposition, are evident, although not to the extent that such events confound phylogenetic inference in most of the fungi investigated to date. Fungi do not show the substantial trafficking in mobile genetic elements seen in Bacteria. Horizontal gene transfer among widely divergent taxa, another means of recombination, has not yet been strongly demonstrated in fungi (Rosewich and Kistler 2000). Under strict clonality, mutations are only transmitted vertically from parent to offspring and such populations might be expected to evolve more slowly than non-clonal populations under conditions where adaptive mutations are limiting. Of course, large population size may make a wide variety of mutations available. Because fungi often reproduce predominantly asexually, their populations may occupy the "grey zone" between panmixia (random mating) and clonality. Milgroom (1996) reviewed the evolutionary significance of recombination and critically examined how frequencies of multilocus genotypes can be used to find evidence of recombination, to test a hypothesis of random mating, and to determine recombination frequency. Clonality in fungi has been reviewed by Anderson and Kohn (1995). More recently, in the context of considering how fungi fit the classical models of population genetics, Anderson and Kohn (1998) provided an overview of the phylogenetic criterion for recombination, also reviewing the evidence for mitochondrial recombination in fungi.

In a review on assessing fitness in fungal populations, Pringle and Taylor (2002) recommended choosing appropriate fitness measures matched to components of often complex life cycles, as well as considering life history and ecological characteristics, such as iterative versus single sporulation. The goals would be to predict or measure the fitness of pathogen genotypes and to determine the effects of specific pathogen genotypes on the fitness of host genotypes (see also: Antonovics and Kareiva 1988; Brunet and Mundt 2000).

McDonald (1997) reviewed genetic markers and sampling designs most suitable for examining population genetic structure. Although isozymes and other electrophoretically based markers continue to be useful, DNA nucleotide sequence is the gold standard because of its high information content and reproducibility. Markers can provide resolution on different temporal scales, for example, nucleotide sequence variation to examine ancient patterns of population divergence (Carbone and Kohn 2001a) and DNA fingerprints to resolve population genetic structure on a more recent time scale (Carbone and Kohn 2001b; our DNA fingerprints are RFLPs, but AFLPs or microsatellites are also expected to evolve rapidly and therefore represent recent evolution). In the absence of any prior knowledge of pathogen population structure, McDonald (1997) has proposed a hierarchical sampling strategy as a starting point. This preliminary sample can be screened with appropriate molecular markers to determine the spatial and temporal scales for further sampling. The extended sample should cover the full range of genetic and phenotypic heterogeneity in the pathogen population. This is important because each isolate provides only one snapshot of genetic and phenotypic variation at a specific time point. This does not imply, however, that we need large sample sizes to detect the full range of genetic and phenotypic heterogeneity. Evolutionary methods can reconstruct genealogical relationships from sampled genotypes and in the process infer missing intermediate or ancestral genotypes. Complex phenotypes can also be inferred by superimposing phenotypic data on mutational networks (described in this review). Fungal genomic data, especially on whole genomes, cannot become available fast enough as questions concerning molecular evolutionary processes, such as gene duplication and paralogy, currently confound routine analyses (see Fig. 1). The increasing availability of genetic and phenotypic data will necessitate the development of a fungal pathogen database that can facilitate the storage, retrieval and comparative analyses of genetic and phenotypic data (Kang *et al.* 2002).

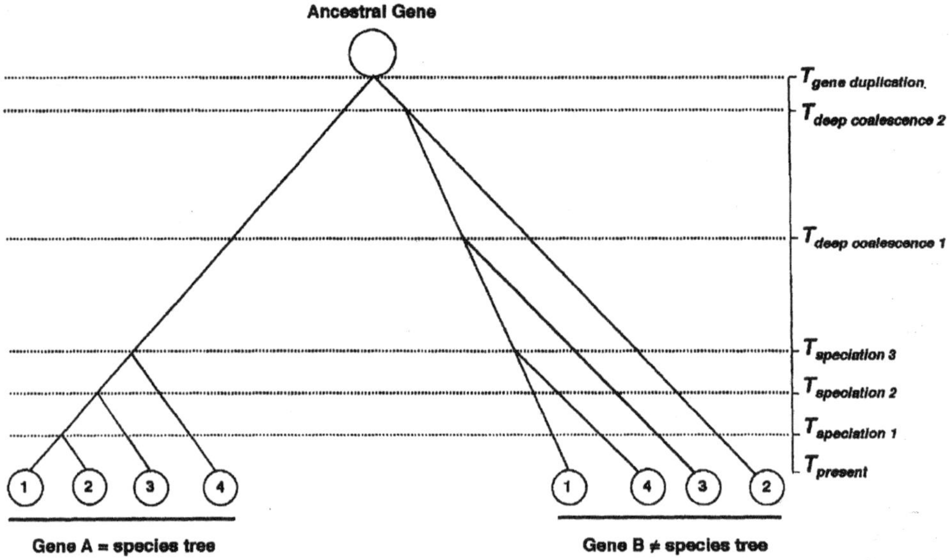

Fig. 1. Gene duplication, paralogy, orthology, deep coalescence and phylogenetic inference. Gene duplication and deep coalescences can result in genealogies that do not track with the species tree. In the figure above, genes A and B were derived from a duplication event in an ancestral gene at $T_{gene\ duplication}$ time units ago. This duplication was followed by two splitting events in gene B ($T_{deep\ coalescence\ 2\ and\ 1}$) and then three speciation events ($T_{speciation\ 3,\ 2,\ and\ 1}$) splitting the ancestral species into four species, designated as 1, 2, 3, and 4. Gene A sequences for species 1, 2, 3 and 4 are derived from the speciation events and are referred to as orthologous gene sequences; gene B sequences for species 1 and 4 are also orthologous. Collectively, the B genes were derived from the duplication event and are referred to as paralogous to the A genes. The inferred phylogeny for the A genes is concordant with the species tree; the B phylogeny is discordant with the species tree. Looking backwards in time from the present ($T_{present}$) only the B genes for species 1 and 4 coalesce on the species branches. The deep coalescence in the B gene for species 1 with species 3 and 2, respectively, does not coincide with the recent splitting of these species. As a result of the deep coalescences, lineages 2 and 3 are sorted into distinct species (*i.e.* are paraphyletic) and, if ignored, can confound our inferences of the underlying species tree and the evolutionary processes in these species.

McDonald and Linde (2002) focused on the evolutionary potential of pathogen populations and proposed a "risk" model that relates the population genetics of plant pathogens to their ability to cause disease. The evolutionary potential is determined by examining the fine-scale genetic structure and evolutionary processes that influence patterns of genetic diversity in pathogen populations. The contribution of each of these processes to population genetic structure predicts whether pathogen populations will evolve rapidly or slowly in response to different control strategies. For example, a recombining population structure would allow mutations for virulence that arise at different loci to be combined into novel and potentially more pathogenic genotypes that can overcome control strategies and thereby allow the pathogen to evolve to a higher level of pathogenicity. According to the risk model, pathogens with the highest evolutionary potential would have large effective population sizes, mixed sexual and asexual reproduction and asexual spores that are widely dispersed. Pathogens with small effective population sizes, that are strictly asexual and that undergo limited gene flow would have the lowest evolutionary potential.

2. GENETIC MARKERS FOR EXAMINING POPULATION GENETIC STRUCTURE AND SPECIES LIMITS FROM POPULATION SAMPLES

Genetic diversity among individuals in populations has been identified using electrophoretically-based markers, such as allozymes, for at least thirty years (Scribner *et al.* 1994). More recently, markers have been developed by means of random amplified polymorphic DNAs (RAPDs), restriction or amplified fragment length polymorphisms (RFLPs or AFLPs) in nuclear or mitochondrial DNA, DNA fingerprints, electrophoretic karyotypes, microsatellites, and minisatellites. A major limitation in the genetic interpretation, as loci and alleles, of electrophoretically-derived markers is that co-migrating bands shared by two individuals do not necessarily reflect descent from a common ancestor; identity by allelic state does not necessarily indicate identity by descent (Lynch 1988). Consequently, these markers are not optimal for phylogenetic reconstruction although they have been useful in systematics for discriminating between species, and in population genetics for typing strains (*e.g.* McEwen *et al.* 2000; Taylor *et al.* 1999a), estimating gene diversity (Keller *et al.* 1997; Linde *et al.* 2002; McDonald *et al.* 1995) and determining genotype diversity (Ceresini *et al.* 2002; Chen and McDonald 1996; Kohn *et al.* 1991; Kumar *et al.* 1999; Milgroom *et al.* 1992). Fortunately, nucleotide sequence data offer the possibility of reconstructing patterns of descent among genotypes within a species, or populations of one or more species. Once polarity is established, ancestral and derived states can be distinguished from sequence data using a combination of coalescent and Bayesian approaches, described later in this chapter.

In selecting loci, several potential complications should be considered (Avise 1998). The first could be allelic variation in single-copy loci from diploids or from haploid heterokaryotic organisms, or in loci belonging to multi-gene families. Although methods have been described for separating individual haplotypes from bi-allelic loci which produce a composite phenotype (Avise 1998), they are not feasible for large population studies. A second complication could arise if all loci have not accumulated sufficient mutations at the intraspecific level, or have undergone extensive intra- and inter-genic recombination. Recombination would scramble genealogical relationships, necessitating inference of phylogenetic networks rather than trees, a breaking area of theoretical research (Bandelt *et al.* 1999; Huber *et al.* 2001; Posada and Crandall 2001b; Strimmer and Moulton 2000; van Nimwegen *et al.* 1999; Wang *et al.* 2001). A further complication could arise when different loci evolve at different rates. In some cases, the locus could have diverged before the population split. As a result, distinct lineages that existed in the ancestral population would be randomly sorted to daughter populations. If not detected, this would result in overestimates of branch lengths and divergence times among populations. Another possibility is introgressive hybridization (Fregene *et al.* 1994; O'Donnell *et al.* 2000; Schardl 2001; Scribner and Avise 1993; Scribner and Avise 1994). Even artifactual "noise" can produce phylogenetic signatures that mimic recombination, rate heterogeneity, etc. and must be distinguished from the real phenomena (Hillis and Huelsenbeck 1992). All of these possibilities influence phylogenetic reconstructions at the population and species levels.

When a phylogenetic tree is inferred from a specific DNA sequence among a group of populations of one or more species, it is a possible species tree in which the populations or species are the operational taxonomic units (OTUs). OTUs can be different individuals within a population, distinct populations, species or any other extant taxa. Species trees are useful tools for estimating the evolutionary relationships among species and for testing hypotheses about the speciation process. When a phylogenetic tree is inferred from a particular DNA sequence within a population, then the tree represents a possible intraspecific gene phylogeny, or gene genealogy, with the DNA sequences themselves as the OTUs. Gene genealogies are powerful tools for examining a variety of population-level processes as discussed later in this review.

It is important to note that the evolutionary pathway of a particular gene genealogy can differ from that of the overall population or species tree in several ways. First, if a tree is thought of as a compilation of many gene genealogies (Avise 1989), then sampling error is possible when reconstructing trees from the sequences of a small number of genes. This sampling error is higher for species that are recently evolved because lineage sorting is not yet complete (see Fig. 1). Lineage sorting is the failure of gene sequences to coalesce to a common ancestor and is also referred to as deep coalescence because the coalescence of ancestral gene copies predates previous speciation events (Maddison 1997). To avoid this type of error, multiple, physically unlinked loci within a species should be used in the reconstruction of the population or species tree. The criterion for whether or not to combine multiple genealogies from different genomic regions (termed loci) should not be measures of the overall concordance among gene genealogies (Carbone *et al.* 1999; Barker and Lutzoni 2002; Darlu and Lecointre 2002). Rather, it can be based on concordance and, most significantly, on increased phylogenetic resolution afforded by combining some, if not all loci for some, if not all, clades. The theory underlying this approach awaits further evaluation in simulations.

When populations or species are well-defined entities it is not difficult to find a genomic region with interspecific variation; many studies in higher eukaryotes have focused on variation in mitochondrial (mt) DNA (Pumo *et al.* 1996; Shaw 1996). This molecule is well-suited for phylogenetic analysis for two major reasons: (i) a rapid rate of evolution, primarily in the form of base substitutions, and (ii) a mostly maternal mode of inheritance with effectively haploid transmission across generations. The chloroplast genome has served much the same function in plants (Gielly and Taberlet 1994; Sang *et al.* 1997). A combination of variation found in the mitochondrial and nuclear large subunit ribosomal RNA genes has been useful in identifying fungal species (Kretzer and Bruns 1999; Taylor and Bruns 1999), however, rate heterogeneity between mt and nuclear genomes often precludes combined interspecific phylogenetic analyses (Moncalvo *et al.* 2000). Intraspecific mt DNA variation in fungi has been useful for testing hypotheses on the evolutionary origins of plant pathogens (O'Donnell *et al.* 1998b; Ristaino *et al.* 2001) and for providing evidence of recombination in the mt genome (Anderson *et al.* 2001; Saville *et al.* 1998).

Many fungi are haploid, and many, though not all, undergo extensive asexual reproduction, with or without periodic sexual reproductive episodes. Studies in intraspecific variation of fungi as well as of species delimitation have been based on variation found in nuclear ribosomal DNA (rDNA) but have more recently utilized a wider range of protein-coding genes (for reviews see Kang *et al.* 2002; Taylor *et al.* 2000). In fungi, ribosomal and protein-encoding mitochondrial genes have been shown to be rife with group I introns. These introns frequently encode maturases and can be very large (ca 2000 bp), much larger than their nuclear counterparts, which are smaller (ca 300 bp) and lack maturases. While universal primer sequences have been developed for fungal mt DNA genes, these regions are frequently sprinkled with introns, which sometimes also fall within the priming sites. This has impeded the use of mitochondrial regions for speciation studies in fungi. The suite of nuclear ribosomal RNA genes is also limiting because these genes frequently lack resolution at the species level, and are more useful for resolution at higher taxonomic levels (O'Donnell *et al.* 1997). More recently, intraspecific studies in fungi have focused on variation in coding and noncoding portions of nuclear protein-encoding genes (Carbone *et al.* 1999; Carbone and Kohn 1999; Carbone and Kohn 2001a; Couch and Kohn 2002; Geiser *et al.* 1998). Because a single gene genealogy represents only one of many tracks toward the true species tree, multiple gene genealogies are a better approximation of the species tree and have been useful for testing hypotheses on species origins and conspecificity (Geiser *et al.* 2001; Kroken and Taylor 2001; O'Donnell 2000; O'Donnell *et al.* 1998a; O'Donnell *et al.* 2000; Shen *et al.* 2002). For population studies of fungi, concordance among multiple genealogies offers the possibility of combining datasets to achieve the best resolution of genotypes. Several studies have combined gene genealogies inferred from several nuclear genes (Carbone *et al.* 1999; Couch and Kohn 2002; O'Donnell *et al.* 2000) and from genealogies inferred from nuclear and mitochondrial small subunit ribosomal RNA genes (O'Donnell *et al.* 1998b; Skovgaard *et al.* 2001). Such combinations of markers have been useful in determining patterns of infection, reproduction and dispersal in populations of plant-pathogens (Carbone *et al.* 1999; Carbone and Kohn 2001b; Kohli and Kohn 1996; Phillips *et al.* 2002; Skovgaard *et al.* 2001; Zhan *et al.* 2002). Combining datasets from multiple genomic regions may further enhance our inference of the species tree by providing finer resolution of deep coalescence events in the history of the species. Lineage sorting (*i.e.* deep coalescence) can result in discordance among gene genealogies and introduce significant errors in our estimates of the species tree. Hybridization and recombination events can also result in incongruencies between gene genealogies and further confound our inference of the species tree. Several methods have been developed that consider the possibility of lineage sorting and recombination when reconstructing a species tree from one or more gene genealogies (Page 1998; Page and Charleston 1997; Taylor et al. 2000).

The phylogenetic approaches that we will discuss in this review require nucleotide sequence data. Not all variation found in coding and noncoding regions is equally informative in reconstructing gene genealogies. There are several advantages in sequencing an entire locus rather than focusing only on SNPs. The potential contamination of SNPs with nonallelic paralogous sequence variation and the possibility of gene conversion between target loci and duplicated regions may introduce, if ignored, significant errors in our estimates of allelic diversity at a locus (Hurles *et al.* 2002). The phylogenetic-compatibility approach we describe below would be useful for detecting and demarcating the putative boundaries of gene conversion events, an essential first step in the utilization of SNP data in phylogenetic reconstructions. In the case of microsatellites and other highly polymorphic sites, caution should be exercised when using these markers in phylogenetic reconstructions because different microsatellite allelic size classes do not always follow a simple stepwise model of evolution (Fisher *et al.* 2000). Although their utility in phylogenetic inference is limited, microsatellites have been useful in examining the geographic partitioning of genetic variation in populations (Fisher *et al.* 2001). One strategy for incorporating variation at microsatellite loci into phylogenetic reconstructions uses variation found in highly polymorphic loci (*e.g.* microsatellites, DNA fingerprints) to extend gene genealogies inferred using nucleotide polymorphisms (Carbone *et al.* 1999; Carbone and Kohn 2001b). Another strategy uses DNA fingerprinting and a Bayesian model to assign recombining individuals of uncertain origin to populations (Fisher *et al.* 2002b).

3. PHYLOGENETIC AND COMPATIBILITY APPROACHES

Gene genealogies are tree-like representations of the history of descent from the ancestral sequence of one or more loci (genomic regions). Single or multi-locus gene genealogies derived by phylogenetic, coalescent or Bayesian approaches, can be explored to estimate the contribution of the key drivers of evolution in populations: mutation, selection, changes in population size, genetic drift, gene flow, genetic exchange and recombination. A variety of methods are available for using gene genealogies to estimate the relative contributions of mutation versus recombination (Burt *et al.* 1996; Carbone and Kohn 2001a; Carbone and Kohn 2001b), to detect selection (Hudson and Kaplan 1995a; Hudson and Kaplan 1995b), and to estimate average levels of gene flow (Hudson *et al.* 1992). Burt *et al.* (1996) described three methods for discriminating between a clonal (mutation alone) versus recombining population structure (see Anderson and Kohn, 1998 for a graphical representation). The first approach is empirical and compares gene genealogies from several different genomic regions for each population sample. It is based on the assumption that if mutation is the dominant evolutionary force giving rise to new genotypes, then clones should be related to each other in clonal lineages and gene genealogies from different genomic regions should be congruent. If recombination is the diversifying force, then gene genealogies should be incongruent (Fig. 2).

The second method infers gene genealogies for a number of loci, and uses likelihood analyses (Felsenstein 1981) to test hypotheses under two models: (i) that all loci have the same topology as would be expected in a clonally evolving organism, and (ii) that loci have different topologies as expected if recombination is an important diversifying force. Under the first model, ln likelihoods are summed over all loci; under the second model, ln likelihoods are determined separately for each locus and then summed across all loci. The model that fits the data best would have the best likelihood estimates. A third approach is to perform a permutation test and to compare the observed tree length with tree lengths from randomized data sets. In a recombining data set the observed tree length should fall within the distribution of tree lengths from randomized data sets. All three methods were used to provide strong evidence for genetic exchange and recombination in *Coccidioides immitis,* an important human pathogen that has been thought to have a strictly asexual life cycle (Burt *et al.* 1996). Although this study showed *C. immitis* to have a highly recombining population structure, it was not possible to determine whether recombination has been a historical or contemporary and ongoing process because the consensus gene genealogy was unresolved.

In subsequent studies, a multiple gene genealogical approach was used to detect geographic differentiation and to identify putative biological species among different populations of *C. immitis* (Fisher *et al.* 2001; Fisher *et al.* 2002b; Koufopanou *et al.* 1997). The rationale was that if recombination occurred within strongly supported clades in all gene genealogies, but not between clades, then these clades defined the boundaries of biological species. This was an extension of the work of Dykhuizen and Green on clonal lineages in the bacterium *Escherichia coli* (Dykhuizen and Green 1991). A further line of evidence that *C. immitis* comprised two reproductively isolated groups was that the splitting of the two groups of isolates in the cladogram was strongly correlated with the geographical origin of isolates (Koufopanou *et al.*

(a)

```
  1234
  ACGT
1 ....
2 GT..
3 GTAC
4 ..AC
```

(b)

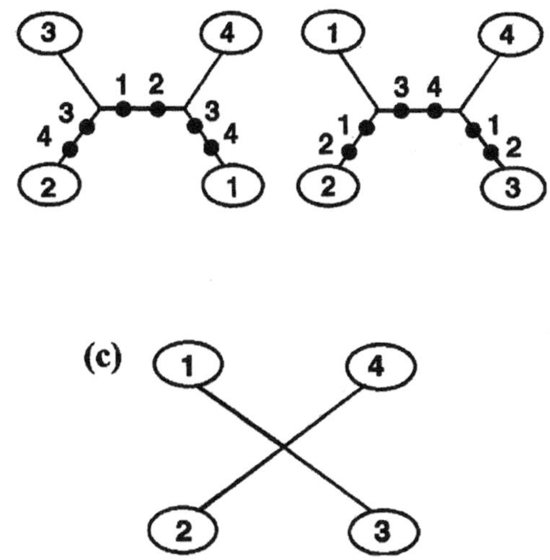

Fig. 2. Phylogenetic inference and detection of recombination. A hypothetical example of the phylogenetic method. Phylogenetic methods compare genealogies inferred from different genomic regions to determine patterns of descent and to detect recombination events in the history of the sample (Anderson and Kohn 1998; Hein 1990; Hein 1993; Posada 2002; Posada and Crandall 2001a; Posada and Crandall 2002; Posada *et al.* 2002; Robertson *et al.* 1995). (a) A multiple DNA sequence alignment for a sample of 4 haplotypes (numbered on left), showing only SNPs (numbered across top row, left to right). Consensus sequence is in second row. In the alignment, dots designate bases matching the consensus sequence. (b) The two equally most parsimonious trees for the data set in (a), each with a consistency index (CI) of 0.6667. The solid circles and the numbers along the branches designate mutations in the sample. In the presence of recombination different DNA regions yield different trees that cannot be reconciled into a single tree without introducing significant errors in branch lengths as a result of parallel mutations or reversals (sites 3 and 4 on one tree and sites 1 and 2 on the other tree). (c) The strict consensus tree for the two trees shown in (b). Because phylogenetic methods test for overall topological concordance among trees they lack inferences into the organization of recombination events (*i.e.*, patterns of recombination) along DNA sequences, the magnitude (*i.e.* number and location of recombination events) along a DNA sequence and the timing and frequency of recombination events (*i.e.* contemporary versus historical). Furthermore, because phylogenetic methods test for overall concordance among trees, they may miss patterns of localized recombination in some but not all clades in the phylogeny (see Figs. 3, 4).

(a)

polymorphic sites (x: sites 1–4; y: sites 5–12; z: sites 13–20)

haplotype	G	T	T	G	T	T	G	G	T	T	C	C	A	A	T	T	C	C	T	T
1	.	.	.	.	.	.	.	.	.	.	.	.	.	.	.	.	.	.	.	.
2	.	.	.	.	.	.	.	.	.	.	.	.	T	.	A	.	.	.	.	.
3	.	.	.	.	A	.	.	C	.	.	.	.	T	.	A	.	.	.	.	.
4	.	.	.	.	A	.	.	C	.	.	.	.	.	.	.	.	.	.	.	.
5	C	A	A	.	.	.	.	.	.	A	.	.	.	.	.	.	.	.	.	A
6	C	.	.	.	.	.	.	.	.	A	.	.	.	.	.	.	G	.	.	.
7	C	.	.	.	.	.	.	.	.	.	.	G	.	.	.	.	G	.	.	.
8	C	A	A	.	.	.	.	.	.	.	.	G	.	.	.	.	.	.	.	A
9	.	.	.	.	.	.	.	.	.	A	.	.	.	.	.	.	.	.	.	.
10	.	.	.	C	.	A	C	.	A	A	.	.	.	.	.	.	.	G	A	.
11	.	.	.	C	.	A	C	.	A	.	.	.	.	.	.	.	.	G	.	.
12	.	.	.	C	.	A	.	.	A	.	G	.	.	T	.	.	.	.	.	.
13	.	.	.	C	.	A	.	.	A	.	G	.	.	T	.	A	.	.	.	.

(b)

(x: sites 1–4; y: sites 5–12; z: sites 13–20)

site	1	2	3	4	5	6	7	8	9	10	11	12	13	14	15	16	17	18	19	20
1	.	.	.	.	.	.	.	.	.	x	.	.	.	.	.	.	.	.	.	.
2	.	.	.	.	.	.	.	.	.	x	.	x	.	.	.	.	.	.	.	.
3	.	.	.	.	.	.	.	.	.	x	.	x	.	.	.	.	.	.	.	.
4	.	.	.	.	.	.	.	.	.	x	.	.	.	.	.	.	.	.	.	.
5	.	.	.	.	.	.	.	.	.	.	.	.	x	.	x	.	.	.	.	.
6	.	.	.	.	.	.	.	.	.	x	.	.	.	.	.	.	.	.	.	.
7	.	.	.	.	.	.	.	.	.	x	.	.	.	.	.	.	.	.	.	.
8	.	.	.	.	.	.	.	.	.	.	.	.	x	.	x	.	.	.	.	.
9	.	.	.	.	.	.	.	.	.	x	.	.	.	.	.	.	.	.	.	.
10	x	x	x	x	.	x	x	.	x	.	.	.	.	.	.	.	x	x	.	x
11	.	.	.	.	.	.	.	.	.	.	.	.	.	.	.	.	.	.	.	.
12	.	x	x	.	.	.	.	.	.	.	.	.	.	.	.	.	x	.	.	x
13	.	.	.	.	x	.	.	x	.	.	.	.	.	.	.	.	.	.	.	.
14	.	.	.	.	.	.	.	.	.	.	.	.	.	.	.	.	.	.	.	.
15	.	.	.	.	x	.	.	x	.	.	.	.	.	.	.	.	.	.	.	.
16	.	.	.	.	.	.	.	.	.	.	.	.	.	.	.	.	.	.	.	.
17	.	.	.	.	.	.	.	.	.	x	.	x	.	.	.	.	.	.	.	.
18	.	.	.	.	.	.	.	.	.	x	.	.	.	.	.	.	.	.	.	.
19	.	.	.	.	.	.	.	.	.	.	.	.	.	.	.	.	.	.	.	.

(c)

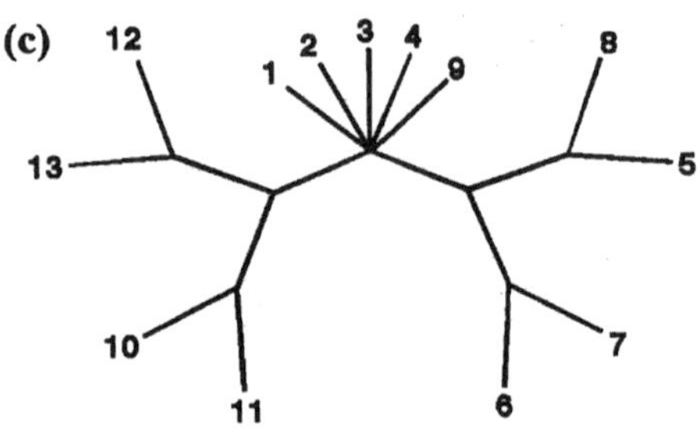

Fig. 3. Compatibility analysis and phylogenetic inference. A hypothetical example of the compatibility method used to assess the support or conflict among individual sites along a DNA sequence alignment. Compatibility methods examine the overall support or conflict among variable sites (*i.e.* SNPs) in a sequence alignment and have been useful in identifying different segments, termed 'partitions', with distinct phylogenetic histories in sequence alignments (Jakobsen *et al.* 1997). **(a)** A multilocus DNA sequence alignment showing only SNPs for a sample of 13 multilocus haplotypes. The 3 loci are designated as x, y and z. The consensus sequence is shown in the second row at the top and a match with a base in the consensus is indicated with a dot. **(b)** The site compatibility matrix for combined loci. The matrix was generated using GENETREE v9.0 (http://www.stats.ox.ac.uk/~griff/software.html). The numbers along the top and side of the matrix are for variable positions in (a). Compatible sites are indicated by '.' and incompatible sites by 'x'. The matrix shows that all sites, with the exception of sites 11, 14, 16 and 19, are incompatible with at least one other site. This conflict yields 8 equally parsimonious trees with a consistency index (CI) of 0.7692. **(c)** The unrooted cladogram (strict consensus) of all trees showing one unresolved fan-shaped polytomy.

1997). This correlation with geography was interpreted as reproductive isolation, but the possibility could not be ruled out that these distinct lineages, though highly divergent and geographically-associated, were still capable of genetic exchange. Carbone and Kohn (2001a) implemented both phylogenetic (Fig. 2) and compatibility approaches (Fig. 3) to reconstruct patterns of mutation and recombination in populations of *Sclerotinia sclerotiorum*.

Compatibility approaches use the principle of site compatibility/incompatibility (Jakobsen *et al.* 1997) to identify non-recombining partitions in the data set. In the presence of recombination, the evolutionary process is more accurately represented in a mutational network that can accommodate reticulations (*i.e.* loops), as well as bifurcations and multifurcations (Posada and Crandall 2001b). One method that has been proposed to resolve phylogenetic relationships within loops is based on the relative frequencies of interior and tip haplotypes (Posada and Crandall 2001b). An alternative approach was described by Carbone and Kohn (2001a). This study used a combined phylogenetic-compatibility approach to identify non-recombining partitions or "recombination blocks" in distinct populations of *S. sclerotiorum* (illustrated schematically in Fig. 4 and the inference of alternative phylogenies for each of the two recombination blocks in each of the two clades with blocks). These alternative phylogenies could then be converted to networks, nested and then extended with DNA fingerprint data (Carbone and Kohn 2001b), providing a robust framework for performing a wide variety of associations of genotype with different phenotypic categories (Phillips *et al.* 2002), as described below. While a combined phylogenetic-compatibility approach was useful for identifying recombination blocks that were converted into networks displaying alternative phylogenetic histories (Fig.1 in Carbone and Kohn 2001a) this analysis on its own could not provide inferences on the ages of the inferred recombination events and the timing of recombination events in the history of *S. sclerotiorum*. These temporal aspects were inferred by means of coalescent approaches, described in this review.

Gene genealogies can be used to discriminate between recurrent and non-recurrent evolutionary processes (Templeton 1995). Non-recurrent processes, such as host jumps and fragmentation events, affect entire populations of individuals simultaneously, creating new evolutionary lineages and potentially new species. If separated for a long time, these new lineages might show a strong host or geographical association because of the accumulation of many host- or geography-restricted mutations. A fragmentation or splitting event could only be detected if the specific DNA locus sampled started to diverge before the fragmentation event; a locus that diverged after fragmentation would provide no resolution of the fragmentation event. Recurrent processes, such as gene flow and population expansion events, operate within evolutionary lineages and affect the structure or pattern of evolution within lineages, but not their pre-divergence history. An expansion event can be detected only if some of the haplotypes are older than the expansion event; haplotypes that arose during or after the expansion event would not provide insight into the expansion event (Templeton 1993). Both recurrent and non-recurrent events can occur throughout the history of the species. A strong geographical association among individuals within an evolutionary lineage may arise from a non-recurrent event affecting population history such as a fragmentation event, or from a recurrent event affecting population structure such as restricted gene flow, or from both non-recurrent and recurrent events.

Recurrent events can be distinguished from non-recurrent events because they predict qualitatively different patterns in the gene genealogy. For example, if restricted gene flow is the reason for the observed geographical association, then (i) new haplotypes within the evolutionary lineage or clade should have a more restricted geographical range and should be positioned at the tips of the clade, (ii) older haplotypes should be located at the interior nodes of the clade and have a broader geographical distribution, and (iii) this pattern should be repeated among many haplotypes within the clade. In contrast, if the geographical association is the result of a fragmentation event, then (i) haplotypes in the fragmented clade with restricted geographical distribution should have ranges that are completely or mostly non-overlapping with haplotypes

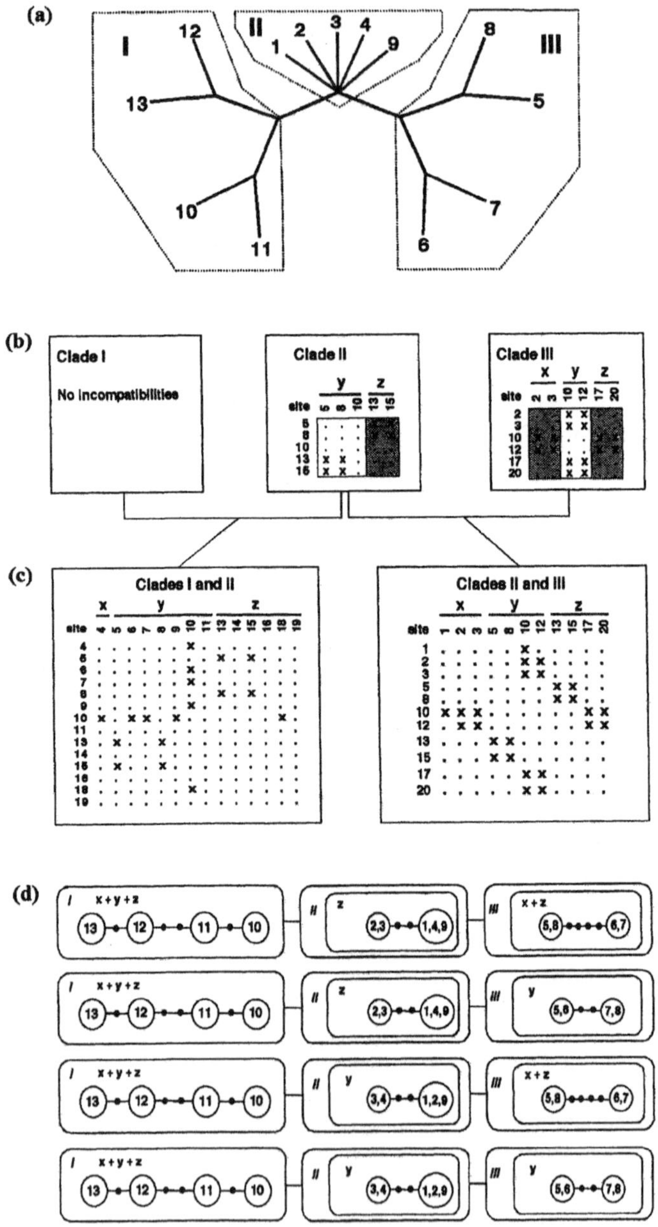

Fig. 4. Phylogenetic-compatibility method for inferring mutational networks. Phylogenetic and compatibility methods are combined to localize recombination events to specific clades in the history of the sample. Follow the three steps described below. This is a continuation of the example in Fig. 3. Step 1. Compatibility matrices are generated for all subsets of haplotypes that share a common ancestry in the strict consensus tree shown in **(a)**. A clade is defined as the largest most inclusive group of two or more haplotypes sharing a specific pattern of compatibility/incompatibility. Each of the three clades enclosed with dashed lines in (a) has a distinct mutation and recombination history as inferred from site compatibility matrices in (b). No incompatibility is detected in clade I and there is no variation in locus x for haplotypes in clade II. If we combine haplotypes in clades I and II or II and III as shown in **(c)**, this would disrupt distinct blocks in clades II and III and introduce homoplasy (*i.e.* incompatibilities) in clade I. Although these patterns are interpreted as arising from reciprocal recombination events, the pattern of recombination in clade III is also consistent with gene conversion whereby variation in locus y is the result of a nonreciprocal cross-over event (see also Wiehe *et al.* 2000). Step 2. The matrices generated for each clade are examined for clusters of two or more identical sites, which define a recombination block, shown as shaded and unshaded rectangles in (b). Within a recombination block all sites are compatible and infer one most parsimonious tree. Step 3. The four unrooted alternative networks showing all possible combinations of marginal networks for clades II and III are shown in (d). Marginal mutational networks are based on recombination blocks identified within clades II and III. There is no recombination within clade I.

found in the ancestral clade, (ii) the fragmented clades should be separated by a large number of mutational steps, and (iii) this pattern should affect only a part of the gene genealogy (for example see Fig. 1 in Carbone and Kohn 2001a).

Results from computer simulations using coalescent theory with several models that include both recurrent and non-recurrent events support the same basic predictions. For example, under a gene flow model, coalescent theory would predict an increase in the geographical distribution of individuals as the evolutionary age of the lineage increases (Hudson 1990). To date, intraspecific phylogeographic methods have relied heavily on the overlay of geography (essentially by eye-ball) on gene genealogies as a method of detecting associations of geography with genotypic variation (Avise 1989; Avise 1994; Avise 1998; O'Donnell *et al.* 1998a; Vilgalys and Sun 1994). Although superimposing geographic distributions on the phylogeny is helpful as an initial step in exploring the data, this approach does not provide (i) a way of testing the null hypothesis of no geographic association, (ii) an assessment of whether sample sizes are sufficient to test among alternative hypotheses, or (iii) a framework for inferring the evolutionary processes that created observed patterns of geographical association.

A powerful method for investigating genotype-phenotype associations within a species, and an entry point to an analytical method for identifying recurrent and non-recurrent population processes, is conversion of each gene genealogy to a nested design (Templeton *et al.* 1987). The first step in the nested design is to convert the gene genealogy into a haplotype network. Templeton *et al.* (1987) have proposed an algorithm for estimating the probability of all non-parsimonious connections among haplotypes to include only those haplotype connections in the haplotype tree with probabilities ≥ 95%. The estimated haplotype network with ambiguities is converted into the nested design using nesting rules (outlined in Crandall 1996; Templeton *et al.* 1987; Templeton and Sing 1993). The nesting procedure involves first grouping together neighboring haplotypes in the network that differ by one mutational step in 1-step clades, followed by clustering of 1-step clades in 2-step clades, and so on, until all individuals are grouped in a nested hierarchy (see Fig. 5).

One advantage of using a nested hierarchical scheme is that even in the absence of a root for the haplotype network, older lineages are usually found at interior nodes or at higher clade levels. This is because older lineages have more mutational derivatives than recent lineages, which are preferentially found on the tips of the tree or at the lowest clade level (Castelloe and

Templeton 1994). While the nested design indicates the relative ages of lineages found at different clade levels, it does not indicate the age-ordering of lineages that belong to the same clade level. For this task, coalescent theory can be applied.

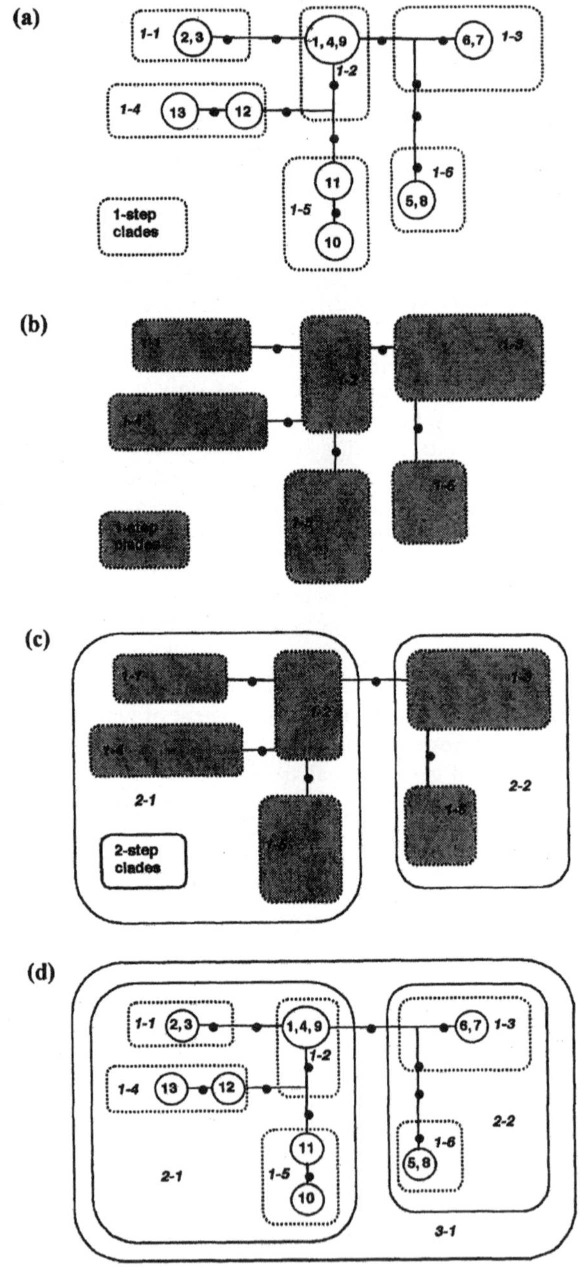

Fig. 5. Conversion of a phylogeny to a nested design for tests of association: host and clade, geographical location and clade. A hypothetical example of the steps for converting a phylogenetic network to a nested design and testing for phenotypic associations. **(a)** Start with the unrooted haplotype network from the example in Fig. 4. In the network, haplotypes (enclosed in circles) are referred to as 0-step clades because all individuals within 0-step clades have identical sequences. The first step in the nesting procedure is to group all haplotypes (0-step clades) that are separated by a single mutation into 1-step clades. The nesting is always performed starting with tip clades and moving toward the interior of the network, following the nesting rules (Crandall 1996; Templeton *et al.* 1987; Templeton and Sing 1993). **(b)** The 0-step clades within each 1-step clade are pooled such that 1-step clades are now the fundamental units for subsequent nesting. The nesting continues by grouping together all 1-step clades that are separated by a single mutation into 2-step clades **(c)** and then grouping 2-step clades into 3-step clades **(d)**. In this example, the entire cladogram is nested into one 3-step clade. The total unrooted nested haplotype network in **(d)** is used for performing nested contingency analysis. Each nesting level provides an independent grouping of clades from the previous level. Consequently, the tests of association performed at each clade level with the different phenotypic categories (*e.g.* geography or host) are also independent from the outcomes at other clade levels. In some cases, 1-step clades contain only one haplotype (*e.g.*, within 1-3 and 1-6) and cannot be tested for significant haplotype-phenotype associations at the 1-step clade level. However, the nested design provides a subsequent grouping of 1-step clades into 2-step clades such that tests of association can be performed at the 2-step clade level (*e.g.*, within 2-2 for clades 1-3 and 1-6).

The nested haplotype network can be used to test for a wide range of associations. For example, any association of haplotypes with geography can be determined using a random, two-way, contingency permutation analysis where geography is treated as a categorical variable. Significant association of geography with haplotype is an indication of restricted gene flow. If a significant geographical association is detected, then geographical distance can be considered. Determining the association between geographical distance and haplotype is a prerequisite for testing alternative hypotheses explaining restricted gene flow by discriminating among short- or long-distance dispersal events (*e.g.*, isolation by distance, range expansion, allopatric fragmentation). Two measures of geographical distance are calculated for sister clades within each nesting level. First, the average clade distance or D_C is calculated for each nested interior or tip clade. This is a measure of the geographical range of each nested sister clade. To calculate D_C, the geographical center of the clade is first calculated by averaging the latitude and longitude (in decimal degrees) for all sampling locations within the nested clade. Then, the distance separating each haplotype within the nested clade from its geographical center is calculated, using the formula for great circle distances. Finally, these haplotype distances are averaged to obtain the D_C for each interior or tip clade. The second geographical measure is the average nested clade distance or D_n calculated between the nested interior or tip clades. This is a measure of the relative geographical distribution of sister clades. This is calculated in a similar fashion to D_C, except that the geographical center is now calculated for all haplotypes within the nesting level and not for each nested sister clade separately.

The null hypothesis of no geographical association of clades can be tested using a random permutation procedure (Roff and Bentzen 1989). For each random permutation of interior and tip clades versus sampling location, the D_C and D_n distances are recalculated and this is repeated to obtain the distributions for D_C and D_n. In this two-way exact contingency test, a minimum of 1000 permutations is required for a 5% level of significance. Given that a significant geographical association has been detected, the next step is to determine whether the pattern of restricted gene flow has arisen from short- or from long-distance dispersal (Templeton *et al.* 1995). Under a model of restricted gene flow, older haplotypes have a wider geographical distribution and are usually interior in the cladogram or network; more recently evolved haplotypes have a more restricted geographical distribution and are usually tips in the cladogram or network (Nath and Griffiths 1996). Interior versus tip contrasts for significant D_C and D_n distance measures are important in discriminating between long- or short-distance movements (Templeton *et al.* 1995). For example, significantly larger values for D_n than for D_C in tip clades indicate long-distance population movement (allopatric fragmentation or range expansion), while concordance between D_C and D_n (*i.e.*, both significantly large or both significantly small, based on the random permutations tests, for tip clades) indicates short distance dispersal (isolation by distance). These distance measures assume that the geographical range of populations has been adequately sampled. With inadequate sampling it is possible to erroneously infer long-distance dispersal instead of isolation by distance (Templeton 1998; Templeton *et al.* 1995). It is important to note that not all nested clade analyses from different loci will yield statistically significant D_C and D_n values. This may be due to insufficient genetic resolution (not enough characters to distinguish haplotypes), small sample size, inadequate geographical sampling, extensive dispersal, or cladogram uncertainty as a result of extensive genetic exchange or recombination. Templeton and co-workers (Templeton *et al.* 1995) have provided an inference key for consistent interpretation of both significant and non-significant distance measures. The nested analysis and in particular the inference key has been criticized for not being statistical (Knowles and Maddison 2002). This limitation can be overcome by integrating the coalescent with nested clade analysis and the inference key (for an example see Carbone and Kohn 2001a). Once a significant geographical association is detected (attributed to restricted gene flow), migration rates can be estimated using methods that make use of the temporal and spatial information in gene trees (described below).

4. COALESCENT APPROACHES FOR EXAMINING GENEALOGICAL PROCESSES AND ESTIMATING POPULATION GENETIC PARAMETERS

In order to use gene genealogies to estimate population parameters and examine population processes, two things must be recognized. First, the genealogy captures the mutational history of genotypes derived relatively recently from a common ancestor. The gene genealogy at population level, unlike the sample of single individuals for each of many species, captures both ancestors and many intermediates in the mutational history of each site of a locus. Second, a sample provides a snapshot of only part of the actual ancestral tree; different samples would produce different ancestries. Although there is no way of observing the underlying ancestry of the sample, the ancestral relationships among a group of individuals can be described mathematically using a stochastic process known as the coalescent (Kingman 1982a; Kingman 1982b; Kingman 1982c).

The coalescent is a mathematical approximation (model) of the actual ancestral structure of a population. Given a gene genealogy showing a particular configuration of variation for a sample of genes, the coalescent process evaluates all possible pathways backwards in time to the ancestral gene of the sample (Fig. 6). According to the coalescent, all extant lineages in the population at time t trace back to one common ancestral lineage at some time in the past, which is the root of the sample of lineages. All that is required to describe the coalescent is the unrooted topology that shows which DNA sequences are closely related and a time scale that determines the rate at which coalescent events occur. In the unrooted mutational network (Fig. 6), the vertices (internodes) represent lineages, and mutations are placed along the paths joining lineages (nodes).

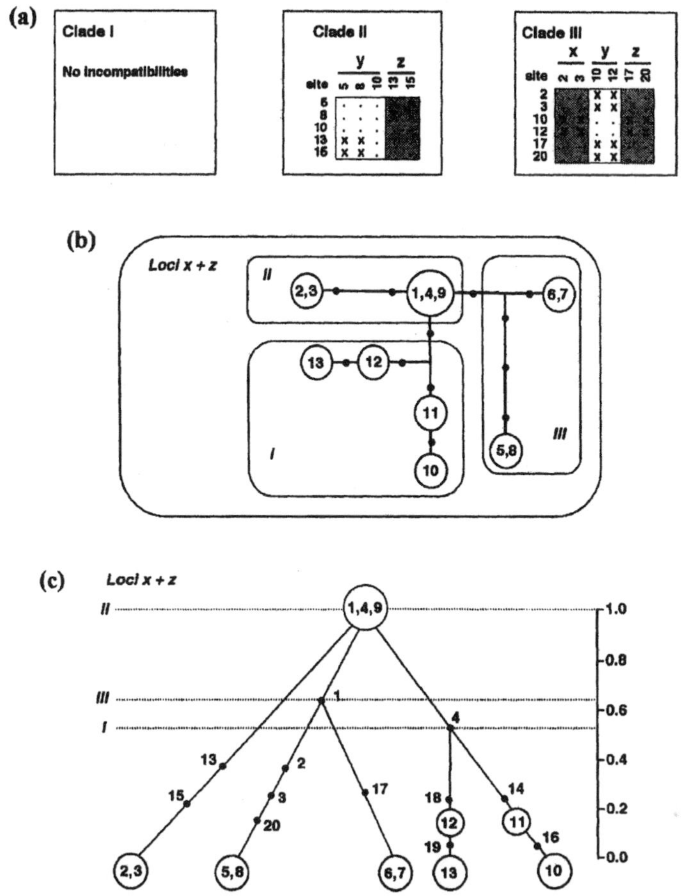

Fig. 6. Genealogical-coalescent inference and estimating ages of clades. Genealogical and coalescent methods can be combined to determine the age of recombination events, ages of mutations, and clades in our sample. First identify compatible blocks (Fig. 4) that link together a locus or loci in all clades in the sample. These blocks represent hierarchical patterns of compatibility in the entire data set. In the matrices shown in **(a)**, loci x and z have compatible histories within each clade and can be combined to infer one most parsimonious mutational network with a consistency index of 1.000 as shown in **(b)**. In the unrooted mutational network, identical haplotypes are enclosed in circles and haplotypes that belong to each of clades I, II and III are boxed. Mutations separating haplotypes are indicated with solid circles along the lines connecting haplotypes. Loci y and z have incompatible histories in clades II and III and cannot be combined without introducing significant phylogenetic conflict as shown in Fig. 4. The relative ages of clades I, II and III in **(b)** can be determined using the coalescent. The coalescent-based gene genealogy with the highest root probability is shown in **(c)**. The inferred genealogy is based on 1 million simulations of the coalescent, an estimate of θ, the population mean mutation rate as $\theta = 3.9$ (Watterson 1975) and constant population sizes and growth rates. The time scale is in coalescent units of effective population size. In the gene genealogy, the direction of divergence is from the top of the genealogy (oldest; *i.e.*, the past) to the bottom (youngest, *i.e.*, the present); coalescence is from the bottom (present) to the top (past). Since the gene genealogy is rooted, all of the mutations (solid circles with numbers) and bifurcations are also time-ordered from top to bottom. The ancestral lineage (haplotypes 1,4,9) is based on likelihood estimations from the coalescent. The configuration of mutations in the ancestral haplotype matches the consensus sequence in this region (Fig. 3). The order of clade divergence is II, III and I.

A key assumption is the infinitely-many-sites model of mutation, where there may be only one mutation at a given site in the sequence – no "multiple hits" (Kimura 1987). Another critical assumption is that the mutation rate is constant and that all mutations are neutral and sampled from a large haploid population of constant effective population size N_e. Furthermore, in the highly simplified model presented here, there can be no recombination and no selection back to the time of coalescence. This is the simplest model for describing how variation has arisen within a specific DNA sequence.

One very useful application of the coalescent is in rooting intraspecific genealogies (Griffiths and Tavaré 1994a; Griffiths and Tavaré 1995). All possible rooted trees can be inferred from any given unrooted tree by placing the root at a vertex (representing a distinct lineage in the unrooted tree) or between mutations (representing potential lineages not in the current sample), and then reading mutation paths between the root and the lineages. All positions in the unrooted tree are evaluated as potential roots for the sample of sequences. The possible roots are the extant lineages in the sample plus all other putative lineages between mutations. For the example in Fig. 6, the sample is comprised of 8 lineages, 12 mutations and 13 possible rooted trees (8 rooted trees for extant lineages plus 5 rooted trees for putative lineages between mutations). The total number of rooted trees can also be determined by adding 1 to the total number of segregating sites (s) in the sample. Since the coalescence times for different lineages within our sample are not known, there exist many topologically different coalescent trees for each rooted tree. Coalescence theory allows us to evaluate statistically all rooted topologies to determine which rooted tree is the best approximation of the true gene genealogy. Here, the assumption is that there are no other forces besides mutation acting on the sequences.

In coalescent analysis, the genealogical process is simulated many times and these simulations provide simultaneous estimates of population parameters and ancestral population processes. Coalescent modeling is particularly useful because it allows for a full likelihood analysis of evolutionary models making it possible to use likelihood ratio tests to evaluate competing phylogeographic hypotheses and to assign confidence intervals to population parameter estimates (Carbone and Kohn 2001a; Knowles and Maddison 2002). The stochastic properties of gene genealogies can be used to estimate population parameters such as rates of mutation, migration, recombination and selection. Although we have presented a simple model to explain basic concepts, to accurately model a genealogy using the coalescent, it may be necessary to consider recombination and the coalescence of lineages (Rosenberg and Nordborg 2002). Depending on the magnitude of recombination it may not be possible to represent the genealogical process as a strictly bifurcating tree, unless the DNA region is first subdivided into non-recombining partitions (Fig. 6). Several coalescent methods have been proposed for identifying recombination events at specific nucleotide positions in a sample of DNA sequences (Griffiths and Marjoram 1996; Kuhner *et al.* 2000). These methods identify non-recombining partitions as DNA segments that coalesce to the same most recent common ancestor in the history of the sample. Once the effects of recombination are removed from the sample, the coalescent can provide additional parameter estimates such as the magnitude and direction of gene flow (Bahlo and Griffiths 2000; Beerli and Felsenstein 1999; Beerli and Felsenstein 2001; Nielsen and Wakeley 2001), effective population sizes (Kuhner *et al.* 1995) and selection (Hudson and Kaplan 1995a; Neuhauser and Krone 1997). Because these coalescent-based approaches assume neutrality and no recombination they are most powerful when used in conjunction with other genealogical methods that can (i) test the neutral mutation hypothesis (Fu 1997; Fu and Li 1993;

Tajima 1989) and (ii) identify potential recombination events in the history of the sample (Fig. 4).

While other methods test for recombination in populations (Burt *et al.* 1996), the coalescence approach can also be applied to estimate the magnitude of recombination and other population processes (Harding *et al.* 1997a; Harding *et al.* 1997b). Coalescence theory can be used to estimate recombination and mutation rates (Griffiths and Marjoram 1996; Griffiths and Tavaré 1994b; Hey and Wakeley 1997; Wakeley and Hey 1997), the times to the most recent common ancestor (TMRCA) of different sequences or haplotypes (Harding *et al.* 1997a; Harding *et al.* 1997b), the ages of mutations, migration rates and effective population sizes (Beerli and Felsenstein 1999; Beerli and Felsenstein 2001), and even the number of recombination events in the ancestry of the sample (Griffiths and Marjoram 1996). In the example shown in Fig. 4 migration estimates could be based on variation segregating in regions that are non-recombining (*i.e.* same recombination block). Regions falling in the same block (loci x and z) can be examined simultaneously and more accurate migration estimates can be obtained by summing over all compatible loci. By adding more sites, the combined analysis provides a more accurate estimate of the genealogy, the underlying migration patterns, and effective population sizes (Beerli and Felsenstein 1999; Beerli and Felsenstein 2001). In simulation studies, migration estimates were closer to their true values when the number of sites per locus was increased or when parameter estimates were obtained by summing over multiple unlinked loci (Beerli and Felsenstein 1999). Regions with different evolutionary histories (*i.e.* different recombination blocks – locus y in Fig. 4) could be treated as independent unlinked loci with recombination between them. This intuitive interpretation requires further testing with empirical and simulated datasets. Although the coalescent has traditionally been used to model the ancestral history in populations, it is not applicable exclusively to population history since populations may have both intra- and interspecific components. This makes the coalescent the tool of choice for studying both population and species-level processes.

In addition to examining the distribution and rates of migration, mutation and recombination in the ancestral histories of populations, the coalescent-based gene genealogies will allow us to examine patterns of divergence at the amino acid level. Although positive selection is necessary for the evolution of novel gene function (Benner and Gaucher 2001; Benner *et al.* 1994; Fukami-Kobayashi *et al.* 2002; Gaucher *et al.* 2001), both drift and negative selection have been reported as important diversifying mechanism in viruses (Kils-Hütten *et al.* 2001; Carbone *et al.* unpublished) and complex gene families (Ohta 2000). Inferences on selective pressures can be based on the ratio of nonsynonymous (r) to synonymous (s) substitutions for different genes, such that a ratio of $r/s = 1$ would suggest selective neutrality, $r/s > 1$ positive selection and $r/s < 1$ negative selection (Ohta 2000). This approach could be used to test the hypothesis that positive selection on a gene is an important mechanism that allows invading genotypes to adapt to a new environment. The alternative hypothesis is negative selection, which can also be explained using a neutral mutation hypothesis whereby deleterious or beneficial mutations arise spontaneously and are then either purged or become fixed in the population. It will be possible to distinguish between these competing hypotheses by examining the age distribution of mutations associated with amino acid changes within a coalescent framework. Replacement substitutions that are located in deep branches of the genealogy are older and possibly not detrimental to gene function; replacement substitutions on terminal branches of the genealogy are recently evolved and may be detrimental or beneficial. It is important to note that the presence of some purifying (negative) selection does not violate the neutral mutation hypothesis and the assumption of neutrality in our coalescent model. These approaches can be used to examine the distribution and rates of selection, in addition to drift and recombination, in pathogen populations – important in estimating the magnitude of directional selection in different agroecosystems (McDonald and Linde 2002b). Furthermore, within a nested statistical framework it will be possible to test whether episodes of positive selection are significantly associated with specific transitions in disease phenotypes. Significant associations may suggest important functional domains that can be further examined using gene disruptions and gene-knock-out mutants.

5. BAYESIAN APPROACHES FOR PHYLOGENETIC INFERENCE AND ESTIMATING POPULATION PARAMETERS

All genealogical methods depend on certain assumptions about the loci on which they are based. Each locus is potentially subjected to a variety of evolutionary forces such as selection and recombination, in addition to stochastic variation. These forces can significantly distort estimates of different population parameters, such as mutation, recombination and migration rates.

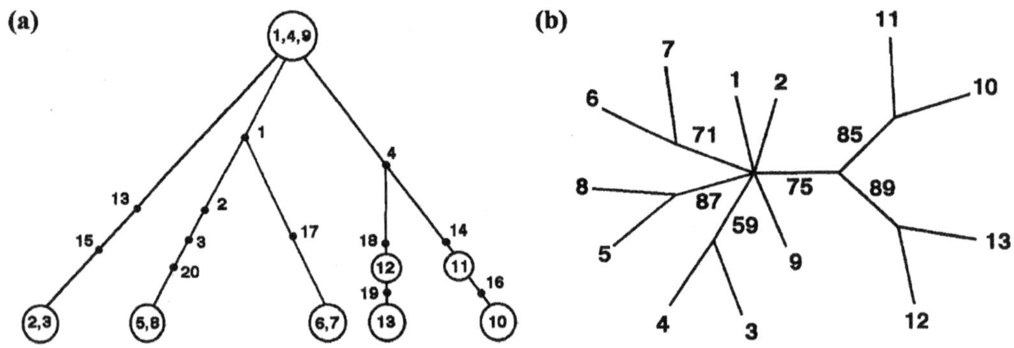

Fig. 7. Bayesian and coalescent inference of phylogeny. **(a)** In the simplest coalescent model (Fig. 6), the ancestral history of the sample was inferred by assuming a constant population mean mutation rate (Watterson's estimate) and no recombination in the history of the sequences. Assuming a starting substitution parameter value of $\theta = 3.9$, the coalescent was used to obtain a maximum likelihood estimate of the tree with the highest root probability, shown in **(a)**, which is our best inference of phylogeny. **(b)** In Bayesian analysis, a substitution model is specified for substitution parameter estimation and a starting number of generations of the Markov chain to initiate the Markov Chain Monte Carlo (MCMC) analysis. MCMC explores the parameter space by sampling trees according to their posterior probabilities (*i.e.* the joint probability density of trees, branch lengths and substitution parameters). The tree with the highest posterior probability, the best phylogenetic inference for the example described in Figs. 3-6, is shown in **(b)**, estimated using the program MRBAYES (Huelsenbeck and Ronquist 2001; http://morphbank.ebc.uu.se/mrbayes/). The substitution parameters were estimated using a time-reversible substitution model (*i.e.* substitution parameters were based on the average frequencies of nucleotides and transitions/transversions over all sequences) and substitution rates distributed equally among sites. Other possible models that could be explored, such as HKY (Hasegawa *et al.* 1985), assume gamma distributed rate variation among-sites, unequal nucleotide frequencies and different transition/transversion rates. The numbers on the interior branches represent the posterior probability of the clades in the tree, analogous to the bootstrap probability in maximum likelihood analysis. These probabilities can potentially be used to provide statistical confidence on the reliability of clades in the gene genealogy, however, the magnitude of posterior probabilities should be interpreted with caution because these estimates can be inflated (Suzuki *et al.* 2002).

Bayesian approaches can deal with multiple sources of phylogenetic uncertainty in phylogenies because they go beyond simple models of evolution (*e.g.* infinite sites) to accommodate complex parameter-rich substitution models (*e.g.* constant or gamma distributed rate variation among-sites, unequal nucleotide frequencies and different transition/transversion rates). What is a gamma distribution? The gamma distribution models site-to-site variation using one parameter, a, that determines the shape of the distribution. In searching for the best tree different gamma shape parameters are evaluated in combination with other parameters in the model (*e.g.* base frequencies, branch length) to determine the combination that maximizes the probability of the tree.

Bayesian methods address phylogenetic uncertainty by averaging inferences of evolutionary processes and parameter estimates over all possible phylogenies, in a manner similar to the coalescent (Huelsenbeck *et al.* 2000). It is important to note that both Bayesian and coalescent methods estimate parameters and accommodate uncertainty in phylogenies using similar mathematical approaches that are conditional on the observed data. The difference between the two methods lies in how the starting parameters for the coalescent process are defined (Fig. 7). The coalescent treats starting parameter estimates (*i.e.,* substitution, migration and population growth rates) as nonrandom variables. In Bayesian inference these starting parameters are modeled as probability distributions and estimated using maximum likelihood. After parameter estimation, Bayesian analysis implements Bayes formula to calculate the posterior probability, defined as the product of the likelihood and the prior probability, *i.e.*, the probability that some hypothesis is true prior to sampling. Instead of calculating likelihoods for all possible outcomes using Markov Chain Monte Carlo (MCMC) as performed in the coalescent, Bayesian inferences uses MCMC to estimate all possible posterior tree probabilities. The posterior probability of a tree can be interpreted as the probability that the estimated tree is the true tree under a particular evolutionary model (Fig. 7).

What is a Markov chain? Within a genealogical framework, a simple example of a Markov chain is an infinite-sites model, where mutations occur randomly along a sequence, but only once at a given site such that the probability of a mutation occurring in a given time interval depends only on the probability of a mutation occurring in the previous time interval. If we assume that the probability of transitioning from one generation to another (*i.e.* successive nodes in a genealogy) follows a Poisson distribution with the mean given by the product of the mutation rate and branch length, then the time between nodes in the genealogy becomes a Markov chain where the probability of the entire genealogy can be estimated by summing the probabilities of one or more successive generations in the tree. For larger samples computing these continuous probability distributions is computationally prohibitive and a combined MCMC method is used instead to estimate the probability of the genealogy. MCMC methods start with the current sample genealogy and perform multiple independent simulations of the genealogy to determine the approximate times between nodes.

In the Bayesian framework, the tree with the maximum posterior probability is interpreted as our best inference of phylogeny. Other applications of Bayesian inference include estimating divergence times of species with or without the assumption of a molecular clock (Huelsenbeck *et al.* 2000) and detecting selection (Nielsen and Huelsenbeck 2002). Some caution should be exercised when using posterior probabilities for assessing the reliability of interior branches (or clades) in phylogenetic trees as the rate of false-positives can be quite high (Suzuki *et al.* 2002). Several Bayesian approaches to estimating population parameters and genealogical history simultaneously have also been proposed (Drummond *et al.* 2002; Nielsen 2000). When individuals are sampled from a population at different time intervals, a combination of Bayesian and coalescent-based methods tend to perform better than using either method on its own (for an example, see Drummond *et al.* 2002).

6. THE POPULATION-SPECIES INTERFACE

From an evolutionary perspective, species cannot be static entities. There is a continuum from genetically-distinct individuals in populations, through populations of phenotypically similar individuals in sibling species, to reproductively isolated and fully diverged species. Since a continuum of genetic variation and group divergence exists, it is difficult to determine exactly when genetically-distinct groups of individuals should be recognized as sibling species and when sibling species should be recognized as species. While the general concept of a species has been widely accepted by biologists as an entity that defines a reproductively isolated and genetically-distinct group of phenotypically similar individuals, the criteria for species delimitation have been a source of controversy (Darwin 1859; Dobzhansky 1951; Mayr 1942; Mayr 1970). In fact, the delimitation of taxonomic species is somewhat at odds with the dynamic process of speciation. Both gene genealogies and species trees provide an historical framework that allows us to study both population and species-level processes. In order to study speciation processes by investigating the population-species interface, phylogenies must span both the population level and the species level. By necessity, species level phylogenies originate from top-down studies informed by taxonomic species concepts. DNA sequences with variation at only one of these levels contain limited information about the genetics of the speciation process. Only DNA sequences that resolve at both levels can be used to infer both population and species-level trees.

When species are well-defined, genetic variation is sufficient to delimit their boundaries. Many studies have sought such defining patterns of genetic variation (flies: Bush 1969; Gleason *et al.* 1998; Schloetterer *et al.* 1994; birds: Avise 1994; Freeman and Zink 1995; plants: Rieseberg *et al.* 1996; fungi: Carbone and Kohn 1993; Craven *et al.* 2001; Fisher *et al.* 2002a; LoBuglio *et al.* 1996; Lutzoni and Vilgalys 1995; O'Donnell 1996; O'Donnell 2000; Skupski *et al.* 1997; Taylor *et al.* 1999b). While this "top-down" approach finds well-defined patterns, it lacks resolution when species are not well-defined and affords limited insight into the speciation process (Templeton 1994). Here, a "bottom-up", micro-evolutionary approach, based on population sampling over the geographical range of the "top-down"-defined species units (Templeton 1994) is warranted. This approach views individuals in a species as sharing adaptations to a locale or niche that are shaped through time and space by specific evolutionary processes, such as gene flow, genetic drift, selection, mutation and recombination. Recent studies have shown that bottom-up approaches are useful for delimiting the boundaries of closely related species and for elucidating the forces driving population divergence and speciation (Routman 1993; Templeton 1994; Templeton 1998; Templeton *et al.* 1995).

Once genetic variation spanning the species-population interface has been identified, the study of the genetics of speciation can begin. When approaching this interface from the species level, it is important to distinguish genetic variation that was involved in the speciation process from other variation responsible for species differences that has evolved since the speciation event. A potential source of difficulty arises when nucleotide sequence variation among species is great. While a high degree of genetic divergence results in species that are phylogenetically well-defined entities, it becomes difficult to trace back the ancestral history of species to infer what polymorphisms were involved in the speciation process. The sharing of polymorphisms and the splitting of ancestral polymorphisms among species can further confound the problem, as evidenced by incongruencies between species trees and gene trees (Avise 1989). At the species level, the ratio of shared to fixed polymorphisms is very small. Looking back in time, this ratio increases as the sibling species level approaches. At this level, the number of fixed polymorphisms is smaller, yet sufficient to define siblings as phylogenetically distinct entities. Further extensions downward to the sibling species-population interface obscure phylogenetic resolution. As the speciation event is approached, the ratio of shared to fixed polymorphisms becomes larger, making it very difficult to focus on the speciation process. When approaching the actual speciation process phylogenetic resolution breaks down entirely because of the paucity of genetic variation. So while this "top-down" macroevolutionary approach is ideally suited to detecting lineages that might be species, it provides few insights into the genetics of speciation. Upward extensions from the population level to the species-population interface should shed light on the speciation process. In this "bottom-up" approach the focus is on using gene genealogies as tools to measure the extent of genetic variation within clonal lineages, genetically isolated populations and sibling species to define the boundaries of a species and to identify the microevolutionary forces driving speciation (Templeton 1994).

This approach was used to study speciation in three closely related fungal species of the genus *Aspergillus* (Geiser *et al.* 1998). Gene genealogies were inferred from eleven protein-encoding loci for thirty-one isolates of *A. flavus*, two isolates of *A. parasiticus* and five isolates of *A. oryzae*. For each locus, isolates of *A. flavus* grouped into two distinct clades, with few shared polymorphisms, resulting in one long evolutionary branch separating the two clades. A long branch between the two groups could indicate a long history of reproductive isolation, and was interpreted here as a cryptic speciation event within *A. flavus*. Although the three species were collected from different geographical areas, all isolates of *A. flavus* were sampled from the same geographical area. Without rejecting geographic divergence among population samples of *A. flavus,* the alternative interpretation cannot be rejected that the low level of shared polmorphisms among the two *A. flavus* groups resulted from a fragmentation event not necessarily followed by reproductive isolation. The two groups could be two geographically separated populations rather than cryptic species. A number of other studies have used a similar approach to detect cryptic speciation within fungal species complexes (Burt *et al.* 1996; Geiser *et al.* 1998; Koufopanou *et al.* 1997; O'Donnell *et al.* 1998a; Steenkamp *et al.* 2002).

More definitive evidence of speciation is the formation of a hybrid zone, an area of contact between geographically contiguous populations where hybridization takes place (Arnold 1997; Brasier *et al.* 1999; Rieseberg *et al.* 1988; Schardl 2001). Even in populations which are today asexual, or in sexual populations of individuals that preferentially self-fertilize, a hybrid zone might exist where historical genetic exchange and recombination have resulted in a decoupling of molecular characters that were completely coupled on either side of the hybrid zone. The existence of such a hybrid zone could mean that speciation has been incomplete. It has been argued that hybrid zones are the result of range expansions following allopatric speciation. Although determining which of these mechanisms created the hybrid zone would be difficult, elucidating the genetic structure of the hybrid zone may be more important in the study of speciation. Arnold (1997) has proposed the Evolutionary Novelty model, which emphasizes the importance of reticulation in hybrid zones, as a mechanism for creating novel evolutionary lineages.

Both species and population-level phylogenies are necessary to examine the evolutionary forces that shaped the present geographical patterns, such as, gene flow, drift (especially bottlenecks), and selection (Harrison 1991; Templeton 1994). The limitations of species trees in examining the speciation process can be overcome by incorporating a bottom-up, nested statistical approach based on population sampling (Templeton 1994; Templeton 1998; Templeton *et al.* 1995). The nesting is dictated by the haplotype network. In the nested analysis, geographical range is treated as a variable character that can change throughout the evolutionary history of the species. With the nested design, it is possible to test for the existence of a geographical pattern by performing a nested contingency analysis in which each geographical location is treated as a categorical variable. By adding geographical distance to the analysis it is possible to discriminate statistically among the alternative geographical processes. Treating geographical location as a dynamic variable acknowledges the possibility that geographical ranges can expand and contract through time, and that these changes can alter geographical patterns and affect the course of speciation. For example, if the geographical ranges of allopatric species expand so that they overlap, or if migration occurs, then gene flow can resume. In sexual populations, the amount of gene flow depends on the ability of individuals in the populations to interbreed. In asexual populations, gene flow may be detected as a past, historical process. The initial fragmentation event could be the defining starting point of speciation in organisms with predominantly asexual life histories. The contributions of specific genetic, morphological, or ecological-demographic adaptations in the speciation process could also be tested using the same nested statistical design that was used for testing for geographical associations. Concordance or discordance among ecological, morphological and molecular data sets provides increased resolution into the mechanisms of speciation. As a result, nested clade analysis becomes a powerful tool for examining both the geographical patterns and evolutionary mechanisms that are responsible for the speciation process.

7. CONCLUSIONS

While fungal genomics data, especially on whole genomes, cannot accrete fast enough to satisfy our needs in more fully parsing out fungal molecular evolutionary processes and their commonalities and unique features compared with other eukaryotes, we are well ahead on the bioinformatic aspects, *i.e.* powerful analytical methods for inferring process as well as pattern. With substantial sequencing of multiple coding and non-coding genomic regions, based on considered sampling of isolates, we have analytical techniques in hand, and new ones nearly in hand, for incisive statistical exploration of the genomic data. In particular, watch for improved models for inferring network (not tree) genealogies that fully incorporate recombination using coalescent approaches, as well as the extensive deployment of Bayesian approaches for hypothesis testing.

Acknowledgements: We thank the Natural Sciences Engineering and Research Council of Canada for continuing research support.

REFERENCES

Anderson JB and Kohn LM (1998). Genotyping, gene genealogies and genomics bring fungal population genetics above ground. Trends Ecol Evol 13:444-449.

Anderson JB, Wickens C, Khan M, Cowen LE, Federspiel N, Jones T, and Kohn LM (2001) Infrequent genetic exchange and recombination in the mitochondrial genome of *Candida albicans*. J Bacteriol 183:865-872.

Antonovics J and Kareiva P (1988) Frequency-dependent selection and competition: Empirical approaches. Philos Trans R Soc Lond B Biol Sci 319:601-614.

Arnold ML (1997). Natural hybridization and evolution. Oxford: Oxford University Press.

Avise JC (1989) Gene trees and organismal histories: A phylogenetic approach to population biology. Evolution 43:1192-1208.

Avise JC (1994) Molecular Markers, Natural History and Evolution. New York: Chapman and Hall.

Avise JC (1998) The history and purview of phylogeography: a personal reflection. Mol Ecol 7:371-379.

Avise JC (2000) Phylogeography : the history and formation of species. Cambridge, MA: Harvard University Press.

Bahlo M and Griffiths RC (2000) Inference from gene trees in a subdivided population. Theor Popul Biol 57:79-95.

Bandelt HJ, Forster P, and Roehl A (1999) Median-joining networks for inferring intraspecific phylogenies. Mol Biol Evol 16:37-48.

Barker FK and Lutzoni FM (2002) The utility of the incongruence length difference test. Syst Biol 51:625-637.

Beerli P and Felsenstein J (1999) Maximum-likelihood estimation of migration rates and effective population numbers in two populations using a coalescent approach. Genetics 152:763-773.

Beerli P and Felsenstein J (2001) Maximum likelihood estimation of a migration matrix and effective population sizes in n subpopulations by using a coalescent approach. Proc Natl Acad Sci USA 98:4563-4568.

Benner SA and Gaucher EA (2001) Evolution, language and analogy in functional genomics. Trends Genet 17:414-418.

Benner SA, Jenny TF, Cohen MA, and Gonnet GH (1994) Predicting the conformation of proteins from sequences. Progress and future progress. Adv Enzyme Regul 34:269-353.

Brasier CM, Cooke DEL, and Duncan JM (1999) Origin of a new *Phytophthora* pathogen through interspecific hybridization. Proc Natl Acad Sci USA:5878-5883.

Brunet J and Mundt CC (2000) Disease, frequency-dependent selection, and genetic polymorphisms: experiments with stripe rust and wheat. Evolution 54:406-415.

Burdon J (1993) The structure of pathogen populations in natural plant communities. Annu Rev Phytopathol 31:305-323.

Burt A, Carter DA, Koenig GL, White TJ, and Taylor JW (1996) Molecular markers reveal cryptic sex in the human pathogen *Coccidioides immitis*. Proc Natl Acad Sci USA 93:770-773.

Bush GL (1969) Sympatric host race formation and speciation in frugivorous flies of the genus *Rhagoletis* (Diptera, Tephritidae). Evolution 23:237-251.

Carbone I, Anderson JB, and Kohn LM (1999) Patterns of descent in clonal lineages and their multilocus fingerprints are resolved with combined gene genealogies. Evolution 53:11-21.

Carbone I and Kohn LM (1993) Ribosomal DNA sequence divergence within internal transcribed spacer 1 of the Sclerotiniaceae. Mycologia 85:415-427.

Carbone I and Kohn LM (1999) A method for designing primer sets for speciation studies in filamentous ascomycetes. Mycologia 91:553-556.

Carbone I and Kohn LM (2001a) A microbial population-species interface: nested cladistic and coalescent inference with multilocus data. Mol Ecol 10:947-964.

Carbone I and Kohn LM (2001b) Multilocus nested haplotype networks extended with DNA fingerprints show common origin and fine-scale, ongoing genetic divergence in a wild microbial metapopulation. Mol Ecol 10:2409-2422.

Castelloe J and Templeton AR (1994) Root probabilities for intraspecific gene trees under neutral coalescent theory. Mol Phyl Evol 3:102-113.

Ceresini PC, Shew HD, Vilgalys RJ, and Cubeta MA (2002) Genetic diversity of *Rhizoctonia solani* AG-3 from potato and tobacco in North Carolina. Mycologia 94:437-449.

Chen RS and McDonald BA (1996) Sexual reproduction plays a major role in the genetic structure of populations of the fungus *Mycosphaerella graminicola*. Genetics 142:1119-1127.

Couch BC and Kohn LM (2002) A multilocus gene genealogy concordant with host preference indicates segregation of a new species, *Magnaporthe oryzae*, from *M. grisea*. Mycologia 94:683-693.

Cowen LE, Nantel A, Whiteway MS, Thomas DY, Tessier DC, Kohn LM, and Anderson JB (2002) Population genomics of drug resistance in *Candida albicans*. Proc Natl Acad Sci USA 99:9284-9289.

Crandall KA (1996) Multiple interspecies transmissions of human and simian T-cell leukemia/lymphoma virus type I sequences. Mol Biol Evol 13:115-131.

Craven KD, Hsiau PTW, Leuchtmann A, Hollin W, and Schardl CL (2001) Multigene phylogeny of *Epichloe* species, fungal symbionts of grasses. Annals of the Missouri Botanical Garden 88:14-34.

Darlu P and Lecointre G (2002) When does the incongruence length difference test fail? Mol Biol Evol 19:432-437.

Darwin C (1859) On the origin of species by means of natural selection or the preservation of favoured races in the struggle for life. London, UK: John Murray.

De Queiroz K (1998) The general lineage concept of species, species criteria, and the process of speciation: a conceptual unification and terminological recommendations. In: DJ Howard, SH Berlocher, ed. Endless Forms: Species and Speciation. New York: Oxford University Press, pp. 57-75.

Dobzhansky T (1951). Genetics and the origin of species. New York: Columbia University Press.
Drummond AJ, Nicholls GK, Rodrigo AG, and Solomon W (2002). Estimating mutation parameters, population history and genealogy simultaneously from temporally spaced sequence data. Genetics 161:1307-1320.
Dykhuizen DE and Green L (1991). Recombination in *Escherichia coli* and the definition of biological species. J Bacteriol 173:7257-7268.
Felsenstein J (1981). Evolutionary trees from DNA sequences: A maximum likelihood approach. J Mol Evol 17:368-376.
Fisher MC, Koenig G, White TJ, and Taylor JW (2000). A test for concordance between the multilocus genealogies of genes and microsatellites in the pathogenic fungus *Coccidioides immitis*. Mol Biol Evol 17:1164-1174.
Fisher MC, Koenig GL, White TJ, San-Blas G, Negroni R, Alvarez IG, Wanke B, and Taylor JW (2001). Biogeographic range expansion into South America by *Coccidioides immitis* mirrors New World patterns of human migration. Proc Natl Acad Sci USA 98:4558-4562.
Fisher MC, Koenig GL, White TJ, and Taylor JW (2002a) Molecular and phenotypic description of *Coccidioides posadasii* sp. nov., previously recognized as the non-California population of *Coccidioides immitis*. Mycologia 94:73-84.
Fisher MC, Rannala B, Chaturvedi V, and Taylor JW (2002b) Disease surveillance in recombining pathogens: multilocus genotypes identify sources of human *Coccidioides* infections. Proc Natl Acad Sci USA 99:9067-9071.
Freeman S and Zink RM (1995) A phylogenetic study of the blackbirds based on variation in mitochondrial DNA restriction sites. Syst Biol 44:409-420.
Fregene MA, Vargas J, Ikea J, Angel F, Tohme J, Asiedu RA, Akorda MO, and Roca WM (1994) Variability of chloroplast DNA and nuclear ribosomal DNA in cassava (*Manihot esculenta* Crantz) and its wild relatives. Theor Appl Genet 89:719-727.
Fu Y-X (1997) Statistical tests of neutrality of mutations against population growth, hitchhiking and background selection. Genetics 147:915-925.
Fu Y-X and Li W-H (1993). Statistical tests of neutrality of mutations. Genetics 133:693-709.
Fukami-Kobayashi K, Schreiber DR, and Benner SA (2002). Detecting compensatory covariation signals in protein evolution using reconstructed ancestral sequences. J Mol Biol 319:729-743.
Gaucher EA, Miyamoto MM, and Benner SA (2001). Function-structure analysis of proteins using covarion-based evolutionary approaches: Elongation factors. Proc Natl Acad Sci U S A 98:548-552.
Geiser DM, Juba JH, Wang B, and Jeffers SN (2001). *Fusarium hostae* sp. nov., a relative of *F. redolens* with a *Gibberella* teleomorph. Mycologia 93:670-678.
Geiser DM, Pitt JI, and Taylor JW (1998). Cryptic speciation and recombination in the aflatoxin-producing fungus *Aspergillus flavus*. Proc Natl Acad Sci USA 95:388-393.
Gielly L and Taberlet P (1994). The use of chloroplast DNA to resolve plant phylogenies: Noncoding versus rbcL sequences. Mol Biol Evol 11:769-777.
Gleason JM, Griffith EC, and Powell JR (1998). A molecular phylogeny of the *Drosophila willistoni* group: Conflicts between species concepts? Evolution 52:1093-1103.
Griffiths RC and Marjoram P (1996) Ancestral inference from samples of DNA sequences with recombination. J Computat Biol 3:479-502.
Griffiths RC and Tavaré S (1994a) Ancestral inference in population genetics. Stat Sci 9:307-319.
Griffiths RC and Tavaré S (1994b) Simulating probability distributions in the coalescent. Theor Popul Biol 46:131-159.
Griffiths RC and Tavaré S (1995) Unrooted genealogical tree probabilities in the infinitely-many-sites model. Math Biosci 127:77-98.
Harding RM, Fullerton SM, Griffiths RC, Bond J, Cox MJ, Schneider JA, Moulin DS, and Clegg JB (1997a) Archaic African and Asian lineages in the genetic ancestry of modern humans. Am J Hum Genet 60:772-789.
Harding RM, Fullerton SM, Griffiths RC, and Clegg JB (1997b) A gene tree for beta-globin sequences from Melanesia. J Mol Evol 1:S133-S138.
Harrison RG (1991) Molecular changes at speciation. Annu Rev Ecol Syst 22:281-308.
Hasegawa M, Kishino H, and Yano T (1985). Dating of the human-ape splitting by a molecular clock of mitochondrial DNA. J Mol Evol 22:160-174.
Hein J (1990). Reconstructing evolution of sequences subject to recombination using parsimony. Math Biosci 98:185-200.
Hein J (1993). A heuristic method to reconstruct the history of sequences subject to recombination. J Mol Evol 36:396-405.

Hey J and Wakeley J (1997). A coalescent estimator of the population recombination rate. Genetics 145:833-846.

Hillis DM and Huelsenbeck JP (1992) Signal, noise, and reliability in molecular phylogenetic analyses. J Hered 83:189-195.

Huber KT, Watson EE, and Hendy MD (2001). An algorithm for constructing local regions in a phylogenetic network. Mol Phyl Evol 19:1-8.

Hudson RR (1990) Gene genealogies and the coalescent process. Oxf Surv Evol Biol 1990:1-44.

Hudson RR and Kaplan NL (1995a) .The coalescent process and background selection. Philos Trans R Soc Lond B Biol Sci 349:19-23.

Hudson RR and Kaplan NL (1995b). Deleterious background selection with recombination. Genetics 141:1605-1617.

Hudson RR, Slatkin M, and Maddison WP (1992). Estimation of levels of gene flow from DNA sequence data. Genetics 132:583-589.

Huelsenbeck JP, Larget B, and Swofford D (2000). A compound poisson process for relaxing the molecular clock. Genetics 154:1879-1892.

Huelsenbeck JP and Ronquist F (2001). MRBAYES: Bayesian inference of phylogenetic trees. Bioinformatics 17:754-755.

Hurles M, Bailey J, and Eichler E (2002). Are 100,000 "SNPs" Useless? Science 298:1509a.

Jakobsen IB, Wilson SR, and Easteal S (1997) The partition matrix: exploring variable phylogenetic signals along nucleotide sequence alignments. Mol Biol Evol 14:474-484.

Kang S, Ayers JE, DeWolf ED, Geiser DM, Kuldau G, Moorman GW, Mullins E, Uddin W, Correll JC, Deckert G, Lee YH, Lee YW, Martin FN, and Subbarao K (2002). The internet-based fungal pathogen database: A proposed model. Phytopathol 92:232-236.

Keller SM, McDermott JM, Pettway RE, Wolfe MS, and McDonald BA (1997). Gene flow and sexual reproduction in the wheat glume blotch pathogen *Phaeosphaeria nodorum* (anamorph *Stagonospora nodorum*). Phytopathol 87:353-358.

Kils-Hütten L, Cheynier R, Wain-Hobson S, and Meyerhans A (2001). Phylogenetic reconstruction of intrapatient evolution of human immunodeficiency virus type 1: predominance of drift and purifying selection. J Gen Virol 82:1621-1627.

Kimura M (1987). Molecular evolutionary clock and the neutral theory. J Mol Evol 26:24-33.

Kingman JFC (1982a). On the genealogy of large populations. J App Prob 19:27-43.

Kingman JFC (1982b). Exchangeability and the evolution of large populations. In: G Koch, F Spizzichino, ed. Exchangeability in Probability and Statistics. Amsterdam: North-Holland, pp. 97-112.

Kingman JFC (1982c). The coalescent. Stoch Processes Appl 13:235-248.

Knowles LL and Maddison WP (2002). Statistical phylogeography. Mol Ecol 11:2623-2635.

Kohli Y and Kohn LM (1996) Mitochondrial haplotypes in populations of the plant-infecting fungus *Sclerotinia sclerotiorum*: wide distribution in agriculture, local distribution in the wild. Mol Ecol 5:773-783.

Kohn LM, Stasovski E, Carbone I, Royer J, and Anderson JB (1991). Mycelial incompatibility and molecular markers identify genetic variability in field populations of *Sclerotinia sclerotiorum*. Phytopathol 81:480-485.

Koufopanou V, Burt A, and Taylor JW (1997). Concordance of gene genealogies reveals reproductive isolation in the pathogenic fungus *Coccidioides immitis*. Proc Natl Acad Sci USA 94:5478-5482.

Kretzer AM and Bruns TD (1999). Use of atp6 in fungal phylogenetics: an example from the boletales. Mol Phylogenet Evol 13:483-492.

Kroken S and Taylor JW (2001). A gene genealogical approach to recognize phylogenetic species boundaries in the lichenized fungus *Letharia*. Mycologia 93:38-53.

Kuhner MK, Yamato J, and Felsenstein J (1995). Estimating effective population size and mutation rate from sequence data using Metropolis-Hastings sampling. Genetics 140:1421-1430.

Kuhner MK, Yamato J, and Felsenstein J (2000) .Maximum likelihood estimation of recombination rates from population data. Genetics 156:1393-1401.

Kumar J, Nelson RJ, and Zeigler RS (1999). Population structure and dynamics of *Magnaporthe grisea* in the Indian Himalayas. Genetics 152:971-984.

Leung H, Nelson RJ, and Leach JE (1993). Population structure of plant pathogenic fungi and bacteria. Adv plant pathol 10:157-205.

Linde CC, Zhan J, and McDonald BA (2002). Population structure of *Mycosphaerella graminicola:* from lesions to continents. Phytopathol 92:946-955.

LoBuglio KF, Berbee ML, and Taylor JW (1996). Phylogenetic origins of the asexual mycorrhizal symbiont *Cenococcum geophilum* Fr. and other mycorrhizal fungi among the ascomycetes. Mol Phyl Evol 6:287-294.

Lutzoni F and Vilgalys R (1995). *Omphalina* (Basidiomycota, Agaricales) as a model system for the study of coevolution in lichens. Cryptogamic Botany 5:71-81.
Lynch M (1988). Estimation of relatedness by DNA fingerprinting. Mol Biol Evol 5:584-599.
Maddison WP (1997). Gene trees in species trees. Syst Biol 46:523-536.
Mayr E (1942). Systematics and the origin of species. New York: Columbia University Press.
Mayr E (1970). Populations, species, and evolution. Cambridge, Massachusetts: Belknap Press.
McDonald BA (1997). The population genetics of fungi: Tools and techniques. Phytopathol 87:448-453.
McDonald BA and Linde C (2002a). Pathogen population genetics, evolutionary potential, and durable resistance. Annu Rev Phytopathol 40:349-379.
McDonald BA and Linde C (2002b). The population genetics of plant pathogens and breeding strategies for durable resistance. Euphytica 124:163-180.
McDonald BA, Pettway RE, Chen RS, Boeger JM, and Martinez JP (1995). The population genetics of *Septoria tritici* (teleomorph *Mycosphaerella graminicola*). Can J Bot 73:S292-S301.
McEwen JG, Taylor JW, Carter D, Xu J, Felipe MS, Vilgalys R, Mitchell TG, Kasuga T, White T, Bui T, and Soares CM (2000). Molecular typing of pathogenic fungi. Med Mycol 38:189-197.
Milgroom MG (1996). Recombination and the multilocus structure of fungal populations. Annu Rev Phytopathol 34:457-477.
Milgroom MG, Lipari SE, and Powell WA (1992). DNA fingerprinting and analysis of population structure in the chestnut blight fungus, *Cryphonectria parasitica*. Genetics 131:297-306.
Moncalvo JM, Drehmel D, and Vilgalys R (2000). Variation in modes and rates of evolution in nuclear and mitochondrial ribosomal DNA in the mushroom genus *Amanita* (Agaricales, Basidiomycota): phylogenetic implications. Mol Phylogenet Evol 16:48-63.
Nath H and Griffiths RC (1996). Estimation in an island model using simulation. Theor Popul Biol 50:227-253.
Nei M (1973). Analysis of gene diversity in subdivided populations. Proc Natl Acad Sci USA 70:3321-3323.
Nei M (1987). Molecular Evolutionary Genetics. New York: Columbia University Press.
Neuhauser C and Krone SM (1997) The genealogy of samples in models with selection. Genetics 145:519-534.
Nielsen R (2000). Estimation of population parameters and recombination rates from single nucleotide polymorphisms. Genetics 154:931-942.
Nielsen R and Huelsenbeck JP (2002). Detecting positively selected amino acid sites using posterior predictive P-values. Pac Symp Biocomput:576-588.
Nielsen R and Wakeley J (2001) Distinguishing migration from isolation: a Markov chain Monte Carlo approach. Genetics 158:885-896.
O'Donnell K (1996). Progress towards a phylogenetic classification of *Fusarium*. Sydowia 48:57-70.
O'Donnell K (2000). Molecular phylogeny of the *Nectria haematococca-Fusarium solani* species complex. Mycologia 92:919-938.
O'Donnell K, Cigelnik E, and Nirenberg HI (1998a). Molecular systematics and phylogeography of the *Gibberella fujikuroi* species complex. Mycologia 90:465-493.
O'Donnell K, Cigelnik E, Weber NS, and Trappe JM (1997). Phylogenetic relationships among ascomycetous truffles and the true and false morels inferred from 18S and 28S ribosomal DNA sequence analysis. Mycologia 89:48-65.
O'Donnell K, Kistler HC, Cigelnik E, and Ploetz RC (1998b). Multiple evolutionary origins of the fungus causing Panama disease of banana: concordant evidence from nuclear and mitochondrial gene genealogies. Proc Natl Acad Sci USA 95:2044-2049.
O'Donnell K, Kistler HC, Tacke BK, and Casper HH (2000) .Gene genealogies reveal global phylogeographic structure and reproductive isolation among lineages of *Fusarium graminearum,* the fungus causing wheat scab. Proc Natl Acad Sci USA 97:7905-7910.
Ohta T (2000) Mechanisms of molecular evolution. Philos Trans R Soc Lond B Biol Sci 355:1623-1626.
Page RDM (1998). GeneTree: Comparing gene and species phylogenies using reconciled trees. Bioinformatics 14:819-820.
Page RDM and Charleston MA (1997) From gene to organismal phylogeny: Reconciled trees and the gene tree/species tree problem. Mol Phyl Evol 7:231-240.
Phillips DV, Carbone I, Gold SE, and Kohn LM (2002). Phylogeography and genotype–symptom associations in early and late season infections of canola by *Sclerotinia sclerotiorum*. Phytopathol 92:785-793.
Posada D (2002). Evaluation of methods for detecting recombination from DNA sequences: empirical data. Mol Biol Evol 19:708-717.

Posada D and Crandall KA (2001a). Evaluation of methods for detecting recombination from DNA sequences: computer simulations. Proc Natl Acad Sci USA 98:13757-13762.
Posada D and Crandall KA (2001b). Intraspecific gene genealogies: trees grafting into networks. Trends Ecol Evol 16:37-45.
Posada D and Crandall KA (2002). The effect of recombination on the accuracy of phylogeny estimation. J Mol Evol 54:396-402.
Posada D, Crandall KA, and Holmes EC (2002). Recombination in evolutionary genomics. Annu Rev Genet 36:75-97.
Pumo DE, Iksoo K, Remsen J, Phillips CJ, and Genoways HH (1996). Molecular systematics of the fruit bat, *Artibeus jamaicensis*: origin of an unusual island population. J Mammal 77:491-503.
Rieseberg LH, Arias DM, Ungerer MC, Linder CR, and Sinervo B (1996) The effects of mating design of introgression between chromosomally divergent sunflower species. Theor Appl Genet 93:633-644.
Rieseberg LH, Soltis DE, and Palmer JD (1988). A molecular reexamination of introgression between *Helianthus annuus* and *Helianthus bolanderi* (Compositae). Evolution 42:227-238.
Ristaino JB, Groves CT, and Parra GR (2001) PCR amplification of the Irish potato famine pathogen from historic specimens. Nature 411:695-697.
Robertson DL, Hahn BH, and Sharp PM (1995). Recombination in AIDS viruses. J Mol Evol 40:249-259.
Roff DA and Bentzen P (1989) The statistical analysis of mitochondrial DNA polymorphisms: C^2 and the problem of small samples. Mol Biol Evol 6:539-545.
Rosenberg NA and Nordborg M (2002). Genealogical trees, coalescent theory and the analysis of genetic polymorphisms. Nat Rev Genet 3:380-390.
Rosewich UL and Kistler HC (2000). Role of horizontal gene transfer in the evolution of fungi. Annu Rev Phytopathol 38:325-363.
Routman E (1993). Population structure and genetic diversity of metamorphic and paedomorphic populations of the tiger salamander, *Ambystoma tigrinum*. J Evol Biol 6:329-357.
Sang T, Crawford DJ, and Stuessy TF (1997). Chloroplast DNA phylogeny, reticulate evolution, and biogeography of *Paeonia* (Paeoniaceae). Am J Bot 84:1120-1136.
Saville BJ, Kohli Y, and Anderson JB (1998). mtDNA recombination in a natural population. Proc Natl Acad Sci USA 95:1331-1335.
Schardl CL (2001). *Epichloë festucae* and related mutualistic symbionts of grasses. Fungal Genet Biol 33:69-82.
Schloetterer C, Hauser M, T., Von H, A., and Tautz D (1994) Comparative evolutionary analysis of rDNA ITS regions in *Drosophila*. Mol Biol Evol 11:513-522.
Scribner KT, Arntzen JW, and Burke T (1994). Comparative analysis of intra- and inter-population genetic diversity in *Bufo bufo*, using allozyme, single-locus microsatellite, minisatellite, and multilocus minisatellite data. Mol Biol Evol 11:737-748.
Scribner KT and Avise JC (1993). Cytonuclear genetic architecture in mosquitofish populations and the possible roles of introgressive hybridization. Mol Ecol 2:139-149.
Scribner KT and Avise JC (1994). Population cage experiments with a vertebrate: the temporal demography and cytonuclear genetics of hybridization in *Gambusia* fishes. Evolution 48:155-171.
Shaw KL (1996). Sequential radiations and patterns of speciation in the Hawaiian cricket genus *Laupala* inferred from DNA sequences. Evolution 50:237-255.
Shen Q, Geiser DM, and Royse DJ (2002). Molecular phylogenetic analysis of *Grifola frondosa* (maitake) reveals a species partition separating eastern North American and Asian isolates. Mycologia 94:472-482.
Skovgaard K, Nirenberg HI, O'Donnell K, and Rosendahl S (2001). Evolution of *Fusarium oxysporum* f. sp. *vasinfectum* races inferred from multigene genealogies. Phytopathol 91:1231-1237.
Skupski MP, Jackson DA, and Natvig DOa (1997). Phylogenetic analysis of heterothallic *Neurospora* species. Fungal Genet Biol 21:153-162.
Steenkamp ET, Wingfield BD, Desjardins AE, Marasas WFO, and Wingfield MJ (2002). Cryptic speciation in *Fusarium subglutinans*. Mycologia 94:1032-1043.
Strimmer K and Moulton V (2000). Likelihood analysis of phylogenetic networks using directed graphical models. Mol Biol Evol 17:875-881.
Suzuki Y, Glazko GV, and Nei M (2002). Overcredibility of molecular phylogenies obtained by Bayesian phylogenetics. Proc Natl Acad Sci USA 99:16138–16143.
Tajima F (1989) Statistical method for testing the neutral mutation hypothesis by DNA polymorphism. Genetics 123:585-595.

Taylor DL and Bruns TD (1999) Community structure of ectomycorrhizal fungi in a *Pinus muricata* forest: minimal overlap between the mature forest and resistant propagule communities. Mol Ecol 8:1837-1850.

Taylor JW, Geiser DM, Burt A, and Koufopanou V (1999a) The evolutionary biology and population genetics underlying fungal strain typing. Clin Microbiol Rev 12:126-146.

Taylor JW, Jacobson DJ, and Fisher MC (1999b) The evolution of asexual fungi: reproduction, speciation and classification. Annu Rev Phytopathol 37:197-246.

Taylor JW, Jacobson DJ, Kroken S, Kasuga T, Geiser DM, Hibbett DS, and Fisher MC (2000) Phylogenetic species recognition and species concepts in fungi. Fungal Genet Biol 31:21-32.

Templeton AR (1993) The "Eve" hypotheses: a genetic critique and reanalysis. Am Anthropol 95:51-72.

Templeton AR (1994) The role of molecular genetics in speciation studies. In: B Schierwater, B Streit, GP Wagner, R DeSalle, ed. Molecular Ecology and Evolution: Approaches and Applications. Basel, Switzerland: Birkhäuser Verlag, pp. 455-477.

Templeton AR (1995) A cladistic analysis of phenotypic associations with haplotypes inferred from restriction endonuclease mapping or DNA sequencing. V. Analysis of case/control sampling designs: Alzheimer's disease and the apoprotein E locus. Genetics 140:403-409.

Templeton AR (1998) Nested clade analyses of phylogeographic data: testing hypotheses about gene flow and population history. Mol Ecol 7:381-397.

Templeton AR, Boerwinkle E, and Sing CF (1987) A cladistic analysis of phenotypic associations with haplotypes inferred from restriction endonuclease mapping. I. Basic theory and an analysis of alcohol dehydrogenase activity in *Drosophila*. Genetics 117:343-351.

Templeton AR, Routman E, and Phillips CA (1995) Separating population structure from population history: a cladistic analysis of the geographical distribution of mitochondrial DNA haplotypes in the tiger salamander, *Ambystoma tigrinum*. Genetics 140:767-782.

Templeton AR and Sing CF (1993) A cladistic analysis of phenotypic associations with haplotypes inferred from restriction endonuclease mapping. IV. Nested analyses with cladogram uncertainty and recombination. Genetics 134:659-669.

van Nimwegen E, Crutchfield JP, and Huynen M (1999) Neutral evolution of mutational robustness. Proc Natl Acad Sci USA 96:9716-9720.

Vilgalys R and Sun BL (1994) Ancient and recent patterns of geographic speciation in the oyster mushroom *Pleurotus* revealed by phylogenetic analysis of ribosomal DNA sequences. Proc Natl Acad Sci USA 91:4599-4603.

Wakeley J and Hey J (1997) Estimating ancestral population parameters. Genetics 145:847-855.

Wang L, Zhang K, and Zhang L (2001) Perfect phylogenetic networks with recombination. J Computat Biol 8:69-78.

Watterson GA (1975) On the number of segregating sites in genetic models without recombination. Theor Popul Biol 7:256-276.

Wiehe T, Mountain J, Parham P, and Slatkin M (2000) Distinguishing recombination and intragenic gene conversion by linkage disequilibrium patterns. Genet Res 75:61-73.

Wright S (1951) The genetical structure of populations. Annals of Eugenics 15:323-354.

Zeyl C (2000) Budding yeast as a model organism for population genetics. Yeast 16:773-784.

Zhan J, Kema GH, Waalwijk C, and McDonald BA (2002) Distribution of mating type alleles in the wheat pathogen *Mycosphaerella graminicola* over spatial scales from lesions to continents. Fungal Genet Biol 36:128-136.

Applied Mycology & Biotechnology
An International Series. Volume 4. Fungal Genomics

3

Molecular and Genetic Basis of Plant-Fungal Pathogen Interactions

Seogchan Kang[1] and **Katherine F. Dobinson**[2]
[1]Department of Plant Pathology, 311 Buckhout, The Pennsylvania State University, University Park, PA 16802, USA (sxk55@psu.edu); [2]Southern Crop Protection and Food Research Centre, Agriculture and Agri-Food Canada, London, Ontario N5V-4T3, Canada, and Departments of Biology, Microbiology and Immunology, The University of Western Ontario, London, ON, Canada (dobinsonk@agr.gc.ca).

Our understanding of the genetics and molecular biology that govern fungal-plant disease interactions has greatly increased during the past decade. This expansion of knowledge has been driven in large part by the development of new tools to investigate pathogenicity and the host response to infection. We present here an overview of the recent genetic and molecular biology research on plant-fungal pathogen interactions with emphasis on the technological advances in the field, and on what we have learned about the specificity of their interactions, and the corresponding host responses. We conclude with comments on the prospects for future research, and its application to disease interactions that are of economic importance.

1. INTRODUCTION

During their evolution, fungi have adapted diverse strategies to meet their nutritional requirements. One such strategy is the intimate association with other organisms. Certain associations are symbiotic (or benign parasitism), such as those between the mycorrhizal or endophytic fungi and their plant hosts, or between the lichen mycobiont and photobiont partners. Many other interactions are not benign; fungi that have evolved the ability to exploit other organisms via pathogenic associations often cause devastating diseases in plants or animals. Fungal diseases are by far the most serious threat to global crop production, and possess the ability to inflict enormous losses that can result in serious socio-economic hardship. For example, the re-appearance of wheat and barley scab (causal agent *Fusarium graminearum*) in North America resulted, in 1993 alone, in yield and quality losses estimated at $1 billion US (McMullen *et al.* 1997). Rice blast disease caused by *Magnaporthe grisea* has been the most explosive and potentially damaging disease of the world's rice crop. The threat of blast disease will continue to increase because agricultural intensification favours development of the disease. *Phytophthora infestans* (causal agent of potato late blight), which was responsible for the Irish potato famine, has again become prevalent due to the emergence of more aggressive, fungicide resistant strains. The appearance of fungicide tolerant pathogens is cause for concern because for many crops only a limited number of alternatives to chemical control are available. In addition to the crop production losses engendered by fungal pathogens, certain groups of pathogens produce toxins in infected crops, and thus directly pose a health hazard to humans and animals (see the chapter by Yu *et al.* in this volume). Global food security therefore depends heavily on both reducing the potential of plant diseases to cause catastrophic crop losses, and preventing the introduction of toxins into the food chain.

Because plant health is affected by complex interactions among plant, pathogen, and environment, a better understanding of the mechanisms that underpin these interactions will expedite our efforts to develop effective means for controlling pathogenic fungi. In parallel with the transition of biological sciences as a whole, the introduction and application of various molecular biological tools has been fundamentally transforming the way we deal with plant diseases, and study plant-fungal pathogen interactions. Genetic engineering and molecular marker-assisted breeding strategies have provided many new opportunities to improve disease resistance in various crop plants. The availability of a large array of molecular markers has also rapidly expanded our knowledge on the evolutionary relationship among plant pathogens and between pathogens and nonpathogenic organisms (pathogen systematics), and on how pathogen populations are structured and change in time and space (population genetics). Lastly, the isolation and characterization of plant and pathogen genes important for defence and pathogenicity, respectively, has significantly increased our understanding of how plant diseases occur.

In this chapter, we review the current status of our knowledge of the molecular and genetic basis of plant-fungal pathogen interactions. Given the volume of information, and the rapid pace at which this knowledge has accumulated, it is not possible to cover all of the salient discoveries in a single chapter. A review of fungal pathogenicity genes can be found in volume 3 of this series, and we present here an introduction to some important concepts relating to plant-fungal pathogen interactions, and highlight the results of recent molecular and genetic studies on specificity and host responses in selected pathosystems. We preface our review with a brief overview of the molecular and cytological tools that are particularly relevant to the study of plant-fungal pathogen interactions. Additional research tools for studying fungal pathogens can be found in a recent review (Gold *et al.* 2001).

2. RESEARCH TOOLS FOR STUDYING PLANT-PATHOGEN INTERACTIONS

The application of molecular and cytological techniques for studying fungal pathogens has lagged somewhat behind their application to the analysis of genetically tractable model systems such as *Saccharomyces cerevisiae*, *Neurospora crassa*, and *Aspergillus nidulans*. The use of these tools in the field of plant pathology has required that they be adapted to circumvent the difficulties associated with working with fungi which lack a sexual stage, and/or which cannot be cultured in the absence of the host plant. Transformation-mediated mutagenesis and complementation analyses, which to date have undoubtedly been the most widely applied method for molecular studies of pathogenicity, have provided considerable information during the past twenty years on gene function in pathogenic fungi (see the chapter by Tudzynski and Sharon in volume 3). The refinement of cytological techniques for tracking pathogen growth, and monitoring gene expression *in planta*, together with the advent of large-scale, high-throughput systems for gene discovery and functional characterization (see Bennett and Arnold 2001;

Sweigard and Ebbole 2001), have given us new and powerful tools to further our understanding of phytopathogenic fungi and their disease interactions.

The most recent addition to our molecular toolbox is genomics. Since the publication in 1995 of the first complete microbial genome sequence, that of the human pathogenic bacterium *Haemophilus influenzae* (Fleischmann *et al.* 1995), the number of microbial genome sequencing projects that have been completed or are in progress has grown exponentially (see the TIGR Microbial Genome Database, http://www.tigr.org/tdb/mdb/mdbcomplete.html); to date, >200 are listed in this database. This information can now be exploited to unveil the evolutionary and genetic basis of different microbial life styles, such as pathogenesis, symbiosis, and growth in different environments. In combination with an array of functional genomic and bioinformatic tools, access to the genetic blueprints of pathogenic microbes has also led to more questions about these processes, and opened new avenues of research for developing effective control measures against human/animal microbial diseases (Rosamond and Allsop 2000; Grandi 2001).

Although plant pathogenic microbes, especially plant pathogenic fungi, are currently underrepresented in the public sequence databases, this situation is improving due to concerted efforts to promote the sequencing of plant-associated microbial genomes. Among the plant pathogenic fungi, *Magnaporthe grisea*, *Aspergillus flavus*, and *Fusarium graminearum* were identified by the advisory committee of the Fungal Genome Initiative (Pennisi 2001) as priorities for initial sequencing. The Plant-Associated Microbe Genomics Initiative (www.apsnet.org/online/feature/microbe/), spearheaded by the American Phytopathological Society, has further expanded the list of organisms to include a diverse range of fungi, based on a number of criteria, including, but not limited to, economic importance and genetic tractability. The *M. grisea* sequencing project, which is now nearing completion, has used a whole genome shotgun sequencing method to obtain more than six-fold sequence coverage of its genome (http://www-genome.wi.mit.edu/annotation/fungi/magnaporthe/). Sequencing of additional fungi, including *F. graminearum*, *Phytophthora sojae*, *Phytophthora ramorum* (the causal agent of sudden oak death), and two soybean rust pathogens (*Phakopsora pachyrhizi* and *P. meibomiae*), is also in progress.

In contrast to the relatively slow start on the plant pathogen side, sequencing of plant genomes has steadily progressed during the past 10 years. The genomes of two model plants, *Arabidopsis thaliana* and rice, have now been published (TAGI 2000; Cantrell and Reeves 2002), and genome sequencing and/or large-scale expressed sequence tag (EST) analysis of many other crop plants are currently underway. Uncovering the genetic design of both plants and their pathogens through genome sequencing not only allows us to identify candidate genes for defence and pathogenicity, respectively, but also provides opportunities to apply functional genomic tools to study the mechanism of their interactions at the genome scale.

2.1 Analysis of Fungal Gene Function by DNA-mediated Transformation

Since the earliest reports of DNA-mediated transformation of *Colletotrichum lindemuthianum*, *Cladosporium fulvum*, and *M. grisea* (Oliver *et al.* 1987; Parsons *et al.* 1987; Rodriguez and Yoder 1987), procedures have been developed for transformation of many phytopathogenic fungi, and applied to their genetic analysis (Mullins and Kang 2001). Stable transformation of filamentous fungi by autonomously replicating plasmid vectors has had only limited success (Lemke and Peng 1995); more typically, transformation relies upon the integration of the transforming DNA and associated selectable marker into the fungal genome, by either illegitimate or homologous recombination. Many cosmid and plasmid transformation vectors bearing genes conferring resistance to hygromycin B, geneticin, benomyl, phleomycin, carbendazim, or bialaphos are now available (Mullins and Kang 2001); a number of these vectors can be obtained through the Fungal Genetics Stock Center (www.fgsc.net/).

2.1.1 Mutagenesis by Random DNA Insertion

Mutant hunts are a standard in genetic studies of microorganisms. DNA-mediated transformation has been widely used for the random mutagenesis of fungal plant pathogens, as an alternative to chemical, UV and X-ray mutagens. A major advantage of transformation-mediated mutagenesis is that the integration into the genome of foreign (plasmid) DNA bearing a selectable marker provides a convenient tag for subsequent identification of the mutated gene (Mullins and Kang 2001). This method has been used with a variety of fungi, including the well-studied species *Ustilago maydis*, *M. grisea*, *Cochliobolus heterostrophus*, to generate collections of transformants, which have subsequently been screened for pathogenicity defects, or mutation of specific pathogenicity genes, or other sequences of interest (see Maier and Schäfer 1999 for review).

Although the typically low transformation efficiency of phytopathogenic fungi renders the generation and selection of transformants a rather laborious process, the frequency of recovery of nonpathogenic mutants can be up to one or two percent (Bölker *et al.* 1995; Sweigard *et al.* 1998; Maier and Schafer 1999; Thon *et al.* 2000). In some systems, the efficiency of transformation may be enhanced by REMI (restriction enzyme-mediated integration), a method first developed for *S. cerevisiae* (Schiestl and Petes 1991). In this method, the transforming plasmid DNA is linearized with a restriction enzyme, and the restriction enzyme is also added to the transformation reaction. As a result, the transforming DNA is preferentially integrated into sites in the fungal genome that have been cleaved by the enzyme (Maier and Schäfer 1999). One potential drawback to this method is that although it generates more single-site insertions than does non-REMI transformation, it also produces a rather high proportion of untagged mutants (up to 50% under some conditions), presumably as a result of faulty DNA repair of the restricted fungal genome (Sweigard *et al.* 1998; Linnemannstöns *et al.* 1999). This particular problem is perhaps not of great concern when dealing with a fungus in which genetic crosses can be used to demonstrate cosegregation of the selectable marker with the mutant phenotype, but it does complicate mutant analysis for those fungi which lack a sexual stage.

An important advance in fungal transformation methodology has been the recent development of *Agrobacterium tumefaciens*-mediated transformation (ATMT). Since the first report of DNA transfer into *S. cerevisiae* by *A. tumefaciens* (Bundock *et al.* 1995), ATMT has been developed for several saprophytic and phytopathogenic fungi (de Groot *et al.* 1998; Chen *et al.* 2000; Covert *et al.* 2001; Mullins *et al.* 2001; Rho *et al.* 2001; Zwiers and De Waard 2001), but it has not yet been widely used for insertional mutagenesis. However, it may in the future become the method of choice; ATMT can be used to transform a variety of tissues, including conidia, mycelia, protoplasts, and even fruiting body tissue (Chen *et al.* 2000). Reports to date also indicate that in certain systems transformation by this method may be more efficient than the classical protoplast transformation, and that it generates relatively high frequency of single-copy insertions (de Groot *et al.* 1998; Mullins *et al.* 2001).

2.1.2 Complementation analysis and targeted mutagenesis via gene replacement

In fungal biology, complementation analysis has been a long-standing tool for genetic analysis. For many phytopathogenic fungi, the introduction of a functional copy of the gene of interest into a mutant strain through classical genetic techniques is problematic, either because the fungus lacks a sexual stage, or because the mutant strains are otherwise genetically incompatible. These limitations can be circumvented by DNA-mediated transformation of the target gene from one strain of a species into a mutant strain of the same species, or even of a different species. In the latter case, the expression of the heterologous gene can be used to demonstrate functional identity or differences between genes that have been predicted, on the basis of sequence information, to have the same function in different species. A good example of the application of this approach is seen in the recent report of transformation of an *M. grisea cpkA* mutant with the *BKA1* gene from the obligate pathogen *Blumeria graminis* (Bindslev *et al.* 2001). Mutations in the *CPKA* gene, which encodes a cyclic AMP-dependent protein kinase, result in delayed, and defective appressorium formation, and consequently are defective in pathogenicity (Xu *et al.* 1997). Bindslev *et al.* (2001) demonstrated that expression of the *BKA1* gene in the *cpkA* mutant complemented the defect in appressorium development, and restored pathogenicity.

Targeted gene disruption is considered a definitive method for determining gene function; most phytopathogenic fungi are haploid, and thus the effects of gene inactivation can be readily assessed. This approach requires: (i) the construction of a gene disruption, or knockout (KO) vector that contains within the gene a dominant selectable marker, such as the hygromycin resistance gene, which allows selection of transformants, (ii) transformation of the wild-type strain with the construct, iii) selection of transformants, and iv) identification of transformants in which the wild-type copy of the gene has been replaced by the KO construct. Construction of gene disruption vectors, often a multi-step, and potentially rate-limiting process, has been facilitated by the development of commercially available *in vitro* transposon (Tn) mutagenesis vectors (Hamer *et al.* 2001). A further modification of this method is the use of a binary plasmid vector that can be propagated in both *E. coli* and *A. tumefaciens* (Mullins *et al.* 2001), and thus used for ATMT.

The recently developed transposon-arrayed gene knockout (TAGKO) system (Hamer *et al.* 2001) combines gene discovery with functional genetic analysis. Genomic DNA libraries are constructed, and subjected to *in vitro* transposon mutagenesis to create a collection of tagged clones; these clones can then be used not only to acquire sequence information, but also for targeted mutagenesis. This method, developed first for *M. grisea*, has the advantages that cosmid, plasmid, or BAC libraries may be used, and that the mutant genes may be used directly for transformation.

2.2 Functional Genomics

The rapid advances in the technological resources dedicated to genome sequencing and post-genomics analyses have already fundamentally transformed the way we study the molecular basis of plant diseases. DNA microarray-based assays make it possible to monitor global gene expression patterns in both plant and pathogen. Since cellular activities are regulated not only at the transcriptional level, proteomic tools have been developed for monitoring downstream changes, at both the translational and post-translational levels. To complement the gene and protein expression analyses, metabolomic (metabolite profiling) tools can be used to simultaneously survey the presence of a large number of metabolites (Fiehn *et al.* 2000a). The judicious application of these functional genomic tools will also assist in systematically identifying genes, biochemical pathways, and global regulatory networks that are critical to pathogenesis and the host defense response, which will in turn facilitate the identification of novel strategies for disease control. Undoubtedly, the utility of genomics for studying the molecular and cellular basis of plant disease will continue to increase, as more novel techniques for utilizing the genome sequence data become available. A recent report summarizing the scope and progress of the National Science Foundation-sponsored genome projects that aim to generate resources for Arabidopsis functional genomics well illustrates the current status of Arabidopsis functional genomics, and may serve as a crystal ball to view the future of studies of agriculturally significant plants (Ausubel 2002).

Table 1. EST collections from phytopathogenic fungi

Organism[1]	Source of material	Reference
Blumeria graminis	Infected plant Conidia	Thomas *et al.* 2001
Botrytis cinerea	Nitrogen starved mycelia	www.genoscope.cns.fr/externe/English/Projets/Projet_W/W.html
Colletotrichum trifolii	Mycelia	D. Samac[2]
Fusarium graminearum	Infected plant Infected plant	J.-R. Xu[3] Krüger *et al.* 2002
Fusarium sporotrichioides	*Tri10* overexpression culture	B. Roe, Q. Ren, A. Peterson, D. Kupfer, H.S. Lai, M. Beremand, A. Peplow, A. Tag[4]
Magnaporthe grisea	Mycelia, Appressoria	R.A. Dean[5], D. Ebbole[6]
Mycosphaerella graminicola	Mycelia, Infected plant	Keon *et al.* 2000
Phytophthora infestans	Mycelia	Kamoun *et al.* 1999b; Waugh *et al.* 2000
Phytophthora sojae	Mycelia Infected plant	Qutob *et al.* 2000; Waugh *et al.* 2000
Verticillium dahliae	Developing microsclerotia, Liquid-grown culture	Neumann and Dobinson 2002

[1]Sequences have been compiled in a searchable database at www.COGEME.ex.ac.uk (Soanes *et al.* 2002); [2]USDA-ARS, Department of Plant Pathology, University of Minnesota; [3]Department of Botany and Plant Pathology, Purdue University; [4]*Fusarium sporotrichioides* cDNA Sequencing Project (www.genome.ou.edu/fsporo.html), The University of Oklahoma, Department of Chemistry and Biochemistry, Norman, Oklahoma, and Texas A&M University, Department of Plant Pathology and Microbiology, College Station, Texas; [5]Fungal Genetics Laboratory, North Carolina State Biotechnology Center; [6]Department of Plant Pathology and Microbiology, Texas A&M University.

2.2.1 Expressed Sequence Tag (EST) Analysis

In the absence of a large-scale, whole genome sequencing effort, which requires considerable financial and scientific resources, EST analysis provides an alternative genomics tool. Even a small-scale analysis can be useful for gene discovery, and for preliminary comparative analyses of gene expression, particularly for those organisms for which genetic information is negligible or completely lacking. EST studies have been undertaken for a broad range of fungal plant pathogens, as well as the oomycete pathogens *P. infestans* and *P. sojae* (Table 1; Soanes *et al.* 2002). An advantage of these analyses is that they may be designed to exploit prior knowledge of the effects of nutritional status or environmental factors on the expression of genes associated with pathogenicity and development (Keon *et al.* 2000; Thomas *et al.* 2001; Neumann and Dobinson 2002). Another consideration is that an EST analysis of infected plant tissue, such as has been done with the *F. graminearum*/wheat and *P. sojae*/soybean interactions (Kruger *et al.* 2002; Qutob *et al.* 2002), provides information not only about pathogen gene expression *in planta*, but also about plant gene expression in response to infection.

The use of EST collections for subsequent functional studies is nicely illustrated by a recent study of *P. sojae* (Qutob *et al.* 2000). Screening of an EST dataset (3035 sequences) by *in silico* methods identified 202 non-redundant sequences predicted to encode either secreted or membrane-associated proteins. Sixteen of these clones, encoding putative secreted proteins, were selected for further analysis, using an *Agrobacterium tumefaciens*/potato virus X-mediated transient expression system was to assay the activity of the candidate proteins *in planta*. Based on the results of these experiments, a protein having necrosis-inducing activity (PsojNIP) was identified; other data indicated that the protein acts during the transition from the biotrophic to necrotrophic phase of the disease, to accelerate plant death (Qutob *et al.* 2002).

2.2.2 Global gene expression analysis

The identification of plant or fungal genes by total genome sequencing and EST analysis is complemented by gene expression analysis using DNA microarrays, which comprise thousands of individual gene fragments or oligonucleotides corresponding to individual genes, printed in a high-density array. The power of microarray transcription profiling is found in its ability to assess gene expression on a whole genome scale (Wisman and Ohlrogge 2000; Zhu and Wang 2000). Such a global analysis will uncover many new genes that are involved in various cellular processes, facilitate the characterization of the molecular basis of mutant phenotypes, and reveal how groups of genes are regulated by different environmental and developmental stimuli. Determining when and where genes of unknown function are expressed, in reference to the expression of genes of known function, will also provide valuable clues about the possible function of the unknowns (i.e. "guilt by association"). For example, a number of studies have employed microarray techniques to investigate the response of Arabidopsis to various stimuli that are known to elicit defense responses (Maleck *et al.* 2000; Schenk *et al.* 2000; Cheong *et al.* 2002). In response to one or more of four stimuli: an avirulent strain of *Alternaria brassicicola*, salicylic acid, methyl jasmonate, and ethylene, 705 of the 2,357 Arabidopsis genes tested were significantly up- or down-regulated (Schenk *et al.* 2000). Of these, 169 were regulated by multiple stimuli, suggesting the existence of substantial cross-talk among the different signalling pathways that control plant defense. Consistent with this supposition, many Arabidopsis genes were also differentially regulated by four or more conditions affecting systemic acquired resistance (Maleck *et al.* 2000). Transcription profiling of Arabidopsis using an oligonucleotide-based array also suggested the existence of novel interactions between wounding, pathogen, abiotic stress, and hormonal responses (Cheong *et al.* 2002). As the data accumulate in other systems on global gene expression in response to pathogenesis/defense-related stimuli, the regulatory networks that control plant defense, and their interconnections at the transcription level, should become apparent. Given the rapid progress in fungal pathogen genome sequencing, global gene expression analysis of fungal pathogens under various pathogenesis-related developmental and growth conditions and in infected plant tissues is likely to be routine in the near future. Large-scale expression profiling, using the SAGE (serial analysis of gene expression) technique, has already been done with the obligate barley mildew fungus *Blumeria graminis* (Thomas *et al.* 2002).

2.2.3 Proteomics

Since cellular activities are regulated not only at the transcriptional level, but also at the translational and post-translational levels, a comprehensive understanding of the nature and mechanisms of cellular activities requires methods for studying the identity and level of cellular proteins. Several tools, based primarily on 2D-gel electrophoresis of proteins, and subsequent identification of individual protein spots on the gel by mass spectrometry (MS), are now available for global analysis of protein expression (Pandey and Mann 2000). In addition to its use for monitoring the levels of "all" proteins expressed under certain conditions, such an analysis can be focused on identifying only those proteins that undergo specific chemical modifications in response to specific stimuli (Peck *et al.* 2001). This particular application is exemplified by the global analysis of protein phosphorylation in Arabidopsis in response to a bacterial elicitor, which resulted in the discovery of a protein that is phosphorylated in response to both bacterial and fungal elicitors (Peck *et al.* 2001). Novel proteomic microarray tools are also being developed to speed up the pace of protein identification (Abbott 1999). The yeast two-hybrid system, for example, which uses the activation of reporter gene expression to detect interaction between two proteins, and which has been widely used to identify protein-protein interactions, is amenable to high-throughput screening (Pandey and Mann 2000). A recently developed protein microarray technique (MacBeath and Schreiber 2000) could also potentially provide a platform for multiple, high-throughput functional analyses of proteins, including protein-protein interactions, and the identification of enzyme substrates, and protein targets of small molecules.

2.2.4 Metabolomics

Metabolomics (or metabolic profiling), the simultaneous surveying of the levels and identities of cellular metabolites, can complement gene and protein expression studies by providing information on biochemical activities and their regulation. Comparison of metabolic profiles among mutants that exhibit similar phenotypes may allow the identification of those metabolites that function in the same biochemical or regulatory pathway. This type of analysis has not yet been reported for phytopathogenic fungi, but its feasibility has been demonstrated with *S. cerevisiae* (Raamsdonk *et al.* 2001); its utility when used in combination with data-mining tools, has also been demonstrated for the characterization of phenotypic changes in plants caused by different genetic backgrounds and environmental conditions (Fiehn *et al.* 2000a; Fiehn *et al.* 2000b; Roessner *et al.* 2000; Roessner *et al.* 2001).

Metabolomics also has the potential to provide valuable clues as to the biological role(s) of those genes for which mutations do not cause any discernable phenotypic alterations, but result in changes in intracellular metabolic activity. In such cases, comparison of the concentrations and types of individual metabolites in the mutants, versus those in wild-type strains, might uncover unique metabolic signatures associated with individual mutations, which could in turn lead to the identification of potential sites of action for those gene products. To date, metabolomics has not been utilized in studies of plant-fungal pathogen interactions. However, considering its potential for providing snapshots of the global networks of metabolic activities, this approach deserves much more attention. It will, for example, be of great value for the analysis of biochemical pathways critical to the outcome of plant-pathogen interactions, including those that direct the synthesis or breakdown of toxic compounds, and secondary metabolites. In parallel with DNA microarray and proteomic analyses, comparative metabolomic analyses of a host plant in response to various pathogens (e.g., compatible vs. incompatible interactions, biotrophic vs. nectrotrophic pathogens, different species of pathogens with similar modes of infection, etc.) may also reveal critical features of the global network that control defense responses.

2.3 Molecular Cytology

Although the genetic and genomic tools described above can be used at the level of whole tissue and/or plant (i.e., averaged changes) to survey global changes in mRNA, proteins, and metabolites in response to fungal pathogen infection, technical limitations make them unsuitable for monitoring molecular changes occurring in individual cells. Considering the intimate and dynamic nature of interaction between fungal pathogens and their hosts throughout the disease cycle, a comprehensive understanding of their interactions requires that we examine the host-pathogen interaction at the cellular level. Rapid advances in cytological tools and techniques, which have been well documented in a number of review articles (Hardham and Mitchell 1998; Heath 2000; Howard 2001; Lorang *et al.* 2001), have made it possible to carry out such detailed studies.

The use as vital markers of fluorescent proteins, particularly green fluorescent protein (GFP) and its spectral variants (red, cyan, etc.), permits the direct visualization by fluorescent imaging of cells or proteins of interest. Certain plants, such as *A. thaliana* and tobacco, have also shown themselves to be readily amenable to GFP expression (Kohler *et al.* 1997; Cutler *et al.* 2000; Kato *et al.* 2002). A number of phylogenetically diverse phytopathogenic fungi have now been transformed with GFP, or its colour variants, as a reporter gene (Spellig *et al.* 1996; van West *et al.* 1999; Bowyer *et al.* 2000; Stephenson *et al.* 2000; Lorang *et al.* 2001; Rohel *et al.* 2001; Sexton and Howlett 2001; Czymmek *et al.* 2002; Lagopodi *et al.* 2002), and an array of novel fluorescent protein genes isolated from reef corals have been successfully expressed in fungal pathogens (Bourett *et al.* 2002). Visualization of both GFP and DsRed (red fluorescent protein; RFP) in dual label experiments (Mas *et al.* 2000) has demonstrated the possibility of using these markers to simultaneously observe both the host and pathogen throughout the infection process. The non-invasive optical sectioning property of confocal microscopy (Czymmek *et al.* 1994) has made an important contribution to these types of analyses; infection by the pathogen, together with certain aspects of the host defense responses, can now be visualized without destroying the infected plants. In addition, three-dimensional, time-resolved data from specific plant-pathogenic infection sites can be obtained (Czymmek *et al.* 2002). In combination with genetic manipulation of a host and its fungal pathogen, these cytological tools promise to have a great impact on studies of the dynamics of plant-pathogen interactions at the cellular and molecular levels.

3. GENETIC BASIS OF HOST SPECIFICITY

Although a very large number of plant pathogens exist in nature, each plant species is resistant to most of these pathogens (general or non-host resistance) (Heath 1991), and susceptible only to the limited number of pathogens that have evolved to overcome its defense systems. In some plant-pathogen systems, cultivars (or varieties) of a host plant also exhibit differential resistance to individual races (or pathotypes) within a pathogen species (race specific resistance). Flor's pioneering work on the genetic basis of race specific resistance in flax against *Melampsora lini* (a rust pathogen) described gene-for-gene relationships between resistance genes in flax cultivars, and corresponding virulence/avirulence genes in *M. lini* (Flor 1971), which determined the outcome of the disease interaction between different cultivars and pathogen races. Since that time, it has been shown that gene-for-gene interactions determine plants' compatibility with many other fungal pathogens, as well as bacteria, viruses, and nematodes. In some plant-fungus interactions, toxic compounds produced by a host or a pathogen determine compatibility (Osbourn 1996; Markham and Hille 2001; Wolpert *et al.* 2002).

3.1 Gene-for-Gene Interactions

The powerful gene-for-gene surveillance system (Flor 1971) is mediated by host recognition of the pathogen, which triggers activation of host defense responses that limit pathogen ingress. In these interactions, a pathogen that carries a specific avirulence (*AVR*) gene is unable to infect those cultivars carrying the complementary resistance (*R*) gene. Since gene-for-gene relationships govern compatibility of plant-pathogen interactions in many pathosystems, the question of how resistance is triggered in the presence of an *AVR* gene-*R* gene pair remains an important biological question.

To date, a large number of *AVR* and *R* genes have been cloned and characterized (Takken and Joosten 2000; Hulbert *et al.* 2001; Leach *et al.* 2001; Luderer and Joosten 2001). *AVR* genes have been identified based on their role in triggering specific *R* gene-mediated resistance, but there is still little understanding of the underlying molecular recognition, and only rarely has the gene/protein sequence provided a clue as to what role, if any, these genes play during normal growth and colonization of the host plant (Leach *et al.* 2001; Luderer and Joosten 2001; Staskawicz *et al.* 2001). Avirulence genes that have been identified encode molecules that may be recognized, directly or indirectly, by the corresponding *R* gene product; others encode enzymes involved in production of small molecule ligands that serve as recognition factors.

We do not yet fully understand in molecular detail why a pathogen might retain an *AVR* gene that prevents it from infecting certain host genotypes, but accumulating evidence suggests a dual role for some *AVR* genes as both gene-for-gene signals, and virulence (or fitness) factors (Leach *et al.* 2001). Although *R* gene-mediated resistance is highly effective once triggered, pathogens can evade this resistance through various mechanisms, including modification of *AVR* gene expression, modification of the structure of the gene product, or deletion of the *AVR* gene from their genome (see below for the *M. grisea AVR-Pita* gene as an example). A better understanding of the biological role(s) of these genes, and of how a pathogen can modify its *AVR* gene to avoid triggering resistance yet preserve virulence/fitness, will provide insight on the durability of the corresponding *R* genes in controlling disease in the field. The following is a summary of gene-for-gene interactions in selected fungal pathogens. The current status of our understanding of the regulatory mechanisms that direct *R* gene-mediated defence responses, gained from studies of *A. thaliana*, is summarized in a separate section (see 4.1).

3.1.1 Rice Blast

Worldwide, rice blast, caused by *M. grisea* (Hebert) Barr. (anamorph, *Pyricularia grisea* Sacc.), is one of the most economically devastating crop diseases. In addition, the broad collective host range of *M. grisea* puts at risk numerous members of the grass family. For example, *M. grisea* strains that infect perennial ryegrass (*Lolium perenne*) have recently become the most destructive of all turfgrass diseases in the US (Viji *et al.* 2001; Farman 2002).

Rice blast is a classical gene-for-gene system, in which genetic analysis has identified *AVR* genes that trigger the hypersensitive resistance response in host plants expressing corresponding *R* genes. Numerous *AVR* genes have been genetically identified in *M. grisea* (Valent and Chumley 1987; Leung *et al.* 1988; Ellingboe *et al.* 1990; Valent *et al.* 1991; Ellingboe 1992; Silué *et al.* 1992a; Silue *et al.* 1992b; Lau *et al.* 1993; Chao and Ellingboe 1997; Dioh *et al.* 2000), and correspondingly, numerous *R* genes against *M. grisea* have been identified in rice (Marchetti *et al.* 1987; Silué *et al.* 1992a; Wang *et al.* 1994). In addition to *AVR* genes, two different classes of genes (*S* and *M* for suppressor and modifier, respectively) have been shown to control compatibility toward certain rice cultivars (Ellingboe 1992; Lau *et al.* 1993; Lau and Ellingboe 1993). The function of the *S* gene is to suppress the expression or function of specific *AVR* gene(s), whereas the *M* gene is required for the expression or function of one or more *AVR* genes. The question as to whether or not all the *AVR* genes in *M. grisea* have corresponding *S* and *M* genes remains to be investigated.

Molecular cloning of several *AVR* genes from *M. grisea* (Kang *et al.* 1995; Sweigard *et al.* 1995; Farman and Leong 1998; Orbach *et al.* 2000; Farman *et al.* 2002) has provided an opportunity to study the molecular basis of their avirulence function. Among these *AVR* genes, *AVR-Pita* is by far the best characterized, and it is unique among the *AVR* genes of cereal fungal pathogens because its corresponding *R* gene, *Pi-ta*, has also been cloned and characterized (Bryan *et al.* 2000; Jia *et al.* 2000; Orbach *et al.* 2000). *AVR-Pita* differs from other fungal *AVR* genes that have been characterized in that it is predicted to encode not a small extracellular polypeptides that function as elicitors of *R* gene associated cell death (Luderer and Joosten 2001), but a 223 amino acid protein with high similarity to fungal zinc metalloproteases of the deuterolysin family (Orbach *et al.* 2000). Although protease activity of AVR-Pita has not been directly demonstrated, it is implicated; Orbach *et al.* (2000) reported that in a spontaneous mutant which had lost avirulence, a single amino acid substitution in which a glycine residue replaced the glutamic acid residue at position 178 that is predicted to be critical for protease function. The results of site-directed mutagenesis of *AVR-Pita* also indicate that the protease signature motif is essential for triggering *Pi-ta*-mediated resistance (G.T. Bryan, L. Farrall and B. Valent, personal communication). It appears that *AVR-Pita* is only expressed during infection, and in a compatible interaction it is highly expressed throughout invasive growth, and colonization of the plant tissue suggesting a role during pathogenesis.

Bryan *et al.* (2000) reported that the *Pi-ta* resistance gene encodes a predicted cytoplasmic receptor with a centrally located nucleotide-binding site, and a carboxyl terminal leucine-rich domain. A biolistic assay was used to demonstrate that transient expression of *AVR-Pita* in rice seedlings carrying the *Pi-ta* gene was sufficient to initiate a *Pi-ta*-mediated defense response (Bryan *et al.* 2000; Jia *et al.* 2000); only the putative mature form, AVR-Pita$_{176}$, induced a hypersensitive response. In a yeast two-hybrid assay, the putative mature protease, but not the intact AVR-Pita protein, binds specifically to the leucine-rich domain at the C-terminus of the Pi-ta resistance protein (Jia *et al.* 2000). Taken together, these results support the hypothesis that the AVR-Pita preproprotein is processed *in vivo* to a mature, active form.

The genome organization and evolution of *AVR-Pita* exhibits many interesting features. Among *M. grisea* strains (>200) from various hosts, the copy number of *AVR-Pita* gene homologues, and their degree of their sequence similarity to *AVR-Pita*, vary substantially, indicating that *AVR-Pita* is a member of a gene family that undergoes frequent deletion/amplification events (C. Khang and S. Kang, unpublished results). This variation, and the frequent occurrence of spontaneous mutations in *AVR-Pita*, including point mutations, deletions (ranging in size from 100 bp to over 12.5 kb), and an insertion by transposon Pot3 (Kang *et al.* 2000; Orbach *et al.* 2000) indicate that the *AVR-Pita* gene family is highly dynamic.

The UAA stop codon of the *AVR-Pita* open reading frame in O-137 is separated from the telomeric repeat sequences by only 48 bp (Orbach *et al.* 2000). The *AVR-Pita* gene in a strain from *Digitaria* is also telomeric (the start codon is separated from the telomere repeat by 1689 bp), although its orientation relative to the telomere repeat is inverted from that in O-137 (C. Khang and S. Kang, unpublished results). Telomere location has been shown in other organisms to be associated with genetic instability (Charron and Michels 1988; Hernandez-Rivas *et al.* 1997; Freitas-Junior *et al.* 2000). The high frequency at which *AVR-Pita* mutates to virulence is thus likely due to its telomeric location. The finding that other *AVR* genes in *M. grisea*, including *AVR1-Ku86*, *AVR1-MedNoï* and *PWL1*, are telomere linked (Kang *et al.* 1995; Dioh *et al.* 2000; Kang 2001), further suggests that the presence of *AVR* genes in the highly dynamic chromosome ends provides an advantage to the fungus by allowing it to rapidly overcome newly deployed *R* genes.

The interfertility of *M. grisea* strains that infect different grass species has allowed the genetic analysis of host specificity at the species level. *PWT1* and *PWT2*, which control compatibility on wheat, have been identified through a genetic cross between a wheat pathogen and a foxtail millet isolate (Murakami *et al.* 2000). Two unlinked genes, *PWL1* and *PWL2*, determining the compatibility to weeping lovegrass have been identified in independent crosses (Valent and Chumley 1987; Valent and Chumley 1991). Browning or auto-fluorescence of weeping lovegrass cells around developing colony margins correlated with the presence of the *PWL1* gene in an invading fungus (Heath *et al.* 1990). The fungus was unable to grow beyond these brown cells, suggesting that weeping lovegrass recognizes the product or by-product of *PWL1*, and subsequently initiates a successful defense response.

The number of genes homologous to *PWL2*, and their degree of sequence homology are highly variable, even among the isolates from the same host species (Kang *et al.* 1995; Sweigard *et al.* 1995), suggesting that, like *AVR-Pita*, *PWL2* is a member of a gene family. Numerous *PWL* genes and their alleles have been cloned from strains isolated from diverse hosts (Kang *et al.* 1995). The *PWL1* gene encodes a protein with 75% identity to the *PWL2* gene product. *PWL3*, cloned from a finger millet pathogen, is allelic to the *PWL4* gene isolated from a weeping lovegrass pathogen. Neither gene can trigger the defense response in weeping lovegrass. However, that the *PWL4* ORF is functional has been demonstrated by placing it under the control of either the *PWL1* or *PWL2* promoter (Kang *et al.* 1995). The products encoded by members of the *PWL* gene family range in size from 137 to 147 amino acids, and have the following common characteristics: (i) the amino terminus has features characteristic of eukaryotic signal peptides, suggesting that the proteins may be secreted; (ii) several glycine residues are well conserved and evenly distributed throughout the protein, and (iii) the proteins are highly hydrophilic with many charged residues.

3.1.2 Tomato-*Cladosporium fulvum*

Like *M. grisea*, the biotrophic fungus *Cladosporium fulvum* interacts with its host in a gene-for-gene fashion. The *C. fulvum*-tomato interaction is arguably the best characterized of the fungal gene-for-gene systems (see Rivas and Thomas (2002) for review), in terms of our understanding at the molecular and cellular levels, and studies of this system have provided much information about pathogen recognition and the induction of plant defences during incompatible interactions.

Of the *C. fulvum AVR* genes cloned to date (see Luderer and Joosten 2001 for review), the expression of two that confer race-specific resistance, *Avr9* and *Avr4*, and of the interactions between the encoded proteins and their corresponding resistance proteins have been studied in detail at the molecular level. Both genes encode elicitors, first identified as low molecular weight proteins that were secreted into the plant apoplastic fluid, which induce hypersensitive responses (HR) in tomato varieties containing the *R* genes *Cf-9* and *Cf-4*, respectively (Van Kan *et al.* 1991; Joosten *et al.* 1994). AVR9 is synthesized as an inactive, 63 amino acid precursor; after removal of the signal peptide, the protein undergoes two or more additional proteolytic cleavages by plant and fungal proteases to produce the active elicitor (Van den Ackerveken *et al.* 1993b). The *Avr9* gene is present only in avirulent strains (Van Kan *et al.* 1991); its expression is highly induced *in planta* during pathogenesis (Van Kan *et al.* 1991), and can also be induced *in vitro* by nitrogen starvation (Van den Ackerveken *et al.* 1994). Using a reporter gene construct with the *Avr9* promoter fused to the *uidA* (GUS) gene, it has been shown that expression of *Avr9* is likely to be controlled by a GATA-type transcription factor, similar to AREA of *Aspergillus nidulans* (Snoeijers *et al.* 1999).

Like *Avr9*, the *Avr4* gene is highly expressed *in planta* during pathogenesis (Joosten *et al.* 1997), and encodes a precursor (135 amino acid) that is processed to yield the active elicitor (Joosten *et al.* 1994). In contrast to *Avr9*, however, *Avr4* is present in virulent strains of *C. fulvum*; many of the alleles in these strains have been shown to contain a single point mutation that results in an amino acid substitution in the encoded protein (Joosten *et al.* 1997). Although *in planta* transcription levels of these alleles are comparable to that of the wild-type allele, the AVR4 protein is detectable only in plants infected by strains carrying the wild-type allele. Over-expression of the mutant alleles using the potato virus X (PVX)-based expression system showed that some of the mutant proteins can confer avirulence to tomato plants containing *Cf-4*, suggesting that instability of these proteins underlies the loss of avirulence function (Joosten *et al.* 1997).

C. fulvum secretes several other low molecular weight, extracellular proteins (ECPs) that exhibit elicitor activity on particular lines of *Lycopersicon esculentum* or accessions of its wild relative *L. pimpinefolium* (Van den Ackerveken *et al.* 1993a; Laugé *et al.* 1998; Laugé *et al.* 2000); ECP2 has also been shown to elicit an HR on *Nicotiana paniculata* (Laugé *et al.* 2000). Although the HR-associated recognition of the ECPs, like that of AVR4 and AVR9, is conferred by single, dominant resistant genes, the recognition is not race specific; the ECPs appear to be produced by all *C. fulvum* strains (Laugé and De Wit 1998; Joosten and De Wit 1999). ECP1 and ECP2 have further been shown to function as virulence factors, presumably by suppressing host defense responses (Laugé *et al.* 1997).

Sequence analysis of the *AVR*s and *ECP*s has shown that there is no significant similarity between the elicitors at the amino acid sequence level, but all contain four, six, or eight cysteine residues (Van den Ackerveken *et al.* 1993a; Laugé *et al.* 2000); most, but not all, of these residues are essential for elicitor function, suggesting their involvement in the formation of disulphide bridges needed for stability and activity (van den Hooven *et al.* 2001; Luderer *et al.* 2002).

Several *Cf* resistance genes have now been isolated from tomato (reviewed in (Rivas and Thomas 2002)). *Cf-4* and *Cf-9* are members of the *Hcr9* (homologues of *C. fulvum* resistance gene *Cf-9*) gene family, all of which map to the short arm of chromosome 1 in three major clusters (Parniske *et al.* 1997). The results of detailed genetic and molecular studies of this locus suggest that the *Hcr9* genes were generated by duplication and translocation of a common ancestral sequence (Parniske *et al.* 1997). Using *AVR9*-disrupted *C. fulvum* strains, Laugé *et al.* (Laugé *et al.* 1997) demonstrated that one of these genes, now designated *Cf-9B* (Panter *et al.* 2002), confers weak resistance against *C. fulvum*. The results of recent studies have further demonstrated that *Cf-9B* resistance is limited to mature tomato plants, and it has been proposed that this resistance may be due to the additive effects of the weak, race-specific resistance conferred by *Cf-9B*, and the weak, nonspecific resistance of mature plants (Panter *et al.* 2002).

All of the *Cf* genes analysed to date are predicted to encode membrane-anchored, glycoproteins containing leucine-rich repeats (LRRs); Cf-9 and Cf-4 share more than 90% identity (Jones *et al.* 1994; Dixon *et al.* 1996; Thomas *et al.* 1997; Dixon *et al.* 1998). Although the sequences of the Cf proteins suggested a direct role in elicitor binding, the results of ligand-binding assays indicate that Cf-9 does not interact directly with AVR9 (Luderer and Joosten 2001), suggesting that a third party, perhaps the previously identified membrane-associated high affinity binding site (Kooman-Gersmann *et al.* 1996), is required in the signalling interaction. More recently, tagged, functional Cf-9 and Cf-4 proteins were used to demonstrate the presence of these proteins in membrane-associated complexes of approx. 400 kDa (Rivas *et al.* 2002; Rivas and Thomas 2002); the protein partners in this complex have not yet been identified.

3.1.3 *AVR* Genes in other fungal pathogens

Proteins that act as elicitors of the hypersensitive response have been identified in the barley leaf scald pathogen *Rhynchosporium secalis*, and also in species of the fungal-like pathogens *Phytophthora* and *Pythium*. The *Nip1* gene in *R. secalis* has a dual role similar to that of *Ecp2* in *C. fulvum* (Rohe *et al.* 1995; Knogge 1996). The *Nip1* gene encodes an 82 amino acid precursor that is processed to a 60 amino acid mature protein that triggers host defense responses in barley varieties carrying the *Rrs1* resistance gene. The NIP1 protein also functions as a toxin, or virulence factor, inducing necrosis in barley cultivars that lack *Rrs1*, as well as in other monocots, and also dicots (Wevelsiep *et al.* 1991); NIP1 appears to exert its virulence activity by stimulating the plasma membrane H^+-ATPase (Wevelsiep *et al.* 1993).

The family of low molecular weight, extracellular elicitor proteins (elicitins) that is secreted by various *Phytophthora* and *Pythium* species (Huet *et al.* 1995; Yu 1995; Mao and Tyler 1996; Kamoun *et al.* 1997) includes at least five distinct classes of proteins (Kamoun *et al.* 1997). The elicitins have been proposed to be species-specific avirulence factors (Bonnet *et al.* 1994; Yu 1995; Kamoun *et al.* 1997), based in part on a strong negative correlation between the elicitin production by *Pythophthora* strains, and their virulence on tobacco; wild-type strains of *P. infestans* are avirulent on *Nicotiana* species, while strains in which the elicitin-encoding gene *INF1* has been silenced are able to complete the disease cycle in *N. benthamiana*. The finding that expression of the *INF*1 gene from a potato virus X (PVX) vector confers avirulence to PVX in tobacco, and concomitantly causes an HR (Kamoun *et al.* 1999a) strengthens this hypothesis.

3.2 Fungal Toxins as Host Specificity Determinants

Certain fungal pathogens, mostly those in the genera *Cochliobolus* and *Alternaria*, produce host-selective toxins (HSTs) that cause plant cell death, or suppress the defense responses in susceptible hosts and thus allow colonization by the pathogen. There is now much evidence to support the roles of HSTs as major virulence and host specificity factors. Application of some of the HSTs in the absence of corresponding pathogen causes the expression of disease symptoms that mimic those caused by fungal infection, and also enables nonpathogenic fungi to colonize the treated plant (Comstock and Scheffer 1973; Xiao *et al.* 1991); mutants that do not produce the toxins are nonpathogenic or significantly less aggressive than their wild-type counterparts (Yang *et al.* 1996; Akamatsu *et al.* 1997). Here, we will briefly review a few examples of host compatibility determined by HSTs, and discuss in some detail the role of host cell death in the resulting resistance/susceptibility. Readers who are interested in the details of the synthesis and modes of actions of HSTs are referred to two recent, and comprehensive reviews (Markham and Hille 2001; Wolpert *et al.* 2002).

Table 2. Selected producers of HSTs and the chemical structures of HSTs [a]

Pathogen	Host	Toxin	Chemical structure
Alternaria alternata	Tomato	AAL-toxin	Aminopentol ester
Alternaria alternata	Apple	AM-toxin	Cyclic tetrapeptide
Alternaria alternata	Tangerine	ACT-toxin	Epoxy-decatrienoic ester
Bipolaris sacchari	Sugarcane	HS-toxin	Glycosylated sesquiterpene
Cochliobolus carbonum	Corn	HC-toxin	Cyclic tetrapeptide
Cochliobolus heterostrophus	Corn	T-toxin	Polyketide
Cochliobolus miyabeanus	Rice	Ophiobolins	C_{25}-terpenoid
Cochliobolus victoriae	Oats	Victorin	Cyclic pentapeptide
Pyrenophora tritici-repentis	Wheat	ToxA and ToxB	Small mw protein

[a]Most of the data summarized in this table are adapted from two recent reviews (Markham and Hille 2001; Wolpert *et al.* 2002). Information for *C. miyabeanus* is from Xiao *et al.* (1991).

Host plants that are resistant to HST-producing fungi either lack the target of the toxin, can detoxify the HST (Multani *et al.* 1996), or are otherwise able to compensate for the inhibition by the HST of a target enzyme activity (Brandwagt *et al.* 2000); resistance is typically conferred by single genes. Although most HSTs are secondary metabolites (see Table 2 for examples) that are synthesized through the combined action of multiple genes (Markham and Hille 2001; Wolpert *et al.* 2002), the HSTs of *Pyrenophora tritici-repentis*, the causal agent of tan spot of wheat, are a notable exception. These are encoded by single genes (Ciuffetti *et al.* 1997; Martinez *et al.* 2001), as demonstrated by the ability of the *P. tritici-repentis ToxA* gene, which encodes a 13.2 kDa HST protein, to confer pathogenicity when transferred into a nonpathogenic strain (Ciuffetti *et al.* 1997), and of the purified ToxA and ToxB proteins to elicit cultivar-specific necrosis in a manner consistent with the compatibility between the fungus and the cultivars tested (Tuori *et al.* 2000; Martinez *et al.* 2001).

3.2.1 Interactions between *Cochliobolus* spp. and their hosts

The diversity of fungal HSTs is exemplified by those produced by members of the genus *Cochliobolus*. In the *C. carbonum* /maize disease interaction, the cyclic tetrapeptide HC-toxin is an important virulence and specificity factor (Panaccione *et al.* 1992). Although the results of early genetic analysis suggested that toxin production was conferred by a single gene, *TOX2*, detailed molecular studies have since demonstrated that the *TOX2* locus, covering more than 500 kb, is in fact comprised of multiple copies of several genes involved in the regulation, biosynthesis, and export of the toxin (Ahn and Walton 1996; Ahn *et al.* 2002). Non-HC-toxin producing races (races 2 and 3) lack the *TOX2* locus, and are less virulent than the toxin-producing race 1. HC-toxin is a potent inhibitor of histone deacetylase, an enzyme that along with histone acetylase, controls the level of acetylation of histones (Brosch *et al.* 1995), and hence can affect the regulation of gene expression. HC-toxin may therefore act to repress the expression of host defense-related genes, rendering the host defenceless against invading race 1 strains.

Resistance to HC-toxin producing strains of *C. carbonum* is conferred by the *Hm1* and *Hm2* loci, which encode NADPH-dependent reductases that convert the toxin to an inactive form (Multani *et al.* 1996). The interactions between toxin-producing or non-producing strains of *C. carbonum*, and resistant or susceptible maize varieties can be depicted in a quadratic check similar to those used to portray *AVR/R* gene interactions (Table 3). However, in the *AVR/R* gene interactions, in which avirulence and resistance are dominant, an incompatible reaction results only if the host and pathogen carry the complementary *R* and *AVR* genes, respectively; all other interactions result in disease. In contrast, in the *C. carbonum*/maize interaction, severe disease occurs only if the fungus produces the HC-toxin, and the maize variety does not contain the *HM* genes; all other interactions lead to minor disease.

Table 3. *C. carbonum*/maize compatibility.

	Maize[b]	
C. carbonum[a]	R	r
T	-	+
t	-	-

[a]T = dominant virulence, conferred by production of HC toxin; [b]R = dominant resistance, conferred by production of HC toxin reductase; Presence and absence of severe disease are indicated by + and -, respectively.

High virulence on maize of the causal agent of Southern corn leaf blight, *C. heterostrophus*, depends on both production by the fungus of the polyketide T-toxin, and the presence in maize of a specific mitochondrial genotype (Dewey *et al.* 1988; Yang *et al.* 1996; Rose *et al.* 2002). Two genes, *PKS1* (encoding a polyketide synthase) and *DEC1* (encoding a decarboxylase), have been shown to be necessary for T-toxin synthesis (Yang *et al.* 1996; Rose *et al.* 2002), and additional, yet-to-be-identified genes are likely to be involved in its synthesis (Rose *et al.* 2002). The finding that both genes, located at two unlinked loci (*Tox1A* and *Tox1B*), are only present in T-toxin producing strains (race T), together with other evidence (Rose *et al.* 2002) support the hypothesis that the genes involved in the synthesis of T-toxin have been horizontally transferred to the *C. heterostrophus* genome as a large contiguous gene cluster and have subsequently undergone a chromosomal translocation.

Unlike the *C. carbonum*/maize compatibility, resistance of maize to *C. heterostrophus* is determined by the lack of a target for T-toxin. Maize lines carrying Texas cytoplasm for male sterility (T-cytoplasm) are highly susceptible to race T strains due to the presence of the mitochondrial gene, *T-urf13*, (Dewey *et al.* 1986). Upon binding of T-toxin, the *T-urf13* gene product (13 kDa) disrupts mitochondrial function by creating pores in the mitochondrial membrane, causing cell death (Rhoads *et al.* 1998). Maize susceptibility to *Mycosphaerella zeae-maydis* is also determined by the presence of the *T-urf13* gene, because the fungus produces a group of toxins (PM-toxin) that are structurally similar to T-toxin and also bind to the *T-urf13* gene product (Dewey *et al.* 1987).

3.3 Fungal Detoxification of Host Toxins as a Host Range Determinant

In addition to an array of inducible defense responses that occur upon pathogen infection, plants also rely on preformed structural and chemical defense barriers to protect themselves from many potential pathogens. A variety of antifungal compounds, including phenols and phenolic glycosides, unsaturated lactones, sulphur compounds, saponins, cyanogenic glycosides, and glucosinolates, may act as chemical defense barriers (Osbourn 1996). Due to the ability of saponins (glycosylated triterpenoids, steroids, or steroidal alkaloids) to bind sterols and thus disrupt the integrity of fungal membrane, fungal pathogens that infect saponin-containing plants employ a number of mechanisms to counteract this chemical defense (Osbourn 1996). These mechanisms include suppression of the saponin membranolytic action by lowering the pH (e.g., *Alternaria solani*), tolerance to the compounds due to the lack of sterols in their membrane (e.g., the oomycete pathogens), and enzymatic degradation of saponins to inactive forms (e.g., *Gaeumannomyces graminis*).

In the interaction between the root infecting pathogen *G. graminis* (causal agent of take-all disease) and its cereal host oat, compatibility is determined by the presence/absence in the plant of avenacin A-1, a saponin, and the fungus' ability to produce avenacinase, which detoxifies avenacin A-1 (Bowyer *et al.* 1995). Isolates of *G. graminis* var. *tritici* (*Ggt*) are sensitive to avenacin A-1 and can infect wheat (a non-saponin containing host), but not the saponin-producing oat species. Isolates of *G. graminis* var. *avenae* (*Gga*) are able to infect both hosts, and are insensitive to avenacin A-1 due to the presence of a gene encoding avenacinase. Disruption of the avenacinase gene via gene replacement converted a *Gga* strain sensitive to avenacin A-1 (Bowyer *et al.* 1995). While the resulting mutant remained pathogenic to wheat, it failed to infect oats, indicating that the ability to detoxify avenacin A-1 is a major host range determinant. Mutants of a diploid oat species, *Avena strigosa*, which are defective in the production of avenacin A-1, became susceptible to *Ggt* and other fungi that are nonpathogenic to the wild-type oat variety, further supporting the role of avenacin A-1 as a chemical defense barrier against potential pathogens (Papadopoulou *et al.* 1999).

In certain fungal pathogens the ability to detoxify saponins might function primarily in an offensive, rather than defensive, role (Martin-Hernandez *et al.* 2000; Bouarab *et al.* 2002). The tomato pathogen *Septoria lycopersici*, for example, produces tomatinase, which hydrolyses α-tomatine, a steroidal glycoalkaloid (Martin-Hernandez *et al.* 2000). Although a mutant of *S. lycopersici* with a disrupted tomatinase gene remains pathogenic to tomato, tomato leaves infected with this mutant exhibit enhanced cell death, and elevated expression of defense-related genes relative to that of plants infected with a wild-type strain, suggesting that in addition to conferring resistance to α-tomatine, tomatinase might be important for suppressing HR (Martin-Hernandez *et al.* 2000). In *Nicotiana benthamiana*, wild-type strains of *S. lycopersici* caused spreading disease lesions, while tomatinase-deficient mutants failed to proliferate in the leaf tissue. At the cellular and molecular levels, the tomatinase-deficient mutants induced stronger defense responses in the infected host than did the wild-type strain; gene silencing of a key component in the *N. benthamiana* defense signalling pathways allowed the mutants to cause normal disease symptoms, indicating that their reduced virulence was not simply due to an increased sensitivity to α-tomatine (Bouarab *et al.* 2002). Treatment of leaves prior to fungal infection with either purified tomatinase, or β_2-tomatine (a breakdown product, produced by tomatinase, of α-tomatine), but not water or α-tomatine, also restored virulence of the mutants. The same treatments resulted in attenuation of the HR in leaf tissues of plants infected with an avirulent bacterial pathogen, or transiently expressing the bacterial avirulence gene, suggesting that tomatinase suppresses host defense systems by converting anti-fungal saponins to compounds inhibitory to the proper activation of HR (Bouarab *et al.* 2002).

4. REGULATION OF PLANT DEFENSE AGAINST FUNGAL PATHOGENS

Due to its rapid generation time, ease of handling and to the availability of a vast amount of genetic resources, *Arabidopsis thaliana* has been extensively utilized in the study of the nature and regulation of plant defense mechanisms. A number of fungi and fungal-like organisms have been shown to be pathogenic to *A. thaliana* and have been used to study the mechanisms of defense responses (Crute *et al.* 1994; Mauch-Mani and Slusarenko 1994; Adam and Somerville 1996; Penninckx *et al.* 1996; Staswick *et al.* 1998; Vijayan *et al.* 1998; Schulze-Lefert and Vogel 2000; Roetschi *et al.* 2001; Tierens *et al.* 2002). We will focus here primarily on how *A. thaliana* regulates its defense responses against various fungal pathogens, with an emphasis on the signalling pathways involved in this process. We do note, however, that *A. thaliana* has not to date been widely used to study the disease interaction from the pathogen perspective, and that given the increasing value of *A. thaliana* as a model system, more attention to those fungal pathogens to which it is susceptible, and to the molecular analysis of their pathogenicity on *A. thaliana*, is certainly warranted.

4.1 Signalling Pathways Controlling Defense

The analysis of Arabidopsis mutants that are altered in their defense responses against various pathogens/pests has revealed the involvement of multiple, interacting signalling pathways that control plant defense responses (Maleck and Dietrich 1999; Feys and Parker 2000; Dangl and Jones 2001). The activation and coordination of these signalling pathways involves one or more of at least three chemical signals, including salicylic acid (SA), jasmonic acid (JA), and ethylene (ETH). Although these signals appear to control unique signalling pathways, and to be important for resistance to different groups of pathogens, they do not function independently. Instead, the pathways are interconnected via both the negative and positive regulatory mechanisms, permitting the plant to fine-tune its defense responses (Fig. 1).

4.1.1 Salicylic acid-dependent signalling

Local infection by a pathogen induces systemic acquired resistance (SAR) throughout the plant, augmenting its ability to defend against subsequent attacks by a broad spectrum of pathogens (Mauch-Mani and Métraux 1997; Maleck and Dietrich 1999). The onset of SAR is strongly correlated with transcriptional induction of several pathogenesis-related (*PR*) genes (Uknes *et al.* 1992). Transgenic (NahG) plants that express a bacterial enzyme which can degrade SA to a biologically inactive form cannot accumulate SA, fails to develop SAR, and exhibits increased susceptibility to many pathogens, suggesting that the induction of SAR requires SA, and that the SA-dependent signalling pathway is important for defense (Gaffney *et al.* 1993). Consistent with this hypothesis, the application of SA analogs 2,6-dichloroisonicotinic acid (INA) and benzothiadiazole (BTH) induces SAR (Lawton *et al.* 1996), and the overproduction of SA in tobacco enhances its resistance to both viral and fungal pathogens (Verberne *et al.* 2000).

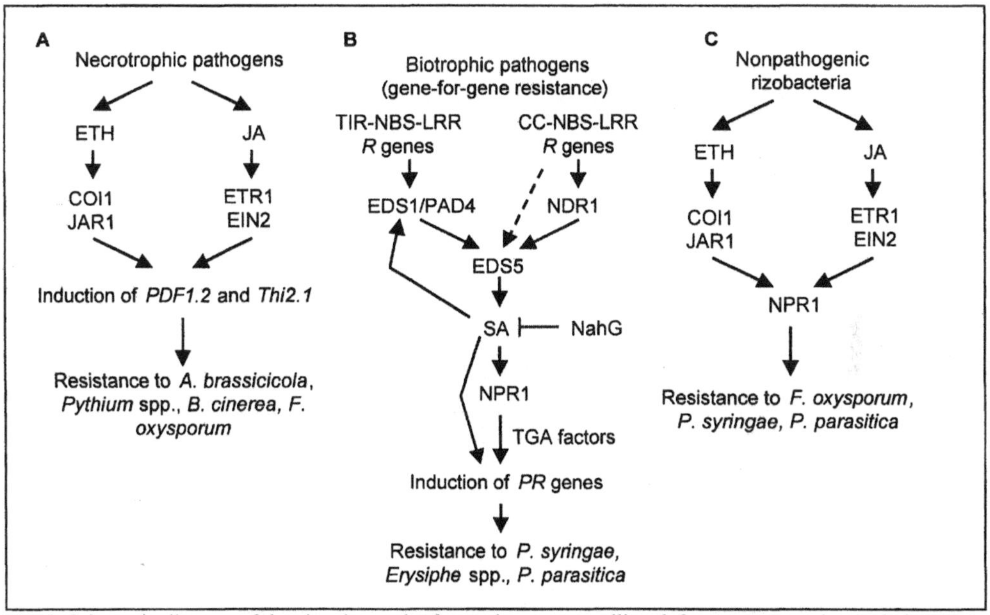

Fig. 1. Schematic diagram of the signal transduction pathways controlling defense responses in *A. thaliana*. A. The jasmonic acid (JA)/ethylene (ETH)-dependent signaling pathway is typically associated with resistance to insects and necrotrophic/hemibiotrophic fungal pathogens. B. Resistance to biotrophic pathogens mainly relies on the salicylic acid (SA)-dependent signaling pathway. Certain *R* genes in the CC-NBD-LRR class function independently of both EDS/PAD4 and NDR1. NahG denotes a bacterial enzyme degrading SA to an inactive form. Some of the SA-mediated responses do not depend on NPR1. C. Induced systemic resistance (ISR), which is triggered by nonpathogenic rizobacteria (mainly fluorescent *Pseudomonas* spp.), depends on components of both the SA- and JA/ETH-dependent signaling pathways.

Structural and regulatory components of the SA-dependent signalling pathway have been identified by mutant analysis. The *NPR1* (non-expresser of *PR*) gene encodes one of the key regulatory factors; *npr1* mutants of *A. thaliana* exhibit increased susceptibility to certain bacterial and fungal pathogens, and are unable to activate SAR in response to pathogen infection or treatment with SA, suggesting that this gene functions downstream from SA (Cao *et al.* 1994; Delaney *et al.* 1995; Shah *et al.* 1997). Overexpression of *NPR1* in transgenic Arabidopsis confers increased resistance to *Pseudomonas syringae* and *Peronospora parasitica* without negatively affecting the growth or development of the transgenic plants (Cao *et al.* 1998). *PR* gene expression is induced in the *NPR1* transgenic lines only in the presence of SAR-activating signals (Cao *et al.* 1998), suggesting that NPR1 activity is controlled by those signals.

NPR1 accumulates in the nucleus in responses to SA or INA, and its nuclear localization is required for activation of *PR* gene expression (Kinkema *et al.* 2000). Consistent with its nuclear localization, and the presence in the protein of two domains that are involved in protein-protein interactions, NPR1 has been shown to interact with a number of basic domain/leucine zipper transcription factors that bind to the *PR1* promoter (Zhang *et al.* 1999; Zhou *et al.* 2000). Thus, NPR1 appears to indirectly regulate the expression of SAR associated genes, including *PR* genes.

Homologues of NPR1 have been identified in diverse plants, including wheat and rice (Kinkema *et al.* 2000; Chern *et al.* 2001). Overexpression of *A. thaliana NPR1* in rice increases its resistance to a bacterial pathogen, *Xanthomonas oryzae* pv. *oryzae* (Chern *et al.* 2001), and NPR1 has been shown to interact with several rice basic domain/leucine zipper transcription factors (Chern *et al.* 2001), suggesting that an NPR1-regulated signalling pathway is conserved in plants. However, although NPR1 is a major regulator in the SA-dependent signalling pathway, there is accumulating evidence that certain aspects of defense are controlled by SA-dependent, NPR1-independent signalling pathway(s) (Clarke *et al.* 1998; Clarke *et al.* 2000; Kachroo *et al.* 2000; Shah *et al.* 2001; Murray *et al.* 2002).

A putative negative regulator of SAR, SNI1 (suppressor of *npr1-1*, *i*nducible 1), was identified through a screen for mutants that suppressed expression of the phenotype exhibited by *npr1-1* mutants (Li *et al.* 1999). In the *npr1-1* background, the recessive *sni1* mutation restores *PR* gene induction by SA, SAR, and disease resistance, and in the *NPR1* background, renders the plant more sensitive to SAR signals. The SNI1 protein appears to be a nuclear protein with limited similarity to the mouse retinoblastoma protein, a negative transcription regulator. Another suppressor mutant of *npr1-1*, *snc1* (suppressor of *npr1-1*, constitutive 1) differs from *sni1* in that it constitutively produces levels of SA approximately 15- to 21-fold higher than those in wild-type plants (Li *et al.* 2001). It appears that the mutation activates an *R* gene-mediated, SA-requiring, but NPR1-independent, resistance pathway.

Several Arabidopsis mutants that are defective in the synthesis of SA and/or the regulation of its synthesis have also been isolated (Glazebrook *et al.* 1997; Nawrath and Métraux 1999; Wildermuth *et al.* 2001). One group of mutants, including *acd2, lsd1-7, cpr1, cpr5, cpr6, cim2, cim3, cep* and *ssi1*, constitutively accumulate high levels of SA, express *PR* genes and confer enhanced resistance to a broad range of pathogens (Dietrich *et al.* 1994; Weymann *et al.* 1995; Bowling *et al.* 1997; Clarke *et al.* 1998; Shah *et al.* 1999; Silva *et al.* 1999). Some of these mutants, including *lsd1-lsd7, acd2, cep, cpr5, ssi1*, exhibit spontaneous HR-like lesions and are referred to as lesion-mimic mutants.

Mutants in the second group, including *eds5* (enhanced disease susceptibility 5), *pad4* (phytoalexin deficient 4), *sid1* (SA induction deficient 1) and *sid2*, exhibit impaired SAR responses, and produces significantly reduced levels of SA in response to biological and chemical agents that are known to trigger defense SAR (Rogers and Ausubel 1997; Zhou *et al.* 1998; Nawrath and Métraux 1999); one of these, *sid1*, is allelic to *eds5*. Although the disease susceptibility of the *sid1* and *sid2* mutants is similar to that of NahG transgenic plants, NahG plants exhibit reduced accumulation of the phytoalexin camalexin, and defective expression of the three SAR-induced *PR-1, PR-2* and *PR-5*, while in the *sid1* and *sid2* mutants, camalexin production is normal, and expression of only *PR-1* is defective.

The *EDS5/SID1* gene encodes one of the more than 50 MATE (multidrug and toxin extrusion) transporters in Arabidopsis, and is presumably involved in transporting organic molecules (Nawrath *et al.* 2002), but the exact biological function of the *EDS5/SID1* gene product, and its role in SA accumulation have not yet been determined. All *sid2* mutations have been localized to *ICS1*, one of two genes that encode isochorismate synthase, and thus result in the blockage of the conversion of chorismate to isochorismate (Wildermuth *et al.* 2001), the immediate precursor to SA. Expression of *ICS1* is induced by *Erysiphe* infection, and appears to be under the negative feedback control of NPR1, but not SA (Figure 1). Although the *sid2* mutants fail express high levels of SA in response to SAR-inducing signals, they still contain low levels of SA, which may be synthesized by the phenylalanine ammonium lyase (PAL) pathway, and/or the second isochorismate synthase (Wildermuth *et al.* 2001).

4.1.2 Jasmonic acid/ethylene-dependent signalling

The plant hormone ethylene (ETH), a major modulator of plant growth and development, and jasmonic acid (JA) and its more volatile methyl ester derivative methyl jasmonate (MeJA), been implicated in plant defense (Mauch-Mani and Métraux 1997; Maleck and Dietrich 1999; Feys and Parker 2000; Seo *et al.* 2001; Turner *et al.* 2002). The JA/ETH-dependent signalling pathway is important for defense against bacterial pathogens such as *Erwinia carotovora* (Norman-Setterbald *et al.* 2000), necrotrophic/hemibiotrophic fungal pathogens such as *Alternaria brassicicola*, *Botrytis cinerea*, and *Pythium irregulare* (Penninckx *et al.* 1996; Penninckx *et al.* 1998; Staswick *et al.* 1998; Thomma *et al.* 1998; Vijayan *et al.* 1998; Pieterse and van Loon 1999; Thomma *et al.* 1999), as well as insects (McConn *et al.* 1997; Staswick and Lehman 1999). Although the ethylene-insensitive mutant *ein2-1* is more susceptible to *B. cinerea* than are wild-type plants (Thomma *et al.* 1999), and mutants in the JA-mediated signalling pathway (*coi1* and *jar1*) are susceptible to *Pythium* spp., *A. brassicicola* and *B. cinerea* (Staswick *et al.* 1998; Thomma *et al.* 1998; Vijayan *et al.* 1998), mutants defective in the SA-dependent signalling pathway, *npr1* and *nahG*, do not exhibit enhanced susceptibility to these fungi (Thomma *et al.* 1998). However, the results of a recent study suggest that both the SA- and JA/ETH-dependent signalling pathways are important for resistance to *B. cinerea* (Zimmerli *et al.* 2001).

The results of the Arabidopsis studies are supported by those from studies of tomato mutants that are defective in the JA/ETH synthesis and signalling pathway (Lund *et al.* 1998; Díaz *et al.* 2002). A transgenic line expressing ACC deaminase, which produces negligible levels of ETH, exhibits higher susceptibility than wild-type plants to *B. cinerea* (Díaz *et al.* 2002). Conversely, the *Nr* (*Never ripe*) mutant, which is severely reduced in ETH sensitivity, exhibits increased resistance to *F. oxysporum* (Lund *et al.* 1998). While *Nr* expressed increased susceptibility to *B. cinerea* in detached leaf assays, when leaves of intact plants were infected, no significant difference between the *Nr* and wild-type plants was observed (Díaz *et al.* 2002). Another mutant, *epinastic* (*epi*), in which components of the ETH response pathway are constitutively active, was more resistant than its wild type to *B. cinerea* (Díaz *et al.* 2002). A JA-deficient mutant *defenseless* (*def1*) was more susceptible than wild-type plants to *B. cinerea* (Díaz *et al.* 2002).

Activation of the JA/ETH-dependent signalling pathway induces the expression of *PDF1.2* (Penninckx *et al.* 1996), a member of a gene family that encodes antifungal proteins called defensins (Terras *et al.* 1993). Transgenic plants that over-express JA carboxyl methyltransferase, which converts JA to MeJA, constitutively express *PDF1.2* and are more resistant to *B. cinerea* (Seo *et al.* 2001). Expression of *PDF1.2* is not affected by mutations in the SA-dependent signalling pathway, but the ethylene-insensitive mutants *etr1* and *ein2,* and a JA-insensitive mutant, *coi1*, block the expression of *PDF1.2*, supporting the involvement of both JA and ETH in its expression (Penninckx *et al.* 1996; Manners *et al.* 1998; Thomma *et al.* 1998);

concomitant activation of the JA- and ETH-controlled signal pathways is necessary for the activation of *PDF1.2* (Penninckx *et al.* 1998).

Isolates of *F. oxysporum* have been identified that exhibit differential virulence on various Arabidopsis ecotypes (Mauch-Mani and Slusarenko 1994; Epple *et al.* 1995; Epple *et al.* 1997; Epple *et al.* 1998; Vignutelli *et al.* 1998). Expression of the Arabidopsis *Thi2.1* gene, which encodes an antifungal thionin, is induced by MeJA, wounding, and by *F. oxysporum* (Bohlmann *et al.* 1998; Vignutelli *et al.* 1998); that induction appears to occur solely through the JA/ETH-dependent signal transduction pathway, since SA had no effect on *Thi2.1* expression (Epple *et al.* 1995). Its expression level directly correlates with resistance to *F. oxysporum* f. sp. *matthiolae* (Epple *et al.* 1998), and over-expression of the gene clearly enhances resistance to that pathogen (Epple *et al.* 1997). Inoculation of *F. oxysporum* on lower rosette leaves significantly increased resistance to a subsequent infection by *P. parasitica* in the other leaves, suggesting that infection also induced the SAR (Mauch-Mani and Slusarenko 1994).

In an analysis of JA synthesis defective mutants, the *fad3-2 fad7-2 fad8* mutant, which is defective in synthesizing the fatty precursors of JA, was shown to be highly susceptible to *Pythium mastophorum*, while the wild-type parent was not (Vijayan *et al.* 1998). Application of MeJA to the mutant restored its resistance to the fungus. Although *opr3* mutants fail to synthesize JA due to the lack of the enzyme that converts 12-oxophytodienoic acid (OPDA) to JA, their resistance to an insect, or to a fungal pathogen (*A. brassicicola*) is indistinguishable from that of wild-type plants (Stintzi *et al.* 2001). In addition, genes under the control of the JA-dependent signal pathway are normally induced by wounding and exogenous application with OPDA. These results suggest that OPDA also functions as a signal molecule in the JA-dependent signal pathway.

4.1.3 Regulation of *R* Gene-mediated responses in Arabidopsis

A number of *R* genes have been identified through genetic and molecular analyses of gene-for-gene interactions between *Arabidopsis* and its fungal pathogens, including the powdery mildew fungus *E. cichoracearum* (Adam and Somerville 1996; Xiao *et al.* 1997), *P. parasitica* (Holub *et al.* 1994; Tör *et al.* 1994; Botella *et al.* 1998; McDowell *et al.* 1998) and *Plasmodiophora brassicae* (Fuchs and Sacristán 1996). Almost all Arabidopsis *R* genes cloned to date belong to a large family of putative cytoplasmic proteins containing a nucleotide-binding site (NBS) and a leucine-rich repeat (LRR) motif (Dangl and Jones 2001; Hulbert *et al.* 2001). Two exceptions to this rule are *RPW8.1* and *RPW8.2*, which are located in tandem at the *RPW8* locus in certain ecotypes and confer broad-spectrum resistance to powdery mildew fungi (Xiao *et al.* 2001). The *RPW8.1* and *RPW8.2* genes encode related proteins (45% identity) that contain an N-terminal transmembrane (or signal anchor) domain, and a coiled-coil (CC) domain. The NBS-LRR *R* gene family can be subdivided into two groups according to their N-terminal structures of the proteins; those in the TIR-NBS-LRR group contain the TIR domain, homologous to that in the Drosophila Toll and mammalian interleukin 1 receptors, while members of the CC-NBS-LRR group contain a CC domain (Dangl and Jones 2001).

The conservation of structural features among *R* gene products that confer resistance to diverse pathogens strongly suggests that Arabidopsis utilizes a small number of common signalling pathways to transmit cues from the different *R* genes. This hypothesis is supported by the results of several genetic and molecular studies of *R* gene-mediated resistance (Dodds and Schwechheimer 2002; Nishimura and Somerville 2002). As summarized in Table 4, there exist several *R* gene-mediated signalling pathways in Arabidopsis (McDowell *et al.* 2000; Bittner-Eddy and Beynon 2001). Three mutants, *eds1* (enhanced disease susceptibility 1), *pad4*, and *ndr1* (non-race specific disease resistance 1), are defective in transmitting gene-for-gene signals triggered by bacterial and fungal pathogens (Century *et al.* 1995; Parker *et al.* 1996). In the *eds1* mutant background, the presence of certain *R* genes belonging to the TIR-NBS-LRR group fails to trigger gene-for-gene defense responses (Dodds and Schwechheimer 2002). The same set of *R* genes also fails to function properly in the *pad4* background, suggesting that both EDS1 and PAD4 are involved in transmitting signals from these *R* genes (Feys *et al.* 2001). The *RPW8.1* and *RPW8.2* genes also require EDS1 for their function (Xiao *et al.* 2001). The role of PAD4 for resistance mediated by *RPW8.1* and *RPW8.2* has not been tested.

SA accumulation in response to infection by fungal pathogens carrying *AVR* genes that correspond to EDS1/PAD4-dependent *R* genes requires both *EDS1* and *PAD4*, which encode lipase-like proteins (Falk *et al.* 1999; Jirage *et al.* 1999), and appears to require direct physical interaction between EDS1 and PAD4 (Feys *et al.* 2001). *R* gene-mediated HR, in contrast, requires only EDS1 (Feys *et al.* 2001).

Table 4. Dependence of Arabidopsis *R* gene function on *EDS1, NDR1, RAR1* and *SGT1*[a]

R GENE[B]	EDS1[c]	NDR1[c]	RAR1[c]	SGT1[c]
RPP4 (TIR)	+	-	+	+
RPP5 (TIR)	+	-	+	+
RPS4 (TIR)	+	-	+	-
RPS2 (CC)	-	+	+	-
RPS5 (CC)	-	+	+	-
RPM1 (CC)	-	+	+	-
RPP2 (TIR)	+	-	-	+
RPP8 (CC)	-	-	-	-
RPP1A (TIR)	+	-	-	-

[a] The data shown in this table are adapted from a recent review (Dodds and Schwechheimer 2002); [b] *RPP* genes confer resistance to *P. parasitica. RPS* and *RPM* genes correspond to *AVR* genes in *P. syringae*, TIR and CC indicate the TIR-NBS-LRR and CC-NBS-LRR families of *R* genes; [c] + (*R* gene function requires the gene); – (*R* gene function does not requires the gene).

In contrast to the resistance conferred by TIR-NBS-LRR *R* genes, resistance by certain members of the CC-NBS-LRR *R* gene family requires the presence of the *NDR1* gene, rather than *EDS1* (Aarts *et al.* 1998). The *NDR1* gene encodes a putative membrane protein (Century *et al.* 1997), and appears to mediate the accumulation of SA in response to reactive oxygen intermediates (ROI) generated during pathogen infection (Shapiro and Zhang 2001).

Two CC-NBS-LRR *R* genes, *RPP8* and *RPP13*, which confer resistance to certain strains of *P. parasitica*, require neither NDR1, nor EDS1/PAD4 (Dodds and Schwechheimer 2002), indicating the existence of additional pathway(s) that transmit signals from these genes (McDowell *et al.* 2000; Bittner-Eddy and Beynon 2001). A series of recent studies also indicates the existence of additional regulatory components for *R* gene-mediated defense responses (Austin *et al.* 2002; Azevedo *et al.* 2002; Muskett *et al.* 2002; Tör *et al.* 2002; Tornero *et al.* 2002). Mutations in *SGT1*, and *AtRAR1*, an ortholog of *HvRAR1*, a gene required for the function of certain powdery mildew *R* genes in barley (Shirasu *et al.* 1999), for example, compromise the function of certain *R* genes (Table 4). Interestingly, some *R* genes whose function is affected by mutations in the *AtRAR1* gene also require EDS1 and NDR1, suggesting that the EDS1/PAD4-

and NDR1-dependent *R* gene signal pathways do not function independently but converge at some point (Muskett *et al.* 2002; Tornero *et al.* 2002).

The *AtRAR1* gene encodes a protein with two zinc-binding domains (Shirasu *et al.* 1999; Muskett *et al.* 2002; Tornero *et al.* 2002), which physically interacts with two highly related proteins (87% identity), AtSGT1a and AtSGT1b (Azevedo *et al.* 2002). Mutations in *AtSGT1b*, but not *AtSGT1b*, cause defects in the transmission of signals from certain *R* genes (Austin *et al.* 2002; Tör *et al.* 2002), and hence increased susceptibility to some avirulent isolates of *P. parasitica*. Both *AtSGT1a* and *AtSGT1b* exhibit significant similarity to yeast *SGT1* and can complement mutations in the yeast gene, suggesting a conserved role for these proteins (Azevedo *et al.* 2002). Further evidence for such conservation comes from the results of biochemical studies which have demonstrated that SGT1 interacts with the SCF (Skp1-cullin-F-box)-type E3 ubiquitin ligase complex in both yeast and barley, and that the SGT1-RAR1 complex interacts with the COP9 signalosome, a protein complex involved in protein degradation (Azevedo *et al.* 2002). Interaction between the COP9 signalosome and the SCF-type E3 ubiquitin ligase complex has been observed in both yeast and Arabidopsis (Lyapina *et al.* 2001; Scvhwechheimer *et al.* 2001). The COI1 protein, a key factor required for transmitting JA signals and defense against insects and pathogens, also physically interacts with components of the SCF-type ubiquitin ligase complex and a histone deacetylase (Devoto *et al.* 2002). These results suggest that ubiquitination-mediated protein degradation is important for controlling *R* gene-mediated defense responses through both the SA- and JA/ETH-dependent signalling pathways.

4.1.4 Additional Arabidopsis genes important for resistance to fungal pathogens

One of the Arabidopsis mutants, *edr1*, which exhibits enhanced resistance to *Pseudomonas syringae*, also exhibits enhanced resistance to the powdery mildew fungus, *E. cichoracearum* (Frye and Innes 1998). The *edr1* mutant is not, however, more resistant to *P. parasitica*, suggesting that the gene is involved in defense against specific pathogens. The enhanced resistance to *E. cichoracearum* is not due to the constitutive expression of *PR1*, *BGL2* or *PR5* genes; infection by *E. cichoracearum* thus appears to induce stronger defense responses in the mutant than in wild-type plants, suggesting that the mutation renders the plant more sensitive to signals triggered by pathogen infection. Four additional loci, termed *pmr1-4* (powdery mildew resistant 1-4), are also involved in defense against this fungus (Vogel and Somerville 2000). Mutations in these loci result in increased resistance to the pathogen, but again, the enhanced resistance does not appear to be a result of constitutive activation of either the SA- or JA/ETH-dependent signalling pathways. Similarly, the data indicate that these genes play a role in the specificity of the defense response; compared with wild-type plants, one of the mutants, *pmr1*, was more susceptible to a closely related powdery mildew fungus, *E. orontii*, but remained susceptible to *P. parasitica* and *P. syringae*, while the *pmr4* mutant was also more resistant to *P. parasitica* but not to *P. syringae*.

Mutations in a newly described *PMR* gene, *PMR6*, also result in increased resistance to two species of powdery mildew fungus, *E. cichoracearum* and *E. orontii*, but do not affect susceptibility to *P. parasitica* and *P. syringae* (Vogel *et al.* 2002). The levels of *PR1* and *PDF1.2* in *pmr6* are indistinguishable from those in wild-type plants, and double mutants between *pmr6* and mutants that affect the SA- or JA/ETH-dependent signalling pathways (*npr1*, *coi1*, *etr1*, and the *NahG* transgene) maintain increased resistance, suggesting that the increased resistance is independent of the activation of the SA- and JA/ETH-dependent signalling pathways. Instead, increased resistance is probably due to the loss of a host factor required for successful colonization by the fungus. The *PMR6* gene encodes a pectate lyase-like protein with a novel C-terminal domain, which is predicted to be located at the exterior surface of the plasma membrane. Consistent with the putative biochemical function of the protein, the composition of the cell wall in the *pmr6* mutant differs from that of wild-type plants (Vogel *et al.* 2002).

A new mutant, *esa1* (enhanced susceptibility to Alternaria 1), shows increased susceptibility to necrotrophic/hemibiotrophic fungi, including *A. brassicicola*, *B. cinerea*, and *Plectosphaerella cucumerina*, but exhibits normal resistance to biotrophic pathogens, *P. syringae* and *P. parasitica* (Tierens *et al.* 2002). In the mutant, accumulation of the phytoalexin camalexin and induction of *PDF1.2* in response to pathogen infection is significantly delayed relative to that of wild-type plants. However, the mutant normally responds normally to JA and ETH. Its increased susceptibility to certain pathogens is probably due to a defect in transmitting defense-associated signals generated by reactive oxygen intermediates (Tierens *et al.* 2002).

4.1.5 Cross-talk between signalling pathways

There is increasing evidence to suggest that the SA- and JA/ETH-dependent signalling pathways do not operate independently (Doares *et al.* 1995; Maleck and Dietrich 1999; Pieterse and van Loon 1999; Feys and Parker 2000; Schenk *et al.* 2000; Zimmerli *et al.* 2001; Devadas *et al.* 2002). Plants seem to utilize cross-talk mechanisms between these pathways in order to coordinate the various defense responses needed to effectively counteract pathogens with different modes of pathogenicity. Induced systemic resistance (ISR), for example, requires the JA/ETH-dependent signalling pathway as well as NPR1, a key regulatory factor in the SA-dependent pathway, while SA is dispensable for ISR (Pieterse and van Loon 1999). In tomato, the application of SA or its functional analog inhibits the production of compounds whose synthesis is regulated by JA, probably through the inhibition of JA synthesis (Doares *et al.* 1995). Conversely, the expression of *PDF1.2*, a gene under the control of JA and ETH, appears to be suppressed by SA (Clarke *et al.* 1998), and is upregulated in mutants defective in the SA-dependent signalling pathway (Zimmerli *et al.* 2001).

In contrast to these antagonistic interactions between the pathways, there also exists evidence for positive interactions. Microarray analyses using 2,375 Arabidopsis genes revealed a number of genes whose expression is coinduced or cosuppressed by SA and MeJA (55 and 28, respectively) (Schenk *et al.* 2000). Moreover, the SA- and JA/ETH-dependent signalling pathways appear to act in concert to modulate *PR* gene expression, cell death, and disease resistance (Devadas *et al.* 2002). However, although the SA-dependent, NPR1-independent signalling pathway(s) appear to positively interact with components of the JA/ETH-dependent signalling for defense against bacterial and fungal pathogens (Clarke *et al.* 2000), this interaction is not required for viral resistance (Kachroo *et al.* 2000). These few examples illustrate the complexity of the positive and negative cross-talk between different signalling pathways; much still remains to be determined about the intricacies of these interactions.

4.2 Programmed Cell Death: a Double-edged Sword?

In the presence of an *AVR* gene-*R* gene pair, HR in and around the initial infection sites prevents extensive ramification of the fungal pathogen. HR is accompanied by the induction of multifaceted defense responses, including: production of reactive oxygen intermediates and phytoalexins, fortification of the cell wall via cross-linking of cell wall proteins, expression of

PR genes, and localized host cell death. HR-associated cell death appears to be a form of programmed cell death (PCD), which is a genetically controlled, cellular suicide mechanism that involves the active participation of host (Shirasu and Schulze-Lefert 2000; Lam *et al.* 2001). Although a compatible disease interaction also often results in host cell death, the process is distinct from HR-associated cell death in that it is slower than HR-associated cell death, and spreads well beyond the original site of infection; it is not clear if the plant plays an active role in directing cell death during compatible interactions. A recent report (Vogel *et al.* 2002) suggests the involvement of a host gene in disease development.

Mutants that spontaneously develop necrotic lesions resembling those caused by HR-associated cell death (i.e., lesion mimic mutants) have been identified and characterized in several plant species, including barley (Buschges *et al.* 1997), *A. thaliana* (Dangl *et al.* 1996), maize (Johal *et al.* 1995), and rice (Takahashi *et al.* 1999; Yin *et al.* 2000), supporting the host involvement in PCD associated with HR. Lesion mimic mutants are defective in negatively controlling the onset of PCD, thus making the mutants hyper-responsive to PCD triggers. Many of the mutants also exhibit increased resistance to certain pathogens (Simmons *et al.* 1999; Takahashi *et al.* 1999; Shirasu and Schulze-Lefert 2000; Wolpert *et al.* 2002), supporting the role of PCD in defense.

Hyper-responsive PCD, or triggering PCD prior to pathogen infection, does not always increase resistance to pathogens (Jarosch *et al.* 1999; Govrin and Levine 2000). On the contrary, it appears that in some cases host PCD facilitates colonization and disease development by necrotrophic fungi (Govrin and Levine 2000; Dickman *et al.* 2001). This hypothesis is supported by the observations that the HSTs secreted by some of *Cochliobolus* spp. and *Alternaria* spp. cause a cell death which exhibits some of the characteristics of PCD (Wolpert *et al.* 2002), and that transgenic expression in tobacco plants of animal genes that inhibit PCD confers increased resistance to *B. cinerea*, *Cercospora nicotianae*, and *Sclerotinia sclerotiorum* (Dickman *et al.* 2001). Similarly, the *A. thaliana dnd1* mutant, which exhibits a highly attenuated HR with no cell death upon infection by pathogens (Yu *et al.* 1998), is more resistant to infection by *B. cinerea* and *S. sclerotiorum* (Govrin and Levine 2000) than is the wild-type plant, and treatment of *A. thaliana* with either reagents that produce reactive oxygen intermediates, or an avirulent strain of *P. syringae* prior to infection with *B. cinerea* or *S. sclerotiorum* causes increased disease severity, suggesting that HR-associated cell death promotes growth of these pathogens. Conversely, these treatments resulted in increased resistance to a virulent strain of *P. syringae* (a biotrophic pathogen).

Germinating conidia of *Cochliobolus miyabeanus* produce a mixture of HSTs called ophiobolins (Xiao *et al.* 1991); when rice leaves were treated with conidial germination fluid (CGF) of *C. miyabeanus*, containing ophiobolins, HR-like lesions developed in cultivars resistant (Tetep) and susceptible (Nakdong) to *C. miyabeanus* (YongHwan Lee, personal communication). CGF treatment protected cultivar Nakdong from infection by a virulent isolate of *M. grisea*; no invasive mycelial growth of *M. grisea* was observed in the CGF-treated leaves. In comparison, untreated plants of cultivar Nakdong showed severe blast symptoms when infected with the same *M. grisea* isolate. At the molecular level, the expression patterns of pathogenesis-related (*PR*) genes in the CGF-treated cultivar Nakdong infected with *M. grisea* resembled those in *M. grisea*-infected cultivar Tetep (resistant to *M. grisea*); maximum transcript accumulation of the *PR1* gene occurred within 24 hours after inoculation with *M. grisea*. In untreated cultivar Nakdong, *PR1* transcripts were not detectable until 2-3 days post inoculation with *M. grisea*. These data suggest that the ophiobolins (and potentially other factors) in the CGF converted a compatible interaction into an incompatible interaction, by triggering HR. Yet CGF treatment had the opposite effect on infection by a strain of *A. alternata* nonpathogenic to rice (Xiao *et al.* 1991); it allowed this necrotrophic fungus to successfully colonize leaf tissues, suggesting that triggering the HR facilitated its infection. Clearly, the role that PCD plays in resistance and disease depends upon the nature of invading pathogen.

5. CONCLUSIONS

Food production and distribution must increase concurrently in order to meet the demands of the world's rapidly expanding population. This increased production must be accomplished in the face of erosion of existing cultivated acreage due to adverse global climatic changes, and/or rapid urbanization in certain parts of the world, and plant diseases that continue to threaten global crop production. The endogenous defense systems of plants have been, and continue to be, widely exploited, primarily through breeding, for the management of many important plant diseases; a better understanding of the disease interaction and how the plant defense systems function will facilitate our efforts to further develop new disease resistant crop plants via breeding, genetic engineering, or both means. Considering the increasing need to employ environmentally friendly and economical means for disease control, we cannot understate the economic importance of accumulating such an understanding.

During their evolution, plant pathogens have developed many different strategies to aid their successful colonization of their host plants (e.g. necrotrophic vs. biotrophic infection, tissue specificity, and diverse mechanisms of penetration and proliferation, etc.). Against this diversity, a one-size-fits-all approach to defense is unlikely to be effective; successful defense is more likely to rely upon the ability of a plant to invoke the appropriate defense responses according to the nature of invading pathogen. The possible involvement of programmed plant cell death in both defense and disease development well illustrates the importance of the triggering appropriate responses for successful defense. As summarized in this chapter, our understanding of the mechanisms underlying fungal-plant disease interactions has greatly advanced during the past decade, aided in large part by the development of new research tools, and a few selected model plant-fungal pathogen systems. The pace of this advance will only quicken as genome sequences of both plants and their fungal pathogens become available; genome information and functional genomic tools will serve as key stepping stones in furthering our understanding of the biology of fungal pathogens of agricultural significance, and the responses of their hosts to infection. Judicious application of genomic tools will make it possible to study plant-fungal pathogen interactions at a global scale, from both sides of the interaction, and to address the many complex questions which seemed in the not-so-distant past to be impractical or impossible to study.

In order to take full advantage of the accumulating knowledge and research tools for the effective management of diverse fungal diseases, we must, however, expand our knowledge to lesser-characterized, but economically important disease interactions, such as those caused by the soil-borne fungal pathogens. Although the threat posed by these fungi is significant, our knowledge of soil-borne diseases at the molecular and cellular level is considerably less than that of the foliar diseases. The current paradigms about the nature and mechanisms of fungal pathogenicity and plant defense have been derived largely from studies of foliar pathogens. Doubtless, certain defense responses will be conserved regardless of the sites of infection and colonization. However, considering the physical, physiological and environmental variation in diverse parts of the plant, it would be expected that there might be some differential regulation of their defense mechanisms in various organs and tissues. From the pathogen perspective, fungi are also likely to have evolved different strategies that allow colonization of specific tissues. Of course, it is highly unlikely that different pathogens would utilize completely different strategies for their pathogenesis. Rather, they are likely to use some of the conserved cellular machinery, and to appropriate additional factors for infecting particular hosts or tissues. Our rapidly expanding collection of research tools will certainly facilitate the identification of such factors.

REFERENCES

Aarts N, Metz M, Holub E, Staskawicz BJ, Daniels MJ, and Parker JE (1998). Different requirements for *EDS1* and *NDR1* by disease resistance genes define at least two *R* gene-mediated signalling pathways in *Arabidopsis*. Proc Natl Acad Sci USA 95: 10306-10311.

Abbott A (1999). A post-genomic challenge; learning to read patterns of protein synthesis. Nature 402: 715-720.

Adam L, and Somerville S (1996). Genetic characterization of five powdery mildew disease resistance loci in *Arabidopsis thaliana*. Plant J 9: 341-356.

Ahn J-H, Cheng Y-Q, and Walton JD (2002). An extended physical map of the *TOX2* locus of *Cochliobolus carbonum* required for biosynthesis of HC-toxin. Fungal Genet Biol 35: 31-38.

Ahn J-H, and Walton JD (1996). Chromosomal organization of *TOX2*, a complex locus required for host-selective toxin biosynthesis in *Cochliobolus carbonum*. Plant Cell 8: 887-897.

Akamatsu H, Itoh Y, Kodama M, Otani H, and Kohmoto K (1997). AAL-toxin-deficient mutants of *Alternaria alternata* tomato pathotype by restriction enzyme-mediated integration. Phytopathology 87: 967-972.

Austin MJ, Muskett P, Kahn K, Feys BJ, Jones JDG, and Parker JE (2002). Regulatory role of *SGT1* in early *R* gene-mediated plant defenses. Science 295: 2077-2080.

Ausubel FM (2002). Summaries of National Science Foundation-sponsored Arabidopsis 2010 plant genome projects that are generating Arabidopsis resources for the community. Plant Physiol 129: 394-437.

Azevedo C, Sandanandom A, Kitagawa K, Freialdenhoven A, Shirasu K, and Schulze-Lefert P (2002). The RAR1 interactor SGT1, an essential component of *R*-gene triggered disease resistance. Science 295: 2073-2076.

Bennett JW, and Arnold J (2001). Genomics for Fungi. In: The Mycota VIII. Biology of the Fungal Cell. Berlin: Howard R, and Gow N. Springer-Verlag, pp 267-297.

Bindslev L, Kershaw MJ, Talbot NJ, and Oliver RP (2001). Complementation of the *Magnaporthe grisea deltacpkA* mutation by the *Blumeria graminis PKA-c* gene: functional genetic analysis for an obligate pathogen. Mol Plant-Microbe Interact 14: 1368-1375.

Bittner-Eddy PD, and Beynon JL (2001). The *Arabidopsis* downy mildew resistance gene, *RPP13-Nd*, functions independently of *NDR1* and *EDS1* and does not require the accumulation of salicylic acid. Mol Plant-Microbe Interact 14: 416-421.

Bohlmann H, Vignutelli A, Hilpert A, Miersch O, Wasternack C, and Apel K (1998). Wounding and chemicals induce expression of the *Arabidopsis thaliana* gene *Thi2.1*, encoding a fungal defense thionin, via the octadecanoid pathway. FEBS Lett 437: 281-286.

Bölker M, Böhnert HU, Braun KH, Görl J, and Kahmann R (1995). Tagging pathogenicity genes in Ustilago maydis by restriction enzyme mediate integration (REMI). Mol Gen Genet 248: 547-552.

Bonnet P, Lacourt I, Venard P, and Ricci P (1994). Diversity in pathogenicity of tobacco and in elicitin production among isolates of *Phytophthora parasitica*. J Phytopathol 141: 25-37.

Botella MA, Parker JE, Frost LN, Bittner-Eddy PD, Beynon JL, Daniels MJ, Holub EB, and Jones JD (1998). Three genes of the *Arabidopsis RPP1* complex resistance locus recognize distinct *Peronospora parasitica* avirulence determinants. Plant Cell 10: 1847-1860.

Bouarab K, Melton R, Peart J, Baulcombe D, and Osbourn A (2002). A saponin-detoxifying enzyme mediates suppression of plant defenses. Nature 418: 889-892.

Bourett TM, Sweigard JA, Czymmek KJ, Carroll AM, and Howard RJ (2002). Reef coral fluorescent proteins for visualizing fungal proteins. Fungal Genet Biol 37: 211-220.

Bowling SA, Clarke JD, Liu Y, Klessig DF, and Dong X (1997). The *cpr5* mutant of Arabidopsis expresses both NPR1-dependent and NPR1-independent resistance. Plant Cell 9: 1573-1584.
Bowyer P, Clarke Br, Lunness P, Daniels MJ, and Osbourn AE (1995). Host range of a plant pathogenic fungus determined by a saponin detoxifying enzyme. Science 267: 371-374.
Bowyer P, Mueller E, and Lucas J (2000). Use of an isocitrate lyase promoter-GFP fusion to monitor carbon metabolism of the plant pathogen *Tapesia yallundae* during infection of wheat. Mol Plant Pathol 1: 253-262.
Brandwagt BF, Mesbah LA, Takken FL, Laurent PL, Kneppers TJ, Hille J, and Nijkamp HJ (2000). A longevity assurance gene homolog of tomato mediates resistance to *Alternaria alternata* f. sp. *lycopersici* toxins and fumonisin B1. Proc Natl Acad Sci USA 97: 4961-4966.
Brosch G, Ransom R, Lechner T, Walton JD, and Loidl P (1995). Inhibition of maize histone deacetylase by HC toxin, the host-selective toxin of *Cochliobolus carbonum*. Plant Cell 7: 1941-1950.
Bryan GT, Wu K, Farrall L, Jia Y, Hershey HP, McAdams SA, Faulk KN, Donaldson GK, Tarchini R, and Valent B (2000). A single amino acid difference distinguishes resistant and susceptible alleles of the rice blast resistance gene *Pi-ta*. Plant Cell 12: 2033-2046.
Bundock P, den Dulk-Ras A, Beijersbergen A, and Hooykaas PJJ (1995). Trans-kingdom T-DNA transfer from *Agrobacterium tumefaciens* to *Saccharomyces cerevisiae*. EMBO J 14: 3206-3214.
Buschges R, Hollricher K, Panstruga R, Simons G, Wolter M, Frijters A, van Daelen R, van der Lee T, Diergaarde P, Groenendijk J, Topsch S, Vos P, Salamini F, and Schulze-Lefert P (1997). The barley *Mlo* gene: a novel control element of plant pathogen resistance. Cell 88: 695-705.
Cantrell RP, and Reeves TG (2002). The cereal of the world's poor takes center stage. Science 296: 53.
Cao H, Bowling SA, Gordon AS, and Dong X (1994). Characterization of an *Arabidopsis* mutant that is nonresponsive to inducers of systemic acquired resistance. Plant Cell 6: 1583-1592.
Cao H, Li X, and Dong X (1998). Generation of broad-spectrum disease resistance by overexpression of an essential regulatory gene in systemic acquired resistance. Proc Natl Acad Sci USA 95: 6531-6536.
Century KS, Holub E, and Staskawicz BJ (1995). *NDR1*, locus of *Arabidopsis thaliana* that is required for disease resistance to both a bacterial and a fungal pathogen. Proc Natl Acad Sci USA 92: 6597-6601.
Century KS, Shapiro AD, Repetti PP, Dahlbeck D, Holub E, and Staskawicz BJ (1997). NDR1, a pathogen-induced component required for Arabidopsis disease resistance. Science 278: 1963-1965.
Chao CT, and Ellingboe AH (1997). Genetic analysis of avirulence/virulence of an isolate of *Magnaporthe grisea* from a rice field in Texas. Phytopathology 87: 71-76.
Charron MJ, and Michels CA (1988). The naturally occurring alleles of *MAL1* in *Saccharomyces* species evolved by various mutagenic processes including chromosomal rearrangement. Genetics 120: 83-93.
Chen X, Stone M, Schlagnhaufer C, and Romaine CP (2000). A fruiting body tissue method for efficient *Agrobacterium*-mediated transformation of *Agaricus bisporus*. Appl Env Microbiol 66: 4510-4513.
Cheong YH, Chang H-S, Gupta R, Wang X, Zhu T, and Luan S (2002). Transcriptional profiling reveals novel interactions between wounding, pathogen, abiotic stress, and hormonal responses in Arabidopsis. Plant Physiol 129: 661-667.
Chern M-S, Fitzgerald HA, Yadav RC, Canlas PE, Dong X, and Ronald PC (2001). Evidence for disease-resistance pathway in rice similar to the *NPR1*-mediated signalling pathway in *Arabidopsis*. Plant J 27: 101-113.
Ciuffetti LM, Tuori RP, and Gaventa JM (1997). A single gene encodes a selective toxin causal to the development of tan spot of wheat. Plant Cell 9: 135-144.
Clarke JD, Liu Y, Klessig DF, and Dong X (1998). Uncoupling *PR* gene expression from NPR1 and bacterial resistance: characterization of the dominant *Arabidopsis cpr6-1* mutant. Plant Cell 10: 557-569.
Clarke JD, Volko SM, Ledford H, Ausubel FM, and Dong X (2000). Roles of salicylic acid, jasmonic acid, and ethylene in *cpr*-induced resistance in Arabidopsis. Plant Cell 12: 2175-2190.
Comstock JC, and Scheffer RP (1973). Role of host-selective toxin in colonization of corn leaves by *Helminthosporium carbonum*. Phytopathology 63: 24-29.
Covert SF, Kapoor P, Lee M, Briley A, and Nairn CJ (2001). *Agrobacterium tumefaciens*-mediated transformation of *Fusarium circinatum*. Mycol Res 105: 259-264.
Crute I, Beynon J, Dangl J, Holub E, Mauch-Mani B, Slusarenko A, Staskawicz B, and Ausubel F (1994). Microbial pathogenesis of *Arabidopsis*. In: Arabidopsis. NY: Meyerowitz E, and Somerville C. Cold Spring Harbor Laboratory Press, pp 705-747.
Cutler SR, Ehrhardt DW, Griffitts JS, and Somerville CR (2000). Random GFP:cDNA fusions enable visualization of subcellular structures in cells of *Arabidopsis* at a high frequency. Proc Natl Acad Sci USA 97: 3718-3723.

Czymmek KJ, Bourett TM, Dezwaan TM, Sweigard JA, and Howard RJ (2002). Cytoplasmic expression of fluorescent proteins in the blast pathogen, *Magnaporthe grisea*, and its utility for *in vivo* fungal imaging during plant pathogenesis. Mycologia 94: 280-289.

Czymmek KJ, Whallon JH, and Klomparens KL (1994). Confocal microscopy in mycological research. Exp Mycology 18: 275-293.

Dangl JL, Dietrich RA, and Richberg MH (1996). Death don't have no mercy: cell death programs in plant-microbe interactions. Plant Cell 8: 1793-1807.

Dangl JL, and Jones JDG (2001). Plant pathogens and integrated defense responses to infection. Nature 411: 826-833.

de Groot MJA, Bundock P, Hooykaas PJJ, and Beijersbergen AGM (1998). *Agrobacterium tumefaciens*-mediated transformation of filamentous fungi. Nature Biotech 16: 839-842.

Delaney TP, Friedrich L, and Ryals JA (1995). Arabidopsis signal transduction mutant defective in chemically and biologically induced disease resistance. Proc Natl Acad Sci USA 92: 6602-6606.

Devadas SK, Enyedi A, and Raina R (2002). The *Arabidopsis hrl1* mutation reveals novel overlapping roles for salicylic acid, jasmonic acid and ethylene signaling in cell death and defense against pathogens. Plant J 30: 467-480.

Devoto A, Nieto-Rostro M, Xie D, Ellis C, Harmston R, Patrick E, Davis J, Sherrat L, Coleman M, and Turner JG (2002). COI1 links jasmonate signalling and fertility to the SCF ubiquitin-ligase complex in *Arabidopsis*. Plant J 32: 457-466.

Dewey RE, Levings III CS, and Timothy DH (1986). Novel recombinations in the maize mitochondrial genome produce a unique transcriptional unit in the Texas male-sterile cytoplasm. Cell 44: 439-449.

Dewey RE, Siedow JN, Timothy DH, and Levings III CS (1988). A 13-kilodalton maize mitochondrial protein in *E. coli* confers sensitivity to *Bipolaris maydis* toxin. Science 239: 293-295.

Dewey RE, Timothy DH, and Levings III CS (1987). A mitochondrial protein associated with cytoplasmic male sterility in the T cytoplasm of maize. Proc Natl Acad Sci USA 84: 5374-5378.

Díaz J, ten Have A, and van Kan JAL (2002). The role of ethylene and wound signaling in resistance of tomato to *Botrytis cinerea*. Plant Physiol 129: 1341-1351.

Dickman MB, Park YK, Oltersdorf T, Li W, Clemente T, and French R (2001). Abrogation of disease development in plants expressing animal antiapoptotic genes. Proc Natl Acad Sci USA 98: 6957-6962.

Dietrich RA, Delaney TP, Uknes SJ, Ward ER, Ryals JA, and Dangl JL (1994). Arabidopsis mutants simulating disease resistance response. Cell 77: 565-577.

Dioh W, Tharreau D, Notteghem JL, Orbach MJ, and Lebrun MH (2000). Mapping of avirulence genes in the rice blast fungus, *Magnaporthe grisea,* with RFLP and RAPD markers. Mol Plant-Microbe Interact 13: 217-227.

Dixon MS, Hatzixanthis K, Jones DA, Harrison K, and Jones JDG (1998). The tomato *Cf-5* disease resistance gene and six homologs show pronounced allelic variation in leucine-rich repeat copy number. Plant Cell 10: 1915-1925.

Dixon MS, Jones DA, Keddie JS, Thomas CM, Harrison K, and Jones JDG (1996). The tomato *Cf-2* disease resistance locus comprises two functional genes encoding leucine-rich repeat proteins. Cell 84: 451-459.

Doares SH, Narvaez-Vasquez J, Conconi A, and Ryan CA (1995). Salicylic acid inhibits synthesis of proteinase inhibitors in tomato leaves induced by systemin and jasmonic acid. Plant Physiol 108: 1741-1746.

Dodds PN, and Schwechheimer C (2002). A breakdown in defense signaling. Plant Cell 14: S5-S8.

Ellingboe AH (1992). Segregation of avirulence/virulence on three rice cultivars in 16 crosses of *Magnaporthe grisea*. Phytopathology 82: 597-601.

Ellingboe AH, Wu BC, and Robertson W (1990). Inheritance of avirulence/virulence in a cross of two isolates of *Magnaporthe grisea* pathogenic to rice. Phytopathology 80: 108-111.

Epple P, Apel K, and Bohlmann H (1995). An *Arabidopsis thaliana* thionin gene is inducible via a signal transduction pathway different from that for pathogenesis-related proteins. Plant Physiol 109: 813-820.

Epple P, Apel K, and Bohlmann H (1997). Overexpression of an endogenous thionin enhances resistance of *Arabidopsis* against *Fusarium oxysporum*. Plant Cell 9: 509-520.

Epple P, Vignutelli A, Apel K, and Bohlmann H (1998). Differential induction of the *Arabidopsis thaliana Thi2.1* gene by *Fusarium oxysporum* f. sp. *matthiolae*. Mol Plant-Microbe Interact 11: 523-529.

Falk A, Feys BJ, Frost LN, Jones JDG, Daniels MJ, and Parker JE (1999). *ESD1*, an essential component of *R* gene-mediated disease resistance in *Arabidopsis* has homology to eukaryotic lipases. Proc Natl Acad Sci USA 96: 3292-3297.

Farman ML (2002). *Pyricularia grisea* isolates causing gray leaf spot on perennial ryegrass (*Lolium perenne*) in the United States: Relationship to *P. grisea* isolates from other host plants. Phytopathology 92: 245-254.
Farman ML, Eto Y, Nakao T, Tosa Y, Nakayashiki H, Mayama S, and Leong SA (2002). Analysis of the structure of the *AVR1-CO39* avirulence locus in virulent rice-infecting isolates of *Magnaporthe grisea*. Mol Plant-Microbe Interact 15: 6-16.
Farman ML, and Leong SA (1998). Chromosome walking to the *AVR1-CO39* avirulence gene of *Magnaporthe grisea*: Discrepancy between the physical and genetic maps. Genetics 150: 1049-1058.
Feys BJ, Moisan LJ, Newman M-A, and Parker JE (2001). Direct interaction between the *Arabidopsis* disease resistance signaling proteins, EDS1 and PAD4. EMBO J 20: 5400-5411.
Feys BJ, and Parker JE (2000). Interplay of signaling pathways in plant disease resistance. Trends Genet 16: 449-455.
Fiehn O, Kopka J, Dörmann P, Altmann T, Trethewey RN, and Willmitzer L (2000a). Metabolite profiling for plant functional genomics. Nature Biotech 18: 1157-1161.
Fiehn O, Kopka J, Trethewey RN, and Willmitzer L (2000b). Identification of uncommon plant metabolites based on calculation of elemental compositions using gas chromatography and quadrupole mass spectrometry. Anal Chem 72: 3573-3580.
Fleischmann RD, Adams MD, White O, Clayton RA, Kirkness EF, Kerlavage AR, Bult CJ, Tomb JF, Dougherty BA, Merrick JM, and al. e (1995). Whole-genome random sequencing and assembly of *Haemophilus influenzae*. Science 269: 496-512.
Flor HH (1971). Current status of the gene-for-gene concept. Ann Rev Phytopathol 9: 275-296.
Freitas-Junior LH, Bottius E, Pirrit LA, Deitsch KW, Scheidig C, Guinet F, Nehrbass U, Wellems TE, and Scherf A (2000). Frequent ectopic recombination of virulence factor genes in telomeric chromosome clusters of *P. falciparum*. Nature 407: 1018-1022.
Frye CA, and Innes RW (1998). An Arabidopsis mutant with enhanced resistance to powdery mildew. Plant Cell 10: 947-956.
Fuchs H, and Sacristán MD (1996). Identification of a gene in *Arabidopsis thaliana* controlling resistance to clubroot (*Plasmodiophora brassicae*) and characterization of the resistance response. Mol Plant-Microbe Interact 9: 91-97.
Gaffney T, Friedrich L, Vernooij B, Negrotto D, Nye G, Uknes S, Ward E, Kessmann H, and Ryals J (1993). Requirement of salicylic acid for the induction of systemic acquired resistance. Science 261: 754-756.
Glazebrook J, Rogers EE, and Ausubel FM (1997). Use of Arabidopsis for genetic dissection of plant defense responses. Annu Rev Genet 31: 547-569.
Gold SE, García-Pedrajas MD, and Martinez-Espinoza AD (2001). New (and used) approaches to the study of fungal pathogenicity. Annu Rev Phytopathol 39: 337-365.
Govrin EM, and Levine A (2000). The hypersensitive response facilitates plant infection by the necrotrophic pathogen *Botrytis cinerea*. Curr Biol 10: 751-757.
Grandi G (2001). Antibacterial vaccine design using genomics and proteomics. Trends Biotech 19: 181-188.
Hamer L, Adachi K, Montenegro-Chamorro MV, Tanzer MM, Mahanty SK, Lo C, Tarpey RW, Skalchunes AR, Heiniger RW, Frank SA, Darveaux BA, Lampe DJ, Slater TM, Ramamurthy L, DeZwaan TM, Nelson GH, Shuster JR, Woessner J, and Hamer JE (2001). Gene discovery and gene function assignment in filamentous fungi. Proc Natl Acad Sci USA 98: 5110-5115.
Hardham AR, and Mitchell HJ (1998). Use of molecular cytology to study the structure and biology of phytopathogenic and mycorrhizal fungi. Fungal Genet Biol 24: 252-284.
Heath MC (1991). The role of gene-for-gene interactions in the determination of host species specificity. Phytopathology 81: 127-130.
Heath MC (2000). Advances in imaging the cell biology of plant-microbe interactions. Annu Rev Phytopathol 38: 443-459.
Heath MC, Valent B, Howard RJ, and Chumley FG (1990). Correlations between cytologically detected plant-fungal interactions and pathogenicity of *Magnaporthe grisea* toward weeping lovegrass. Phytopathology 80: 1382-1386.
Hernandez-Rivas R, Mattei D, Sterkers Y, Peterson DS, Wellems TE, and Scherf A (1997). Expressed *var* genes are found in *Plasmodium falciparum* subtelomeric regions. Mol Cell Biol 17: 604-611.
Holub EB, Beynon JL, and Crute IR (1994). Phenotypic and genetic characterization of interactions between isolates of *Peronospora parasitica* and accessions of *Arabidopsis thaliana*. Mol Plant-Microbe Interact 7: 223-239.
Howard RJ (2001). Cytology of fungal pathogens and plant-host interactions. Curr Opin Microbiol 67: 1987-1994.

Huet JC, Le Caer JP, Nespoulous C, and Pernollet JC (1995). The relationships between the toxicity and the primary and secondary structures of elicitin-like protein elicitors secreted by the phytopathogenic fungus *Pythium vexans*. Mol Plant-Microbe Interact 8: 302-310.

Hulbert SH, Webb CA, Smith SM, and Sun Q (2001). Resistance gene complexes: evolution and utilization. Ann Rev Phytopathol 39: 285-312.

Jarosch B, Kogel K-H, and Schaffrath U (1999). The ambivalence of the barley *Mlo* locus: mutations conferring resistance against powdery mildew (*Blumeria graminis* f. sp. *hordei*) enhances susceptibility to the rice blast fungus *Magnaporthe grisea*. Mol Plant-Microbe Interact 12: 508-514.

Jia Y, McAdams SA, Bryan GT, Hershey HP, and Valent B (2000). Direct interaction of resistance gene and avirulence gene products confers rice blast resistance. EMBO J 19: 4004-4014.

Jirage D, Tootle TL, Reuber TL, Frost LN, Feys BJ, Parker JE, Ausubel FM, and Glazebrook J (1999). *Arabidopsis thaliana PAD4* encodes a lipase-like gene that is important for salicylic acid signaling. Proc Natl Acad Sci USA 96: 13583-13588.

Johal GS, Hulbert SH, and Briggs SP (1995). Disease lesion mimics of maize. BioEssays 17: 685-692.

Jones DA, Thomas CM, Hammond-Kosack KE, Balint-Kurti PJ, and Jones JDG (1994). Isolation of the tomato *Cf-9* gene for resistance to *Cladosporium fulvum* by transposon tagging. Science 266: 789-793.

Joosten MH, Vogelsang R, Cozijnsen TJ, Verberne MC, and De Wit PJ (1997). The biotrophic fungus *Cladosporium fulvum* circumvents *Cf-4*-mediated resistance by producing unstable AVR4 elicitors. Plant Cell 9: 367-379.

Joosten MHAJ, Cozijnsen TJ, and de Wit PJGM (1994). Host resistance to a fungal tomato pathogen lost by a single base-pair change in an avirulence gene. Nature 367: 384-386.

Joosten MHAJ, and De Wit PJGM (1999). The tomato-*Cladosporium fulvum* interaction: a versatile experimental system to study plant-pathogen interactions. Annu Rev Phytopathol 37: 335-367.

Kachroo P, Yoshioka K, Shah J, Dooner H, and Klessig DF (2000). Resistance to turnip crinkle virus in Arabidopsis is regulated by two host genes and is salicylic acid dependent but *NPR1*, ethylene, and jasmonate independent. Plant Cell 12: 677-690.

Kamoun S, Honée G, Weide R, Laugé R, Kooman-Gersmann M, de Groot KE, Govers F, and de Wit PJGM (1999a). The fungal gene *Avr9* and the oomycete gene *infl* confer avirulence to potato virus X on tobacco. Mol Plant-Microbe Interact 12: 459-462.

Kamoun S, Hraber P, Sobral B, Nuss D, and Govers F (1999b). Initial assessment of gene diversity for the oomycete pathogen *Phytophthora infestans* based on expressed sequences. Fungal Genet Biol 28: 94-106.

Kamoun S, Lindqvist H, and Govers F (1997). A novel class of elicitin-like genes from *Phytophthora infestans*. Mol Plant-Microbe Interact 10: 1028-1030.

Kamoun S, van West P, de Jong AJ, de Groot KE, Vleeshouwers VGAA, and Govers F (1997b). A gene encoding a protein elicitor of *Phytophthora infestans* is down-regulated during infection of potato. Mol Plant-Microbe Interact 10: 13-20.

Kang S (2001). Organization and distribution pattern of MGLR-3, a novel retrotransposon in the rice blast fungus *Magnaporthe grisea*. Fungal Genet Biol 32: 11-19.

Kang S, Mullins ED, DeZwann TM, and Orbach MJ (2000). Pathogenesis and genome organization of the rice blast fungus. In: Fungal Pathology. Boston: Kronstad JW. Kluwer Academic Press, pp 195-235.

Kang S, Sweigard JA, and Valent B (1995). The *PWL* host specificity gene family in the rice blast fungus *Magnaporthe grisea*. Mol Plant-Microbe Interact 8: 939-948.

Kato N, Pontier D, and Lam E (2002). Spectral profiling for the simultaneous observation of four distinct fluorescent proteins and detection of protein-protein interaction via fluorescence resonance energy transfer in tobacco leaf nuclei. Plant Physiol 129: 931-942.

Keon J, Bailey A, and Hargreaves J (2000). A group of expressed cDNA sequences from the wheat fungal leaf blotch pathogen, *Mycosphaerella graminicola* (*Septoria tritici*). Fungal Genet Biol 29: 118-133.

Kinkema M, Fan W, and Dong X (2000). Nuclear localization of NPR1 is required for activation of *PR* gene expression. Plant Cell 12: 2339-2350.

Knogge W (1996). Fungal infection of plants. Plant Cell 8: 1711-1722.

Kohler RH, Zipfel WR, Webb WW, and Hanson MR (1997). The green fluorescent protein as a marker to visualize plant mitochondria *in vivo*. Plant J 11: 613-621.

Kooman-Gersmann M, Honee G, Bonnema G, and De Wit PJGM (1996). A high affinity binding site for the Avr9 peptide elicitor of *Cladosporium fulvum* is present on plasma membranes of tomato and other solanaceous plants. Plant Cell 8: 929-938.

Kruger WM, Pritsch C, Chao S, and Muehlbauer GJ (2002). Functional and comparative bioinformatic analysis of expressed genes from wheat spikes infected with *Fusarium graminearum*. Mol Plant-Microbe Interact 5: 445-456.

Lagopodi AL, Ram AF, Lamers GE, Punt PJ, Van den Hondel CAMJJ, Lugtenberg BJJ, and Bloemberg GV (2002). Novel aspects of tomato root colonization and infection by *Fusarium oxysporum* f. sp. *radicis-lycopersici* revealed by confocal laser scanning microscopic analysis using the green fluorescent protein as a marker. Mol Plant -Microbe Interact 15: 172-179.

Lam E, Kato N, and Lawton M (2001). Programmed cell death, mitochondria and the plant hypersensitive response. Nature 411: 848-853.

Lau GW, Chao CT, and Ellingboe AH (1993). Interaction of genes controlling avirulence/virulence of *Magnaporthe grisea* on rice cultivar Katy. Phytopathology 83: 375-382.

Lau GW, and Ellingboe AH (1993). Genetic analysis of mutations to increased virulence in *Magnaporthe grisea.* Phytopathology 83: 1093-1096.

Laugé R, and De Wit PJGM (1998). Fungal avirulence genes: structure and possible functions. Fungal Genet Biol 24: 285-297.

Laugé R, Goodwin PH, de Wit PJGM, and Joosten MHAJ (2000). Specific HR-associated recognition of secreted proteins from *Cladosporium fulvum* occurs in both host and non-host plants. Plant J 23: 735-745.

Laugé R, Joosten MHA, Haanstra JPW, Goodwin PH, Lindhout P, and de Wit PJGM (1998). Successful search for a resistance gene in tomato targeted against a virulence factor of a fungal pathogen. Proc Natl Acad Sci USA 95: 9014-9018.

Laugé R, Joosten MHAJ, Van den Ackerveken GFJM, Van den Broek HWJ, and de Wit PJGM (1997). The in planta-produced extracellular proteins ECP1 and ECP2 of *Cladosporium fulvum* are virulence factors. Mol Plant-Microbe Interact 10: 725-734.

Lawton KA, Friedrich L, Hunt M, Weymann K, Delaney T, Kessmann H, Staub T, and Ryals J (1996). Benzothiadiazole induces disease resistance in Arabidopsis by activation of the systemic acquired resistance signal transduction pathway. Plant J 10: 71-82.

Leach JE, Vera Cruz CM, Bai J, and Leung H (2001). Pathogen fitness penalty as a predictor of durability of disease resistance genes. Annu Rev Phytopathol 39: 187-224.

Lemke PA, and Peng M (1995). Genetic manipulation of fungi by DNA-mediated transformation. In: The Mycota II Genetics and Biotechnology. Berlin: Kuck U. Springer-Verlag, pp 109-139.

Leung H, Borromeo ES, Bernardo MA, and Notteghem JL (1988). Genetic analysis of virulence in the rice blast fungus *Magnaporthe grisea*. Phytopathology 78: 1227-1233.

Li X, Clarke JD, Zhang Y, and Dong X (2001). Activation of an EDS1-mediated *R*-gene pathway in the *snc1* mutant leads to constitutive, NPR1-independent pathogen resistance. Mol Plant -Microbe Interact 14: 1131-1139.

Li X, Zhang Y, Clarke JD, Li Y, and Dong X (1999). Identification and cloning of a negative regulator of systemic acquired resistance, SNI1, through a screen for suppressors of *npr1-1*. Cell 98: 329-339.

Linnemannstöns P, Vob T, Hedden P, Gaskin P, and Tudzynski B (1999). Deletions in the gibberellin biosynthesis gene cluster of *Gibberella fujikuroi* by restriction enzyme-mediated integration and conventional transformation-mediated mutagenesis. Appl Env Microbiol 65: 2558-2564.

Lorang JM, Tuori RP, Martinez JP, Sawyer TL, Redman RS, Rollins JA, Wolpert TJ, Johnson KB, Rodriguez RJ, Dickman MB, and Ciuffetti LM (2001). Green fluorescent protein is lighting up fungal biology. Appl Environ Microbiol 67: 1987-1994.

Luderer R, De Kock MJD, Dees RHL, De wit PJGM, and Joosten MHAJ (2002). Functional analysis of cysteine resides of ECP elicitor proteins of the fungal tomato pathogen *Cladosporium fulvum*. Mol Plant Pathol 3: 91-95.

Luderer R, and Joosten MHAJ (2001). Avirulence proteins of plant pathogens: determinants of victory and defeat. Mol Plant Pathol 2: 355-364.

Lund ST, Stall RE, and Klee HJ (1998). Ethylene regulates the susceptible response to pathogen infection in tomato. Plant Cell 10: 371-382.

Lyapina S, Cope G, Shevchenko A, Serino G, Tsuge T, Zhou C, Wolf DA, Wei N, Shevchenko A, and Deshaies RJ (2001). Promotion of NEDD8-CUL1 conjugate cleavage by COP9 signalosome. Science 292: 1382-1385.

MacBeath G, and Schreiber SL (2000). Printing proteins as microarrays for high-throughput function determination. Science 289: 1760-1763.

Maier FJ, and Schäfer W (1999). Mutagenesis via insertional or restriction enzyme-mediated-integration (REMI) as a tool to tag pathogenicity related genes in plant pathogenic fungi. Biol Chem 380: 855-864.

Maleck K, and Dietrich RA (1999). Defense on multiple fronts: how do plants cope with diverse enemies? Trends Plant Sci 4: 215-219.

Maleck K, Levine A, Eulgem T, Morgan A, Schmid J, Lawton KA, Dangl JA, and Dietrich RA (2000). The transcriptome of Arabidopsis thaliana during systemic acquired resistance. Nature Genet 26: 403-409.

Manners JM, Penninckx IA, Vermaere K, Kazan K, Brown RL, Morgan A, Maclean DJ, Curtis MD, Cammue BP, and Broekaert WF (1998). The promoter of the plant defensin gene *PDF1.2* from *Arabidopsis* is systemically activated by fungal pathogens and responds to methyl jasmonate but not to salicylic acid. Plant Mol Biol 38: 1071-1080.

Mao Y, and Tyler BM (1996). Cloning and sequence analysis of elicitin genes of *Phytophthora sojae*. Fungal Genet Biol 20: 169-172.

Marchetti MA, Lai X, and Bollich CN (1987). Inheritance of resistance to *Pyricularia oryzae* in rice cultivars grown in the United States. Phytopathology 77: 799-804.

Markham JE, and Hille J (2001). Host-selective toxins as agents of cell death in plant-fungus interactions. Mol Plant Pathol 2: 229-239.

Martinez JP, Ottum SA, Ali S, Francl LJ, and Ciuffetti LM (2001). Characterization of the *ToxB* gene from *Pyrenophora tritici-repentis*. Mol Plant -Microbe Interact 14: 675-677.

Martin-Hernandez AM, Dufresne M, Hugouvieux V, Melton RE, and Osbourn A (2000). Effects of targeted replacement of the tomatinase gene on the interaction of *Septoria lycopersici* with tomato plants. Mol Plant-Microbe Interact 13: 1301-1311.

Mas P, Devlin PF, Panda S, and Kay SA (2000). Functional interaction of phytochrome B and cryptochrome 2. Nature 408: 207-211.

Mauch-Mani B, and Métraux JP (1997). Systemic acquired resistance. Annu Rev Phytopathol 35: 235-270.

Mauch-Mani B, and Slusarenko AJ (1994). Systemic acquired resistance in *Arabidopsis thaliana* induced by a predisposing infection with a pathogenic isolate of *Fusarium oxysporum*. Mol Plant-Microbe Interact 7: 378-383.

McConn M, Creelman RA, Bell E, Mullet JE, and Browse J (1997). Jasmonate is essential for insect defense in *Arabidopsis*. Proc Natl Acad Sci USA 94: 5473-5477.

McDowell JM, Cuzick A, Can C, Beynon J, Dangl JL, and Holub EB (2000). Downey mildew (*Peronospora parasitica*) resistance genes in Arabidopsis vary in functional requirements for *NDR1*, *EDS1*, *NPR1*, and salicylic acid accumulation. Plant J 22: 523-529.

McDowell JM, Dhandaydham M, Long TA, Aarts MG, Goff S, Holub EB, and Dangl JL (1998). Intragenic recombination and diversifying selection contribute to the evolution of downy mildew resistance at the *RPP8* locus of *Arabidopsis*. Plant Cell 10: 1861-1874.

McMullen M, Jones R, and Gallenberg D (1997). Scab of wheat and barley: A re-emerging disease of devastating impact. Plant Dis 81: 1340-1348.

Mullins E, Romaine CP, Chen X, Geiser D, Raina R, and Kang S (2001). *Agrobacterium tumefaciens*-mediated transformation of *Fusarium oxysporum*: An efficient tool for insertional mutagenesis and gene transfer. Phytopathology 91: 173-180.

Mullins ED, and Kang S (2001). Transformation: a tool for studying fungal pathogens of plants. Cell Mol Life Sci 58: 2043-2052.

Multani DS, Meeley RB, Paterson AH, Gray J, Briggs SP, and Johal GS (1996). Plant-pathogen microevolution: molecular basis for the origin of a fungal disease in maize. Proc Natl Acad Sci USA 95: 1686-1691.

Murakami J, Tosa Y, Kataoka T, Tomita R, Kawasaki J, Chuma I, Sesumi Y, Kusaba M, Nakayashiki H, and Mayama S (2000). Analysis of host species specificity of *Magnaporthe grisea* toward wheat using a genetic cross between isolates from wheat and foxtail millet. Phytopathology 90: 1060-1067.

Murray SL, Thomson C, Chini A, Read ND, and Loake GJ (2002). Characterization of a novel, defense-related *Arabidopsis* mutant, *cir1*, isolated by luciferase imaging. Mol Plant-Microbe Interact 15: 557-566.

Muskett PR, Khan K, Austin MJ, Moisan LJ, Sadanandom A, Shirasu K, Jones JDG, and Parker JE (2002). Arabidopsis *RAR1* exerts rate-limiting control of *R* gene-mediated defenses against multiple pathogens. Plant Cell 14: 979-992.

Nawrath C, Heck S, Parinthawong N, and Métraux J-P (2002). ESD5, an essential component of salicylic acid-dependent signaling for disease resistance in Arabidopsis, is a member of the MATE transporter family. Plant Cell 14: 275-286.

Nawrath C, and Métraux JP (1999). Salicylic acid induction-deficient mutants of *Arabidopsis* express *PR-2* and *PR-5* and accumulate high levels of camalexin after pathogen inoculation. Plant Cell 11: 1393-1404.

Neumann MJ, and Dobinson KF (2002). Sequence tag analysis of gene expression during pathogenic growth and microsclerotia development in the vascular wilt pathogen *Verticillium dahliae*. Fungal Genet Biol in press.

Nishimura M, and Somerville S (2002). Resisting attack. Science 295: 2032-2033.

Norman-Setterbald C, Vidal S, and Palva ET (2000). Interacting signal pathways control defense gene expression in Arabidopsis in response to cell wall-degrading enzymes from *Erwinia carotovora*. Mol Plant-Microbe Interact 13: 430-438.

Oliver RP, Roberts IN, Harding R, Kenyon L, Punt PJ, Dingemanse MA, and van den Hondel CAMJJ (1987). Transformation of *Fulvia fulva*, a fungal pathogen of tomato, to hygromycin B resistance. Curr Genet 12: 231-233.

Orbach MJ, Farrall L, Sweigard JA, Chumley FG, and Valent B (2000). A telomeric avirulence gene determines efficacy for rice blast resistance gene *Pi-ta*. Plant Cell 12: 2019-2032.

Osbourn AE (1996). Preformed antimicrobial compounds and plant defense against fungal attack. Plant Cell 8: 1821-1831.

Panaccione DG, Scott-Craig JS, Pocard J-A, and Walton JD (1992). A cyclic peptide synthetase gene required for pathogenicity of the fungus *Cochliobolus carbonum* on maize. Proc Natl Acad Sci USA 89: 6590-6594.

Pandey A, and Mann M (2000). Proteomics to study genes and genomics. Nature 405: 837-846.

Panter SN, Hammond-Kosack KE, Harrison K, Jones JDG, and Jones DA (2002). Developmental control of promoter activity is not responsible for mature onset of *Cf-9B*-mediated resistance to leaf mold in tomato. Mol Plant -Microbe Interact 15: 1099-1107.

Papadopoulou K, Melton RE, Leggett M, Daniels MJ, and Osbourn AE (1999). Compromised disease resistance in saponin-deficient plants. Proc Natl Acad USA 96: 12923-12928.

Parker JE, Holub EB, Frost LN, Falk A, Gunn ND, and Daniels MJ (1996). Characterization of *eds1*, a mutation in Arabidopsis suppressing resistance to *Peronospora parasitica* specified by several different *RPP* genes. Plant Cell 8: 2033-2046.

Parniske M, Hammond-Kosack KE, Golstein C, Thomas CM, Jones DA, Harrison K, Wulff BBH, and Jones JDG (1997). Novel disease resistance specificities result from sequence exchange between tandemly repeated genes at the *Cf-4/9* locus of tomato. Cell 91: 821-832.

Parsons KA, Chumley FG, and Valent B (1987). Genetic transformation of the fungal pathogen responsible for rice blast disease. Proc Natl Acad Sci USA 84: 4161-4165.

Peck SC, Nuhse TS, Hess D, Iglesias A, Meins F, and Boller T (2001). Directed proteomics identifies a plant-specific protein rapidly phosphorylated in response to bacterial and fungal elicitors. Plant Cell 13: 1467-1475.

Penninckx IAMA, Eggermont K, Terras FR, Thomma BP, De Samblanx GW, Buchala A, Metraux JP, Manners JM, and Broekaert WF (1996). Pathogen-induced systemic activation of a plant defensin gene in Arabidopsis follows a salicylic acid-independent pathway. Plant Cell 8: 2309-2323.

Penninckx IAMA, Thomma BP, Buchala A, Metraux JP, and Broekaert WF (1998). Concomitant activation of jasmonate and ethylene response pathways is required for induction of a plant defensin gene in Arabidopsis. Plant Cell 10: 2103-2114.

Pennisi E (2001). The push to pit genomics against fungal genomics. Science 292: 2273-2274.

Pieterse CMJ, and van Loon LC (1999). Salicylic acid-independent plant defense pathways. Trends Plant Sci 4: 52-58.

Qutob D, Hraber PT, Sobral BWS, and Gijzen M (2000). Comparative analysis of expressed sequences in *Phytophthora sojae*. Plant Physiol 123: 243-253.

Qutob D, Kamoun S, and Gijzen M (2002). Expression of a *Phytophthora sojae* necrosis inducing protein occurs during transition from biotrophy to necrotrophy. Plant J 32: 361-373.

Raamsdonk LM, Teusink B, Broadhurst D, Zhang N, Hayes A, Walsh MC, Berden JA, Brindle KM, Kell DB, Rowland JJ, Westerhoff HV, van Dam K, and Oliver SG (2001). A functional genomics strategy that uses metabolome data to reveal the phenotype of silent mutations. Nature Biotech 19: 45-50.

Rho H, Kang S, and Lee Y (2001). *Agrobacterium tumefaciens* mediated transformation of the plant pathogenic fungus *Magnaporthe grisea*. Mol Cells 12: 407-411.

Rhoads DM, Brunner-Neuenschwander B, Levings CS, and Siedow JN (1998). Characterization of the interaction between fungal phytotoxins and Urf13, the cms-T maize mitochondrial T-toxin receptor. In: Molecular Genetics of Host-Specific Toxins in Plant Disease. Dordrecht, Netherlands: Kohmoto K, and Yoder OC. Kluwer, pp 355-365.

Rivas S, Mucyn T, van den Burg HA, Vervoort J, and Jones JDG (2002). An ~400 kDa membrane-associated complex that contains one molecule of the resistance protein Cf-4. Plant J 29: 783-796.

Rivas S, and Thomas CM (2002). Recent advances in the study of tomato *Cf* resistance genes. Mol Plant Pathol 3: 277-282.

Rodriguez RJ, and Yoder OC (1987). Selectable genes for transformation of the fungal plant pathogen *Glomerella cingulata* f.sp. *phaseoli* (*Colletotrichum lindemuthianum*). Gene 54: 73-81.

Roessner U, Luedemann A, Brust D, Fiehn O, Linke T, and Willmitzer L (2001). Metabolic profiling allows comprehensive phenotyping of genetically or environmentally modified plant systems. Plant Cell 13: 11-29.

Roessner U, Wagner C, Kopka J, Trethewey RN, and Willmitzer L (2000). Simultaneous analysis of metabolites in potato tuber by gas chromatography-mass spectrometry. Plant J 23: 131-142.

Roetschi A, Si-Ammour A, Belbahri L, Mauch F, and Mauch-Mani B (2001). Characterization of an *Arabidopsis-Phytophthora* pathosystem: resistance requires a functional *PAD2* gene and is independent of salicylic acid, ethylene and jasmonic acid signalling. Plant J 28: 293-305.

Rogers EE, and Ausubel FM (1997). Arabidopsis enhanced disease susceptibility mutants exhibit enhanced susceptibility to several bacterial pathogens and alterations in *PR-1* gene expression. Plant Cell 9: 305-316.

Rohe M, Gierlich A, Hermann H, Hahn M, Schmidt B, Rosahl S, and Knogge W (1995). The race-specific elicitor, NIP1, from the barley pathogen, *Rhynchosporium secalis*, determines avirulence on host plants of the *Rrs1* resistance genotype. EMBO J 14: 4168-4177.

Rohel EA, Payne AC, Fraaije BA, and Hollomon DW (2001). Exploring infection of wheat and carbohydrate metabolism in *Mycosphaerella graminicola* transformants with differentially regulated green fluorescent protein expression. Mol Plant -Microbe Interact 14: 156-163.

Rosamond J, and Allsop A (2000). Harnessing the power of the genomics in the search for new antibiotics. Science 287: 1973-1976.

Rose MS, Yun S-H, Asvarak T, Lu S-W, Yoder OC, and Turgeon BG (2002). A decarboxylase encoded at the *Cochliobolus heterostrophus* translocation-associated *Tox1B* locus is required for polyketide (T-toxin) biosynthesis and high virulence on T-cytoplasm maize. Mol Plant-Microbe Interact 15: 883-893.

Schenk PM, Kazan K, Wilson I, Anderson JP, Richmond T, Somerville SC, and Manners JM (2000). Coordinated plant defense responses in *Arabidopsis* revealed by microarray analysis. Proc Natl Acad Sci USA 97: 11655-11660.

Schiestl RH, and Petes TD (1991). Integration of DNA fragments by illegitimate recombination in *Saccharomyces cerevisiae*. Proc Natl Acad Sci USA 88: 7585-7589.

Schulze-Lefert P, and Vogel J (2000). Closing the ranks to attack by powdery mildew. Trends Plant Sci 5: 343-348.

Scvhwechheimer C, Serino G, Callis J, Crosby WL, Lyapina S, Deshaies RJ, Gray WM, Estelle M, and Deng X-W (2001). Interactions of the COP9 signalosome with the E3 ubiquitin ligase SCF^{TIR3} in mediating auxin response. Science 292: 1379-1382.

Seo HS, Song JT, Cheong J-J, Lee Y-H, Lee Y-W, Hwang I, Lee JS, and Choi YD (2001). Jasmonic acid carboxyl methyltransferase: a key enzyme for jasmonate-regulated plant responses. Proc Natl Acad Sci USA 98: 4788-4793.

Sexton AC, and Howlett BJ (2001). Green fluorescent protein as a reporter in the *Brassica-Leptosphaeria maculans* interaction. Physiol Mol Plant Pathol 58: 13-21.

Shah J, Kachroo P, and Klessig DF (1999). The Arabidopsis *ssi1* mutation restores pathogenesis-related gene expression in *npr1* plants and renders defensin gene expression salicylic acid dependent. Plant Cell 11: 191-206.

Shah J, Kachroo P, Nandi A, and Klessig DF (2001). A recessive mutation in the *Arabidopsis SSI2* gene confers SA- and *NPR1*-independent expression of *PR* genes and resistance against bacterial and oomycete pathogens. Plant J 25: 563-574.

Shah J, Tsui F, and Klessig DF (1997). Characterization of a salicylic acid-insensitive mutant (sai1) of *Arabidopsis thaliana*, identified in a selective screen utilizing the SA-inducible expression of the *tms2* gene. Mol Plant-Microbe Interact 10: 69-78.

Shapiro AD, and Zhang C (2001). The role of *NDR1* in avirulence gene-directed signaling and control of programmed cell death in Arabidopsis. Plant Physiol 127: 1089-1101.

Shirasu K, Lahaye T, Tan M-W, Zhou F, Azevedo C, and Schulze-Lefert P (1999). A novel class of eukaryotic zinc binding proteins is required for disease resistance signaling in barley and development in *C. elegans*. Cell 99: 355-366.

Shirasu K, and Schulze-Lefert P (2000). Regulators of cell death in disease resistance. Plant Mol Biol 44: 371-385.

Silué D, Notteghem JL, and Tharreau D (1992a). Evidence for a gene-for-gene relationship in the *Oryza sativa-Magnaporthe grisea* pathosystem. Phytopathology 82: 577-580.

Silué D, Tharreau D, and Notteghem JL (1992b). Identification of *Magnaporthe grisea* avirulence genes to seven rice cultivars. Phytopathology 82: 1462-1467.
Silva H, Yoshioka K, Dooner HK, and Klessig DF (1999). Characterization of a new Arabidopsis mutant exhibiting enhanced disease resistance. Mol Plant-Microbe Interact 12: 1053-1063.
Simmons C, Hantke S, Grant S, Johal GS, and Briggs SP (1999). The maize lethal leaf spot 1 mutant has elevated resistance to fungal infection at the leaf epidermis. Mol Plant-Microbe Interact 11: 1110-1118.
Snoeijers SS, Vossen P, Goosen T, Van den Broek HWJ, and de Wit PJGM (1999). Transcription of the avirulence gene *Avr9* of the fungal tomato pathogen *Cladosporium fulvum* is regulated by a GATA-type transcription factor in *Aspergillus nidulans*. Mol Gen Genet 261: 653-659.
Soanes DM, Skinner W, Keon J, Hargreaves J, and Talbot NJ (2002). Genomics of phytopathogenic fungi and the development of bioinformatic resources. Mol Plant-Microbe Interact 15: 421-427.
Spellig T, Bottin A, and Kahmann R (1996). Green fluorescent protein (GFP) as a new vital marker in the phytopathogenic fungus *Ustilago maydis*. Mol Gen Genet 252: 503-509.
Staskawicz BJ, Mudgett MB, Dangl JL, and Galan JE (2001). Common and contrasting themes of plant and animal diseases. Science 292: 1945-1952.
Staswick PE, and Lehman CC (1999). Jasmonic acid-signaled responses in plants. In: Induced Plant Defenses against Pathogens and Herbivores. St. Paul: Agrawal AA, Tuzun S, and Bent E. APS Press, pp 117-136.
Staswick PE, Yuen GY, and Lehman CC (1998). Jasmonate signaling mutants of *Arabidopsis* are susceptible to the soil fungus *Pythium irregulare*. Plant J 15: 747-754.
Stephenson S, Hatfield J, Rusu A, Maclean DJ, and Manners JM (2000). CgDN3: An essential pathogenicity gene of *Colletotrichum gloeosporioides* necessary to avert a hypersensitive-like response in the host *Stylosanthes guianensis*. Mol Plant-Microbe Interact 13: 929-941.
Stintzi A, Weber H, Reymond P, Browse J, and Farmer EE (2001). Plant defense in the absence of jasmonic acid: the role of cyclopentenones. Proc Natl Acad Sci USA 98: 12837-12842.
Sweigard JA, Carroll AM, Farrall L, Chumley FG, and Valent B (1998). *Magnaporthe grisea* pathogenicity genes obtained through insertional mutagenesis. Mol Plant-Microbe Interact 11: 404-412.
Sweigard JA, Carroll AM, Kang S, Farrall L, Chumley FG, and Valent B (1995). Identification, cloning, characterization of *PWL2*, a gene for host-species specificity in the rice blast fungus. Plant Cell 7: 1221-1233.
Sweigard JA, and Ebbole DJ (2001). Functional analysis of pathogenicity genes in a genomics world. Curr Opin Microbiol 4: 387-392.
Takahashi A, Kawasaki T, Henmi K, Shii K, Kodama O, Satoh H, and Shimamoto K (1999). Lesion mimic mutants of rice with alterations in early signaling events of defense. Plant J 17: 535-545.
Takken FLW, and Joosten MHAJ (2000). Plant resistance genes: Their structure, function, and evolution. Eur J Plant Pathol 106: 699-713.
Terras FRG, Torrekens S, Van Leuven F, Osborn RW, Vanderleyden J, Cammue BPA, and Broekaert WF (1993). A new family of basic cysteine-rich plant antifungal proteins from Brassicaceae species. FEBS Lett 316: 233-240.
TAGI (The Arabidopsis Genome Initiative) (2000). Analysis of the genome sequence of the flowering plant *Arabidopsis thaliana*. Nature 408: 796-815.
Thomas CM, Jones DA, Parniske M, Harrison K, Balint-Kurti PJ, Hatzixanthis K, and G. JJD (1997). Characterization of the tomato *Cf-4* gene for resistance to *Cladosporium fulvum* identifies sequences that determine recognitional specificity in Cf-4 and Cf-9. Plant Cell 9: 2209-2224.
Thomas SW, Glaring MA, Rasmussen SW, Kinane JT, and Oliver RP (2002). Transcription profiling in the barley mildew pathogen Blumeria graminis by serial analysis of gene expression (SAGE). Mol Plant-Microbe Interact 15: 847-856.
Thomas SW, Rasmussen SW, Glaring MA, Rouster JA, Christiansen SK, and Oliver RP (2001). Gene identification in the obligate fungal pathogen *Blumeria graminis* by expressed sequence tag analysis. Fungal Genet Biol 33: 195-211.
Thomma BPHJ, Eggermont K, Penninckx IAMA, Mauch-Mani B, Vogelsang R, Cammue BPA, and Broekaert WF (1998). Separate jasmonate-dependent and salicylate-dependent defense-response pathways in Arabidopsis are essential for resistance to distinct microbial pathogens. Proc Natl Acad Sci USA 95: 15107-15111.
Thomma BPHJ, Eggermont K, Tierens KFM-J, and Broekaert WF (1999). Requirement of functional *ethylene-insensitive 2* gene for efficient resistance of Arabidopsis to infection by *Botrytis cinerea*. Plant Physiol 121: 1093-1101.
Thon MR, Nuckles EM, and Vaillancourt LJ (2000). Restriction enzyme-mediated integration used to produce pathogenicity mutants of *Colletotrichum graminicola*. Mol Plant-Microbe Interact 13: 1356-1365.

Tierens KFM-J, Thomma BPHJ, Bari RP, Garmier M, Eggermont K, Brouwer M, Penninckx IAMA, Broekaert WF, and Cammue BPA (2002). *Esa1*, an *Arabidopsis* mutant with enhanced susceptibility to a range of necrotrophic fungal pathogens, shows a distorted induction of defense responses by reactive oxygen generating compounds. Plant J 29: 131-140.

Tör M, Gordon P, Cuzick A, Eulgem T, Sinapidou E, Mert F, Can C, Dangl JL, and Holub EB (2002). Arabidopsis SGT1b is required for defense signaling conferred by several downy mildew resistance genes. Plant Cell 14: 993-1003.

Tör M, Holub EB, Brose E, Musker R, Gunn N, Can C, Crute IR, and Beynon JL (1994). Map positions of three loci in *Arabidopsis thaliana* associated with isolate-specific recognition of *Peronospora parasitica* (downy mildew). Mol Plant-Microbe Interact 7: 214-222.

Tornero P, Merritt P, Sadanandom A, Shirasu K, Innes RW, and Dangl JL (2002). *RAR1* and *NDR1* contribute quantitatively to disease resistance in Arabidopsis, and their relative contributions are dependent on the *R* genes assayed. Plant Cell 14: 1005-1015.

Tuori RP, Wolpert TJ, and Ciuffetti LM (2000). Heterologous expression of functional Ptr ToxA. Mol Plant-Microbe Interact 13: 456-464.

Turner JG, Ellis C, and Devoto A (2002). The jasmonate signal pathway. Plant Cell 14: S131-S151.

Uknes S, Mauch-Mani B, Moyer M, Potter S, Williams S, Dincher S, Chandler D, Slusarenko A, Ward E, and Ryals J (1992). Acquired resistance in Arabidopsis. Plant Cell 4: 645-656.

Valent B, and Chumley FG (1987). Genetic analysis of host species specificity in *Magnaporthe grisea*. In: Molecular Strategies for Crop Protection. New York: Arntzen CJ, and Ryan C. Alan R. Liss, pp 83-93.

Valent B, and Chumley FG (1991). Molecular genetic analysis of the rice blast fungus, *Magnaporthe grisea*. Annu Rev Phytopathol 29: 443-467.

Valent B, Farrall L, and Chumley FG (1991). *Magnaporthe grisea* genes for pathogenicity and virulence identified through a series of backcrosses. Genetics 127: 87-101.

Van den Ackerveken GFJM, Dunn RM, Cozijnsen TJ, Vossen P, Van den Broek HWJ, and de Wit PJGM (1994). Nitrogen limitation induces expression of the avirulence gene *avr9* in the tomato pathogen *Cladosporium fulvum*. Mol Gen Genet 243: 277-285.

Van den Ackerveken GFJM, Van Kan JAL, Joosten MHAJ, Muisers JM, Verbakel HM, and de Wit PJGM (1993a). Characterization of two putative pathogenicity genes of the fungal tomato pathogen *Cladosporium fulvum*. Mol Plant-Microbe Interact 6: 210-215.

Van den Ackerveken GFJM, Vossen P, and de Wit PJGM (1993b). The AVR9 race-specific elicitor of *Cladosporium fulvum* is processed by endogenous and plant proteases. Plant Physiol 103: 91-96.

van den Hooven HW, van den Burg HA, Vossen P, Boeren S, de Wit PJGM, and Vervoort J (2001). Disulfide bond structure of the AVR9 elicitor of the tomato pathogen Cladosporium fulvum: Evidence for a cystine knot. Biochem 40: 3458-3466.

Van Kan JAL, Van den Ackerveken GFJM, and de Wit PJGM (1991). Cloning and characterization of cDNA of avirulence gene avr9 of the fungal pathogen *Cladosporium fulvum*, causal agent of tomato leaf mold. Mol Plant-Microbe Interact 4: 52-59.

van West P, Reid B, Campbell TA, Sandrock RW, Fry WE, Kamoun S, and Gow NAR (1999). Green fluorescent protein (GFP) as a reporter gene for the plant pathogenic oomycete *Phytophthora palmivora*. FEMS Microbiol Lett 178: 71-80.

Verberne MC, Verpoorte R, Bol JF, Mercado-Blanco J, and Linthorst HJM (2000). Overproduction of salicylic acid in plants by bacterial transgenes enhances pathogen resistance. Nat Biotech 18: 779-783.

Vignutelli A, Wasternack C, Apel K, and Bohlmann H (1998). Systemic and local induction of an *Arabidopsis* thionin gene by wounding and pathogens. Plant J 14: 285-295.

Vijayan P, Shockey J, Levesque CA, Cook RJ, and Browse J (1998). A role for jasmonate in pathogen defense of *Arabidopsis*. Proc Natl Acad Sci USA 95: 7209-7214.

Viji GA, Wu B, Kang S, Uddin W, and Huff DR (2001). *Pyricularia grisea* causing gray leaf spot (blast) of perennial ryegrass turf: Population structure and host specificity. Plant Dis 85: 817-826.

Vogel J, Raab TK, Schiff C, and Somerville SC (2002). *PMR6*, a pectate lyase-like gene required for powdery mildew susceptibility in Arabidopsis. Plant Cell 14: 2095-2106.

Vogel J, and Somerville S (2000). Isolation and characterization of powdery mildew-resistant *Arabidopsis* mutants. Proc Natl Acad Sci USA 97: 1897-1902.

Wang G-L, Mackill DJ, Bonman JM, McCouch SR, Champox MC, and Nelson RJ (1994). RFLP mapping of genes conferring complete and partial resistance to blast in a durably resistant rice cultivar. Genetics 136: 1421-1434.

Waugh M, Hraber P, Weller J, Wu Y, Chen G, Inman J, Kiphart D, and Sobral B (2000). The Phytophthora genome initiative database: informatics and analysis for distributed pathogenomic research. Nucl Acids Res 28: 87-90.

Wevelsiep L, Kogel KH, and Knogge W (1991). Purification and characterization of peptides from *Rhynchosporium secalis* inducing necrosis in barley. Physiol Mol Plant Pathol 39: 471-482.

Wevelsiep L, Rüpping E, and Knogge W (1993). Stimulation of barley plasmalemma H^+-ATPase by phytotoxic peptides from the fungal pathogen *Rhynchosporium secalis*. Plant Physiol 101: 297-301.

Weymann K, Hunt M, Uknes S, Neuenschwander U, Lawton K, Steiner H-Y, and Ryals J (1995). Suppression and restoration of lesion formation in Arabidopsis *lsd* mutants. Plant Cell 7: 2013-2022.

Wildermuth MC, Dewdney J, Wu G, and Ausubel FM (2001). Isochorismate synthase is required to synthesize salicylic acid for plant defense. Nature 414: 562-565.

Wisman E, and Ohlrogge J (2000). Arabidopsis microarray service facilities. Plant Physiology 124: 1468-1471.

Wolpert TJ, Dunkle LD, and Ciuffetti LM (2002). Host-selective toxins and avirulence determinants: what's in a name? Annu Rev Phytopathol 40: 251-285.

Xiao JZ, Tsuda M, Doke N, and Nishimura S (1991). Phytotoxins produced by germinating spores of *Bipolaris oryzae*. Phytopathology 81: 58-64.

Xiao X, Ellwood S, Calis O, Patrick E, Li T, Coleman M, and Turner JG (2001). Broad-spectrum mildew resistance in *Arabidopsis thaliana* mediated by *RPW8*. Science 291: 118-120.

Xiao X, Ellwood S, Findlay K, Oliver RP, and Turner JG (1997). Characterization of three loci controlling resistance of *Arabidopsis thaliana* accession Ms-0 to two powdery mildew diseases. Plant J 12: 757-768.

Xu J, Urban M, Sweigard JA, and Hamer JE (1997). The *CPKA* gene of *Magnaporthe grisea* is essential for appressorial penetration. Mol Plant-Microbe Interact 10: 187-194.

Yang G, Rose MS, Turgeon BG, and Yoder OC (1996). A polyketide synthase is required for fungal virulence and production of the polyketide T-toxin. Plant Cell 8: 2139-2150.

Yin Z, Chen J, Zeng L, Goh M, Leung H, Kuhush GV, and Wang G-L (2000). Characterizing rice lesion mimic mutants and identifying a mutant with broad-spectrum resistance to rice blast and bacterial blight. Mol Plant-Microbe Interact 13: 869-876.

Yu I, Parker J, and Bent AF (1998). Gene-for-gene disease resistance without the hypersensitive response in *Arabidopsis dnd1* mutant. Proc Natl Acad Sci USA 95: 7819-7824.

Yu LM (1995). Elicitins from *Phytophthora* and basic resistance in tobacco. Proc Natl Acad Sci USA 92: 4088-4094.

Zhang Y, Fan W, Kinkema M, Li X, and Dong X (1999). Interaction of NPR1 with basic leucine zipper protein transcription factors that bind sequences required for salicylic acid induction of the *PR-1* gene. Proc Natl Acad Sci USA 96: 6523-6528.

Zhou J-M, Trifa Y, Silva H, Pontier D, Lam E, Shah J, and Klessig DF (2000). NPR1 differentially interacts with members of the TGA/OBF family of transcription factors that bind an element of the PR-1 gene required for induction by salicylic acid. Mol Plant-Microbe Interact 13: 191-202.

Zhou N, Tootle TL, Tsui F, Klessig DF, and Glazebrook J (1998). *PAD4* functions upstream from salicylic acid to control defense responses in Arabidopsis. Plant Cell 10: 1021-1030.

Zhu T, and Wang X (2000). Large-scale profiling of the Arabidopsis transcriptome. Plant Physiol 124: 1472-1476.

Zimmerli L, Métraux J-P, and Mauch-Mani B (2001). β-aminobutryric acid-induced protection of Arabidopsis against the necrotrophic fungus *Botrytis cinerea*. Plant Physiol 126: 517-523.

Zwiers L-H, and De Waard MA (2001). Efficient *Agrobacterium tumefaciens*-mediated gene disruption in the phytopathogen *Mycosphaerella graminicola*. Curr Genet 39: 388-393.

Applied Mycology & Biotechnology
An International Series. Volume 4. Fungal Genomics

4

Genomics of *Candida albicans*

Siegfried Salomon, Angelika Felk, and Wilhelm Schäfer
Institute of General Botany, Department of Molecular Phytopathology and Genetics (AMPIII), University of Hamburg, Ohnhorststrasse 18, D-22609 Hamburg, Germany (schäfer@botanik.uni-hamburg.de).

The polymorphic fungus *Candida albicans* is an opportunistic human pathogen of increasing medical importance, especially in immunocompromised individuals. *C. albicans* causes oral and vaginal candidiasis as well as disseminated infections in neutropenic patients. The genetic basis of virulence is assumed to be very complicated and seems to involve a lot of different genes. This fungus is a challenging organism for molecular genetic studies, because it is diploid, the sexual cycle is unknown, and it has a divergent codon usage. This review summarises current knowledge about the genomic and genetic variability of this fungal pathogen and outlines molecular techniques for functional analysis of gene expressions and proteins which are supposed virulence factors.

1. INTRODUCTION

1.1. Occurrence of *Candida albicans*

The number of fungal infections in man has distinctly increased during the last ten years (Fridkin and Jarvis 1996; Garbino *et al.* 2002). The majority of nosocomial mycoses are caused by *Candida*-species (Perea and Patterson 2002). The *Candida* genus is part of the class of blastomycetes and the subdivision of deuteromycotina. These are imperfect fungi, which are characterised by the lack of any kind of sexual cycle. These fungi include more than 200 species, of which, however, merely less than 10 % can cause diseases in man (Müller and Löffler 1982). *Candida albicans* is a thin-walled, gram-positive, capsule-less yeast of oval to round shape with a diploid chromosome set. It multiplies by means of holoblastic, lateral budding. This fungus has septate hyphae, but can also appear in yeast form. This ability of the fungus is referred to as dimorphism (Müller and Löffler 1982). With regards to *Candida*-infections, *Candida albicans* with 90 % of all *Candida*-infections occupies the leading role among other *Candida*-species such as *C. dubliniensis*, *C. tropicalis*, *C. parapsilosis*, *C. glabrata* and *C. krusei*.

Candida albicans is an opportunistic fungus, meaning it can only cause an infection in a host with predisposing factors. According to Brandis and Otte (1984), 50% of the world population are colonised by *Candida*. The fungus is part of the natural microflora of the oral cavity, the gastro-intestinal tract and the vaginal mucosa. The primary infection in humans probably occurs during the passage through the birth channel (Fridkin and Jarvis 1996). The humoral and cellular immuno-defense mechanism prevents the invasive spread of the organism. Whenever the immune system is weakened, by for instance, disease, intake of antibiotics or immunosuppressants, the fungus is now enabled to change from a commensal to a pathogenic state. This means that infections caused by *Candida* are nearly always endogenic. The superficial mycosis attacks skin, nails, and mucosa. The subcutaneous mycosis is characterised by locally infected tissue. The systemic mycosis attacks internal organs, with the causative agent spreading *via* the blood stream (Odds 1994a). Systemic mycoses mainly occur, when patients suffer from a lack of leucocytes and, thus, show a restricted immuno-defense mechanism.

1.2. Virulence Factors

The virulence of *C. albicans* is multifactorial, meaning the fungus expresses a pattern of different virulence factors at different stages of infection. Additionally, the expression varies from strain to strain (Odds 1988). Among the most important virulence factors is the change of phenotypes (yeast-mycelium-transition). This ability of the fast and reversible transition of yeast to mycelium cells enables the organism to adapt to the different micromilieus of the host. The expression pattern changes with this alternation. The fungus, for instance, increasingly expresses hydrophobic molecules for improved adhesion to cell surface and receptors for host cell proteins.

Due to the changes in the cell wall during hyphal growth and the secretion of lytic enzymes, such as proteases, lipases, and phospholipases, hyphae are better equipped to penetrate host cells. Furthermore, hyphae show an increased capacity for agglutination (Calderone and Braun 1991), which causes them to form plaques and local lesions in the host tissue. Another virulence factor is thigmotropism and describes the fungus'ability to orient itself towards surface structures. This could assist the fungus in penetrating the host tissue despite unevenness in the surface or microscopic injuries (Nikawa 1998). Furthermore, one should mention the surface hydrophobicity. *C. albicans* is characterised by the ability to change the hydrophobicity of the cell surface, for instance by means of changes in concentrations and lengths of fibrillas. This leads to an improved adhesion of the organism to host cells and improved agglutination of the cells when building lesions (Glee 1995). As an example for molecular mimicry, one can mention the fibrinogen-binding protein at the cell surface of *C. albicans*. These proteins enable the fungus to bind to thrombocytes and, thus, make itself invisible to the host's defence mechanism. One must also mention the high growth rate of the organism (under optimal growth conditions the generation time of *C. albicans* is one hour, Lay *et al.* 1998). In addition, *C. albicans* distinguishes itself through a high rate of genomic reorganisation. This mainly includes frequent chromosome-translocations (Suzuki *et al.* 1986). The plasticity of the genome supports the fungus in overcoming the immuno-defense mechanism of the host.

2. *CANDIDA ALBICANS* GENETICS

2.1. Sexual Reproduction

The genus *Candida* comprises more than 150 species, whose main common feature is the absence of any sexual form (Odds 1988). The sexual stages of only a few *Candida* species (for examples: *C. krusei, C. famata, and C. kefyr*) have been described (Rinaldi 1993). As known so far *C. albicans* lacks a sexual cycle in nature, is constitutive diploid, and, therefore, the fungus is difficult to analyse genetically.

Sequence similarities to *Saccharomyces cerevisiae* revealed that *C. albicans* possess a mating-type (*MAT*) locus called mating type-like (*MTL*) locus with sexual cycle regulators Mata1p, Matα1p, and Matα2p on chromosome 5 (Hull and Johnson 1999; Odds *et al.* 2000). Like diploid *S. cerevisiae*, which are heterozygous with regard to the *MAT* locus, *MTLa* and *MTLα* genes of *C. albicans* are located separately on homologous chromosomes and are interrupted by introns. The *MTL* locus of *C. albicans* contains further genes encoding proteins similar to poly(A) polymerases, oxysterol binding proteins, and phosphatidylinositol kinases, which are not observed in the *MAT* loci of other fungi. Although the *MTL* locus (8742 bp and 8861 bp) is larger than the *MAT* locus of *S. cerevisiae* (642 bp and 747 bp), the predicted *C.*

albicans proteins show more than 50 % similarity to equivalents of *S. cerevisiae* and the flanking DNA sequences are more than 99 % identical (Hull and Johnson 1999).

Mating of *C. albicans* could be only induced (in mutants) when the parent strains contain either *MTLa* or *MTLα* alleles. In two independent studies the artificial haploid *MTLa* and *MTLα* strains were obtained in different ways. Hull *et al.* (2000) deleted specifically one of *MTL* sequences and demonstrated mating *in vivo*. Magee and Magee (2000) used mutants with the loss of one of the two chromosomes 5 by growth on sorbose. The induced monosomy of chromosome 5 leads to the loss of the *MTLa* or *MTLα* gene. It was shown that *C. albicans* can also mate *in vitro* (Magee and Magee 2000). Both studies demonstrated that mating is a rare event and leads to a tetraploid progeny (Hull *et al.* 2000; Magee and Magee 2000), which underwent spontaneous random chromosome loss (Hull *et al.* 2000). It is not known how the recombinant cells return to the diploid state in general. The diploidy of *C. albicans* could then be restored through meiosis or chromosome loss but meiosis has not yet been demonstrated. Gow *et al.* (2000) also suggested an alternative possibility of natural generation of transient haploid mating partners under favourable conditions. However, the importance of mating in *C. albicans* and the ability of this yeast to undergo meiosis remain unclear.

In a further comparative *in silico* study from Tzung *et al.* (2001) using published sequences of genome projects of *C. albicans*, *S. cerevisiae*, *Caenorhabditis elegans*, and *Drosophila melanogaster*. They analysed over 500 genes, which are important for sexual differentiation in *S. cerevisiae* and found many homologous genes which are implicated in the initiation of meiosis, chromosome recombination, and the formation of synaptonemal complexes. Simultaneously important homologous genes for meiosis were not found, but some *C. albicans* sequences have homologies to meiosis related genes in *C. elegans* and *D. melanogaster* Tzung *et al.* (2001). This observation possibly indicates the existence of alternative mechanisms of genetic exchange by *C. albicans*. Another hint of sexual reproduction in *C. albicans* was found during population genetic analyses of 52 natural isolates (Gräser *et al.* 1996). Although the majority of isolates provides evidence for clonality, direct examination of genotypic frequencies between several loci also indicates that interlocus recombination occurs. Still it remains to be shown that *C. albicans* has a meaningful sexual cycle.

2.2. Ploidy of *Candida albicans*

In the early 1980s it was shown by means of the DNA content (Olaiya and Sogin 1979; Whelan *et al.* 1980; Saracheck *et al.* 1981) and reassociation experiments (Riggsby 1982) that *C. albicans* is a diploid organism. The next evidence for diploidy was supplied by the heterozygosity of several genes of natural isolates which are converted to homozygosity by mitotic crossover (Whelan *et al.* 1980; Whelan and Magee 1981). Using chromosomes separation techniques like clamped homogeneous electric field (CHEF) (Magee *et al.* 1988), field–inversion gel electrophoresis (FIGE), and orthogonal field alternation gel electrophoresis (OFAGE) (Lasker *et al.* 1989), eight chromosomal bands were isolated. The diploid genome of *C. albicans* consists of 16 chromosomes. The chromosomes were numbered from the largest, approximately 3.4 Mbp (chromosome 1), to the smallest approximately 1.1 Mbp (chromosome 7; Fig. 1) (Chu *et al.* 1993; Magee and Chibana 2002). The largest of the chromosomes with variable size of two homologues in different strains and during serial culture was called chromosome R, because it contains ribosomal DNA cistrons. The number of tandem repeats of ribosomal DNA on the homologues of chromosome R can vary due to exchange to sister-chromatid (Iwaguchi *et al.* 1992a; Rustchenko *et al.* 1993).

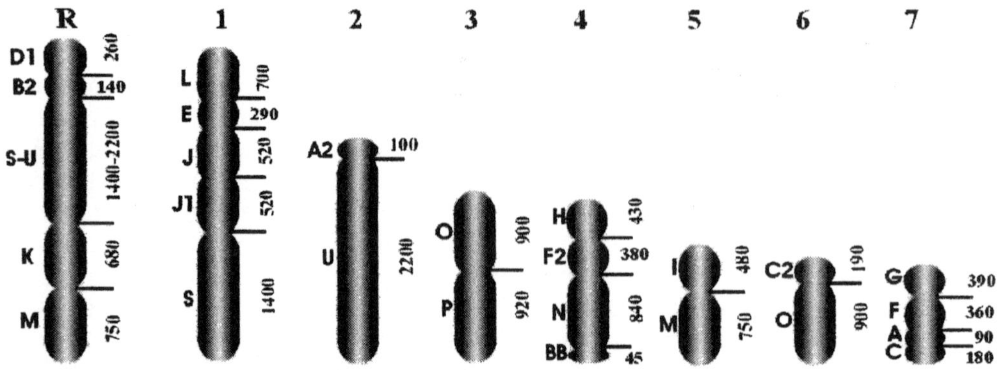

Fig. 1. Physical map of the *C. albicans* chromosomes (*Sfi*I macrorestriction map; left: name of the *Sfi*I fragment, right: fragment sizes in kb; http://alces.med.umn.edu/*Candida*.html).

2.3. Variability of Karyotypes

Despite primer clonal propagation (Lott 1999), the genome of *C. albicans* shows apparent plasticity and characteristic by frequent karyotypic variation (Iwaguchi *et al.* 1990; Rustchenko-Bulgac 1991). Differences in chromosomal banding pattern have been noted by Suzuki *et al.* (1989) and Rustchenko-Bulgac *et al.* (1990) in laboratory strains undergoing spontaneous colony morphology changes and ploidy shifts. Also electrophoretic karyotypes of clinical isolates show a wide variation (Iwaguchi *et al.* 1990; Magee *et al.* 1992) and karyotyping rearrangements seem to increase by conditions of stress associated with new niches in the host or host immune system.

Among the changes that lead to karyotypic diversity in *C. albicans* repetitive sequences play a major role. The major repeat sequences (MRS) of *C. albicans* consist of three elements (Fig. 2): repetitive sequences (RPSs), RB2, and HOK (Iwaguchi *et al.* 1992b; Chibana *et al.* 1994; Chindamporn *et al.* 1998;). HOK and RB2 neighbouring the RPS sequence and all three repetitive sequences are found on every chromosomes except chromosome 3 (Chindamporn *et al.* 1998). Only RB2a, out of one subclone from RB2, could be detected on this chromosome (Chindamporn *et al.* 1998). The MRS consists usually of a few to as many as 50 tandem repeats of RPS (Iwaguchi *et al.* 1992b; Magee and Chibana 2002). The RPSs are composed of 172 bp tandem repeats, called alts, the sequences of which can vary between different kinds of alts. The flanking DNA of a repeating alt region is highly homologue to the different RPS units. The size of RPS depends on the number of alt repeats within it. The alts are followed by highly conserved sequence, COM29, which contains an 8-bp restriction site of *Sfi*I (Magee and Chibana 2002). The *Sfi*I sites are highly conserved in the RPS sequences of different chromosomes (Chibana *et al.* 1994). The unique occurrence of *Sfi*I in only RPSs was used in a genomic project for detection of localisation of markers on chromosomes (chapter 4). The fully mapped chromosome 7 in *C. albicans* strain 1006 consists of four *Sfi*I fragments: 7C, 7A, 7F, and 7G, and two copies of the MRS. The repeats form the borders between *Sfi*I fragments 7A-7F and 7F-7G are oriented in an inverted repeat configuration (Chibana *et al.* 2000).

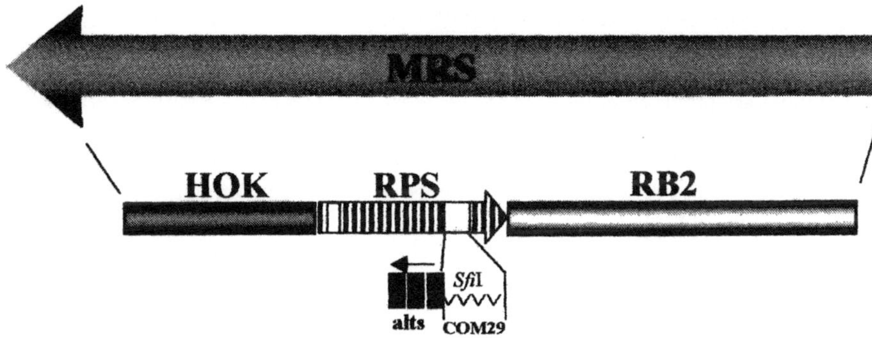

Fig. 2. Composition of MRS element on chromosome 7. The arrow shows the defined direction of the MRS. The RPS sequence contains several *Sfi*I sites and alt sequences. (Chibana *et al.* 1994; Chindamporn *et al.* 1998).

Other described repeat sequences in *C. albicans's* are the repetitive elements CARE-1 (*C. albicans* repetitive element-1; Lasker *et al.* 1991) and CARE-2 (Lasker *et al.* 1992; Thrash-Bingham and Gorman 1993; Magee and Chibana 2002). They are dispersed on several chromosomes. The middle repetitive element CARE-1 consists of 467 bp and contains a total of 27 perfectly directed repeats 5-10 bp in length and scattered stretches of A or T residues. CARE-1 elements are reiterated with a frequency of about two to twelve copies per haploid genome (Lasker *et al.* 1991). The 600 bp region of CARE-2 is characterised by six perfectly direct repeats in 6 bp in length and was shown to be present with a minimum of 10-14 copies per strain. CARE-2 repeats are highly variable in different strains but show low frequency of new polymorphism by higher temperature (Lasker *et al.* 1992). Moderately repetitive 223 bp DNA element Rel-1 is specific to the genome of *C. albicans* and contains small subrepeats (Thrash-Bingham and Gorman 1993). The 2789-bp Rel-2 sequence contains repetitive and unique DNA sequences. The repetitive DNA is present on most *C. albicans* chromosomes, whereas the unique sequence maps to chromosome 7; however, in some strains, it is also present on additional chromosomes (Thrash-Bingham and Gorman 1993). Chibana *et al.* (2000) pointed out four events that contribute to generate the genetic diversity of *C. albicans*: (1) chromosome length polymorphism (CLP), (2) reciprocal translocation occurring at the MRS loci, (3) chromosomal deletion, and (4) trisomy of individual chromosomes.

2.4. Chromosome Length Polymorphism

The chromosome length polymorphisms (CLP) are due to either differences in the size of the MRS on different homologues (Chibana *et al.* 2000) or the variation in the size of the telomeres or to small translocations (Magee and Chibana 2002). Tandem repeated RPS elements in MRS can change the size of chromosome homologue due to changes in the number of copies of RPS. About 10% of chromosome 7a in strain NUM114 is composed of the MRSs with several tandem repeats of RPS (Chibana *et al.* 2000). McEachern and Hicks (1993) reported on telomere length polymorphism in *C. albicans*. They observed an extension of the telomeres at higher temperature. The telomere of *C. albicans* is more complex than that of most fungi and consists of tandem copies of a 23 bp sequence, which are identical in a number of *C. albicans* strains. It was shown that telomere extension was accompanied by increases in the number of the 23 bp repeats. Also CARE-2 and Rel-2 repeated elements were assigned to the subtelomeric regions on chromosome 7 fragments (Chibana *et al.* 1998), which can influence the size variation in telomere or in subtelomeric repeats (Chibana *et al.* 2000). In *S. cerevisiae* seven differently repeated sequence families were found, some of which are highly variable in copy number in homologous chromosomes and location between strains (Louis *et al.* 1994). CARE-2 and Rel-2 contain long terminal repeat (LTR) fragments of retrotransposons (Goodwin and Pouter 1998) and many of these subtelomeric repeats lie adjacent to a region rich in retroansposon fragments. In the genome of *C. albicans*, more than 350 transposon insertions were identified, their insertion or excision can also evoke CLP (Goodwin and Poulter 2000).

2.5. Translocations

Thrash-Bingham and Gorman (1992) reported for the first time on the contribution of DNA translocations to chromosome length polymorphisms. Magee and Chibana (2002) described translocations of *Sfi*I fragments between chromosomes 1 and 5, 4 and 7, 5 and 6 in different strains (Magee and Chibana 2002). Only chromosome 3, which does not contain one complete copy of MRS, is not involved in translocations. These translocations are located around a repeated element MRS, which includes RPS with several *Sfi*I sites (Iwaguchi *et al.* 1992b; Chibana 1994). Thus, it was reported that reciprocal translocation of *Sfi*I fragments between chromosomes 7 and 4 in strains WO-1 (Chu *et al.* 1993) and NUM1000 (Chibana *et al.* 2000). The recombinations occurred between the MRSs of *Sfi*I fragments 4F-4N and 7F-7G in WO-1 and between *Sfi*I fragments 4F-4N and 7A-7F in NUM1000. Navarro-Garcia *et al.* (1995) observed translocations of *Sfi*I fragments 7G and 2A in strain 1001. Translocations can also build an extra chromosome as in strain WO-1 composed of *Sfi*I fragments 4H, 7F, 7G (Iwaguchi *et al.* 2000; Chibana *et al.* 2000).

2.6. Aneuploidy

Using *C. albicans* with multiple linked and unlinked heterozygosities in diploid and tetraploid strains, Hilton *et al.* (1985) concluded that the recovery of auxotrophic markers suggested that heat shock can induce the loss of entire chromosomes. Under stress situations like heat shock (Hilton *et al.* 1985), growth on non-utilizable carbon sources (Jarbon *et al.* 1998), or selection for fluconazole resistance (Perepnikhatka *et al.* 1999) occurred with the loss of one part or the whole homologue of the chromosome.

As a spontaneous derivative of the white-opaque switching strain WO-1, *C. albicans* strain WO-2 was isolated. WO-2 has lost two chromosomes compared to parental strain WO-1: the translocation product which contains *Sfi*I fragments 6C-5M and one intact homologue of chromosome 7. The missed chromosomes amounted to 10% of the genome. The aneuploidy of WO-2 was shown to be stable and affected the growth characteristics (Magee and Magee 1997). Although strains with different karyotypes show highly conserved *Sfi*I restriction maps (Chu *et al.* 1993) and *Sfi*I fragments of chromosome 7 are mostly conserved, Chibana *et al.* (1998) reported that strain NUM1000 has lost the *Sfi*I fragment 7C and two-thirds of *Sfi*I fragment A on chromosome 7.

Barton and Gull (1992) observed slow growth of *ade2* (gene of the *C. albicans* phosphoribosylaminoimidazole carboxylase) heterozygotic cells in the presence of methylbenzimidazol-2-yl carbamate, which was a result of the loss of one chromosome 3 homologue. Large red colonies, which grew much faster, may be a result of the re-establishment of the diploid complement, possibly *via* a second non-disjunction event. Both studies showed that *C. albicans* can tolerate aneuploidy and it can contribute to gene regulation. The spontaneously occurring mutants of *C. albicans* with the ability to assimilate certain carbon sources like D-arabinose and L-sorbose, that are not utilized by the parental strain, show specific changes in chromosome numbers (Rustchenko-Bulgac *et al.* 1994). The growth of *C. albicans* on non-utilizable sugar L-sorbose generates after about a week colonies which lack chromosome 5, are able to metabolise L-sorbose, and grow slowly on glucose medium (Janbon *et al.* 1998). After prolonged incubation of these strains on glucose-containing medium, they lose the ability to grow on L-sorbose and revert to the diploid state. The gene for sorbose utilization (*SOU1*) which is required for L-sorbose assimilation in *C. albicans,* is located on chromosome 4. Since *SOU1* does not reside on chromosome 5, Janbon *et al.* (1998) suggested that chromosome 5 contains a repressor that inhibits the expression of *SOU1* in dependence on gene dosage. The loss of one copy of chromosome 5 decreases the concentration of repressor molecules and leads to insufficient inhibition of the expression of the *SOU1* gene with subsequent ability to metabolise L-sorbose. Further indication for the influence of changing the chromosome copy number on regulation of physiologically important genes was shown by Perepnikhatka *et al.* (1999). They reported that fluconazole-resistant mutants obtained by selection on fluconazole medium, exhibit two different changes in karyotype of *C. albicans*. Firstly, aneuploidy of chromosome 4 after incubation of the mutants for seven days and secondly, trisomy of chromosome 3 after incubation for 35 or 40 days (Perepnikhatka *et al.* 1999).

2.7. Genetic Typing Methods

Strain typing is an important part of *Candida* research. This allows identifying of particular strains in outbreaks of infection, their relative abundance and origin due to novel infecting strain or reinfection by the original strain, monitoring the dynamics of yeast populations and development of drug resistance (Sullivan and Coleman 2002). The diagnostic methods are based on natural genetic variations, which have been observed by chromosomal length polymorphism, isoenzymes, restriction fragment length polymorphism (RFLP), DNA fingerprinting, PCR fingerprinting, and randomly amplified polymorphic DNA (RAPD).

Analysis of karyotype by means of pulsed-field gel electrophoresis (PFGE; Schwartz and Cantor 1984) is most widely used in *Candida* epidemiology. The size and number of separated chromosomes can widely vary in different strains. The method was extended by infrequent cleavage of genomic DNA with *Sfi*I and *Not*I before electrophoretic separation (Doi *et al.* 1994; Pontieri *et al.* 1996; Riederer *et al.* 1998). The identification of MRS restriction patterns by restriction endonuclease digestion, usually with *Eco*RI, and probing with specific fingerprinting probes 27A (Scherer and Stevens, 1987), Ca3 (Sadhu *et al.* 1991), RPS (Chindamporn *et al.* 1998), and CARE- 2 (Lasker *et al.* 1992) by Southern blot has simplified the identification of specific strains. The probes 27A and Ca3 are included in the MRS, and a part of each overlap with RPS. Most of the B fragment of Ca3 is included in HOK, and C1 fragment contains portions of both RPS and HOK (Chindamporn *et al.* 1998). This method, called DNA fingerprinting, reduced the number of bands, some of which are highly polymorphic. Simultaneously, others are nonvariable. The composition of MRS is dynamic both *in vivo* and *in vitro* and yields differences in the restriction patterns (Pujol *et al.* 1999).

For the analysis of *Candida* populations the randomly amplified polymorphic DNA technique (RAPD) is quick and easy to perform (Bostock *et al.* 1993). This PCR-based method utilize a single 10 to 15 bp primer of arbitrary sequence by amplification at low stringency amplification reaction conditions (Welsh and McClelland 1990). The obtained fingerprinting consists of three to six bands. The variability of different strains is caused by nucleotide variations at the primer annealing sites. The usage of primers homologous to microsatellite sequences with relatively high annealing temperature improved the reproducibility of RAPD (Weising *et al.* 1995).

With multilocus enzyme electrophoresis (MLEE) strain differences can be analysed on protein level. This technique involves separating isoenzymes by starch gel electrophoresis and subsequent visualization of specific enzymes using appropriate staining reactions. Differences in DNA sequences correlate with the number of substitutions of amino acids in proteins, which influences the charge of protein and its mobility (Caugant and Sandven 1993; Pujol *et al.* 1993; Rosa *et al.* 2000). Pujol *et al.* (1997) tested 23 different enzyme profiles for 29 *Candida* isolates and showed, that MLEE is an effective method for assessing the genetic relatedness of *C. albicans* isolates.

2.8. Retrotransposons

Retrotransposons are mobile genetic elements capable of autonomous transposition *via* RNA intermediates (Boeke *et al.*1985). They are widespread in eukaryotes and often make up a large proportion of their genomes. In maize, for example, it has been estimated that at least 50% of the genome is derived from retrotransposons (SanMiguel *et al.* 1996). Such retrotransposons carry two long terminal repeats (LTRs), typically 250-600 bp in length, flanking an internal protein-coding domain 5-7 kb long. The LTRs contain signals for promoting and terminating transcription and are also required in the reverse transcription process. The internal domain usually contains two open reading frames (ORF) known as *gag* and *pol*. The *gag* ORF encodes proteins that make up the major structural component of a cytoplasmic particle in which reverse transcription occurs. The *pol* ORF encodes enzymes (reverse transcriptase, ribonuclease H, protease, and integrase) which are involved in making a double-stranded DNA copy of the retrotransposon mRNA and inserting it into the host genome.

Several LTR retrotransposons present in different yeasts have been shown to be active, for example Tyl, Ty2, and Ty3 of *Saccharomyces cerevisiae*, Ty5 of *Saccharomyces paradoxus*, and Tfl and Tf2 of *Schizosaccharomyces pombe* (Curcio *et al.* 1988; Zou *et al.* 1996; Behrens *et al.* 2000). Matthews *et al.* (1997) first described a retrotransposon from the pathogenic yeast *Candida albicans*. This retrotransposon (Tca2) was originally discovered due to its appearance as an abundant, extrachromosomal linear, double-stranded DNA molecule (known as pCal) in the *C. albicans* strain h0G1042. Holton *et al.* (2001) showed that Tca2 is widespread in *C. albicans*, but that different *C. albicans* strains vary greatly in the amount of extrachromosomal Tca2 DNA that they produce and that Tca2 DNA levels are also strongly temperature dependent. In the 6426 bp long sequence Tca2 contains two long ORFs. These are similar to the gag and pol ORFs of other retrotransposons, which are flanked by 280-bp LTRs. Tca2 is present at an estimated 50 copies per cell. No other retrotransposons are known to produce such abundant extrachromosomal DNA copies. It was suggested that retrotransposons like Tca2 could be used to perform a system for efficient random insertional mutagenesis of *C. albicans* (Matthews *et al.* 1997).

2.9. Gene Duplication and Gene Families

Sequence duplication and the correlated emergence of gene families are believed to play a major role in molecular evolution. After duplication, the different copies of a gene can diverge and/or acquire novel regulations that may eventually lead to novel functions (Ohono 1970; Skrabanek and Wolfe 1998; Hughes 1999). For *C. albicans* several gene families are known and characterised.

2.9.1. Secreted aspartat proteases gene family

C. albicans produces secreted enzymes which disturb cells and intercellular matrix and, as a result, provoke tissue lesions. The most observed enzymes are secreted aspartic proteases (*SAP*) that are encoded by a gene family with at least ten genes (Table 1). All ten SAP genes encode preproenzymes approximately 60-200 amino acids longer than the mature enzyme. The N-terminal secretion signal is cleaved by a signal peptidase in the endoplasmic reticulum (ER). The propeptide is removed to activate the proteinases by the subtilisin-like Kex2 proteinase in the Golgi before being transported *via* vesicles to the cell surface for secretion or glycosylphosphatidylinositol (GPI)-anchoring (Hube and Naglik 2001). The mature enzyme contains sequence motifs typical for all aspartic proteinases, including the two conserved aspartate residues of the active site. Conserved cysteine residues are probably implicated in maintaining the three-dimensional structure of the enzyme (Hube 1996). Unlike *SAP*1-8, *SAP*9 and *SAP*10 both have C-terminal consensus sequences typical for GPI proteins (Hube and Naglik 2001).

Table 1. SAP genes of *C. albicans*. Size of the open reading frame, deduced protein size, NCBI accession number, location on chromosome and *Sfi*I fragment (if data available).

Gene	ORF [bp]	NCBI Accession	Chromosome	Protein [AS]	References
SAP1	1173	X56867 P28872	6(O)	391	Hube *et al.* 1991 Magee *et al.* 1993
SAP2	1194	A45280	R(U,S)	398	Wright *et al.* 1992 Magee *et al.* 1993
SAP3	1194	L22358	3(Z)	398	White *et al.* 1993
SAP4	1251	L12452	6(O)	417	Miyasaki *et al.* 1994
SAP5	1254	P43094	6	418	Monod *et al.* 1994
SAP6	1254	P43095	6	418	Monod *et al.* 1994
SAP7	1764	P43096	1	588	Monod *et al.* 1994
SAP8	1215	042778	3(O)	405	Monod *et al.* 1998
SAP9	1632	042779	3	544	Monod *et al.* 1998
SAP10	1323	AF146440	4	441	Felk *et al.* (direct submission, 2000)

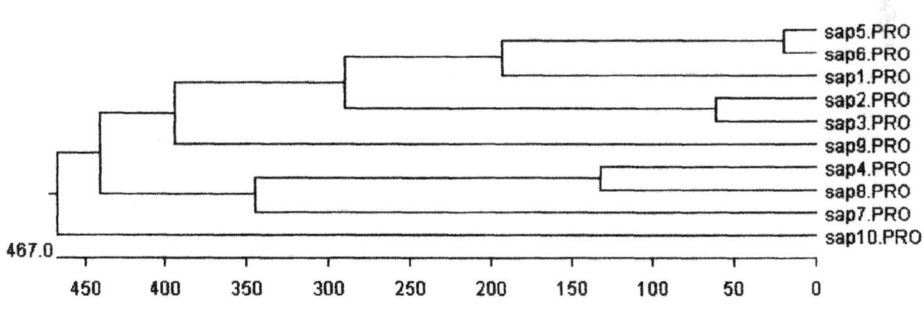

Fig. 3. Dendrogram of the Sap enzyme family of *C. albicans* (Sap1–Sap10 based on amino acid sequences). The length of each pair of branches represents the distance between sequence pairs, while the units at the bottom of the tree indicate the number of amino acid substitution events.

2.9.2. Agglutinin-like sequence gene family

An other large gene family in *C. albicans* is called *ALS* (agglutinin-like sequence) due to the resemblance of domains of its encoded proteins to α-agglutinin, a cell-surface adhesion glycoprotein in *Saccharomyces cerevisiae* (Lipke *et al.* 1989; Hoyer *et al.* 1995). Here, alpha-agglutinin, encoded by AG alpha 1, has the function of an adhesion glycoprotein that mediates mating of haploid cells (Hoyer *et al.* 1995). The presence of this gene(s) in *C. albicans* is curious, since the organism has not been observed to undergo meiosis. Presently, nine genes of the *ALS* family have been reported in the literature (Table 2), although a small number of additional genes are found in the *C. albicans* genome (Hoyer *et al.* 1995, 1998a,b; Gaur and Klotz 1997; Hoyer and Hecht 2000, 2001). There are different alleles for each of the genes due to changes of simple repeat numbers. *ALS* genes conform to a basic three-domain structure that includes a relatively conserved 59 domain of 1299 to 1308 nucleotides (433 to 436 amino acids), a central domain of variable length consisting entirely of a tandemly repeated 108-bp motif, and a 39 domain of variable length and sequence that encodes a serine-threonine-rich protein (Hoyer *et al.* 1998b).

Table 2. *ALS* genes of *C. albicans*. Size of the open reading frame, deduced protein size, NCBI accession number, location on chromosome and *Sfi*I fragment (if data available).

Name	DNA sequence [bp]	NCBI Accession	Chromosome	Protein [AS]	References
ALS1	3783	L25902	6(O)	1260	Hoyer *et al.* 1995
ALS2	1404	AF024580	6(C)	468	Hoyer *et al.* 1998b
ALS3	3360	U87956	R	1119	Hoyer *et al.* 1998a
ALS4	1407	AH006930	6(C)	469	Hoyer *et al.* 1998b
ALS5	3813	AF068866	3	1270	Hoyer and Hecht 2001
ALS6	4332	AF075293 AAD42033	3	1443	Hoyer and Hecht 2000
ALS7	6897	AF201684 AF075294	3	433	Hoyer and Hecht 2000
ALS8	4383	AF051313 AAD02580	R	1047	Leng *et al.* (direct submission, 1999)
ALS9	1404	AH010248 AF229990	6(O)	468	Hoyer *et al.* (direct submission, 2000)

Phylogenetic analysis of the *ALS* and *SAP* gene families show that the *ALS* family is younger than the *SAP* family. *ALS* genes in *C. albicans*, *C. dubliniensis*, and *C. tropicalis* tend to be located on chromosomes that also encode genes from the *SAP* family, yet the two families have unexpectedly different evolutionary histories (Figure 4). Homologous recombination between the tandem repeat sequences present in *ALS* genes could explain the different histories for co-localized genes in a predominantly clonal organism like *C. albicans* (Hoyer *et al.* 2001).

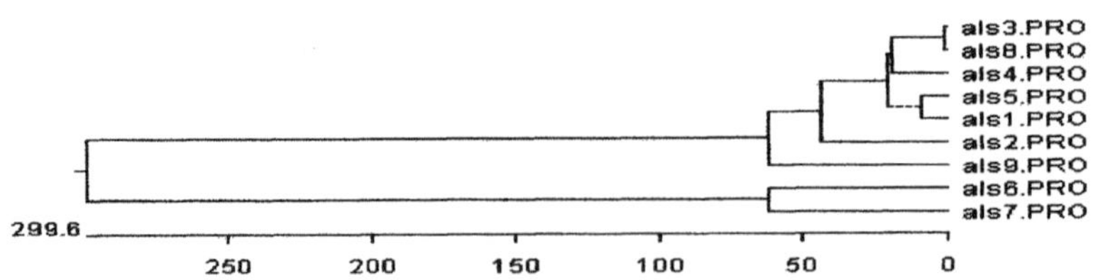

Fig. 4. Dendrogram of the *ALS* gene family of *C. albicans* (*ALS*1–*ALS*9) based on amino acid sequences. Two distinct groups are clustered within the family (*ALS*6, 7 and *ALS*1, 2, 3, 4, 5, 8, 9). The length of each pair of branches represents the distance between sequence pairs, while the units at the bottom of the tree indicate the number of amino acid substitution events.

Heterologous expression of *ALS* genes in *S. cerevisiae* confers to an adherence phenotype on the organism, suggesting Als proteins function in adhesion to host surfaces, a property that is positively correlated with *Candida* pathogenesis (Calderone and Braun 1991; Gaur and Klotz 1997; Fu *et al.* 1998). The *ALS1* gene encodes an adhesion protein that mediates attachment to endothelial cells (Hoyer 2001). Kamai *et al.* (2002) found that the *ALS1* gene product is important for the adherence of the organism to the oral mucosa during the early stage of infection. *ALS* genes exhibit several levels of variability including strain- and allele-specific size differences for the same gene, strain-specific differences in gene regulation, the absence of particular *ALS* genes in certain isolates, and additional *ALS* coding regions in others (Hoyer *et al.* 2001).

Similar to the *SAP* family *ALS* genes are differentially expressed under a variety of conditions that include morphological form, growth medium composition, growth phase, and strain of *C. albicans* (Hube *et al.* 1994; Hoyer *et al.* 1998a,b). The differential regulation and genetic variability of the *ALS* genes result in a diverse cell-surface ALS protein profile that is also affected by growth conditions (Hoyer *et al.* 2001). *ALS* genes are also found in other *Candida* species that are isolated from cases of clinical disease. The *ALS* genes are one example of a gene family associated with pathogenicity mechanisms in *C. albicans* and other *Candida* species (Hoyer 2001).

2.9.3. Lipase multigene family

Extracellular lipolytic activity enables the pathogen *Candida albicans* to grow on lipids as the sole source of carbon. Ten members of a lipase gene family (LIP1–LIP10) are cloned and characterised. The ORFs of all ten lipase genes are between 1281 and 1416 bp long (Table 3) and encode highly similar proteins with up to 80% identical amino acid sequences (Fig. 4). Each deduced lipase sequence has conserved lipase motifs, four conserved cysteine residues, conserved putative N-glycosylation sites and similar hydrophobicity profiles. All LIP genes, except LIP7, also encode an N-terminal signal sequence. The Lip isoenzyme family could be divided into two subgroups (Fig. 5). Lip4p, Lip5p, Lip8p, and Lip9p, which are more than 73% identical to each other, and Lip1p, Lip2p, Lip3p, Lip6p, and Lip10p, which are at least 54% identical to each other. Lip7p was the most divergent lipase in this isoenzyme family (Hube *et al.* 2000).

Table 3. Sequence data of all ten *C. albicans* lipase genes and their deduced proteins. Preprotein size of the deduced protein, signal peptide size of the signal peptide, GXSXG position of the serine residue in the lipase motif of the catalytic triad, YHG position of the histidine residue of the catalytic triad (FHS for Lip7), and localization on chromosome (determined for known sequences linked to the lipase genes, Hube *et al.* 2000). No signal sequence was identified for Lip7p.

Gene	*LIP1*	*LIP2*	*LIP3*	*LIP4*	*LIP5*	*LIP6*	*LIP7*	*LIP8*	*LIP9*	*LIP10*
ORF (bp)	1407	1401	1416	1380	1392	1392	1281	1380	1362	1398
Pre-protein (amino acids)	468	466	471	459	463	463	426	460	453	465
Signal peptide (amino acids)	16	16	16	14	14	16	–	14	14	16
kDa	50.8	50.4	51.4	49.6	50.0	50.5	47.9	49.7	49.2	50.4
GXSXG	196	196	196	194	194	196	190	194	194	196
YHG	344	344	344	339	339	344	321 (FHS)	339	339	344
Chromosomal location	1	n.d.	1	6	7	1	R 7	n.d.	1	1
Accession number	AF 188894	AF 189152	AF 191316	AF 191317	AF 191318	AF 191319	AF 191320	AF 191321	AF 191322	AF 191323

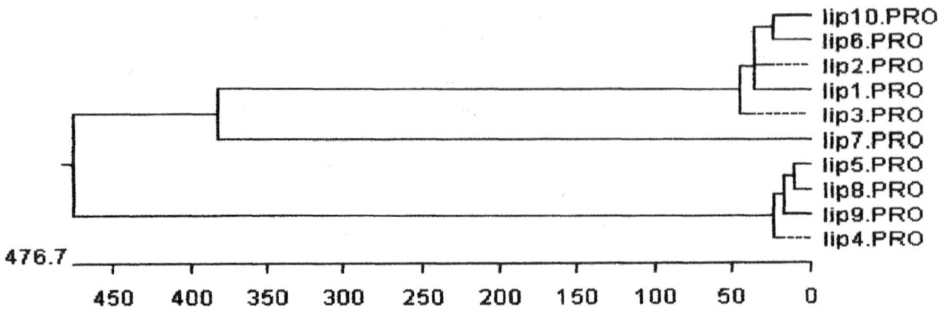

Fig. 5. Dendrogram of the lipase isoenzyme family of *C. albicans* (*Lip1-Lip10*). The length of each pair of branches represents the distance between sequence pairs, while the units at the bottom of the tree indicate the number of amino acid substitution events.

2.9.4. Drug resistance multigene family

Several lines of evidence suggest that multidrug transporter genes of the class of ABC transporters may be members of a larger multigene family in *C. albicans*. Besides the cloning of *C. albicans* drug resistance genes *CDR1* and *CDR2*, preliminary data from screening of a *C. albicans* genomic library with a conserved multidrug ABC transporter gene probe resulted in the isolation of at least two distinct genes with similarity to the two genes mentioned above isolated by functional screening (Sanglard *et al.* 1997). It will be interesting to resolve the complexity of this gene family in the future and to determine whether only specific members of this multigene family are involved in resistance to azole antifungal agents in *C. albicans* (see also Chapter 3.4).

2.10. Codon Usage

During the last 20 years, a number of alterations in the standard genetic code have been found in both prokaryotes and eukaryotes. In prokaryotes genetic code changes take place by the reassignment of the UGA stop codon to tryptophan in *Mycoplasma spp.* and *Spiroplasma spp.* (reviewed by Osawa 1995). Ten species of the eukaryotic genus *Candida* decode standard leucine CUG codon as serine (Santos *et al.* 1995; Santos *et al.* 1997). By using an *in vitro* cell-free translation system Sugita and Nakase (1999) identified the amino acid assignment of codon CUG in 78 species and 7 varieties of galactose-lacking *Candida* species equipped with Q9 as the major ubiquinone. Of these, only 11 species used codon CUG as a leucine codon. The remaining species decoded CUG as serine. Their result suggests that non-universal decoding is not a rare event, and that it is widely distributed in the genus *Candida*. Lee *et al.* (2002) found 17 non-universal serine codons (CUG) in the *Candida rugosa* LIP2 gene, and Tang *et al.* (2001) describe 19 CUG codons in the LIP4 ORF of *C. rugosa*, wereas lipase genes of *C. albicans* contain only around one to three non-universal serine codons. The code change from the standard leucine CUG codon to serine is mediated by a novel ser-tRNA(CAG), which induces aberrant mRNA decoding *in vitro* (Moura *et al.* 2002). The data from Santos *et al.* (1999) show hat genetic code ambiguity allows for increased tolerance to sudden and severe environmental changes and for growth under conditions that are lethal to parent cells not expressing CUG ambiguity. However, such codon usage is rare in *C. albicans*, and in contrast to *C. rugosa*.

The usage of a non-standard genetic code leads to problems by using a heterologous expression system like *S. cerevisiae* or *Pichia pastoris*. In the case of the exo-ß-1,3-glucanase from the pathogenic fungus *C. albicans,* there are two such translational events, one in the prepro-leader sequence and the other at residue 64 (Cutfield *et al.* 2000). Overexpression of active mature enzyme in a yeast host indicated that these two positions are tolerant to substitution. By comparing the crystal structure of the recombinant protein with that of the native, it was shown how either serine or leucine can be accommodated at position 64. Examination of the relatively few solved protein structures from *C. albicans* indicates that other CUG encoded serines are also found at non-essential surface sites (Cutfield *et al.* 2000). The non-universal serine codons (CUG) in *Candida spp.* can be converted into universal serine codons (UCU) by overlap extension PCR-based multiple site-directed mutagenesis (Tang *et al.* 2001; Lee *et al.* 2002). This step is necessary for a detailed biochemical characterisation of the proteins after heterologous expression.

2.11. Genome Project of *C. albicans*

The genome project was initiated in the early 1990s for efficient discovery of genes, which encode, firstly, virulence factors and, secondly, metabolic enzymes not present in mammalian cells. This project was started in two directions: developement of a physical map of the *C. albicans* genome at the University of Minnesota and sequencing of whole *C. albicans* genome at the Stanford University. The *C. albicans* genome is very plastic, with a variable karyotype. The strain selected for sequencing (SC5314) contains 8 chromosome pairs, 7 of which with constant size, and one polymorphic chromosome (R), with size ranging from 3.2 to 4.0 Mb. The presence of tandemly arrayed genes for the ribosomal RNAs accounts for size variation. The genome size of this strain is approximately 16 Mb (http://www.ncbi.nlm.nih.gov/cgi-bin/Entrez/map00?taxid=5476).

For construction of a physical map a fosmid library was produced by M. Strathmann at Lawrence Berkeley Laboratory at the University of California. It contains overlapping clones by partial digestion of genomic DNA with restriction enzyme *Sau*3A and subsequent ligation of fragments of about 40 kb into cloning site of the fosmid vector. In contrast to cosmid clones, fosmids propagate at low copy number in *E. coli*. The genome of *C. albicans* contains a lot of middle tandemly repeated DNA sequences. Only the use of fosmids, which are present as one or two copies per cell, could solve a problem of recombination of these repeats in bacteria (Magee and Scherer 1998). The fosmid library includes 3,840 clones, which consists of overlapping clones und contains about 10 haploid genomic equivalents of DNA. In the addition to the fosmid library, several smaller libraries of *Hin*dIII- or *Eco*RI-digested DNA were made. These small libraries were partially sequenced and the sequences were compared to the entire genomic data base. Sequences which show homology to genes in the database were used as markers and served to make probes. The markers were detected by hybridisation on the filters containing replicas of the 1,920 clones of the first half of the fosmid library. Simultaneously, by hybridisation to pulse-field electrophoresis separations of both the chromosomes and a *Sfi*I digest of the chromosomes established localisation of markers on chromosomes. The probes were hybridised to the several overlapping fosmids, which are grouped to one contig and determinated to chromosome region.

The large scale genome shotgun sequencing project by the Stanford Genome Technology Center is finished with the sequencing of 10.4 haploid genome equivalents (Tzung *et al.* 2002). The genomic sequence of *C. albicans* is available *via* public database http://sequence-www.stanford.edu/group/*Candida*. The approximately sixteen million base pairs large genome of *C. albicans* is subdivided into more than 1,680 contig DNA sequences (May 2000 data release). The contigs range from 2 to 151 kb and result in total 16.2 Mb. Preliminary studies estimate, that the *C. albicans* genome contains approximately 8000 open reading frames (ORFs) by consideration of translated protein sequences larger then 100 amino acids.

The complete genome of *Candida albicans* mitochondrion was sequenced (total 40420 bp, Genome Technology Center, Stanford University, 855 California Ave, Palo Alto, CA 94304, USA) and published in the Genbank under accession NC_002653 (Anderson *et al.* 2001).

2.12. Comparative Genetics

Comparison of gene order among genomes can be used for two purposes: inferring the phylogenetic relationships of species, and estimating the number and type of genomic rearrangements that have occurred, since two genomes last shared a common ancestor. Three mechanisms of rearrangement are usually considered: inversion, transposition, and reciprocal translocation (Nadeau and Taylor 1984; Sankoff 1993; Blanchette *et al.* 1996). Sequence comparison in computational molecular biology is a powerful tool for deriving evolutionary and functional relationships between genes. However, classical alignment algorithms handle only local mutations (i.e. insertion, deletion, and substitution of nucleotides) and ignore global rearrangements like inversions and transpositions of long fragments. Gene order comparisons have been made on sequenced organelle and viral genomes (Sankoff *et al.* 1992; Boore *et al.* 1995; Hannenhalli *et al.* 1995), and on more sparsely mapped mammalian and plant nuclear chromosomes (Nadeau and Taylor 1984; Bafna and Pevzner 1995; Paterson *et al.* 1996).

The extent of gene order conservation among ascomycete fungi has previously been estimated by comparing the *S. cerevisiae* genome sequence (Goffeau *et al.* 1997) to DNA sequences from other species, using either random ''genome survey'' sequences from both ends of small clones (Altmann-Jöhl and Philippsen 1996; Ozier-Kalogeropoulos *et al.* 1998; Hartung *et al.* 1998) or existing European Molecular Biology Laboratory database sequences (Keogh *et al.* 1998). With relates to *S. cerevisiae* and *C. albicans*, two species separated by 140–330 million years (Berbee and Taylor 1993; Pesole *et al.* 1995), gene order is substantially different between these two yeasts. Before the sequencing projects of both fungi were started, only one example of conserved gene order and orientation has been reported (STE6–UBA1; Raymond *et al.* 1998). Based on computer simulations Seoighe *et al.* (2000) found that 3,188 pairs of genes, that appear to be adjacent in *C. albicans,* do not have *S. cerevisiae* orthologues. For 298 gene pairs (9%), which are adjacent in one species, being conserved as adjacent in the other. The inversion of small segments of DNA, less than 10 genes long, has been a major cause of rearrangement, which means that even where a pair of genes has been conserved as adjacent, the transcriptional orientations of the two genes relative to one another are often different. Seoighe *et al.* (2000) estimate that about 1,100 single-gene inversions have occurred since the divergence between these species. Other genes, being adjacent in one species, are in the same neighbourhood in the other, but their precise arrangement has been disrupted, probably by multiple successive multigene inversions. A more accurate description of the size distribution is clearly needed, but will require comparisons between more closely related yeast species or other pathogenic fungi like *Ustilago maydis*. This basidiomycete smut pathogen of maize grows as a yeast when haploid, and filamentous when the cells are dikaryotic.

3. FITNESS AND VIRULENCE GENES

In comparison to the other pathogenic organisms, which are toxins producing as a single pathogen factor; the pathogenicity of *C. albicans* is more complex and multifactorial. The ability of *C. albicans* to switch between yeast and hypha cell form (called dimorphism), surface virulence molecules and hydrophobicity, adherence to the host cells, molecular mimicry, phenotypic switching, and secreted hydrolytic enzymes like aspartat proteases, phospholipases and lipases are examples of pathogenic factors (Odds 1988; Cutler 1991; Odds 1994a; 1994b; San-Blas *et al.* 2000). For each step of infection, several virulence factors are essential. According to Odds (1994a) the development of *Candida* infection was divided into four steps: (1) asymptomatic colonisation of skin or mucosa, (2) surface mycosis with invasion into the upper layer of skin or mucosa, (3) deep mycosis with penetration of fungal cells into deep tissue and into blood vessels, and (4) life-threatening systemic mycosis with dissemination of *Candida* cells over the whole body.

3.1. Dimorphism

The transition between yeast form (YF) growth (round to ellipsoid cells) and filamentous form (FF) growth (pseudohypae and true hyphae) of *C. albicans* (Figure 6) is postulated to contribute to the virulence of the organism (Madhani and Fink 1998; Brown and Gow 1999). The filamentation of *C. albicans* can be stimulated under different conditions. The strongest inducer is serum (Shepherd *et al.* 1980), in liquid serum medium the hyphal form arise at 37°C within minutes. The change of pH in the culture medium together with a simultaneously increasement of temperature to 37°C promote germination (Buffo *et al.* 1984). Also proline is known as an inducer of filamentous growth of *C. albicans* (Dabrowa *et al.* 1976; Ernst 2000).

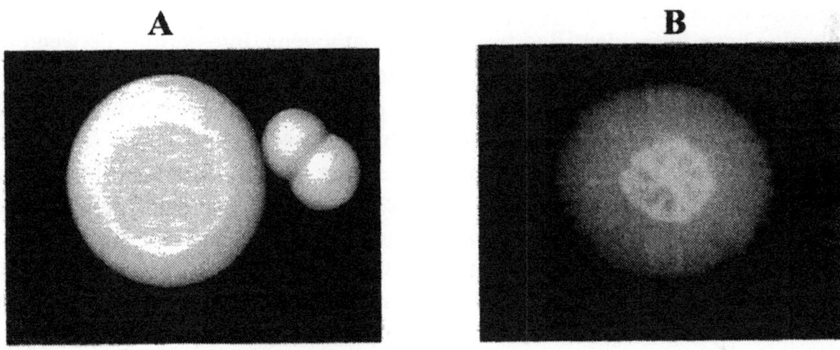

Fig. 6. Transition between growth forms like yeast cells and filamentous cells: A) colonies of *C. albicans* strain SC 5314 (Gillum *et al.* 1984) on normal growth media YPG (1% (w/v) yeast extract, 2% (w/v) peptone, 2% (w/v) glucose, 1,35% (w/v) Agar). B) hyphal induction on Spider-Medium (1% (w/v) mannitol, 1 % (w/v) nutrient broth, 0,2 % (w/v) K_2HPO_4, 1,35 % (w/v) Agar; Liu *et al.* (1994); Photograph by N. Borchert).

3.2. Phenotyping Switching

Different switching systems were described for *C. albicans* (Soll *et al.* 1987): some strains alternates between phenotypes distinguished by several colony morphologies (conversion from the original smooth to other variant colony morphologies), other switch between colonies with and without dense myceliation and a third system is typified by the patient isolate WO-1 and alternates between white (W) hemispherical colonies and gray flat colonies, designated opaque (O). White / opaque phenotypic switching affects the shape and size of cells, their ability to form hyphae, their surface properties (adhesion, permeability), membrane composition, range of secretory products, sensitivity to neutrophils and oxidants, antigenicity, and drug susceptibility (Soll 2001; 2002).

The *C. albicans* strains are able to spontaneously and reversibly switch phenotypes at high frequency (Slutsky *et al.* 1987; Soll 2002). Phenotypic switching is discussed as virulence factor and reflects a mechanism of adaptation to the changing environment by spontaneously generation of several phenotypes (Soll 2002). During phenotypic switching, dramatic reversible physiological and morphological changes are observed (Slutsky *et al.* 1987;

Kennedy *et al.* 1988; Anderson *et al.* 1990; Kolotila and Diamond 1990; Lan *et al.* 2002). Therefore, switching in *C. albicans* exerts pleiotropic effect on phenotype. Soll (2002) provided the first simple model for phase-regulated gene expression during switching, in which white and opaque phase-specific gene expression is controlled by a single white and a single opaque phase-specific activation factor. In this model, the promoters of genes coordinately expressed in each phase share common *cis*-regulatory sequences (Soll 2002).

Lan *et al.* (2002) report an application of a custom Affymetrix GeneChip representative of the entire *C. albicans* genome and assay the global expression profiles of white and opaque switch phenotypes of the WO-1 strain. Of 13,025 probe sets examined, 221 ORFs were expressed at a level higher in opaque cells than in white cells and 152 were more highly expressed in white cells. The differential expressed sequences represent functions as metabolism, adhesion, cell surface composition, stress response, signalling, mating type, and virulence genes.

3.3. Adhesine and Cell Wall

Microbial adherence is one of the most important determinants of pathogenesis, yet very few adhesins have been identified from fungal pathogens. Interactions between pathogens and their mammalian host are typically mediated by molecules that are either secreted or associated with cell wall. The cell wall of *C. albicans* is dynamic, which can cause dramatic changes in protein composition under different growth conditions (Chaffin *et al.* 1998). Several hyphal-specific cell wall proteins have been characterised and, while some have been assigned roles in cell wall integrity or adhesiveness to host cells, others have no known function. In addition to their functional characteristics, the accessibility of cell wall proteins makes them useful targets for sensitive diagnostic tests and some simple therapeutic approaches. Braun and Johnson (1997) show that a deletion of gene *TUP1* in *C. albicans* causes filamentous growth in the absence of any environmentally induced signal. *TUP1* has an orthologue in *S. cerevisiae.* Here, the gene has been studied extensively and it encodes a part of a transcriptional repression complex. It is brought to promoters by regulated DNA-binding proteins (Keleher *et al.*1992). Since remodelling of the cell wall appears to be a major aspect of the transition between blastospore and filamentous growth, Braun *et al.* (2000) showed that *TUP1* directly represses many of these cell wall protein-encoding genes. The *Δtup1* strain, therefore, could be used as a useful tool to study the expression of various genes during filamentation in *C. albicans*.

Four structurally related adhesins, Hwp1p, Ala1p/Als5p, Als1p from *Candida albicans* and Epa1p from *Candida glabrata*, are members of a class of proteins termed glycosylphosphatidylinositol-dependent cell wall proteins (GPI-CWP) (Sundstrom 2002). These proteins have N-terminal signal peptides and C-terminal domains that mediate glycosylphosphatidylinositol (GPI) membrane anchor addition, as well as other determinants leading to attachment to cell wall glucan. While common signal GPI motifs facilitate cell surface presence, unique features mediate ligand binding specificities of adhesins. The first glimpse of structural features of putative adhesins was made possibly by biophysical characterisations of the N-terminal domain of Als5p. One protein, Int1p, not belonging to GPI-CWP, initially described as an adhesin, has recently been shown to be similar in primary amino acid sequence to a gene of *S. cerevisiae* (Bud4p) which is involved in the budding pattern (Sundstrom 2002).

3.4. Multi-Drug Resistance

Yeast, which is an ideal eukaryotic model, has also been shown to have multidrug resistance (MDR) or pleiotropic drug resistance (PDR) genes. Several *pdr* mutants have been reported in *S. cerevisiae* and at least PDR1, PDR2, PDR3, PDR4, PDR5, and PDR6 loci of PDR in *S. cerevisiae* have been identified, which are located on different chromosomes (Dexter *et al.* 1994).

The widespread use of prolonged fluconazole therapy increased the incidence of treatment failure due to fluconazole-resistant *C. albicans* (Boschman *et al.* 1989; Fox *et al.* 1991; Kitchen *et al.* 1991; Ruhnke *et al.* 1994; Van den Bossche *et al.* 1994b). A number of studies identified the major azole resistance mechanisms (Albertson *et al.* 1996; Joseph-Horne *et al.* 1997; Sanglard *et al.* 1995; Van den Bossche *et al.* 1994b; White 1997; White *et al.* 1998). These include overexpression of or mutations in the drug target, sterol demethylase, mutations in other parts of the sterol biosynthesis pathway, and, most commonly, overexpression of drug efflux proteins. The fact that *C. albicans* is resistant to many metabolic inhibitors e.g. benomyl, a tubulin binding agent and methotrexate, a dihydrofolate reductase inhibitor, has prevented the development of effective drugs. However, in contrast to *S. cerevisiae* and other yeasts, there is not sufficient knowledge regarding the basis of natural or acquired resistance in *Candida* against various drugs.

Candida albicans possesses transporters such as Cdr1p and Cdr2p (*Candida* drug resistance proteins) with homology to proteins of the ATP-binding cassette (ABC) family (Higgins 1992; 1995; Ouellette *et al.* 1994; Jenkinson 1996; Decottignies *et al.* 1998), as well as Benp, which has homology to the major facilitator superfamily (MFS) class of drug proton antiport efflux pumps (Albertson *et al.* 1996; Sanglard *et al.* 1996; Cannon *et al.* 1998; White *et al.* 1998). The *BEN* gene encodes a transporter associated with resistance to benomyl and methotrexate, when it is expressed in *S. cerevisiae*. The gene responsible for benomyl and methotrexate resistance has been isolated and analysed to show that both phenotypes were encoded by the same DNA fragment. Since benomyl and methotrexate are structurally and functionally unrelated, a potential MDR element was suggested. However, the work of Fling *et al.* (1991) revealed that the protein encoded by the ORF had no sequence similarity to any known protein including P-glycoprotein of the MDR family. The resistance gene has been detected in several *C. albicans* strains and *C. stellatoidea.*

The mechanism of this gene in *C. albicans* remains to be determined. In certain instances polyenes resistance arises naturally in clinical material. Hitchcock *et al.* (1987) described a *C. albicans* strain with an ergosterol deficiency that arose spontaneously in laboratory culture, is resistant not only to polyenes but also to azole antifungals. Among antifungal agents available to treat infections caused by *C. albicans*, fuconazole is by far the most commonly used compound (Powderly 1994). The chance of such multiple resistance occurring is unknown, but the study referred to prove that the possibility cannot be denied. Six different genes have been identified, which are involved in resistance to polyenes and all are shown to affect ergosterol biosynthesis. The C-5 sterol desaturase gene (ERG3), essential for yeast ergosterol biosynthesis, was cloned and sequenced from *C. albicans* by homology with the *S. cerevisiae* ERG3. After transformation of *S. cerevisiae* ergosterol deficient mutants with the ERG3 gene from *C. albicans* the ergosterol synthesis (Miyazaki *et al.* 1999) was restored.

By functional complementation of a *Δpdr5* null mutant of *S. cerevisiae*, Prasad *et al.* (1995) cloned and sequenced the multidrug-resistance gene CDR1 of *C. albicans*. Transformation of a PDR5-disrupted host hypersensitive to cycloheximide and chloramphenicol with CDR1 resulted in resistance to cycloheximide, chloramphenicol, and other drugs, such as the antifungal miconazole, with collateral hypersensitivity to oligomycin, nystatin and 2,4-dinitrophenol. The nucleotide sequence of CDR1 revealed that, like PDR5, it encodes a putative membrane pump belonging to the ABC (ATP-binding cassette) superfamily

(Prasad *et al.* 1995; Nakamura *et al.* 2001). CDR1 encodes a 1501-residue protein of 169.9 kDa, whose predicted structural organisation is characterised by two homologous parts, each comprising a hydrophobic region with a set of six transmembrane stretches, preceded by a hydrophilic nucleotide binding fold (Prasad *et al.* 1995).

To isolate additional factors potentially responsible for resistance to azole antifungal agents in *C. albicans*, the hypersusceptibility of a *S. cerevisiae* multidrug transporter mutant, Δ*pdr5*, to these agents was complemented with a *C. albicans* genomic library. Several new genes were isolated, one of which was the ABC transporter gene called *CDR2* (Sanglard *et al.* 1997). The protein Cdr2p encoded by this gene exhibited 84% identity with Cdr1p and could confer resistance to azole antifungal agents, to other antifungals (terbinafine, amorolfine) and to a variety of metabolic inhibitors. The disruption of *CDR2* in the *C. albicans* strain CAF4-2 did not render cells more susceptible to these substances. When the disruption of *CDR2* was performed in the background of a mutant in which *CDR1* was deleted, the resulting double Δ*cdr1*/Δ*cdr2* mutant was more susceptible to these agents than the single Δ*cdr1* mutant (Sanglard *et al.* 1997).

Wirsching *et al.* (2000) shows, that in clinical *C. albicans* strains, fluconazole resistance frequently correlates with an constitutive activation of the *MDR1* gene. *MDR1* encodes a membrane transport protein of the major facilitator superfamily. Homozygous Δ*mdr1* mutants have an enhanced susceptibility against fluconazole. The result, that the disruption of both alleles of the *MDR1* gene have no effect on the susceptibility to benomyl, was interpreted as *C. albicans* having other resistance mechanisms that compensate the loss of the *MDR1* gene (Morschhäuser 2002). An constitutive high level *MDR1* expression still causes resistance to other toxic compounds in addition to fluconazole (Wirsching *et al.* 2000; Morschhäuser 2002).

3.5. Secreted Enzymes

3.5.1. Secreted aspartic proteases

C. albicans produces secreted enzymes which disturb the host cell and intercellular matrix and, as a result, provoke tissue lesions. Most enzymes are secreted aspartic proteases (*SAP*) that are encoded by a gene family. The production of *SAP*s is increased when proteins are presented as sole nitrogen source (Crandall and Edwards 1987). *SAP*s can also degrade various proteins like collagen, keratin, mucin, laminin, fibronectin, serum albumins, immunoglobulin IgA, α_2-macroglobuline, and haemoglobin (Ray and Payne 1990; 1991; Goldman *et al.* 1995; Colina *et al.* 1996; Rüchel 1981; 1986; Rüchel and Böning 1983). Moreover, it was shown that different functions of the host can be influenced by secreted proteases (Tab. 4). Rüchel (1983) and Kaminishi *et al.* (1994) show activation of blood clotting factors II (prothrombin), X (Stuart factor), and XII (Hageman factor) by *C. albicans* proteases.

Some *SAP* expression studies and investigations of particular Δ*sap* mutants of *C. albicans* showed differential usage of these proteases during different infection types (the data of these studies are summarised in Table 5; reviewed by Felk 2002; Hube and Naglik 2002). Deleted single genes *SAP1*, *SAP2*, and *SAP3* as well as both deleted *SAP1* and *SAP3* impair virulence of *C. albicans* by infection of reconstituted human epithelia (RHE) (Schaller *et al.* 1999). De Bernardis *et al.* (1999) reported on reduced virulence of Δ*sap1* and Δ*sap3* single mutants during vaginal infections of rat (Table 4). The Δ*sap2* mutant was almost avirulent in the same infection model. The transcripts of *SAP1-SAP3* are also presented by systemic mycoses of *intra-peritoneally* (i.p.) infected mice. During the dynamic development of

Table 4. Relevance of SAP isoenzyme subfamilies during infections of *C. albicans*.

Subfamilies	Infection types	References
Sap1p-Sap3p	- skin / mucous infections	Schaller *et al.* 1999
		Naglik *et al.* 1999
		De Repentigny *et al.* 2000
		Felk *et al.* 2002
		Kretschmar *et al.* 2002
	- vaginal infections	De Bernardis *et al.* 1999
Sap4p-Sap6p	- systemic mycosis	Sanglard *et al.* 1997
		Kretschmar *et al.* 1999
		Staib *et al.* 2000
		Felk *et al.* 2002
	- skin / mucous infections	Schaller *et al.* 1999
	- oesophagus infections	Staib *et al.* 2000
		Kretschmar *et al.* 2002
	- colonisation of surfaces	Naglik *et al.* 1999
Sap7p	- oral infections	Naglik *et al.* 1999
Sap8p	- skin / mucous infections	Schaller *et al.* 1999
	- systemic mycosis	Felk 2002
Sap9p-Sap10p	- postulated processing function	Hube and Naglik 2001
		Felk 2002

Candida infection *SAP2* transcripts and, less often, *SAP1* transcripts were detected at all investigated infection stages (Felk 2002; Felk *et al.* 2002). The expression of *SAP3* was, at a later point of time (72 h after initiation of infection), not detectable. The *Δsap4/5/6* triple and double mutants (*Δsap4/5, Δsap5/6* and *Δsap4/6*) showed significantly reduced virulence in mice model of i.p. infection (Kretschmar *et al.* 1999; Felk *et al.* 2002). Sanglard *et al.* (1997) postulated that Sap4p-Sap6p isoenzymes are important for the normal progression of systemic infection by *C. albicans*, because the survival time of with *Δsap4/5/6* triple mutant *intra-venous* (i.v.) infected guinea pigs was significantly longer than that of control animals infected with the wild type strain. Sap6p is probably the most essential protease for systemic infections in Sap4p-Sap6p isoenzyme subfamily (Felk *et al.* 2002). Hube *et al.* (1994) demonstrated at first that the expression of *SAP4-6* genes is restricted to the hyphal growth form. Later studies showed strong expression of these genes during i.p. and oral infections of mice (Staib *et al.* 2000; Felk *et al.* 2002; Felk 2002). Also in all oral patient probes from *Candida* carrier, including HIV-positive patients, *SAP4-6* subfamily was predominant expressed (Naglik *et al.* 1999). In contrast to these observations no expression of *SAP4-6* was detected in vaginal model of rat (De Bernardis *et al.* 1995; 2001). This emphasises the distinct role of *SAP4-6* subfamily in the various form of candidiasis.

3.5.2. Extracellular phospholipases (PL)

The secretion of extracellular phospholipases in *C. albicans* was firstly reported in the 1960s by growing the yeast on solid media containing egg yolk or lecithin and analysing the lipid breakdown products (Costa *et al.* 1968; Ghannoum 2000). Phospholipase activity was found in many pathogenic *C. albicans* strains by using media containing blood serum and sheep erythrocytes (Costa *et al.* 1968). Banno *et al.* (1985) suggested that *C. albicans* secreted three types of phospholipases: lysophospholipase (Lyso-PL), lysophospholipase-transacylase (LPTA), and a phospholipase B (PLB). Barrett-Bee *et al.* (1985) were the first to evaluate the role of extracellular phospholipases in virulence by using a murine model of candidiasis. When phospholipase activity was measured in six yeasts (four strains of *C. albicans* and a single strain each of *C. parapsilosis* and *S. cerevisiae*), a correlation was found between phospholipase activity and two potential parameters of pathogenicity. The *C. albicans* isolates which adhered most strongly to buccal epithelial cells and were most pathogenic in mice had the highest phospholipase activities. Less pathogenic isolates of *C. albicans*, *C. parapsilosis*, and *S. cerevisiae* were less adherent to epithelial cells and less lethal to mice and had lower phospholipase activities (Barrett-Bee *et al.* 1985). These findings suggest a correlation between phospholipase activity and *Candidal* virulence.

Table 5. Phenotypes of *C. albicans* *Δsap*-mutants (RHE: reconstituted human epidermis).

Mutant	Phenotype	References
Δsap1	- reduced virulence in the RHE model and premature expression of *SAP2* and *SAP8*	Schaller *et al.* 1999
	In vivo: reduced virulence in the model of i. v. infections	Hube *et al.* 1997
	- reduced virulence in the rat model of vaginal infections	De Bernardis *et al.* 1999
Δsap2	- reduced virulence in the RHE model	Schaller *et al.* 1999
	In vitro: aggregation of cells during the growth in minimal medium	Hube *et al.* 1997
	- reduced growth in medium with protein as a sole nitrogen source and reduced proteolytical activity in the culture supernatant	Hube *et al.* 1997
	- reduced ability to damage the endothelial cells and to activate the expression of E-selectin	Ibrahim *et al.* 1998
	In vivo: reduced virulence in the model of i. v. infections	Hube *et al.* 1997
	- strongly reduced virulence in the rat model of vaginal infections	De Bernardis *et al.* 1999
Δsap3	- reduced virulence in the RHE model	Schaller *et al.* 1999
	In vivo: reduced virulence in the model of i. v. infections	Hube *et al.* 1997
	- reduced virulence in the rat model of vaginal infections	De Bernardis *et al.* 1999
Δsap1 / 3	- strongly reduced virulence in the RHE model	Schaller *et al.* 1999
	- premature expression of *SAP2* and *SAP8* and additional expression of *SAP5* in the RHE-model	Schaller *et al.* 1999
Δsap6 *Δsap4 / 6* *Δsap5 / 6* *Δsap4 / 5*	*In vivo*: significantly reduced virulence in the model of i.p. infections	Felk *et al.* 2002
Δsap4 / 5 / 6	*In vitro*: reduced growth in medium with protein as a sole nitrogen source and reduced proteolytical activity in the culture supernatant	Sanglard *et al.* 1997
	In vivo: significantly reduced virulence in the model of i.v. and i.p. infections	Sanglard *et al.* 1997 Kretschmar *et al.* 1999

3.5.3. Secreted lipases

Lipases (EC 3.1.1.3) are triacylglycerolester hydrolases that catalyse the hydrolysis of triglycerides at the interface between the insoluble substrate and water. The most obvious function of lipases is the hydrolysis of lipids in order to use fatty acids and/or glycerol as substrates. *C. albicans* is able to grow on media containing lipids as the sole source of carbon. Although this growth is slow compared to growth in other media, the fungus is able to hydrolyse the lipids and transport the hydrolytic products into the cell. Possible lipid substrates may be found on human tissue such as the skin or the intestinal tract (Hube *et al.* 2000). The high number of *LIP* genes may provide an adaptive advantage to persist on these surfaces, even in the absence of carbohydrates, and may assist *C. albicans* in competing with the normal microbial flora.

3.6. Gene Regulation and DNA Array Studies of Virulence Factors

DNA microarrays will become a widely used commodity in biological and biomedical research. The principle of the microarray technology is a reverse dot blot. The power and universality of microarrays derives from the specificity and affinity of complementary base-pairing. A few hundred up to hundreds of thousands of DNA fragments (considered as probes) are spotted onto various supports, such as nylon membranes, activated glass slides, or silicon. The samples to be analysed, DNA or RNA, can be labelled either radioactively, fluorescently or conjugated with aptens and then be used to hybridize the array. For many applications, PCR products are attached to the support as probe molecules. For global analyses a highly complex array with many thousands of probe molecules are required. Especially the preparation of PCR fragments is a major part of the microarray production regarding to time and cost involvement (Diehl *et al.* 2002). The success of the microarray technique is due to the possibility to study for the first time an incredibly high number of different parameters in parallel in a single experiment. In addition, the microarray technique can be used for either studying the transcription of a certain number of genes under specific conditions or to study DNA with the application being identification and genotyping of polymorphisms or mutations. A DNA microarray with 51–86 % coverage of the *C. albicans* genome (3,609–5,669 ORFs) is available (Tessier 2001) and will be used to measure drug resistance in *C. albicans* (Cowen *et al.* 2001). Other DNA microarray-based studies of gene expression in yeasts have focused on the immediate effects of development (Chu *et al.* 1998), regulators (DeRisi *et al.* 2000), antifungal drugs (Bammert and Fostel 2000; DeBacker *et al.* 2001), and stressful conditions (Gasch *et al.* 2000).

Nantel *et al.* (2002) used DNA microarrays to investigate the transcription profiles of 6333 predicted ORFs in cells undergoing yeast-to-hyphal transition and their responses to changes in temperature and culture medium. They could identify several genes whose transcriptional profiles are similar to those of known virulence factors that are modulated by the switch to hyphal growth caused by addition of serum and a 37°C growth temperature.

It has been shown that yeast-to-hypha transition is regulated by Efg1p and Cph1p (Liu *et al.* 1994; Stoldt *et al.* 1997; Nantel *et al.* 2002). Both proteins are transcription factors that are regulated by cAMP and mitogen-activated protein kinase (MAPK) signalling pathways and are homologous to Phd1p and Ste12p of *S. cerevisiae* (Gimeno and Fink 1994). The deletion of transcription factor Efg1p led to the drastic block in hyphal formation under standard conditions, using serum or GlcNAc as inducers (Lo *et al.* 1997; Stold, 1997). Ernst (2000) proposed that Efg1p is depending on environmental cues an activator as well as repressor of morphogenesis. Under "microaerophilic" and anaerobic conditions the *Δefg1* mutants are not defective in the hyphal formation and appear rather stimulated (Brown *et al.* 1999; Sonneborn *et al.* 1999a). The expression of *EFG1* also appears to be inactivated when the hyphal formation has been initiated (Stoldt *et al.* 1997). The lack of *EFG1* expression in strains CAI8 and WO-1 derivative favours a rod-like, pseudohyphal phenotype (Sonneborn *et al.* 1999b). The transcription factor Cph1p of the MAPK signalling cascade has a minor contribution to hyphal growth. The deletion of Cph1p blocked the hyphal development only on certain solid starvation media Spider or SLAD (Liu *et al.* 1994; Köhler and Fink 1996; Leberer *et al.* 1996; Csank *et al.* 1998). *C. albicans Cph1/Cph1 EFG1/EFG1* double mutants are defective in filamentous growth, unable to form hyphae or pseudohyphae in response to many stimuli, including serum and avirulent in mouse model infections (Lo *et al.* 1997).

Obviously, not only morphogenesis is regulated *via* signalling pathways, but also other cellular events. Sonneborn *et al.* (1999b) presented evidence that white-opaque phenotypic switching in the strain WO-1 is regulated by Efg1p. The gene *EFG1* is expressed in white, but not in opaque cells (Sonneborn *et al.* 1999b). The ectopic expression of *EFG1* in opaque cells leads immediately to switching to the white phase (Sonneborn *et al.* 1999b).

Also, different pHs that are present in diverse host niches influence morphological changes. Acidic pH favours the yeast form and pH near neutral assists filamentation of *C. albicans*. pH-regulated glucosidase genes *PHR1* and *PHR2* were identified (Saporito-Irwin *et al.* 1995; Mühlschlegel and Fonzi 1997; Fonzi 1999). The expression of *PHR1* is only detected when the ambient pH is above 5.5 (Saporito-Irwin *et al.* 1995). An inverse expression pattern showed *PHR2* with maximal level by pH below 5 (Mühlschlegel and Fonzi 1997). Both Phr1p and Phr2p are cell surface glycosylphosphatidylinositol-anchored proteins that cross-link cell wall ß-1,6- and ß-1,3-glucans and are important for both morphology and growth of *C. albicans*. In animal models of vaginal and haematogenous infections, which correspond to acidic and alkaline environments, pathogenicity of *Δphr1* and *Δphr2* mutants was differently attenuated. Mutants lacking *PHR1* are compromised to cause systemic mycoses but retain virulence in vaginal infection model (De Bernardis *et al.* 1998). Opposited to that the *Δphr2* mutants showed an attenuated ability to cause vaginal, but not systemic infections (De Bernardis *et al.* 1998). Davis *et al.* (2000) showed that both *PHR1* and *PHR2* genes and alkaline induced filamentation are controlled by Rim101p signalling pathway. The identified transcription factor Rim101p (called also Prr2p) controls pH-regulated genes (Ramon *et al.* 1999; Davis *et al.* 2000). In acidic pH, Rim101p exists as full length protein. Under alkaline conditions Rim101p regulates gene expression. A carboxy-terminal truncation of Rim101p causes the alteration of protein conformation (El Barkani *et al.* 2000). The short form of Rim101p is required for induction of gene expression as it was shown for *PHR1* and *PRA1*, and simultaneously, for repression of acidicly activated genes like *PHR2* (Davis *et al.* 2000).

In the current model MAPK and cAMP signalling pathways (Fig. 7) are triggered separately by various environmental signals *via* sensors, and a threshold level of signalling may be transduced by single strongly induced or several weakly stimulated pathways (Ernst, 2000). Obviously *C. albicans* possess Cph1p- and Efg1p-independent signalling pathways promoting hyphal formation, because it was recently shown that *in vitro* non-filamentous and low virulent *Δefg1 / Δcph1* double mutant was able to build hyphae in the tongue of orally infected immunosupressed piglets (Riggle *et al.* 1999). The transcription factor Czf1p appears to belong to Efg1p and Cph1p separate signalling pathways. The Czf1p pathway is activated only under mechanical stress, for example when the cells are hidden in a matrix, because inactivation of *CZF1* leads to delayed filamentation of embedded cells, by overexpressing *CZF1* it was shown that embedded cells display enhanced filamentation (Brown *et al.* 1999).

In a different signalling pathway the transcription factor CaTup1p (Braun and Johnson 1997; Braun and Jonson 2000) gets more important. In combination with CaMig1p and CaNrg1p, CaTup1p negatively regulates the morphogenesis of *C. albicans* (Braun and Johnson 1997). Inactivation of CaTup1p repressor caused exclusive constitutive pseudohyphal growth of *C. albicans* and depression of hypha-specific genes (Braun and Johnson 1997; Braun and Jonson 2000; Braun *et al.* 2000). Murad *et al.* (2001) showed that CaMig1p and CaNrg1p in combination with CaTup1p regulate a different set of genes (Tab. 6). Like other aforementioned signalling pathways, the transcription factors of the Tup1 signalling pathway regulate not only morphology of *C. albicans*, but also other cell functions. The transcript profiling with an array carrying 2002 genes, which represented about one-quarter of *C. albicans* genome (www.pasteur.fr/recherche/unites/RIF/transcriptdata/), revealed that CaNrg1p-regulated genes were enriched for the functional category energy and CaMig1p-regulated genes for cell rescue functions (Bailey *et al.* 1996; Hoyer *et al.* 1998b; Staab and Sundstrom 1998; Birse *et al.* 1993; Murad *et al.* 2001). Most of the hypha-specific genes are controlled by multiple pathways. In further studies using microarray technique for the screening of 7000 genes and cluster analysis of 637 genes it was shown that expression of a common set of differentially expressed genes (Table 6) is regulated by transcription factors Cph1p, Efg1p, and newly discovered Cph2p (Lane *et al.* 2001a; 2001b).

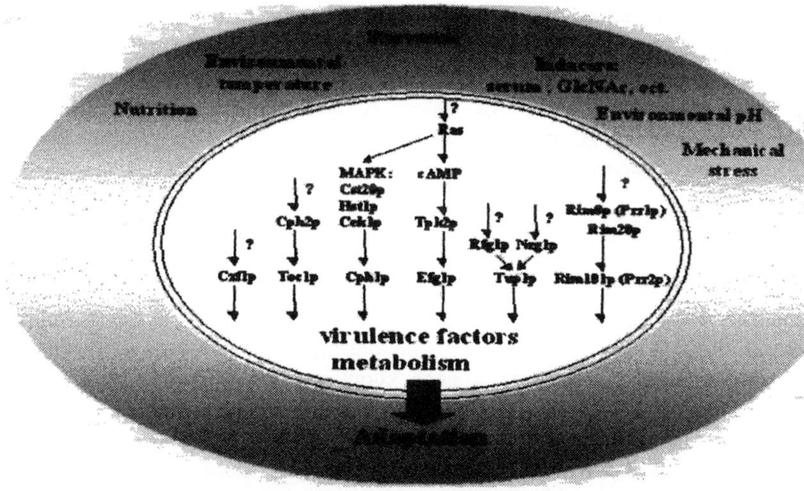

Fig. 7. In several studies it was shown that the Cph1p-, Cph2p-, Czf1p-, Efg1p-, Tup1p-, and pH-modulated signalling pathways of *C. albicans* (Ernst 2000; Davis *et al.* 2000; Murad *et al.* 2001; Lane *et al.* 2001a; Brown *et al.* 1999) coregulate cell physiology, morphogenetic changes, and other virulence factors. The changed expression patterns influence the adaptation of the fungus to the new environmental state.

Table 6. Genes that are regulated by multiple signalling pathways (Lane *et al.* 2001a;b; Murad *et al.* 2001).

Gene	Function of Gene	Regulated by Transcription Factor	References
ALS8	Agglutinin-like cell surface glycoprotein	CaTup1p; CaNrg1p	Hoyer *et al.* 1998
DDR48	Stress-response-protein	Cph1p; Cph2p; Efg1p	Treger and McEntee 1990
ECE1	Protein associated with hyphal growth	CaTup1p; CaNrg1p Cph1p; Cph2p; Efg1p	Birse *et al.* 1993
ECE99 (RBT1)	Cell wall protein	Cph1p; Cph2p; Efg1p	Braun *et al.* 2000
HSP12	12kd heat-shock protein	Cph1p; Cph2p; Efg1p	Stone *et al.* 1990
HWP1	Hyphal-specific cell wall protein	CaTup1p; CaNrg1p Cph1p; Cph2p; Efg1p	Staab and Sundstrom 1998
HYR1	Yeast cell wall protein	CaTup1p; CaNrg1p Cph1p; Cph2p; Efg1p	Bailey *et al.* 1996
PCK1	Phosphoenolpyruvate carboxykinase	CaTup1p; CaMig1p	Bailey *et al.* 1996
SAP5 / SAP6	Secreted aspartat protease 5 / 6	Cph1p; Cph2p; Efg1p	Monod *et al.* 1994
WH11	White colony-forming gene	Cph1p; Cph2p; Efg1p	Srikantha and Soll 1993
YJL79 (RBT4)	Cell wall protein	Cph1p; Cph2p; Efg1p	Braun *et al.* 2000

Although the expression of most of these genes is associated with morphogenesis, some of them like *ECE99*, *WH11*, *HSP12*, and *DDR48*, did not show a strict correlation with morphogenesis (Lane *et al.* 2001b). These experiments serve as an evidence that morphology of *C. albicans* is coordinated with other virulence factors. Braun and Johnson (2000) supposed two models of the regulatory circuit leading to filamentous growth, termed central control and network control. The authors tend to the second model, that reflects that the gene responds individually to various pathways.

4. GENETIC MANIPULATIONS OF *CANDIDA ALBICANS*

Genetic manipulations in the diploid opportunistic human fungal pathogen *C. albicans* require efficient transformation methods. Lithium acetate (Ito *et al.* 1983) and spheroplast transformation (Hinnen *et al.* 1978) are widely used methods (for review see DeBacker *et al.*

2000). Electroporation can also be used as a simple method for transformation (Thompson *et al.* 1998). Integrative transformation is an important tool to generate stable transformants for gene overexpression and gene disruptions. Since *C. albicans* is resistant to a number of selective drugs commonly used for selection after transformations, such as G418, cycloheximide, hygromycin B, methotrexate, and methionin sulfoxide (DeBacker *et al.* 2000), the use of the auxotrophic marker *URA*3 is the most popular system.

4.1. *URA*-Blaster Technique

Gene disruptions in the human fungal pathogen *C. albicans* are usually created using multiple rounds of targeted integration called the *URA*-blaster method (Fonzi and Irwin 1993). The *URA*-blaster protocol was before described for gene disruption in *Saccharomyces* (Alani and Kleckner 1987). Since *C. albicans* is a diploid organism, two consecutive steps of gene disruption are required to generate a gene knock-out. In principle, the same disruption cassette, consisting of the *C. albicans URA3* gene (encoding orotidine 5'-monophosphate [OMP] decarboxylase) flanked by *Salmonella typhimurium hisG* direct repeats, can be used for disruption of both alleles of the gene. This is possible, because, after the first round of disruption, homologous recombination between the direct repeats flanking the *URA*3 marker leads to loss of the marker gene (Figure 8). The subsequent counterselection on 5-fluoroorotic acid (5-FOA) containing medium allows the efficient recovery of Ura^- revertants. Resulting heterozygous and homozygous null mutants can be auxotrophic (Ura^-) or prototrophic (Ura^+) for uracil biosynthesis. The *URA*-blaster method was successfully applied to disrupt many genes of *C. albicans* (Gow *et al.* 1994; Buurman *et al.* 1998; Lussier *et al.* 1998; Viaene *et al.* 2000; Sanyal and Carbon 2002).

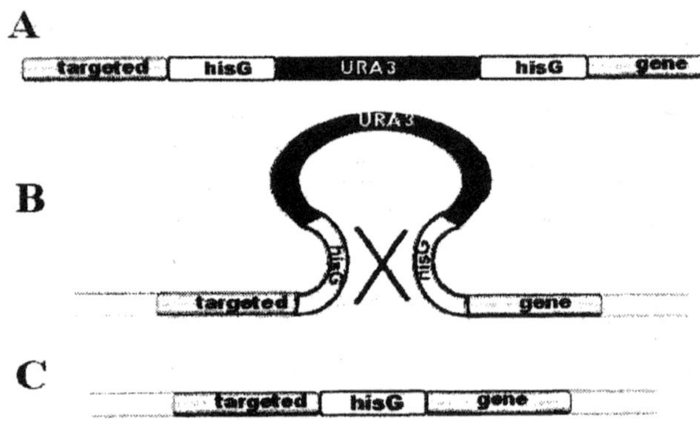

Fig. 8. *URA*-Blaster Strategy: gene disruption based on the use of the *URA3* gene for selection and recovery of the disruption cassette by intrachromosomal recombination between direct repeats of the hisG gene (Fonzi and Irwin 1993). A: disruption cassette for homologous recombination based gene targeting. B: mitotic recombination between the hisG repeat sequences results in a return to uridine auxotrophy. C: gene locus after insertion of the disruption cassette and excision of the *URA3* selectionmarker. hisG direct repeats of the *S. typhimurium hisG* gene; URA3 *C. albicans URA3* gene as a selection marker.

Recently, using the *URA*-blaster method, one improvement was the development of PCR-mediated cloning methods to avoid any time consuming cloning steps (DeBacker *et al.* 2000; De Hoogt *et al.* 2000). The original *URA*-blaster disruption cassette cannot be used in a PCR-based disruption approach, because the first *hisG* repeat flanking the *URA*3 gene can anneal to the second *hisG* repeat during PCR and prevent, therefore, the amplification of the complete cassette. This can be avoided by using the split-maker recombination strategy to get rid of the

URA3 gene disruption cassette (Fairhead *et al.* 1996; DeBacker *et al.* 2000). De Hoogt *et al.* (2000) used this technique to disrupt the *C. albicans* FAL1 gene (ATP-dependent RNA helicase). They amplified 5' and 3' *FAL1*-specific regions *via* PCR and ligated these fragments into an *URA*-blaster cassette. The resulting ligation reactions were used separately as templates to generate two FAL1 disruption cassettes with overlapping *URA3* marker regions. Simultaneous transformation with both overlapping disruption cassettes yielded efficient disruption of one FAL1 allele.

Another problem was that the *URA*-blaster disruption cassette is too large for an efficient PCR amplification, so it cannot be used for PCR product-directed gene disruptions. Wilson *et al.* (2000) described a gene disruption cassette without the *hisG* repeats to create gene knock-outs through a PCR product-directed disruption.

The application of the *URA*-blaster disruption cassette can lead to several problems: in the first transformation round the recovery of the initial wild-type genotype through gene conversion or mitotic recombination events can occur (DeBacker *et al.* 2000). Bain *et al.* (2001) showed that the *URA*-status of otherwise isogenic mutants affect the adhesion of *C. albicans* cells. Moreover, the effect of the *URA*-status on adhesion was also dependent on the null mutant background. Their results demonstrate that the *URA*-status is not neutral in determining adhesive properties of *C. albicans* mutants that are generated *via* the *URA*-blaster protocol.

4.2. *URA*-Flipper Strategy or Flp-Recombinase Mediated Site Specific Recombination

A refinement of the *URA*-blaster strategy, the so-called *URA*-flipper strategy, use the FLP recombinase of *S.cerevisiae* for the excision of the transformation marker (Sadowski 1986). Here a *URA3* marker flanked by short sequences which act as recognition sites for the Flp recombinase will be used for the gene disruption cassette (Figure 9). The Flp recombinase activity is thereby controlled by the *SAP2* promoter of *C. albicans* (Morschhäuser *et al.* 1999). After homologous recombination of the desired target gene, excision of the marker is accomplished by cultivation of the mutant cells in growth media with low nitrogen content that stimulate the *SAP2* promotor (DeBacker *et al.* 2000). Since this strategy is conceptually similar to the *URA*-blaster strategy, it has some of the same limitations.

Fig. 9. *URA*-Flipper Strategy: gene disruption based on the use of the *URA3* gene for selection and recovery of the disruption cassette by Flp-mediated, site specific recombination (Morschhäuser *et al.* 1999). A: disruption cassette for homologous recombination based gene targeting. B: gene locus after insertion of the disruption cassette and Flp recombinase mediated excision of the *URA3* selectionmarker. FRT recombinase recognition site; P *SAP2* promoter; FLP site-specific recombinase.

4.3. IMP Dehydrogenase Gene as a Dominant Selectable Marker

The identification of the inosine-5-monophosphate dehydrogenase (*IMH3*) gene in *C. albicans* lead Köhler *et al.* (1997) to the suggestion that mycophenolic acid (MPA), a specific inhibitor of IMP dehydrogenases, might be used to develop a dominant selectable marker system in *Candida*. In other eukaryotic organisms, it has been shown that amplification of an IMP dehydrogenase gene can confer MPA resistance in a gene-dosage-dependent manner

(Collart and Huberman 1987; Hedstrom and Wang 1990). IMP dehydrogenase (EC 1.1.1.205) is the key enzyme in the *de novo* biosynthesis of GMP, catalyzing the NAD-dependent oxidation of IMP to XMP. This conversion of IMP to XMP is the rate-limiting step in the biosynthetic pathway of guanine nucleotides (Jackson *et al.*1975). Köhler *et al.* (1997) showed that increasing the gene dosage by introducing plasmids carrying the *IMH3* ORF renders the recipient *C. albicans* cells resistant to elevated concentrations of MPA due to overexpression of the drug target enzyme. The *IMH3* gene can be used in a similar manner as the *URA*3 gene in the original *URA*-blaster disruption cassette, or better with the advantage of the FLP-mediated site-specific recombination. Wirsching *et al.* (2001) used a MPA(R)-flipping strategy for the disruption of the *CdMDR1* drug resistance gene in *C. dubliniensis.*

4.4. Visual Selection Marker

A progressive way to develop selection systems for *C. albicans* is the usage of visual selection markers. First trails with visual selection markers were done by Morschhäuser *et al.* (1998). They suggested that the use of the green fluorescent protein (*GFP*) reporter gene would allow gene disruption and subsequent selection of transformants by fluorescence-activated cell sorting. Since the cells still contain the green fluorescent protein from the first disruption step, the selection after disruption of the second allele would pose a problem. The same GFP marker can not to be used for a second time (DeBacker *et al.* 2000).

Viaene *et al.* (2000) studied the suitability of the *MET15* gene as a visual selection marker. They generated a methionine(*met15*)-deficient strain of *C. albicans* by using the *URA*-blaster technique. The deficient strain showed a brown colony colour on Pb(2+)-containing medium. The transformation with an integrative or replicative plasmid containing one of the *C. albicans* MET15 alleles leads to a complementation of the deficient strain. Transfomants may be identified by white colonies on Pb(2+)-containing medium (Viaene *et al.* 2000).

5. CONCLUSIONS

The past few years have brought a continuous increase in knowledge of the genetic basis of *C. albicans* virulence factors. The improvements in techniques to study gene functions by inactivation (disruption) and gene overexpression gives an insight to different molecular mechanisms of the fungus. Some virulence factors are well characterised, a multidrug transporter gene family and several secreted enzymes are known. Different transformation methods are available to enable gene-targeting in *C. albicans* and study gene functions in this organism. The genome project and comparative genetics allow a greater understanding of gene functions in shorter times than was conceivable a decade ago. Micro arrays of immobilised DNA (DNA chips) have been successfully applied for detecting simultaneously the expression of thousands of genes. Large-scale gene transcript analysis allows to study the interconnectiveness of gene regulation in response to environmental stimuli. Analysis of the expression of all proteins in an organism is much more difficult to conduct. Proteomic technologies, such as mass spectrometry (MS) and yeast two-hybrid assays could investigate the overall protein expression and protein-protein interaction in this pathogen, but this techniques requires time-consuming adaptations. One alternative approach to study gene expression is the construction of promoter/GFP fusions for every gene in the genome. Although this method has so far been developed only for *S. cerevisiae*, it should be applicable to *C. albicans.* Application of all these techniques to *C. albicans* will open up new avenues in the functional analysis of genes that are important for surviving and pathogenesis of this organism.

Acknowledgements: We thank Cornelia M. Gregel and Brigitte Witt for critical reading of the manuscript.

REFERENCES

Albertson GD, Niimi M, Cannon RD, and Jenkinson HF (1996).Multiple efflux mechanisms are involved in *Candida albicans* fluconazole resistance. Antimicrob Agents Chemother. 40:2835–2841.

Alani E, Cao L, Kleckner N. (1987). A method for gene disruption that allows repeated use of URA3 selection in the construction of multiply disrupted yeast strains. Genetics 116(4):541-545.

Altmann-Jöhl R, and Philippsen P (1996). AgTHR4, a new selection marker for transformation of the filamentous fungus *Ashbya gossypii*, maps in a four-gene cluster that is conserved between A. gossypii and Saccharomyces cerevisiae. Mol Gen Genet 250:69–80.

Anderson JB, Wickens C, Khan M, Cowen LE, Federspiel N, Jones T, and Kohn LM (2001). Infrequent genetic exchange and recombination in the mitochondrial genome of *Candida albicans.* J Bacteriol 183(3):865-872

Anderson J, Mihalik K, and Soll DR (1990). Ultrastructure and antigenicity of the unique cell and pimple of the *Candida* opaque phenotype. J Bacteriol 172:224-235.

Bafna V and Pevzner PA (1995). Sorting by Reversals: Genome Rearrangements in Plant Organelles and Evolutionary History of X Chromosome Mol. Biol. Evol. 12:239–246.

Bailey DA, Feldmann PJ, Bovey M, Gow NA, and Brown AJ (1996). The *Candida albicans* HYR1 gene, which is activated in response to hyphal development, belongs to a gene family encoding yeast cell wall proteins. J Bacteriol 178:5353-5360.

Bain JM, Stubberfield C, and Gow NA (2001). Ura-status-dependent adhesion of *Candida albicans* mutants. FEMS Microbiol Lett 204(2):323-328.

Bammert GF and Fostel JM (2000). Genome-wide expression patterns in Saccharomyces cerevisiae: comparison of drug treatments and genetic alterations affecting biosynthesis of ergosterol. Antimicrob. Agents Chemother. 44:1255–1265.

Banno Y, Yamada T, and Nozawa Y (1985). Secreted phospholipases of the dimorphic fungus, *Candida albicans*; separation of three enzymes and some biological properties. Sabouraudia 23:47–54.

Barrett-Bee K, Hayes Y, Wilson RG, and Ryley JF (1985). A comparison of phospholipase activity, cellular adherence and pathogenicity of yeasts. J. Gen. Microbiol. 131:1217–1221.

Barton RC and Gull K (1992). Isolation, characterization, and genetic analysis of monosomic, aneuploid mutants of *Candida albicans*. Mol Microbiol 6:171-177.

Beck-Sague CM and Jarvis WR (1993). Secular trends in the epidemiology of nosocomial fungal infections in the United States, 1980-1990. J Infect Dis 49:183-215.

Behrens R, Hayles J, and Nurse P (2000). Fission yeast retrotransposon Tfl integration is targeted to 5' ends of open reading frames. Nucleic Acids Res 28:4709-4716.

Berbee, ML and Taylor JW (1993). In: Reynolds DR and Taylor JW. eds. The Fungal Holomorph: Mitotic, Meiotic and Pleomorphic Speciation in Fungal Systematics Wallingford,UK: CAB International. pp. 67–78.

Birse CE, Irvin MY, Fonzi WA, and Sypherd PS (1993). Cloning and characterization of ECE1, a gene expressed in association with cell elongation of the dimorphic pathogen *Candida albicans*. Infect Immun 61:3648-3655.

Blanchette, M., Kunisawa, T. and Sankoff, D. (1996). Parametric genome rearrangement. Gene 172:GC11–GC17.

Boeke JD, Garfinkel DJ, Styles CA, and Fink GR (1985). Ty elements transpose through an RNA intermediate. Cell 40:491-500.

Boore JL, Collins TM, Stanton D, Daehler LL and Brown WM (1995). Deducing the pattern of arthropod phylogeny from mitochondrial DNA rearrangements. Nature (London) 376:163–165.

Boschman CR, Bodnar UR, Tornatore MA, Obias AA, Noskin GA, Englund K, Postelnick MA, Suriano T, and Peterson LR (1998). Thirteen-year evolution of azole resistance in yeast isolates and prevalence of resistant strains carried by cancer patients at a large medical center. Antimicrob. Agents Chemother. 42:734–738.

Bostock A, Khattak MN, Matthews R, and Burnie J (1993). Comparison of PCR fingerprinting, by random amplification of polymorphic DNA, with other molecular typing methods for *Candida albicans*. J Gen Microbiol 139:2179-2184.

Brandis, H and Otte HJ (1984). Lehrbuch der medizinischen Mikrobiologie (5 th Edition) Stuttgart, Germany, Gustav Fischer Verlag

Braun BR and Johnson AD (1997). Control of filament formation in *Candida albicans* by the transcriptional repressor TUP1. Science 277:105–109.

Braun BR and Johnson AD (2000). TUP1, CPH1 and EFG1 make independent contributions to filamentation in *Candida albicans*. Genetics 155:57-67.

Braun BR, Head WS, Wang MX, and Johnson AD (2000). Identification and characterization of TUP1-regulated genes in *Candida albicans*. Genetics 156:31-44.

Brown AJP (2002). Expression of growth form-specific factors during morphogenesis in *Candida albicans*. In: RA Calderone, ed. *Candida* and Candidiasis. Washington DC: ASM Press, pp 87-106.

Brown AJP and Gow NAR (1999). Regulatory networks controlling *Candida albicans* morphogenesis. Trends Microbiol 7(8):333-338.

Brown DH Jr, Giusani AD, Chen X, and Kumamoto C (1999). Filamentous growth of *Candida albicans* in response to physical environmental cues and its regulation by the unique CZF1 gene. Mol Microbiol 34:651-662.

Buffo J and Herman MA., Soll DR (1984). A characterization of pH-regulated dimorphism in *Candida albicans*. Mycopathologia 85:21-30.

Buurman ET, Westwater C, Hube B, Brown AJP, Odds FC, and Gow NAR (1998). Molecular analysis of CaMnt1p, a mannosyl transferase important for adhesion and virulence of *Candida albicans*. Proc Natl Acad Sci USA 95:7670-7675

Calderone RA and Braun PC (1991). Adherence and receptor relationships of *Candida albicans*. Microbiol Rev 55:1–20.

Cannon RD, Fischer FJ, Niimi K, Niimi M, and Arisawa M (1998). Drug pumping mechanisms in *Candida albicans*. Jpn. J. Med. Mycol. 39:73–78.

Caugant DA and Sandven P (1993). Epidemiological analysis of *Candida albicans* strains by multilocus enzyme electrophoresis. J Clin Microbiol 31:215–220.

Chaffin WL, Lopez-Ribot JL, Casanova M, Gozalbo D, and Martinez JP (1998). Cell wall and secreted proteins of *Candida albicans*: identification, function, and expression. Microbiol Mol Biol Rev 62:130–180.

Chibana H, Beckerman JL, and Magee PT (2000). Fine-resolution physical mapping of genomic diversity *in Candida albicans*. Genome Res 10:1865-1877.

Chibana H, Iwaguchi S, Homma M, Chindamporn A, Nakagawa Y, and Tanaka K (1994). Diversity of tandemly repetitive sequences due to short periodic repetitions in the chromosomes of *Candida albicans*. J Bacteriol 176:3851-3858.

Chibana H, Magee BB, Grindle S, Ran Y, Scherer S, and Magee PT (1998). A physical map of chromosome 7 of *Candida albicans*. Genetics 149:1739-1752.

Chindamporn A, Nakagawa Y, Mizuguchi I, Chibana H, Doi M, and Tanaka K (1998). Repetitive sequences (RPSs) in the chromosomes of *Candida albicans* are sandwiched between two novel stretches, HOK and RB2, common to each chromosome. Microbiol. 144: 849-857.

Chu S, DeRisi J, Eisen M, Mulholland J, Botstein D, Brown PO, and Herskowitz I (1998). The transcriptional program of sporulation in budding yeast. Science 282:699–705.

Chu W-S, Magee BB, and Magee PT (1993). Construction of an SfiI macrorestriction map of the *Candida albicans* genome. J Bacteriology 175:6637-6651.

Colina AR, Aumont F, Deslauriers N, Belhumeur P, and De Repentigny L (1996). Evidence for degradation of gastrointestinal mucin by *Candida albicans* secretory aspartyl proteinase. Infect Immun 64:4514-4519.

Collart FR and Huberman E (1987). Amplification of the IMP dehydrogenase gene in Chinese hamster cells resistant to mycophenolic acid. Mol Cell Biol 7:3328–3331.

Costa A, Costa C, Misefari A, and Amato A (1968). On the enzymatic activity of certain fungi. VII. Phosphatidase activity on media containing sheep's blood of pathogenic strains of *Candida albicans*. Atti Soc Sci Fis Mat Nat XIV:93–101.

Cowen LE, Kohn LM, and Anderson JB (2001). Divergence in Fitness and Evolution of Drug Resistance in Experimental Populations of *Candida albicans*. J Bacteriol, 183:2971-2978.

Crandall M and Edwards JE Jr (1987). Segregation of proteinase-negative mutants from heterozygous *Candida albicans*. J Gen Microbiol 133 (Pt 10):2817-2824.

Csank C, Schröppel K, Leberer E, Harcus D, Mohamed O, Meloche S, Thomas DY, and Whiteway M (1998). Roles of the *Candida albicans* mitogen-activated protein kinase homolog, Cek1p, in hyphal development and systemic candidiasis. Infect Immun 66:2713-2721.

Curcio MJ, Sanders NJ, and Gärfinkel DJ (1988). Transpositional competence and transcription of endogenous Ty elements in *Saccharomyces cerevisiae*: implications for regulation of transposition. Mol Cell Biol 8:3571-3581.

Cutfield JF, Sullivan PA, and Cutfield SM (2000). Minor structural consequences of alternative CUG codon usage (Ser for Leu) in *Candida albicans* exoglucanase. Protein Eng 13:735 - 738

Cutler JE (1991). Putative virulence factors of *Candida albicans*. Annu Rev Microbiol 45:187-218.

Dabrowa N, Taxer, SS. and Howard, DH (1976). Germination of *Candida albicans* induced by proline. Infect Immun 13(3):830-835.

Davis D, Wilson RB, and Mitchell AP (2000). RIM101-dependent and –independent pathways govern pH responses in *Candida albicans*. Mol Cell Biol 20:971-978.

De Backer MD, Ilyina T, Ma XJ, Vandoninck S, Luyten WH and Vanden Bossche H (2001). Genomic profiling of the response of *Candida albicans* to itraconazole treatment using a DNA microarray. Antimicrob. Agents Chemother. 45:1660–1670.

De Backer MD, Magee PT, and Pla J (2000). Recent developments in molecular genetics of *Candida albicans*. Annu Rev Microbiol 54:463-498.

De Bernardis F, Arancia S, Morelli L, Hube B, Sanglard D, Schäfer W, and Cassone A (1999). Evidence that members of the secretory aspartic proteinase gene family, in particular SAP2, are virulence factors for *Candida* vaginitis. J Infect Dis 179:201-208.

De Bernardis F, Cassone A, Sturtevant J, and Calderone R (1995). Expression of *Candida albicans* SAP1 and SAP2 in experimental vaginitis. Infect Immun 63:1887-1892.

De Bernardis F, Sullivan PA., and Cassone A (2001). Aspartyl proteinases of *Candida albicans* and their role in pathogenecity. Med Microbial 39:303-313.

De Hoogt R, Luyten WH, Contreras R, and De Backer MD (2000). PCR- and ligation-mediated synthesis of split-marker cassettes with long flanking homology regions for gene disruption in *Candida albicans*. Biotechniques 28(6):1112-1116.

De Repentigny L, Aumount F, Bernard K, and Belhumeur P (2000). Characterization of binding of *Candida albicans* to small intestinal mucin and its role in adherence to mucosal epithelial cells. Infect Immun 68:3172-3179.

De Viragh PA, Sanglard D, Togni G, Falchetto R, and Monod M (1993). Cloning and sequencing of two *Candida* parapsilosis genes encoding acid proteases. J Gen Microbiol 139:335-342.

Decottignies A, Grant AM, Nichols JW, de Wet H, McIntosh DB, and Goffeau A (1998). ATPase and multidrug transport activities of the overexpressed yeast ABC protein Yor1p. J Biol Chem 273(20):12612-12622.

DeRisi J, van den Hazel B, Marc P, Balzi E, Brown P, Jacq C, and Goffeau A (2000). Genome microarray analysis of transcriptional activation in multidrug resistance yeast mutants. FEBS Lett. 470:156–160.

Dexter D, Moye-Rowley W S, Wu AL, and Golin J (1994). Mutations in the Yeast PDR3, PDR4, PDR7 and PDR9 Pleiotropic (Multiple) Drug Resistance Loci Affect the Transcript Level of an ATP Binding Cassette Transporter Encoding Gene, PDR5. Genetics 136:505-515.

Diehl F, Beckmann B, Kellner N, Hauser NC, Diehl S, and Hoheisel JD (2002). Manufacturing DNA microarrays from unpurified PCR products. Nucleic Acids Research 30(16):e79.

Doi M, Homma M, Iwaguchi S, Horibe K, and Tanaka K (1994). Strain relatedness of *Candida albicans* strains isolated from children with leukemia and their bedside parents. J Clin Microbiol 32:2253-2259.

El Barkani A, Kurzai O, and Fonzi W A, Ramon A, Porta A, Frosch M, and Mühlschlegel FA (2000). Dominant active alleles of RIM101 (PRR2) bypass the pH restriction on filamentation of *Candida albicans*. Mol Cell Biol 20:4635-4647.

Ernst JF (2000). Transcription factors in *Candida albicans* – environmental control of morphogenesis. Microbiol 146:1763-1774.

Fairhead C, Llorente B, Denis F, Soler M, and Dujon B. (1996). New vectors for combinatorial deletions in yeast chromosomes and for gap-repair cloning using 'split-marker' recombination. Yeast 12(14):1439-1457

Felk A (2002). Secreted aspartat proteases and their importance as virulence factor PhD dissertation, FU University, Berlin.

Felk A, Kretschmar M, Albrecht A, Schaller M, Beinhauer S, Nichterlein T, Sanglard D, Korting HC, Schäfer W, and Hube B (2002). *Candida albicans* hyphal formation and the expression of the Efg1-regulated proteinases Sap4-6 are required for the invasion of parenchymal organs. Infect Immun 70:3689-3700.

Fling ME, Kopf J, Tamarkin A, Gorman JA, Smith HA, and Koltin Y (1991). Analysis of a *Candida albicans* gene that encodes a novel mechanism for resistance to benomyl and methotrexate. Mol Gen Genet. 227(2):318-329.

Fonzi WA (1999). PHR1 and PHR2 of *Candida albicans* encode putative glycosidases required for proper cross-linking of ß-1,3- and ß-1,6-glucans. J Bacteriol 181:7070-7079.

Fonzi WA and Irwin MY (1993). Isogenic strain construction and gene mapping in *Candida albicans*. Genetics 134:717-728.

Fox R, Neal KR, Leen CLS, Ellis ME, and Mandal BK (1991).Fluconazole resistant *Candida* in AIDS. J. Infect. 22:201–204.

Fridkin SK and Jarvis WR (1996). Epidemiology of nosocomial fungal infections. Clin Microbiol Rev 9:499-511.

Fu Y, Rieg G, Fonzi WA, Belander PH, Edwards JE, and Fillerr SG (1998). Expression of the *Candida albicans* gene ALS1 in *Saccharomyces cerevisiae* induces adherence to endothelial and epithelial cells. Infect Immun 66:1783–1786.

Garbino J, Kolarova L, Rohner P, Lew D, Pichna P, Pittet D (2002). Secular trends of candidemia over 12 years in adult patients at a tertiary care hospital. Medicine (Baltimore) 81(6):425-33.

Gasch AP, Spellman PT, Kao CM, Carmel HO, Eisen MB, Storz G, Botstein D and Brown PO (2000). Genomic expression programs in the response of yeast cells to environmental changes Mol. Biol. Cell 11:4241–4257.

Gaur NK and Klotz SA (1997). Expression, cloning, and characterization of a *Candida albicans* gene, ALA1, that confers adherence properties upon *Saccharomyces cerevisiae* for extracellular matrix proteins. Infect Immun 65:5289–5294.

Ghannoum MA (2000). Potential Role of Phospholipases in Virulence and Fungal Pathogenesis Clinical Microbiology Reviews 13:122–143.

Gillum AM, Tsay, EY, and Kirsch DR (1984). Isolation of the *Candida albicans* gene for orotidine-5'-phosphate decarboxylase by complementation of *S. cerevisiae* ura3 and *E. coli* pyrF mutations. Mol Gen Genet 198(1):179-182.

Gimeno CJ and Fink GR (1994). Induction of pseudohyphal growth by overexpression of PHD1, a Saccharomyces serevisiae gene related to transcriptional regulators of fungal development. Mol Cell Biol 14:2100-2112.

Glee PM, Sundstrom P, and Hazen KC (1995). Expression of surface hydrophobic proteins by *Candida albicans in vivo*. Infect Immun. 63(4):1373-1379.

Goffeau A, Aert R, Agostini-Carbone ML, Ahmed A, Aigle M, Alberghina L, Allen E, Alt-Moörbe J, Andre B, Andrews S, *et al.* (1997) The Yeast Genome Directory. Nature (London) 387:Suppl., 5–105.

Goldman RC, Frost DJ, Capobianco JO, Kadam S, Rasmussen RR, and Abad-Zapatero C (1995). Antifungal drug targets: *Candida* secreted aspartyl protease and fungal wall beta-glucan synthesis. Infect Agents Dis 4:228-247.

Goodwin TJD and Poulter RTM (1998). The CARE-2 and rel-2 repetitive elements of *Candida albicans* contain LTR fragments of a new retrotransposons. Gene 218:85-93.

Goodwin TJD and Poulter RTM (2000). Multiple LTR-retransposon families in the asexual yeast *Candida albicans*. Genome Res 10:174-191.

Gow NAR, Brown AJP, and Odds FC (2000). *Candida*'s arranged marriage. Science 289:256-257.

Gow NAR, Robbins PW, Lester JW, Brown AJP, Fonzi WA, Chapman T, and Kinsman OS (1994). A hyphal-specific chitin synthase gene (CHS2) is not essential for growth, dimorphism, or virulence of *Candida albicans.* Proc Natl Acad Sci USA 91:6216-6220.

Gräser Y, Volovsek M, Arrington J, Schonian G, Presber W, Mitchell TG, Vilgalys R (1996). Molecular markers reveal that population structure of the human pathogen *Candida albicans* exhibits both clonality and recombination. Proc Natl Acad Sci USA 93(22):12473-12477.

Hannenhalli S, Chappey C, Koonin EV, and Pevzner PA (1995). Genome sequence comparison and scenarios for gene rearrangements: a test case. Genomics 30:299–311.

Hartung K, Frishman D, Hinnen A, and Wolfl S (1998). Single-read sequence tags of a limited number of genomic DNA fragments provide an inexpensive tool for comparative genome analysis. Yeast 14:1327–1332.

Hedstrom L and Wang CC (1990). Mycophenolic acid and thiazole adenine dinucleotide inhibition of *Tritrichomonas foetus* inosine 59-monophosphate dehydrogenase: implications on enzyme mechanism. Biochemistry 29:849–854.

Higgins CF (1992). ABC transporters: from microorganisms to man. Annu. Rev. Cell Biol. 8:67–113.

Higgins CF (1995). The ABC of channel regulation. Cell 82:693–696.

Hilton C, Markie D, Corner B, Rikkerink E, and Poulter R (1985). Heat shock induces chromosome loss in the yeast *Candida albicans*. Mol Gen Genet 200:162-168.

Hinnen A, Hicks JB, and Fink GR (1978). Transformation of yeast. Proc Natl Acad Sci USA 75:1929-1933

Hitchcock CA, Russell NJ, Barrett-Bee KJ. (1987). Sterols in *Candida albicans* mutants resistant to polyene or azole antifungals, and of a double mutant *C. albicans* 6.4. Crit Rev Microbiol. (1):111-115.

Holton NJ, Goodwin TJD, Butler MI, and Poulter RTM (2001). An active retrotransposon in *Candida albicans*. Nucleic Acids Research 29:4014-4024.

Hoyer LL (2001). The ALS gene family of *Candida albicans*. Trends Microbiol 9(4):176-180.

Hoyer LL, and Hecht JE (2000). The ALS6 and ALS7 genes of *Candida albicans*. Yeast 16: 847–855.

Hoyer LL, and Hecht JE (2001). The ALS5 gene of *Candida albicans* and analysis of the Als5p N-terminal domain. Yeast 18:49–60.

Hoyer LL, Fundyga R, Hecht JE, Kapteyn JC, Klis FM, and Arnold J (2001). Characterization of Agglutinin-like Sequence Genes From Non-*albicans Candida* and Phylogenetic Analysis of the ALS Family. Genetics 157:1555 - 1567.

Hoyer LL, Payne TL, and Hecht JE (1998b) Identification of *Candida albicans* ALS2 and ALS4 and localization of ALS proteins to the fungal cell surface. J Bacteriol 180:5334–5343.

Hoyer LL, Payne TL, Bell M, Myers AM, and Scherer S (1998a). *Candida albicans* ALS3 and insights of the ALS gene family. Curr Genet 33:451-459.

Hoyer LL, Scherer S, Shatzman AR, and Livi GP (1995). *Candida albicans* ALS1: domains related to a *Saccharomyces cerevisiae* sexual agglutinin separated by a repeating motif. Mol Microbiol 15:39–54.

Hube B (1996). *Candida albicans* secreted aspartic proteinases. Curr Top Med Mycol 7:55-69.

Hube B, Monod M, Schofield DA, Brown AJP, and Gow NAR (1994). Expression of seven members of the gene family encoding secretory aspartyl proteinases in *Candida albicans*. Mol Microbiol 14:87-99.

Hube B and Naglik J (2001). *Candida albicans* proteinases: resolving the mystery of a gene family. Microbiology 147:1997 – 2005

Hube B and Naglik J (2002). Extracellular hydrolases. In: RA Calderone, ed. *Candida* and Candidiasis. Washington DC: ASM Press, pp 107-122.

Hube B, Sanglard D, Odds FC, Hess D, Monod M, Schäfer W, Brown AJP, and Gow NAR (1997). Gene disruption of each of the secretory aspartyl proteinase genes SAP1, SAP2 and SAP3 in *Candida albicans* attenuates virulence. Infect Immun 65:3529-3538.

Hube B, Stehr F, Bossenz M, Mazur A, Kretschmar M, Schäfer W (2000). Secreted lipases of *Candida albicans*: cloning, characterisation and expression analysis of a new gene family with at least ten members. Arch Microbiol 174(5):362-74.

Hube B, Turver CJ, Odds FC, Eiffert H, Boulnois GJ, Kochel H, and Ruchel R (1991). Sequence of the *Candida albicans* gene encoding the secretory aspartate proteinase. J Med Vet Mycol 29 (2):129-132

Hughes AL (1999). Adaptive Evolution of Genes and Genomes. New York: Oxford University Press. pp. 143-179.

Hull CM and Johnson A.D (1999). Identification of mating type-like locus in the asexual pathogenic yeast *Candida albicans*. Science 285:1271-1275.

Hull CM, Raisner RM, and Johnson AD (2000). Evidence for mating of the "asexual" yeast *Candida albicans* in a mammalian host. Science 289:307-309.

Ibrahim AS, Filler SG, Sanglard D, Edwards JE Jr, and Hube B (1998). Secreted aspartyl proteinases and interaction of *Candida albicans* with human endothelial cells. Infect Immun 66:3003-3005.

Ito H, Fukuda Y, Murata K, and Kimura A (1983). Transformation of intact yeast cells treated with alkali cations. J Bacteriol 153:163-168

Iwaguchi S, Homma M, and Tanaka K (1992 a). Clonal variation of chromosome size derived from the rDNA cluster region in *Candida albicans*. J Gen Microbiol 138:1177-1184.

Iwaguchi S, Homma M, and Tanaka K 1990. Variation in the electrophoretic karyotype analysed by the assignment of DNA probes in *Candida albicans*. J Gen Microbiol 136:2433-2442.

Iwaguchi S, Homma M, Chibana H, and Tanaka K (1992 b). Isolation and characterization of repeated sequence (RPS1) of *Candida albicans*. J Gen Microbiol 138:1893-1900.

Iwaguchi S-I, Kanbe T, Tohne T, Magee PT, and Suzuki T (2000). High-frequency occurrence of chromosome translocation in a mutant strain of *Candida albicans* by a suppressor mutation of ploidy shift. Yeast 16:411-422.

Jackson RC, Weber G, and Weber HG (1975). Inosine monophosphate dehydrogenase: an enzyme linked with proliferation and malignancy. Nature 256:331–333.

Jarbon G, Sherman F, and Rustchenko E (1998). Monosomy of a specific chromosome determines L-sorbose utilisation: a novel regulatory mechanism in *Candida albicans*. Proc Natl Acad Sci USA 95:5150-5155.

Jenkinson HF (1996). Ins and outs of antimicrobial resistance: era of the drug pumps. J. Dent. Res. 75:736–742.

Joseph-Horne T, and Hollomon DW (1997). Molecular mechanisms of azole resistance in fungi. FEMS Microbiol Lett 149:141–149.

Kamai Y, Kubota M, Kamai Y, Hosokawa T, Fukuoka T, and Filler SG (2002). Contribution of *Candida albicans* ALS1 to the pathogenesis of experimental oropharyngeal Candidiasis. Infect Immun 70:5256 – 5258

Kaminishi H, Hamatake H, Cho T, Tamaki T, Suenaga N, Fujii T, Hagihara Y, and Maeda H (1994). Activation of blood clotting factors by microbial proteinases. FEMS Microbiol Lett 121:327-332.

Keleher CA, Redd MJ, Schultz J, Carlson M, and Johnson AD (1992). Ssn6-Tup1 is a general repressor of transcription in yeast. Cell 68:709–719.

Kennedy MJ, Rogers AL, Hanselman LR, Soll DR, and Yancey RJ (1988). Variation in adhesion and cell surface hydrophobicity in *Candida albicans* white and opaque phenotypes. Mycopathologia 102:149-156.

Keogh RS, Seoighe C, and Wolfe KH (1998). Evolution of gene order and chromosome number in Saccharomyces, Kluyveromyces and related fungi. Yeast 14:443–457.

Kitchen VS, Savage M, and Harris JRW (1991). *Candida albicans* resistance in AIDS. J. Infect. 22:204–205.

Köhler GA, White TC, Agabian N (1997). Overexpression of a cloned IMP dehydrogenase gene of *Candida albicans* confers resistance to the specific inhibitor mycophenolic acid. J Bacteriol 179(7):2331-2338

Köhler JR and Fink GR (1996). *Candida albicans* strains heterozygous and homozygous in mitogen-activated protein kinase signalling components have defects in hyphal development. Proc Natl Acad Sci USA 93: 13223-13228.

Kolotila MP and Diamond RD (1990). Effects of neutrophils and in vitro oxidants on survival and phenotypic switching of *Candida albicans* WO-I. Infect Immun 58:1174-1179.

Kretschmar M, Felk A, Staib P, Schaller M, Hess D, Callapina M, Morschhäuser J, Schäfer W, Korting HC, Hof H, Hube B, and Nichterlein T (2002). Individual acid aspartic proteinases (Saps) 1-6 of *Candida albicans* are not essential for invasion and colonization of the gastrointestinal tract in mice. Microb Pathog 32:61-70.

Kretschmar M, Hube B, Bertsch T, Sanglard D, Merker R, Schröder M, Hof H, and Nichterlein T (1999). Germ tubes and proteinase activity contribute to virulence of *Candida albicans* in murine peritonitis. Infect Immun 67:6637-6642.

Lan C-Y, Newport G, Murillo LA, Jones T, Scherer S, Davis RW, and Agabian N (2002). Metabolic specialization associated with phenotypic switching in *Candida albicans*. Proc Natl Acad Sci USA 99:14907-14912.

Lane S, Zhou S, Pan T, Dai Q, and Liu H (2001 a). The basic helix-loop-helix transcription factor Cph2 regulates hyphal development in *Candida albicans* partly via Tec1. Mol Cell Biol 21:6418-6428.

Lane S, Birse C, Zhou S, Matson R, and Liu H (2001 b). DNA array studies demonstrate convergent regulation of virulence factors by Chp1, Cph2, and Efg1 in *Candida albicans*. J Biol Chem 276:48988-48996.

Lasker BA, Carle GF, Kobayashi GS, and Medoff GS (1989). Comparison of separation of *Candida albicans* chromosome-sized DNA by pulsed-field gel electrophoresis techniques. Nucl Acids Res 17:3783-3793.

Lasker BA, Page LS, Lott TJ, and Kobayashi GS (1992). Isolation, characterization, and sequencing of *Candida albicans* repetitive element 2. Gene 116:51-57.

Lasker BA, Page LS, Lott TJ, Kobayashi GS, and Medoff G (1991). Characterization of CARE-1: *Candida albicans* repetitive element-1. Gene 102:45-50.

Lay J, Henry LK, Clifford J, Koltin Y, Bulawa CE, and Becker JM (1998). Altered expression of selectable marker URA3 in gene-disrupted *Candida albicans* strains complicates interpretation of virulence studies. Infect Immun 66(11):5301-5306.

Leberer E, Harcus D, Broadbent ID, Clark KL, Dignard D, Ziegelbauer K, Schmidt A, Gow NAR, Brown AJP, and Thomas DY (1996). Signal transduction through homologs of the Ste20p and Ste7p protein kinases can trigger hyphal formation in the pathogenic fungus *Candida albicans*. Proc Natl Acad Sci USA 93:13217-13222.

Lee GC, Lee LC, Sava V, and Shaw JF (2002). Multiple mutagenesis of non-universal serine codons of the *Candida rugosa* LIP2 gene and biochemical characterization of purified recombinant LIP2 lipase overexpressed *in Pichia pastoris*. Biochem J 366(Pt 2):603-611

Lipke PN, Wojciechowicz D, and Kurjan J (1989). AGa1 is the structural gene for the *Saccharomyces cerevisiae* a-agglutinin, a cell surface glycoprotein involved in cell-cell interactions during mating. Mol Cell Biol 9:3155–3165.

Liu H, Köhler JR, and Fink GR (1994). Supression of hyphal formation in *Candida albicans* by mutation of a STE12 homolog. Science 266:1723-1725.

Lo HJ, Köhler JR, Didomenico B, Loebenberg D, Cacciapuoti A, and Fink GR (1997). Nonfilamentous *C. albicans* mutants are avirulent. Cell 90:939-949.

Lott TJ, Holloway BP, Logan DA, Fundyga R, and Arnold J (1999). Towards understanding the evolution of the human commensal yeast *Candida albicans*. Microbiol 145:1137-1143.

Louis EJ, Naumova ES, Lee A, Naumov G, and Haber JE (1994). The chomosome end in yeast: ist mosaic nature and influence on recombinational dynamics. Genetics 136:789-802.

Lussier M, Sdicu A-M, Shahinian S, and Bussey H (1998). The *Candida albicans* KRE9 gene is required for cell wall beta -1,6-glucan synthesis and is essential for growth on glucose. Proc Natl Acad Sci USA 95:9825-9830.

Madhani HD and Fink GR (1998). The control of filamentous differentiation and virulence in fungi. Trends Cell Biol 8(9):348-353.

Magee BB and Magee PT (1997). WO-2, a stable aneuploid derivative of *Candida albicans* strain WO-1, can switch from white to opaque and form hyphae. Microbiol 143:289-295.

Magee BB, Hube B, Wright RJ, Sullivan PJ, and Magee PT (1993). The genes encoding the secreted aspartyl proteinases of *Candida albicans* constitute a family with at least three members. Infect Immun 61:3240-3243.

Magee BB, Koltin Y, Gorman JA., and Magee PT (1988). Assignment of cloned genes to the seven electrophoretically separated *Candida albicans* chromosomes. Mol Cell Biol 8:4721-4726.

Magee PT and Chibana H (2002). The genome of *Candida albicans* and other *Candida* species. In: RA Calderone, ed. *Candida* and Candidiasis. Washington DC: ASM Press. pp 293-304.

Magee PT and Scherer S (1998). Genom mapping and gene discovery in *Candida albicans*. ASM News 64:505-511.

Magee PT, Bowdin L, and Staudinger J (1992). Comparison of molecular typing methods for *Candida albicans*. J Clin Microbiol 30:2674-2679.

Magee BB and Magee PT (2000). Induction of mating in *Candida albicans* by construction of MTLa and MTLα*Δ* strains. Science 289:310-312.

Matthews GD, Goodwin TJ, Butler MI, Berryman TA, and Poulter RT (1997). pCal, a highly unusual Tyl/copia retrotransposon from the pathogenic yeast *Candida albicans*. J Bacteriol 179:7118-7128.

McEachern MJ and Hicks JB (1993). Unusually large telomeric repeats in the yeast *Candida albicans*. Mol Cell Biol 13:551-560.

Miyasaki SH, White TC, and Agabian N (1994). A fourth secreted aspartyl proteinase gene (*SAP*4) and a CARE2 repetitive element are located upstream of the *SAP*1 gene in *Candida albicans*. J Bacteriol 176:1702-1710.

Miyazaki Y, Geber A, Miyazaki H, Falconer D, Parkinson T, Hitchcock C, Grimberg B, Nyswaner K, and Bennett JE (1999). Cloning, sequencing, expression and allelic sequence diversity of ERG3 (C-5 sterol desaturase gene) in *Candida albicans*. Gene 236(1):43-51.

Monod M, Hube B, Hess D, and Sanglard D (1998). Differential regulation of *SAP*8 and *SAP*9, which encode two new members of the secreted aspartic proteinase family in *Candida albicans*. Microbiol 144:2731-2737.

Monod M, Togni G, Hube B, and Sanglard D (1994). Multiplicity of genes encoding secreted aspartic proteinases in *Candida* species. Mol Microbiol 13:357-368.

Morschhäuser J (2002).The genetic basis of fluconazole resistance development *in Candida albicans*. Biochim Biophys Acta 1587(2-3): 240-248.

Morschhäuser J, Michel S, and Staib P (1999). Sequential gene disruption in *Candida albicans* by FLP-mediated site-specific recombination. Mol Microbiol32(3):547-556.

Morschhäuser J, Michel S, Hacker J (1998). Expression of a chromosomally integrated, single-copy GFP gene in *Candida albicans*, and its use as a reporter of gene regulation. Mol Gen Genet 257(4):412-420.

Moura G, Miranda I, Cheesman C, Tuite MF, and Santos MA (2002). Stop codon decoding in *Candida albicans*: from non-standard back to standard. Yeast 19(9):727-733.

Mühlschlegel FA and Fonzi WA (1997). PHR2 of *Candida albicans* encodes a functional homolog of the pH-regulated gene PHR1 with an inverted pattern of expression. Mol Cell Biol 17:5960-5967.

Müller E and Loeffler W (1982). Mykologie (4 th Edition) Stuttgart, Germany, Thieme Verlag.

Murad AM, d'Enfert C, Gaillardin C, Tournu H, Tekaia F, Talibi D, Marechal D, Marchais V, Cottin J, and Brown AJ (2001). Transcript profiling in *Candida albicans* reveals new cellular functions for the transcriptional repressors CaTup1, CaMig1 and CaNrg1. Mol Microbiol 42:981-993.

Nadeau JH and Taylor BA (1984). Lengths of chromosomal segments conserved since divergence of man and mouse. Proc Natl Acad Sci USA 81:814–818.

Naglik JR, Newport G, White TC, Fernandes-Naglik LL, Greenspan JS, Greenspan D, Sweet SP, Challacombe SJ, and Agabian N (1999). *In vivo* analysis of secreted aspartyl proteinase expression in human oral candidiasis. Infect Immun 67:2482-2490.

Nakamura K, Niimi M, Niimi K, Holmes AR, Yates JE, Decottignies A, Monk BC, Goffeau A, and Cannon RD (2001). Functional expression of *Candida albicans* drug efflux pump Cdr1p in a *Saccharomyces cerevisiae* strain deficient in membrane transporters. Antimicrob Agents Chemother 45(12):3366-3374.

Nantel A, Dignard D, Bachewich C, Harcus D, Marcil A, Bouin A-P, Sensen CW, Hogues H, van het Hoog M, Gordon P, Rigby T, Benoit F, Tessier DC, Thomas DY, and Whiteway M (2002). Transcription Profiling of *Candida albicans* Cells Undergoing the Yeast-to-Hyphal Transition. Mol Biol Cell 13:3452 - 3465.

Navarro-Garcia F, Perez-Diaz RM, Magee BB, Pla J, Nombela C, and Magee PT (1995). Chromosome reorganization in *Candida albicans* 1001 strain. J Met Vet Mycol 33:361-366.

Nikawa H, Nishimura H, Hamada T, Makihira S, and Samaranayake LP (1998). Relationship between thigmotropism and *Candida* biofilm formation *in vitro*. Mycopathologia 144(3):125-129.

Odds FC, Brown AJP, and Gow NAR (2000). Might *Candida albicans* be made to mate after all? Trends Microbiol. 8:4-6.

Odds FC (1988). *Candida* and Candidosis. 2nd ed. London: Bailliere Tindal.

Odds FC (1994 a). *Candida* species and virulence. ASM News 60:313-318.

Odds FC (1994 b). Pathogenesis of *Candida* infections. J Am Acad Dermatol 31:2-5.

Ohno S (1970). Evolution by gene duplication. London: Allen and Unwin.

Olaiya AF and Sogin SJ (1979). Ploidy determination of *Candida albicans*. J Bacteriol 140:1043-1049.

Osawa S (1995). Evolution of the Genetic Code. New York: Oxford University Press.

Ouellette M, Legare D, and Papadopoulou B (1994). Microbial multidrugresistance ABC transporters. Trends Microbiol. 2:407–411.

Ozier-Kalogeropoulos O, Malpertuy A, Boyer J, Tekaia F, and Dujon B (1998). Random exploration of the Kluyveromyces lactis genome and comparison with that of Saccharomyces cerevisiae. Nucleic Acids Res. 26:5511–5524.

Paterson AH, Lan TH, Reischmann KP, Chang C, Lin YR, Liu SC, Burow MD, Kowalski SP, Katsar CS, DelMonte TA, Feldmann KA, Schertz KF, Wendel JF (1996). Toward a unified genetic map of higher plants, transcending the monocot-dicot divergence. Nat Genet. 14(4):380-382.

Perea S and Patterson TF (2002). Antifungal resistance in pathogenic fungi. Clin Infect Dis 35(9):1073-1080.

Perepnikhatka V, Fischer FJ, Niimi M, Baker RA, Cannon RD, Wang Y-K, Sherman F, and Rustchenko E (1999). Specific chromosome alterations in fluconazole-resistant mutants of *Candida albicans*. J Bacteriol 181: 4041-4049.

Pesole G, Lotti M, Alberghina L, and Saccone C (1995). Evolutionary origin of nonuniversal CUGSer codon in some *Candida* species as inferred from a molecular phylogeny. Genetics 141:903–907.

Pontieri E, Gregori L, Gennarelli M, Ceddia T, Novelli G, Dallapiccola B, De Bernardis F, and Carruba G (1996). Correlation of SfiI macrorestriction endonuclease fingerprint analysis of *Candida* parapsilosis isolates with source of isolation. J Med Microbiol 45:173-178.

Powderly WG (1994). Resistant candidiasis. AIDS Res Hum Retroviruses 10:925-929.

Prasad R, De Wergifosse P, Goffeau A, Balzi E (1995). Molecular cloning and characterization of a novel gene of *Candida albicans*, CDR1, conferring multiple resistance to drugs and antifungals. Curr Genet. 27(4):320-329.

Pujol C, Joly S, Lockhart SR, Noel S, Tibayrenc M, and Soll DR (1997). Parity among the randomly amplified polymorphic DNA method, multilocus enzyme electrophoresis, and Southern blot hybridization with the moderately repetitive DNA probe Ca3 for fingerprinting *Candida albicans.* J Clin Microbiol 35:2348-2358.

Pujol C, Joly S, Nolan B, Srikantha T, and Soll DR (1999). Microevolutionary changes in *Candida albicans* identified by the complex Ca3 fingerprinting probe involve insertions and deletions of the full-length repetitive sequence RPS at specific genomic sites. Microbiol 145:2635-2646.

Pujol C, Reynes J, Renaud F, Mallie′ M, and Bastide J-M (1993). Analyse ge′ne′tique de souches de *Candida albicans* par e′lectrophore`se des isoenzymes. J Mycol Med 3:14–19.

Ramon AM, Porta A, and Fonzi WA (1999). Effect of environmental pH on morphological development of *Candida albicans* is mediated via the PacC-related transcription factor encoded by Prr2. J Bacteriol 181:7524-7530.

Ray TL and Payne CD (1990). Comparative production and rapid purification of *Candida* acid proteinase from protein-supplemented cultures. Infect Immun 58(2): 508-514.

Ray TL and Payne CD (1991). Degradation of basement membrane matrix proteins laminin, fibronectin, and type IV collagen by *Candida* acid proteinase. J Invest Dermatol 96:606A.

Raymond M, Dignard D, Alarco AM, Mainville N, Magee BB, and Thomas DY (1998). A Ste6p/P-glycoprotein homologue from the asexual yeast *Candida albicans* transports the a-factor mating pheromone in Saccharomyces cerevisiae. Mol. Microbiol. 27:587–598.

Riederer K, Fozo P, and Khatib R (1998). Typing of *Candida albicans* and *Candida* parapsilosis: species-related limitations of electrophoretic karyotyping and restriction endonuclease analysis of genomic DNA. Mycoses 41:397-402.

Riggle PJ, Andrutis KA, Chen X, Tzipori SR, and Kumamoto CA (1999). Invasive lesions containing filamentous forms produced by a *Candida albicans* mutant that is defective in filamentous growth in culture. Infect Immun 67:3649-3652.

Riggsby WS, Torrey-Banza IJ, Wills JW, and Townes TM (1982). DNA content, genetic complexity and the ploidy question in *Candida albicans*. Mol Cell Biol 2:853-862.

Rinaldi MG (1993). Biology and pathogenicity of *Candida* species. In: GP Bodey ed. Candidiasis: pathogenesis, diagnosis and treatment. New York. Raven Press. Ltd. pp. 1-20.

Rosa EAR, Pereira CV, Rosa RT, and Hofling JF (2000). Grouping oral *Candida* species by multilocus enzyme electrophoresis. Int J Syst Evol Microbiol 50:1343 – 1349.

Rüchel R (1981). Properties of a purified proteinase from the yeast *Candida albicans*. Biochem Biophys Acta 659:99-113.

Rüchel R (1983). On the renin-like activity of *Candida* proteinase and activation of blood coagulation in vitro. Zbl Bakt I Abt Orig A 255:368-379.

Rüchel R (1986). Cleavage of immunoglobulins by pathogenic yeast of the genus *Candida*. Microbiol Sci 3:316-319.

Rüchel R and Böning B (1983). Detection of *Candida* protease by enzyme immunoassay and interaction of the enzyme with α-a-macroglobulin. J Immunol Methods 61:107-116.

Ruhnke M, Eigler A, Tennagen I, Geiseler B, Engelmann E, and Trautmann M (1994). Emergence of fluconazole-resistant strains of *Candida albicans* in patients with recurrent oropharyngeal candidosis and human immunodeficiency virus infection. J. Clin. Microbiol. 32:2092–2098.

Rustchenko EP, Curran TM, and Sherman F (1993). Variations in the number of ribosomal DNA units in morphological mutants and normal strains of *Candida albicans* and in normal strains of Saccharomyces cerevisiae. J Bacteriol 175:7189-7199.

Rustchenko-Bulgac EP (1991). Variation of *Candida albicans* electrophoretic karyotypes. J Bacteriol 173:6586-6596.

Rustchenko-Bulgac EP, Howard DH, and Sherman F (1994). Chromosomal alterations of *Candida albicans* are associated with the gain and loss of assimilating functions. J Bacteriol 176:3221-3241.

Rustchenko-Bulgac EP, Sherman F, and Hicks JB (1990). Chromosomal rearrangements associated with morphological mutants provide a means for genetic variation of *Candida albicans*. J Bacteriol 172:1276-1283.

Sadhu C, McEachern MJ, Rustchenko-Bulgac EP, Schmid J, Soll DR, and Hicks JB (1991). Telomeric and dispersed repeat sequences in *Candida* yeasts and their use in strain identification. J Bacteriol 173:842-850.

Sadowski P (1986). Site-specific recombinases: changing partners and doing the twist. J Bacteriol 165(2):341-347.

San-Blas G, Travassos LR, Fries BC, Goldman DL, Casadevall A, Carmona AK, Barros TF, Puccia R, Hostetter MK, Shanks SG, Copping VMS, Knox Y, and Gow NAR (2000). Fungal morphogenesis and virulence. Med. Mycol. 38 (Suppl. I): 79-86.

Sanglard D, Ischer F, Monod M, and Bille J (1996). Susceptibilities of *Candida albicans* multidrug transporter mutants to various antifungal agents and other metabolic inhibitors. Antimicrob. Agents Chemother. 40:2300–2305.

Sanglard D, Ischer F, Monod M, and Bille J (1997). Cloning of *Candida albicans* genes conferring resistance to azole antifungal agents: characterization of CDR2, a new multidrug ABC transporter gene. Microbiology. 143 (Pt2):405-416.

Sanglard D, Kuchler K, Ischer F, Pagani JL, Monod M, and Bille J (1995). Mechanisms of resistance to azole antifungal agents in *Candida albicans* isolates from AIDS patients involve specific multidrug transporters. Antimicrob. Agents Chemother. 39:2378–2386.

Sankoff D (1993). Analytical approaches to genomic evolution. Biochimie 75:409–413.

Sankoff D, Leduc G, Antoine N, Paquin B, Lang BF, and Cedergren R (1992). Gene order comparisons for phylogenetic inference: evolution of the mitochondrial genome. Proc Natl Acad Sci USA 89:6575–6579.

SanMiguel P, Tikhonov A, Jin Y-K, Motchoulskaia N, Zakharov D, Melake-Berhan A, Springer PS, Edwards KJ, Lee M, Avramova Z, and Bennetzen JL (1996). Nested retrotransposons in the intergenic regions of the maize genome. Science 274:765-768.

Santos MA and Tuite MF (1995). The CUG codon is decoded in vivo as serine and not leucine in *Candida albicans.* Nucleic Acids Res 23:1481 - 1486

Santos MAS, Cheesman C, Costa V, Moradas-Ferreira P, and Tuite MF (1999). Selective advantages created by codon ambiguity allowed for the evolution of an alternative genetic code in *Candida spp*. Molecular Microbiology 31(3):937–947.

Santos MAS, Ueda T, Kimitsuna W, and Tuite MF (1997). The non-standard genetic code of *Candida spp*. an evolving genetic code or a novel mechanism for adaptation? Mol Microbiol 26: 423–431.

Sanyal K and Carbon J (2002). The CENP-A homolog CaCse4p in the pathogenic yeast *Candida albicans* is a centromere protein essential for chromosome transmission. . Proc Natl Acad Sci USA 99:12969-12974.

Saporito-Irwin SM, Birse CE, Sypherd PS, and Fonzi WA (1995). PHR1, a pH-regulated gene of *Candida albicans*, is required for morphogenesis. Mol Cell Biol 15:601-613.

Sarachek A, Rhoads DD, and Schwarzhoff RH (1981). Hybridization of *Candida albicans* through fusion of protoplasts. Arch Microbiol 129:1-8.

Schaller M, Korting HC, Schäfer W, Bastert J, Chen WC, Hube B (1999). Secreted aspartic proteinase (Sap) activity contributes to tissue damage in a model of human oral candidosis. Mol Microbiol 34: 169-180.

Scherer S and Stevens DA (1987). Application of DNA typing methods to epidemiology and taxonomy of *Candida* species. J Clin Microbiol 25:675-679.

Schwartz DC and Cantor CR (1984). Separation of yeast chromosome-sized DNAs by pulsed field gradient gel electrophoresis. Cell 37:67-75.

Seoighe S, Federspiel N, Jones T, Hansen N, Bivolarovic V, Surzycki R, Tamse R, Komp C, Huizar L, Davis RW, Scherer S, Tait E, Duncan J. Shaw DJ, Harrisi D, Murphyi L, Oliveri K, Taylori K, Rajandreami M-A, Barrelli BG, and Wolfe KH (2000). Prevalence of small inversions in yeast gene order evolution. Proc Natl Acad Sci USA 97:14433–14437

Shepherd MG, Yin CY, Ram SP, and Sullivan PA (1980). Germ tube induction in *Candida albicans.* Can J Microbiol 26:21-26.

Skrabanek L and Wolfe KH (1998). Eukaryote genome duplication — where's the evidence? Curr Opin Genet Devel 8: 694-700.

Slutsky B, Staebell M, Anderson J, Risen L, Pfaller M, and Soll DR (1987). „White-opaque-transition“: a second high–frequency switching system in *Candida albicans.* J Bacteriol 169:189-197.

Soll DR (2001). In: Calderone RA and Cihlar RL eds. Fungal Pathogenesis: Principles and Clinical Applications. New York: Marcel Dekker Vol. 14. pp. 161–182.

Soll DR (2002). Phenotypic Switching. In: RA Calderone, ed. *Candida* and Candidiasis. Washington DC: ASM Press. pp 123-142.

Soll DR, Langtimm CJ, McDowell J, Hicks J, Galask R (1987). High-frequency switching in *Candida* strains isolated from vaginitis patients. J Clin Microbiol 25(9):1611-1622.

Sonneborn A, Bockmühl D, and Ernst JF (1999 a). Chlamydospore formation in *Candida albicans* requires the Efg1p morphogenetic regulator. Infect Immun 67:5514-5517.

Sonneborn A, Tebarth B, and Ernst JF (1999 b). Control of white-opaque phenotypic switching in *Candida albicans* by the Efg1p morphogenetic regulator. Infect Immun 67:4655-4660.

Srikantha T and Soll DR (1993). A white-specific gene in the white-opaque switching system of *Candida albicans*. Gene 131:53-60.

Staab JF and Sundstrom P (1998). Genetic organization and sequence analysis of the hypha-specific cell wall protein gene HWP1 of *Candida albicans*. Yeast 14:681-686.

Staib P, Kretschmar M, Nichterlein T, Hof H, and Morschhäuser J (2000). Differential activation of a *Candida albicans* virulence gene family during infection. Proc Natl Acad Sci USA 97: 6102-6107.

Stoldt VR, Sonneborn A, Leuker C, and Ernst JF (1997). Efg1 an essential regulator of morphogenesis of the human pathogen *Candida albicans*, is a member of a conserved class of bHLH proteins regulating morphogenetic processes in fungi. EMBO 16:1982-1991.

Stone RL, Matarese V, Magee BB, Magee PT, and Bernlohr DA (1990). Cloning, sequencing and chromosomal assignment of a gene from Saccharomyces cerevisiae which is negatively regulated by glucose and positively by lipids. Gene 96: 171-176.

Sugita T and Nakase T (1999). Non-universal usage of the leucine CUG codon and the molecular phylogeny of the genus *Candida*. Syst Appl Microbiol 22(1):79-86.

Sullivan DJ and Coleman DC (2002). The genome of *Candida albicans* and other *Candida* species. In: RA Calderone, ed. *Candida* and Candidiasis. Washington DC: ASM Press, pp 427-441.

Sundstrom P (2002). Adhesion in *Candida spp.* Cell Microbiol 4 (8) 461-469.

Suzuki T, Kanbe T, Kuroiwa T, and Tanaka K (1986). Occurrence of ploidy shift in a strain of the imperfect yeast *Candida albicans*. J Gen Microbiol 132 (Pt 2):443-53.

Suzuki T, Kobayashi I, Kanbe T, and Tanaka K (1989). High frequency variation of colony morphology and chromosome reorganization in the pathogenic yeast *Candida albicans*. J Gen Microbiol 135:425-434.

Tang SJ, Shaw JF, Sun KH, Sun GH, Chang TY, Lin CK, Lo YC, and Lee GC (2001). Recombinant expression and characterization of the *Candida rugosa* lip4 lipase in *Pichia pastoris*: comparison of glycosylation, activity, and stability.Arch Biochem Biophys 387(1):93-98.

Tessier DC., Thomas DY, and Brousseau R (2001). In H-J Rehm and P Stadler, ed. Biotechnology., eds. Rehm, H.-J. and Stadler, P. (Wiley-VCH, Weinheim, Germany), Vol. 5b, Chapter 9. Weinheim, Germany: Wiley-VCH.

Thompson JR, Register E, Curotto J, Kurtz M, and Kelly R (1998). An improved protocol for the preparation of yeast cells for transformation by electroporation. Yeast 14:565-571.

Thrash-Bingham C, Gorman JA (1993). Identification, characterization and sequence of *Candida albicans* repetitive DNAs Rel-1 and Rel-2. Curr Genet 23(5-6):455-462.

Thrash-Birgham C and Gorman JA (1992). DNA translocations contribute to chromosome length polymorphisms in *Candida albicans*. Curr Genet 22:93-100.

Treger JM and McEntee K (1990). Structure of the DNA damage-inducible gene DDR48 and evidence for its role in mutagenesis in Saccharomyces cerevisiae. Mol Cell Biol 10:3174-3184.

Tzung K-W, Williams RM, Scherer S, Federspiel N, Jones T, Hansen N, Bivolarevic V, Huizar L, Komp C, Surzycki R, Tamse R, Davis RW, and Agabian N (2001). Genomic evidence for complete sexual cycle in *Candida albicans*. Proc Natl Acad Sci USA 98:3249-3253.

Van den Bossche H, Marichal P, and Odds FC. (1994b). Molecular mechanisms of drug resistance in fungi. Trends Microbiol 2:393-400.

Van den Bossche H, Warnock DW, Dupont B, Kerridge D, Sen GS, Improvisi L, Marichal P, Odds FC, Provost F, and Ronin O (1994a). Mechanisms and clinical impact of antifungal drug resistence. J Med Vet Mycol 32:189-202.

Viaene J, Tiels P, Logghe M, Dewaele S, Martinet W, and Contreras R (2000). MET15 as a visual selection marker for *Candida albicans*. Yeast 16(13):1205-1215.

Weising K, Atkinson R, and Gardner R (1995). Genomic fingerprinting by microsatellite-primed PCR: a critical evaluation. PCR Methods Applic 4:249-255.

Welsh J and McClelland M (1990). Fingerprinting genomes using PCR with arbitrary primers. Nucleic Acids Res 18:7213-7224.

Whelan WL, Magee PT (1981). Natural heterozygosity in *Candida albicans*. J Bacteriol 145(2):896-903.

Whelan WL, Partridge RM, and Magee PT (1980). Heterozygosity and segregation in *Candida albicans*. Mol Gen Genet 180:107-113.

White TC (1997). Increased mRNA levels of ERG16, CDR, and MDR1 correlate with increases in azole resistance in *Candida albicans* isolates from a patient infected with human immunodeficiency virus. Antimicrob. Agents Chemother. 41:1482–1487.

White TC, Marr KA, and Bowden RA (1998). Clinical, cellular, and molecular factors that contribute to antifungal drug resistance. Clin. Microbiol. Rev. 11:382–402.

White TC, Miyasaki SH, and Agabian N (1993). Three distinct secreted aspartyl proteinases in *Candida albicans*. J Bacteriol 175:6126-6133.

White TC and Agabian N (1995). *Candida albicans* secreted aspartyl proteinases: isoenzyme pattern is determined by cell type, and levels are determined by environmental factors. J Bacteriol 177:5215-5221.

Wilson RB, Davis D, Enloe BM, Mitchell AP (2000). A recyclable *Candida albicans* URA3 cassette for PCR product-directed gene disruptions.Yeast 16(1):65-70

Wirsching S, Michel S, and Morschhäuser J (2000). Targeted gene disruption in *Candida albicans* wild-type strains: the role of the MDR1 gene in fluconazole resistance of clinical *Candida albicans* isolates.Mol Microbiol 36(4):856-865.

Wirsching S, Moran GP, Sullivan DJ, Coleman DC, Morschhäuser J (2001). MDR1-mediated drug resistance in *Candida dubliniensis*. Antimicrob Agents Chemother 45(12):3416-3421.
Wright RJ, Carne A, Hieber AD, Lamont IL, Emerson GW, Sullivan PA (1992). A second gene for a secreted aspartate proteinase in *Candida albicans*. J Bacteriol 174:7848-7853.
Zou S, Ke N, Kim JM, and Voytas DF (1996). The *Saccharomyces* retrotransposon Ty5 integrates preferentially into regions of silent chromatin at the telomeres and mating loci. Genes Der 10:634-645.

Applied Mycology & Biotechnology
An International Series. Volume 4. Fungal Genomics

5

Molecular Genetics and Genomics of *Phytophthora*

Susan J. Assinder
School of Biological Sciences, University of Wales, Bangor, Gwynedd, LL57 2UW, Wales, UK.
(s.assinder@bangor.ac.uk).

Members of the genus *Phytophthora* are destructive plant pathogens, causing significant economic losses worldwide. Extensive background knowledge exists on the transmission biology of the genus, but it is only in the last decade that the mechanisms underlying the success of the *Phytophthoras* as plant pathogens have begun to be elucidated at the cellular and molecular level. Using molecular genetic tools, substantial progress has been made in identifying key molecules involved in the pathogen infection cycle. Genomic approaches are now opening up new routes to gene discovery and promise to have a significant impact upon our understanding of the biology and pathology of this important group of organisms. This article reviews the tools and resources available for molecular genetic and genomic studies in *Phytophthora* and discusses the insights that these studies are giving into the molecular basis of the plant-pathogen relationship.

1. INTRODUCTION

Members of the genus *Phytophthora* cause some of the most destructive plant diseases in the world. Concern about the economic impacts of these diseases, and their general recalcitrance to control, has led to a concerted effort over the last decade to develop tools with which to dissect the causal organisms at the molecular level. Technical advances, such as DNA transformation, use of reporter genes, and genetic manipulation using gene silencing have allowed significant progress towards identifying key genes involved in the infection process. More recent efforts in structural and functional genomics have opened up new routes to gene discovery and functional analysis. With the means at hand to generate low-cost high-throughput sequence information, the bioinformatic tools to interpret these data and an expanding array of functional assays, the opportunity now exists to elucidate the molecular mechanisms underlying the associations of *Phytophthora* species with their plant hosts. This article reviews the biology of the *Phytophthoras*, concentrating on those aspects of the life cycle that contribute to their success as pathogens. The progress made towards dissecting these processes at the molecular level using conventional cloning strategies is discussed, along with the recent impact of novel gene discovery tools based on the exploitation of genomic data.

2. THE BIOLOGY OF THE *PHYTOPHTHORA*

2.1 Evolutionary Position

The oomycete genus *Phytophthora* was historically included on morphological and physiological grounds within the Kingdom Fungi. However, modern molecular (Förster *et al.* 1990; Cooke *et al.* 2000) and biochemical (Pfyffer *et al.* 1990) analyses suggest that the oomycetes share little taxonomic affinity with fungi. Sequence analysis of 18S ribosomal RNA and of conserved genes such as actin and tubulin has revealed that the oomycetes, which also include the closely related genus *Pythium* and the biotrophic downy mildews, are in fact more closely related to diatoms and brown algae. Therefore, they are now classified as Stramenopiles, a Kingdom that includes heterokont algae and diatoms (Sogin and Silberman 1998). It has been suggested (Tyler 2001) that the features common to both oomycetes and fungi, such as filamentous growth and heterotrophy, arose through convergent evolution driven by the requirements of a pathogenic lifestyle.

2.2 *Phytophthoras* as Agents of Disease

Most of the sixty or so known species within the genus *Phytophthora* are highly destructive and economically significant plant pathogens, damaging to both important crop species and natural ecosystems. Virtually all dicotyledonous plants are affected by one or more species of *Phytophthora* (Erwin *et al.* 1983; Erwin and Ribiero 1996). The distinct physiology of the *Phytophthoras* has been a major factor in the success of these organisms as pathogens, since many chemicals effective against fungi are ineffectual against *Phytophthora* infections. *Phytophthora* species also display an unusually high degree of genetic flexibility and adapt rapidly both to chemical control measures (Gisi and Cohen 1996) and to genetic resistance bred into plant hosts (Fry 1982).

Probably the best known oomycete pathogen is *Phytophthora infestans*, the causal agent of late blight of potato and tomato, which causes more losses than any other member of the genus. *P. infestans* notably destroyed the Irish potato crop in 1845 and 1846. The resulting famine led to the death and displacement of millions of people and spurred immigration from Ireland to North America. The threat has continued to the present day, with a new wave of late blight epidemics in the 1990s due to the emergence of highly aggressive and fungicide-insensitive isolates in North America and Europe (Fry and Goodwin 1997 a,b; Schiermeier 2001; Smart and Fry 2001). Although chemicals targeting the pathogen can provide some level of control, late blight is still a damaging disease, particularly since the crops infected by *P. infestans* are grown in every state in the US. Crop losses and control measures are estimated to cost several billion dollars annually worldwide (Duncan 1999; Schiermeier 2001).

Other significant pathogens are *Phytophthora sojae*, responsible for soybean root rot, *Phytophthora palmivora* and *Phytophthora megakarya*, which cause black pod of cocoa and represent a significant threat to the chocolate industry and *Phytophthora cinnamomi*, which can infect nearly 2000 different plant species. The latest addition to the list is *Phytophthora ramorum*, the causal agent of sudden oak death syndrome. This recently described species was previously known only in Germany and the Netherlands, where it had been observed on Rhododendron and Viburnum species. The disease has reached epidemic proportions in forests along approximately 300 km of the central coast of California (Rizzo *et al.* 2002). There are fears that it might be spreading to other regions in North America and also expanding its host range to include other trees, such as redwoods (Knight 2002).

One species, *Phytophthora brassicae* (formerly *Phytophthora porri*) infects the model plant *Arabidopsis thaliana* (Roetschi *et al.* 2001). A typical facultative biotrophic interaction is observed, strongly reminiscent of that seen in agronomically important diseases caused by *Phytophthora* species. This pathosystem thus has significant potential as a model for elucidating the molecular mechanisms underlying the interactions of the pathogenic *Phytophthoras* with their hosts.

2.3 The *Phytophthora* Life Cycle

In general, *Phytophthora* species spread through an infection cycle involving the production of sexual and/or asexual spores, dispersal of spores, and infection of a new host. The new infection serves as a source of spores to re-commence the cycle, and transport of infected host tissue can introduce *Phytophthora* diseases to new geographic areas. The majority of *Phytophthora* species are root pathogens but a few, including *P. infestans* and *P. ramorum*, have evolved to attack the aerial parts of the plant.

For most *Phytophthora* species, contact with a potential host plant is made by motile biflagellate zoospores (Duniway 1983). Zoospores are able to swim actively in water surfaces on plant tissues and in the soil, enabling the pathogen to select its infection site. Some species (e.g. *P. infestans*) produce wind-dispersed sporangia, which can either germinate directly or release zoospores by cytoplasmic cleavage when they land on a host plant. Under laboratory conditions, sporangia of *P. infestans* will release zoospores when incubated below 12°C, but germinate by hyphal outgrowth at higher temperatures. While both sporangia and zoospores can act as inocula for plant infection, it is thought that the latter are responsible for rapid infections because of their greater numbers and more efficient germination. Thus late blight epidemics spread fastest when the weather is cool, since zoospores are then the predominant infectious particles.

When the zoospore reaches a host leaf or root surface, it settles, encysts and germinates (Figure 1). The tip of the germ tube swells to form an appressorium or appressorium-like structure that facilitates adhesion to and penetration of the plant surface. In root-infecting species, the germ tube can also penetrate directly without the aid of an appressorium. Deriving nutrients from the host, the hyphae spread through the plant tissue. For species employing air-borne dispersal, sporangiophores develop on the underside of the leaves of the dying plant and these release sporangia to effect aerial spread of the pathogen and re-entry into the disease cycle. Soilborne *Phytophthora* species form sporangia on the root surface, which release zoospores by cytokinesis. The asexual cycle is rapid and can be repeated several times during the lifetime of a plant.

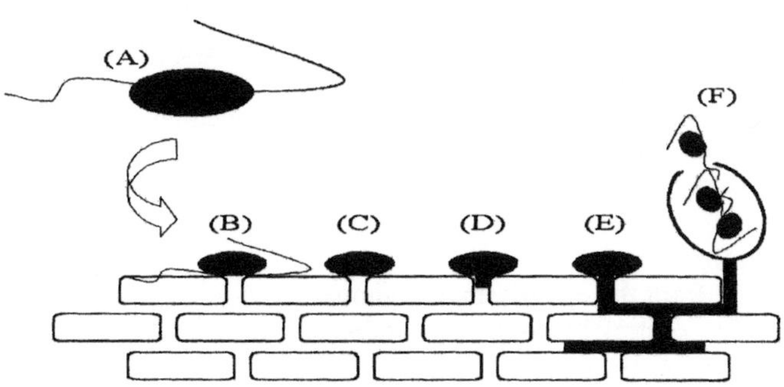

Fig. 1. Generalised schematic diagram showing the *Phytophthora* infection cycle. (A) The biflagellate motile zoospore is attracted to the host plant surface through chemotactic and electrotactic factors. (B) The zoospore settles with the ventral surface against the plant surface. (C) The zoospore encysts, with loss of the flagella and secretion of adhesive. (D) The cyst germinates and (E) the germ tube penetrates into the host plant. (F) Multinucleate sporangia develop on the plant surface and cleave to form uninucleate zoospores, which can re-initiate the infection cycle.

The sexual cycle also plays an important role in the biology of the *Phytophthoras*, as it contributes to genetic variation and produces thick-walled oospores that are adapted for over-wintering and survival under adverse environmental conditions. About half the species of *Phytophthora* (e.g. *P. sojae*) are homothallic and readily form oospores after colonisation of plant tissue, whilst the others (e.g. *P. infestans*, *P. palmivora*) are heterothallic and require the presence of opposite mating types, known as A1 and A2, for sexual reproduction. The development of male and female gametangia (antheridia and oogonia respectively) is hormonally stimulated within a mating zone where asexual growth and sporulation are inhibited. Fertilisation involves emptying of some of the contents of the antheridium into the oogonium and fusion of the haploid nuclei. Oospores germinate under suitable conditions to produce single or multiple germ tubes, which can then form sporangia, allowing the pathogen to re-enter the asexual cycle.

3. MOLECULAR BASIS OF THE *PHYTOPHTHORA* INFECTION CYCLE

The infection cycle comprises all the processes that are necessary for the pathogen to establish and maintain a successful infection. A *Phytophthora* zoospore is essentially a pre-programmed cell requiring only the appropriate molecular signals to progress through a complex sequence of events culminating in infection. In the 30-40 minutes between a zoospore first encountering a host plant and an infection being established, it undergoes remarkable changes in organisation to transform from a wall-less motile cell into a walled cyst, which then gives rise to a new hyphal apex. It is becoming clear that several features contribute to the rapid spread of these infective propagules and successful disease establishment (reviewed in Hardham 2001).

3.1 Formation of Zoospores

Cold shock induces cytokinesis in the multinucleate sporangium and leads to compartmentalisation of single nuclei within each zoospore. Divalent cations, in particular Ca^{2+}, have been shown to play an important role in this process. Jackson and Hardham (1996) demonstrated that two increases in Ca^{2+} concentration always occurred in sporangia of *P. cinnamomi* that underwent cytokinesis in response to cold shock. There was a rapid and transient rise within the first minute of cold shock, followed by a gradual and more prolonged rise accompanying cell division. Judelson and Roberts (2002) further demonstrated that the calcium channel blocker verapamil and the calmodulin antagonist trifluoroperazine inhibit zoosporogenesis and encystment in *P. infestans*.

Since zoospores are wall-less cells whose outer surface is the plasma membrane, osmoregulation is a key issue during their formation. The cytoplasm of developing *Phytophthora nicotianae* zoospores contains a high concentration of proline, which is believed to counterbalance the hyper-osmotic conditions in the sporangial lumen (Ambikapathy *et al.* 2002). Proline is expelled rapidly from the zoospores upon their release from the sporangium to prevent osmotic rupture of the zoospore membrane. Once released, zoospores maintain their water balance by means of a contractile water expulsion vacuole, which takes up water from the cytoplasm and pumps it out of the cell. The water expulsion vacuole consists of a central bladder, surrounded by a vesicular and tubular spongiome (reviewed in Patterson 1980). Water expulsion is believed to be powered by proton-pumping ATPases on the spongiome membranes, as evidenced by the fact that treatment of *P. nicotianae* zoospores with the ATPase inhibitor potassium nitrate slowed the pulse rate of the vacuole and led to premature encystment (Mitchell and Hardham 1999).

Recent work has given some insights into the molecules and pathways involved in zoosporogenesis. A survey of expression patterns of putative protein kinase genes identified one for which mRNA accumulation was first detected under conditions that induce sporangial cleavage (Judelson and Roberts 2002). The transcript persisted in motile zoospores and in germinated cysts but was not detected in other tissues. Consistent with the calcium-dependency of zoosporogenesis, the structure of the predicted protein resembled that of Ca^{2+}- and calmodulin-regulated serine/threonine protein kinases, but with an unusually short calmodulin-binding region.

Zoospore formation has also been shown to involve signalling via a G-protein pathway. Using a gene-silencing approach (Section 5.3), Latijnhouwers *et al.* (2003) constructed mutants of *P. infestans* that were deficient in the *Pigpa1* gene, which encodes a G-protein α subunit. Zoospore release was reduced to an average of 66% of wildtype level and there was a marked increase in the number of unusually large zoospores due to incomplete cleavage of the sporangial cytoplasm. These observations suggest that *Pigpa1* plays a role in the cytoplasmic cleavage that precedes zoospore formation.

Zoospores of *Phytophthora* species contain several characteristic types of peripheral vesicles. The largest of these have been proposed to act as nutrient stores and in *P. cinnamomi* have been shown to contain three immunologically related high molecular weight novel proteins, designated LPVs (Marshall *et al.* 2001). Northern blot analysis revealed that hyphae which had been induced to form sporangia contained three large transcripts corresponding to three *Lpv* genes, whereas these transcripts were absent from uninduced hyphae. The transcripts accumulated in a co-ordinate fashion four to six hours after induction, suggesting a role in sporangia formation.

3.2 Zoospore Motility

Phytophthora zoospores swim in a corkscrew fashion, changing direction when they meet obstacles or by spontaneous random turns. Most of the directional thrust comes from the anterior flagellum, with the posterior flagellum serving mainly as a rudder (reviewed by Carlile 1983). The swimming pattern is strongly influenced by calcium (Irving and Grant 1984; Reid *et al.* 1995). Further to its role in zoosporogenesis, G-protein signalling also appears to be involved in regulating swimming behaviour. Latijnhouwers *et al.* (2003) showed that PiGPA1-deficient mutants of *P. infestans* swam abnormally with more frequent turning and thus covered less distance than wildtype zoospores. These observations suggest that a functional G-protein pathway is normally needed to suppress turning. The mutants also failed to autoaggregate, a phenomenon normally observed in zoospore suspensions at high concentrations and presumed to be important in the life cycle as a means of intensifying the pressure at sites of infection (Ko and Chase 1973; Reid *et al.* 1995).

The mechanism by which zoospores are attracted to their hosts is not fully elucidated but may involve several factors. Zoospores exhibit directional swimming responses with respect to chemical, nutrient, ionic or electrical gradients. Using capillary tube assays, a number of common components of root exudates have been shown to be chemo-attractants, including sugars, amino acids, alcohols and phenolic compounds (e.g. Carlile 1983; Cahill and Hardham 1994; reviewed in van West *et al.* 2002; Tyler 2002). Some examples of host-specific attractants have been reported; for example, *P. sojae*, a host-specialised pathogen of soybean roots, is attracted to the soybean flavonoids daidzein and genestein (Morris and Ward 1992). Chemotaxis towards glutamic and aspartic acids is impaired in PiGPA1-deficient mutants of *P. infestans*, implicating PiGPA1 in gradient sensing ((Latijnhouwers *et al.* 2003).

Although numerous chemotactic responses have been demonstrated for *Phytophthora* zoospores, questions have been raised about their significance in the infection process due to the problem of distinguishing *bone fide* chemotaxis from zoospore immobilisation and/or autoaggregation in capillary model systems (van West *et al.* 2003). In addition to releasing chemical exudates, plant roots generate external electric currents and fields due to the flow of protons in and out of growing and wounded regions. Electrotaxis has been shown to also be an important guidance cue for *Phytophthora* zoospores in close proximity to a root (Morris *et al.* 1992; Morris and Gow 1993), and van West *et al.* (2003) propose that the endogenous electrical field profile of a root is the dominant predictor of zoospore accumulation sites.

3.3 Encystment and Adhesion

Upon encountering a host plant, zoospores rapidly form walled cysts that germinate and infect the host tissue. This involves detachment of the flagella, secretion of an adhesive material and formation of a cellulosic cell wall. The spore typically orientates, settles and adheres with the ventral surface, bearing the two flagella, towards the host surface. Little is known about the molecular signalling processes that achieve zoospore orientation but it is clearly advantageous for subsequent infection steps.

Within the first two minutes of encystment, a proteinaceous adhesive is released from small vesicles underlying the ventral surface (Hardham and Gubler 1990). Cyst coat glycoprotein is released by exocytosis of small dorsal vesicles and a microfibrillar cyst wall is synthesised beneath the coat. Firm adhesion to the host surface is a key part of the infection cycle, since it prevents the zoospore from being dislodged and allows it to exert physical force in order to penetrate the plant surface (Hardham 2001). The exact molecular nature of the adhesive has yet to be elucidated, but it is known from antibody labelling experiments to have a molecular mass of around 200 kDa (Hardham 2001) and to require Ca^{2+} for its function (Gubler *et al.* 1989). Gaulin *et al.* (2002) showed using gene silencing that the CBEL cellulose-binding glycoprotein of *Phytophthora parasitica* is essential for adhesion to cellulosic substrates (Section 8.1).

Encystment of *Phytophthora* zoospores can be triggered by mechanical agitation or by a variety of chemical treatments (Byrt *et al.* 1982; Grant *et al.* 1985). Calcium-modulating drugs induce encystment in several *Phytophthora* species (Irving and Grant 1984; Reid *et al.* 1995) and the soybean isoflavone, daidzein, was shown to trigger a calcium influx leading to encystment of zoospores of *P. sojae* (Connolly *et al.* 1999). Recent work has begun to elucidate the molecular signalling processes involved in the regulation of encystment. It was established some years ago that phosphatidic acid (PA) is a potent inducer of encystment in *P. palmivora* (Zhang *et al.* 1992). Building on this observation, Latijnhouwers *et al.* (2002) showed that encystment of *P. infestans* zoospores is triggered not only by PA but also by the G-protein activators mastoparan, and n- and sec-butanol. The mastoporan-induced increase was shown to be due to stimulation of phospholipase-D leading to the accumulation of PA. Thus the regulation of encystment in *P. infestans* involves a mastoporan- and butanol-inducible phospholipase-D pathway.

To identify other molecules that could promote encystment, Bishop-Hurley *et al.* 2002 used a phage-display protocol to select bioactive peptides that could bind to the surface molecules of *Phytophthora capsici* zoospores *in vitro*. The selected peptides contained an abundance of polar amino acids and proline, but were otherwise not conserved. Around half of the phage species selected for their ability to bind to the zoospore surface also induced encystment, and these effects could be reproduced by the corresponding peptides in free solution. This approach has the potential for exploitation in the development of novel plant disease resistance strategies.

3.4 Penetration and Colonization

Phytophthora cysts germinate within 20-30 minutes of zoospore encystment. The point of emergence of the germ tube is pre-determined and occurs at what was the centre of the ventral surface of the zoospore (Hardham and Gubler 1990). As is the case for much of the infection process, the germination of zoospore cysts is dependent on the availability of external calcium. Germination can be prevented by the application of substances that interfere with calcium-mediated processes, including calcium channel blockers, calcium ionophores, calcium chelators, calmodulin inhibitors and compounds such as TMB-8 that block the release of calcium from intracellular stores. The infection sequence from zoospore encystment through to cyst germination and host penetration in *P. parasitica* has been shown to involve sequential calcium fluxes across the cell membrane (Warburton and Deacon 1998). There is a rapid net influx of Ca^{2+} coincident with encystment, which is followed by a larger efflux associated with germination, consistent with release of Ca^{2+} from intracellular stores.

In root-infecting *Phytophthora* species, the germ tube may penetrate intercellularly along the anticlinal walls between the epidermal cells immediately below the spore. An alternative mode of infection is the formation of an appressorium-like structure that enables direct cell penetration through the periclinal wall (Hardham 2001). In leaf-infecting species, the formation of appressoria is the more usual mode of penetration. Appressoria in *Phytophthora* species are generally smaller than those produced by fungi and lack a cross-wall separating the appressorial swelling from the germ tube. As in fungi, ridges and grooves on the host surface have been shown to induce appressorium formation in both *P. infestans* (Bircher and Hohl 1997) and *P. sojae* (Morris *et al.* 1998).

Much work remains to elucidate the molecular processes involved in the germination of cysts and the formation of appressoria in *Phytophthora*. Using a subtractive hybridisation approach, Gornhardt *et al.* (2000) cloned three *P. infestans* genes that were up-regulated in germinating cysts shortly before onset of invasion of the host tissue. The genes were clustered and belonged to a small polymorphic gene family. The deduced proteins were distinguished by the presence of an internal octapeptide repeat rich in acidic amino acids and were designated as *Car* (for cyst-germination-specific acidic repeat) proteins. *Car* genes are transiently expressed during cyst germination and formation of appressoria. The proteins exhibit homology to mammalian mucins and are localised on the surface of the pre-infection structures. It is proposed that Car proteins are components of the mucous layer that protects the germling from desiccation, physical damage, or the adverse effects of the plant defence response and that they may also assist in adhesion to the leaf surface.

Several studies have demonstrated that G-protein signalling plays an important role in fungal pathogenicity (reviewed in Lengeler *et al.* 2000). Recent work has shown that the same is true of *P. infestans* (Laxalt *et al.* 2002). The genes encoding the α (*Pigpa1*) and β (*Pigpb1*) G protein subunits were found to be differentially expressed during asexual development of *P. infestans*, with the highest mRNA levels being observed in sporangia. Both genes were expressed in cysts and germinating cysts, but no *Pigpa1* mRNA was detected in mycelium. As described in the previous sections, silencing of the *Pigpa1* gene led to a variety of abnormalities in the production and behaviour of zoospores. The mutants were also defective in later stages of infection, producing fewer appressoria and many of these were deformed with only small tip swellings. Similarly, silencing of *Pigpb1* was shown to cause defective sporulation and aberrant vegetative growth (Latijnhouwers and Govers 2003).

4. UNDERSTANDING THE PLANT-PATHOGEN INTERACTION

Interaction between a *Phytophthora* pathogen and a plant can result in successful infection with the development of disease symptoms (compatible reaction) or the inhibition of infection through a mounting of a plant defence response (incompatible reaction). As detailed above, several developmental processes are required for a *Phytophthora* pathogen to invade a plant host. However, the pathogenicity of an organism does not rely solely on its ability to form infection structures. Pathogenicity determinants also include elicitors of plant defence responses, molecules involved in overcoming physical and chemical barriers to infection (e.g. cell wall degrading enzymes), and molecules that are toxic or which can detoxify antimicrobial substances produced by the plant.

4.1 Elicitors of Plant Defence Responses

Race-specific resistance is characterised by interactions between the products of dominant resistant (*R*) gene alleles in the host and corresponding avirulence (*Avr*) gene alleles in the pathogen, the so called 'gene-for-gene' hypothesis. The plant defence response is a form of localised programmed cell death known as the hypersensitive response (HR) which prevents further spread of the pathogen. This has been best studied in the reaction between *P. infestans* and the potato, where at least eleven *R* genes have been identified (Wastie 1991). Specificity towards a particular *R* gene rests with a single dominant *Avr* gene for most interactions (Al-Kherb *et al.* 1995).

A major effort has been underway for several years aimed at cloning *Phytophthora Avr* genes. The rationale behind these efforts is that this should give some insight into both how *Phytophthora* pathogens attack plants and also how plants defend themselves against the pathogens. The construction of detailed genetic linkage maps has pinpointed several *Avr* genes for positional cloning in *P. infestans* and *P. sojae* (Section 6.1).

A number of other *Phytophthora* genes have been described whose products are able to elicit plant defence responses. Many of these are members of the elicitin family. Elicitins are cysteine-rich extracellular proteins secreted by many *Phytophthora* species, which were originally identified by their ability to evoke the HR when infiltrated into tobacco leaves. Most *Phytophthoras* infect only a limited number of plant species, implying that non-host plants possess a mechanism by which they can overcome infection. The major elicitin in *P. infestans* is encoded by the *inf1* gene (Kamoun *et al.* 1997 a). This is highly expressed in cultured mycelium, but not in sporangia, zoospores or cysts. It is down-regulated in the early stages of infection, but is highly expressed in the later stages when sporulation and necrosis occur (Kamoun *et al.* 1997 b). Convincing evidence for the involvement of elicitins in this defence response was obtained using a gene silencing strategy to inhibit production of INF1 (Section 8.2). INF1-silenced strains of *P. infestans* induced disease lesions when inoculated on the wild tobacco species *Nicotiana benthamiana*, suggesting that INF1 functions as an avirulence factor in this interaction (Kamoun *et al.* 1998 b). Further support for this view comes from the analysis of isolates of *P. parasitica* that are naturally able to infect tobacco to cause black shank disease. Colas *et al.* (2001) showed that these isolates evade tobacco recognition by either not producing elicitins or by down-regulating expression *in planta* of the main elicitin-encoding gene *parA1*.

Although the functions of most elicitins remain to be determined, it has been proposed that the elicitin domain functions in binding a variety of lipid or lipid-like molecules (reviewed in Blein *et al.* 2002). Two out of eight genes shown to be up-regulated during sexual development in *P. infestans* encoded elicitins (Fabritius *et al.* 2002). Qutob *et al.* (2003) searched databases of *P. sojae* ESTs for transcripts encoding proteins with elicitin-like domains. They concluded that the gene family is large and diverse, comprising at least nine genes with varying patterns of expression and HR-inducing activity.

Other molecules able to elicit plant defence responses are now coming to light. A recent example is the identification in *P. parasitica* of a 24 kDa cell wall protein dubbed NPP1 (for necrosis-inducing *Phytophthora* protein 1) which triggers plant defense responses in parsley and *Arabidopsis* (Fellbrich *et al.* 2002). Infiltration of NPP1 into leaves of *Arabidopsis* plants resulted in HR-like cell death. Qutob *et al.* (2002) identified a *P. sojae* orthologue of NPP1 by screening proteins encoded by ESTs for secretory peptide leader sequences (Section 6.3) and testing selected candidates using a heterologous expression assay. Orthologues of NPP1/PsojNIP are known to occur in other *Phytophthora* species, including *P. infestans*, and

Phytophthora medicaginis, but not in plants. It is suggested that these proteins facilitate the colonisation of host tissues during the necrotrophic phase of growth.

A recent study in *P. infestans* using a DNA sequence data mining approach (Section 6.3) identified two novel necrosis-inducing genes, *crn1* and *crn2*, which encode extracellular proteins belonging to a large and complex family (Torto *et al.* 2003). Both genes are expressed constitutively during colonisation of tomato by *P. infestans* and *crn2* was shown to induce tomato defense response genes. The *crn* genes showed different responses compared to *inf1* and *PsojNIP* in heterologous functional assays in *N. benthamiana* (Section 8.2). Necrosis was delayed for the *crn* genes and the necrotic lesions differed in appearance, possibly due to the induction of secondary metabolite pathways. These observations suggest that the CRN proteins induce different defence pathways in plants than those induced by INF1 and PsojNIP.

4.2 Cell-wall Degrading Enzymes

Progress is being made in characterising the cell-wall degrading enzymes involved in entry into the host plant. Intercellular penetration requires the advancing germ tube to degrade the pectin component of the middle lamella. Endopolygalacturonases (endoPGs) are pectinases that have been implicated in the invasion of plant tissue by many pathogenic microbes. A gene (*Pipg1*) encoding a fungal-like extracellular endoPG has been identified in *P. infestans* (Torto *et al.* 2002). The *Pipg1* gene was expressed during pre-infection and infection stages, suggesting that the enzyme plays a role in the initial stages of infection. Germlings of *P. cinnamomi* secrete polygalacturonases within an hour of spore encystment on the root surface (Gotesson *et al.* 2002). The *P. cinnamomi* genome has been shown to contain a large polygalacturonase gene family of 19 members, organised into three clusters. The encoded enzymes fall into subgroups according to the extent of N-glycosylation, isoelectric point and N- and C-terminal structure. Three putative exo-1,3-β-glucanase genes, one endo-1,3-β-glucanase gene and one endo-1,3:1,4-β-glucanase gene have been cloned and characterised from *P. infestans* (McLeod *et al.* 2003). Brunner *et al.* (2002) cloned a *P. infestans* gene (*bgx1*) encoding a beta-glucosidase/xylosidase; the *bgx1* gene was present in various *Phytophthora* species, but is apparently absent in species of the related genus, *Pythium*.

Plants in turn secrete enzymes directed against the pathogen cell wall. As a counter-measure, *P. sojae* secretes a group of Glucanase Inhibiting Proteins (GIPs). GIP1 specifically inhibits EgaseA, the soybean endo-β-1,3-glucanase that is required for release of oligoglucoside elicitors from the pathogen's cell wall. It thus acts as a novel counterdefensive weapon that suppresses the plant defense response by slowing down early recognition of the host (Rose *et al.* 2002).

4.3 Detoxifying Compounds and Toxins

Plants synthesise antimicrobial secondary metabolites such as phytoalexins to inhibit pathogen proliferation, and many fungi produce enzymes that detoxify these compounds (reviewed in Morrissey and Osbourn 1999). Although this is an area that has not been explored extensively in *Phytophthora* species, it is not unreasonable to expect that similar mechanisms will be present. Plants also produce reactive oxygen species (ROS) in response to infection, which are involved in limiting pathogen growth. A number of genes encoding ROS-inactivating enzymes have been cloned from fungi (Mayer *et al.* 2001) and corresponding sequences have been identified in *Phytophthora* EST databases, suggesting that they may also defend themselves in this way (Kamoun *et al.* 1999 b).

4.4 Towards a Global Understanding of the Plant-Pathogen Interaction

Studies of the *Phytophthora* infection cycle to date have largely concentrated on identifying individual genes or elucidating the involvement of specific molecular pathways. We are now entering an era where new technologies will allow us to take a more global perspective of the whole infection process.

The basic tenet of this strategy is that genes with a key role in either host defence or pathogenicity would be expected to be up-regulated during the pathogen-host interaction. As an early validation of the approach, Pieterse *et al.* (1994) isolated two *in planta*-induced (*ipi*) genes by differential screening of a *P. infestans* genomic library with cDNA prepared from the pathogen growing both on potato leaves and on artificial medium. Both genes were expressed at high levels in the early phases of the pathogenic interaction, suggesting that their gene products have a function in the early stages of the infection process. Subsequent experiments by van West *et al.* (1998) addressed the question of whether *ipiO* was expressed during various developmental stages of the pathogen. The *ipiO* mRNA was detected in zoospores, cysts, germinating cysts, and young mycelia, but not in sporangia or in old mycelia grown *in vitro*.

Extending this idea in more general terms, Birch and Kamoun (2002) coined the term 'interaction transcriptome' to describe the sum of transcripts, from both host and pathogen, which are produced during their association. One strategy for characterising the interaction transcriptome is to construct cDNA libraries from a variety of developmental and infection stages of the pathogen and to then identify Expressed Sequence Tags (ESTs) by high-throughput sequencing. An alternative approach is to use amplified fragment length polymorphism-based mRNA fingerprinting (c-DNA-AFLP). This method involves restriction digestion of cDNAs with two enzymes, one recognising a 6bp site and the other a 4bp site. Following ligation of adaptors, subsets of cDNA populations are PCR-amplified and the products compared on polyacrylamide gels. Bands derived from up-regulated cDNAs can be excised, re-amplified and sequenced. This approach has been used recently for expression profiling in *P. infestans* and identified 64 transcripts up-regulated in germinating zoospore cysts but not in vegetative mycelium (Avrora *et al.* 2003). Real-time RT-PCR (Wang and Brown 1999) was then used to quantify expression of a selection of cyst transcripts, the first time that this method has been applied to a plant-pathogen interaction.

Another key technology in this context is proteomics, which combines high-resolution 2-D gel electrophoresis with mass spectrometry of tryptic digestion products to give a fast and effective method for identifying stage-specific proteins (Blackstock and Weir 1999; Pandey and Mann 2000). As the previous sections have described, *Phytophthora* pathogens produce several different cell types prior to and during infection. Since these various cell types can be isolated independent of the host plant, we have the opportunity to elucidate the molecular processes that underlie differentiation through the various stages of the *Phytophthora* life cycle. Mitchell *et al.* (2002) characterised *P. nicotianae* zoospore and cyst membrane proteins using one- and two-dimensional gel electrophoresis. Their comparison showed that at least 53 proteins were specific to, or occurred preferentially in, one or other spore type. Similarly, Shepherd *et al.* (2003) used a proteomics-based approach to identify newly synthesised proteins from germinated cysts and appressoria of *P. infestans*. They showed that approximately 1% of proteins are specific for each of the mycelial, sporangial, zoospore, cyst and germinated cyst stages of the life cycle. Profiles of *P. palmivora* and *P. infestans* contained equivalent numbers of protein spots, with 30% of them having similar or identical positions in 2-D gels, suggesting that *P. infestans* proteome databases could be a useful predictor of proteins in P. *palmivora*. Although proteomics approaches are most effective when they can be used in conjunction with genome sequence databases, these studies demonstrated that this may not be an absolute requirement for *Phytophthora* species. Hence, the current *P. infestans* and *P. sojae* genome sequencing projects (Section 6.2) could facilitate proteomic studies not only in these two species, but also in other *Phytophthoras* that are unlikely to be sequenced for some years.

5. TOOLS FOR MOLECULAR GENETICS

Despite their early identification as significant pathogens and their acknowledged continuing importance, *Phytophthora* species have somewhat lagged behind model higher fungi in the development of molecular tools for genetic analysis. The conclusion that the oomycetes have an independent evolutionary history has awoken researchers to the realisation that our understanding of their biology and pathology will not necessarily be advanced by studies of ascomycetes and basidiomycetes. This has led to development in recent years of an expanding toolbox for the molecular analysis of the *Phytophthoras*.

5.1 Transformation Systems

Transformation systems have been described for several *Phytophthora* species using a variety of technologies (reviewed in Judelson 1997). Strategies used in filamentous fungi proved to be non-transferable to oomycetes and successful transformation required the development of novel vectors and experimental protocols. Promoters from the oomycete downy-mildew *Bremia lactucae* showed strong activity in transient transformation assays of *P. infestans* protoplasts and were subsequently incorporated into transformation vectors (Judelson and Michelmore 1991). The first reported transformation was of *P. infestans* by Judelson *et al.* (1991), using liposome-PEG mediated transformation of protoplasts generated from germinated sporangia using the enzyme Novozym 234, followed by regeneration and antibiotic selection on agar medium. However, Novozym 234 is no longer marketed and the protoplast approach has largely been superseded by alternative techniques. Cvitanich and Judelson (2003) used microprojectile bombardment to transform sporangia, zoospores and mycelia of *P. infestans* to G418-resistance. The numbers of transformants were similar to the maximum rates obtained using protoplasting enzymes. The method was also highly reproducible (compared to the use of Novozym 234 which gave notoriously variable results between enzyme batches) and suitable for high-throughput experiments. Another recent advance has been the development for *P. infestans* of an *Agrobacterium tumefaciens* mediated transformation system (Vijn and Govers 2003). Between ten and thirty transformants were obtained per 10^7 zoospores, most of which contained a single T-DNA copy randomly integrated at a chromosomal locus. Electroporation of zoospores has also been reported (Tyler 2001).

5.2 Reporter Genes

Reporter genes have proven to be versatile tools to address a variety of biological questions in a range of *Phytophthora* species. Using oomycete promoter and terminator sequences, a plant-adapted green fluorescent protein (GFP) has been used as a vital reporter in *P. parasitica* (Bottin *et al.* 1999) and *P. palmivora* (van West *et al.* 1999a). The gene has been useful both as a quantitative reporter of gene expression and as a vital marker allowing the study of development of *Phytophthora in vitro* and in the host plant. The β-glucuronidase (GUS) reporter gene has also been used for studies in several *Phytophthora* species. In *P. infestans,* it has been used to analyse sexual preference during mating interactions (Judelson 1997), to quantify late blight resistance of potato (Kamoun *et al.* 1998 a) and to determine the location and time of expression *in planta* of the *ipiO* gene using a transcriptional fusion with the *ipiO* promoter (van West *et al.* 1998). van West *et al.* (1999a) compared the use of GUS and GFP for molecular and physiological studies of *P. palmivora* development. It was concluded that GFP was better visualised in high magnification microscopic studies, whereas

GUS was superior for macroscopic analyses.

5.3 Reverse Genetics

Reverse genetics in *Phytophthora* species has until fairly recently been hampered by the diploid nature of the organisms and difficulties in transformation and regeneration. Conventional gene disruption has not been successful due to a low frequency of homologous recombination of knockout constructs. An exciting and promising alternative technology is post-transcriptional gene silencing (van West *et al.* 1999 b). This phenomenon was discovered first for the *inf1* gene, which codes for the most abundant secreted elicitor protein in culture filtrates (Kamoun *et al.* 1993). Transformation of *P. infestans* with antisense, sense, and promoter-less constructs of *inf1* led to stable silencing of the endogenous gene in up to 20% of the transformants. Silencing was accompanied by the complete absence of *inf1* mRNA and INF1 protein and proved stable over repeated vegetative culture of the pathogen both *in vitro* and *in planta.* Heterokaryons obtained by somatic fusion of an *inf1*-silenced transgenic strain and a wild type strain were also stably silent. The *inf1* gene also remained silenced in non-transgenic homokaryotic single zoospore isolates derived from the silenced heterokaryons, showing that the presence of nuclear transgenic sequences is not essential to maintain silencing. These observations rule out DNA:DNA interactions as the basis for the phenomenon and suggest instead the involvement of a diffusible *trans*-acting silencing factor. The exact mechanism of silencing remains to be elucidated. Nuclear run-on assays have shown that *inf1* gene silencing is regulated at the transcriptional level, but DNA hypermethylation, often associated with transcriptionally regulated silencing, was not detected in the transgenic or endogenous *inf1* genes (van West *et al.* 1999 b).

The internuclear transfer of gene silencing is of particular interest from the perspective of functional genomics, since the induction of silencing in a limited number of nuclei in a coenocytic hypha appears to be sufficient to silence the whole hypha. This approach has now been exploited in functional analyses in a number of *Phytophthora* species (Section 8.1).

6 STRUCTURAL GENOMICS OF *PHYTOPHTHORA*

Genome sequence information is the key to gene discovery and the first step towards learning the function of each gene in an organism. In recent years, large-scale DNA sequencing approaches have accelerated the genetic characterisation of *Phytophthora* species. The primary goal is to identify novel virulence and avirulence genes, thereby elucidating the molecular basis of pathogenicity and host specificity and revealing putative targets for chemical control (reviewed in Birch and Whisson 2001).

6.1 Map-Based Cloning

Detailed maps have been constructed for both *P. infestans* and *P. sojae*. The first *P. infestans* map was based on the analysis of AFLP markers in F1 progeny from a cross between two diploid isolates (van der Lee *et al.* 1997). The map covered 827 cM, and was composed of 10 major and 7 minor linkage groups containing 183 AFLP markers, 7 RFLP markers and the mating type locus. Van der Lee *et al.* (2001) used bulked segregant analysis to select AFLP markers linked to *Avr* genes. These were used to construct two high-density linkage maps, one containing *Avr4* (located on linkage group A2-a) and the other containing a cluster of three tightly linked genes, *Avr3*, *Avr10*, and *Avr11* (on linkage group VIII). The data also allowed positioning of Avr1 on linkage group IV and Avr2 on linkage group VI.

The first *P. sojae* map was constructed using over one hundred F2 individuals from two crosses between different races using one race as a common parent in both crosses (Whisson *et al.* 1995). The map located 22 RFLP markers, 228 RAPD markers, and 7 avirulence genes into 10 major and 12 minor linkage groups over a total distance of 830.5 cM. May *et al.*

(2002) extended this analysis using two new crosses and produced a revised map comprising 35 RFLP, 236 RAPD, and 105 AFLP markers, as well as ten avirulence genes. The map is composed of 21 major and seven minor linkage groups and covers a total map distance of 1640.4 cM.

To date, seven avirulence genes have been placed on the *P. infestans* map and ten on the *P. sojae* map (reviewed in Tyler 2001). Two bacterial artificial chromosome (BAC) libraries have been generated from *P. infestans* to enable map-based cloning of genes and studies of genome structure and organisation. Randall and Judelson (1999) constructed a library with fourfold coverage of the genome and an average insert size of 75 kb. They also reported the transformation of entire BAC clones into *P. infestans*, thereby making it possible to determine the presence of a gene within a specific BAC clone. Whisson *et al.* (2001) subsequently made a larger BAC library with 10-fold genome coverage and an average insert size of 98 kb. They then used a three-dimensional pooling strategy to screen the library for AFLP markers. The pools used to construct a contig of 11 BAC clones in a region of the genome containing the cluster of three avirulence genes on linkage group VIII. The BAC contig is predicted to include *Avr11* but additional mapping is required to determine whether or not it also encompasses *Avr3* and *Avr10*. Using similar approaches, *Avr1a* of *P. sojae* has been localised to a 114 kb region (MacGregor *et al.* 2002) and Tyler (2001) reports the identification of a BAC contig spanning the *P. sojae Avr1b*/*Avr1k* locus.

6.2 Sequencing Projects

Pilot cDNA sequencing projects were carried out for both *P. infestans* and *P. sojae* within the framework of the *Phytophthora* Genome Initiative (PGI) (Waugh *et al.* 2000). These resulted in 2000-3000 expressed sequence tags (ESTs) for each species, which are compiled in searchable databases at the website of the National Centre for Genome Resources (http://www.ncgr.org/pgc). Kamoun *et al.* (1999) analysed 760 unique EST sequences obtained by random sequencing of clones from a cDNA library constructed from *P. infestans* mycelial mRNA. Over 60% of the clones showed matches to sequences in public databases and could be ascribed a putative cellular role. Similarly, Qutob *et al.* (2000) classified over 2000 cDNA transcripts from *P. sojae*. The transcripts were derived from three cDNA libraries using mRNA isolated from axenically grown mycelium and zoospores and from infected plant tissue. *P. infestans* and *P. sojae* transcribed sequences are GC-rich (58%) compared to plants (42%). On this basis, Qutob *et al.* (2000) estimated that two-thirds of the ESTs from the infected plant library were derived from *P. sojae* transcripts.

Large-scale sequencing projects are now in progress for both species in the expectation that understanding the genetic make-up of these economically important organisms will lead to novel approaches for disease management. Efforts are being co-ordinated and promoted under the auspices of an expanded and improved version of the PGI, known as the Phytophthora Genome Consortium. Approaches include the sequencing of ESTs from various developmental and infection stages, together with sequencing of selected contigs of BAC clones. The US Department of Agriculture is funding the sequencing of a further 41 000 *P. sojae* ESTs and 14 000 *P. infestans* ESTs. An additional 35 000 *P. infestans* ESTs developed by an international consortium funded by Syngenta are expected to become publicly available in the near future (Lam 2001).

The major goal of a complete *Phytophthora* genome sequence is now within reach. Although *P. infestans* is arguably the most important member of the genus on economic grounds, it unfortunately has an unusually large genome for an oomycete (250 Mb), over four times the size of that of *P. sojae* (62 Mb). Hence the latter is the favoured candidate to be the first fully sequenced *Phytophthora*. A complete sequence of *P. sojae* will greatly aid research into other *Phytophthora* species through comparative genomics. In expectation of funding becoming available, efforts are already in progress to assemble the *P. sojae* BAC library into contigs in preparation for genome sequencing.

6.3 Data Mining

Once sequence data has been generated, identifying candidate genes for functional assays presents a whole new set of challenges. Kamoun *et al.* (2002) proposed a series of specific criteria for genes likely to be involved in virulence or avirulence. These include genes encoding degradative enzymes since these are likely to be involved in host penetration (Section 4.2), those that are up-regulated during preinfection and infection stages (Section 4.4), those that encode extracellular proteins and those that are conserved among several pathogenic oomycetes. It is suggested that such genes represent potential targets for fungicides and should thus be prioritised as candidates for functional assays.

Torto *et al.* (2003) have developed a strategy to identify putative extracellular proteins in *P. infestans*, the rationale being that molecules which either promote infection or trigger defense responses are typically secreted. Using an algorithm (PexFinder) that can predict the presence of signal peptides and their cleavage sites, they screened over 2000 ESTs in the *P. infestans* database. They identified 142 non-redundant Pex (*Phytophthora* extracellular protein) cDNAs, of which 78 had no homologues in public databases. They selected 63 cDNAs for functional expression in plants using a viral vector system (Section 8.2). Qutob *et al.* (2002) used the same approach in *P. sojae*. A PexFinder analysis of over 3000 ESTs originating from mycelium, zoospore, and infected soybean tissues yielded 176 putative secreted proteins. Sixteen clones were selected for expression analysis, one of which was found to encode PsojNIP as a powerful inducer of necrosis and cell death (Section 4.1).

Using a similar rationale, Bos *et al.* (2003) report the validation and preliminary application of an approach that combines data mining with comparative genomics as a rapid alternative to positional cloning of *Avr* genes. Their strategy is based on the logic that *Avr* genes defining cultivar-specific resistance would be expected to exhibit significant intraspecific variation in DNA sequence. Candidate *Avr* genes selected using the criteria explained above (Kamoun *et al.* 2002) are amplified from a panel of characterised *Phytophthora* races. Polymorphic genes are identified and the extent of linkage disequilibrium determined between the polymorphic marker and the avirulence phenotype. The genes are then tested in functional assays. The advantage of the approach is that it has the potential to identify 'orphan' *Avr* genes that do not correspond to known R genes.

These exciting approaches of combining computational tools for data mining of ESTs with a high throughput functional assay have great potential for advancing functional genomic studies in *Phytophthora*.

7. *PHYTOPHTHORA* GENOME ORGANISATION

Although genome sequencing in the *Phytophthoras* is in its infancy, some common themes are already emerging regarding genome organisation and complexity. We are also now in a position to start identifying the control regions of individual genes and thereby draw conclusions about the regulation of gene expression.

7.1 Genome Size

The *Phytophthoras* exhibit a wide range of genome sizes. Cytophotometric determination of the nuclear DNA content of around 40 isolates of *P. infestans* estimated the genome size in this species to be approximately 237 Mb (Tooley and Therrien 1987). In contrast, the size of the *P. sojae* genome was estimated by nuclear Feulgen staining and image analysis to be 62 to 98 Mb (Voglmayr and Greilhuber 1998). Microscopic estimates place the haploid chromosome counts of the two species at 8-10 (Sansome and Brasier 1973) and 10-13 (Sansome and Brasier 1974) respectively.

Aneuploidy is common in the *Phytophthoras*. Whittaker *et al.* (1991) showed that *P. infestans* isolates from selected countries exhibited a range of DNA contents from a basic 2C level (presumed diploid) through intermediate quantities to 4C (presumed tetraploid). Similarly, DNA content measurements of 35 isolates collected in Ireland revealed the presence of putative diploid, triploid, and tetraploid individuals in the population (Tooley *et al.* 1993). Polyploids can also be selected *in vitro* on double-drug media due to aberrant sexual development and zoospore-mediated heterokaryosis (Judelson and Yang 1998).

7.2 Repetitive Sequences

Repetitive sequences are common in *Phytophthora* genomes. Judelson and Randall (1998) identified 33 distinct families of both tandemly repeated and dispersed repeated sequence in *P. infestans* following screening of genomic libraries. Together these comprised around 51% of the genome, with copy numbers ranging from 70 to 8,400. Some of the elements were found only in *P. infestans* and the closely related species *Phytophthora mirabilis,* but others also occurred in more distantly related *Phytophthora* species. Reverse transcriptase motifs were detected in seven of the repeat families that were widely distributed throughout the genus. Judelson and Tooley (2000) used primers based on the sequences of two of the species-specific repeats to develop a PCR-based method for detecting and quantifying *P. infestans* in plants. This highly sensitive assay has a detection limit of 10 fg of *P. infestans* DNA (equivalent to 0.02 nuclei), which is around 100 times more sensitive than methods based on amplification of internal transcribed spacer (ITS) of ribosomal DNA.

In *P. sojae*, repetitive sequences are estimated to constitute around 50% of the genome. Mao and Tyler (1996) identified five families of tandemly repeated sequences following genomic subtraction of chromosomal DNA from different isolates of the pathogen. These sequences varied in copy number between isolates. Five of the repetitive sequences were tandemly repeated and were localised on single chromosomes, whereas the sixth corresponded to the ribosomal RNA genes. These data suggest that gene amplification could contribute to the generation of genetic diversity in *P. sojae*. Panabieres and Le Berre (1999) identified a family of repetitive sequences in the genome of *P. cryptogea*. The elements ranged in size from 100 to 1200 bp and were deduced to represent 3' terminal fragments of a larger element. The sequences of the elements were highly conserved and most copies were flanked by terminal direct repeats, suggesting that their distribution within the genome involves insertion events. Hybridisation studies with 20 other *Phytophthora* species detected a few copies of the element in *P. cinnamomi* but not in other members of the genus.

Various lines of evidence suggest that transposable elements have been a major force in the shaping of *Phytophthora* genomes. Tooley and Garfinkel (1996) identified multiple copies of Ty1-copia group retrotransposon sequences in *P. infestans* by PCR amplification using degenerate primers for reverse transcriptase coding sequences. Some elements contained complete open reading frames while others were truncated, indicating the presence of potentially active as well as inactive elements. Liou *et al.* (2002) cloned a partial sequence (designated G2Ty-1) of the reverse transcriptase gene of a Ty1-copia retrotransposon from *P. parasitica.* It was demonstrated that G2Ty-1 existed as a moderately repetitive sequence in the genome of *P. parasitica,* where it exhibits a high degree of mitotic stability. Judelson (2002) showed by PCR using specific and degenerate primers that Gypsy retroelements are widely distributed throughout the genus *Phytophthora.* Out of 37 species tested, 29 showed the presence of the elements, with the abundance ranging from 10 to 10,000 copies per genome. However, each of 12 family members sequenced contained defects that would inactivate its protein product. DNA transposable elements of the *mariner* class have also been identified in EST databases of *P. infestans* (Kamoun *et al.* 1999 b) and *P. sojae* (Qutob *et al.* 2000). It is possible that transposition events underlie at least some of the phenotypic instability observed in many *Phytophthora* species both in the field and in culture (Caten and Jinks 1968; Judelson 2002).

Some *Phytophthora* repeat sequences are highly polymorphic and have been invaluable as multilocus probes in population and phylogenetic studies. For example, RG57 is a 1.2 kb genomic fragment of *P. infestans* that hybridises to 25 or more genomic *Eco*RI fragments (Goodwin *et al.* 1992). It has been used extensively to investigate the genetic diversity of *P. infestans* populations in North America (Forbes *et al.* 1998) and Europe (Sujkowski *et al.* 1994; Lebreton & Andrivon 1998; Zwankhuizen *et al.* 2000; Purvis *et al.* 2001). Similarly, DNA fingerprint analysis using G2Ty-1 (Liou *et al.* 2002) showed that this retrotransposon could be used as a genetic marker to study phylogeny in *P. parasitica*. Banding patterns of G2Ty-1 were highly polymorphic and strains displayed host-specific banding patterns.

7.3 Coding Sequences

The genome sequencing carried out so far in *Phytophthora* species is suggesting a tendency for functional genes to be located in high-density gene islands dispersed among clusters of repetitive sequences. In a 200 kb *P. sojae* BAC contig spanning two avirulence genes, the average spacing of genes was less than 300 base pairs, with three pairs of overlapping genes. Likewise, Qutob *et al.* (2002) reported that seven genes, including three genes encoding necrosis-inducing proteins, were tightly clustered in a 10.8 kb region of the *P. sojae* genome, separated by an average of less than 0.5 kb. The clustering of avirulence genes suggests that, as in many pathogens of plants and animals, genes involved in infection might be clustered in pathogenicity islands (Tyler 2001).

A similar high gene density was described for the two *in planta*-induced (*ipi*) genes of *P. infestans* identified by Pieterse *et al.* (1994). The *ipiB* gene belongs to a small gene family consisting of at least three genes, *ipiB1*, *ipiB2* and *ipiB3*, which are clustered in a head-to-tail arrangement. There are also two very similar *ipiO* genes, *ipiO1* and *ipiO2*, which are closely linked and arranged in an inverted orientation.

Early studies aimed at developing transformation systems for *Phytophthora* (Section 5.1) were frustrated by the fact that promoters from non-oomycete species do not function in *Phytophthora* (Judelson and Michelmore 1991; Judelson *et al.* 1992). Although typical eukarytotic consensus sequences such as TATA-boxes are uncommon, a conserved 16 bp sequence (GCTCATTYYNCAWTTT) has been detected in seven out of eight oomycetous genes for which the transcription start point has been determined, suggesting that oomycetes have a sequence preference for transcription initiation (Pieterse *et al.* 1994). Specialised databases of sequences and functional elements of 5' and 3' untranslated regions of eukaryotic mRNAs have shown that the untranslated regions of oomycete genes are amongst the shortest recorded for eukaryotes (Pesole *et al.* 2000). An analysis of 63 *Phytophthora* sequences in GenBank Release 131.0 (September 2002) showed that 21 genes contained introns, ranging in size from 26-172 bp (Kamoun 2003). Introns contain typical conserved sequences at the exon-intron junctions (5'-GTRNGT...YAG-3') and a motif (CTAAC) believed to be important for lariat formation during splicing has also been identified (Karlovsky and Prell 1991; Pieterse *et al.* 1995). There is conservation around the translation start codon (ACCATGA) typical of the eukaryotic consensus (Kamoun 2003) and there is a bias toward GC-rich codons (Unkles *et al.* 1991; http://www.oardc.ohiostate.edu/phytophthora/codon.htm).

7.4 The Extrachromosomal Genome

The circular 33 kb mitochondrial (mt) DNA of *P. infestans* has been completely sequenced (Lang and Forget 1992). It is very A+T-rich (76%) with a high gene density and the organisation of the genes more closely resembles that of plants than fungi. Most of the genes in the mtDNA of *P. infestans* are not usually found in the mitochondrial genomes of animals and plants, including eleven small- and six large-subunit ribosomal protein genes and three subunits of the NADH dehydrogenase complex.

The mtDNA in *P. infestans* is inherited from one parent only (probably through the oogonium) and is not re-arranged by sex (Whittaker *et al.* 1994). Hence, mtDNA markers have proven valuable in tracing lineages over time and space (Ristaino *et al.* 2001). The original RFLP markers detected by PCR define only four haplotypes (Griffith and Shaw 1998). We have recently identified new polymorphisms in inter-genic mitochondrial DNA spacers for *P. infestans* and four related taxa, which offer higher resolution in the definition of variation in mtDNA (Wattier *et al.* 2003). Both intra- and inter-taxon variation was observed, showing potential for the exploitation of these markers in both molecular ecology and molecular systematics.

Judelson and Fabritius (2000) discovered an unusual RNA element in an isolate of *P. infestans*. The RNA element was not widely distributed throughout *P. infestans* and had little effect on growth or pathogenicity. It existed mainly as linear single-stranded molecules of around 625 nucleotides in length and appeared to be autonomous since cross-hybridising sequences were not found in *P. infestans* DNA. There was no evidence for encapsulation of the RNA into a form of viral particle, and it was therefore classified as a linear RNA plasmid. Two-thirds of the elements co-purified with nuclei, whilst the remainder were cytoplasmic in location, but not mitochondrial. Maternal inheritance was usually observed in sexual crosses.

8. FUNCTIONAL GENOMICS OF *PHYTOPHTHORA*

Linking a gene sequence to a phenotype is arguably the most challenging aspect of genomics research. The situation is made more difficult in the *Phytophthoras* since, being diploid, it is not possible to apply the same strategies for gene knockouts as are used routinely in haploid fungi. Homologous functional assays have been developed that exploit the gene silencing phenomenon, whilst assays in heterologous species have also proven useful in elucidating the function of cloned *Phytophthora* genes.

8.1 Homologous Assays

The stability and efficiency of the silencing phenomenon provides a novel route for functional analysis. Since promoterless constructs can be used without modification to generate silenced strains (van West *et al.* 1999 b), it provides a rapid approach for simultaneously assaying pools of cDNA clones, and also offers the prospect of co-silencing members of conserved gene families. Following the discovery of silencing using the *infl* elicitin-encoding gene, the phenomenon has been successfully exploited to investigate the functions of a number of other *P. infestans* genes. Latijnhowers *et al.* (2003) silenced the *Pigpal* gene encoding an G-protein subunit and demonstrated pleiotropic effects on zoospore release, zoospore motility, chemotaxis and appressorium formation and sporulation (Section 3). Ah Fong and Judelson (2003) used a similar approach to silence the *P. infestans* gene *piCdc14*, which encodes a homologue of the *Saccharomyces cerevisiae* cell cycle regulator Cdc14. The phenotypes of silenced transformants suggested that *piCdc14* may function to synchronise nuclear behaviour during sporulation and maintain dormancy in spores until germination.

The technique has also been applied in other *Phytophthora* species. For example, Gaulin *et al.* (2002) investigated the biological role of CBEL in *P. parasitica* by generating transgenic strains silenced for CBEL expression (Section 3.3). Phenotypic characterisation of these strains showed that they were impaired in their ability to attach to cellophane membranes, but were not greatly affected in pathogenicity towards tobacco plants. The absence of CBEL also led to abnormalities in cell wall structure, suggesting that the protein is involved *in vivo* both in cell wall deposition and adhesion to cellulosic substrates.

Despite its potential, the universality of the silencing approach remains to be proven. The technology has so far been applied to only a small of genes and its molecular basis remains unclear. Other approaches in development include ribozyme-mediated disruption of gene expression, gene disruption through homologous recombination and transposon mutagenesis (Kamoun *et al.* 2002).

8.2 Heterologous Assays

Heterologous assays can be of two forms, 'complementation' or 'gain-of-function'. In a complementation assay, a candidate gene of interest is transformed into a heterologous host known to lack a functional copy of the gene. For example, the *P. infestans Piypt* gene was shown to have a role in vesicle transport and secretion on the basis of its ability to complement a *S. cerevisiae* strain carrying a mutation in the equivalent gene (Chen and Roxby 1996). This strategy is conceptually simple and experimentally straightforward, but is constrained by the availability of appropriate yeast mutants and the fact that it can be applied only with highly conserved genes.

Gain-of-function assays are particularly useful in the context of identifying genes encoding virulence and avirulence functions. One approach is to use a heterologous *Phytophthora* species as a host for the gene under test. For example, Panabieres *et al.* (1998) demonstrated functional expression in *P. infestans* of the gene encoding the basic elicitin cryptogein from *Phytophthora cryptogea*. The secreted cryptogein B from *P. infestans* co-transformants increased their ability to cause a HR-like necrosis of tobacco leaves.

An alternative strategy is to use a plant host for functional assays of *Phytophthora* genes. A number of suitable methods are available (reviewed in Kamoun *et al.* 2002). Ectopic expression of *Phytophthora* avirulence genes in plant cells containing the matching resistance gene has been shown to lead to phenotypic effects. For example, functional expression of a *P. infestans inf* elicitin gene in *N. benthamiana* using Potato Virus X (PVX) resulted in specific induction of the hypersensitive response (Kamoun *et al.* 1999 a). *Agrobacterium tumefaciens*-based assays such as agroinfiltration have also been used (e.g. Van der Hoorn *et al.* 2000). An approach that is proving particularly valuable is agroinfection using a binary PVX- *Agrobacterium* vector (Takken *et al.* 2000). This approach allows the high throughput rate demanded by large-scale gene discovery programmes and has been exploited in a number of recent studies (Qutob *et al.* 2002; Bos *et al.* 2003; Qutob *et al.* 2003; Torto *et al.* 2003). However, PVX-based systems cannot be used for assaying inserts greater than 2 kb in size, which places a limitation on their usefulness. To overcome this problem, Kamoun *et al.* (2003) have developed a new strategy of agrosuppression, that involves inoculation of plants with a mixture of *A. tumefaciens* strains carrying a binary plasmid containing the candidate gene plus a tumour-inducing T-DNA. Upon induction of the HR, tumour-formation is suppressed, giving an easily scorable phenotype. This approach has considerable potential for rapid high throughput functional screening of *Phytophthora* genes.

9. CONCLUSIONS

The *Phytophthoras* remain a relatively poorly studied group of organisms despite their indisputable commercial significance. However, it is clear that the tide is turning. Robust and reliable molecular tools are now in place to facilitate rapid progress and the increased availability of genome sequence data is allowing the development of powerful new approaches for gene discovery. Analyses of *Phytophthora* genomes and proteomes, coupled with high throughput functional assays, promise to accelerate the identification of proteins of relevance to the plant-pathogen relationship. There is every reason to be optimistic that the next few years will witness a significant increase in our understanding of the molecular basis of pathogenesis, which will aid the development of specific and sustainable control measures for *Phytophthora* infections.

Acknowledgements: The author acknowledges the many researchers in the *Phytophthora* community who communicated unpublished results for this review.

REFERENCES

Ah Fong AMV and Judelson HS (2003). Cell cycle regulator Cdc14 is expressed during sporulation but hot hyphal growth in the fungus-like oomycete *Phytophthora infestans*. Mol Microbiol (In press).

Al-Kherb SM, Fininsa C, Shattock RC, and Shaw DS (1995). The inheritance of virulence of *Phytophthora infestans* to potato. Plant Pathol. 44: 552–562.

Ambikapathy J, Marshall JS, Hocart CH, and Hardham AR (2002). The role of proline in osmoregulation in *Phytophthora nicotianae*. Fungal Genet Biol 35: 287-299.

Avrora AO, Venter E, Birch PRJ, and Whisson SC (2003). Profiling and quantifying differential gene transcription in *Phytophthora infestans* prior to and during the early stages of potato infection. Fungal Genet Biol (in press).

Birch PRJ and Whisson SC (2001). *Phytophthora infestans* enters the genomic era. Mol Plant Pathol 2: 257-263.

Bircher U and Hohl HR (1997). Environmental signalling during induction of appressorium formation in *Phytophthora*. Mycol Res 101: 395-402.

Bishop-Hurley SL, Mounter SA, Laskey J, Morris RO, Elder J, Roop P, Rouse C, Schmidt FJ, and English JT (2002). Phage-displayed peptides as developmental agonists for *Phytophthora capsici* zoospores. Appl Environ Microbiol 68: 3315-3320.

Blackstock WP and Weir MP (1999). Proteomics: quantitative and physical mapping of cellular proteins. Trends Biotech 17: 121-127.

Blein, JP, Coutos-Thévenot P, Marion D, and Ponchet M (2002). From elicitins to lipid-transfer proteins: a new insight in cell signalling involved in plant defence mechanisms. Trends Plant Sci 7: 293–296.

Bos JIB, Armstrong M, Whisson SC, Torto T, Ochwo M, Birch PRJ, and Kamoun S (2003). Intraspecific comparative genomics to identify avirulence genes from *Phytophthora*. New Phytol 159: 63-72.

Bottin A, Larche L, Villalba F, Gaulin E, Esquerre-Tugaye MT, and Rickauer M (1999). Green fluorescent protein (GFP) as gene expression reporter and vital marker for studying development and microbe-plant interaction in the tobacco pathogen *Phytophthora parasitica* var. *nicotianae*. FEMS Microbiol Letts 176: 51-56.

Brunner F, Wirtz W, Rose JKC, Darvill AG, Govers F, Scheel D, and Nurnberger T (2002). A β glucosidase/xylosidase from the phytopathogenic oomycete, *Phytophthora infestans*. Phytochemistry 59: 689-696.

Byrt PN, Irving HR, and Grant BR (1982). The effect of organic compounds on the encystment, viability and germination of zoospores of *Phytophthora cinnamomi*. J Gen Microbiol 128: 1189-1198.

Cahill DM and Hardham AR (1994). Exploitation of zoospore taxis in the development of a novel dipstick immunoassay for the specific detection of *Phytophthora cinnamomi*. Phytopath 84: 193-200.

Carlile MJ (1983). Motility, taxis and tropism in *Phytophthora*. In: DC Erwin, S Bartnicki-Garcia and PH Tsao, Eds. *Phytophthora*: Its Biology, Taxonomy, Ecology and Pathology. St Paul, MN: APS Press, pp 95-107.

Caten CE and Jinks JL (1968). Spontaneous variability of single isolates of *Phytophthora infestans*. I. Cultural variation. Can J Bot 46: 329-347.

Chen Y and Roxby R (1996). Characterization of a *Phytophthora infestans* gene involved in vesicle transport. Gene 181: 89-94.

Colas V, Conrod S, Venard P, Keller H, Ricci P, and Panabieres F (2001). Elicitin genes expressed *in vitro* by certain tobacco isolates of *Phytophthora parasitica* are down regulated during compatible interactions. Molec Plant-Microbe Interact 14: 326-335.

Connolly MS, Williams N, Heckman CA, and Morris PF (1999). Soybean isoflavones trigger a calcium influx in *Phytophthora sojae*. Fungal Genet Biol 28: 6–11.

Cooke DEL, Drenth A, Duncan JM, Wagels G, and Brasier CM (2000). A molecular phylogeny of *Phytophthora* and related oomycetes. Fungal Genet Biol 30: 17–32.

Cvitanich C and Judelson HS (2001). Transformation of the oomycete *Phytophthora infestans* using microprojectile bombardment. Fungal Genet Newslet 48 (Suppl.): 129.

Duncan J (1999). *Phytophthora* – an abiding threat to our crops. Microbiol Today 26: 114-116.

Duniway JM (1983). Role of physical factors in the development of *Phytophthora* diseases. In: DC Erwin, S Bartnicki-Garcia and PH Tsao, Eds. *Phytophthora*: Its Biology, Taxonomy, Ecology and Pathology. St Paul, MN: APS Press, pp 175-187.

Erwin DC, Bartnicki-Garcia S, and Tsao PH (1983). *Phytophthora*: Its Biology, Taxonomy, Ecology, and Pathology. St Paul, MN: APS Press.

Erwin DC and Ribeiro OK (1996). *Phytophthora* Diseases Worldwide. St Paul, MN: APS Press.

Fabritius AL, Cvitanich C, and Judelson HS (2002). Stage-specific gene expression during sexual development in *Phytophthora infestans.* Molec Microbiol 45: 1057-1066.

Fellbrich G, Romanski A, Varet A, Blume B, Brunner F, Engelhardt S, Felix G, Kemmerling B, Krzymowska M, and Nurnberger T (2002). NPP1, a *Phytophthora*-associated trigger of plant defense in parsley and *Arabidopsis.* Plant J 32: 375-390.

Forbes GA, Goodwin SB, Drenth A, Oyarzun P, Ordonez ME, and Fry WE (1998). A global marker database for *Phytophthora infestans*. Plant Dis 82: 811-818.

Förster H, Coffey MD, Elwood H, and Sogin ML (1990). Sequence analysis of the small subunit ribosomal RNAs of three zoosporic fungi and implications for fungal evolution. Mycologia 82: 306–312.

Fry WE (1982). Principles of plant disease management. Academic Press.

Fry WE and Goodwin SB (1997a). Re-emergence of potato and tomato late blight in the United States. Plant Dis 81:1349-1357.

Fry WE and Goodwin SB (1997b). Resurgence of the Irish potato famine fungus. Bioscience 47: 363-371.

Gaulin E, Jauneau A, Villalba F, Rickauer M, Esquerre-Tugaye MT, and Bottin A (2002). The CBEL glycoprotein of *Phytophthora parasitica* var. *nicotianae* is involved in cell wall deposition and adhesion to cellulosic substrates. J Cell Sci 115: 4565-4575.

Gisi U and Cohen Y (1996). Resistance to phenylamide fungicides: a case study with *Phytophthora infestans* involving mating type and race structure. Ann Rev Phytopathol 34: 549–572.

Goodwin SB, Drenth A, and Fry WE (1992). Cloning and genetic analyses of 2 highly polymorphic, moderately repetitive nuclear DNAs from *Phytophthora infestans*. Curr Genet 22: 107-115.

Gornhardt B, Rouhara I, and Schmelzer E (2000). Cyst germination proteins of the potato pathogen *Phytophthora infestans* share homology with human mucins. Molec Plant-Microbe Interact 13: 32-42.

Gotesson A, Marshall JS, Jones DA, and Hardham AR (2002). Characterization and evolutionary analysis of a large polygalacturonase gene family in the oomycete plant pathogen *Phytophthora cinnamomi*. Mol Plant-Microbe Interact 15: 907-921.

Grant BR, Irving HR, and Radda M (1985). The effect of pectin and related compounds on encystment and germination of *Phytophthora palmivora* zoospores. J Gen Microbiol 131: 669-676.

Griffith GW and Shaw DS (1998). Polymorphisms in *Phytophthora infestans*: four mitochondrial haplotypes are detected after PCR amplification of DNA from pure cultures or from host lesions. Appl Environ Microbiol 64: 4007–4014.

Gubler F, Hardham AR, and Duniec J (1989). Characterising adhesiveness of *Phytophthora cinnamoni* zoospores during encystment. Protoplasma 149: 24-30.

Hardham AR (2001). The cell biology behind *Phytophthora* pathogenicity. Australas Plant Pathol 30: 91-98.

Hardham AR and Gubler F (1990). Polarity of attachment of zoospores of a root pathogen and pre-alignment of the emerging germ tube. Cell Biol Int Reports 14: 947-956.

Irving HR and Grant BR (1984). The effect of calcium on zoospore differentiation in *Phytophthora cinnamomi.* J Gen Microbiol 130: 1569-1576.

Jackson SL and Hardham AR (1996). A transient rise in cytoplasmic free calcium is required to induce cytokinesis in zoosporangia of *Phytophthora cinnamomi.* Eur J Cell Biol 69: 180-188.

Judelson HS (1997). The genetics and biology of *Phytophthora infestans*: modern approaches to a historical challenge. Fungal Genet Biol 22: 65-76.

Judelson HS (2002) Sequence variation and genomic amplification of a family of gypsy-like elements in the oomycete genus *Phytophthora.* Mol Biol Evol 19: 1313-1322.

Judelson HS and Fabritius AL (2000). A linear RNA replicon from the oomycete *Phytophthora infestans*. Mol Gen Genet 263: 395-403.

Judelson HS and Michelmore RW (1991). Transient expression of genes in the oomycete *Phytophthora infestans* using *Bremia lactucae* regulatory sequences. Curr Genet 19:453-459.

Judelson HS and Randall TA (1998). Families of repeated DNA in the oomycete *Phytophthora infestans* and their distribution within the genus. Genome 41:605-615.

Judelson HS and Roberts S (2002). Novel protein kinase induced during sporangial cleavage in the oomycete *Phytophthora infestans.* Euk Cell 1: 687-695.

Judelson HS and Tooley PW (2000). Enhanced polymerase chain reaction methods for detecting and quantifying *Phytophthora infestans* in plants. Phytopath 90: 1112-1119.

Judelson HS, Tyler BM, and Michelmore RW (1991). Transformation of the oomycete pathogen, *Phytophthora infestans*. Mol Plant-Microbe Interact 4: 602-607.

Judelson HJ, Tyler BM, and Michelmore RW (1992). Regulatory sequences for expressing genes in oomycete fungi. Mol Gen Genet 234: 138-146.

Judelson HS and Yang G (1998). Recombination pathways in *Phytophthora infestans*: polyploidy resulting from aberrant sexual development and zoospore-mediated heterokaryosis. Mycol Res 102: 1245-1253.

Kamoun S (2003). Molecular genetics of pathogenic oomycetes. Euk Cell 2: 191-199.

Kamoun S, Dong S, Hamada W, Huitema E, Kinney D, Morgan WR, Styer A, Testa A, and Torto TA (2002). From sequence to phenotype: functional genomics of *Phytophthora*. Can J Plant Pathol 24: 6-9.

Kamoun S, Hamada W, and Huitema E (2003). Agrosuppression: A bioassay for the hypersensitive response suited to high-throughput screening. Mol Plant-Microbe Interact 16: 7-13.

Kamoun S, Honee G, Weide R, Lauge R, Kooman-Gersmann M, de Groot K, Govers F, and de Wit PJGM (1999 a). The fungal gene *AVR9* and the oomycete gene *infl* confer avirulence to potato virus X on tobacco. Mol Plant-Microbe Interact 12: 459-462.

Kamoun S, Hraber P, Sobral B, Nuss D, and Govers F (1999b). Initial assessment of gene diversity for the oomycete pathogen *Phytophthora infestans* based on expressed sequences. Fungal Genet Biol 28: 94 -106.

Kamoun S, Lindqvist H, and Govers F (1997 a). A novel class of elicitin-like genes from Phytophthora infestans. Molec Plant-Microbe Interact 10: 1028-1030.

Kamoun S, van West P, de Jong AJ, de Groot KE, Vieeshouwers VGAA and Govers F (1997 b). A gene encoding a protein elicitor of *Phytophthora infestans* is down-regulated during infection of potato. Molec Plant-Microbe Interact 10: 13-20.

Kamoun S, van West P, and Govers F (1998 a). Quantification of late blight resistance of potato using transgenic *Phytophthora infestans* expressing beta-glucuronidase. Eur J Plant Path 104: 521-525.

Kamoun S, van West P, Vleeshouwers VG, de Groot KE, and Govers F (1998 b). Resistance of *Nicotiana benthamiana* to *Phytophthora infestans* is mediated by the recognition of the elicitor protein INF1. Plant Cell 10: 1413-1426.

Kamoun S, Young M, Glascock C, and Tyler BM (1993). Extracellular protein elicitors from *Phytophthora*: Host-specificity and induction of resistance to fungal and bacterial phytopathogens. Molec Plant-Microbe Interact 6: 15-25.

Karlovsky P and Prell HH (1991). The *TRP1* gene of *Phytophthora parasitica* encoding indole-3-glycerolphosphate synthase-N-(5'-phosphoribosyl)anthranilate isomerase: structure and evolutionary distance from homologous fungal genes. Gene 109: 161-165.

Knight J (2002). Fears mount as oak blight infects redwoods. Nature 415: 251.

Ko W and Chase L (1973). Aggregation of zoospores of *Phytophthora palmivora*. J Gen Microbiol 78: 79-82.

Lam ST (2001). Phytophthora genomics consortium. Phytopath 91 (Suppl.): S158 (Abstr.).

Lang BF and Forget L (1992). The mitochondrial genome of *Phytophthora infestans*. In: SJ O'Brien, ed. Genetic Maps, 6th ed. Cold Spring Harbor, New York: Cold Spring Harbor Laboratory Press. pp 3133–3135.

Latijnhouwers M and Govers F (2003). A *Phytophthora infestans* G-protein β subunit is involved in sporangia formation. (In press).

Latijnhouwers M, Ligterink W, Vleeshouwers VGAA, van West P, and Govers F (2003). A G_α subunit controls zoospore motility, appressorium formation and virulence in the potato late blight pathogen *Phytophthora infestans*. (In press).

Latijnhouwers M, Munnik T, and Govers F (2002). Phospholipase D in *Phytophthora infestans* and its role in zoospore encystment. Molec Plant-Microbe Interact 15: 939-946.

Laxalt MA, Latijnhouwers M, van Hulten M, and Govers F (2002). Differential expression of G protein α and β subunit genes during development of *Phytophthora infestans*. Fungal Genet Biol 36: 137-146.

Lebreton L and Andrivon D (1998). French isolates of *Phytophthora infestans* from potato and tomato differ in phenotype and genotype. Eur J Plant Path 104: 583-594.

Lengeler KB, Davidson RC, D'Souza C, Harashima T, Shen WC, Wang P, Pan X, Waugh M, and Heitman J (2000). Signal transduction cascades regulating fungal development and virulence. Microbial Mol Biol Rev 64: 746-785.

Liou, RF, Lee, JT, Lee, HC, and Ann, PJ (2002). Analysis of *Phytophthora parasitica* by retrotransposon-derived DNA fingerprinting. Bot Bul Acad Sin 43: 21–29.

MacGregor T, Bhattacharyya M, Tyler B, Bhat R, Schmitthenner AF, and Gijzen M (2002). Genetic and physical mapping of *Avr1a* in *Phytophthora sojae*. Genetics 160: 949-959.

Mao Y and Tyler BM (1996). The *Phytophthora sojae* genome contains tandem repeat sequences which vary from strain to strain. Fungal Genet Biol 20:43-51.

Marshall JS, Wilkinson JM, Moore T, and Hardham AR (2001). Structure and expression of the genes encoding proteins resident in large peripheral vesicles of *Phytophthora cinnamomi* zoospores. *Protoplasma* 215: 226-239.

May KJ, Whisson SC, Zwart RS, Searle IR, Irwin JAG, Maclean DJ, Carroll BJ, and Drenth A (2002). Inheritance and mapping of 11 avirulence genes in *Phytophthora sojae*. Fungal Genet Biol 37: 1-12.

Mayer AM, Staples RC, and Gil-Ad NL (2001). Mechanisms of survival of necrotrophic fungal plant pathogens in hosts expressing the hypersensitive response. Phytochem 58: 33-41.

McLeod A, Smart CD, and Fry WE (2003). Characterization of 1,3-β-glucanase and 1,3:1,4-β-glucanase genes from *Phytophthora infestans*. Fungal Genet Biol 38: 250-263.

Mitchell HJ and Hardham AR (1999). Characterisation of the water expulsion vacuole in *Phytophthora nicotianae* zoospores. Protoplasma 206: 118-130.

Mitchell HJ, Kovac KA, and Hardham AR (2002). Characterisation of *Phytophthora nicotianae* zoospore and cyst membrane proteins. Mycol Res 106: 1211-1223.

Morris BM and Gow NAR (1993). Mechanism of electrotaxis of zoospores of phytopathogenic fungi. Phytopath 83: 877-882.

Morris BM, Reid B, and Gow NAR (1992). Electrotaxis of zoospores of *Phytophthora palmivora* at physiologically relevant field strengths. Plant Cell Environ 15: 345-353.

Morris PF, Bone E, and Tyler BM (1998). Chemotropic and contact responses of *Phytophthora sojae* hyphae to soybean isoflavonoids and artificial substrates. Plant Physiol 117: 1171-1178.

Morris PF and Ward EWB (1992). Chemoattraction of zoospores of the soybean pathogen *Phytophthora sojae*, by isoflavones. Physiol. Molec Plant Path 40: 17-22.

Morrissey JP and Osbourn AE (1999). Fungal resistance to plant antibiotics as a mechanism of pathogenesis. Microbiol Mol Biol Rev 63: 708-724.

Panabieres F and Le Berre JY (1999). A family of repeated DNA in the genome of the oomycete plant pathogen *Phytophthora cryptogea*. Curr Genet 36:105-112.

Panabieres F, Birch PR, Unkles SE, Ponchet M, Lacourt I, Venard P, Keller H, Allasia V, Ricci P, and Duncan JM (1998). Heterologous expression of a basic elicitin from *Phytophthora cryptogea* in *Phytophthora infestans* increases its ability to cause leaf necrosis in tobacco. Microbiol 144: 3343-3349.

Pandey A and Mann M (2000). Proteomics to study genes and genomes. Nature 405: 837-846.

Patterson DJ (1980). Contractile vacuoles and associated structures: their organization and function. Biol Rev 55: 1-46.

Pesole G, Liuni S, Grillo G, Licciulli F, Larizza A, Makalowski W, and Saccone C (2000). UTRdb and UTRsite: specialized databases of sequences and functional elements of 5' and 3' untranslated regions of eukaryotic mRNAs. Nucl Acids Res 28: 193-196.

Pieterse CM, van't Klooster J, van den Berg-Velthuis GC, and Govers F (1995). *NiaA*, the structural nitrate reductase gene of *Phytophthora infestans*: isolation, characterization and expression analysis in *Aspergillus nidulans*. Curr Genet 27: 359-366.

Pieterse CMJ, van West P, Verbakel HM, Brasse PWHM, van den Berg-Velthuis GCM, and Govers F (1994). Structure and genomic organization of the ipiB and ipiO gene clusters of *Phytophthora infestans*. Gene 138: 67-77.

Pfyffer G, Borashi-Gaia, E, Weber B, Hoesch L, Orpin CG, and Rast D (1990). A further report on the occurrence of acylic sugar alcohols in fungi. Mycol Res 92: 219–222.

Purvis AI, Pipe ND, Day JP, Shattock RC, Shaw DS and Assinder SJ (2001). AFLP and RFLP (RG57) fingerprints can give conflicting evidence about the relatedness of isolates of *Phytophthora infestans*. Mycol Res 105: 1321-1330.

Qutob D, Hraber PT, Sobral BW, and Gijzen M (2000). Comparative analysis of expressed sequences in *Phytophthora sojae*. Plant Physiol 123: 243-254.

Qutob D, Huitema E, Gijzen M, and Kamoun S (2003). Variation in structure and activity among elicitins from *Phytophthora sojae*. Molec Plant Path 4: 119-124.

Qutob D, Kamoun S, and Gijzen M (2002). Expression of a *Phytophthora sojae* necrosis-inducing protein occurs during transition from biotrophy to necrotrophy. Plant J 32: 361-373.

Randall TA and Judelson HS (1999). Construction of a bacterial artificial chromosome library of *Phytophthora infestans* and transformation of clones into *P. infestans*. Fungal Genet Biol 28: 160-170.

Reid B, Morris B, and Gow NAR (1995). Calcium-dependent genus-specific autoaggregation of zoospores of phytopathogenic fungi. Exp Mycol 19: 202-213.

Ristaino JB, Groves CT, and Parra GR (2001). PCR amplification of the Irish potato famine pathogen from historic specimens. Nature 411: 695-697.

Rizzo DM, Garbelotto M, Davidson JM, Slaughter GW, and Koike ST (2002). *Phytophthora ramorum* as the cause of extensive mortality of *Quercus* spp. and *Lithocarpus densiflorus* in California. Plant Dis 86: 205-214.

Roetschi A, Si-Ammour A, Belbahri L, Mauch F, and Mauch-Mani B (2001). Characterization of an *Arabidopsis-Phytophthora* pathosystem: resistance requires a functional *PAD2* gene and is independent of salicylic acid, ethylene and jasmonic acid signalling. Plant J 28: 293-305.

Rose JK, Ham KS, Darvill AG, and Albersheim P (2002). Molecular cloning and characterization of glucanase inhibitor proteins: coevolution of a counterdefense mechanism by plant pathogens. Plant Cell 14: 1329-1345.

Rowley A, Choudhary JS, Marzioch M, Ward MA, Weir M, Solari RCE, and Blackstock WP (2000). Applications of protein mass spectrometry in cell biology. Methods 20: 383-397.

Sansome E and Brasier CM (1973). Diploidy and chromosomal structural hybridity in *Phytophthora infestans*. Nature 241: 344-345.

Sansome E and Brasier CM (1974). Polyploidy associated with varietal differentiation in the *megasperma* complex of *Phytophthora*. Trans Br Mycol Soc 63: 461-467.

Schiermeier Q (2001). Russia needs help to fend off potato famine, researchers warn. Nature 410: 1011.

Shepherd SJ, van west P, and Gow NAR (2003). Proteomic analysis of asexual development of *Phytophthora palmivora*. Mycol Res 107: 395-400.

Smart CD and Fry WE (2001). Invasions by the late blight pathogen: renewed sex and enhanced fitness. Biol Invasions 3: 235-243.

Sogin ML and Silberman JD (1998). Evolution of the protists and protistan parasites from the perspective of molecular systematics. Int J Parasitol 28: 11–20.

Sujkowski LS, Goodwin SB, Dyer AT, and Fry WE (1994). Increased genotypic diversity via migration and possible occurrence of sexual reproduction of *Phytophthora infestans* in Poland. Phytopath 84: 201-207.

Takken FLW, Luderer R, Gabriels SHEJ, Westerink N, Lu R, de Witt PJGM, and Joosten MHAJ (2000). A functional cloning strategy, based on a binary PVX-expression vector, to isolate HR-inducing genes of plant pathogens. Plant J 24: 275-283.

Tooley PW and Garfinkel DJ (1996). Presence of Ty1-copia group retrotransposon sequences in the potato late blight pathogen *Phytophthora infestans*. Mol Plant-Microbe Interact 9: 305-309.

Tooley PW and Therrien CD (1987). Cytophotometric determination of the nuclear DNA content of 23 Mexican and 18 non-Mexican isolates of *Phytophthora infestans*. Exp Mycol 11:19-26.

Tooley PW, Therrien CD, Sim JH, O'Sullivan E, and Dowley LJ (1993). Mating-type, nuclear-DNA content and isozyme genotypes of Irish isolates of *Phytophthora infestans*. Mycol Res 97:1131-1134.

Torto TA, Rauser L, and Kamoun S (2002). The *pipg 1* gene of the oomycete *Phytophthora infestans* encodes a fungal-like endopolygalacturonase. Curr Genet 40: 385-390.

Torto TA, Li SA, Styer A, Huitema E, Testa A, Gow NAR, van West P, and Kamoun S (2003). EST mining and functional expression assays identify extracellular effector proteins from *Phytophthora*. Genome Research 13: 1675-1685.

Tyler BM (2001). Genetics and genomics of the oomycete-host interface. Trends Genet 17: 611-614.

Tyler BM (2002). Molecular basis of recognition between *Phytophthora* pathogens and their hosts. Annu Rev Phytopath 40: 137-167.

Unkles SE, Moon RP, Hawkins AR, Duncan JM, and Kinghorn JR (1991). Actin in the oomycetous fungus *Phytophthora infestans* is the product of several genes. Gene 100: 105-112.

Van der Hoorn RAL, Laurent F, Roth R, and De wit PJ (2000). Agroinfiltration is a versatile tool that facilitates comparative analyses of *Avr9/Cf-9*-induced and *Avr4/Cf-4*-induced necrosis. Molec Plant-Microbe Interact 13: 439-446.

van der Lee T, De Witte I, Drenth A, Alfonso C, and Govers F (1997). AFLP linkage map of the oomycete *Phytophthora infestans*. Fungal Genet Biol 21: 278-291.

van der Lee T, Robold A, Testa A, van't Klooster JW, and Govers F (2001). Mapping of avirulence genes in *Phytophthora infestans* with amplified fragment length polymorphism markers selected by bulked segregant analysis. Genetics 157: 949-956.

van West P, Appiah AA, and Gow NAR (2003). Advances in research on root pathogenic oomycetes. Physiol Molec Plant Pathol. 62: 99-113.

van West P, de Jong AJ, Judelson HS, Emons AMC, and Govers F (1998). The *ipiO* gene of *Phytophthora infestans* is highly expressed in invading hyphae during infection. Fungal Genet Biol 23: 126-138.

van West P, Kamoun S, van't Klooster JW, and Govers F (1999 b) Internuclear gene silencing in *Phytophthora infestans*. Mol Cell 3: 339-348.

van West P, Morris BM, Reid B, Appiah AA, Osborne MC, Campbell TA, Shepherd SJ, and Gow NAR (2002). Oomycete plant pathogens use electric fields to target roots. Molec Plant-Microbe Interact 15: 790-798.

van West P, Reid B, Campbell TA, Sandrock RW, Fry WE, Kamoun S, and Gow NAR (1999 a) Green fluorescent protein (GFP) as a reporter gene for the plant pathogenic oomycete *Phytophthora palmivora*. FEMS Microbiol Lett 178: 71-80.

Vijn I and Govers F (2003). *Agrobacterium tumefaciens* mediated transformation of the oomycete plant pathogen *Phytophthora infestans*. Mol Plant Path (In press).

Voglmayr H and Greilhuber J (1998). Genome size determination in peronosporales (Oomycota) by Feulgen image analysis. Fungal Genet Biol 25: 181-195.

Wang T and Brown MJ (1999). MRNA quantification by real time TaqMan polymerase chain reaction: validation and comparison with RNase protection. Anal Biochem 269: 198-201.

Warburton AJ and Deacon JW (1998). Transmembrane Ca^{2+} fluxes associated with zoospore encystment and cyst germination by the phytopathogen *Phytophthora parasitica*. Fungal Genet Biol 25: 54 –62.

Wastie RL (1991). Breeding for resistance. In: D.S. Ingram and P.H. Williams, Eds. Advances in Plant Pathology, Vol. 7: *Phytophthora infestans*, the cause of late blight of potato. London: Academic Press. pp 193–223.

Waugh M, Hraber P, Weller J, Wu Y, Chen G, Inman J, Kiphart D, and Sobral B (2000). The *Phytophthora* Genome Initiative database: informatics and analysis for distributed pathogenomic research. Nucl Acids Res 28: 87-90.

Wattier RAM, Gathercole LL, Assinder SJ, Gliddon CJ, Deahl KL, Shaw DS, and Mills DI (2003). Sequence variation of inter-genic mitochondrial DNA spacers (mtDNA-IGS) of *Phytophthora infestans* (Oomycetes) and related species. Molec Ecol 3: 136-138.

Whisson SC, Drenth A, Maclean DJ, and Irwin JA (1995). *Phytophthora sojae* avirulence genes, RAPD, and RFLP markers used to construct a detailed genetic linkage map. Molec Plant-Microbe Interact 8: 988-995.

Whisson SC, van der Lee T, Bryan GJ, Waugh R, Govers F, and Birch PR (2001). Physical mapping across an avirulence locus of *Phytophthora infestans* using a highly representative, large-insert bacterial artificial chromosome library. Mol Genet Genomics 266:289-295.

Whittaker SL, Shattock RC, and Shaw DS (1991). Variation in DNA content of nuclei of *Phytophthora infestans* as measured by a microfluorimetric method using the fluorochrome DAPI. Mycol Res 95: 602-610.

Whittaker SL, Assinder SJ, and Shaw DS (1994). Inheritance of mitochondrial DNA in *Phytophthora infestans*. Mycol Res 98: 569-575.

Zhang Q, Griffith JM, and Grant BR (1992). Role of phosphatidic acid during differentiation of *Phytophthora palmivora* zoospores. J Gen Microbiol 138: 451-459.

Zwankhuizen MJ, Govers F, and Zadoks JC (2000). Inoculum sources and genotypic diversity of *Phytophthora infestans* in Southern Flevoland, the Netherlands. Eur J Plant Path 106: 667-680.

Applied Mycology & Biotechnology
An International Series. Volume 4. Fungal Genomics

6

Genomics of Phytopathogenic *Fusarium*

Haruhisa Suga[1] and Mitsuro Hyakumachi[2]
[1]Molecular Genetics Research Center, Gifu University, Gifu 501-1193 Japan, (suga@cc.gifu-u.ac.jp);
[2]Laboratory of Plant Pathology, Faculty of Agriculture, Gifu University, Gifu 501-1193 Japan.

Recent genomic studies have elucidated many aspects of phytopathogenic *Fusarium* species. Molecular phylogenetic analyses have helped to clarify ambiguities in traditional classification systems of *Fusarium*. In *Fusarium oxysporum* and *F. redolens*, for example, phylogenetic analyses have revealed that pathogenicity factors have had multiple evolutionary origins. Genome organization and molecular mechanisms of pathogenicity are still not well understood in many *Fusarium* species. The advent of pulsed field gel electrophoresis has opened a new avenue to study chromosomes of these fungi. Also, using molecular approaches, tolerance against antimicrobial compounds, signal transduction systems and some secondary metabolites have been shown to be involved in the pathogenicity of some *Fusarium* species. Whole genome sequencing of *F. graminearum* is in progress. Genomic studies are improving the understanding of the biology of *Fusarium* and elucidation of the pathogenicity on the molecular level will help to control *Fusarium* incited disease.

1. INTRODUCTION

Fusarium is a large cosmopolitan genus of imperfect fungi and is of interest primarily because numerous species are important plant pathogens (Nelson *et al.* 1981), produce of a wide range of secondary metabolites, and/or cause opportunistic mycoses in humans (Austwick 1982; Michniewicz 1989; Vesonder 1989). Although *Fusarium* research over the past 100 years has advanced our understanding of this important group of fungi, many aspects of its biology still need to be addressed. Traditional classification of *Fusarium* has been based exclusively on morphology. However, this system of classification has been controversial for many years because it has resulted in the description of markedly different numbers of sections and species by different taxonomists (Matuo 1980; Aoki 1998). In addition, morphological plasticity and the general paucity of morphological characters have made delineation of some *Fusarium* species difficult and have resulted in species that consist of strains with markedly different physiological characters, such as host-plant specificity. Snyder and Hansen (1940; 1941; 1945) integrated this diversity of host specificity within a single species into the 'forma specialis' (f. sp.) system; each forma specialis within in a species exhibited high level of virulence on a particular host species (or group of species) but not on others.

In traditional *Fusarium* taxonomy, each section, species or forma specialis was considered to be a monophyletic unit. However, recent molecular phylogenetic analyses have revealed that this is not the case, at least for some classification units of *Fusarium* (O'Donnell 1996; O'Donnell *et al.* 1998a; O'Donnell 2000; Baayen *et al.* 2001; Kistler 2001).

Sexual states (teleomorphs) have been described for some *Fusarium* species. All known *Fusarium* teleomorphs are included in the Ascomycota order Hypocreales (Samuels *et al.* 2001) but have been classified into several different genera (e.g *Gibberella, Nectria*). Both homothallic and heterothallic species have been described (Puhalla 1981; Samuels *et al.* 2001). The chromosomes of some of these species have been observed during meiosis by light microscopy (Puhalla 1981). However, like other fungi, *Fusarium* has small chromosomes compared to plants and animals, and light microscopy is not satisfactory for a precise determination of chromosome number or for karyotyping (Taga *et al.* 1998). As a result pulsed field gel electrophoresis (PFGE) has become widely used for electrophoretic karyotyping (EK) of *Fusarium* as well as for fungi in general (Millus and McCluskey 1990; Zolan 1995). PFGE is highly effective at resolving chromosomes of many fungi and at revealing EK variations within and among species (Zolan 1995). PFGE is also useful for gene mapping (Tzeng *et al.* 1992; Masel *et al.* 1993; Zolan 1995; Xu and Leslie 1996; Akamatsu *et al.* 1999; Suga *et al.* 2002; Zhong *et al.* 2002), and it can detect genome alterations, such as chromosome loss (VanEtten *et al.* 1998), translocations (Orbach *et al.* 1988; Tzeng *et al.* 1992), and large-scale deletions (Miao *et al.* 1991a; Sone *et al.* 1997). Therefore, PFGE is a powerful tool for studying fungal genome structure (Millus and McCluskey 1990; Zolan 1995).

Amplified fragment length polymorphism (AFLP) is another tool that has been used to study fungal genomes by facilitating the construction of high-density genetic linkage maps (Kema *et al.* 2002; Zhong *et al.* 2002). Recently, AFLP was used to construct dense linkage maps of *Fusarium verticillioides* (teleomorph *Gibberella moniliformis*) (Jurgenson *et al.* 2002b) and *Fusarium graminearum* (teleomorph *Gibberella zeae*) (Jurgenson *et al.* 2002a).

Fusarium species cause diseases of a wide range of plant species. Most *Fusarium* diseases are soil-borne and, in comparison to pathogens that infect aerial parts of plants, the processes by which *Fusarium* infects its hosts are not well understood. For example, although *F. oxysporum* may show minor modifications in their morphology during infection, structures specific for infection, such as appresoria or infection pegs have not been observed (Mendgen *et al.* 1996; Lagopodi *et al.* 2002). Also, host-specific toxins like those produced by *Alternaria* and *Cochliobolus* species are unknown in *Fusarium*. Nevertheless, insights into the pathogenicity of *Fusarium* have been gained even before the advent of advanced molecular techniques. For example, tolerance of plant antimicrobial compounds (e.g. phytoalexins) in *Nectria haematococca* mating population (MP) VI (anamorph: *Fusarium solani*) (VanEtten *et al.* 1980; Tegtmeier and VanEtten 1982; Kistler and VanEtten 1984; VanEtten *et al.* 1989), and mycotoxin production by *F. graminearum* (Casale and Hart 1988; Desjardins and Hohn 1997) are thought to be involved in pathogenicity. The lack of a teleomorphs precludes a classical genetic approach for identification of pathogenicity genes in many *Fusarium* species. Transformation-mediated gene disruption, however, can be used to identify genes in both sexual and asexual fungi as well as sexual fungi and has been applied to study pathogenicity genes in *Fusarium*.

Genomic analyses have overcome some of the limitations encountered in some of the more traditional techniques used to study *Fusarium*. This review focuses on recent discoveries in the fields of classification, pathogenicity and genome structure of *Fusarium*. In addition, technologies used in fungal genomics are also discussed.

2. IMPACT OF MOLECULAR PHYLOGENETIC ANALYSES ON THE CLASSIFICATION OF PHYTOPATOGENIC *FUSARIUM*

Several species concepts, including the morphological, biological and phylogenetic species concepts, have been applied to *Fusarium* taxonomy. The biological species concept is based on sexual mating compatibility; strains within one biological species are sexually compatible with one another but not with strains in other biological species. This concept is more natural and testable than the morphological species concept and has been applied to two major groups within *Fusarium*: the *N. haematococca*-species complex (NHSC), which consists of strains that are morphologically *F. solani* (Section *Martiellela*) (Matuo and Snyder 1973) and the *Gibberella fujikuroi*-species complex (GFSC), which consists of strains that are morphologically classified as a number of different species including *F. moniliforme*, *F. proliferatum* and *F. subglutinans* (Section *Liseola*) (Leslie 1991; Leslie 1995; Samuels *et al.* 2001). The biological species concept is, however, of limited practical use in *Fusarium* classification because of the absence of teleomorphic stages in many members of *Fusarium* and because female sterility, which is frequently observed in *Fusarium*, can confound placement of isolates in a given biological species (Burnett 1983; Leslie 1995). In addition, complete homothallism may prevent mating with other isolates (Taylor *et al.* 2000). Furthermore, mating compatibility only assures the potentiality of gene flow (Taylor *et al.* 2000). Gene flow could be expected in a whole biological species but it may be confined to geographically or genetically isolated groups. For example, outcrossing under experimental conditions was observed in the homothallic *G. zeae* (Bowden and Leslie 1999). However, the presence of eight genetically distinct lineages in *G. zeae* suggests that gene flow in nature is more confined than what would be expected from their mating competency (O'Donnell *et al.* 2000; Ward *et al.* 2002).

The phylogenetic species concept is based on the similarity of nucleic acids and is applicable to members of *Fusarium* with or without teleomorphic stage can provide evidence for evolutionary relationships between groups of species. In the initial stages of the development of the phylogenetic species concept for *Fusarium*, Szécsi and Dobrovolszky (1985a; 1985b) generated *Fusarium* phylogenies based on thermal reassociation of genomic DNA. More recently, O'Donnell and coworkers have refined the phylogenetic species concept in *Fusarium* using nucleotide sequence data. The species definition according to the phylogenetic species concept is "the smallest aggregation of populations (sexual) or lineages (asexual) diagnosable by a unique combination of character states in comparable individuals (semaphoronts)" (Nixon and Wheeler 1990).

2.1 Sections and Species

Ribosomal DNA (rDNA) has been widely used for molecular phylogenetic analyses of organisms. Initial molecular systematic studies in *Fusarium* employed the nuclear large subunit (LSU) 28S ribosomal DNA (Peterson and Logrieco 1991) or the nuclear rDNA internal transcribed spacer (ITS) region (O'Donnell 1992). However, it is important to include multiple unrelated genes in molecular phylogentic analyses because analyses based on a single gene may not reflect the true evolutionary history of the individuals/species being compared (Taylor *et al.* 2000). Nonorthologous genes (genes that have not result from a speciation event) lead to erroneous phylogenetic trees (O'Donnell and Cigelnik 1997). Two types of nonorthologous rDNA ITS2 are present in the GFSC and in the closely related *F. oxysporum*-species complex (FOSC), although only one type of rDNA ITS2 (the major ITS2) is discernable by PCR amplifications and by direct sequencing with conserved primers (O'Donnell and Cigelnik 1997). The major ITS2 has been switched between these nonorthologous ITS2 types at least three times during GFSC evolution (O'Donnell and Cigelnik 1997). Nonorthologous regions have also been found in rDNA intergenic spacers (IGS) of *F. oxysporum* (Appel and Gordon 1996) and β-tubulin genes of NHSC (O'Donnell 1996). For combinational phylogenetic analyses of *Fusarium*, sequence information from LSU 28S rDNA, β-tubulin gene, mitochondrial small subunit (mtSSU) rDNA, rDNA ITS, nuclear SSU 18S rDNA, translation elongation factor (EF-1α) gene, phosphate permase gene is now frequently being used (O'Donnell and Cigelnik 1997; O'Donnell *et al.* 1998a; O'Donnell *et al.* 1998b; O'Donnell 2000; O'Donnell *et al.* 2000 ; Baayen *et al.* 2000; Baayen *et al.* 2001; Skovgaard *et al.* 2001).

In some cases molecular phylogentic analyses have confirmed lineages defined by other species concepts. For example, in molecular phylogenetic analyses of GFSC and NHSC all biological species (or mating populations) within these two species complexes could be resolved as distinct lineages (O'Donnell *et al.* 1998a; O'Donnell 2000). This complete congruence between biological and phylogenetical species concepts indicates that both concepts can reveal and represent fundamental taxonomic units (O'Donnell 1996). PFGE analyses of chromosome also support these lineages. Distinct electrophoretic karyotypes were described for each mating population of GFSC (Xu *et al.* 1995) and NHSC (Suga *et al.* 2002).

Molecular phylogenetic analyses have also helped to resolve multiple species that were not previously resolved using morphological characters. For example, twenty-three species wtihin the GFSC that were not previously resolved by morphological characters were differentiated by molecular phylogentic analysis (O'Donnell *et al.* 1998a). In addition, *F. graminearum* group 1 isolates are morphologically indistinguishable from group 2 isolates but were resolved into two distinct species by molecular phylogentic analysis (Group I was re-classified as *F. pseudograminearum*) (Aoki and O'Donnell 1999). Likewise, the species now classified as *F. redolens* was previously classified as *F. oxysporum* based on morphological characters (Snyder and Hansen 1940; Nelson *et al.* 1983; Booth 1971). However, molecular phylogentic analyses showed that *F. redolens* isolates represent a lineage that is distinct from *F. oxysporum* and not even be a sister group to FOSC (O'Donnell *et al.* 1998a; Baayen *et al.* 2001). Recently, *Fusarium* isolates that are pathogenic to cultivated species in the plant genus *Hosta* and share some morphological characters with *F. oxysporum* were discovered to be a new species, *F. hostae*, that is phylogenetically closely related with *F. redolens* (Baayen *et al.* 2001; Geiser *et al.* 2001). Also, it was shown that *Neocosmospora vasinfecta,* which does not produce macroconidia belongs to NHSC by molecular phylogenetic analyses (O'Donnell and Gray 1995; O'Donnell 2000). Another major discovery using nucleotide sequence information for classification purposes is that chlamydospore formation and several mycotoxins are the plesiomorphic (ancestral) state in GFSC, indicating a limited usefulness of the morphological species concept, particularly in this group (O'Donnell *et al.* 1998a). Specifically, section *Liseola* excluded chlamydospore-forming species, such as *F. dlamini*, *F. nygamai* and *F.napiforme* by definition. Furthermore, *F. udum* which had been classified in the section *Elegans,* was found to be nested within the GFSC, which includes only *Fusarium* species in section *Liseola*. These discoveries indicate that sections *Liseola* and *Elegans* are non-monophyletic and therefore artificial (O'Donnell *et al.* 1998a). Nonmonophyly has also been determined for sections *Discolor* and *Dlaminia* (O'Donnell and Cigelnik 1997; O'Donnell *et al.* 1998a).

2.2 Forma Specialis, Race and Vegetative Compatibility Group

Forma specialis and race are typically used to designate pathogenic specificity of fungi on plant species and cultivars, respectively. However, these designations need some attention because they can be misleading and can result in inaccurate inferences concerning the pathogenicity of inidividual isolates. For example, classification of *Fusarium solani* into formae speciales presents two types of problems. First, although *F. solani* f. sp. *pisi* is defined by its specific pathogenicity to pea, some isolates of f. sp. *pisi* are also pathogenic to chickpea (VanEtten and Kistler 1988) (Fig. 1). Therefore, host specificity is not limited to pea in this forma specialis. Second, formae speciales of *F. solani* were equated with MPs of *Nectria haematococca* in an early study (Matuo and Snyder 1973). Consequently, *Nectria haematococca* MP VI is generally considered to be the teleomorph of *F. solani* f. sp. *pisi*. However, some strains of *N. haematococca* MP VI are pathogenic to different plant species (VanEtten and Kistler 1988) (Fig. 1) while other strains are nonpathogenic to pea (VanEtten 1978). Whether such problems exist in other formae speciales of *F. solani* is not known. Nevertheless, the problems in *F. solani* f. sp. *pisi* indicate that the concept of formae speciales may need revision in this species. Race designations can also be problematic in *Fusarium* because they are not defined by cultivar-level specificity. For example, race 1 and race 2 of *F. solani* f. sp. *cucurbitae* are defined by their pathogenesis on different tissues of cucurbits rather than variations in pathogenicity on different cultivars: race 1 is pathogenic on roots, stems and fruits and race 2 is pathogenic only on fruit (Tousson and Snyder 1961). Some races of *F. oxysporum* f. sp. *conglutinans* are also not defined by cultivar-level specificity (Kistler 1997a). Consequently, the concept of forma specialis and race in these and potentially other *Fusarium* species may need to be reconsidered. On the other hand, molecular phylogenetic analyses have contributed significantly to the understanding of the evolution of pathogenic specificity in *Fusarium*. All eight formae speciales of *F. solani* and, race 1 and race 2 of *F. solani* f. sp. *cucurbitae* could be resolved as distinct lineages in the NHSC (O'Donnell 2000). This result indicates that pathogenic specificity may have evolved independently in each lineage in the NHSC, including within the lineages corresponding to race 1 and race 2 in *F. solani* f. sp. *cucurbitae*.

Since Puhalla (1985) established vegetative compatibility as a basis for subdivision within formae speciales of *F. oxysporum*, relationships between vegetative compatibility group (VCG) and pathogenic specificity have been studied extensively in the species (Leslie 1996; Kistler 1997a). Pairs of isolates that can undergo hyphal fusion to produce a heterokaryons belong to the same VCG. Vegetative compatibility is thought to result from action of alleles at several distinct loci. To be vegetatively compatible a pair of isolates needs to have the same alleles at all loci governing compatibility (Leslie 1993). Consequently, members of same VCG are considered to be genetically identical or closely related. Molecular genetic analyses have confirmed this hypothesis. In *F. oxysporum*, isolates belonging to the same VCG always shared an identical sequence for EF-1α and mtSSU rDNA (Baayen *et al.* 2000). Only minimal variation among isolates belonging to the same VCG was apparent by restriction fragment length polymorphism (RFLP) or amplified fragment length polymorphism (AFLP) analyses in *F. oxysporum* (Manicom *et al.* 1990; Elias *et al.* 1993; Bentley *et al.* 1998). Because each forma specialis has a unique VCG or a set of unique VCGs (Kistler 1997a), it has been assumed that all VCGs of a forma specialis originated from a common ancestor and that therefore, each forma specialis has a monophyletic origin. Recently, this assumption was examined by molecular phylogenetic analyses and it was determined that some formae speciales of *F. oxysporum* (f. sp. *cubense*, *lycopersici*, *radicis-lycopersici*, *melonis* (O'Donnell *et al.* 1998b), *asparagi*, *dianthi*, *gladioli*, *lini* (Baayen *et al.* 2000), and *vasinfectum* (Skovgaard *et al.* 2001)) were polyphyletic. Similarly, *F. redolens* f. sp. *asparagi* and *dianthi* were shown to be non-monophyletic (Baayen *et al.* 2001). These findings suggest that host specificity was derived independently in these distinct lineages. However, two other hypotheses can also explain this observation without evoking an independent origin of host specificity. First, host-specificity may be the ancestral state that has been maintained only in a small number of lineages or second, host-specificity may have been transferred horizontally between lineages (Kistler, 2001).

VCG analysis has also been used to study race relationships within specific formae speciales of *F. oxysporum* (Jacobson and Gordon 1988; Ploetz and Correll 1988; Elias and Schneider 1991). While VCG and race were generally correlated in *F. oxysporum* f. sp. *cubense* (Ploetz and Correll 1988), *pisi* (Correll *et al.* 1985), *apii* (Ireland and Lacy 1986) and *dianthi* (Aloi and Baayen 1993), no correlation was observed in *F. oxysporum* f. sp. *melonis* (Jacobson and Gordon 1988) or f. sp. *lycopersici* (Elias and Schneider 1991). Since VCGs represent phylogenetically distinct lineages in *F. oxysporum* (Bentley *et al.* 1998; Baayen *et al.* 2000; Kistler 2001), cultivar-level pathogenic specificity also may have been derived independently in distinct lineages within a forma specialis. Also, some VCGs of *F. oxysporum* contain multiple races (Correll *et al.* 1985; Jacobson and Gordon 1988; Ploetz and Correll 1988; Elias and Schneider 1991; Aloi and Baayen 1993). These results have been interpreted to mean that race determinants may be simple and/or highly mutable (Kistler 1997a).

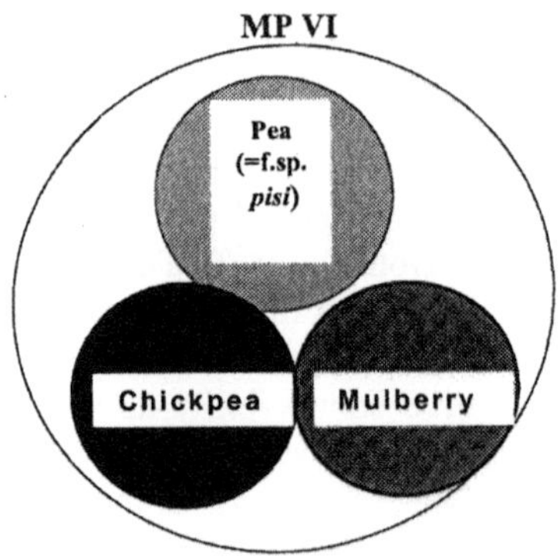

Fig. 1. Diagrammatic representation of pathogenicity of *Nectria haematococca* MP VI. The largest circle indicates all strains belonging to *N. Haematococca* MP VI. Each small circle indicates a group that is pathogenic to the denoted plant. Although the figure shows Pathogenicity to only three plant species, Pathogenicity of MPVI has been verified on nine plant species (VanEtten and Kistler, 1988).

3. GENOME STRUCTURE AND EVOLUTION

Although chromosomes have been observed during meiotic division in *Fusarium* by conventional light microscopy (Puhalla 1981), the use of microscopy to study chromosomes has been limited by their small size. In addition, because many *Fusarium* species lack a meiotic state, light microscopic observations of condensed chromosomes is not feasible. The advent of PFGE opened a new avenue to the study of fungal chromosomes (Zolan 1995). While only four chromosomes were observed in *G. fujikuroi* by conventional microscopy (Howson *et al.* 1963), PFGE differentiated 12 chromosomes (Xu and Leslie 1996). In the last decade, genome structure has been analyzed in some *Fusarium* species with the help of PFGE.

3.1 *Fusarium oxysporum*

PFGE has been widely applied in *F. oxysporum* (Momol *et al.* 1990; Kim *et al.* 1993; Migheli *et al.* 1993; Boehm *et al.* 1994; Kistler *et al.* 1995; To-Anun *et al.* 1995). As a result, chromosome length polymorphisms (CLPs) have been described within and among formae speciales (Migheli *et al.* 1993), races (Kim *et al.* 1993), and VCGs (Boehm *et al.* 1994). It should be noted, however, that a reduced amount of CLPs was observed within VCGs (Boehm *et al.* 1994; Alves-Santos *et al.* 1999). Chromosome numbers (5-15) and estimated genome size (ca 15-50 Mb) of *F. oxysporum* were highly variable throughout these studies. Differences in chromosome number may, in part, be attributable to some limitations of chromosome separation by PFGE (Zolan 1995). Although intensively stained bands suggest co-migration of multiple chromosomes, estimating the number of chromosomes within a given band with higher intensity is often difficult. Despite this technical limitation, there is no doubt that extensive CLPs exist within *F. oxysporum*. Electrophoretic karyotypes (EKs) derived from different subcultures and sample preparations of a given strain are generally highly reproducible (Kim *et al.* 1993; Migheli *et al.* 1993; Boehm *et al.* 1994). However, marked EK alterations have been observed in serial cultures from a single isolate of *F. oxysporum* f. sp. *melonis* that exhibited genetic instability (Davière *et al.* 2001). rDNA and single-copy DNA sequences hybridized to different sized-chromosomes in some, but not all isolates of *F. oxysporum* f. sp. *cubense* (Boehm *et al.* 1994; Kistler *et al.* 1995). A wide variety of repeated DNA sequences, including many transposable elements, have been detected in *F. oxysporum* (Kistler *et al.* 1991; Elias *et al.* 1993; Kim *et al.* 1993; Daboussi and Langin 1994; Gómez-Gómez *et al.* 1999; Hua-Van *et al.* 2000). A positive correlation between CLPs and density of repetitive DNA sequences was shown for *F. oxysporum* (Davière *et al.* 2001). Repetitive DNA sequences might increase the frequency of ectopic recombination events both within and between chromosomes (Zolan 1995). Asexual fungi such as *F. oxysporum* may lack some or all processes needed to maintain genome integrity, such as pre-meiotic deletion of tandem duplications (Selker 1990). Consequently, duplications resulting from translocations or aneuploidy may occur more frequently in asexual than sexual fungi (Kistler and Miao 1992; Kistler *et al.* 1995). It is worth mentioning at this point that although a teleomorph for *F. oxysporum* has never been described, mating-type idiomorphs, MAT-1 and the MAT-2, were found in *F. oxysporum* (Yun *et al.* 2000) and the genes within these idiomorphs are expressed. Therefore, the lack of sexual reproduction in *F. oxysporum* can not been attributed to the abscense of *MAT* genes or to the lack of their expression (Yun *et al.* 2000). Although the basis for the assumed lack of sexual reproduction and the molecular mechanisms to generate genomic diversity are not fully elucidated in *F. oxysporum*, studies that address these issues should provide insight into genome evolution in asexual fungi.

3.2 The *Nectria haematococca* Species Complex

The NHSC includes both asexual and sexual (homothallic and heterothallic) species (O'Donnell 2000). Phylogenetic analyses suggest that meiosis was lost independently at least twice during the evolutionary history of the complex (O'Donnell 2000). Homothallic species appear to be polyphyletic and therefore are likely to have been derived independently from heterothallic species (O'Donnell 2000). Partial DNA sequences of the MAT-1 (GenBank accession No. AY040737) and MAT-2 (Arie *et al.* 1997) idiomorphs have been determined in heterothallic *N. haematococca.* Although homothallic species in the NHSC are assumed to contain at least partial copies of both MAT idiomorphs, idiomorph structures in these species remain unknown (Yun *et al.* 1999; Yun *et al.* 2000).

PFGE generally resolves five to seven chromosomes in the seven formae speciales of *F. solani.* Several genes encoding putative pathogenicity factors (kievitone hydratase, pectate lyase A, pisatin demethylase) have been mapped onto separate chromosomes (Suga *et al.* 2002). Distinct genome organization in the various mating populations of this species complex was evident from both molecular phylogenetic analyses (O'Donnell 2000) and EK analysis (Suga *et al.* 2002). Genome structure analyses of the NHSC have focused on one mating population, *N. haematococca* MP VI (Miao *et al.* 1991a; Miao *et al.* 1991b; Kistler and Benny 1992; Kim *et al.* 1995; Taga *et al.* 1998; Taga *et al.* 1999; Enkerli *et al.* 2000). Ten to 17 chromosomes with a total genome size of ca 40 Mb have been resolved among isolates of *N. haematococca* MP VI (Miao *et al.* 1991b; Kim *et al.* 1995; Taga *et al.* 1998). An approximately 1.6-Mb chromosome was found to be dispensable (Miao *et al.* 1991a; Kistler and Benny 1992; VanEtten *et al.* 1998). Members of *N. haematococca* MP VI with high virulence on pea have a cluster of genes (named PEP for Pea Pathogenicity genes) on the dispensable chromosome (designated the *PDA1* dispensable chromosome) that includes the pea phytoalexin detoxification gene, *PDA1* (Han *et al.* 2001). Isolates with high virulence to chickpea have a chickpea phytoalexin detoxification gene, *MAK1*, on the 1.6-Mb dispensable chromosome (*MAK1* dispensable chromosome) that also bears a second pea phytoalexin detoxification gene *PDA6-1* (Miao and VanEtten 1992b; Covert *et al.* 1996; Enkerli *et al.* 1998; Enkerli *et al.* 2000). The *PDA1* and the *MAK1* dispensable chromosomes share homologous regions but also have unique regions (Funnell and VanEtten 2002). Several types of transposable elements have also been identified on these two dispensable chromosomes. The *MAK1* dispensable chromosome has a class I copia-like transposable element, *Nht2* (Shiflett *et al.* 2002) and a class II *Fot1/Pogo*-like transposable element, *Nht1* (Enkerli *et al.* 1997; Enkerli *et al.* 2000). The *PDA1* dispensable chromosome has a class I SINE-like transposable element, *Nrs1* (Kim *et al.* 1995) and three class II transposable elements, *Fot1/Pogo*-like, *Tc1/Mariner*-like and *Ac*-like (*hAT* family) (Han *et al.* 2001). The *MAK1* dispensable chromosome can become truncated during meiosis (Miao *et al.* 1991a), and the truncated region has been physically mapped (Enkerli *et al.* 2000). It remains unclear whether any of the transposable elements are associated with the truncation (Enkerli *et al.* 1997; Shiflett *et al.* 2002). However, the truncation may be analogous to the positive correlation of CLPs and the density of repetitive DNA sequences in *F. oxysporum* (Davière *et al.* 2001).

A homologous sequence of the pea phytoalexin detoxification gene was also detected on a 1.6-Mb chromosome in the pea pathogen *F. oxysporum* f. sp. *pisi* but not in other *F. oxysporum* formae speciales (Kistler 2001). This suggests that part or all of the dispensible chromosome could have been transfered between species via horizontal gene transfer. This hypothesis is supported by the similarity of the PEP cluster on the dispensible chromosome and bacterial pathogenicity islands, which have been shown to be acquired through horizontal gene transfer (Hacker *et al.* 1997). Similarities of the PEP gene cluster to pathogenicity islands include the presence of transposable elements and multiple pathogenicity genes as well as distinct codon usage and GC content (Han *et al.* 2001). Horizontal transmission of an entire 2-Mb chromosome has been experimentally demonstrated between vegetatively incompatible biotypes of the anthracnose pathogen, *Colletotrichum gloeosporioides* (He *et al.* 1998). Such direct observation of horizontal gene transfer is rare in filamentous fungi and the molecular mechanisms by which it occurs are not known (Rosewich and Kistler 2000). Nevertheless, horizontal transfer of genetic material may play an important role in fungal evolution, including in the the acquisition of phytopathogenicity factors. For example, horizontal gene transfer has been suggested to be responsible for race T evolution of the Southern corn leaf blight pathogen, *Cochliobolus heterostrophus* (Turgeon and Lu 2000).

3.3 *Gibberella* Species

The GFSC also includes both asexual and sexual (heterothallic only) species (O'Donnell *et al.* 1998a). All GFSC MPs (A to F) examined by PFGE appeared to have 12 chromosomes and a total genome size of approximately 45-55 Mb (Xu *et al.* 1995). Because each GFSC MP has a distinct EK, it was suggested that each MP has a distinct genome organization (Xu *et al.* 1995). Phylogenetic gene trees generated from MAT idiomorphs of these MPs are congruent with those generated from several other independent genes, although MP B appeared to have a peculiar placement (O'Donnell *et al.* 1998a; Steenkamp *et al.* 2000). This observation indicates parallel evolution of MAT idiomophs and speciation in GFSC, with the exception of MP B.

Genetic linkage groups based on RFLP and AFLP markers were generated for *G. moniliformis* (synonym *G. fujikuroi* MP A, anamorph: *F. verticillioides*, synonym *F. moniliforme*) and assigned to the 12 chromosomes resolved by PFGE (Xu and Leslie 1996; Jurgenson *et al.* 2002b). Two auxotrophic loci (*arg1*, *nic1*), the mating type (*MAT*) locus, the female sterility locus (*ste1*), the spore killer (*SK*) locus and the fumonisin biosynthetic locus corresponding to *FUM5* (Proctor *et al.* 1999) were also mapped to the linkage groups. The smallest chromosome (ca 0.7 Mb) of *G. moniliformis* was dispensable (Xu and Leslie 1996). In field isolates of other GFSC MPs, considerable variation in length was observed for all chromosomes less than 1 Mb (Xu *et al.* 1995). This observation lead to the speculation that these small chromosomes in the other MPs are also dispensable (Xu *et al.* 1995).

Gibberella zeae (anamorph *F. graminearum*) is a homothallic species that is a pathogen of multiple crop plants including maize, wheat and barley. Like certain homothallic *Cochliobolus* species, self-fertility in *G. zeae* results from the presence of both MAT-1 and MAT-2 idiomorphs within the genome of individual strains (Yun *et al.* 1999, 2000). Outcrossing between different *G. zeae* strains can be accomplished under experimental conditions (Bowden and Leslie 1999) and this has facilitated genetic analysis that has resolved nine linkage groups in the fungus (Jurgenson *et al.* 2002a). Loci controlling amount and type of the trichothecene toxins were mapped to linkage groups 4 and 1, respectively (Jurgenson *et al.* 2002a). The locus governing the type of trichothecene (deoxynivalenol or nivalenol) produced probably corresponds to *Tri13* as described below. The nine *G. zeae* linkage groups have not yet been assigned to chromosomes, because PFGE data has not been generated to date.

Eight phylogenetic lineages have so far been described in *G. zeae* based on combined sequence analysis of six independent nuclear genes (O'Donnell *et al.* 2000; Ward *et al.* 2002). These results indicate that gene flow between lineages is extremely limited or absent in nature, although one putative recombinant between two lineages was found (O'Donnell *et al.* 2000). Lineage phylogeny did not correlate with the trichothecene chemotypes initially characterized by Miller *et al.* (1991). Trichothecene chemotypes of *G. zeae* are the NIV (nivalenol and acetylnivalenolproducing) type, the 3-ADON (deoxynivalenol (DON) and 3-acetyldeoxynivalenol-producing) type, and the 15-ADON (DON and 15-acetyldeoxynivalenol-producing) type. It was determined that specific lineages may contain two or three chemotypes (O'Donnell *et al.* 2000; Ward *et al.* 2002). Recently, it was revealed that a defective *Tri13* in the trichothecene gene cluster is responsible for the switch from NIV to DON production in *G. zeae* (Lee *et al.* 2002). Phylograms inferred from eight genes in the trichothecene gene cluster (Ward *et al.* 2002) and cladograms inferred from *Tri13* (Brown *et al.* 2002) showed congruence of chemotypes. Ward *et al.* (2002) interpreted these results to mean that the various trichothecene chemotypes were maintained within lineages through multiple speciation events by balancing selection. Balancing selection is also thought to be responsible for maintaining allelic diversity at the heterokaryon incompatibility-*c* locus of *Neurospora crassa* (Wu *et al.* 1998) and mating compatibility *b1* locus of *Coprinus cinereus* (May *et al.* 1999).

3.4 Other Species

F. sporotrichioides and related species have been examined by PFGE and similar EK with two intensively stained bands larger than 4.6 Mb and one or two bands smaller than 2 Mb have been observed among isolates of *F. sporotrichioides*, *F. poae*, *F. fusarioides* and *F. acuminatum* subspecies *armeniacum* (Fekete *et al.* 1993; Nagy and Hornok 1994, Nagy *et al.* 1995). Several chromosomes apparently co-migrate within the second largest band (4.6-6.1 Mb) in these species. The *TRI5* gene, which is located in the *F. sporotrichioides* trichothecene biosynthetic gene cluster (Brown *et al.* 2001), hybridized to the comigrating chromosome band in the different species examined except for *F. fusarioides* (Fekete *et al.* 1993; Nagy and Hornok 1994). As is the case for other *Fusarium* species (Miao *et al.* 1991a; Kim *et al.* 1993; Xu *et al.* 1995), a high degree of CLP was observed among chromosomes smaller than 2 Mb in these species (Nagy and Hornok 1994; Nagy *et al.* 1995; Hornok *et al.* 1996). More detailed analysis of a small chromosome of *F. sporotrichioides* identified unique sequences as well as sequences that were also present on larger chromosomes (Nagy *et al.* 1995).

4. PATHOGENICITY GENES

While many *Fusarium* species can cause disease on a wide range of plant species and are also soil saprophytes (Booth 1971), individual strains of *Fusarium* are generally pathogenic only on a narrow range of species (Snyder and Hansen 1940; Snyder and Hansen 1941; Snyder and Hansen 1945). Molecular mechanisms of *Fusarium* pathogenicity are generally poorly understood. However, hypotheses have been put forward concerning *Fusarium* pathogenicity and recently developed molecular techniques, such as gene disruption, are providing means to test these hypotheses. Results of gene disruption studies on *Fusarium* pathogenicity are presented in Table1.

4.1 Cell Wall Degrading Enzymes

Cell wall degrading enzymes are thought to help pathogens directly penetrate host root tissue, and therefore, may be essential for pathogenicity. Studies using antibodies to cutinase (Maiti and Kolattukudy 1979) or pectate lyase (Crawford and Kolattukudy 1987) suggested that cell wall degrading enzymes are important for pathogenicity of *F. solani* f. sp. *pisi*. Transformation of a cutinase gene from *F. solani* into a strain of *Mycosphaerella* that could infect papaya fruit only through wounds enabled tranformants to infect unwounded papaya (Dickman *et al.* 1989).

However, gene disruption studies raised questions about the validity of this concept. A cutinase disruption mutant of *F. solani* f. sp. *pisi* exhibited no difference in pathogenicity compared to a wild-type isolate from which it was derived (Stahl and Schäfer 1992). Similarly, a pectate lyase A disruption mutant of *F. solani* f. sp. *pisi* and an exo-α1,4-polygalacturonase (*pgx4*) disruption mutant of *F. oxysporum* f. sp. *lycopersici* were not decreased in pathogenicity (Guo *et al.* 1996; García-Maceira *et al.* 2000). Genes encoding other cell wall degrading enzymes have also been cloned. The gene coding for *endo*polygalacturonase was cloned from *F. moniliforme* (Caprari *et al.* 1993) and *F. oxysporum* f. sp. *lycopersici* (Arie *et al.* 1998; Di Pietro and Roncero 1998), and the . pectate lyase gene (*pl1*) (Huertas-González *et al.* 1999) and two types of xylanase genes (*xyl2* and *xyl3*) have been cloned from *F. oxysporum* f. sp. *lycopersici* (Ruiz-Roldán *et al.* 1999).

Table 1 Effect of gene disruption on *Fusarium* pathogenicity.

Species	Gene	Gene product	Test plants	Pathogenicity[a]	References
F. avenaceum	*esyn 1*	enniatin synthetase	Potato	R	Herrmann *et al.* 1996a
F. oxysporum f. sp. *lycopersici*	*fmk1*	mitogen activated protein kinase	Tomato	R	Di Pietro *et al.* 2001a
	pgx4	exo-α1,4-polygalacturonase	Tomato	N	García-Maceira *et al.* 2000
	prt1	subtilase	Tomato	N	Di Pietro *et al.* 2001b
f. sp. *melonis*	*ARG1*	argininosuccinate lyase	Melon	R	Namiki *et al.* 2001
Gibberella pulicaris	*Gpabc1*	ATP-binding cassette transporter	Potato	R	Fleißner *et al.* 2002
	Tri5	trichodiene synthase	Parsnip	R	Desjardins *et al.* 1992
			Potato	N	Desjardins *et al.* 1992
Gibberella zeae	*MGV1*	mitogen activated protein kinase	Wheat	R	Hou *et al.* 2002
	Tri5	trichodiene synthase	Corn	N	Proctor *et al.* 1995
			Rye	R	Proctor *et al.* 1995
			Wheat	R	Desjardins *et al.* 1996
Nectria haematococca MPVI	*CTF1α*	cutinase transcription factor	Pea	R	Li *et al.* 2002
	cutinase gene	cutinase	Pea	N	Stahl and Schäfer 1992
			Pea	R	Rogers *et al.* 1994
	MAK1	maackiain monooxygenase	Chickpea	R	Enkerli *et al.* 1998
	PDA1	pisatin demethylase	Pea	R	Wasmann and VanEtten 1996
	pelA	pectate lyase	Pea	N	Guo *et al.* 1996

[a] R: gene disruption resulted reduction of pathogenicity, N: gene disruption did not change pathogenicity

Involvement of the *endo*polygalacturonase gene *pg1* in pathogenicity was examined by transformation of *pg1* into *F. oxysporum* f. sp. *melonis* isolates deficient in *endo*polygalacturonase 1 (Di Pietro and Roncero 1998). Transformants with *pg1* did not exhibit increased pathogenicity on muskmelon (Di Pietro and Roncero 1998). Also, *prt1*, which encodes subtilase in *F. oxysporum* f. sp. *lycopersici*, is assumed to be involved in degradation of protein components in plant cell walls (Di Pietro *et al.* 2001b). However, a *prt1* disruption mutant of the fungus exhibited no detectable difference in their pathogenicity on tomato compared to their wild-type progenitor (Di Pietro *et al.* 2001b). These results would suggest that cell wall degrading enzymes might not be involved in pathogenicity of these *Fusarium* species. However, the involvement of these enzymes in pathogenicity cannot completely be ruled out. For example, a meticulous re-evaluation of a cutinase gdr mutant of *F. solani* f. sp. *pisi* showed reduction of pathogenicity to pea (Rogers *et al.* 1994). A potential problem for evaluation of cell wall degrading enzymes by using gdr mutant may be that functionally similar genes could be present in the genome. To date, three types of cutinase genes (Li *et al.* 2002) and four types of pectate lyase genes (González-Candelas and Kolattukudy 1992; Guo *et al.* 1995b; Guo *et al.* 1995a; Guo *et al.* 1996) have been characterized in *F. solani* f. sp. *pisi*. Normal polygalacturonase activity in *pgx4* gdr mutant of *F. oxysporum* f. sp. *lycopersici* suggests that activity of genes other than *pgx4* contribute to total polygalacturonase activity (García-Maceira *et al.* 2000). In addition to the effect of known or yet undetected genes, synergistic action among cell wall degrading enzymes or other gene products may contribute to pathogenicity. For example, a cutinase gene of *F. solani* f. sp. *pisi* had no detectable effect on pathogenicity of *Cochliobolus heterostrophus* to pea (a pathogen of corn foliage, but not of pea), whereas combinational expression of the cutinase gene and the pisatin demethylase gene (*PDA*), which function in pea phytoalexin detoxification, in *C. heterostrophus* showed increased virulence on pea (Oeser and Yoder 1994).

4.2 Tolerance of Antimicrobial Compounds (Phytoalexins, Phytoanticipins)

Meiotic analysis of *N. haematococca* MP VI revealed a positive correlation between detoxificationof the pea phytoalexin pisatin by the enzyme pisatin demethylase and virulence on pea (VanEtten *et al.* 1980; Tegtmeier and VanEtten 1982; Kistler and VanEtten 1984; VanEtten *et al.* 1989). Two types of pisatin demethylase have been characterized in the fungus: one, PDAL, has low levels of activity and is produced after long exposure to pisatin and the other, PDAH, has moderate to high levels of activity and is produced after only a short exposures to pisatin (Wasmann and VanEtten 1996; VanEtten *et al.* 1989). Two PDAH-encoding genes (*PDAT9*, *PDA1*) and one PDAL-encoding gene (*PDA6-1*) have been cloned (Maloney and VanEtten 1994; Straney and VanEtten 1994; Reimmann and VanEtten 1994). The involvement of pisatin demethylase in pea pathogenicity was confirmed with two experiments. First, the virulence on pea of weakly virulent strains of *N. haematococca* MP VI that did not detoxify pisatin was increased by transformation with *PDAT9* (Ciuffetti and VanEtten 1996), and second, the virulence on pea of a highly virulent strain of the fungus was reduced by disruption of the *PDA1* gene (Wasmann and VanEtten 1996). However, changes in virulence confered by the addition or disruption of a single PDAH gene in these experiments was less than expected based on earlier meiotic analyses of pisatin demethylase activities. As a result VanEtten and coworkers hypothesized that additional pea pathogenicity (*PEP*) genes were present in the *N. haematococca* MP VI genome (Ciuffetti and VanEtten 1996; Wasmann and VanEtten 1996). *PEP* genes were localized to within a 100-kb region of the the 1.6-Mb dispensable chromosome that also carries the *PDA1* gene by a telomere-mediated chromosome breakage approach (Kistler *et al.* 1996, Miao *et al.* 1991b). More recently the *PEP* genes were localized to a gene cluster that includes four putative open reading frames (ORFs) and five genes that are expressed during pea infection (Han *et al.* 2001). At least three genes, *PEP1*, *PEP2* and *PEP5*, contribute to pea pathogenicity (Han *et al.* 2001). Nucleotide sequence analyses conducted in August 2002 did not identify significant sequence similarities between *PEP1* and other genes in the database. In contrast, the predicted *PEP2* protein includes a putative RNP-1 motif found in the conserved RNA-binding domain of polyadenylate-binding and other RNA-binding and the predicted *PEP5* protein has sequence similarity major facilitator superfamily (MFS) transport proteins. These latter proteins are involved in transport of molecules across the cell membrane (Han *et al.* 2001). A nondegrative mechanism for pisatin tolerance in *N. haematococca* MP VI was reported by Denny and VanEtten (1987) and may be explained by the activity of a transport protein such as that encoded by *PEP5* (Flei β ner *et al.* 2002). Transport proteins can contribute to fungal pathogenicity by protection against endogenous (Callahan *et al.* 1999) and/or exogenous toxic compounds (Morrissey and Osbourn 1999; Urban *et al.* 1999). For example, ATP-binding cassette (ABC) transporter genes that confer tolerance to phytoalexins contribute to the pathogenicity of *Gibberella pulicaris* (anamorph: *Fusarium sambucinum*) on potato (Fleiβner *et al.* 2002) and pathogenicity of *Botrytis cinerea* on grapevines (Schoobeek *et al.* 2001). In addition, an ABC transporter gene is necessary for high levels of virulence of *Magnapotha grisea* on rice, but what compounds this transporter transports remains to be determined (Urban *et al.* 1999).

Some isolates of *N. haematococca* MP VI cause disease on chickpea as well as on pea and a positive correlation between virulence on chickpea and tolerance of the chickpea phytoalexin, maackiain (Miao and VanEtten 1992a). Tolerance to maackiain is conferred by the *MAK1* gene, which encodes a cytochrome P450 monooxygenase that detoxifies the phytoalexin. *MAK1* was cloned and localized to the 1.6-Mb dispensable chromosome that also carries the PDAL gene, *PDA6-1* (Miao and VanEtten 1992b; Covert *et al.* 1996). Contribution of *MAK1* to chickpea pathogenicity was examined first by transformation of *MAK1* into a *N. haematococca* MP VI strain that did not detoxify maackiain and was had low virulence on chickpea, and second by disruption of *MAK1* in a strain that was highly virulent on chickpea (Enkerli *et al.* 1998). *MAK1* disruption mutant exhibited lower than expected reductions in virulence on chickpea based on earlier meiotic analyses. Together, these results suggested the presence of additional genes necessary for full virulence on chickpea (Enkerli *et al.* 1998). However, unlike the pea pathogenicity genes, these additional genes were not located on the 1.6-Mb dispensable chromosome (Enkerli *et al.* 1998). Furthermore, Funnell and VanEtten (2002) showed that genes for the pathogenicity of *N. haematococca* on carrot and ripe tomato are also not located on the dispensable chromosome.

The enzyme kievitone hydratase detoxifies the bean phytoalexin kievitone and has long been considered to contribute to the pathogenicity of *F. solani* f. sp. *phaseoli* on bean (Smith and Cleveland 1982; Li *et al.* 1995). Mutants with lowered kievitone hydratase activity were generated by the chemical mutagen, N-methyl-N'-nitro-N-nitrosoguanidine (Smith *et al.* 1984). Although the mutants exhibited reduced virulence on bean, mutational effects other than lowered kievitone hydratase activity could not be ruled out, as mutants also exhibited morphological changes (Smith *et al.* 1984).

α-tomatine is a saponin synthesized by *Lycopersicon* and other solanaceous species (Roddick 1974) and is considered to be a phytoanticipin (VanEtten *et al.* 1995). Many fungal pathogens of tomato have tolerance to α-tomatine (Sandrock and VanEtten 1998) and some such as *F. oxysporum* f. sp. *lycopersici* (Lairini *et al.* 1996) and *F. solani* (Lairini and Ruiz-Rubio 1998) produce the tomatine detoxifying enzyme, tomatinase. Recently, the tomatinase gene (*FoTom1*) was cloned from *F. oxysporum* f. sp. *lycopersici* and its expression followed during tomato infection (Roldán-Arjona *et al.* 1999). The significance of tomatinase for tolerance to tomatine and for tomato pathogenicity was investigated in the tomato leaf spot pathogen *Septoria lycopersici* (Martin-Hernandez *et al.* 2000). Tomatinase gene disruption mutants of this fungus exhibited lowered tolerance to tomatine but were not affected in their ability to produce macroscopic lesion on tomato leaves (Martin-Hernandez *et al.* 2000). A lack of corrletion between phytoanticipin tolerance and virulence has been reported in other host-pathogen systmes. The corn pathogen *G. moniliformis* can detoxify and is tolerant of the corn antimicrobial compounds, 6-methoxy-2-benzoxazolinone (MBOA) and 2-benzoxazolinone (BOA) (Richardson and Bacon 1995; Yue *et al.* 1998). However, BOA-sensitive and BOA-tolerant strains exhibited the same levels of virulence on corn seedlings (Glenn *et al.* 2002). Some plant species produce the phytoanticipin cyanide, and at least some pathogens of these plants can detoxify cyanide via cyanide hydratase. Genes encoding this enzyme have been cloned from *Gloeocercospora sorghi* (Wang and VanEtten 1992) and *F. lateritium* (Cluness *et al.* 1993), however disruption of the gene in *G. sorghi* did not alter virulence of the fungus on cyanogenic sorghum (Wang *et al.* 1999).

4.3 Secondary Metabolites

Species of *Fusarium* produce a variety of secondary metabolites including mycotoxins , which are defined as low molecular weight fungal metabolites that are toxic to vertebrates (Desjardins and Hohn 1997; Vesonder and Golinski 1989). Trichothecene mycotoxins inhibit protein synthesis in animals and yeast and they are also toxic to plants (Casale and Hart 1988; Wakulinski 1989). Consequently, trichothecenes were thought to contribute to the phytopathogenicity of *Fusarium* species in relatively early studies of these compounds (Wakulinski 1989; Desjardins and Hohn 1997). UV irradiation-induced mutants of *F. sporotrichioides* that were unable to produce the trichothecene T-2 toxin and exhibited reduced virulence on parsnip roots provided the first direct evidence for the role of trichothecenes in plant pathogenesis (Desjardins *et al.* 1989). Mutants of *G. pulicaris* and *G. zeae* carrying a disrupted *TRI5* gene (formerly *Tox5*), which encodes the enzyme, trichothecene synthase, that catalyzes the first committed step in trichothecene biosynthesis, provided further evidence for the role of trichothecenes in pathogenesis (Desjardins *et al.* 1992; Hohn and Desjardins 1992; Proctor *et al.* 1995; Desjardins *et al.* 1996). *TRI5* disruption mutants of *G. pulicaris* no longer produced 4, 15-diacetoxyscirpenol and exhibited reduced virulence on parsnip root, but retained full virulence on potato tubers (Desjardins *et al.* 1992). *TRI5* disruption mutants of *G. zeae* no longer produce deoxynivalenol and exhibited reduced virulence on wheat (Desjardins *et al.* 1996) and rye, but retain full virulence on corn seedlings (Proctor *et al.* 1995). Although physiological changes in the disruption mutants other than the loss of trichothecene production could not be ruled out, these studies strongly suggest that trichothecenes are virulence factors in some host plant-pathogen combinations (Proctor *et al.* 1995). Similar to trichothecenes, fumonisin is involved in phytotoxicity (Van Asch *et al.* 1992; Lamprecht *et al.* 1994). An association between fumonisin production and pathogenicity on corn seedlings was identified in a genetic analysis of *G. moniliformis* (Desjardins *et al.* 1995). However, more recent studies of disruption mutants of the *FUM1* gene, which encodes a polyketide synthase gene required for fumonisin production, indicate that fumonisin production does not make a major contribution to the virulence of *G. moniliformis* on corn ears (Proctor *et al.* 1999; Seo *et al.* 2001; Desjardins *et al.* 2002).

Phytotoxins such as fusaric acid (5-n-butylpicolinic acid) and enniatin (N-methylated cyclic hexadepsipeptides) are also produced by *Fusarium* species (Vesonder and Golinski 1989). A positive correlation between fusaric acid content and pathogenicity was observed in *F. oxysporum* f. sp. *niveum*, although no correlation was observed for several other formae speciales (Davis 1969). The phytotoxin enniatin is synthesized by a nonribosomal thiotemplate mechanism catalyzed by the multifunctional enzyme, enniatin synthetase (Zocher and Kleinkauf 1978; Zocher *et al.* 1982; Zocher *et al.* 1983). The gene encoding the enzyme (*esyn1*) has been cloned from *F. scirpi* (Haese *et al.* 1993) and *F. avenaceum* (Herrmann *et al.* 1996a). An *esyn1* disruption mutant of *F. avenaceum* did not product enniatins and exhibited reduced virulence on potato tuber (Herrmann *et al.* 1996a). A 17-kDa protenaceous phytotoxin is also produced by *F. solani* f. sp. *glycines*, which causes the sudden death syndrome on soybean (Jin *et al.* 1996). Its role in the pathogenesis of this fungus has yet to be determined. Together, these studies indicate that secondary metabolites produced by *Fusarium* can function as virulence factors.

4.4 Signal Transduction

Mitogen activated protein (MAP) kinases are involved in the transduction of various extracellular signals (Schaeffer and Weber 1999). Three types of MAP kinases have been characterized (Xu 2000). Fus3/Kss1, Hog1 and Slt2 type MAP kinases are involved in mating and filamentation, growth under high osmotic conditions, and cell integrity, respectively, in *Saccharomyces cerevisiae* (Xu 2000). Recently, the importance of MAP kinases in pathogenicity was shown for several fungal pathogens (Xu and Hamer 1996; Lev *et al.* 1999; Takano *et al.* 2000). In *F. oxysporum* f. sp. *lycopersici*, disruptionof the *fmk1* gene, which encodes a Fus3/Kss1 type MAP kinase, impaired attachment to tomato roots and markedly reduced pathogenicity on tomato (Di Pietro *et al.* 2001a). In *F. graminearum*, disruption of the *MGV1* gene, which encodes a Slt2 type MAP kinase resulted in defective cell walls and markedly reduced virulence on wheat heads (Hou *et al.* 2002). Several protein kinase genes, including Fus3/Kss1 type MAP kinase (*FsMAPK*), have also been cloned from *F. solani* f. sp. *pisi* (Kolattukudy *et al.* 1995; Li *et al.* 1997). Inhibition of cAMP-dependent protein kinase prevented flavonoid-responsive germination of *F. solani* f. sp. *pisi* (Ruan *et al.* 1995).

4.5 Others

Restriction enzyme-mediated integration (REMI) mutagenesis has been used to isolate genes involved in pathogenicity of *F. oxysporum* f. sp. *melonis* (Inoue *et al.* 2001; Namiki *et al.* 2001). *ARG1*, which encodes the enzyme that catalyzes the last step in arginine biosynthesis, was indicated to be involved in pathogenicity on melon (Namiki *et al.* 2001). Mutants impaired in pathogenicity, which require methionine or partially require histidine for growth have also been identified in REMI mutants of *M. grisea* (Sweigard *et al.* 1998; Balhadère *et al.* 1999).

Mutational race change has been examined for *F. oxysporum* f. spp. *melonis* (Burnett 1983) and *lycopersici* (Mes *et al.* 1999a). It is generally assumed that *F. oxysporum* f. sp. *lycopersici* race 1 carries the avirulence gene *avr1* corresponding to the tomato *Fusarium* resistance gene *I* (or *I-1*) and race 2 carries at least *avrI-2* corresponding to the resistance gene *I-2* (Mes *et al.* 1999a). One mutant that showed virulence on tomato with *I-2* was generated from *F. oxysporum* f. sp. *lycopersici* race 2 by gamma-irradiation (Mes *et al.* 1999a). Although a 3.75-Mb chromosomal translocation was detected in this mutant, *avrI-2* remains to be cloned (Mes *et al.* 1999b).

Gene expression factors have been studied mainly for cutinase genes (Li and Kolattukudy 1995; Li and Kolattukudy 1997; Li *et al.* 2002) and *PDA* (Straney and VanEtten 1994) of *F.*

solani. f. sp. *pisi*. CTF1α containing a Zn(II)2Cys6 binuclear cluster DNA binding motif is one of the cutinase transcription factors in *F. solani* f. sp. *pisi* (Li and Kolattukudy 1997). The Zn(II)2Cys6 binuclear cluster DNA binding motif was characterized in fungal transcriptional regulator proteins by Todd and Andrianopoulos (1997). Transcription factors with the Zn(II)2Cys6 binuclear cluster motif appear to be involved in mycelial melanin biosynthesis in the cucumber anthracnose pathogen, *Colletotrichum lagenarium*, and in *M. grisea* (Tsuji *et al.* 2000). Disruption of CTF1α in *F. solani* f. sp. *pisi* resulted in no detectable change in growth rate but eliminated pathogenicity, although the reduction in cutinase production resulting from the CTF1α disruption can not fully explain the loss of virulence (Li *et al.* 2002).

5. MODERN TECHNOLOGY FOR GENOMIC ANALYSES

New technologies for genomic research are developing rapidly. High performance DNA sequence technologies enable whole genome sequencing s(Pennisi 2001). Genome sequences of *Neurospora crassa* (ca 38 Mb) and *M. grisea* (ca 38 Mb) are already available on the worldwide web (http://www-genome.wi.mit.edu/annotation/). The *G. zeae* international genomics consortium has obtained funding for whole genome sequencing of *G. zeae* (http://www.crl.umn.edu/scab/gz-consort.html). Development of DNA microarray technologies facilitates monitoring of gene expression on a genome-wide scale. As described above, targeted gene disruption has significantly contributed to our understanding of *Fusarium* pathogenicity. Gene-tagging techniques, such as REMI and transposon tagging are powerful tools to discover new genes associated with pathogenicity because no prior knowledge of gene products is needed. In future, these technologies, together with genome-wide analyses promise to advance molecular studies of *Fusarium*.

5.1 Functional Genomics

Transformation systems are critical tools for genomic studies. Protoplasts are generally needed for *Fusarium* transformation (Kistler and Benny 1988; Richey *et al.* 1989; Powell and Kistler 1990; Hohn and Desjardins 1992; Tudzynski *et al.* 1996), though efficient protoplast formation can be a cumbersome task or is not even attained for some interesting species or isolates. Recently, an *Agrobacterium tumefaciens*-mediated transformation system that is not based on protoplasts, has been developed for *F. circinatum* (Covert *et al.* 2001) and *F. oxysporum* (Mullins *et al.* 2001).

The targeted gene disruption method has been successfully applied to *Fusarium* species (Desjardins *et al.* 1992; Stahl and Schäfer 1992; Proctor *et al.* 1995; Herrmann *et al.* 1996b; García-Maceira *et al.* 2000). However, infrequent targeted integration and the need for independent construction of gene disruption vectors are impeding high throughput (large-scale) gene disruption systems in filamentous fungi (Sweigard and Ebbole 2001). Although targeted integration frequencies seem to be somewhat locus dependent, frequencies can generally be raised by increasing the DNA fragment size in the disruption vector (Hamer *et al.* 2001a). Infrequent targeting may also be improved by inducing expression of the target gene during transformation. When protoplasts were produced from germlings under conditions that induce expression of the polygalacturonase gene (*pgx4*) and then were used for *pgx4* disruption, high-rate targeting was observed in *F. oxysporum* (García-Maceira *et al.* 2000). Utilization of *in vitro* transposition systems may facilitate time-consuming vector construction (Hamer *et al.* 2001a).

Transposon-arrayed gene knockouts (TAGKO) were successful in *M. grisea* and *Mycosphaerella graminicola* (Hamer *et al.* 2001a).

REMI has been used for tagging fungal pathogenicity genes (Bölker *et al.* 1995; Akamatsu *et al.* 1997; Sweigard *et al.* 1998; Namiki *et al.* 2001). However, REMI mutants are not always tagged by the transforming DNAs (Sweigard *et al.* 1998). In *G. fujikuroi* (synonym *G. fujikuroi* MP C, anamorph: *Fusarium fujikuroi*), the transformation process itself lead to deletions and chromosome translocations (Linnemannstöns *et al.* 1999). This is a more significant problem in imperfect fungi than perfect fungi, the reason being that although mutation tagging can be confirmed by cosegregation of the mutant phenotype and the transforming DNA after crossing (Sweigard *et al.* 1998), this is not possible for fungi without a sexual stage. In *A. tumefaciens*-mediated transformations, T-DNA was randomly inserted into the *F. oxysporum* genome (Mullins *et al.* 2001). When *A. tumefaciens* growing in the absence of acetosyringone was used for transformation of *F. oxysporum*, over 80% transformants had a single T-DNA insert (Mullins *et al.* 2001). Class II transposons such as *impala* from *F. oxysporum* have been considered as gene tagging tool (Daboussi and Langin 1994; Kempken and Kück 1998). The successful use of *impala* for tagging pathogenicity genes was demonstrated for *F. oxysporum* f. sp. *melonis* (Migheli *et al.* 2000). The *ORP1* gene involved in pathogenicity in *M. grisea* was recently found by using *impala* for gene tagging (Villalba *et al.* 2001). In addition to *F. oxysporum*, the *impala* was shown to transpose autonomously in *F. moniliforme* (Hua-Van *et al.* 2001). T-DNA or transposon tagging might be able to reduce mutants that are not tagged by the transforming DNAs.

While the disruption and tagging methods described above target small stretches of DNA, usually a single gene, other methods can be used for larger size-DNA manipulation in fungi. A 100 kb deletion from a chromosomal end was directed by transformation with a plasmid containing the telomere repeat sequence in *N. haematococca* MP VI (Kistler *et al.* 1996). Exposure to benomyl (methyl 1-(butylcarbamoyl) benzimidazol-2-yl carbamate) that is known to destroy astral microtubules (Aist and Bayles 1991) induced chromosome loss in *N. haematococca* MP VI (VanEtten *et al.* 1998). Large DNA fragments can also be directly transformed into fungi. A BAC clone of at least 113 kb was transformed into *M. grisea* (Diaz-Perez *et al.* 1996). Surviving colonies that were obtained by telomere-mediated chromosome breakage or benomyl treatment contained deletions only in dispensable terminal regions or chromosomes. However, *in vivo* manipulation of large size DNA, as well as small size DNA, may be useful tool for genomic analysis as 100 kb deletion in *N. haematococca* MP VI revealed the clustering of PEP and its chromosomal location (Kistler 1997b; VanEtten *et al.* 1998).

5.2 Bioinformatics

Expressed sequence tags (EST) and/or genomic sequencing projects are not an end in itself, but can be tied directly to functional analyses (Sweigard and Ebbole 2001). The development of EST data provides substantial gene inventories in the absence of gene information (Soanes *et al.* 2002). EST databases have recently been generated for two *Fusarium* species, *G. zeae* and *F. sporotrichioides* (http://www.genome.ou.edu/fsporo.html). The *G. zeae* libraries were generated from mycelia under carbon or nitrogen starvation and from maturing perithecia (http://www.msu.edu/~trail/index1.htm). Also, Kruger *et al.* (2002) generated 4,838 ESTs from a cDNA library prepared from wheat spikes of the partially resistant cultivar Sumai 3, infected with *G. zeae* and found sixteen genes that might be specifically expressed in *G. zeae* during infection. Yoder and Turgeon (2001) compared the genome sequences of *Neurospora*, *Cochliobolus*, *Fusarium*, *Botrytis*, and yeast and found major differences in the GC content of these fungi (from a low of 40 % in yeast to a high of 56 % in *Neurospora*). They also noted that generally, pathogens seem to carry more genes dedicated to secondary metabolism than saprophytes. The pathogenic fungi *Cochliobolus, Fusarium,* and *Botrytis* were rich in nonribosomal peptide synthetases and polyketide synthases, whereas the genomes of the saprophytes *Neurospora, Ashbya,* and *Saccharomyces* encoded few or none of these proteins (Yoder and Turgeon 2001). Skinner *et al.* (2001) showed the presence of homologues of pathogenicity genes in various fungal species.

Comparative analyses of genome structure should reveal regions of synteny among *Fusarium* species or with other fungal species. Microsynteny was observed between *M. grisea* and *N. crassa* (Hamer *et al.* 2001b). The muskmelon pathogen, *Myrothecium roridum*, has been known to produce a macrocyclic trichothecene, roridin, and contains a gene cluster, which is apparently homologous to the trichothecene gene clusters of *G. zeae* and *F. sporotrichioides* (Trapp *et al.* 1998). Whereas the gene organization in the cluster was highly conserved in *G. zeae* and *F. sporotrichioides* (Brown *et al.* 2001), the cluster of *M. roridum* showed organizational differences. For example, the distance between *Tri4* and *Tri5* was 40 kb in *M. roridum*, but only 8 kb in *F. sporotrichioides*, the relative gene orientations of *Tri4*, *Tri5* and *Tri6* were different from that of *G. zeae* and *F. sporotrichioides* (Trapp *et al.* 1998).

International nucleotide sequence databases, such as GenBank, have become indispensable tools for genome research (Kang *et al.* 2002). An internet-based database including phenotypic data, such as toxin production, and molecular genetical data, such as fingerprinting pattern, also has been proposed for fungal pathogens (Kang *et al.* 2002). In *F. oxysporum*, an international numbering system for vegetative compatibility groups is being administrated through the worldwide web (http://www.cdl.umn.edu/scab/vcg.html) (Kistler *et al.* 1998). *Fusarium* species are ubiquitous fungi. Strain information annotated with biological and molecular genetical data, published as an internet-based database would be an invaluable tool to study genetic diversity in *Fusarium* species world-wide.

6. CONCLUSION

Recent genomic studies have made a great contribution to the understanding of many aspects of phytopathogenic *Fusarium*. Molecular phylogenetic analyses can assist in excluding morphological characters that are not correlated with genome evolution and therefore may have lead to controversial classifications in the past. Given the instability of morphology and the general paucity of distinct morphological characters in *Fusarium*, molecular phylogenetic analyses are of considerable importance for both the improvement of current classification and also for species identification of isolates of interest. Although biological species are a good indicator of genetically isolated groups, taking into account that many species lack a teleomorphic stage, the application of molecular phylogenetics is essential for the correct classification of this diverse group. In addition, molecular phylogenetic studies also contribute to the study of pathogenicity evolution. Polyphyly of some formae speciales of *F. oxysporum* and *F. redolens* suggest multiple origin of their pathogenicity to the same host. Additional analyses of pea pathogenicity genes in *F. oxysporum* f. sp. *pisi* and *N. haematococca* MP VI may support the hypothesis that some pathogenicity genes have their origin through a horizontal transmission event. Although direct observation of horizontal gene transfer under experimental conditions is difficult, horizontal gene transfer may be an important mechanisms for acquisition of pathogenicity.

Recently developed genetic manipulation techniques are useful to reveal pathogenicity genes. These studies have indicated that tolerance against antimicrobial compounds, signal transduction systems and some secondary metabolites are involved in the pathogenesis of some *Fusarium* species. However, *Fusarium* pathogenicity is still not well understood. Why are some isolates of *Fusarium* parasitic, while many others are not? Why are some *Fusarium* pathogenic to only a limited range of host plant species? The studies reviewed here are only a glimmer of discoveries to come.

ADDENDUM

Three research papers pertinent to this review were published after completion of the review. Jain *et al.* (Curr Genet 2002. 41: 407-413) and Inoue *et al.* (Plant Cell 2002. 14: 1869-1883) demonstrated the reduction of pathogenicity in *F. oxysporum* by disruption of *fga1*, which encodes a G protein α subunit and *FOW1*, which encodes a mitochondrial carrier protein, respectively. Trail *et al.* (Fungal Genet Biol 2002. in press) analyzed 7,996 ESTs generated from *G. zeae* mycelia under carbon or nitrogen starvation and from maturing perithecia (http://www.msu.edu/~trail/index1.htm).

Acknowledgement: We thank L Rosewich Gale and HC Kistler for critically reading this manuscript and for valuable comments.

REFERENCES

Aist JR and Bayles CJ (1991). Organelle motility within mitotic asters of the fungus *Nectria haematococca*. Eur J Cell Biol 56: 358-363.

Akamatsu H, Itoh Y, Kodama M, Otani H and Kohmoto K (1997). AAL-toxin-deficient mutants of *Alternaria alternata* tomato pathotype by restriction enzyme-mediated integration. Phytopathology 87: 967-972.

Akamatsu H, Taga M, Kodama M, Johnson R, Otani H and Kohmoto K (1999). Molecular karyotypes for *Alternaria* plant pathogens known to produce host-specific toxins. Curr Genet 35: 647-656.

Alexander NJ, McCormick SP and Hohn TM (1999). *Tri12*, a trichothecene efflux pump from *Fusarium sporotrichioides*: gene isolation and expression in yeast. Mol Gen Genet 261: 977-984.

Aloi C and Baayen RP (1993). Examination of the relationships between vegetative compatibility groups and races in *Fusarium oxysporum* f. sp. *dianthi*. Plant Pathol 42: 839-850.

Alves-Santos FM, Benito EP, Eslava AP and Díaz-Mínguez JM (1999). Genetic diversity of *Fusarium oxysporum* strains from common bean fields in Spain. Appl Environ Microbiol 65: 3335-3340.

Aoki T (1998). Current situation of the taxonomy of the genus *Fusarium*. Soil Microorganisms 52: 73-83.

Aoki T and O'Donnell K (1999). Morphological and molecular characterization of *Fusarium pseudograminearum* sp. nov., formerly recognized as the Group 1 population of *F. graminearum*. Mycologia 91: 597-609.

Appel DJ and Gordon TR (1996). Relationships among pathogenic and nonpathogenic isolates of *Fusarium oxysporum* based in the partial sequence of the intergenic spacer region of the ribosomal DNA. Mol Plant-Microbe Interact 9: 125-138.

Arie T, Christiansen SK, Yorder OC and Turgeon BG (1997). Efficient cloning of Ascomycete mating type genes by PCR amplification of the conserved *MAT* HMG box. Fungal Genet Biol 21: 118-130.

Arie T, Gouthu S, Shimazaki S, Kamakura T, Kimura M, Inoue M, Takio K, Ozaki A, Yoneyama K and Yamaguchi I (1998). Immunological detection of *endo*polygalacturonase secretion by *Fusarium oxysporum* in plant tissue and sequencing of its encoding gene. Ann Phytopathol Soc Jpn 64: 7-15.

Austwick PKC (1982). *Fusarium* infections in man and animals. In: MO Moss and JE Smith, ed. The applied mycology of *Fusarium*. Cambridge: Cambridge University Press, pp 129-140.

Baayen RP, O'Donnell K, Bonants PJ, Cigelnik E, Kroon LPNM, Roebroeck E and Waalwijk C (2000). Gene genealogies and AFLP analyses in the *Fusarium oxysporum* complex identify monophyletic and nonmonophyletic formae speciales causing wilt and rot disease. Phytopathology 90: 891-900.

Baayen RP, O'Donnell K, Breeuwsma S, Geiser DM and Waalwijk C (2001). Molecular relationships of fungi within the *Fusarium redolens-F. hostae* clade. Phytopathology 91: 1037-1044.

Balhadère PV, Foster AJ and Talbot NJ (1999). Identification of pathogenicity mutants of the rice blast fungus *Magnaporthe grisea* by insertional mutagenesis. Mol Plant-Microbe Interact 12: 129-142.

Bentley S, Pegg KG, Moore NY, Davis RD and Buddenhagen IW (1998). Genetic variation among vegetative compatibility groups of *Fusarium oxysporum* f. sp. *cubense* analyzed by DNA fingerprinting. Phytopathology 88: 1283-1293.

Boehm EWA, Ploetz RC and Kistler HC (1994). Statistical analysis of electrophoretic karyotype variation among vegetative compatibility groups of *Fusarium oxysporum* f. sp. *cubense*. Mol Plant-Microbe Interact 7: 196-207.

Bölker M, Böhnert HU, Braun KH, Görl J and Kahmann R (1995). Tagging pathogenicity genes in *Ustilago maydis* by restriction enzyme-mediated integration (REMI). Mol Gen Genet 248: 547-552.

Booth C (1971). The Genus *Fusarium*. Commonwealth Mycological Institute, Kew, Surrey, UK pp 1-237.

Bowden RL and Leslie JF (1999). Sexual recombination in *Gibberella zeae*. Phytopathology 89: 182-188.

Brown DW, McCormick SP, Alexander NJ, Proctor RH and Desjardins AE (2001). A genetic and biochemical approach to study trichothecene diversity in *Fusarium sporotrichioides* and *Fusarium graminearum*. Fungal Genet Biol 32: 121-133.

Brown DW, McCormick SP, Alexander NJ, Proctor RH and Desjardins AE (2002). Inactivation of a cytochrome P-450 is a determinant of trichothecene diversity in *Fusarium* species. Fungal Genet Biol 36: 224-233.

Burnett JH (1983). Aspect of *Fusarium* genetics. In: MO Moss and JE Smith, ed. The applied mycology of *Fusarium*. Cambridge: Cambridge University Press, pp 39-69.

Callahan TM, Rose MS, Meade MJ, Ehrenshaft M and Upchurch RG (1999). *CFP*, the putative cercosporin transporter of *Cercospora kikuchii* is required for wild type cercosporin production resistance, and virulence on soybean. Mol Plant-Microbe Interact 12: 901-910.

Caprari C, Richter A, Bergmann C, Lo Cicero S, Salvi G, Cervone F and De Lorenzo G (1993). Cloning and characterization of a gene encoding the *endo*polygalacturonase of *Fusarium moniliforme*. Mycol Res 97: 497-505.

Casale WL and Hart LP (1988). Inhibition of 3H-leucine incorporation by trichothecene mycotoxins in maize and wheat tissue. Phytopathology 78: 1673-1677.

Ciuffetti LM and VanEtten HD (1996). Virulence of a pisatin demethylase-deficient *Nectria haematococca* MP VI isolate is increased by transformation with a pisatin demethylase gene. Mol Plant-Microbe Interact 9: 787-792.

Cluness MJ, Turner PD, Clements E, Brown DT and O'Reilly C (1993). Purification and properties of cyanide hydratase from *Fusarium lateritium* and analysis of the corresponding *chy*1 gene. J Gen Microbiol 139: 1807-1815.

Correll JC, Puhalla JE, Schneider RW and Kraft JM (1985). Differentiating races of *Fusarium oxysporum* f. sp. *pisi* based on vegetative compatibility. Phytopathology 75: 1347 (abstract).

Covert SF, Enkerli J, Miao VPW and VanEtten HD (1996). A gene for maackiain detoxification from a dispensable chromosome of *Nectria haematococca*. Mol Gen Genet 251: 397-406.

Covert SF, Kapoor P, Lee M-h, Briley A and Nairn CJ (2001). *Agrobacterium tumefaciens*-mediated transformation of *Fusarium circinatum*. Mycol Res 105: 259-264.

Crawford MS and Kolattukudy PE (1987). Pectate lyase from *Fusarium solani* f. sp. *pisi*: purification, characterization, *in vitro* translation of the mRNA, and involvement in pathogenicity. Arch Biochem Biophys 258: 196-205.

Daboussi MJ and Langin T (1994). Transposable element in the fungal plant pathogen *Fusarium oxysporum*. Genetica 93: 49-59.

Davière J-M, Langin T and Daboussi M-J (2001). Potential role of transposable elements in the rapid reorganization of the *Fusarium oxysporum* genome. Fungal Genet Biol 34: 177-192.

Davis D (1969). Fusaric acid in selective pathogenicity of *Fusarium oxysporum*. Phytopathology 59: 1391-1395.

Denny TP and VanEtten HD (1987). A possible mechanism of nondegradative tolerance of pisatin in *Nectria haematococca* MP VI. Physiol Plant Pathol 30: 93-107.

Desjardins AE and Hohn TM (1997). Mycotoxins in plant pathogenesis. Mol Plant-Microbe Interact 10: 147-152.

Desjardins AE, Hohn TM and McCormick SP (1992). Effect of gene disruption of trichodiene synthase on the virulence of *Gibberella pulicaris*. Mol Plant-Microbe Interact 5: 214-222.

Desjardins AE, Munkvold GP, Plattner RD, Proctor RH (2002). *FUM1*—a gene required for fumonisin biosynthesis but not for maize ear rot and ear infection by *Gibberella moniliformis* in field tests. Mol Plant-Microbe Interact. 15: 1157-1164.

Desjardins AE, Plattner RD, Nelsen TC and Leslie JF (1995). Genetic analysis of fumonisin production and virulence of *Gibberella fujikuroi* mating population A (*Fusarium moniliforme*) on maize (*Zea mays*) seedlings. Appl Environ Microbiol 61: 79-86.

Desjardins AE, Proctor RH, Bai G, McCormick SP, Shaner G, Buechley G and Hohn TM (1996). Reduced virulence of trichothecene-nonproducing mutants of *Gibberella zeae* in wheat field tests. Mol Plant-Microbe Interact 9: 775-781.

Desjardins AE, Spencer GF, Plattner RD and Beremand MN (1989). Furanocoumarin phytoalexins, trichothecene toxins, and infection of *Pastinaca sativa* by *Fusarium sporotrichioides*. Phytopathology 79: 170-175.

Di Pietro A, García-Maceira FI, Méglecz E, Gutierrez-Corona JF and Roncero MIG (2001a). A MAP kinase of the vascular wilt fungus *Fusarium oxysporum* is essential for root penetration and pathogenesis. Mol Microbiol 39: 1140-1152.

Di Pietro A, Huertas-González MD, Gutierrez-Corona JF, Martínez-Cadena G, Méglecz E and Roncero MIG (2001b). Molecular characterization of a subtilase from the vascular wilt fungus *Fusarium oxysporum*. Mol Plant-Microbe Interact 14: 653-662.

Di Pietro A and Roncero MIG (1998). Cloning, expression, and role in pathogenicity of *pg1* encoding the major extracellular endopolygalacturonase of the vascular wilt pathogen *Fusarium oxysporum*. Mol Plant-Microbe Interact 11: 91-98.

Diaz-Perez SV, Crouch VW and Orbach MJ (1996). Construction and characterization of a *Magnaporthe grisea* bacterial artificial chromosome library. Fungal Genet Biol 20: 280-288.

Dickman MB, Podila GK and Kolattukudy PE (1989). Insertion of cutinase gene into a wound pathogen enables it to infect intact host. Nature 342: 446-448.

Elias KS and Schneider RW (1991). Vegetative compatibility group in *Fusarium oxysporum* f. sp. *lycopersici*. Phytopathology 81: 159-162.

Elias KS, Zamir D, Lichtman-Pleban T and Katan T (1993). Population structure of *Fusarium oxysporum* f. sp. *lycopersici*: restriction fragment length polymorphisms provide genetic evidence that vegetative compatibility group is an indicator of evolutionary origin. Mol Plant-Microbe Interact 6: 565-572.

Enkerli J, Bhatt G and Covert SF (1997). *Nht1*, a transposable element cloned from a dispensable chromosome in *Nectria haematococca*. Mol Plant-Microbe Interact 10: 742-749.

Enkerli J, Bhatt G and Covert SF (1998). Maackiain detoxification contributes to the virulence of *Nectria haematococca* MP VI on chickpea. Mol Plant-Microbe Interact 11: 317-326.

Enkerli J, Reed H, Briley A, Bhatt G and Covert SF (2000). Physical map of a conditionally dispensable chromosome in *Nectria haematococca* mating population VI and location of chromosome breakpoints. Genetics 155: 1083-1094.

Fekete C, Nagy R, Debets AJM and Hornok L (1993). Electrophoretic karyotypes and gene mapping in eight species of the *Fusarium* sections *Arthrosporiella* and *Sporotrichiella*. Curr Genet 24: 500-504.

Fleißner A, Sopalla C and Weltring K-M (2002). An ATP-binding cassette multidrug-resistance transporter is necessary for tolerance of *Gibberella pulicaris* to phytoalexins and virulence on potato tubers. Mol Plant-Microbe Interact 15: 102-108.

Funnell DL and VanEtten HD (2002). Pisatin demethylase genes are on dispensable chromosomes while genes for pathogenicity on carrot and ripe tomato are on other chromosomes in *Nectria haematococca*. Mol Plant-Microbe Interact 15: 840-846.

García-Maceira FI, Di Pietro A and Roncero IG (2000). Cloning and disruption of *pgx4* encoding an in planta expressed exopolygalacturonase from *Fusarium oxysporum*. Mol Plant-Microbe Interact 13: 359-365.

Geiser DM, Juba JH, Wang B and Jeffers SN (2001). *Fusarium hostae* sp. nov., a relative of *F. redolens* with a *Gibberella* teleomorph. Mycologia 93: 670-678.

Glenn AE, Gold SE and Bacon CW (2002). *Fdb1* and *Fdb2, Fusarium verticillioides* loci necessary for detoxification of preformed antimicrobials from corn. Mol Plant-Microbe Interact 15: 91-101.

Gómez-Gómez E, Anaya N, Roncero MIG and Hera C (1999). *Folyt1*, a new member of the hAT family, is active in the genome of the plant pathogen *Fusarium oxysporum*. Fungal Genet Biol 27: 67-76.

González-Candelas L and Kolattukudy PE (1992). Isolation and analysis of a novel inducible pectate lyase gene from the phytopathogenic fungus *Fusarium solani* f. sp. *pisi* (*Nectria haematococca*, mating population VI). J Bacteriol 174: 6343-6349.

Guo W, González-Candelas L and Kolattukudy PE (1995a). Cloning of a new pectate lyase gene *pelC* from *Fusarium solani* f. sp. *pisi* (*Nectria haematococca*, mating Type VI) and characterization of the gene product expressed in *Pichia pastoris*. Arch Biochem Biophys 323: 352-360.

Guo W, González-Candelas L and Kolattukudy PE (1995b). Cloning of a novel constitutively expressed pectate lyase gene *pelB* from *Fusarium solani* f. sp. *pisi* (*Nectria haematococca*, mating type IV) and characterization of the gene product expressed in *Pichia pastoris*. J Bacteriol 177: 7070-7077.

Guo W, González-Candelas L and Kolattukudy PE (1996). Identification of a novel *pelD* gene expressed uniquely in planta by *Fusarium solani* f. sp. *pisi* (*Nectria haematococca*, Mating Type VI) and characterization of its protein product as an endo-pectate lyase. Arch Biochem Biophys 332: 305-312.

Hacker J, Blum-Oehler G, Mühldorfer I and Tschäpe H (1997). Pathogenicity islands of virulent bacteria: structure, function and impact on microbial evolution. Mol Microbiol 23: 1089-1097.

Haese A, Schubert M, Herrmann M and Zocher R (1993). Molecular characterization of the enniatin synthetase gene encoding a multifunctional enzyme catalysing *N*-methyldepsipeptide formation in *Fusarium scirpi*. Mol Microbiol 7: 905-914.

Hamer L, Adachi K, Montenegro-Chamorro MV, Tanzer MM, Mahanty SK, Lo C, Tarpey RW, Skalchunes AR, Heiniger RW, Frank SA, Darveaux BA, Lampe DJ, Slater TM, Ramamurthy L, DeZwaan TM, Nelson GH, Shuster JR, Woessner J and Hamer JE (2001a). Gene discovery and gene function assignment in filamentous fungi. Proc Natl Acad Sci USA 98: 5110-5115.

Hamer L, Pan H, Adachi K, Orback MJ, Page A, Ramamurthy L and Woessner JP (2001b). Regions of microsynteny in *Magnaporthe grisea* and *Neurospora crassa*. Fungal Genet Biol 33: 137-143.

Han Y, Liu X, Benny U, Kistler HC and VanEtten HD (2001). Genes determining pathogenicity to pea are clustered on a supernumerary chromosome in the fungal plant pathogen *Nectria haematococca*. Plant J 25: 305-314.

He C, Rusu AG, Poplawski AM, Irwin JAG and Manners JM (1998). Transfer of a supernumerary chromosome between vegetatively incompatible biotypes of the fungus *Colletotrichum gloeosporioides*. Genetics 150: 1459-1466.

Herrmann M, Zocher R and Haese A (1996a). Effect of disruption of the enniatin synthetase gene on the virulence of *Fusarium avenaceum*. Mol Plant-Microbe Interact 9: 226-232.

Herrmann M, Zocher R and Haese A (1996b). Enniatin production by *Fusarium* strains and its effect on potato tuber tissue. Appl Environ Microbiol 62: 393-398.

Hohn TM and Desjardins AE (1992). Isolation and gene disruption of the *Tox5* gene encoding trichodiene synthase in *Gibberella pulicaris*. Mol Plant-Microbe Interact 5: 249-256.

Hornok L, Fekete C and Giczey G (1996). Molecular characterization of *Fusarium poae*. Sydowia 48: 23-31.

Hou Z, Xue C, Peng Y, Katan T, Kistler HC and Xu J-R (2002). A map kinase gene (*MGV1*) in *Fusarium graminearum* is required for female fertility, heterokaryon formation and plant infection. Mol Plant-Microbe Interact in press.

Howson WT, McGinnis RC and Gordon WL (1963). Cytological studies on the perfect stages of some species of *Fusarium*. Can J Genet Cytol 5: 60-64.

Hua-Van A, Davière J-M, Kaper F, Langin T and Daboussi M-J (2000). Genome organization in *Fusarium oxysporum*: clusters of class II transposons. Curr Genet 37: 339-347.

Hua-Van A, Pamphile JA, Langin T and Daboussi M-J (2001). Transposition of autonomous and engineered *impala* transposons in *Fusarium oxysporum* and a related species. Mol Gen Genet 264: 724-731.

Huertas-González MD, Ruiz-Roldán MC, Maceira FIG, Roncero MIG and Di Pietro A (1999). Cloning and characterization of *pl1* encoding an in planta-secreted pectate lyase of *Fusarium oxysporum*. Curr Genet 35: 36-40.

Inoue I, Ohara T, Namiki F and Tsuge T (2001). Isolation of pathogenicity mutants of *Fusarium oxysporum* f. sp. *melonis* by insertional mutagenesis. J Gen Plant Pathol 67: 191-199.

Ireland KF and Lacy ML (1986). Laboratory identification of pathogenic isolates of *Fusarium oxysporum* f. sp. *apii* race 2. Phytopathology 76: 956 (abstract).

Jacobson DJ and Gordon TR (1988). Vegetative compatibility and self-incompatibility within *Fusarium oxysporum* f. sp. *melonis*. Phytopathology 78: 668-672.

Jin H, Hartman GL, Nickell CD and Widholm JM (1996). Characterization and purification of a phytotoxin produced by *Fusarium solani*, the causal agent of soybean sudden death syndrome. Phytopathology 86: 277-282.

Jurgenson JE, Bowden RL, Zeller KA, Leslie JF, Alexander NJ and Plattner RD (2002a). A genetic map of *Gibberella zeae* (*Fusarium graminearum*). Genetics 160: 1451-1460.

Jurgenson JE, Zeller KA and Leslie JF (2002b). Expanded genetic map of *Gibberella moniliformis* (*Fusarium verticillioides*). Appl Environ Microbiol 68: 1972-1979.

Kang S, Ayers JE, DeWolf ED, Geiser DM, Kuldau G, Moorman GW, Mullins E, Uddin W, Correll JC, Deckert G, Lee Y-H, Lee Y-W, Martin FN and Subbarao K (2002). The internet-based fungal pathogen database: a proposed model. Phytopathology 92: 232-236.

Kawamura C, Tsjimoto T and Tsuge T (1999). Targeted disruption of a melanin biosynthesis gene affects conidial development and UV tolerance in the Japanese pear pathotype of *Alternaria alternata*. Mol Plant-Microbe Interact 12: 59-63.

Kema GHJ, Goodwin SB, Hamza S, Verstappen ECP, Cavaletto JR, Van der Lee TAJ, de Weerdt M, Bonants PJM and Waalwijk C (2002). A combined amplified fragment length polymorphism and randomly amplified polymorphism DNA genetic linkage map of *Mycosphaerella graminicola*, the Septoria tritici leaf blotch pathogen of wheat. Genetics 161: 1497-1505.

Kempken F and Kück U (1998). Transposons in filamentous fungi-facts and perspective. Bioessays 20: 652-659.

Kim DH, Martyn RD and Magill CW (1993). Chromosomal polymorphism in *Fusarium oxysporum* f. sp. *niveum*. Phytopathology 83: 1209-1216.

Kim HG, Meinhardt LW, Benny U and Kistler HC (1995). Nrs1, a repetitive element linked to pisatin demethylase gene on a dispensable chromosome of *Nectria haematococca*. Mol Plant-Microbe Interact 8: 524-531.

Kistler HC (1997a). Genetic diversity in the plant-pathogenic fungus *Fusarium oxysporum*. Phytopathology 87: 474-479.

Kistler HC (1997b). Targeted chromosome breakage in filamentous fungi. Fungal Genet Biol 22: 13-18.

Kistler HC (2001). Evolution of host specificity in *Fusarium oxysporum*. In: B Summerell, ed. Paul E Nelson Memorial Symposium. St. Paul, MN: American Phytopathological Society Press, pp 70-82.

Kistler HC, Alabouvette C, Baayen RP, Bentley S, Brayford D, Coddington A, Correll J, Daboussi M-J, Elias K, Fernandez D, Gordon TR, Katan T, Kim HG, Leslie JF, Martyn RD, Migheli Q, Moore NY, O'Donnell K, Ploetz RC, Rutherford MA, Summerell B, Waalwijk C and Woo S (1998). Systematic numbering of vegetative compatibility groups in the plant pathogenic fungus *Fusarium oxysporum*. Phytopathology 88: 30-32.

Kistler HC and Benny U (1992). Autonomously replicating plasmids and chromosome rearrangement during transformation of *Nectria haematococca*. Gene 117: 81-89.

Kistler HC, Benny U, Boehm EWA and Katan T (1995). Genetic duplication in *Fusarium oxysporum*. Curr Genet 28: 173-176.

Kistler HC and Benny UK (1988). Genetic transformation of the fungal plant wilt pathogen, *Fusarium oxysporum*. Curr Genet 13: 145-149.

Kistler HC, Meinhardt LW and Benny U (1996). Mutants of *Nectria haematococca* created by a site-directed chromosome breakage are greatly reduced in virulence toward pea. Mol Plant-Microbe Interact 9: 804-809.

Kistler HC and Miao VPW (1992). New modes of genetic change in filamentous fungi. Annu Rev Phytopathol 30: 131-152.

Kistler HC, Momol EA and Benny U (1991). Repetitive genomic sequences for determining relatedness among strains of *Fusarium oxysporum*. Phytopathology 81: 331-336.

Kistler HC and VanEtten HD (1984). Regulation of pisatin demethylation in *Nectria haematococca* and its influence on pisatin tolerance and virulence. J Gen Microbiol 130: 2605-2613.

Kolattukudy PE, Rogers LM, Li D, Hwang C-S and Flaishman MA (1995). Surface signaling in pathogenesis. Proc Natl Acad Sci USA 92: 4080-4087.

Kruger WM, Pritsch C, Chao S and Muehlbauer GJ (2002). Functional and comparative bioinformatic analysis of expressed genes from wheat spikes infected with *Fusarium graminearum*. Mol Plant-Microbe Interact 15: 445-455.

Lagopodi AL, Ram AF, Lamers GEM, Punt PJ, Van den Hondel CAM, Lugtenberg BJJ and Bloemberg GV (2002). Novel aspects of tomato root colonization and infection by *Fusarium oxysporum* f. sp. *radicis-lycopersici* revealed by confocal laser scanning microscopic analysis using the green fluorescent protein as a marker. Mol Plant-Microbe Interact 15: 172-179.

Lairini K, Perez-Espinosa A, Pineda M and Ruiz-Rubio M (1996). Purification and characterization of tomatinase from *Fusarium oxysporum* f. sp. *lycopersici*. Appl Environ Microbiol 62: 1604-1609.

Lairini K and Ruiz-Rubio M (1998). Detoxification of α-tomatine by *Fusarium solani*. Mycol Res 11: 1375-1380.

Lamprecht SC, Marasas WFO, Alberts JF, Cawood ME, Gelderblom WCA, Shephard GS, Thiel PG and Calitz FJ (1994). Phytotoxicity of fumonisins and TA-toxin to corn and tomato. Phytopathology 84: 383-391.

Lee T, Han Y-K, Kim K-H, Yun S-H and Lee Y-W (2002). *Tri13* and *Tri7* determine deoxynivalenol- and nivalenol-producing chemotypes of *Gibberella zeae*. Appl Environ Microbiol 68: 2148-2154.
Leslie JF (1991). Mating population in *Gibberella fujikuroi* (*Fusarium* section *Liseola*). Phytopathology 81: 1058-1060.
Leslie JF (1993). Fungal vegetative compatibility. Annu Rev Phytopathol 31: 127-150.
Leslie JF (1995). *Gibberella fujikuroi*: available populations and variable traits. Can J Bot 73(Suppl. 1): S282-S291.
Leslie JF (1996). Fungal vegetative compatibility-promises and prospects. Phytoparasitica 24: 3-6.
Lev S, Sharon A, Hadar R, Ma H and Horwitz BA (1999). A mitogen-activated protein kinase of the corn leaf pathogen *Cochliobolus heterostrophus* is involved in conidiation, appressorium formation, and pathogenicity: diverse role for mitogen-activated protein kinase homologs in foliar pathogens. Proc Natl Acad Sci USA 96: 13542-13547.
Li D, Chung K-R, Smith DA and Schardl CL (1995). The *Fusarium solani* gene encoding kievitone hydratase, a secreted enzyme that catalyzes detoxification of a bean phytoalexin. Mol Plant-Microbe Interact 8: 388-397.
Li D and Kolattukudy PE (1995). Cloning and expression of cDNA encoding a protein that binds a palindromic promoter element essential for induction of fungal cutinase by plant cutin. J Biol Chem 270: 11753-11756.
Li D and Kolattukudy PE (1997). Cloning of cutinase transcription factor 1, a transactivating protein containing Cys6Zn2 binuclear DNA-binding motif. J Biol Chem 272: 12462-12467.
Li D, Rogers L and Kolattukudy PE (1997). Cloning and expression of cDNA encoding a mitogen-activated protein kinase from a phytopathogenic filamentous fungus. Gene 195: 161-166.
Li D, Sirakova T, Rogers L, Ettinger WF and Kolattukudy PE (2002). Regulation of constitutively expressed and induced cutinase genes by different zinc finger transcription factors in *Fusarium solani* f. sp. *pisi* (*Nectria haematococca*). J Biol Chem 277: 7905-7912.
Linnemannstöns P, Voß T, Hedden P, Gaskin P and Tudzynski B (1999). Deletion in the Gibberellin biosynthesis gene cluster of *Gibberella fujikuroi* by restriction enzyme-mediated integration and conventional transformation-mediated mutagenesis. Appl Environ Microbiol 65: 2558-2564.
Maiti IB and Kolattukudy PE (1979). Prevention of fungal infection of plants by specific inhibition of cutinase. Science 205: 507-508.
Maloney AP and VanEtten HD (1994). A gene from the fungal plant pathogen *Nectria haematococca* that encodes the phytoalexin-detoxifying enzyme pisatin demethylase defines a new cytochrome P450 family. Mol Gen Genet 243: 506-514.
Manicom BQ, Bar-Joseph M, Kotze JM and Becker MM (1990). A restriction fragment length polymorphism probe relating vegetative compatibility groups and pathogenicity in *Fusarium oxysporum* f. sp. *dianthi*. Phytopathology 80: 336-339.
Martin-Hernandez AM, Dufresne M, Hugouvieux V, Melton R and Osbourn A (2000). Effects of targeted replacement of the tomatinase gene on the interaction of *Septoria lycopersici* with tomato plants. Mol Plant-Microbe Interact 13: 1301-1311.
Masel AM, Irwin JAG and Manners JM (1993). DNA addition or deletion is associated with a major karyotype polymorphism in the fungal phytopathogen *Colletotrichum gloeosporioides*. Mol Gen Genet 237: 73-80.
Matuo T (1980). Chapter 1. In: T Matuo, H Komada, A Matsda ed. *Fusarium* disease of cultivated plants. Tokyo: Zenkoku Noson Kyokai Publishing, pp 17-59.
Matuo T and Snyder WC (1973). Use of morphology and mating populations in the identification of formae speciales in *Fusarium solani*. Phytopathology 63: 562-565.
May G, Shaw F, Badrane H and Vekemans X (1999). The signature of balancing selection: fungal mating compatibility gene evolution. Proc Natl Acad Sci USA 96: 9172-9177.
Mendgen K, Hahn M and Desing H (1996). Morphogenesis and mechanisms of penetration by plant pathogenic fungi. Annu Rev Phytopathol 34: 367-386.
Mes JJ, Weststeijn EA, Herlaar F, Lambalk JJM, Wijbrandi J, Haring MA and Cornelissen BJC (1999a). Biological and molecular characterization of *Fusarium oxysporum* f. sp. *lycopersici* divides race 1 isolates into separative virulence groups. Phytopathology 89: 156-160.
Mes JJ, Wit R, Testerink CS, De Groot F, Haring MA and Cornelissen BJC (1999b). Loss of avirulence and reduced pathogenicity of a gamma-irradiated mutant of *Fusarium oxysporum* f. sp. *lycopersici*. Phytopathology 89: 1131-1137.
Miao VP, Covert SF and VanEtten HD (1991a). A fungal gene for antibiotic resistance on a dispensable ("B") chromosome. Science 254: 1773-1776.

Miao VPW, Mattews DE and VanEtten HD (1991b). Identification and chromosomal locations of a family of cytochrome P-450 genes for pisatin detoxification in the fungus *Nectria haematococca*. Mol Gen Genet 226: 214-223.

Miao VPW and VanEtten HD (1992a). Genetic analysis of the role of phytoalexin detoxification in virulence of the fungus *Nectria haematococca* on chickpea (*Cicer arietinum*). Appl Environ Microbiol 58: 809-814.

Miao VPW and VanEtten HD (1992b). Three genes for metabolism of the phytoalexin maackiain in the plant pathogen *Nectria haematococca*: meiotic instability and relationship to a new gene for pisatin demethylase. Appl Environ Microbiol 58: 801-808.

Michniewicz M (1989). Growth regulators formed by *Fusaria*: Their significance for fungus growth, sporulation and pathogenicity towards the host plant. In: J Chelkowski, ed. *Fusarium*, Mycotoxins, Taxonomy and Pathogenicity. Amsterdam: Elsevier Scientific Publishers, pp 227-241.

Migheli Q, Berio T and Gullino ML (1993). Electrophoretic karyotypes of *Fusarium* spp. Exp Mycol 17: 329-337.

Migheli Q, Steinberg C, Davière J-M, Olivain C, Gerlinger C, Gautheron N, Alabouvette C and Daboussi M-J (2000). Recovery of mutants impaired in pathogenicity after transposition of Impala in *Fusarium oxysporum* f. sp. *melonis*. Phytopathology 90: 1279-1284.

Miller JD, Greenhalgh R, Wang Y and Lu M (1991). Trichothecene chemotypes of three *Fusarium* species. Mycologia 83: 121-130.

Millus D and McCluskey K (1990). Electrophoretic karyotypes of fungi: the new cytology. Mol Plant-Microbe Interact 3: 351-357.

Momol EA, Martin FN, Kimbrough JW and Kistler HC (1990). Analysis of chromosome-sized DNA molecules in strains of two formae speciales of *Fusarium oxysporum* and protoplast fusion products. Phytopathology 80: 993 (Abstract).

Morrissey JP and Osbourn AE (1999). Fungal resistance to plant antibiotics as a mechanism of pathogenesis. Microbiol Mol Biol Rev 63: 708-724.

Mullins ED, Chen X, Romaine P, Raina R, Geiser DM and Kang S (2001). Agrobacterium-mediated transformation of *Fusarium oxysporum*: an efficient tool for insertional mutagenesis and gene transfer. Phytopathology 91: 173-180.

Nagy R and Hornok L (1994). Electrophoretic karyotype differences between two subspecies of *Fusarium acuminatum*. Mycologia 86: 203-208.

Nagy R, Táborhegyi E, Wittner A and Hornok L (1995). Mini-chromosomes in *Fusarium sporotrichioides* are mosaics of dispersed repeats and unique sequences. Microbiology 141: 713-719.

Namiki F, Matsunaga M, Okuda M, Inoue I, Nishi K, Fujita Y and Tsuge T (2001). Mutation of an arginine biosynthesis gene causes reduced pathogenicity in *Fusarium oxysporum* f. sp. *melonis*. Mol Plant-Microbe Interact 14: 580-584.

Nelson PE, Toussoun TA and Cook RJ (ed.) (1981). *Fusarium*: Diseases, Biology, and Taxonomy. University Park: The Pennsylvania State University Press.

Nelson PE, Toussoun TA and Marasas WFO (1983). *Fusarium* species: an illustrated manual for identification. University Park: The Pennsylvania State University Press.

Nixon KC and Wheeler QD (1990). An amplification of the phylogenetic species concept. Cladistics 6: 211-223.

O'Donnell K (1992). Ribosomal DNA internal transcribed spacers are highly divergent in the phytopathogenic ascomycete *Fusarium sambucinum* (*Gibberella pulicaris*). Curr Genet 22: 213-220.

O'Donnell K (1996). Progress towards a phylogenetic classification of *Fusarium*. Sydowia 48: 57-70.

O'Donnell K (2000). Molecular phylogeny of the *Nectria haematococca-Fusarium solani* species complex. Mycologia 92: 919-938.

O'Donnell K and Cigelnik E (1997). Two divergent intragenomic rDNA ITS2 types within a monophyletic lineage of the fungus *Fusarium* are nonorthologous. Mol Phylogenet Evol 7: 103-116.

O'Donnell K, Cigelnik E and Nirenberg HI (1998a). Molecular systematics and phylogeography of the *Gibberella fujikuroi* species complex. Mycologia 90: 465-493.

O'Donnell K and Gray LE (1995). Phylogenetic relationships of the soybean sudden death syndrome pathogen *Fusarium solani* f. sp. *phaseoli* inferred from rDNA sequence data and PCR primers for its identification. Mol Plant-Microbe Interact 8: 709-716.

O'Donnell K, Kistler HC, Cigelnik E and Ploetz RC (1998b). Multiple evolutionary origins of the fungus causing Panama disease of banana: concordant evidence from nuclear and mitochondrial gene genealogies. Proc Natl Acad Sci USA 95: 2044-2049.

O'Donnell K, Kistler HC, Tacke BK and Casper HH (2000). Gene genealogies reveal global phylogeographic structure and reproductive isolation among lineages of *Fusarium graminearum*, the fungus causing wheat scab. Proc Natl Acad Sci USA 97: 7905-7910.

Oeser B and Yoder OC (1994). Pathogenesis by *Cochliobolus heterostrophus* transformants expressing a cutinase-encoding gene from *Nectria haematococca*. Mol Plant-Microbe Interact 7: 282-288.

Orbach MJ, Vollrath D, Davis RW and Yanofsky C (1988). An electrophoretic karyotype of *Neurospora crassa*. Mol Cell Biol 8: 1469-1473.

Pennisi E (2001). The push to pit genomics against fungal pathogens. Science 292: 2273-2274.

Peterson SW and Logrieco A (1991). Ribosomal RNA sequence variation among interfertile strains of some *Gibberella* species. Mycologia 83: 397-402.

Ploetz RC and Correll JC (1988). Vegetative compatibility among races of *Fusarium oxysporum* f. sp. *cubense*. Plant Dis 72: 325-328.

Powell WA and Kistler HC (1990). *In vivo* rearrangement of foreign DNA by *Fusarium oxysporum* produces linear self-replicating plasmid. J Bacteriol 172: 3163-3171.

Proctor RH, Desjardins AE, Plattner RD and Hohn TM (1999). A polyketide synthase gene required for biosynthesis of fumonisin mycotoxins in *Gibberella fujikuroi* mating population A. Fungal Genet Biol 27: 100-112.

Proctor RH, Hohn TM and McCormick SP (1995). Reduced virulence of *Gibberella zeae* caused by disruption of a trichothecene toxin biosynthetic gene. Mol Plant-Microbe Interact 8: 593-601.

Puhalla JE (1981). Genetic considerations of the genus *Fusarium*. In: PE Nelson, TA Toussoun and RJ Cook ed. *Fusarium*: Diseases, Biology, and Taxonomy. University Park: The Pennsylvania State University Press, pp 291-305.

Puhalla JE (1985). Classification of strains of *Fusarium oxysporum* on the basis of vegetative compatibility. Can J Bot 63: 179-183.

Reimmann C and VanEtten HD (1994). Cloning and characterization of the *PDA6-1* gene encoding a fungal cytochrome *P*-450 which detoxifies the phytoalexin pisatin from garden pea. Gene 146: 221-226.

Richardson MD and Bacon CW (1995). Catabolism of 6-methoxy-benzoxazolinone and 2-benzoxazolinone by *Fusarium moniliforme*. Mycologia 87: 510-517.

Richey MG, Marek ET, Schardl CL and Smith DA (1989). Transformation of filamentous fungi with plasmid DNA by electroporation. Phytopathology 79: 844-847.

Roddick JG (1974). The steroidal glycoalkaloid α-tomatine. Phytochemistry 13: 9-25.

Rogers LM, Flaishman MA and Kolattukudy PE (1994). Cutinase gene disruption in *Fusarium solani* f. sp. *pisi* decreases its virulence on pea. Plant Cell 6: 935-945.

Roldán-Arjona T, Pérez-Espinosa A and Ruiz-Rubio M (1999). Tomatinase from *Fusarium oxysporum* f. sp. *lycopersici* defines a new class of saponinases. Mol Plant-Microbe Interact 12: 852-861.

Rosewich UL and Kistler HC (2000). Role of horizontal gene transfer in the evolution of fungi. Annu Rev Phytopathol 38: 325-363.

Ruan Y, Kotraiah V and Straney DC (1995). Flavonoids stimulate spore germination in *Fusarium solani* pathogenic on legumes in a manner sensitive to inhibitors of cAMP-dependent protein kinase. Mol Plant-Microbe Interact 8: 929-938.

Ruiz-Roldán MC, Di Pietro A, Huertas-González MD and Roncero MIG (1999). Two xylanase genes of the vascular wilt pathogen *Fusarium oxysporum* are differentially expressed during infection of tomato plants. Mol Gen Genet 261: 530-536.

Samuels GJ, Nirenberg HI and Seifert KA (2001). Perithecial species of *Fusarium*. In: B Summerell, ed. Paul E Nelson Memorial Symposium. St. Paul, MN: American Phytopathological Society Press, pp 1-14.

Sandrock RW and VanEtten HD (1998). Fungal sensitivity and enzyme degradation of the phytoanticipin α-tomatine. Phytopathology 88: 137-143.

Schaeffer HJ and Weber MJ (1999). Mitogen-activated protein kinases: specific messages from ubiquitous messengers. Mol Cell Biol 19: 2435-2444.

Schoobeek H, Sorbo GD and De Waard MA (2001). The ABC transporter BcatrB affects the sensitivity of *Botrytis cinerea* to phytoalexin resveratrol and fungicide fenpiclonil. Mol Plant-Microbe Interact 14: 562-571.

Selker EU (1990). Premeiotic instability of repeated sequences in *Neurospora crassa*. Annu Rev Genet 24: 579-613.

Seo J-A, Proctor RH and Plattner RD (2001). Characterization of four clustered and coregulated genes associated with fumonisin biosynthesis in *Fusarium verticillioides*. Fungal Genet Biol 34: 155-165.

Shiflett AM, Enkerli J and Covert SF (2002). *Nht2*, a copia LTR retrotransposon from a conditionally dispensable chromosome in *Nectria haematococca*. Curr Genet 41: 99-106.

Skinner W, Keon J and Hargreves J (2001). Gene information for fungal plant pathogens from expressed sequences. Curr Opin Microbiol 4: 381-386.

Skovgaard K, Nirenberg HI, O'Donnell K and Rosendahl S (2001). Evolution of *Fusarium oxysporum* f. sp. *vasinfectum* races inferred from multigene genealogies. Phytopathology 91: 1231-1237.

Smith DA and Cleveland TE (1982). Relation between production of extracellular kievitone hydratase by isolates of *Fusarium* and their pathogenicity on *Phaseolus vulgaris*. Phytopathology 72: 1319-1323.

Smith DA, Wheeler HE, Banks SW and Cleveland TE (1984). Association between lowered kievitone hydratase activity and reduced virulence to bean in variants of *Fusarium solani* f. sp. *phaseoli*. Physiol Plant Pathol 25: 135-147.

Snyder WC and Hansen HN (1940). The species concept in *Fusarium*. Amer Jour Bot 27: 64-67.

Snyder WC and Hansen HN (1941). The species concept in *Fusarium* with reference to section *Martiella*. Amer Jour Bot 28: 738-742.

Snyder WC and Hansen HN (1945). The species concept in *Fusarium* with reference to *Discolor* and other sections. Amer Jour Bot 32: 657-666.

Soanes DM, Skinner W, Keon J, Hargreaves J and Talbot NJ (2002). Genomics of phytopathogenic fungi and the development of bioinformatic resources. Mol Plant-Microbe Interact 15: 421-427.

Sone T, Abe T, Suto M and Tomita F (1997). Identification and characterization of a karyotypic mutation in *Magnaporthe grisea*. Biosci Biotech Biochem 61: 81-86.

Stahl DJ and Schäfer W (1992). Cutinase is not required for fungal pathogenicity on pea. Plant Cell 4: 621-629.

Steenkamp ET, Wingfield BD, Coutinho TA, Zeller KA, Wingfield MJ, Marasas WFO and Leslie JF (2000). PCR-based identification of *Mat-1* and *Mat-2* in the *Gibberella fujikuroi* species complex. Appl Environ Microbiol 66: 4378-4382.

Straney DC and VanEtten HD (1994). Characterization of the *PDA1* promoter of *Nectria haematococca* and identification of a region that binds a pisatin-responsive DNA binding factor. Mol Plant-Microbe Interact 7: 256-266.

Suga H, Ikeda S, Taga M, Kageyama K and Hyakumachi M (2002). Electrophoretic karyotyping and gene mapping of seven formae speciales in *Fusarium solani*. Curr Genet 41: 254-260.

Sweigard JA, Carroll AM, Farrall L, Chumley FG and Valent B (1998). *Magnaporthe grisea* pathogenicity genes obtained through insertional mutagenesis. Mol Plant-Microbe Interact 11: 404-412.

Sweigard JA and Ebbole D (2001). Functional analysis of pathogenicity genes in a genomics world. Curr Opin Microbiol 4: 387-392.

Szécsi Á and Dobrovolszky A (1985a). Genetic distance in fungus genus *Fusarium* measured by comparative computer analysis of DNA thermal denaturation profiles. Mycopathologia 89: 95-100.

Szécsi Á and Dobrovolszky A (1985b). Phylogenetic relationships among *Fusarium* species measured by DNA reassociation. Mycopathologia 89: 89-94.

Taga M, Murata M and Saito H (1998). Comparison of different karyotyping methods in filamentous ascomycetes - a case study of *Nectria haematococca*. Mycol Res 102: 1355-1364.

Taga M, Murata M and VanEtten HD (1999). Visualization of a conditionally dispensable chromosome in the filamentous ascomycete *Nectria haematococca* by fluorescence *in situ* hybridization. Fungal Genet Biol 26: 169-177.

Takano Y, Kikuchi T, Kubo Y, Hamer JE, Mise K and Furusawa I (2000). The *Colletotrichum lagenarium* MAP kinase gene CMK1 regulates diverse aspects of fungal pathogenesis. Mol Plant-Microbe Interact 13: 374-383.

Taylor JW, Jacobson DJ, Kroken S, Kasuga T, Geiser DM, Hibbett DS and Fisher MC (2000). Phylogenetic species recognition and species concepts in fungi. Fungal Genet Biol 31: 21-32.

Tegtmeier KJ and VanEtten HD (1982). The role of pisatin tolerance and degradation in the virulence of *Nectria haematococca* on peas: a genetic analysis. Phytopathology 72: 608-612.

To-Anun C, Nelson H and Ouchi S (1995). Electrophoretic karyotyping of *Fusarium oxysporum*. Ann Phytopathol Soc Jpn 61: 350-356.

Todd RB and Andrianopoulos A (1997). Evolution of a fungal regulatory gene family: the Zn (II) 2 Cys 6 binuclear cluster DNA binding motif. Fungal Genet Biol 21: 388-405.

Tousson TA and Snyder WC (1961). The pathogenicity, distribution, and control of two races of *Fusarium* (*Hypomyces*) *solani* f. *cucurbitae*. Phytopathology 51: 17-22.

Trapp SC, Hohn TM, McCormick S and Jarvis BB (1998). Characterization of the gene cluster for biosynthesis of macrocyclic trichothecene in *Myrothecium roridum*. Mol Gen Genet 257: 421-432.

Tsuji G, Kenmochi Y, Takano Y, Sweigard J, Farrall L, Furusawa I, Horino O and Kubo Y (2000). Novel fungal transcriptional activators, Cmr1p of *Colletotrichum lagenarium* and Pig1p of *Magnaporthe grisea*, contain Cys2His2 zinc finger and Zn(II)2Cys6 binuclear cluster DNA-binding motifs and regulate transcription of melanin biosynthesis genes in a developmentally specific manner. Mol Microbiol 38: 940-954.

Tudzynski B, Mende K, Weltring K-M, Kinghorn JR and Unkles SE (1996). The *Gibberella fujikuroi niaD* gene encoding nitrate reductase: isolation, sequence, homologous transformation and electrophoretic karyotype location. Microbiology 142: 533-539.

Turgeon BG and Lu S-W (2000). Evolution of host specific virulence in *Cochliobolus heterostrophus*. In: JW Kronstad ed. Fungal pathology. Dordrecht: Kluwer Academic Publishers, pp 93-126.

Tzeng T-H, Lyngholm LK, Ford CF and Bronson CR (1992). A restriction fragment length polymorphism map and electrophoretic karyotype of the fungal maize pathogen *Cochliobolus heterostrophus*. Genetics 130: 81-96.

Urban M, Bhargava T and Hamer JE (1999). An ATP-driven efflux pump is a novel pathogenicity factor in rice blast disease. EMBO J 18: 512-521.

Van Asch MAJ, Rijkenberg FHJ and Coutinho TA (1992). Phytotoxicity of fumonisin B1, moniliformin, and T-2 toxin to corn callus cultures. Phytopathology 82: 1330-1332.

VanEtten H, Jorgensen S, Enkerli J and Covert SF (1998). Inducing the loss of conditionally dispensable chromosomes in *Nectria haematococca* during vegetative growth. Curr Genet 33: 299-303.

VanEtten HD (1978). Identification of additional habitats of *Nectria haematococca* mating population VI. Phytopathology 68: 1552-1556.

VanEtten HD and Kistler HC (1988). *Nectria haematococca*, mating population I and VI. Adv Plant Pathol 6: 189-206.

VanEtten HD, Matthews DE and Matthews PS (1989). Phytoalexin detoxification: importance for pathogenicity and practical implications. Annu Rev Phytopathol 27: 143-164.

VanEtten HD, Matthews PS, Tegtmeier KJ, Dietert MF and Stein JI (1980). The association of pisatin tolerance and demethylation with virulence on pea in *Nectria haematococca*. Physiol Plant Pathol 16: 257-268.

VanEtten HD, Sandrock RW, Wasmann CC, Soby SD, McCluskey K and Wang P (1995). Detoxification of phytoanticipins and phytoalexins by phytopathogenic fungi. Can J Bot 73 (Supplement 1): S518-S525.

Vesonder RF and Golinski P (1989). Metabolite of *Fusarium*. In: J Chelkowski, ed. *Fusarium*, Mycotoxins, Taxonomy and Pathogenicity. Amsterdam: Elsevier Scientific Publishers. pp 1-39.

Villalba F, Lebrun M-H, Hua-Van A, Daboussi M-J and Grosjean-Cournoyer M-C (2001). Transposon *impala*, a novel tool for gene tagging in the rice blast fungus *Magnaporthe grisea*. Mol Plant-Microbe Interact 14: 308-315.

Wakulinski W (1989). Phytotoxicity of *Fusarium* metabolites in relation to pathogenicity. In: J Chelkowski, ed. *Fusarium*, Mycotoxins, Taxonomy and Pathogenicity. Amsterdam: Elsevier Scientific Publishers. pp 257-268.

Wang P, Sandrock RW and VanEtten HD (1999). Disruption of the cyanide hydratase gene in *Gloeocercospora sorghi* increases its sensitivity to the phytoanticipin cyanide but does not affect its pathogenicity on the cyanogenic plant sorghum. Fungal Genet Biol 28: 126-134.

Wang P and VanEtten HD (1992). Cloning and properties of a cyanide hydratase gene from the phytopathogenic fungus *Gloeocercospora sorghi*. Biochem Biophys Res Commun 187: 1048-1054.

Ward TJ, Bielawski JP, Kistler CH, Sullivan E and O'Donnell K (2002). Ancestral polymorphism and adaptive evolution in the trichothecene mycotoxin gene cluster of phytopathogenic *Fusarium*. Proc Natl Acad Sci USA 99: 9278-9283.

Wasmann CC and VanEtten HD (1996). Transformation-mediated chromosome loss and disruption of a gene for pisatin demethylase decrease the virulence of *Nectria haematococca* on pea. Mol Plant-Microbe Interact 9: 793-803.

Wu J, Saupe SJ and Glass NL (1998). Evidence for balancing selection operating at the *het*-c heterokaryon incompatibility locus in a group of filamentous fungi. Proc Natl Acad Sci USA 95: 12398-12403.

Wuchiyama J, Kimura M and Yamaguchi I (2000). A trichothecene efflux pump encoded by *Tri102* in the biosynthetic gene cluster of *Fusarium graminearum*. J Antibiot 53: 196-200.

Xu J-R (2000). MAP kinase in fungal pathogens. Fungal Genet Biol 31: 137-152.

Xu J-R and Hamer JE (1996). MAP kinase and cAMP signaling regulate infection structure formation and pathogenic growth in the rice blast fungus *Magnaporthe grisea*. Genes Dev 10: 2696-2706.

Xu J-R and Leslie JF (1996). A genetic map of *Gibberella fujikuroi* mating population A (*Fusarium moniliforme*). Genetics 143: 175-189.

Xu J-R, Yan K, Dickman MB and Leslie JF (1995). Electrophoretic karyotypes distinguish the biological species of *Gibberella fujikuroi* (*Fusarium* section *Liseola*). Mol Plant-Microbe Interact 8: 74-84.

Yoder OC and Turgeon BG (2001). Fungal genomics and pathogenicity. Curr Opin Plant Biol 4: 315-321.

Yue Q, Bacon CW and Richardson MD (1998). Biotransformation of 2-benzoxazolinone and 6-methoxy-benzoxazolinone by *Fusarium moniliforme*. Phytochemistry 48: 451-454.

Yun S-H, Arie T, Kaneko I, Yoder OC and Turgeon BG (2000). Molecular organization of mating type loci in heterothallic, homothallic, and asexual *Gibberella/Fusarium* species. Fungal Genet Biol 31: 7-20.

Yun S-H, Berbee ML, Yoder OC and Turgeon BG (1999). Evolution of the fungal self-fertile reproductive life style from self-sterile ancestors. Proc Natl Acad Sci USA 96: 5592-5597.

Zhong S, Steffenson BJ, Martinez JP and Ciuffetti LM (2002). A molecular genetic map and electrophoretic karyotype of the plant pathogenic fungus *Cochliobolus sativus*. Mol Plant-Microbe Interact 15: 481-492.

Zocher R, Keller U and Kleinkauf H (1982). Enniatin synthetase, a novel type of multifunctional enzyme catalyzing depsipeptide synthesis in *Fusarium oxysporum*. Biochemistry 21: 43-48.

Zocher R, Keller U and Kleinkauf H (1983). Mechanism of depsipeptide formation catalyzed by enniatin synthetase. Biochem Biophys Res Commun 110: 292-299.

Zocher R and Kleinkauf H (1978). Biosynthesis of enniatin B: partial purification and characterization of the synthesizing enzyme and studies of the biosynthesis. Biochem Biophys Res Commun 81: 1162-1167.

Zolan ME (1995). Chromosome-length polymorphism in fungi. Microbiol Rev 59: 686-698.

Applied Mycology & Biotechnology
An International Series. Volume 4. Fungal Genomics

7

Genomics of *Fusarium venenatum:* An Alternative Fungal Host for Making Enzymes

Randy M. Berka, Beth A. Nelson, Elizabeth J. Zaretsky, Wendy T. Yoder and Michael W. Rey
Novozymes Biotech, Inc., 1445 Drew Avenue, Davis, California 95616-4880, USA
(Rambo@novozymesbiotech.com).

Fusarium venenatum A3/5 (formerly *F. graminearum* Schwabe A3/5) has been used since 1985 as the commercial source of Quorn™ mycoprotein, a processed form of fungal mycelia applied in several human food products to simulate chunks of chicken or beef. Regulatory approval of the organism for human consumption made it an attractive candidate to consider as a host for the production of industrial and food grade enzymes. Systems for genetic manipulation and transformation of *F. venenatum* cells have been developed together with several strong promoters and selectable markers for the introduction and expression of heterologous genes. Recent marketing of a heterologous xylanase and a fungal trypsin have provided a "proof of concept" for *F. venenatum* as a useful alternative to more traditional fungal hosts such as *Aspergillus niger* or *A. oryzae*. However, compared to the latter organisms and well-studied model fungi such as *Neurospora crassa* and *A. nidulans*, information regarding the genomics of *F. venenatum* is inadequate. This chapter provides one of the first overviews of *F. venenatum* genomic information based on a compilation of expressed sequence tags and chromosomal gene sequences to initiate momentum for more comprehensive genome sequencing efforts.

1. INTRODUCTION

The filamentous fungus *Fusarium venenatum* A3/5 is used commercially to produce a high-protein product for human consumption known as Quorn™ mycoprotein (http://www.quorn.com). This product is sold in Europe and the United States as a meat/poultry substitute in burgers, sausages, fillets, chicken-style nuggets, and in several pasta dishes. For a more complete documentary describing the development, production, and safety assessment of Quorn™ mycoprotein, the reader is directed to the recent review article by Wiebe (2002).

Largely because of its history of safe use in human food and its adaptability to large scale fermentation schemes, *F. venenatum* has also been developed as a host for the production of industrial enzymes (Royer *et al.*, 1995; Wiebe *et al.*, 1997). Recently, GRAS status was declared for heterologous xylanase produced in *F. venenatum* (US Food & Drug Administration, Agency Response Letter GRAS Notice No. GRN 000054), and a recombinant fungal trypsin was commercialized from this species (San Diego Metropolitan Daily Business Report, November 7,

2002). This underscores the utility of *F. venenatum* as an alternative to more traditional fungal hosts such as *Aspergillus* species. Detailed descriptions of host strain development and the tools available for genetic manipulation and heterologous gene expression in *F. venenatum* have been described in a topical review by Yoder and Lehmbeck (2003).

Given that *F. venenatum* is now accepted as a source of both edible protein and enzymes for various biotechnological applications, studies directed toward understanding the genome composition and organization are defensible to further enhance and accelerate future improvements in the products that are derived from the fungus. This chapter provides an overview of the first steps taken toward exposing the genetic complement of *F. venenatum* based on analysis of its karyotype, expressed sequence tags (ESTs) and a collection of genomic DNA sequences.

2. CHROMOSOMES

The chromosomes of filamentous fungi are typically small (ranging from 1 to <10 Mbp) compared to those of mammalian and plant species, and they are difficult to visualize microscopically. However, the karyotypes of several fungi including two *Fusarium* species have been resolved using CHEF (contour-clamped homogeneous electric field) gel electrophoresis. For example, the chromosomes of *Gibberella fujikori* (Xu *et al.*, 1995) and *F. oxysporum* (Kistler and Momol, reported in Skinner *et al.* 1991) were successfully separated by electrophoretic methods; however chromosomes from several other species remain recalcitrant to this type of analysis. Other investigators have been unable to resolve the chromosomes from *F. sambucinum*, the closest relative to *F. venenatum* (Weltring and Corby Kistler, personal communications). Since it would be advantageous to enumerate the number and sizes of the *F. venenatum* chromosomes for tracking strain lineages, making chromosome-specific libraries, and mapping genes of interest, we sought to elucidate the electrophoretic karyotype of this fungus.

Attempts were made to separate the chromosomes for several wild type and mutant strains of *F. venenatum* using CHEF (contour clamped homogeneous electric field) electrophoresis. However, under conditions which resulted in successful chromosome separation for the closely related species *F. torulosum* and *F. graminearum* and the more distantly related *F. oxysporum*, none of the *F. venenatum* chromosomes could be resolved. The following hypotheses were proposed to explain these results: (i) *F. venenatum* chromatin is organized in such a way that it prevents resolution of the individual chromosomes on CHEF gels; (ii) chromosomes of *F. venenatum* are large (*i.e.*, greater than 10 Mbp) and do not separate on CHEF gels; (iii) *F. venenatum* has similar numbers of chromosomes compared to other *Fusarium* species reported in the literature, but some chromosomes are linked (end-to-end) or have become circularized.

In relation to the first hypothesis, we discounted the possibility that the chromosomal DNA is intricately complexed with protein, which prevents the DNA from resolving on CHEF gels, by employing an S1 nuclease treatment and pre-treating agarose-embedded DNA with very low doses of gamma irradiation. In both cases we observed subsequent resolution of the broken DNA on CHEF gels. Three observations supported the second hypothesis: (i) By using a telomere-specific oligonucleotide probe, based on the published *F. oxysporum* telomeric consensus sequence ($(TTAGGG)_6$, Southern blots of genomic DNA revealed eight bands, probably representing four discrete chromosomes (Fig. 1); (ii) In *BamHI or Hind*III digests of *Bal*31-restricted *F. venenatum* DNA the fragments hybridizing to the telomeric probe disappeared; (iii) We employed the germ tube burst method (Sato *et al.*, 1998) in combination with fluorescence microscopy to visualize *F. venenatum* chromosomes directly (Fig. 2). At least one of the nuclei appears to show four discrete chromosomes, thereby supporting the hypothesis that the *F. venenatum* karyotype is composed of four large (>10 Mbp) chromosomes. Although it may be

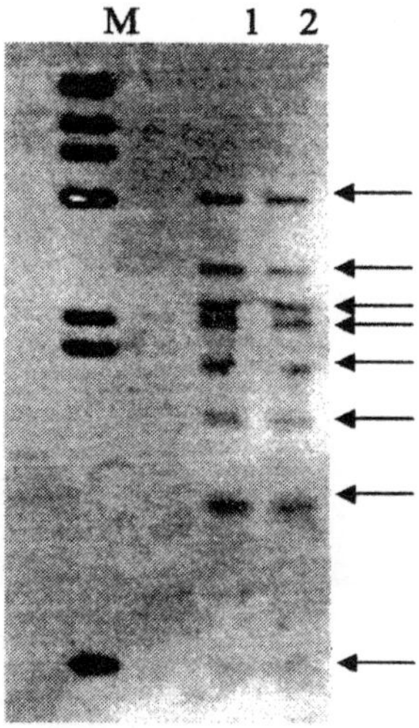

Fig. 1. Southern blot showing the hybridization pattern of a telomeric probe $(TTAGGG)_6$ hybridizing to *F. venenatum* genomic DNA digested with *Hin*dIII. M denotes Boehringer Mannheim size marker #2; Lanes 1 and 2, *F. venenatum* ATCC 20334 *Hind III*-cut genomic DNA. Arrows at the right denote the position of eight hybridization signals (bands) that may correspond to the telomeric repeats in four large chromosomes of *F. venenatum*.

unusual for *Fusarium* species to have such similarly-sized chromosomes, this could be a reflection of the fact that only those species which have been resolved have been reported in the literature. It is quite possible that some species with fewer, equivalently-sized chromosomes exist but that these have not been reported because their large chromosomes have not been amenable to electrophoretic resolution.

The likelihood that the *F. venenatum* genome is comprised of four large chromosomes significantly reduces the chances for constructing chromosome-specific libraries that would be preferred for a genome sequencing effort. Consequently, if such a project were undertaken, the nucleotide sequence information would most likely be gathered using shotgun sequencing strategies. Regardless of the approach used for data collection, the prediction of open reading frames and intron-exon boundaries would be facilitated to a great extent by the availability of EST data. The following section describes an initiative to collect and assemble *F. venenatum* ESTs that have been useful for the development of *F. venenatum* as a host for enzyme production, gene discovery, and for the construction of cDNA microarrays.

3. EXPRESSED SEQUENCE TAGS (ESTs)

Experimental studies of gene expression have entered an era of unprecedented advancement empowered by the rapid development of efficient high-throughput technologies and automated genetic analysis instruments. Large-scale isolation and partial sequencing of anonymous cDNA clones to generate expressed sequence tags (ESTs) (Adams *et al.*, 1991; Matsubara and Okubo,

1993) has enabled the identification of previously unknown genes in a variety of organisms and some measure of their expression levels. Furthermore, analysis of EST libraries provides a "snapshot" of cellular physiology and metabolism at a given point during growth of the organism. Among the filamentous fungi, large-scale EST projects were initiated for *Neurospora crassa* (Nelson *et al.*, 1997) and *Aspergillus nidulans* (Roe *et al.*, 2002) as antecedents to their respective genome sequencing endeavors. Lastly, ESTs from several species have been effective for the development of cDNA and oligonucleotide microarrays (Schena *et al.*, 1995; Marshall

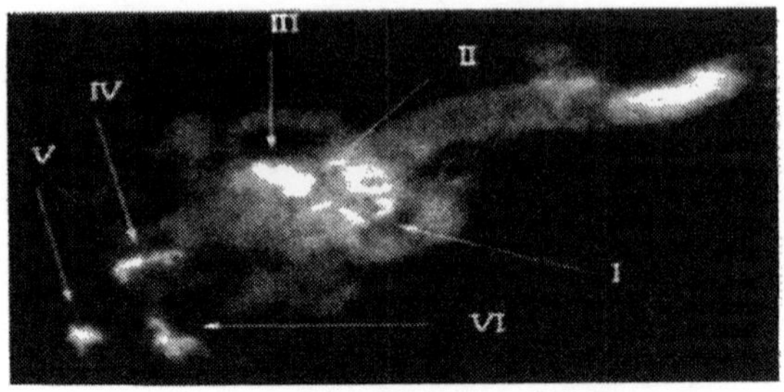

Fig. 2. Chromosomes from six nuclei of a burst hyphal tip of *F. venenatum* (DAPI stain). Nucleus I has the most clearly separated chromosomes, which appear to number four. Nuclei II and III are very close and their chromosomes are not clearly distinguished. The chromosomes of nuclei IV through VI are visible to varying degrees.

and Hodgson, 1998) that are valuable for monitoring global gene expression, gene discovery, detection of mutations and polymorphisms, as well as gene mapping.

This section describes the construction and analysis of directional cDNA (EST) libraries from *Fusarium venenatum* in our laboratories at Novozymes Biotech, Inc. (Rey *et al.*, 2000). The nucleotide sequence information we obtained from these EST libraries was used to generate a cDNA database for *F. venenatum* with aims of (a) identifying abundant cDNAs representing new and potentially strong promoters, (b) classifying ESTs encoding unwanted enzyme activities that could be removed by gene knockouts, (c) discovering sequences encoding new selectable markers for transformation of *F. venenatum*, (d) finding cDNAs encoding new enzyme candidates, and (e) initiating a database of the *F. venenatum* genome/proteome for use in constructing cDNA microarrays.

Several directional cDNA libraries were constructed from *F. venenatum* mRNA (Table 1). Synthesis of cDNA involved a cloning strategy in which the 5'-ends were blunt and the 3'-ends contained a *Not*I site (by virtue of an oligo-dT:*Not*I primer). These asymmetric cDNA molecules were cloned into a pZErO-2 vector (Invitrogen, La Jolla, CA) that had been cleaved with *Eco*RV plus *Not*I. A significant advantage of this vector is the positive selection for inserts that is conferred by the *ccdB* gene, thereby minimizing the background caused by re-ligation of the empty vector during library construction. The mRNA used for cDNA synthesis was extracted from mycelia grown either in two-liter laboratory fermentors or in shake flask cultures grown on various carbon sources and laboratory media. For example, the first library, designated as "Library A," was made using mRNA from cells that were harvested at four days post inoculum from a laboratory fermentor grown on maltodextrin as the principal carbon source, and a second, termed "Library B," was assembled from a six-day sample of the same fermentation. From each unamplified library, we first picked 1192 clones at random, and nucleotide sequence information was obtained from the 5'-end of each cDNA insert. After removal of low quality data, the resulting expressed sequence tags (ESTs) were compared to a non-redundant database (NRDB) using a modified Smith-Waterman (SW) search algorithm. From the 2010 sequences that were compiled with the CAP2 assembler program, an estimated 1206 unique *F. venenatum* genes were represented. From all of the libraries that were constructed (Table 1), we obtained nucleotide sequence information from more than 18,000 clones assembled in contigs and singletons, representing about 4,000 unique *F. venenatum* gene sequences. If we assume that the *F. venenatum* genome is similar in size to that of *N. crassa* (38-42 Mbp encoding approximately 10,000 genes), then it is possible that we have ESTs clones corresponding to more than one-third of the genome. Most of the assembled EST sequences are publicly available in the GENESEQN database (Derwent Information, Alexandria, VA). Approximately 54% of these ESTs showed similarity to known sequences in the NRDB with SW scores of 100 or greater.

3.1 New Promoters

Since neither library A nor B was amplified, the relative abundance of each cDNA clone in the libraries ostensibly reflects the amount of the corresponding mRNA species in the starting material. Thus, the most abundant cDNAs in each collection may represent either mRNAs that are the most stable, or those that are transcribed from strong promoters. The most abundant ESTs in Libraries A and B corresponded to a secreted amyloglucosidase (AMG), a putative vacuolar associated protein subunit (VAPS) and two unknown gene products, each of which comprised about 1-2% of the clones in the libraries (Table 2). Additionally, cDNAs encoding transcription elongation factor 1 (*tef1*), a putative DNA damage response protein (DDRP), and α-amylase were also encountered at a relatively high frequency. Owing to their relative abundance, the corresponding chromosomal genes encoding these products could represent potentially strong promoters for use in *F. venenatum*. To this end, genomic DNA segments encoding several these products were isolated, and the efficacy of their promoters was tested using a lipase reporter gene (Berka *et al.*, 2002). The AMG, VAPS, and Unknown 1 promoter regions used for these constructs were 2.1, 3.1, and 0.9 kb, respectively. We elected to employ around 1 kb or more to ensure that any transcription enhancer elements would be present in the expression constructs. All of these vectors were used to transform an *F. venenatum* recipient strain, and the resulting transformants were tested in shake flask cultures (25 ml) for the ability to express extracellular lipase (Royer *et al.* 1995). Fifteen transformants derived with each vector were analyzed. An untransformed *F. venenatum* strain serving as the negative control produced less than one lipase unit per milliliter (< 1 LU/ml). In contrast, the titers among transformants generated with the AMG-lipase reporter construct ranged from 473 to 2731 LU/ml (mean = 1158 LU/ml; median = 1036 LU/ml), those obtained with the Unknown 1-lipase vector ranged from 184-1967 LU/ml (mean = 894 LU/ml; median = 875 LU/ml), and the yields derived from the VAPS-lipase transformants varied from 125-1990 LU/ml (mean = 337 LU/ml; median = 180 LU/ml). A comparison of the highest yielding transformant containing the AMG-lipase vector to a lipase transformant derived using the *F. oxysporum* trypsin promoter (Royer *et al.*, 1995) demonstrated that higher levels of lipase activity were achieved with the AMG promoter (Berka *et al.* 2002a).

It was not anticipated that AMG mRNA would be the most prevalent species identified. Instead, mRNAs coding for important housekeeping functions were expected to be the most predominant. For example, Nakari *et al.* (1993) found *tef1* cDNA to be the most abundant species in glucose-grown cells of *Trichoderma reesei*. The *tef1* transcript was also abundantly represented in *F. venenatum* libraries A and B, but the frequency at which it was encountered was less than half that of the glucoamylase cDNA. Perhaps the most surprising observation among the *F. venenatum* ESTs was that from the four most prevalent cDNAs, two encoded proteins with unknown functions (Table 2).

3.2 Selectable Marker Genes

Repeated genetic manipulations in fungi often require multiple selectable markers with the appropriate host cell backgrounds for introducing, deleting, disrupting or amplifying specific gene sequences. A source of metabolic genes that confer the ability to grow prototrophically without added amino acids or vitamins can provide the elements for additional vector components that are necessary for genetic transformation of specific auxotrophic mutants. Among the collection of ESTs generated in this work, we identified cDNAs representing a number of potentially useful selectable marker genes. These included genes involved in the biosynthesis of amino acids (e.g., *trpG, met7, proC, arg4, cys3, ilv3, ilv5, trp1*), purines/pyrimidines (e.g., *pur7, pur8, fur1, allA*), and vitamins/cofactors (e.g., *thi4*). The utility of these markers will depend upon the availability of the corresponding auxotrophic mutant strains. However, with the genes in hand it should be possible to generate suitable auxotrophs via gene knock-outs.

3.3 New Enzymes

Sequencing of randomly selected cDNA clones can generate an inventory of genes that are expressed in a culture at the moment the cells are harvested and disrupted. When sequences are analyzed from cells grown on a variety of nutrient sources, it is possible to discover and catalogue cellular enzymes that are produced on those nutrients. At least some of these enzymes, particularly those that are secreted into the medium, may be of interest for various industrial and biotechnological applications.

Among our EST collection we identified several interesting new enzyme genes from *F. venenatum*. These included a constellation of lipolytic activities such as a lysophospholipase (phospholipase B) (Berka *et al.* 2000c) and a triacylglycerol lipase (Tsutsumi *et al.* 2002) having potential applications in baking and for the processing of oils. Both of these genes have been expressed in recombinant strains of *F. venenatum*, and the corresponding enzymes are currently the subjects of additional analysis and characterization.

Several cDNA clones encoding a novel lactonohydrolase were also revealed by random sequencing (Berka *et al.* 2000b). Lactonohydrolase is a type of carboxylic esterase that reversibly catalyzes the hydrolysis of lactones to hydroxy acids, thereby mediating the interconversion between the lactone and acid forms of hydroxyl carboxylic acids. Enzymes of this type may be useful for conversion of aldonate or aromatic lactones as well as for the asymmetric hydrolysis of D-pantoyl lactone which can be employed as a chiral building block for the synthesis of D-pantothenate (Shimazu and Kataoka 1996).

Additionally, ESTs that code for galactose oxidase were discovered among the indiscriminate sequences collected (Golightly *et al.* 2000). Galactose oxidases have been identified in several fungal species, and the corresponding genes have been cloned from *Dactylium dendroides* (McPherson *et al.*, 1992) and *Stigmatella aurantiaca* (Silakowski *et al.* 1998). In the presence of molecular oxygen galactose oxidase catalyzes the oxidation of D-galactose to D-galacto-hexodialdose with the concurrent production of hydrogen peroxide. In addition to D-galactose, the enzyme will oxidize a number of other sugars including dihydroxyacetone, glycerol, raffinose, methyl-α-D-galactopyranose, and methyl-β-D-galactopyranose.

Monogastric animals are unable to hydrolyze the phosphate esters of phytic acid (myo-inositol- hexakisphosphate), and as a result they excrete organic phosphates in their waste, contributing to ground water pollution. Fungal acid phosphatases have been shown to provide synergy with phytase (EC 3.1.3.8) in liberating phosphate from animal feed (Zyla *et al.* 1995 a, b). Acid phosphatase has also shown to be synergistic with glucoamylase in the saccharification of potato maltodextrin, particularly at high substrate concentrations (Zyla 1990). The addition of acid phosphatase may increase the rate of saccharification, possibly allowing for shorter and more efficient hydrolysis reactions. Coincidentally, clones encoding a secreted acid phosphatase were isolated from our inventory *F. venenatum* ESTs (Yaver *et al.* 2001). The corresponding enzyme was produced in recombinant strains for *Aspergillus oryzae* as well as *F. venenatum*.

In the future, agricultural crops may be improved by the incorporation of DNA sequences encoding fungal enzymes into their genomes. For example, head blight caused by *F. graminearum* is a global disease of wheat, maize, barley and other cereals. Devastating losses to wheat and barley crops occurred in the United States and Canada from 1991 to 1997. Clearly, innovative approaches to prevent this disease could prevent severe economic stresses caused by widespread infestations. One possible approach involves the expression of antifungal proteins in cereal crops. Interestingly, several cDNA clones encoding chitinolytic enzymes were revealed in our *F. venenatum* EST collection. These included sequences that direct the synthesis of both endo- and exo-chitinase as well as β-glucanase that most likely function cooperatively to hydrolyze cell wall components during mycelial growth and hyphal extension. An appealing application of these enzymes involved expression of the corresponding cDNAs in hexaploid wheat (Okubara *et al.* 2000, 2001 a, b). The resulting transgenic wheat lines not only transcribed the heterologous fungal sequences but were also more resistant to a challenge infection with *F. graminearum* than their parental lines.

By our arbitrary criteria, more than 45% of the *F. venenatum* ESTs have no significant similarity to sequences in the NRDB. This figure is remarkably consistent with the fraction of known genes in other EST programs and to the percentage of hits obtained when the ORFs of completely sequenced genomes are compared to publicly available databases. The work presented here is by no means exhaustive. Further computational analyses using sequence profiles and motif searching tools may provide added insight about the function of unknown ESTs in this collection.

4. GENOMIC SEQUENCES AND FULL-LENGTH cDNA CLONES

As a direct result of the EST project mentioned above at least 15 full-length cDNA clones (containing at least the translation start region and ATG codon) were identified and sequenced to a high redundancy. In addition, the follow-up work on more than 20 additional ESTs included isolation and nucleotide sequencing of the corresponding genomic DNA segments. Although this subset is by no means extensive, certain trends and observations can be readily deduced about the organization and composition of *F. venenatum* coding sequences in the genome. For example, comparison of the nucleotides immediately upstream of the start codon for these ORFs allows one to derive a consensus sequence for the translation initiation region, commonly referred to as Kozak's box (Kozak 1987) (Table 3). Strikingly, the consensus sequence in this region shows that 94% of the genes have an adenine residue at the -3 position, and the residue at

–5 is a pyrimidine in 91% of the genes. With the exception of positions -6 and -1, there is a general prejudice against G residues in this region. These observations should be considered

Table 1. *F. venenatum* EST libraries constructed for this study.

Library	Amplification	Conditions	Number of clones	% Inserts	Size range (bp)	Avg. size (bp)	# Sequenced
A	N	maltodextrin fermentation	7.5 x 10e4	97	600-2200	1046	958
B	N	maltodextrin fermentation	1.2 x 10e5	97	800-3600	1383	1053
C	Y	maltodextrin fermentation	–	–	–	–	1899
D	Y	maltodextrin fermentation	–	–	–	–	1071
E	Y	Glucose/ammonia ferm.	7.5 x 10e6	90	100-1700	520	480
F	Y	Maltodextrin /ammonia ferm.	<2.5 x 10e6	96	100-2100	672	480
G	Y	Glucose/ammonia ferm.	4.1 x 10e6	92	850-3000	1400	24
H	Y	Maltodextrin /ammonia ferm.	1.2 x 10e7	100	500-3400	1470	24
I	Y	Vogel's + arabinogalactan SF	4.05 x 10e5	96	300-3100	1162	984
J	Y	Vogel's + lactose (α+β) SF	1.1 x 10e6	96	700-3300	1080	1080
K	Y	Vogel's + soyflour SF	6.25 x 10e5	96	500-4000	1303	2219
L	Y	Vogel's + sucrose SF	3.8 x 10e5	88	600-2300	1207	24
M	Y	Sporulation medium SF	2.5 x 10e5	92	550-2500	1108	974
N	Y	Vogel's + pectin SF	6.15 x 10e5	100	700-2800	1200	986
O	Y	Vogel's + xylan SF	8 x 10e5	81	300-1400	720	976
P	Y	Myro medium SF	1.1 x 10e6	100	200-1100	540	1068
Q	Y	Vogel's + PHV SF	3.75 x 10e5	94	500-1400	800	1168
R	Y	Vogel's + soybean oil SF	5 x 10e5	94	300-2300	770	974
S	Y	Vogel's + pannose/maltose SF	3.85 x 10e5	25	300-800	487	1072
T	Y	Vogel's + lignosulfonate SF	1.35 x 10e6	88	350-1950	780	16
U	Y	Vogel's + cellulose/β-lactose SF	2.05 x 10e5	70	700-2400	1100	1074
-	–	–	–	Avg = 89	–	Avg = 987	Total = 18604

SF denotes a culture grown in shake flasks.

Table 2. Most abundant cDNA species found *in F. venenatum* EST libraries A and B.

EST putative identification	Abundance Library A (%)	Abundance Library B (%)
Amyloglucosidase (AMG)	1.8	1.0
Unknown 1	1.8	0.8
Vacuolar associated protein subunit (VAPS)	0.9	1.1
Unknown 2	0.9	0.9
tef1	0.7	0.6
DDRP	0.7	0.6
α-Amylase	0.5	0.4

Table 3. Analysis of nucleotides preceding the start codon in *F. venenatum* genes.

Nucleotide	-10	-9	-8	-7	-6	-5	-4	-3	-2	-1	A	T	G
A	0.38	0.39	0.19	0.43	0.29	0.06	0.19	0.94	0.50	0.25			
C	0.26	0.03	0.44	0.37	0.00	0.35	0.69	0.00	0.33	0.44			
G	0.06	0.11	0.03	0.06	0.31	0.03	0.08	0.06	0.03	0.22			
T	0.29	0.31	0.33	0.14	0.40	0.56	0.03	0.00	0.14	0.08			
Consensus	H	W	H	M	D	Y	M	A	M	V	A	T	G

Ambiguous nucleotides: D (A, G, or T), H (A, C, or T), M (A or C), W, (A or T),V (A, C, or G), and Y (C or T).

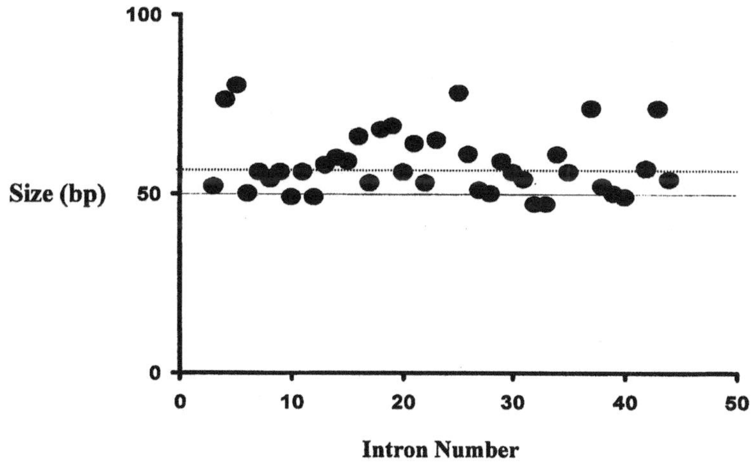

Fig. 3. Scatter plot of intron length in a set of 30 *F. venenatum* genes (45 introns). The median length of 56 bp is indicated by a dotted line.

when designing expression vectors for use in *F. venenatum*. In addition, these statistics might aid in identification of translation start sites in genes that are derived from future genomics investigations.

Most *F. venenatum* genes contain intervening sequences (introns) that punctuate the open reading frames. From the analysis of 45 intron sequences derived from 30 genes and their corresponding cDNAs, several interesting trends emerged. While the sizes of *F. venenatum* introns in our data set varied from 47 bp to 594 bp most introns in this fungus are small with a median length of 56 bp (Fig. 3). The splicing signals in *F. venenatum* are similar to those described for other filamentous fungi (Gurr *et al.* 1987). As shown in Table 4, the splice donor sequences most frequently begin with GTAHGT (where H = A, C, or T). Approximately 98% of the introns in this organism end with the sequence YAG (where Y = C or T). Near the 3' ends of each intron a lariat sequence conforming to the consensus RCTRAC (where R = A or G) is found. Most intron-exon predicting software programs are woefully unreliable. Specific information on the consensus splicing signals in *F. venenatum* may prove to be more dependable than teaching sets derived from heterologous fungal genes when applied to intron-exon boundary prediction in future genome sequencing and annotation activities.

Synonymous codon usage in *F. venenatum* was investigated using a set of protein-coding ORFs derived from 37 full-length cDNA and genomic clones encompassing nearly 20,000 codons. The overall nucleotide composition of the coding regions (excluding introns) was 52.6% G+C, 47.3% A+T suggesting little overall mutational bias or selective pressure at the "wobble" position. Nevertheless, codon usage among these genes is non-random, and several striking biases were observed (Table 5). First, among those codons for which four or more choices exist, there is a strong preference for those ending with a pyrimidine residue (72%). Secondly, for those codons that terminate in a purine (Gln, Glu, and Lys; excluding Met and Trp), there is a strong bias for G as the "wobble" base (75%). Thirdly, a slight bias (63%) for C

Table 4. Fraction of each nucleotide at various positions within *F. venenatum* introns.

Nucleotide	D1	D2	D3	D4	D5	D6	D7	D8	L1	L2	L3	L4	L5	L6	A1	A2	A3
A	0.00	0.00	0.80	0.62	0.07	0.13	0.33	0.33	0.47	0.02	0.00	0.56	0.98	0.00	0.02	1.00	0.00
C	0.00	0.00	0.00	0.13	0.02	0.13	0.33	0.22	0.00	0.93	0.00	0.04	0.00	0.69	0.47	0.00	0.00
G	1.00	0.00	0.16	0.09	0.91	0.07	0.09	0.16	0.42	0.00	0.00	0.33	0.02	0.04	0.00	0.00	1.00
T	0.00	1.00	0.04	0.16	0.00	0.67	0.24	0.29	0.11	0.04	1.00	0.07	0.00	0.27	0.51	0.00	0.00
Consensus	G	T	A	H	G	T	H	H	R	C	T	R	A	Y	Y	A	G

Column headings starting with D denote splice donor positions (5' end of introns); headings with L represent nucleotide positions within the lariat, and the final three columns headed with an A signify positions of the splice acceptor site (3' end of introns). Ambiguous nucleotides include H (A, C, or T), R (A or G), and Y (C or T).

Table 5. Codon usage pattern in 37 *F. venenatum* genes.

gca	Ala(A)	0.13	cag	Gln(Q)	0.65	uug	Leu(L)	0.12	uaa	Ter (.)	0.54
gcc	Ala(A)	0.36	---	Gln(Q)	709	---	Leu(L)	1561	uag	Ter(.)	0.19
gcg	Ala(A)	0.10	gaa	Glu(E)	0.26	aaa	Lys(K)	0.16	uga	Ter(.)	0.27
gcu	Ala(A)	0.41	gag	Glu(E)	0.74	aag	Lys(K)	0.84	---	Ter(.)	37
---	Ala(A)	1689	---	Glu(E)	1044	---	Lys(K)	984	aca	Thr(T)	0.21
aga	Arg(R)	0.10	gga	Gly(G)	0.21	aug	Met(M)	1.00	acc	Thr(T)	0.38
agg	Arg(R)	0.06	ggc	Gly(G)	0.32	---	Met(M)	407	acg	Thr(T)	0.09
cga	Arg(R)	0.24	ggg	Gly(G)	0.05	uuc	Phe(F)	0.66	acu	Thr(T)	0.33
cgc	Arg(R)	0.26	ggu	Gly(G)	0.42	uuu	Phe(F)	0.34	---	Thr(T)	1243
cgg	Arg(R)	0.05	---	Gly(G)	1543	---	Phe(F)	835	ugg	Trp(W)	1.00
cgu	Arg(R)	0.28	cac	His(H)	0.60	cca	Pro(P)	0.15	---	Trp(W)	318
---	Arg(R)	899	cau	His(H)	0.40	ccc	Pro(P)	0.38	uac	Tyr(Y)	0.75
aac	Asn(N)	0.77	---	His(H)	454	ccg	Pro(P)	0.08	uau	Tyr(Y)	0.25
aau	Asn(N)	0.23	aua	Ile(I)	0.05	ccu	Pro(P)	0.39	---	Tyr(Y)	736
---	Asn(N)	942	auc	Ile(I)	0.60	---	Pro(P)	1003	gua	Val(V)	0.07
gac	Asp(D)	0.55	auu	Ile(I)	0.35	agc	Ser(S)	0.19	guc	Val(V)	0.48
gau	Asp(D)	0.45	---	Ile(I)	1039	agu	Ser(S)	0.10	gug	Val(V)	0.12
---	Asp(D)	1258	cua	Leu(L)	0.06	uca	Ser(S)	0.14	guu	Val(V)	0.33
ugc	Cys(C)	0.60	cuc	Leu(L)	0.37	ucc	Ser(S)	0.22	---	Val(V)	1312
ugu	Cys(C)	0.40	cug	Leu(L)	0.16	ucg	Ser(S)	0.09	nnn	???(X)	0
---	Cys(C)	277	cuu	Leu(L)	0.27	ucu	Ser(S)	0.26	TOTAL		19798
caa	Gln(Q)	0.35	uua	Leu(L)	0.02	---	Ser(S)	1508			

in the terminal position was observed among the codons that terminate in a pyrimidine (Asp, Cys, His, Phe, and Tyr). Lastly, codons CGG and AGG (Arg) as well as GGG (Gly), AUA (Ile), GUA (Val), CUA and UUA (Leu) are seldom used in *F. venenatum* genes. Codon usage patterns may be of value in predicting exons and determining correct reading frames in genomic sequence data. Moreover, the use of optimal codons could be important when attempting to express heterologous genes in *F. venenatum*. In this regard several published reports describe the importance of using preferred codons for expression of foreign genes in *Aspergillus* species (Fernandez-Abalos *et al.*, 1998; Villanueva *et al.*, 1999; Moralejo *et al.*, 2000; Krasevec *et al.*, 2000). The stop codon UAA was utilized in approximately half of the *F. venenatum* genes we examined. Interestingly, a purine is preferred (72%) as the nucleotide immediately following a stop codon (Table 6). Additionally, there is a slight but consistent bias for purines in the next three nucleotides as well (61% at each position). Tate *et al.* (1996) observed that translation termination efficiency is determined by the nucleotide that follows the stop codon by compiling a database of natural termination sites from a wide variety of organisms. They observed a marked

Table 6. Nucleotides immediately following the stop codon in *F. venenatum* genes.

Nucleotide	+1	+2	+3	+4	+5	+6
A	0.36	0.25	0.25	0.28	0.22	0.17
C	0.08	0.17	0.14	0.11	0.25	0.06
G	0.36	0.36	0.36	0.33	0.28	0.36
T	0.19	0.22	0.25	0.28	0.25	0.42
Consensus	R_{72}	R_{61}	R_{61}	R_{61}	N	D

Ambiguous nucleotides include D (A, G, or T), R (A or G), and N (A, C, G, or T). Subscript numbers refer to the percent occurrence of a purine at each position.

bias in the nucleotide immediately following the stop codon in mammalian genes. Statistical analysis showed that purines were favored in this position.

5. cDNA MICROARRAYS

Microarray technologies offer a massively parallel approach to monitor global changes in gene expression, genotyping, and comparative genomics (Eisen and Brown, 1999; Harrington *et al.*, 2000; Schena *et al.*, 1998). One adaptation of this technology employs glass microscope slides onto which hundreds to thousands of cDNA clones are immobilized. Fluorescently labeled RNA or cDNA probes prepared from cellular mRNA samples are then used to query the array of immobilized targets by hybridization as in a northern blot. As noted above, approximately 4000 unique *F. venenatum* cDNA sequences were identified following assembly of more than 18,000 ESTs. We purified plasmid DNA samples from each member of this "uniset" using a robotics-assisted procedure, and the resulting samples were spotted onto glass microarray slides (Eisen and Brown, 1999). In one experiment, arrays were probed with fluorescently labeled cDNA derived from cells grown on either maltodextrin or glucose. Among the transcripts that appeared to be induced in maltodextrin-grown cells were several cDNAs that we previously enumerated in a collection of ESTs that were derived from cells fermented in maltodextrin (*e.g.*, amyloglucosidase, α-glucosidase, amylase) and a number of genes of unknown function. These microarrays are currently being used in our laboratory as a valuable tool for functional genomics and gene discovery. Since nearly all of the commercially useful enzymes produced by filamentous fungi are extracellular proteins, the power of cDNA microarrays for enzyme discovery might be further enhanced by applying the techniques described by Diehn *et al.* (2000) in which mRNA from membrane-bound polysomes is used to selectively identify membrane-associated and secreted proteins.

6. CONCLUSIONS

The filamentous fungus *F. venenatum* is both an edible source of protein in human food and a useful organism for the production of enzymes by large-scale fermentation. Thus, it can be viewed as a valuable component in our modern society with certain health and economic benefits. Clearly, additional studies of its physiology, biochemistry, and genetics are warranted for continued improvements in the strains that are used industrially. Undoubtedly these studies would be accelerated by a thorough analysis of the *F. venenatum* genome. Our preliminary investigations of chromosome number and gene repertoire can be seen as a first step toward a more comprehensive genomic effort. It is clear from our observations that genome information is extremely helpful for gene discovery and gene expression profiling studies. Future endeavors to marry EST information with proteomic analysis could provide additional synergies and insights about *F. venenatum* physiology, particularly with regard to the enumeration and classification of proteins. Developing a cohesive strategy to utilize the information from genome data to generate improved enzyme production strains poses an exciting challenge at this time.

REFERENCES

Adams MD, Kelley JM, Gocayne JD, *et al.* (1991). Complementary DNA sequencing: expressed sequence tags and the human genome project. Science 252: 1651-1656.

Berka, RM, Rey MW, Brown K, and Brown SH (2002a). Promoters for expressing genes in a fungal cell. US Patent 6,361,973.

Berka RM, Rey MW (2002b). Polypeptides having lactonohydrolase activity and nucleic acid sequences encoding same. US Patent 6,395,529.

Berka RM, Rey MW, Byun T, Itami R, Tsutsumi N, and Klotz A (2000c). Polypeptides having lysophospholipase activity and nucleic acids encoding same. PCT Patent Application WO 00/28044.

Diehn M, Eisen MB, Botstein D, and Brown PO (2000). Large-scale identification of secreted and membrane-associated gene products using DNA microarrays. Nature Genet 25: 58-62.

Eisen MB, and Brown PO (1999). DNA arrays for analysis of gene expression. Methods Enzymol 303: 179-205.

Fernanadez-Abalos JM, Fox H, Pitt C, Wells B, and Doonan JH (1998) Plant-adapted green fluorescent protein is a versatile vital reporter for gene expression, protein localization and mitosis in the filamentous fungus *Aspergillus nidulans*. Mol Microbiol 27: 121-130.

Golightly EJ, Berka RM, and Rey MW (2000). Polypeptides having galactose oxidase activity and nucleic acids encoding same. US Patent 6,090,604.

Gurr SJ, Unkles SE, and Kinghorn JR (1987). The structure and organization of nuclear genes in filamentous fungi. pp. 93-139, In (JR Kinghorn, ed.) Gene Structure in Eukaryotic Microbes. IRL Press, Oxford.

Harrington CA, Rosenow C, and Retief J (2000). Monitoring gene expression using DNA microarrays. Curr Opinion Microbiol 3: 285-291.

Kozak M (1987). An analysis of 5'-noncoding sequences from 699 vertebrate messenger RNAs. Nucl Acids Res 15: 8125-9148.

Krasevec N, van den Hondel CAMJJ, and Komel R (2000). Expression of human lymphotoxin alpha in *Aspergillus niger*. Pflugers Archiv Eur J Physiol 440: R83.

Lockhart DJ, Dong H, Byrne MC, *et al.* (1996). Expression monitoring by hybridization to high-density oligonucleotide arrays. Nature Biotechnol 14: 1675-1680.

Marshall A and Hodgson J (1998). DNA chips: An array of possibilities. Nature Biotechnol 16: 27-31.

Matsubara K and Okubo K (1993) Identification of new genes by systematic analysis of cDNAs and database construction. Curr Opin Biotechnol 4: 672-677.

McPherson MJ, Ogel ZB, Stevens C, Yadav KD, Keen JN, and Knowles PF (1992) Galactose oxidase of *Dactylium dendroides*. Gene cloning and sequence analysis. J Biol Chem 267: 8146-8152.

Moralejo FJ, Cardoza RE, Gutierrez S, Sisniega H, Faus I, and Martin JF (2000) Overexpression and lack of degradation of thaumatin in an aspergillopepsin A-defective mutant of *Aspergillus awamori* containing an insertion in the *pepA* gene. Appl Microbiol Biotechnol 54: 772-777.

Nakari T, Alatalo E, and Penttila ME (1993). Isolation of *Trichoderma reesei* genes highly expressed on glucose-containing media: characterization of the *tef1* gene encoding translation elongation factor 1 alpha. Gene 136: 313-318.

Nelson MA, Kang S, Braun EL, *et al.* (1997). Expressed sequences from conidial, mycelial, and sexual stages of *Neurospora crassa*. Fungal Genet Biol 21: 348-363.

Okubara P, Hohn T, Berka R, Anderson O, and Blechl A (2000). Expression of candidate anti-*Fusarium* protein genes in hexaploid wheat. Phytopathol 90: (6 Suppl) p. S57.

Okubara PA, Hohn TM, Blechl AE, and Berka RM (2001a). Nucleic acid sequences encoding cell wall-degrading enzymes and use to engineer resistance to *Fusarium* and other pathogens. PCT Patent Application WO 01/16353.

Okubara PA, Dill-Macky R, McCormick SP, Hohn TM, Berka RM, Alexander NA, Lin J, and Blech AE (2001b). Expression of two different candidate anti-*Fusarium* protein genes affords partial protection against the spread of *Fusarium graminearum*. National Scab Forum, Cincinnati, OH.

Rey MW, Nelson BA, Bernauer S, and Berka RM (2000) The *Fusarium venenatum* EST Project: Analysis of 8,000 ESTs. Abstr Fifth European Congress on Fungal Genetics, XIo-3. Arcachon, France.

Roe B, Kupfer D, Clifton S, and Prade R (2002) *Aspergillus nidulans* cosmid and cDNA sequencing. http://www.genome.ou.edu/asper.html.

Royer JC, Moyer DL, Reiwitch SG, Madden MS, Jensen EB, Brown SH, Yonkers CC, Johnstone JA, Golightly EJ, Yoder WT, and Shuster JR (1995) *Fusarium graminearum* A3/5 as a novel host for heterologous protein production. Bio/Technology 13: 1479-1483.

Sato T, Taga M, Saitoh H, Nakayama T, and Takehara T (1998) International Congress of Plant Pathology, Edinburgh, Scotland. August. Paper number 2.2.48.

Schena M, Heller RA, Theriault TP, Konrad K, Lachenmeier E, and Davis RW (1998). Microarrays: biotechnology's discovery platform for functional genomics. Trends Biotechnol 16: 301-306.

Schena M, Shalon D, Davis RW, and Brown PO (1995). Quantitative monitoring of gene expression patterns with a complementary DNA microarray. Science 270: 467-470.

Shimazu S and Kataoka M (1996) Optical resolution of pantolactone by a novel fungal enzyme, lactonohydrolase. Annals New York Acad Sci 799: 650-658.

Silakowski B, Ehret H, and Schairer HU (1998) *fbfB*, a gene encoding a putative galactose oxidase, is involved in *Stigmatella aurantiaca* fruiting body formation. J Bacteriol 180:1241-1247.

Skinner DZ, Budde AD, and Leong SA (1991). Molecular karyotype analysis of fungi. pp. 86-103, In (JW Bennett and LL Lasure, eds.) More Gene Manipulations in Fungi. Academic Press, San Diego.

Southern EM (1996). DNA chips: Analyzing sequence by hybridization to oligonucleotides on a large scale. Trends Genet 12: 110-115.

Tate WP, Poole ES, Dalphin ME, Major LL, Crawford DJG, and Mannering SA (1996). The translational stop signal: Codon with a context or extended factor recognition element? Biochemie 78: 945-952.

Tsutsumi N, Sasaki Y, Rey MW, Zaretsky E, Spendler T, and Vind J (2002) Lipolytic enzyme. PCT Patent Application WO 02/00852.

Villanueva A, MacCabe PA, Buesa J, and Ramon D (1999). Apparent mRNA instability in *Aspergillus nidulans* and *Aspergillus terreus* of a heterologous cDNA encoding the major capsid antigen of Rotavirus. Revista Iberoamericana de Micologia 16: 130-135.

Wiebe MG (2002). Myco-protein from *Fusarium venenatum*: a well-established product for human consumption. Appl Microbiol Biotechnol 58: 421-427.

Wiebe MG, Nováková M, Miller L, Blakebrough ML, Robson GD, Punt PJ, and Trinci APJ (1997). Protoplast production and transformation of morphological mutants of the Quorn® myco-protein fungus, *Fusarium graminearum* A3/5 using the hygromycin B resistance plasmid pAN7-1. Mycol Res 101:871-877.

Xu J-R, Yan K, Dickman MB, and Leslie JF (1995). Electrophoretic karyotypes distinguish the biological species of *Gibberella fujikuroi* (*Fusarium* section *liseola*). Mol Plant-Microbe Interac 8: 74-84.

Yaver DS, Berka RM, and Rey MW (2001). Polypeptides having acid phosphatase activity and nucleic acids encoding same. US Patent 6,309,869.

Yoder WT and Lehmbeck J (2003). Heterologous expression and protein secretion in filamentous fungi. In (J Tkacz and L Lange, eds.). Advances in Fungal Biotechnology for Industry, Agriculture, and Medicine. Kluwer Academic/Plenum Publishers, New York.

Zyla K (1990) Influence of acid phosphatase activity on the saccharification of potato maltodextrins by *Aspergillus niger* glucoamylase. Acta Biotechnol 10:445-449.

Zyla K, Ledoux DR, and Veum TL (1995a). Complete enzymatic dephosphorylation of corn-soybean meal feed under simulated intestinal conditions of the turkey. J Agric Food Chem 43: 288-294.

Zyla K, Ledoux DR, Carcia A, and Veum TL (1995b). An *in vitro* procedure for studying enzymatic dephosphorylation of phytate in maize-soybean feeds for turkey poults. Br J Nutr 74: 3-17.

Applied Mycology & Biotechnology
An International Series. Volume 4. Fungal Genomics

8

Molecular Characterization of *Rhizoctonia solani*

Mette Lübeck
Department of Plant Biology, Plant Pathology Section, The Royal Veterinary and Agricultural University, 40 Thorvaldsensvej, DK-1871 Frederiksberg C, Denmark. (mette.lubeck@plbio.kvl.dk).

The basidiomycete *Rhizoctonia solani* Kühn (teleomorph: *Thanatephorus cucumeris* (Frank) Donk) is a species complex composed of divergent populations. The genetics of *R. solani* is multifaceted and is currently not well understood. The main attention in genetic studies of *R. solani* has been devoted to characterization of populations and recognition of subgroups within the complex, traditionally focusing on the anastomosis reaction of isolates. From hyphal anastomosis reactions, isolates are divided into anastomosis groups (AGs). The AGs themselves do not necessarily give information on the genetic variation and taxonomic relationships within and between AGs. Currently, a growing number of other means for characterization and grouping of isolates are reported, in particular RFLPs, isozymes, PCR fingerprinting, DNA/DNA hybridization and sequence analysis. In general, these methods have supported the AG concept and revealed that the AGs consist of distinct phylogenetic entities. The methods have also revealed that some of the AGs is further genetically isolated in subgroups. Recently, an universally primed PCR (UP-PCR) cross hybridization assay has been developed for rapid identification of isolates into correct AG subgroup. The assay is based on the fact that UP-PCR products of isolates within AG subgroups cross hybridize strongly; whereas there is little or no cross hybridization of UP-PCR products from isolates belonging to different AG subgroups. The UP-PCR product cross hybridization assay represents a macro-array technique and has potential for routine use in diagnostics of *R. solani*. Even though *R. solani* is causing a wide range of economically important diseases there is a lack of knowledge concerning genes and their function in relation to pathogenicity. At present, a sequence survey shows that GeneBank contains close to 400 DNA sequences derived from *R. solani*. However, most of them are sequences of the nuclear encoded rRNA genes, which mainly are used for taxonomic purposes. The few other sequences are from genes encoding laccases, glyceraldehydes-3-phosphate-dehydrogenases, and part of an ABC transporter, respectively. However, the actual role of these few genes has not been studied. Extending the current focus on genomics to include *R. solani* would be of great utility for building up knowledge on genes and gene expression from this important plant pathogen.

1. INTRODUCTION

The basidiomycete *Rhizoctonia solani* Kühn (teleomorph: *Thanatephorus cucumeris* (Frank) Donk) is well-known as a plant pathogenic fungus causing economically important diseases to many agricultural and horticultural crops worldwide (Ogoshi 1987; Sneh *et al.* 1991; 1996). Species in *Rhizoctonia* in general are not well understood due to the absence of distinct morphological features and because of the lack of information about mating and mating relationships (Vilgalys and Cubeta 1994). Based on hyphal anastomosis reactions, isolates of *R. solani* are divided into anastomosis groups (AGs), of which 14 AGs have been described (Sneh *et al.* 1991, 1996; Carling *et al.* 1999a,b, 2002). The AGs seem to be genetically isolated and *R. solani* is regarded a large species complex composed of many genetically distinct groups with very diverse life histories, likely comprising several species (Anderson 1982; Kuninaga 1996; Vilgalys and Cubeta 1994; Cubeta and Vilgalys 1997). However, the groups cannot be distinguished by their sexual stages because of the inability of many isolates to fruit in culture, which is the reason why most researchers recognize *R. solani* as a single species divided into AGs and AG subgroups (Ogoshi 1987).

Anastomosis grouping is a convenient method for classification but sometimes anastomosis reactions can be difficult to identify making a demand for other methods for identification (Carling 1996; Carling *et al.* 1999b, 2002b). A variety of molecular methods has been used for characterization and grouping of *R. solani* isolates such as DNA/DNA hybridization, analysis of ribosomal DNA by restriction fragment analysis (RFLP) or by sequencing (Kuninaga and Yokosawa 1985a,b; Vilgalys 1988; Carling and Kuninaga 1990; Vilgalys and Gonzalez 1990; Liu and Sinclair 1993; Kuninaga 1996; Schneider *et al.* 1997a; Salazar *et al.* 1999, 2000; Lübeck and Poulsen 2001; Gonzalez *et al.* 2001; Carling *et al.* 2002a,b). Also PCR fingerprinting methods have been employed, *e.g.* random amplified polymorphic DNA (RAPD-PCR), PCR of enterobacterial repetitive intergenic consensus sequences (ERIC-PCR) and repetitive extragenic palindromic sequences (REP-PCR), universally primed PCR (UP-PCR) and amplified fragment length polymorphism (AFLP) (Duncan *et al.* 1993; Toda *et al.* 1999; Julián *et al.* 1999; Lübeck and Poulsen 2001; Fenille *et al.* 2002). These methods have shown large genetic variation and low sequence homology among the different AGs.

The present review focuses on the AG system, the application of different molecular methods, and genetic diversity and population structure in *R. solani*. The usefulness of a diagnostic assay based on hybridization of UP-PCR products for routine identification of *R. solani* isolates into AG subgroups is discussed.

2. HYPHAL ANASTOMOSIS GROUPS

Anastomosis is a common method for gene exchange in fungi, and the anastomosis abilities of *R. solani* isolates facilitate their distribution into specific groups. Anastomosis in *R. solani* is defined as somatic, or vegetative, incompatibility between hyphae of different but related isolates (Anderson 1982). Thus, the AG is a collection of closely related isolates grouped together based on their capability to anastomose with each other (Carling 1996; Cubeta and Vilgalys 1997). Pairing of isolates from the same AG results in interaction of the hyphae between mycelial tip cells and fusion of the hyphae. The process of anastomosis and the genetic factors controlling it is not well known. The hyphae have some way of recognizing each other which make them direct their growth towards other hyphae. McCabe *et al.* (1999) have described anastomosis by video microscopy. They found the process to be limited to the hyphal tip cells of side branches. Within a culture intense anastomosing activity was limited to small areas whereas large areas were found to be without anastomosing cells.

When identical isolates are paired, hyphal fusion is complete and cytoplasmic mixing occurs. This is recognized as the "C3 reaction" (compatible or perfect fusion). However, pairing of non-identical isolates from the same AG results in the "C2 reaction" (incompatible or imperfect fusion). Upon fusion, 5-6 cells on either side of the fused cells become vacuolated and die. When isolates of different AGs are paired, the hyphae grow over and under each other but never make contact or interact with each other, the "C0 reaction" (Carling 1996; McCabe *et al.* 1999). Based solely on anastomosis reaction, each AG can be considered as a genetically isolated population since there is no interaction between groups (McCabe *et al.* 1999).

One more type of reaction is recognized, the "C1 reaction", also known as the bridging reaction. Similar to the C2 reaciton, the C1 is attempts to anastomosis with hyphal contacts but with no fusion of walls. In contrast to the C2 reaction, only occasionally one or both anastomosing cells and adjacent cells die in the bridging reaction (Carling 1996). The concept of bridging was originally defined in terms of fusion frequency and indicated an anastomosis relationship between isolates belonging to different AGs (Carling *et al.* 2002b). Especially anastomosis relationship of the bridging group, AG-BI, to certain isolates of AG2, AG3, AG6 and AG8 was recognized (Sneh *et al.* 1991).

It has been found that most isolates will participate in tuft formation when confronted with other isolates (MacNish *et al.* 1997). Tufts are areas of distinct demarcation in the confrontation zone between isolates, occupied by a band of hyphae raised above the general level of mycelium. Depending on the vegetative compatibility of the isolates, confrontation may result in tuft formation, no reaction or an intermediate reaction. Thus, isolates able to anastomose with a C3 reaction have no tuft formation, while isolates showing C2 reaction always are found to participate in tuft formation. Isolates with C0 or C1 reaction differ in tuft formation (intermediate reactions) (MacNish *et al.* 1997).

As mentioned 14 AGs have been reported, including AG1-AG13 and AG-BI (Ogoshi 1987; Carling *et al.* 1994; 1999a,b, 2002a). Seven of the 14 AGs (AG1, AG2, AG3, AG4, AG6, AG8 and AG9) have been further divided into subgroups to reflect differences observed in culture appearance, morphology, host range, pathogenicity, thiamine requirements, and hyphal fusion frequency. Also molecular studies have revealed further division of some of the AGs (Ogoshi 1987; MacNish *et al.* 1993; Kuninaga *et al.* 1997; Gonzalez *et al.* 2001; Carling *et al.* 2002a, 2002b). The current subdivision of the AGs varies because the subgroups have been identified and defined using very different criteria. Therefore, the genetic relationships are more or less evident for the different subgroups and some of the subgroups are further divided into groups or types. Subdivision of AG1 is originally based on differences in pathogenicity, while subdivision of AG2 is originally based on hyphal fusion frequency (Ogoshi 1987). In addition, the subgroups of AG2 are further divided into types based on pathogenicity (Ogoshi 1987; Schneider *et al.* 1997b; Hyakumachi *et al.* 1998; Carling *et al.* 2002b). Subdivision of AG4, AG6 and AG9 is originally based on DNA/DNA hybridization data (Kuninaga and Yokosawa 1984a, 1984b; Vilgalys 1988; Kuninaga 1996), while pectic zymograms have been used to identify subgroups of AG8 (MacNish *et al.* 1993). The known AGs and AG subgroups including information on genetic variation revealed by different molecular methods are listed in Table 1.

Sometimes isolates fail to anastomose because of genetic instability, environmental or nutrient conditions. Also the bridging reaction is difficult to interpret. In these instances, anastomosis grouping may not always be the best indicator of genetic relatedness (Carling 1996; Vilgalys and Cubeta 1994; Cubeta and Vilgalys 1997). For identification, additional methods to supplement anastomosis reactions are recommended.

3. MOLECULAR METHODS USED FOR CHARACTERIZATION OF *R. SOLANI*

3.1 DNA-DNA Hybridization

Genetic relatedness can be assessed through DNA base sequence complementary analyses by measuring the degree of re-association (hybridization) between dual DNA molecules. DNA-DNA hybridization can be carried out in two ways: either in a free solution or by

Table 1. List of *Rhizoctonia solani* anastomosis groups (AGs) and AG subgroups including genetic variation revealed by different molecular methods.

AG	Genetic variation in AG[#]	Subgroup	Genetic variation in Subgroup[#]
AG1	rDNA-RFLP: multiple RFLP patterns (Vilgalys and Gonzales 1990); three RFLP patterns (Jabaji-Hare *et al.* 1990); six groups (Liu and Sinclair 1993) DNA/DNA hybridization: three groups 45-60%* (Kuninaga 1996) rDNA-ITS sequences: three groups 67-91%* (Kuninaga *et al.* 1997; Gonzalez *et al.* 2001)	AG1-1A	rDNA-RFLP: single RFLP pattern (Jabaji-Hare *et al.* 1990) DNA/DNA hybridization: one group (Kuninaga 1996)
		AG1-1B	rDNA-RFLP: single RFLP pattern (Jabaji-Hare *et al.* 1990) DNA/DNA hybridization: one group (Kuninaga 1996)
		AG1-1C	rDNA-RFLP: single RFLP pattern (Jabaji-Hare *et al.* 1990) DNA/DNA hybridization: one group (Kuninaga 1996) REP and RAPD-PCR: high heterogeneity (Toda *et al.* 1999)
		AG1-1D	rDNA-RFLP, RAPD (Priyatmojo *et al.* 2001)
AG2**	rDNA-RFLP: multiple RFLP patterns (Vilgalys and Gonzales 1990; Jabaji-Hare *et al.* 1990); five groups (Liu and Sinclair 1992) DNA/DNA hybridization: four groups 45% (Kuninaga 1996) rDNA-ITS sequences: six groups (Salazar *et al.* 1999)	AG2-1 (AG2t, 2Nt)	DNA/DNA hybridization: single group (Kuninaga 1996) rDNA-ITS sequences: 92-100%* (Kuninaga *et al.* 1997), two groups of 2-1 (Salazar *et al.* 1999), AG2t and AG2Nt are regarded as members of AG2-1 (Carling *et al.* 2002)
		AG2-2 (AG2-2IIIB, 2-2IV, 2-2LP)	DNA/DNA hybridization: variation in AG2-2 70%* (Kuninaga 1996) rDNA-ITS sequences: 93-94.5%* between AG2-2IIIB and AG2-2IV (Kuninaga *et al.* 1997), 3 groups of AG2-2 (Salazar *et al.* 2000) rDNA-RFLP: 3 groups of AG2-2 (Hyakumachi *et al.* 1998)
		AG2-3	(Naito and Kanematsu 1994) rDNA-ITS sequences (Carling *et al.* 2002a)
		AG2-4	rDNA-ITS sequences (Carling *et al.* 2002a, 2002b)
AG3	rDNA-RFLP: single RFLP pattern (Vilgalys and Gonzales 1990; Jabaji-Hare *et al.* 1990), two groups (Liu *et al.* 1993), two groups (Ceresini *et al.* 1999) DNA/DNA hybridization: single group (Kuninaga 1996) rDNA-ITS sequences: two groups 91-95%*	AG3 PT	AFLP: 32 phenotypes of 32 isolates (Ceresini *et al.* 2002a) Multilocus PCR-RFLP (Ceresini *et al.* 2002b, 2003)

	(Kuninaga *et al.* 1997; 2000, Gonzalez *et al.* 2001) REP, RAPD-PCR and AFLP: high heterogeneity (Toda *et al.* 1999; Ceresini *et al.* 2002a)		
		AG3 TB	AFLP: 28 phenotypes of 36 isolates (Ceresini *et al.* 2002a)
AG4	rDNA-RFLP: single RFLP pattern (Vilgalys and Gonzales 1990); two RFLP patterns (Jabaji-Hare *et al.* 1990; Liu *et al.* 1993) ITS1-5.8S-ITS2 sequences: three groups*** (Boysen *et al.* 1995) DNA/DNA hybridization: two groups 30-47%* (Kuninaga and Yokosawa 1984; Vilgalys 1988), three groups rDNA-ITS sequences: two groups 92-97%*(Kuninaga *et al.* 1997); 3 groups (Gonzalez *et al.* 2001, Kuramae *et al.* 2003) REP and RAPD-PCR: high heterogeneity (Toda *et al.* 1999)	AG4 HGI	rDNA-RFLP: single RFLP pattern (Jabaji-Hare *et al.* 1990); DNA/DNA hybridization: one group (Kuninaga and Yokosawa 1984; Vilgalys 1988)
		AG4 HGII	rDNA-RFLP: single RFLP pattern (Jabaji-Hare et al. 1990); DNA/DNA hybridization: one group (Kuninaga and Yokosawa 1984; Vilgalys 1988)
		AG4HGII	rDNA-RFLP, rDNA-LSU (Gonzalez *et al.* 2001)
AG5	rDNA-RFLP: multiple RFLP patterns (Vilgalys and Gonzales 1990; Jabaji-Hare *et al.* 1990); three groups (Liu *et al.* 1993) DNA/DNA hybridization: single group (Kuninaga 1996)		
AG6	rDNA-RFLP: multiple RFLP patterns (Vilgalys and Gonzales 1990; Jabaji-Hare *et al.* 1990) DNA/DNA hybridization: two groups 47-62% (Kuninaga and Yokosawa 1984) rDNA ITS sequences: 96-97.4%* (Kuninaga *et al.* 1997; Gonzalez *et al.* 2001)	AG6-I	DNA/DNA hybridization: one group (Kuninaga and Yokosawa 1984; Vilgalys 1988)
		AG6-V	DNA/DNA hybridization: heterogeneous group 55-66% (Kuninaga and Yokosawa 1984; Vilgalys 1988)
AG7	rDNA-RFLP: single RFLP pattern (Vilgalys and Gonzales 1990) * DNA/DNA hybridization: single group (Kuninaga 1996)		
AG8	rDNA-RFLP: single RFLP pattern (Vilgalys and Gonzales 1990) * DNA/DNA hybridization: single group (Kuninaga 1996) Pectic zymogram: five groups (MacNish *et al.* 1993) rDNA ITS sequences: heterogeneity (Gonzalez *et al.* 2001)		
AG9	rDNA-RFLP: multiple RFLP patterns (Jabaji-Hare *et al.* 1990; Liu *et al.* 1990) DNA/DNA hybridization: variation (? groups) (Kuninaga 1996)	AG9TP	DNA/DNA hybridization: one group (Kuninaga 1996)

	rDNA-ITS sequences: one group 99.6-100%* (Kuninaga *et al.* 1997)
AG9TX	DNA/DNA hybridization: one group (Kuninaga 1996)
AG10	rDNA-RFLP: two groups (Liu *et al.* 1993) rDNA-ITS sequences (Carling *et al.* 2002a)
AG11	(Carling *et al.* 1994) rDNA-ITS sequences (Carling *et al.* 2002a)
AG12	(Carling *et al.* 1999b) rDNA-ITS sequences (Carling *et al.* 2002a)
AG13	rDNA-ITS sequences (Carling *et al.* 2002a)
AG-BI***	rDNA-RFLP: single RFLP pattern (Vilgalys and Gonzales 1990) DNA/DNA hybridization: single group (Kuninaga 1996) REP and RAPD-PCR: high heterogeneity (Toda *et al.* 1999)

results for genetic variation vary not only because of method but also depending on how many isolates and subgroups were included in the different studies; *indicate percent similarity between subgroups. A single DNA/DNA hybridization group indicates that there are > 90% hybridization values among isolates; **AG2-2 has recently been revised by Carling *et al.* (2002b). On the basis of anastomosis reactions between AG-BI and AG2-1 and AG2-2, it is proposed that AG-BI is included as a subset of AG2. Furthermore, anastomosis reactions of AG2-3 and AG2-4 show very weak bridging-type anastomosis reactions with all other AG-2 members and might be regarded as separate independent AGs. The phylogenetic relationship of AG2 with all known AGs is presented in Carling *et al.* (2002a); ***two isolates representing one AG4 group described by Boysen *et al.* (1996) were later identified as anamorphs of *Ceratobasidium* (binucleate *Rhizoctonia*) (Carling *et al.* 2002a) based on analysis of the rDNA ITS region by Kuninaga *et al.* (1997) and Salazar *et al.* (1999).

immobilization of one of the DNA species on a filter. When hybridizing in a free solution, the kinetics of duplex formation (hybridization) of DNA is monitored spectrophotometrically, whereas for the other method radioactivity labeling is used. DNA-DNA hybridization studies carried out by Kuninaga and Yokosawa (1985a, 1985b), Vilgalys (1988) and Carling and Kuninaga (1990) on an assortment of AGs of *R. solani* confirmed that DNA-DNA hybridization groups support the groupings based on anastomosis reactions (Kuninaga 1996). Between the different AGs, the hybridization value is mostly less than 15%, indicating very little genetic relationships. At the same time, it was found that substantial genetic differences between subgroups of certain AGs exist, e.g. subgroups of AG1, AG2, AG4, AG6 and AG9. The hybridization value of subgroups within an AG is often less than 60%. In contrast, within AG subgroups most members have very high DNA-DNA hybridization values (> 90%), indicating a high degree of genetic relatedness among isolates (Kuninaga 1996). DNA-DNA hybridization carried out by Kuninaga and Yokosawa (1984a, 1984b) and Vilgalys (1988) revealed previously unknown heterogeneity within AG4 and AG6 (Table 1). The hybridization values between members of AG4 subgroups HGI and HGII were 30 – 47%, while the hybridization values between members of AG6 subgroups HGI and GV were 55 – 66% (Kuninaga 1996).

From DNA-DNA hybridization studies of filamentous fungi and yeast, it has been found that high levels of cross hybridization seem to correspond with ability to mate (biological species), and that cross hybridization between even closely related species is exceedingly low (Bruns *et al.* 1991, Kuninaga 1996, Naumova *et al.* 2001). According to Kuninaga (1996) and Naumova *et al.* (2001), DNA-DNA hybridization values less than 50-65% and below seem to represent isolates of different species. Among different AGs and some of the AG subgroups, the DNA-DNA hybridization values are less than 45% and, as mentioned above, often less than 15% (Kuninga 1996). The interpretation of DNA relatedness in the *R. solani* complex thus suggests that different AGs and even some of the AG subgroups obviously represent distinct evolutionary units, corresponding to separate species, which Kuninga (1996) termed as "genomic species".

3.2 Restriction Fragment Length Polymorphism

Examination of the nuclear encoded ribosomal DNA (rDNA) repeat unit and restriction digestion patterns (RFLPs) has been found useful for distinguishing isolates of *R. solani* into groups, using Southern blotting methods or PCR (Jabajii-Hare *et al.* 1990; Vilgalys and Gonzales 1990; Liu *et al.* 1992; Liu and Sinclair 1992; 1993, Cubeta *et al.* 1996; Schneider *et al.* 1997a; Ceresini *et al.* 1999; Lübeck and Poulsen 2001). Vilgays and Gonzalez (1990) used Southern blotting with a probe containing the rDNA repeat from an AG4 strain. They found that isolates within AG3, AG4, AG7, AG8 and AGBI possessed a single RFLP pattern, while isolates from AG1, AG2, AG5 and AG6 possessed multiple RFLP patterns. Within the subgroups of AG1 and AG2 they found considerable genetic variability. In a similar study conducted by Jabaji-Hare *et al.* (1990), using a probe containing the rDNA repeat from *Armillaria ostoyae*, somewhat different results were obtained. They found that isolates from AG4, AG5, AG7, AG8, AG9, and AGBI possessed a variety of RFLP patterns some of which were not unique. Thus, comparison of the results obtained from different groups may vary because of the use of different isolates and different probes. Therefore, it is recommended to use some kind of reference isolates for these studies similar to the so-called "tester isolates" used for evaluation of anastomosis reactions.

Due to the advantage of PCR compared with Southern blotting, i.e. less labor intensive (less DNA required, no radioactivity or bioluminescence needed), PCR amplification of the rDNA combined with restriction digestion has become popular. In particular RFLP of amplified internal transcribed sequence (ITS) regions of rDNA (ITS rDNA polymorphism) has been used. The improvements in DNA sequencing and the easy access to sequencing facilities with lower costs have recently expanded the analysis of rDNA as described in section 3.4.

In general, the RFLP variation obtained either by Southern blotting or by PCR revealed similar patterns in each AG subgroup, thus again confirming the genetic similarity of isolates within an AG subgroup (Table 1). However, as described individual isolates can sometimes possess more than one rDNA length variant, and some AG subgroups possess several discrete rDNA-RFLP patterns (Vilgalys and Gonzales 1990, Liu and Sinclair 1992, 1993). For example, sequence analysis of the ITS region of AG1 isolates revealed a very high level of variation (nucleotide substitutions and deletions/insertions), suggesting that genetic estimation from restriction banding patterns alone would be difficult in this AG (Kuninaga *et al.* 1997). Thus infra-group variation in rDNA polymorphism prevents the use of these methods for routine identification. However, rDNA polymorphic patterns are useful as a coarse grouping of isolates and as a supplement to other tests. For example, we used ITS rDNA polymorphism on isolates derived from diseased sugar beets and potato for comparison with and validation of identification by PCR fingerprinting (see below) and found a nice correlation (Lübeck and Poulsen 2001). Similarly, a good correlation between results obtained from ITS rDNA polymorphisms and pectic zymograms of isolates associated with tulips was found by Schneider *et al.* (1997a). In any case, caution should be taken in directly using ITS rDNA polymorphisms for identification without using additional criteria.

Very recently, single-copy multilocus RFLP markers have been developed and used to study the genetics of AG1-1A and AG3 isolates (Rosewich *et al.* 1999; Ceresini *et al.* 2002b; 2003). These RFLPs are neutral co-dominant genetic markers, which unambiguously distinguish between homo- and heterozygotes (Rosewich *et al.* 1999) in contrast to the PCR fingerprinting techniques described below. The application of this technique for population genetics is discussed later in section 5.

3.3 PCR Fingerprinting Techniques

Genetic differences among AG subgroups have also been assessed through PCR fingerprinting methods, which have become versatile approaches partly because of their relative ease and the fast obtaining of rather detailed results. Since the first methods, random amplified polymorphic DNA (RAPD) (Williams *et al.* 1990) and arbitrarily primed PCR (AP-PCR) (Welsh and McClelland 1990), were described, a vast number of variants have been developed. Common for the mentioned PCR fingerprinting methods is that it is possible to amplify DNA from any organism without previous knowledge of any DNA sequences. Basically, the methods require primers which, used singly, provide amplification of DNA fragments randomly distributed throughout the genome, thereby creating genomic 'fingerprints'. Using a given primer, polymorphisms among strains are detected as differences in any of the fragments in the DNA banding profiles.

In *R. solani*, RAPD markers facilitated discrimination between Australian AG subgroups and, in some cases, between different isolates within some AG subgroups (Duncan *et al.* 1993). Also Toda *et al.* (1999) found that RAPD-PCR of 41 isolates belonging to 11 AGs showed considerable variation. Similar to Duncan *et al.* (1993) they found variable fingerprint patterns of isolates within some of the AG subgroups, especially 1-1C, 3, 4, BI (Table 1). Also Fenille *et al.* (2002) used RAPD to characterize isolates of *R. solani* associated with soybean in Brazil and found that they clustered together with AG1-1A, AG2-2IIIB and AG4HGI.

Similarly, another PCR fingerprinting technique, UP-PCR (Bulat *et al.* 1998), revealed considerable variability among the isolates belonging to different AGs and different subgroups (Fig. 1; Lübeck and Poulsen 2001). UP-PCR resembles the RAPD technique, the main difference being the use of specific designed UP primers that are relatively long (15-18 bp). UP primers are used at relatively high annealing temperature and primarily target intergenic, more variable areas of the genome. For this reason the method is suitable for generation of species-specific bands and for detection of infra-specific variation (Bulat *et al.* 1998; Lübeck and Lübeck 2003a, b).

Visual comparison of aligned UP-PCR banding profiles of reference isolates with *R. solani* isolates obtained from diseased sugar beets and potatoes in many cases enabled their identification (Lübeck and Poulsen 2001). This was facilitated by the fact that many bands were shared between isolates within an AG subgroup (Fig. 1). The UP-PCR technique combined with hybridization for subgroup identification is described further in section 6.

Also the AFLP technique has been applied to *R. solani* isolates (Ceresini *et al.* 2002a). AFLP differs from the above fingerprinting techniques in several respects. Prior to amplification, the genomic DNA is digested simultaneously with two restriction enzymes, and biotinylated double stranded DNA based linkers are ligated to the fragments. Amplification is then carried out using primers matching the linker sequences (Vos *et al.* 1995). The amplification products are separated on denaturing polyacrylamide sequencing gels and visualized by exposing to autoradiographic films. The complexity of the banding pattern is usually up to 50-70 fragments (Majer *et al.* 1996). The advantages of the method ascribe to the reproducibility, and the proportion of the genome being analyzed per reaction (Majer *et al.* 1996).

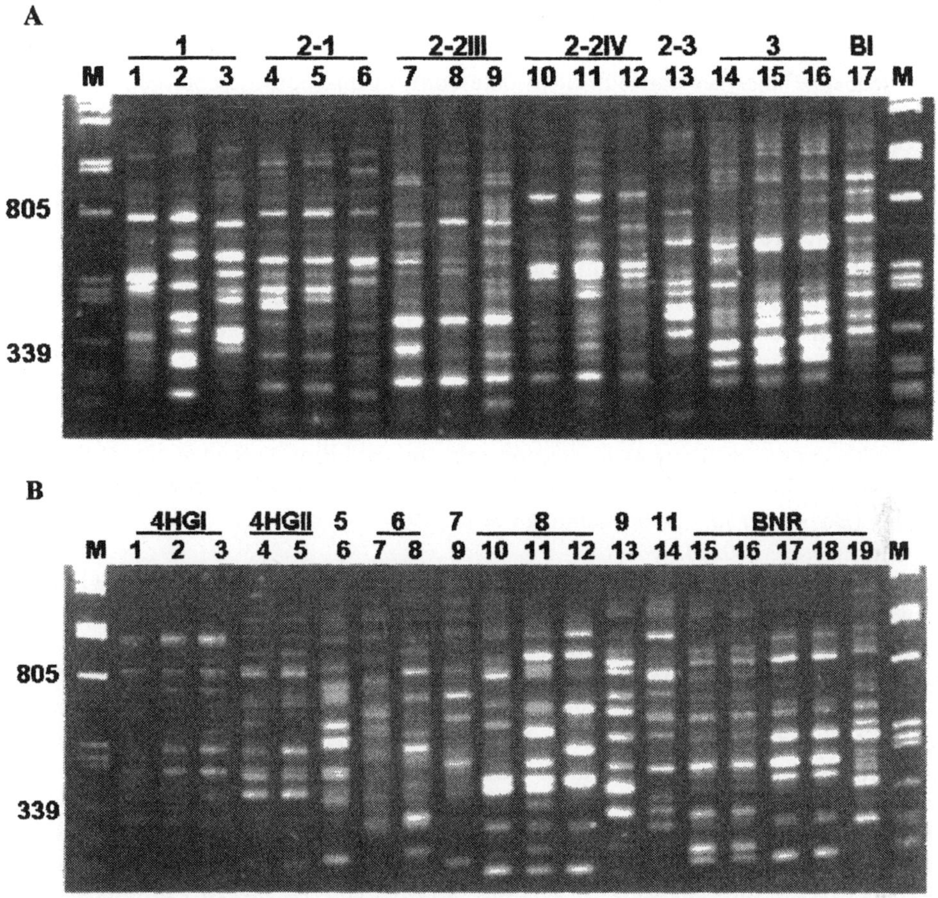

Fig. 1. UP-PCR banding profiles of *Rhizoctonia* isolates. Lane M. molecular size markers (λ phage DNA digested with *Pst*1). AGs are indicated above the lanes. The three AG1 isolates belong to AG1-1A, 1-1B, 1-1C respectively. The AG2-1 isolates no. 4 and 5 belong to AG2-1 and 2t, while the two AG6 isolates belong to AG6HGI and 6GV, respectively. The three AG8 isolates belong to three different zymogram groups. In A, isolate no. 6, 15, 16 and B, isolate no. 2, 3, 5 and all BNR (binucleate *Rhizoctonia* isolates) are all isolated from potato or sugar beet. Reprinted from Lübeck and Poulsen (2001) with permission from Elsevier.

AFLP analysis has been applied to AG-3 isolates from potato (AG-3 PT) and tobacco (AG-3 TB) (Ceresini *et al.* 2002a). The analysis showed that each of in total 32 AG-3 PT isolates analyzed had a distinct AFLP phenotype, whereas 28 AFLP phenotypes were found among 36 isolates of AG-3 TB (Ceresini *et al.* 2002a). Thus, it seems like the AFLP technique has a very high discriminatory ability facilitating infra-group variation.

3.4 Phylogenetic Analysis Using Ribosomal DNA Sequences

Comparative analysis of nuclear encoded rRNA genes is extensively used to examine taxonomic and phylogenetic relationships of fungi (Taylor *et al.* 2000). Database search of *R. solani* sequences showed that a majority of the nearly 400 sequences currently present are rDNA sequences, indicating that considerable efforts have been devoted to phylogenetic analysis in *R. solani*. More than 300 of the sequences are the ITS region (ITS1-5.8S-ITS2) and the availability of such a large ITS database facilitate expanded phylogenetic analysis of

R. solani. Also sequences of the 18S subunit and the large (28S) subunit (LSU) regions are available for *R. solani* isolates.

The sequence analyses carried out to date confirm that considerable genetic variability exists both within an AG and sometimes also within an AG subgroup. Furthermore, the analyses confirm that the AGs and AG subgroups of *R. solani* are genetically distinct. Thus, the analysis of rRNA genes largely supports anastomosis grouping similar to results obtained by DNA/DNA hybridization, RFLP analysis, PCR fingerprinting and other methods. All current data have revealed a completely conserved 5.8S region, but show great variation in both internal transcribed spacers (ITS) regions (Boysen *et al.* 1996; Kuninaga *et al.* 1997; Salazar *et al.* 1999, 2000; Gonzalez *et al.* 2001; Carling *et al.* 2002a, 2002b). Gonzalez *et al.* (2001) reported that the ITS region was very difficult to align and exhibited more homoplasy than the LSU region. Although their alignment revealed many differences in the ITS region, they found that most were concentrated in six highly variable regions (Gonzalez *et al.* 2001). The variation in size and nucleotide sequence, in general, is proportionately greater within isolates with greater differences in anastomosis compatibility, pathogenicity and habitat. It has been shown that differences in rDNA ITS regions correlate with differences in biological properties (Boysen *et al.* 1996; Kuninaga *et al.* 1997; Salazar *et al.* 1999; Gonzalez *et al.* 2001; Carling *et al.* 2002b).

An example of phylogenetic analysis is described by Kuninaga *et al.* (1997), who examined the sequence of the ITS regions of 45 isolates, representing 11 AGs and 11 subgroups. Phylogenetic analysis based on the sequence data revealed that isolates of the different AGs formed a distinct cluster. They showed that the sequence was above 96% for isolates of the same subgroup within an AG, 66-100% for isolates of different subgroups within an AG, and 55-96% for isolates of different AGs. The studies have also indicated that the ITS1 region tend to be more variable than the ITS2 region.

When grown on nutrient medium, many *R. solani* isolates resembles *Rhizoctonia* isolates belonging to other species *e.g.* isolates of *Ceratobasidium* and *Thanatephorus* (Ogoshi 1987). However, examination of their nuclear condition has shown that *Rhizoctonia* isolates of *Thanatephorus* are multinucleate. In contrast, *Rhizoctonia* isolates of *Ceratobasidium* only have two nuclei per hyphal cell, and are known as binucleate *Rhizoctonia*. These have also been shown to consist of AGs. However, there are reports that express uncertainty about the taxonomic placement of certain AGs of *Ceratobasidium* and *Thanatephorus*, *e.g.* hyphal fusion has been observed among isolates of AG6 (*Thanatephorus*) with AG-F (*Ceratobasidium*).

Gonzalez *et al.* (2001) tested the hypothesis that *Ceratobasidium* and *Thanatephorus* represent distinct evolutionary lineages of fungi with *Rhizoctonia* anamorphs and that anastomosis groups represent the most fundamental evolutionary units within *R. solani*. Phylogenetic analysis of ITS sequences of 122 isolates revealed 31 genetically distinct groups from *Thanatephorus* (21 groups) and *Ceratobasidium* (10 groups) that corresponded well with previously recognized AG or AG subgroups. Of the 122 isolates, 99 belonged to *R. solani*, and the 21 genetically distinct ITS groups revealed corresponded extremely well with known AG subgroups. However, six *Ceratobasidium* AGs clustered within the *Thanatephorus* clade even though it was not supported by high bootstrap values. Statistical evidence from rDNA phylogenies suggests that some isolates currently classified in *Ceratobasidium* based on nuclear condition and hyphal anastomosis reaction might eventually be more correctly classified within *Thanatephorus* (Gonzalez *et al.* 2001).

Results from combined analysis of ITS and LSU sequence data suggest that AG1, AG4, AG6, and AG8 represent well-defined and genetically isolated groups, while AG2 and AG3 are of multiple origins (polyphyletic). Thus experimental results from the phylogenetic studies on the rDNA ITS sequences make several common conclusions: most AG and AG subgroups represent genetically distinct groups supporting AG subgrouping, certain AGs are not monophyletic and there is greater taxonomic support for AG subgroups than AG. The hypothesis that AGs represent the most fundamental evolutionary units within *Thanatephorus* thus was rejected (Gonzalez *et al.* 2001).

Phylogenetic analysis can be used for identification purposes due to the availability of the large amount of ITS sequences in the databases. This was recently done by Kuramae *et al.* (2003) who used ITS sequences as a supplement to identify *R. solani* isolates derived from different vegetables in Brazil.

4. IMPLICATIONS OF MOLECULAR METHODS FOR AG SUBGROUPING

The various molecular methods are useful in defining groups in *R. solani*. Often the methods generate "fixed patterns" of variation (groups with high similarity and low or absent similarity among groups), which very likely is associated with intersterility barriers and other species differences. The AG subgroups have been established as genetically isolated groups and represent highly divergent evolutionary units. There is currently revision of the AGs and subgroups for examination of whether the groups are distinct and well separated.

Some recent examples concern isolates of *R. solani* AG-3, which are frequently associated with diseases of potato (AG-3 PT) and tobacco (AG-3 TB). The two types have been found to differ not only by pathogenicity but also by fatty acid, isozyme composition, ITS sequence analysis and AFLP fingerprint patterns (Laroche *et al.* 1992; Kuninaga *et al.* 1997, 2000; Gonzalez *et al.* 2001; Ceresini *et al.* 2002a). Even though they are somatically related based on anastomosis reaction, their host range does not overlap. This suggests that they represent genetically subdivided populations among closely related species that have evolved a high level of specificity on different hosts (Ceresini *et al.* 2002a). By testing this with AFLP and somatic compatibility, it was found that there was a complete lack of common AFLP phenotypes and compatible somatic interactions among potato and tobacco isolates supporting the subdivision (Ceresini *et al.* 2002a).

Carling *et al.* (2002b) examined a range of AG2 isolates belonging to all known subgroups of this AG by hyphal anastomosis reactions, rDNA-ITS sequences and virulence levels. They concluded that anastomosis reaction is unreliable and not sufficient for subgroup identification within AG2. Anastomosis reaction based on hyphal fusion frequency is difficult for placement of an isolate into a subset of AG2, and it was found that anastomosis reactions between AG-BI and the original subsets of AG2 (AG2-1 and AG2-2) are very strong (Carling *et al.* 2002b). Phylogenetic evidence suggests that AG-BI cluster together with AG2 isolates, and Carling *et al.* (2002b) propose that it should be included as a subset of AG2 with the designation AG2 BI, reducing the number of AGs to 13. On the other hand, the subgroups AG2-3 and AG2-4 show very weak bridging-type anastomosis reactions with all other subgroups of AG2 and each of them might represent an AG (Carling *et al.* 2002b). The independent AG status of AG2-3 and AG2-4 has recently been further supported by phylogenetic analysis (Gonzalez *et al.* 2001; Carling *et al.* 2002b). While grouping based on anastomosis reactions and rDNA ITS sequences is consistent with each other in AG2, grouping based on pathogenicity does not conform established grouping patterns (Carling *et al.* 2002b), in contrast to what is found in AG3. Virulence and host range seem not to be suitable as a general group-defining criterion.

According to Carling *et al.* (2002b), the concept of bridging has changed from describing the anastomosis relationships between AG-BI and certain other AGs to including description of relationships among subsets of AG2. If, as proposed by Carling *et al.* (2002b), AG-BI is included as a subgroup of AG2, the main inter-AG bridging reactions include AG2-AG3 bridging and AG2-AG8 bridging (Carling 1996).

In AG2-1, a subgroup pathogenic to tulips has been found, AG2t (Schneider et al. 1997b), but the isolates has not been found to be different from the Japanese AG2-1 tester isolates with ITS-RFLP, pectic zymography (Schneider et al. 1997a) and UP-PCR (Lübeck and Poulsen 2001). In addition, a subgroup of 2-1 "Italian Nt isolates" has been identified (Nicoletti et al. 1999, Kuninaga et al. 2000) as well as 2-1 isolates have been identified from Alaska and Australia differing from Japanese 2-1 isolates (Salazar et al. 1999, Carling et al. 2002b). Salazar et al. (1999) and Carling et al. (2002b) agree that 2-1 is heterogeneous regarding virulence and ITS sequence differences. Carling et al. (2002b) consider AG2t and Nt isolates to be included in AG2-1. As mentioned, AG2-2 is further divided into three ecological types: AG2-2IIIB (rush type), AG2-2IV (root rot type) (Ogoshi 1987), and AG2-2LP (Hyakumachi et al. 1998). This subdivision of AG2-2 into the three ecological types is not clear based on ITS sequence differences (Carling et al. 2002b).

Molecular systematic evidence indicates that many AGs and AG subgroups represent good biological species. In spite of this, only a few studies have explored the relationship of AG and subgroups to species or other taxonomic units. Boidin *et al.* (1998) recognized four species: (1) *T. microsclerotiums* (Weber) Boidin, Mugmier & Canales including AG1-1B, (2) *T. sasaki* (Shirai) Tu & Kimbrough including AG1-1A and AG1-1C, (3) *T. practicola* (Kotila) Flentje including AG4, and (4) *T. cucumeris* including AG2, 3, 5, 6, 8, 9 and BI. AG7, AG10-13 was not included in any of the above species. Also Mordue *et al.* (1989) recognized AG4 as a distinct species, *T. practicola*, but the remaining subgroups were assigned to *T. cucumeris*. Gonzalez *et al.* (2001) support that AG4 is a separate species based on their combined analysis of ITS and LSU sequence data. This also seems apparent in a single most parsimonious tree based on ITS sequences presented in Carling *et al.* (2002a).

These examples clearly demonstrate the utility of combined methods for subgrouping in *R. solani*. However, while regarding the subgroups as genetically distinct groups there is also a need for understanding the populations within these subgroups and their genetic diversity.

5. GENETIC DIVERSITY AND POPULATION STRUCTURE

A genetic structure of a population can broadly be defined as the amount and distribution of genetic variation within and between populations. There is a lack of knowledge of the genetic basis for the variation of *R. solani*, making it very difficult to understand the population structure of the fungus. Understanding the genetics is necessary to understand how the flow of genetic information within and between populations takes place. Genetic studies are more complicated for *R. solani* than for other fungal species due to several characteristics of the species, such as the multinucleate hyphal cells, the lack of clamp connections to differentiate homokaryons from heterokaryons, the absence of vegetative spores and the difficulty in obtaining sexual spores in vitro (Julián *et al.* 1999, Rosewich *et al.* 1999).

In general, *R. solani* are assumed to be primarily asexual, surviving as mycelium and sclerotia in soil and on plant material. The fungus is disseminated by the asexual production of sclerotia and vegetative mycelium (Adams 1996). Basidiospores of *T. cucumeris* are monokaryotic and assumed to have limited dispersal as they are very fragile. They are believed to have little or no importance in the disease cycle. However, they may play an important role in contributing to the genetic diversity and structure of field populations. Field isolates are generally assumed to be heterokaryotic. If recombination contributes to population structure, heterokaryotisation via somatic fusion of the mycelium from two basidiospores must occur prior to infection, (Rosewich *et al.* 1999). The dynamics of asexual and sexual interactions are poorly understood in *R. solani* (Cubeta and Vilgalys 1997). Only within an AG and presumably within an AG subgroup is sexual compatibility possible.

Several groups of *R. solani* are believed to be homothallic because of their ability to produce sexual spores that directly give rise to fertile cultures. AG1-1A, AG1-1C, AG4 and AG8 have been crossed in the laboratory and possess a heterothallic (outcrossing) mating system (Vilgalys and Cubeta 1994, Cubeta and Vilgalys 1997, Rosewich *et al.* 1999). For *e.g.* AG2 and AG3, it has been suggested that the mating system differs from that of AG1 and AG4 (Vilgalys and Cubeta 1994). Very little information is available on the genetic condition, heterokaryotic or homokaryotic, of field isolates and the nature of individuals of pathogenic *R. solani*, even though many field isolates are reported to be heterokaryotic (Julián *et al.* 1999, Rosewich *et al.* 1999). In *R. solani* AG4, it has been observed that heterokaryons sometimes spontaneously revert to their component heterokaryons, giving rise to a new pool of homokaryons that can intercross among themselves or with other colonies (Vilgalys and Cubeta 1994).

Progeny generated via asexual or homothallic sexual reproduction will be genetically identical or nearly identical to the parental isolate and all parts of the genome will have the same evolutionary history because of the lack of recombination (Taylor *et al.* 1999, Ceresini *et al.* 2002a). In contrast, progeny generated via heterothallic sexual reproduction will be genetically different than parental isolates as a result of recombination of genetically different nuclei. Different regions of the genome of the progeny will have different evolutionary history (Taylor *et al.* 1999). At the extremes of models of population structure, populations may be panmictic or strictly clonal. However, an intermediate model may be appropriate for many fungi, which are characterized by frequent recombination and by the occurrence of one or a few successful individual genotypes that reproduce clonally and may increase to high frequencies.

Population genetics would resolve many unanswered questions concerning the genetics of *R. solani*, as it would allow estimation of outbreeding, inbreeding, population structure, spatial orientation, and genetic variation (Adams 1996). The first step is to develop neutral genetic markers, which can distinguish between homo- and heterozygotes. The genetic isolation of AGs and potential differences in sexuality requires separate genetic analysis for each AG (Julián *et al.* 1999). Only recently, such studies have been carried out in *R. solani*.

One example is described in Rosewich *et al.* (1999), who used co-dominant markers, single-copy RFLPs, to investigate the genetics of 182 AG1-1A isolates from six commercial rice fields in Texas at a population level. Visual examination of the allelic information from the single-locus RFLP probes allowed an easy separation into heterozygote and homozygote classes at seven loci. Isolates indicative of a heterokaryon-homokaryon mating were not observed. The study indicated that novel genotypes in AG1-1A are produced by sexual recombination, that the AG is heterothallic and that some clones appear to be geographically more widespread and prevalent than others. The population of AG1-1A from rice displays high genetic diversity, coupled with efficient mechanisms of migration, indicating a high degree of gene flow. Novel genotypes apparently are generated by sexual recombination. Once favorable gene combinations are formed, selection can act upon successful individuals and increase them by asexual reproduction to high frequency (Rosewich *et al.* 1999).

In studies of AG3, recombination plays a role since it reduces the one-to-one strict association between independent characters and results in non-clonal population structure (Ceresini *et al.* 2002a). Using polymorphic co-dominant single-locus PCR-RFLP markers a model of population structure for *R. solani* AG3 PT (potato type) that includes both recombination and clonality was revealed (Ceresini *et al.* 2002b). From the RFLP results, it was evident that AG3 PT isolates had one or more heterozygous loci and thus seem to be heterokaryotic similar to what was found by Rosewich *et al.* (1999) for AG1-1A isolates.

6. UP-PCR-A MICRO-ARRAY BASED DIAGNOSTIC TOOL

Precise identification and early detection of plant pathogens is important for disease management in most crops. Many plant pathogens including *R. solani* are difficult to identify using morphological criteria, which can be time consuming and require extensive knowledge in taxonomy. Among methods that have gained considerably interests for species identification, detection and diagnostics are PCR based methods. In the following, UP-PCR combined with hybridization for identification of R. solani into correct AG subgroup is described.

UP-PCR generates complex banding profiles primarily from intergenic more variable areas of the genome and is a powerful technique with several applications (Bulat *et al.* 1998; Lübeck and Lübeck 2003a, 2003b). One of the applications is a cross hybridization assay of UP-PCR products generated from different fungal strains. UP-PCR product cross hybridization allows grouping of the strains into genetic entities which has been shown to correspond with species. The power of species identification based on strong UP-PCR product cross hybridization has been shown for several fungi, including sibling species of different yeast (reviewed in Lübeck and Lübeck 2003a). UP-PCR product cross hybridization resembles the DNA-DNA re-association techniques, facilitating investigation of sequence complementarity. The UP-PCR product cross hybridization is, however, much faster facilitating multiple hybridization analysis of each isolate in contrast to the time consuming pair-wise comparison in conventional DNA/DNA hybridization experiments.

Lübeck and Poulsen (2001) applied the UP-PCR product cross hybridization assay to *R. solani* isolates to test whether the assay could be used for identification. Twenty-one anastomosis group (AG) tester isolates belonging to 11 AGs were amplified with a single UP primer, which generated multiple PCR fragments for each isolate (Fig. 1). The amplified products were spotted onto a filter, immobilized and used for cross hybridization against amplification products from the different isolates. Isolates within AG subgroups cross hybridized strongly, whereas between different AGs little or no cross hybridization occurred. Sixteen *Rhizoctonia* isolates from diseased sugar beets and potatoes were identified using the assay. The results were supported by RFLP analysis of the ITS1-5.8S-ITS2 region.

The results using UP-PCR product cross hybridization assay were compared with results obtained in DNA/DNA hybridization experiments on *R. solani* carried out by Kuninaga and Yokosawa (1985a, 1985b) and Vilgalys (1988). The UP-PCR hybridization data appeared to be in agreement with total DNA/DNA hybridization. The percentage of genome similarity (expressed as DNA hybridization value) was estimated from the intensity of signals in the UP-PCR assay. Strong UP-PCR product hybridization signal seems to correspond with a DNA hybridization value of more than 75%. Significant signal corresponds with a DNA hybridization value of approximately 60-75%, and weak signal corresponds with a DNA hybridization value of approximately 40-60%. No UP-PCR product hybridization signal is obtained with a DNA hybridization value less than 40% (Lübeck and Poulsen 2001). As most AG subgroups have hybridization values less than 60% with each other (Kuninaga 1996), the assay showed that *R. solani* isolates readily can be identified using the UP-PCR product cross hybridization assay. The relationship of UP-PCR product cross hybridization, DNA/DNA hybridization and AGs is shown in Table 2.

DNA hybridization array technology, e.g. micro-arrays and DNA chips (Freeman *et al.* 2000), is believed to have great applications in diagnostics and lead to multiplexing, *i.e.* many pathogens per assay, with fast and highly accurate detection capabilities (Wilgenbus and Lichter 1999, Lévesque 2001). Micro-array technology is based on the classical DNA hybridization principle, with the main difference that multiple specific probes are attached to

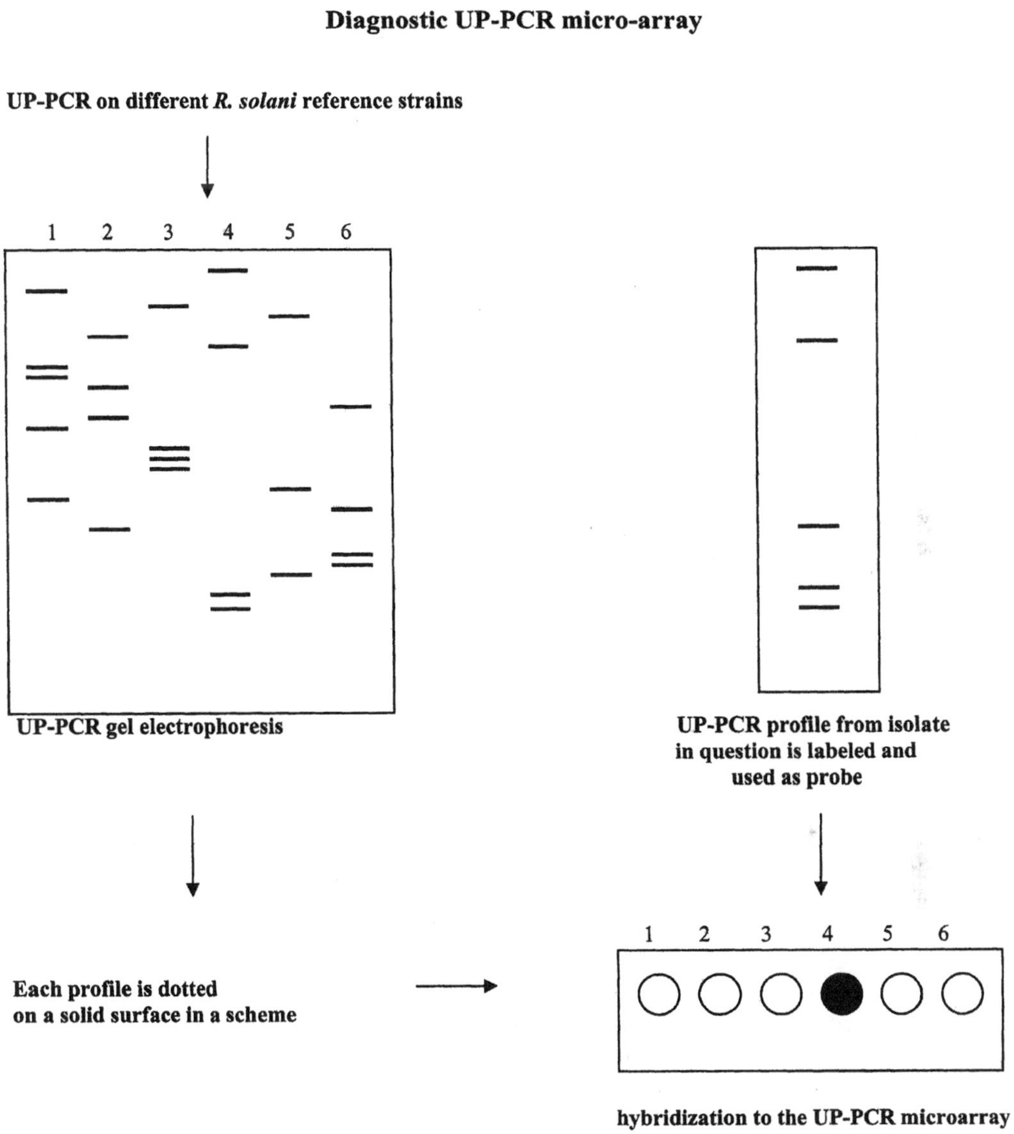

Fig. 2. Flowdiagram showing the UP-PCR product cross hybridization assay for identification of *R. solani*. UP-PCR banding profiles for reference isolates (1, 2, 3, 4, 5, 6) belonging to the different anastomosis subgroups are shown schematically. All amplification products for each isolate are then attached to a solid surface, *e.g.* a glass slide or membrane, which is the "UP-PCR product based microarray". The circles in the array each represent amplification products from a single reference isolate. For identification the isolate in question is labeled and used as probe. A strong signal indicates the correct anastomosis subgroup of the isolate in question.

a solid surface. The advantage is that it allows the simultaneous analysis of large numbers of nucleic acid hybridization experiments. The potential of the UP-PCR product cross hybridization technique as a micro-array based diagnostic tool for routine use has newly been discussed by Lübeck and Lübeck (2003b), focusing on the identification of plant pathogen species.

Table 2. Estimated scale of DNA hybridization values for UP-PCR product cross hybridization signals and implications for *R. solani* AGs and AG subgroups.

UP-PCR product cross hybridization signal[#]	DNA hybridization values## (% DNA relatedness)	Implications for anastomosis group relationship (pairwise comparison)
No signal	< 40 %	Isolates belong to different AGs
Weak signal	40-60 %	Isolates belong to different AG subgroups (in AG1, AG4 and AG6)
Significant signal	60-75 %	Isolates belong to different subgroups in AG2-2
Strong signal	> 75 %	Isolates belong to same subgroup

[#]Based on results obtained in Lübeck and Poulsen (2001); [##]Estimated values by comparison of hybridization signals with results from DNA/DNA hybridization experiments from Kuninaga and Yokosawa (1985a, 1985b) and Vilgalys (1988).

Since DNA hybridization arrays essential is "multiplex hybridization arrays using different types of array platforms, including traditional membranes" (Freeman *et al.* 2000), the UP-PCR product cross hybridization assay itself represents a macro-array technique. All amplification products for each strain are regarded as a single probe, which are transferred and attached to a solid surface, the membrane, in a systematic scheme (Fig. 2).

UP-PCR product based micro-arrays have the potential to be developed for many different purposes and with different complexity (Lübeck and Lübeck 2003b). As the simple array with isolates of *R. solani*, representing many of the different AGs and AG subgroups, facilitated an easy determination to correct subgroup of non-characterized *R. solani* isolates (Lübeck and Poulsen 2001), a commercial *R. solani*-specific micro-array consisting of amplification products from representative isolates from the various AGs and AG subgroups, could be constructed.

7. CONCLUSIONS

The genetic studies in *R. solani* have mainly concerned with the characterization of populations and recognition of subgroups within the complex. Questions whether subgroups represent independent evolutionary units have started to be unraveled. The recent phylogenetic studies support that most subgroups are genetically distinct evolutionary units. However, no final taxonomic decisions about the phylogenetic groupings have been made even though the molecular data strongly indicate that many groups do represent species. Regarding the subgroups as distinct species, population studies can be carried out within each subgroup. Recently, population studies have been carried out using co-dominant single-locus RFLP markers in some of the subgroups. This has for example revealed a model of population structure for *R. solani* AG3-PT and AG1-1A that include both recombination and clonality. These kinds of population studies are expected in the future for the different subgroups and will help clarifying genetic diversity, the impact of mating and gene flow, which may differ for the different groups.

Up till now, there have not been sequenced and characterized genes involved in pathogenicity and no gene expression analysis has been reported. In addition to rRNA genes, the only sequenced genes derived from *R. solani* are four different laccase genes (Wahleithner et al. 1996), four sequences coding for glyceraldehydes-3-phosphate-dehydrogenase and part of an ABC reporter according to GeneBank. Furthermore, the actual role of these few genes has not been studied. Even though DNA hybridization array technology, *e.g.* micro-arrays and DNA chips, has been developed in recent years for automated rapid screening of gene expression and sequence variation of large number of samples, there is only a single report on the use of such principles in *R. solani* research. This report is concerning development of a diagnostic approach, UP-PCR product cross hybridization assay, for subgroup identification, in accordance with the current attention on subgrouping in *R. solani*. Focus on the DNA-array technology for expression analysis of genes, *e.g.* expressed sequence tags (ESTs) would be very important for building up knowledge on genes and gene expression of importance for pathogenicity. This would enable gaining an overview of the cascade of gene expression involved in pathogenicity and would have great impact in our understanding of the success of *R. solani* as a plant pathogen. Unraveling the most important steps for pathogenicity could help identification of better control strategies. However, because of the large genetic diversity within *R. solani* such studies likely have to be carried out separately for each subgroup.

Acknowledgements: I thank Dr. Peter Stephensen Lübeck (The Danish Plant Directorate, Sorgenfri, Denmark) for constructing Fig. 2 and Dr. Dan Funck Jensen (The Royal Veterinary and Agricultural University, Frederiksberg, Denmark) for his valuable comments.

REFERENCES

Adams GC (1996). Genetics of *Rhizoctonia* species. In: *Rhizoctonia* Species: Taxonomy, Molecular Biology, Ecology, Pathology and Disease Control. (Sneh B, Jabaji-Hare S, Neate S, and Dijst G Eds.), Kluwer Academic Publishers, Dordrecht, The Netherlands, pp 101-106.

Anderson NA (1982). The genetics and pathology of *Rhizoctonia solani*. Annu Rev Phytopathol 20:329-347.

Boidin J, Mugnier J, and Canales R (1998). Taxonomic moleculaire des Aphyllophorales. Mycotaxon 66:445-491.

Boysen M, Borja M, del Moral C, Salazar O, and Rubio V (1995). Identification at strain level of *Rhizoctonia solani* AG4 isolates by direct sequence of asymmetric PCR products of the ITS region. Curr Genet 29:174-181.

Bruns TD, White TJ, and Taylor JW (1991). Fungal molecular systematics. Annu Rev Ecol Syst 22, 525-564.

Bulat SA, Lübeck M, Mironenko NV, Jensen DF, and Lübeck PS (1998). UP-PCR analysis and ITS1 ribotyping of *Trichoderma* and *Gliocladium* fungi. Mycol Res 102:933-943.

Carling DE (1996). Grouping in *Rhizoctonia solani* by hyphal anastomosis reaction. In: *Rhizoctonia* Species: Taxonomy, Molecular Biology, Ecology, Pathology and Disease Control. (Sneh B, Jabaji-Hare S, Neate S, and Dijst G Eds.), Kluwer Academic Publishers, Dordrecht, The Netherlands, pp 37-47.

Carling DE, and Kuninaga S (1990). DNA base-sequence homology in *Rhizoctonia solani* Kuhn: inter- and intra-group relatedness of anastomosis group-9. Phytopathology 80:1362-1364.

Carling DE, Baird RE, Kuninaga S, and Gitaitis RD (1999a). Characterization of anastomosis group (AG)-13 of *Rhizoctonia solani*. Phytopathology 89:S11.

Carling DE, Baird RE, Gitaitis RD, Brainard KA, and Kuninaga S (2002a). Characterization of AG-13, a newly reported anastomosis group of *Rhizoctonia solani*. Phytopathology 92:893-899.

Carling DE, Kuninaga S, and Brainard KA (2002b). Hyphal anastomosis reactions, rDNA-internal transcribed spacer sequences, and virulence levels among subsets of *Rhizoctonia solani* anastomosis group 2 (AG-2) and AG-BI. Phytopathology 92:43-50.

Carling DE, Pope EJ, Brainard KA, and Carter DA (1999b). Characterization of mycorrhizal isolates of *Rhizoctonia solani* from an orchid, including AG-12, a new anastomosis group. Phytopathology 89:942-946.

Carling DE, Rothrock CS, MacNish GC, Sweetingham MW, Brainard KA, and Winters SW (1994). Characterization of anastomosis group 11 (AG11) of *Rhizoctonia solani*. Phytopathology 84:1387-1393.

Ceresini PC, Shew HD, and Cubeta MA (1999). RFLP analysis of the PCR amplified ribosomal DNA regions ITS and IGS indicated that isolates of *Rhizoctonia solani* from potato and tobacco represent distinct groups within the anastomosis group 3. Phytopathology 89:S12.

Ceresini PC, Shew HD, Vilgalys RJ, and Cubeta MA (2002a). Genetic diversity of *Rhizoctonia solani* AG-3 from potato and tobacco in North Carolina. Mycologia 94:437-449.

Ceresini PC, Shew HD, Vilgalys RJ, Rosewich UL, and Cubeta MA (2002b). Genetic structure of populations of *Rhizoctonia solani* AG-3 on potato in eastern North Carolina. Mycologia 94:450-460.

Ceresini PC, Shew HD, Vilgalys RJ, Rosewich Gale L, and Cubeta MA (2003). Detecting migrants in populations of *Rhizoctonia solani* anastomosis group 3 from potato in North Carolina using multilocus genotype probabilities. Phytopathology 93:610-615.

Cubata MA, Vilgalys R, and Conzales D (1996). Molecular analysis of ribosomal RNA genes in *Rhizoctonia* fungi. In: *Rhizoctonia* Species: Taxonomy, Molecular Biology, Ecology, Pathology and Disease Control. (Sneh B, Jabaji-Hare S, Neate S, and Dijst G Eds.), Kluwer Academic Publishers, Dordrecht, The Netherlands, pp . 81-86

Cubeta MA, and Vilgalys R (1997). Population biology of the *Rhizoctonia solani* complex. Phytopathology 87:480-484.
Duncan S, Barton JE, and O'Brien PA (1993). Analysis of variation in isolates of *Rhizoctonia solani* by random amplified polymorphic DNA assay. Mycol Res 97:1075-1082.
Fenille RC, de Souza NL, and Kuarmae EE (2002). Characterization of *Rhizoctonia solani* associated with soybean in Brazil. Eur J Plant Pathol 108:783-792.
Freeman, W.M., Robertson, D.J., and Vrana, K.E. (2000). Fundamentals of DNA hybridization arrays for gene expression analysis. BioTechniques 29, 1042-1055.
Gonzalez D, Carling DE, Kuninaga S, Vilgalys R, and Cubeta MA (2001). Ribosomal DNA systematics of *Ceratobasidium* and *Thanatephorus* with *Rhizoctonia* anamorphs. Mycologia 93:1138-1150.
Hyakumachi M, Mushika T, Ogiso Y, Toda T, Kageyama K and Tsuge T (1998). Characterization of a new cultural type (LP) of *Rhizoctonia solani* AG2-2 isolated from warm-season turfgrasses, and its genetic differentiation from other cultural types. Plant Pathol 47:1-0.
Jabaji-Hare SH, Meller Y, Gill S, and Charest PM (1990). Investigation of genetic relatedness among anastomosis groups of *Rhizoctonia solani* using cloned DNA probes. Can J Plant Pathol 12:393-404.
Julián MC, Acero J, Salazar O, Keijer J, and Rubio V (1999). Mating type-correlated molecular markers and demonstration of heterokaryosis in the phytopathogenic fungus *Thanatephorus cucumeris* (*Rhizoctonia solani*) AG 1-1C by AFLP DNA fingerprinting analysis. J Biotech 67:49-56.
Kuninaga S (1996). DNA base sequence relatedness. In: *Rhizoctonia* Species: Taxonomy, Molecular Biology, Ecology, Pathology and Disease Control. (Sneh B, Jabaji-Hare S, Neate S, and Dijst G Eds.), Kluwer Academic Publishers, Dordrecht, The Netherlands, pp 74-78.
Kuninaga S, Carling DE, Takeuchi T, and Yokosawa R (2000). Comparison of rDNA ITS sequences between potato and tobacco strains in Rhizoctonia solani AG-3. J Gen Plant Pathol 66:2-11.
Kuninaga S, Natsuaki T, Takeuchi T, and Yokosawa R (1997). Sequence variation of the rDNA ITS regions within and between anastomosis groups in *Rhizoctonia solani*. Curr Genet 32:237-243.
Kuninaga S, Nicoletti R, Lahoz E, and Naito S (2000). Ascription of Nt-isolates of *Rhizoctonia solani* to anastomosis group 2-1 (AG-2-1) on account of rDNA-ITS sequence similarity. J Plant Pathol 82:61-64.
Kuninaga S, and Yokosawa R (1984a). DNA base sequence homology in *Rhizoctonia solani* Kühn. IV. Genetic relatedness within AG4. Ann Phytopathol Soc Jpn 50:322-330.
Kuninaga S, and Yokosawa R (1984b). DNA base sequence homology in *Rhizoctonia solani* Kühn. V. Genetic relatedness within AG6. Ann Phytopathol Soc Jpn 50:346-352.
Kuninaga S, and Yokosawa R (1985a). DNA base sequence homology in *Rhizoctonia solani* Kühn. VI. Genetic relatedness among seven anastomosis groups. Ann Phytopathol Soc Jpn 51:127-132.
Kuninaga S, and Yokosawa R (1985b). DNA base sequence homology in *Rhizoctonia solani* Kühn. VII. Genetic relatedness between AG-BI and other anastomosis groups. Ann Phytopathol Soc Jpn 51:133-138.
Kuramae, EE, Buzeto AL, Ciampi MB, and Souza NL (2003). Identification of *Rhizoctonia solani* AG 1-1B in lettuce, AG 4 HG-I in tomato and melon, and AG 4 HG-III in brócoli and spinach, in Brazil. Eur J Plant Pathol 109: 391-395.
Laroche JP, Jabaji-Hare SH, and Carhest PM (1992). Differentiation of two anastomosis groups of *Rhizoctonia solani* by isozyme analysis. Phytopathology 82:1387-1393.
Lévesque C.A. (2001). Molecular methods for detection of plant pathogens – what is the future? Can J Plant Pathol 24:333-336.
Liu ZL, Domier LL, and Sinclair JB (1993). ISG-specific ribosomal DNA polymorphism of the *Rhizoctonia solani* species complex. Mycologia 85:795-800.
Liu ZL, and Sinclair JB (1992). Genetic diversity of *Rhizoctonia solani* anastomosis group 2. Phytopathology 82:778-787.
Liu ZL, and Sinclair JB (1993). Differentiation of intraspecific groups within anastomosis group 1 of *Rhizoctonia solani* using ribosomal DNA internal transcribed spacer and isozyme comparisons. Can J Plant Pathol 15:272-280.
Lübeck M, Alekhina IA, Lübeck PS, Jensen DF, Bulat SA (1999). Delineation of *Trichoderma harzianum* Rifai into two different genetic entities by a highly robust fingerprinting method, UP-PCR. Mycol Res 103:289-298.
Lübeck M, Poulsen SK, Lübeck PS, Jensen DF and Thrane U (2000). Identification of *Trichoderma* strains from buildings materials by ITS1 ribotyping, UP-PCR fingerprinting and UP-PCR cross hybridization. FEMS Microbiol. Lett 185:129-134.
Lübeck M, and Poulsen H (2001). UP-PCR and UP-PCR cross blot hybridization as a tool for studying anastomosis group relationship in the *Rhizoctonia solani* complex, FEMS Microbiol. Lett. 201, 83-89.
Lübeck M, and Lübeck PS (2003a). Universally Primed PCR (UP-PCR) and its applications in mycology. In: Current Trends in Mycological Research (Rai M, and Deshmukh SK, Eds.). Science Publisher Inc., USA. In

press.
Lübeck M, and Lübeck PS (2003b). The potential of universally primed PCR (UP-PCR) as a micro-array based diagnostic tool. In: Emerging Concepts in Plant Health Management (Caesar A, and Lartey RT). Research Signpost, Kerala, India. In press.
MacNish GC, Carling DE, and Brainard KA (1993). Characterization of *Rhizoctonia solani* AG8 from bare patches by pectic isozyme (zymogram) and anastomosis techniques. Phytopathology 83:922-927.
MacNish GC, Carling DE, and Brainard KA (1997). Relationship of microscopic and macroscopic vegetative reactions in *Rhizoctonia solani* and the occurrence of vegetatively compatible populations (VCPs) in AG8. Mycol Res 101:61-68.
Majer D, Mithen R, Lewis BG, Vos P and Oliver R (1996). The use of AFLP fingerprinting for the detection of genetic variation in fungi. Mycol Res 100:1107-1111.
McCabe PM, Gallagher MP, and Deacon JW (1999). Microscopic observation of perfect hyphal fusion in *Rhizoctonia solani*. Mycol Res 103:487-490.
Mordue JEM, Currah RS, and Bridge PD (1989). An integrated approach to *Rhizoctonia* taxonomy: cultural, biochemical and numerical techniques. Mycol Res 92:78-90.
Naito S, and Kanematsu S (1994). Characterization and pathogenicity of a new anastomosis subgroup AG2-3 og *Rhizoctonia solani* Kühn isolated from leaves of soybean. Ann Phytopathol Soc Jpn 60:681-690.
Naumova ES, Smith MTh, Boekhout T, de Hoog GS, and Naumov GI (2001). Molecular differentiation of sibling species in the *Galactomyces geotrichum* complex. Antonie van Leeuwenhoek Int. J. Gen. Mol. Microbiol. 80:263-273.
Nicoletti R, Lahoz E, Kanematsu S, Naito S, and Contillo R (1999). Characterization of *Rhizoctonia solani* isolates from tobacco fields related to anastomosis groups 2-1 and BI (AG2-1 and AG-BI). J Phytopathol 147:71-77.
Ogoshi A (1987). Ecology and pathogenicity of anastomosis and intraspecific groups of *Rhizoctonia solani* Kühn. Annu Rev Phytopathol 25:125-143.
Priyatmojo A, Escoplao VE, Tangonan NG, Pascual CB, Suga H, Kageyama K, and Hykumachi M (2001). Characterization of a new subgroup of *Rhizoctonia solani* anastomosis group 1 (AG-1-1D), causal agent of a necrotic leaf spot on coffee. Phytopathology 91:1054-1061.
Rosewich UL, Pettway RE, McDonald BA, and Kistler HC (1999). High levels of gene flow and heterozygote excess characterize *Rhizoctonia solani* AG1-1A (*Thanatephorus cucumeris*) from Texas. Fungal Gen Biol 28:148-159.
Salazar O, Schneider JHM, Julián MC, Keijer J, and Rubio V (1999). Phylogenetic subgrouping of *Rhizoctonia solani* AG2 isolates based on ribosomal ITS sequences. Mycologia 91:459-467.
Salazar O, Julián MC, Hyakumachi M, and Rubio V (1999). Phylogenetic grouping of cultural types of *Rhizoctonia solani* AG2-2 based on ribosomal ITS sequences. Mycologia 92:505-509.
Schneider JHM, Salazar O, Rubio V, and Keijer J (1997a). Identification of *Rhizoctonia solani* associated with field-grown tulips using ITS rDNA polymorphism and pectic zymograms. Eur J Plant Pathol 103:607-622.
Schneider JHM, Schilder MT, and Dijst G (1997b). Characterization of *Rhizoctonia solani* AG2 isolates causing bare patch in field grown tulips in the Netherlands. Eur J Plant Pathol 103:265-279.
Sneh B, Burpee LL, and Ogoshi A (1991). Identification of *Rhizoctonia* speices. The American Phytopathological Society, St. Paul, Minnesota.
Sneh B, Jabaji-Hare S, Neate S, and Dijst G (1996). *Rhizoctonia* Species: Taxonomy, Molecular Biology, Ecology, Pathology and Disease Control, Kluwer Academic Publishers, Dordrecht, The Netherlands.
Taylor JW, Jacobsen DJ, and Fisher MC (1999). The evolution of asexual fungi: reproduction, speciation and classification. Annu Rev Phytopathol 37:197-246.
Taylor JW, Jacobsen DJ, Kroken S, Kasuga T, Geiser DM, Hibbett DS, and Fisher MC (2000). Phylogenetics, species recognition and species concept in fungi, Fungal Genet Biol 31:21-32.
Toda T, Hyakumachi M, and Arora DK (1999). Genetic relatedness among and within different *Rhizoctonia solani* anastomosis groups as assessed by RAPD, ERIC and REP-PCR. Microbiol Res 154:247-258.
Vilgalys R (1988). Genetic relatedness among anastomosis groups in *Rhizoctonia* as measured by DNA/DNA hybridization. Phytopathology 78:698-702.
Vilgalys R, and Gonzalez D (1990). Ribosomal DNA restriction fragment length polymorphisms in *Rhizoctonia solani*. Phytopathology 80:151-158.
Vilgalys R, and Cubeta MA (1994). Molecular systematics and population biology of *Rhizoctonia*. Annu Rev Phytopathol 32:135-155.
Vilgalys R, and Gonzalez D (1990). Ribosomal DNA restriction fragment length polymorphisms in *Rhizoctonia solani*. Phytopathology 78:698-702.
Vos P, Hogers R, Bleeker M, Reijans N, van de Lee T, Hornes M, Frijters A, Pot J, Peleman J, Kuiper M and Zabeau M (1995). AFLP: a new concept for DNA fingerprinting. Nucleic Acids Res 23:4407-4414.

Wahleithner JA, Xu F, Brown KM, Brown SH, Golightly EJ, Halkier T, Kauppinen S, Pederson A, and Schneider P (1996). The identification and characterization of four laccases from the plant pathogenic fungus *Rhizoctonia solani*. Curr Genet 29:395-403.

Welsh J and McClelland M (1990). Fingerprinting genomes using PCR with arbitrary primers. Nucleic Acids Res 18:7213-7218.

Wilgenbus KK, and Lichter P (1999). DNA chip technology ante portas. J Mol Med 77:761-768.

Williams JGK, Kubelik AR, Livak KJ, Rafalski JA and Tingey SV (1990). DNA polymorphisms amplified by arbitrary primers are useful as genetic markers. Nucleic Acids Res 18:6531-6535.

Applied Mycology & Biotechnology
An International Series. Volume 4. Fungal Genomics

9

Genomics of *Trichoderma*

Manuel Rey[1], Antonio Llobell[2], Enrique Monte[3], Felice Scala[4] and Matteo Lorito[4]
[1]Newbiotechnic S.A., Isla de la Cartuja, Seville, Spain, [2]Instituto de Bioquimica Vegetal y Fotosintesis, University of Sevilla/CSIC, Seville, Spain, [3]Centro Hispano Luso de Investigaciones Agrarias, University of Salamanca, Salamanca, Spain, [4]Dipartimento Ar.Bo.Pa.Ve., SEZIONE DI Patologia Vegetale, Laboratori di Biocontrollo, Università di Napoli Federico II, Via Università, 100, 80055 Portici (Napoli) Italy (lorito@unina.it, http://cds.unina.it/%7Elorito/).

Trichoderma strains are considered to be among the most useful fungi in industrial enzyme production, agriculture and bioremediation. Much less developed is the idea of looking at these micromycetes as model microorganisms to study and improve the understanding of important microbial interactions, for instance with plants and pests. In fact, the use of genomic approaches to study the complex and fascinating mechanisms that permit *Trichoderma* to produce large amount of heterologous proteins (this aspect will be reviewed in detail in another chapter of this book), control pathogens and effect plant metabolism and physiology is still in its infancy. Very little has been accomplished to date, even though the need and interest in sustaining both structural and functional genomic projects is widely recognized and has led to funding and start up of several new initiatives. This chapter presents the rationale used for investing in a wide genome effort on these fungi, based on diversity and potential utility of their biological characteristics, and discusses the problems related to the taxonomical discrimination of the various species and strains within this genus. It also describes the methods that have been used to date to obtain useful genomic data especially on *T. reesei* and a few other *Trichoderma* species. Finally, we describe the strategy and the preliminary results of a functional genomic project recently undertaken by an International Consortium comprised of academic institutions and small enterprises, which was purposely formed to follow this initiative. This program aims at exploring genetic biodiversity within the *Trichoderma* genus, providing basic knowledge on the biology of these complex microorganisms, and developing commercial applications in agriculture, food industry and medicine.

1. INTRODUCTION

1.1. *Trichoderma*

Fungal strains within the genus *Trichoderma* include a wide spectrum of evolutionary solutions that range from very effective soil colonizers with high biodegradation potential, to non-strict plant symbionts that colonize the rhizosphere. Some groups of biotypes within this conglomerate are able to antagonize phytopathogenic fungi by using substrate colonization,

Corresponding author: Matteo Lorito

antibiosis and/or mycoparasitism as the main mechanisms. This antagonistic potential is the base for effective applications of different *Trichoderma* strains as an alternative to the chemical control against a wide set of fungal plant pathogens (Chet 1987; Harman and Björkman 1998). As a consequence of the variety of activities displayed by the *Trichoderma* strain conglomerate, a large range of applications have been developed: the antagonistic potential is the basis for the effective control of a wide set of phytopathogenic fungi and the biodegradative capacity is a source of useful enzymes in different industrial sectors (Harman and Kubicek 1998).

Trichoderma biocontrol strains have evolved numerous mechanisms for both attacking other fungi and enhancing plant and root growth (Harman 2000). The colonization of the root system by rhizosphere competent strains of *Trichoderma* results in increased development of root and/or aerial systems and crop yields (Harman and Kubicek 1998). Other activities, like the induction of plant systemic resistance and antagonistic effects on plant pathogenic nematodes (Sharon *et al.* 2001), have also been described. These facts strongly suggest that during the plant-*Trichoderma* interactions, the fungus participates actively in protecting and improving its ecological niche.

Strains of *Trichoderma* may also be aggressive biodegraders (Wardle *et al.* 1993) and act as competitors fungal pathogens in their saprofitic phases, especially when nutrients are a limiting factor (Simon and Sivasithamparam 1989). Strains have been reported as promoting activities of non-pathogenic bacteria (Vrany *et al.* 1990) and mycorrhizal fungi (Calvet *et al.* 1993). In the 1990s, the ability of *Trichoderma* strains to synthesize substances inducing SAR-like responses in plants was shown (Elad 1996; Enkerli *et al.* 1999). For example, work on the harpin protein from *Erwinia amylovora* led to the development of a natural compound, with great commercial expectations in the USA, as the first systemic fungicide of biological origin having a defined molecular structure (Grisham 2000). Molecules produced by *Trichoderma* and/or its metabolic activity also have potential for promoting plant growth (Yedidia *et al.* 1999). Application of the species *T. harzianum* to plants resulted in improved seed germination, increased plant size, and augment of leaf area and weight (Altomare *et al.* 1999; Inbar *et al.* 1994). The scenario of combined systemic biofungicides and plant growth promoters has great market potential if the molecular basis of the activities can be identified.

The strong biodegradation and substrate colonization performances of *Trichoderma* strains is the result of an amazing metabolic versatility and a high secretory potential which leads to the production of a complex set of hydrolytic enzymes. Similarly, the mycoparasitic process is based on the secretion of a rich cocktail of cell wall degrading enzymes (CWDEs) able to hydrolize the cell wall of various hosts (Kubicek *et al.* 2001). Among others, chitinases (de la Cruz *et al.* 1992), β-1,3- glucanases (de la Cruz *et al.* 1995b; Lorito *et al.* 1994a; Noronha and Ulhoa 1996), β-1,6-glucanases (de la Cruz *et al.* 1995a; de la Cruz and Llobell 1999), α-1,3-glucanases (Ait-Lahsen *et al.* 2001) and proteases (Geremia *et al.* 1993; Suárez 2001) have been described as important components of the multi enzymatic system of *Trichoderma* strains. Some of these proteins display strong antifungal activities when are applied *in vitro*, alone and/or combined, against plant pathogens (Harman 2000). Some lytic enzymes can be involved in both antagonistic and saprophytic processes providing an evolutionary advantage to strains with both biodegrading and antagonistic potential, for the efficient colonization of different ecological niches in soil. A principal role in mycoparasitism has been attributed to chitinases (Lorito 1998) and glucanases (Benítez *et al.* 1998). However, fungal proteases may also be significantly involved in cell wall degradation, since fungal cell walls contain chitin and glucan polymers embedded in and covalently linked to a protein matrix (Kapteyn *et al.* 1996).

The production of secondary metabolites by *Trichoderma* strains also shows great variety and application potential. *Trichoderma* strains seem to be an inexhaustible source of antibiotics, from the acetaldehydes gliotoxin and viridin (Dennis and Webster 1971), to alpha-pyrones (Keszler *et al.* 2000), terpenes, polyketides, isocyanide derivatives, piperacines, and complex families of peptaibols (Sivasithamparam and Ghisalberti 1998). All these compounds produce synergistic effects in combination with CWDEs, with strong inhibitory activity on many fungal plant pathogens (Lorito *et al.* 1996a; Schirmböck *et al.* 1994). The potential of genes involved in biosynthetic pathways of antibiotics [e.g. polyketides, Sherman (2002) and peptaibols (Wiest *et al.* 2002)] with applications in human and veterinary medicine not been explored yet.

Several methods for applying both biocontrol and plant growth promotion exerted by *Trichoderma* strains have recently been demonstrated and it is now clear that hundreds of separate genes and gene products are involved in the processes of mycoparasitism, antibiosis, competition for nutrients or space, tolerance to stress through enhanced root and plant development, solubilization and sequestration of inorganic nutrients, induced resistance and inactivation of enzymes produced by pathogens (Monte 2001). Some of these genes have been identified, patented and used to transgenically increase plant disease resistance (Lorito *et al.* 1998), but most of them are still unexploited for developing new biotechnologies.

1.2. "Omics" and The New Era of Science

The sequencing of the first genome, the bacteriophage ΦX174, opened the Genomics era (Sanger *et al.* 1978). The term "genomics" was coined by Thomas Roderick in 1986, to refer to a new scientific discipline of mapping, sequencing, and analyzing genomes. Ten years ago, genomics referred to the quantitative studies about genes, regulatory and non-coding sequences, and genomes. Genomics is now undergoing a transition or expansion from the mapping and sequencing of genomes to an emphasis on genome function. To reflect this shift, genome analysis may now be divided into "structural genomics" (the generation and analysis of information about genes) and "functional genomics" (the generation and analysis of information about what genes do). The aim of functional genomics is to derive as much information about as many genes as possible and as quickly as possible. In a broad sense, genomics is described by Brent (2000) as the generation of information about living things by systematic approaches that can be performed on a large scale.

The term proteome, coined in 1994, as a linguistic equivalent to the concept of genome, is used to describe the complete set of proteins that is expressed and modified by the entire genome in the lifetime of a cell. It is also used in a less universal sense to describe the complement of proteins expressed by a cell at a specific time. More recently, Weinstein (1998) coined the term "omics" and extended the concepts to related research areas such as "proteomics" (protein expression), "transcriptomics" (RNA and gene expression), and "metabolomics" (metabolites and metabolic networks). Experimental approaches to elucidate the complex interplay of gene products and regulatory sequences from a more global perspective become more and more crucial.

Proteomics refers to the study of the proteome using technologies of large-scale protein separation and identification to obtain a global, integrated view of cellular processes including expression levels, post-translational modifications, protein interactions and location. Proteomics catalogues and characterizes cellular proteins, investigates variations in their levels of expression under different physiological conditions, studies their interactions (Pandey and Mann 2000), then identifies and relates their function to cellular metabolic pathways (Hellman 2000; Grishin *et al.* 2000). Thus, it is a fundamental tool for understanding all biological processes (Rossignol 2001). Proteomic technology has powerful applications since the effect of genetic and/or environmental changes can be visualized on two-dimensional electrophoresis (2-DE) gels. The separated proteins can be analyzed by fragmentation and peptide mass fingerprinting or electro-spray mass spectrometry, and/or N-terminal Edman micro-sequencing (Grishin *et al.* 2000; Stasyk *et al.* 2001).

The techniques available to identify mRNA from actively transcribed genes are referred to as transcriptomics, as these are based on the process of transcription, and the complement of mRNAs transcribed from a cell genome is called the transcriptome. The analysis of transcriptomes is a key area since the development and differentiation of a cell or an organism, as well as the adaptation to variable conditions, is determined in a large part by the profile of gene expression. DNA array technology (Bowtell and Sambrook 2003), where thousands of different DNA sequences are arrayed at a high density and with high precision (Maier *et al.* 1997) in a defined matrix on different supports (e.g. nylon or glass) (Maskos and Southern 1992), is rapidly becoming the method of choice for gene expression profiling. The applications of microarrays extend beyond the boundaries of basic research into environmental monitoring, pharmacology, agricultural biotechnology and diagnostics. DNA-microarray technology has overcome the initial stage and emerged as an important tool to meet many of these challenges for instance in the study of gene expression (Schena *et al.* 1995) and identification of transcription factor binding targets (Iyer *et al.* 2001). More recently, protein microarrays have been used for the large-scale analysis of protein (Zhu *et al.* 2001) and peptide (Houseman *et al.* 2002) activities and functions; and carbochips (carbohydrate microarrays) have been developed to quantitatively analyze protein-carbohydrate interactions (Wang *et al.* 2002).

The power of transcriptomics and proteomics can be harnessed in either "mechanistic" or "predictive" modes of analysis. In the "mechanistic" mode, these techniques are used to implicate specific genes or proteins in the mechanism of action. Such research leads to further investigation by using conventional techniques, in order to clarify the roles, if any, played by them. Alternatively, the technologies can be utilized in a "predictive" context, with the hope that biological responses induced by gene products can be described based on comparison of global patterns of gene and protein expression. In this way, the mode of action of a novel protein or metabolite may be identified by comparing the expression pattern they elicit with established expression "fingerprints" of reference gene products where mechanisms are understood. The application of these expression profiling methods in mechanistic studies has already met with some success. However, it remains to be seen whether the predictive capacity of these methods can enhance our ability to detect compounds with biotechnological value.

The term 'metabolome' refers to the entire complement of all the small molecular weight metabolites inside a cell suspension (or other sample) of interest. The techniques available to identify the presence and concentrations of metabolites in a biological sample are known as metabolomics. It is likely that measurement of the metabolome in different physiological states will in fact be much more discriminating for the purposes of functional genomics (Raamsdonk *et al.* 2001). This task can be facilitated by rapid analytical methods such as Pyrolysis mass spectrometry, Fourier-Transform Infrared Spectrometry, Raman spectrometry, GC-MS, and most recently, LC-Electrospray and cap-LC-tandem-electrospray mass spectrometry.

In the last few years, some biotech companies are using the term "phenomics" to describe the technology of automated functional analysis of proteins. The word derives from phenotype (the observable characteristics conferred by a gene). Particular applications to drug discovery are also know as "pharmacogenomics" (Evans and Relling 1999). Some enthusiastic researchers in the field are even starting to refer to the whole operation of molecular analysis of a cell, extending from DNA through RNA to protein, as "operomics".

Array technology (Bowtell and Sambrook 2003) is an essential tool for genomics projects (Celis *et al.* 2000; Chee *et al.* 1996). Many types exist, depending of the number and type of clones (bacteria, plasmids, PCR products, oligonucleotides), substrate (crystal, nylon, nitrocellulose, PVDF etc.) used for support, and type of marker for hybridizations (Ekins and Chu 1999; Leming 1999; Soundy *et al.* 2001). Commercially marketed arrays contain complete or partial ORFs of distinct organisms, e.g. *Saccharomyces cerevisiae* (Research Genetics, Inc.) or humans (Affimetrix, 2000). Applications are numerous, ranging from detection of single-nucleotide polymorphisms (SNPs) to functional genomics, and include gene expression analysis, genotyping and genetic mapping, comparative genomic studies of different organisms, as well as of populations or races of the same species (Hoheisel 2002). For instance, the macroarrays that we have developed derive from cDNA libraries of *Trichoderma* obtained under different conditions (i.e. mycoparasitism) and by using various normalization and subtraction methods it is possible to characterize the clones and study the gene regulation in *Trichoderma* (Bonaldo *et al.* 1996). Results from these type of experiments are treated with data mining techniques included in the bioinformatics package GeneSpring (www.silicongenetics.com) and other less sophisticated programs (e.g. J-Express, www.molmine.com), which permit representation and grouping of expression data in an intelligible manner.

The rapid and massive accumulation of data derived from the different omics approaches (genomics, proteomics, transcriptomics, phenomics, physiomics, metabolomics, etc.) represents the basis of a new type of biological science and of the so-called post-genomic era. In 1993, Walter Gilbert directly speculated on the nature of biology in the post-genome era : "The emerging new paradigm is that all genes will be sequenced and made resident in huge databases available electronically, and thus the starting point of a biological investigation will be theoretical" (Lenoir 2000).

This new era arises from the advances in several different areas of analytical chemistry, analytical biochemistry, image analysis, robotics and process automation which have provided an abundance of effective approaches for the very large scale of the tasks involved in such projects. These technical innovations point to the crucial role that engineering and physical sciences will play in the development of biology during the post-genome era. Clearly, a special role will be covered by informatics, since the inevitable consequence of the scale of research in structural and functional genomics is the obtaining of huge data set arranged in a "phone book- like" manner and that are not easy to examine or understand. Given the sequence of the human genome, for example, it is a difficult task to identify individual genes: similar problems are very common in both structural and functional genomics. In fact, one of the consequences anticipated by Gilbert in 1993 is that databases themselves will become the main target especially at the initial stages of a biological research.

The acquisition of relational databases, as well as the development of efficient methods for searching and viewing these data, represents *de facto* a new discipline called "bioinformatics." In a broader view, bioinformatics contains computational or algorithmic approaches to analyze the information obtained from large amounts of biological data, and this might include predictions on protein structure, the dynamic modelling of complex physiological systems, or the statistical treatment of quantitative traits in populations in order to determine the genetic basis for these traits. Unquestionably, bioinformatics will be an essential component of all research activities that use high-throughput methods for generation of data, and is particularly important as an adjunct to genomics and proteomics projects (Ouzounis *et al.* 1996). Otherwise, the large quantity of data generated from sequence projects and array analysis, also considering the speed with which they are acquired, cannot be effectively harnessed (Rashidi 2000). Data can be stored in a manner which allows easy access and management, and gene expression can be studied on a relational basis using simple interfaces, as in GeneCards (http://bioinfo.weizmann.ac.il/cards/). This, and similar databases, provide easy access for researchers to the results as they appear, and make their dissemination process to the whole scientific community more efficient than by traditional publications. Software packages are available to characterize new sequences and identify homology with others of known function (Misener and Krawetz 2000), including Wisconsin GCG version 10 (Butler 1998).

2. FUNGAL GENOMICS and *TRICHODERMA* GENOMICS

Fungi constitute a group of organisms of paramount importance in the study of basic biological processes, human health and commercial applications. As pointed out in the Fungal Genome Initiative White Paper (Birren *et al.* 2002), within fungal genomes lies the evolutionary history of the origins of many important biological processes found in higher eukaryotes. In addition, experimental tractability of fungi makes them among the most useful model systems in cell biology. Fungal cellular physiology and genetics share key components with animal cells, such as multi-cellularity, development and differentiation, cell cycle, intercellular signaling, or programmed cell death. The shared origins of the genes responsible for these fundamental biological functions between humans and fungi continue to make the history and function of these fungal genes of vital interest to human biology. Fungi, as pathogens, have a heavy negative impact on human health and agriculture. Identifying effective therapies against human opportunistic pathogenic fungi has been more difficult than for bacterial pathogens. Most of the existing drugs produce serious side effects and developed resistance to these compounds is an increasing problem. In agriculture, crop losses in the field and post-harvest due to phytopathogenic fungi exceed 200 billion Euro annually, and in the USA alone over 600 million dollars are spent annually on agricultural fungicides.

Finally, fungi are the target of many biotechnological applications from the development and production of noteworthy pharmaceuticals (penicillins, cephalosporins, etc.) and industrial enzymes (cellulases, lipases, etc.), to the use as systems for homologous and heterologous gene over-expression. Despite their importance, the resources applied to studies of filamentous fungi are very limited in comparison to the total genomics effort (Yoder and Turgeon 2001). With regards to *Trichoderma*, although it is a fungal genus of demonstrable biotechnological value, its genome has been poorly surveyed compared to other microorganisms. This is due to the large diversity of species, the absence of optimized systems for its exploration, and the great variety of genes expressed under a wide range of ambient conditions (Lorito *et al.* 1998; Monte 2001). Due to their ubiquity and rapid substrate colonization, *Trichoderma* species have been commonly used as biocontrol organisms in agriculture, and their enzyme systems are widely used in industry (Papavizas 1985; Samuels 1996; Harman and Kubicek 1998). However, there is a clear need to explore beyond the phenotype and the biocontrol applications, but to exploit the underlying genetic systems. According to present methods, the genomic information available in data bases about *Trichoderma* could be divided in two main groups as follow.

2.1. Classical *ad hoc* Approaches and Digital Genomics

Audic and Claveirie (1997) were the first to apply the term "digital" to the study of gene expression based on absolute count of randomly generated tags from large condition, stage, organ, or tissue-specific cDNA populations presents in database. In a broad sense, the term digital genomics could be considered as the extraction of knowledge from the sequences accumulated in data bases coming from either high-throughput or independent classical ad hoc experiments (Table 1).

An analysis applied to *Trichoderma* digital genomics data reveals the following: i) A glance at the literature clearly indicates that the genera *Trichoderma* is the target of active investigation, and only a few other fungi present more registers in PUBMED than

Trichoderma. Among the 15 fungi proposed by the Fungal International Initiative for initial sequencing based on the relevance in medical, commercial or evolution and fungal diversity areas, none of them show higher citation index than *T. reesei* or *T. harzianum*. Despite this fact, the number of genomic-related molecular biology studies are scarce, as can be deduced from the relatively low number of sequences present in database. Most of the work based on *Trichoderma* has focused on commercial and practical applications and the studies focus on physiological, microbiological and biochemical aspects rather than the molecular basis of these processes.

ii) A deeper analysis of the sequences available reveals that taxonomy studies of the genus *Trichoderma* are an active field. In fact, 1.016 out of the 1319 sequences obtained by classical approaches correspond to internal transcribed sequences (Bruns *et al.* 1991). In addition, there are 64 sequences of *chit42* and 32 sequences of *Tef1*, also obtained for taxonomical purposes as described above. However, the complexity of the genus *Trichoderma* has generated numerous controversies in the understanding of the basic biology of these organisms and their commercial use. Thus, great efforts have been made to clarify this area, also by using the more powerful approaches. As a result, the molecular taxonomy of *Trichoderma* has become one of the most studied among the filamentous fungi (Table 2).

Table 1. Total number of *Trichoderma* sequences available in database until dec. 2002

Type of sequence	Methods	Number of sequences	Publicly available
EST (Genencor)	High-throughput	11050	Partially
EST (NewBioTechnic)	High- throughput	1300*	No
EST (USP[a])	High-throughput	1200*	Yes
ITS[o]	Classic	1016	Yes
Patented	Classic	99	Yes
Polymer degrading enzymes	Classic	148	Yes
Others	Classic	56	Yes

*Unigenes,[a] University Sao Paulo,[o] Internal Transcribed Sequence

Table 2: Number of Internal Transcribed Sequences described for different fungi

Organism/Genus	Number of sequences
Fungi	42.402
Fusarium	1.714
Aspergillus	1.260
Trichoderma	1.016
Neurospora	580
Colletotrichum	527
Penicillium	482
Verticillium	331
Ustilago	63
Rhizopus	61
Botrytis	52

Another hot topic in the *Trichoderma* molecular genetic studies is clearly related to its ability to degrade polymers (Table 3). A fundamental part of the *Trichoderma* antifungal capability consists of a series of genes encoding for a surprising variety of extracellular lytic enzymes, including endochitinases, β-N-acetylhexosaminidase (N-acetyl-β-D-glucosaminidase), chitin-1,4-β-chitobiosidases, proteases, endo- and exo-β-1,3-glucanases, endo β-1,6-glucanases, lipases, xylanases, mananases, pectinases, pectin lyases, amylases, phospholipases, RNAses, DNAses, etc. (Benítez *et al.* 1998; Lorito 1998). The chitinolytic and glucanolytic enzymes are especially valuable for their CWDE activity on fungal plant pathogens, by hydrolyzing polymers not present in plant tissues. Each of these classes of enzymes contains diverse sets of proteins with distinct enzymatic activities, and some have been purified and characterized and their genes cloned (Ait-Lahsen *et al.* 2001; de la Cruz *et al.* 1992; 1995a; 1995b; García *et al.* 1994; Limón *et al.* 1995; Lora *et al.* 1995; Lorito *et al.* 1993; 1994a; Montero 2001; Peterbauer *et al.* 1996; Suárez 2001; Viterbo *et al.* 2001; 2002).

Table 3: Polymer degrading enzymes, their genes and relative entrances in public data bases

Polymer	Gene / Enzyme	Nº E. in DB*
Chitin	*Chit42* (endochitinase)	64
	Chit33 (endochitinase)	2
	Nag1 (chitobiosidase)	2
	Nag2 (chitobiosidase)	2
	Exc1 (exochitinase)	1
	Exc2 (exochitinase)	1
	Ech2	2
	Ech3	2
	Ech3B	2
	Cht1	3
	Cht2	2
	Chit33 (endochitinase)	2
	Chit36Y	2
	Chit-VIRI	1
	Chit-P	1
	Chit-HAM	1
	Chit-HAR1	1
	Chit-HAR2	1
	Chit-HAR3	1
	Chit-G1	1
	Chit-G2	1
Glucan	*Bgn1* (endo-β-1,3-glucanase)	1
	Bgn2 (endo-β-1,3-glucanase)	1
	Bgn3 (endo-β-1,6-glucanase)	2
	Gluc78 (exo-β-1,3-glucanase)	2
	B16-1 (endo-β-1,6-glucanase)	2
	B16-2 (endo-β-1,6-glucanase)	2
	B16-3 (endo-β-1,6-glucanase)	2
Cellulase	*Cbh1* (β-1,4-glucan-cellobiohydrolase)	7
	Cbh2 (β-1,4-glucan-cellobiohydrolase)	6
	Cel2A (β-1,4-endoglucanase)	2
	EglI (β-1,4-endoglucanase)	1
	EglII (β-1,4-endoglucanase)	1
	EglIII (β-1,4-endoglucanase)	1
	EglIV (β-1,4-endoglucanase)	1
Mutan	α-1,3-glucanase	3
Xylan	*Xyl1* (β-xylanase)	2
	Xyn2 (β-xylanase)	2
	Xyn3 (β-xylanase)	1
	Endo-β-1,4-xylanase	1
	α-glucuronidase	1
	α-L-arabinofuranosidase	1

*Number of entrances in public data bases

Once purified, most enzymes have been shown to have a strong antifungal activity, which is synergistic when used in combinations. Research over the past ten years has demonstrated that CWDEs from *Trichoderma* have great potential in agriculture as active components for novel fungicides (Harman 2000). In particular, β-1,6-glucanases (which are not present in plants) have the capacity to degrade β-1,6-glucans, important structural components of fungal walls which link the cell wall chitins to β-1,3-glucans (Kapteyn *et al.* 1996, 1997). In addition, *Trichoderma* enzymes have a high diversity of structural and kinetic properties, which increase the probability of evading inhibitory mechanisms (Ham *et al.* 1997), and have been demonstrated to be synergistic with PR proteins of plants (Lorito *et al.* 1996b), other natural compounds such as antibiotics, as well as xenobiotic compounds (Lorito *et al.* 1998). *T. harzianum* CWDEs do not have an effect on humans and animals, as confirmed by EPA tests used for registration of biocontrol agents in the USA, and they degrade innocuously in the environment. Single or mixed combinations of CWDEs with elevated antifungal effects, obtained from culture filtrates or through heterologous expression, could be included in commercial formulations since they are easily characterized, resist desiccation, are stable at temperatures up to 60° C, and are active over a wide range of pH and temperatures. The inhibitory effect of chemical fungicides can be substantially improved by the addition of minute quantities (10-20 ppm) of *Trichoderma* CWDEs (Lorito *et al.* 1994b), and CWDE genes can be expressed in plants to improve plant disease resistence (Lorito *et al.* 1998).

Many purified *Trichoderma* CWDEs are of interest to the agro-food industry. The enormous potential of the β-(1,4)-endoglucanase produced by *T. longibrachiatum* and *T. reesei* has been used to solve filtration problems associated with the presence of β-glucans during beer production. The addition of this enzyme is a frequent practice in this industrial sector, and genes coding for its production have been incorporated in transgenic yeasts for making beer (Linko *et al.* 1998). Furthermore, β-(1,4)-endoglucanase from *T. longibrachiatum* promotes the liberation of aromatic terpene precursors in grape that leads to the final fruity aroma of wines (Pérez-González *et al.* 1993). Thirdly, in a very distinct example, *Trichoderma* cellulases and hemicellulases have been used for years as an additive to chicken feed formulations to improve digestibility and fecal production.

The rest of the *Trichoderma* sequences (118) are distributed in many functional categories based on similarities to known sequences found in established data bases (Fig. 1).

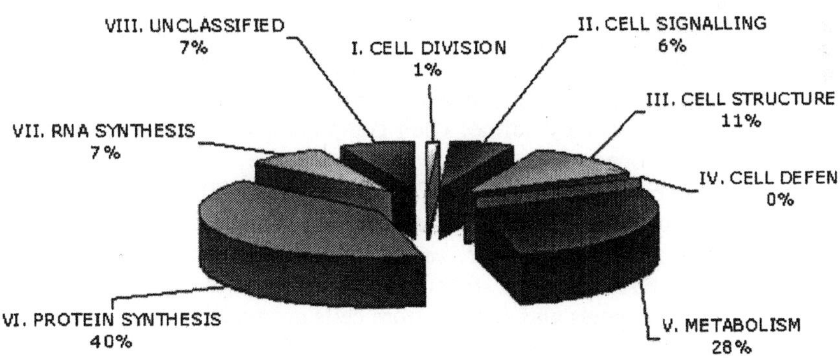

Fig. 1. Functional assignment of *Trichoderma* sequences available in data base according to the classification developed at the Institute for Genomic Research (TIGR, Rockville, MD). Unclassified are those ESTs that show similarity to a sequence with known function or do not fall into any of the classification schemes utilized.

Although it is difficult to extract any conclusions from this miscellaneous group of genes, it is noteworthy that there is an increasing interest by many research groups in genes related to signalling and regulation.

2.2 High-Throughput Genomics

Currently, in a collection of 379 genome-project websites, 55 projects are completed and 324 are in progress. Only 16 of these websites (i.e. 4% of the total) involve fungi, and only 12 species of fungi are represented, of which 7 are pathogens of either plants or humans. This list includes projects on plant pathogens, human pathogens and saprophytes. Regarding *Trichoderma*, to date there are only four projects based on genomics technologies, including genome and expressed sequence tag (EST) (Adams *et al.* 1991) sequencing and expression profiling using micro- and macro-array analysis (see below). Three of these projects are funded by "for-profit" commercial organizations (NewBioTechnic in Spain, Genencor in USA and VTT in Finland) which clearly indicates the biotechnological value of these organisms, and of course limits the access of data by the public.

The main initiatives or current developments in this area are: i) Nakari *et al.* (2000), following a macroarray-like procedure, described a method for the identification and cloning of promoters expressed under defined environmental conditions, such as growth in glucose medium. By using this approach, 5 promoters capable of high expression were isolated and their use patented. ii) Chambergo *et al.* (2002), using EST analysis and cDNA microarrays, studied the metabolism of *T. reesei* focusing on the anaerobic and aerobic degradation of glucose. In particular, these authors tried to address the unresolved questions of why most multicellular microorganisms metabolize glucose by respiration rather than fermentation. This is also the case in *T. reesei*, which does not obtain energy from glucose by anaerobic metabolism but produces ATP from glucose by respiration, unlike *S. cerevisiae*. They found that in *T. reesei* the expression of genes encoding the enzymes of the TCA cycle and the proteins of the electron transport chain is programmed in a way that favors the oxidation of pyruvate via TCA cycle rather than ethanol. They indicated that likely the presence of an evolutionary pressure has directed the flow of metabolites into respiration rather than fermentation, and the prevalence of aerobic metabolism of *T. reesei* in the presence of high levels of glucose. Their genomic work has produced so far the sequence of 2385 randomly selected cDNA corresponding to 1151 unique transcripts and the complete mitochondrial genome of *T. reesei*. Putative functions were assigned to 36% of the transcripts, with unknown proteins representing 3% of them. Interestingly, 61% of the ESTs showed no significant similarity to any other sequence in the data bases, and thus they are possibly specific for *T. reesei* or other filamentous fungi (Chambergo *et al.* 2002). However, these authors also indicated that it was possible that no enough fungal sequences have yet been deposited in gene banks, therefore the dissimilarity may be due to non-availability of sequences for the functions that they define. Their complete EST data base and the details on the experimental procedure and results are available on line (http://trichoderma.iq.usp.br/TrEST.html). iii) Chellappan *et al.* (2000), in collaboration with Genencor, are carrying out an EST and BAC end sequencing project for the discovery and development of new gene products from *T. reesei*. In this study, two cDNA libraries were constructed, one with RNA extracted from cells grown in conditions promoting the production of cellulolytic enzymes and another from cells grown in 18 different conditions. A total of 9.792 ESTs were sequenced which resulted in 2.336 singlets. Each of the two libraries showed similar levels of redundancy, while the redundancy for the entire EST library raised up to 46%. This project has been extended into a structural genomics sequencing effort to obtain the complete sequence of the *T. reesei* genome, which is expected to be released by the middle of the year 2003. (iv) A bacterial artificial chromosome library of at least 25X coverage has been constructed for the biocontrol agent *T. virens* strain Tv29-8. This work lead by C.M. Kenerly at Texas A&M University, has been based on the isolation of high molecular weight DNA fragments to produce a BAC library robotically arrayed as 12243 clones with inserts from 10 to 170 kb. The library was successfully used to isolate full and truncated clones of a 62.8 kb peptide synthetase and many other genes including other peptide synthetases, glucanases and signal transduction-involved sequences (Wiest *et al.* 2002). (v) A functional genomic project to develop applications of gene products from antagonistic *Trichoderma* strains in different industrial sectors has been started in year

2000 lead by NewBioTechnic, S.A. in collaboration with several research groups. The project aims to explore the biodiversity within the genus and do not concentrated on only one strain (see below). The initiative has expanded into a consortium of research teams that was started at the end of 2002 and is financed by the European Commission within its 5th framework.

3. GENETIC VARIABILITY

Trichoderma is a fungal genus that was described in 1794, including anamorphic fungi isolated primarily from soil and decomposing organic matter (Persoon 1794). The genera *Hypocrea, Podostroma* and *Sarawakus*, belonging to the family *Hypocreaceae* (Ascomycetes), have also been described as teleomorphs of *Trichoderma* (Rossman *et al.* 1999). Most of the *Trichoderma* species are morphologically very similar and were considered for many years as a single species, *T. viride* (Bisby 1939). A consolidated taxonomical scheme was needed. Rifai (1969) proposed and defined nine morphological species aggregates. Since these aggregates were broad, Bissett (1991) considered more species distributed in five sections and expanded the *Trichoderma* concept to *Hypocrea* anamorphs, and also included some species previously described in the genus *Gliocladium*. However, given the variability within *Trichoderma* and the long time taxonomic confusion of the species, it has been difficult to use the names assigned by Rifai (1969) or Bisset (1991) as these authors used different taxonomic concepts.

DNA methods have brought additional valuable criteria to the taxonomy of *Trichoderma* which are used today for studies that include identification (Hermosa *et al.* 2001; Lübeck *et al.* 2000) and phylogenetic classification (Kullnig-Gradinger *et al.* 2002; Lieckfeldt and Seifert 2000). DNA sequencing and PCR fingerprinting are applied as common tools for phylogenetic studies (Kuhls *et al.* 1996; Kullnig-Gradinger *et al.* 2002; Meyer *et al.* 1992). In this way, rDNA sequence analyses, including internal transcribed sequences (ITS), have been used for taxonomic studies of *T. longibrachiatum* (Kuhls *et al.* 1997), *T. harzianum* (Gams and Meyer 1998), biocontrol strains (Hermosa *et al.* 2000), *T. viride* (Lieckfeldt *et al.* 1999), *T. aureoviride* (Lieckfeldt *et al.* 2001) and biotypes associated with the green mold of mushroom (Muthumeenakshi *et al.* 1994; Ospina-Giraldo *et al.* 1999; Samuels *et al.* 2002). Also, genetic diversity among *Trichoderma* spp. has been demonstrated with random amplified polymorphic DNA (RAPDs) analysis (Arisan-Atac *et al.* 1995; Chen *et al.* 1999; Hermosa *et al.* 2000; Muthumeenakshi *et al.* 1994; Ospina-Giraldo *et al.* 1998; Zimand *et al.* 1994). The application of various DNA approaches has proven to be useful in determining in the species delimitation of Rifai's *T. harzianum* aggregate and the phylogenetic relationships within Bisset's sections based on the sequence analysis of multiple independent loci. Most isolates of the genus *Trichoderma* that were found to act as mycoparasites of many economically important aerial and soil-borne plant pathogens (Chet 1987), have been classified as *T. harzianum* Rifai. Due to the fact that the species *harzianum* is generally considered as a group made of mycoparasitic and biocontrol strains, and there is large morphological plasticity that results in character overlaps with other species, the identification of the species may be difficult. Several authors have reported a large genetic variability among *T. harzianum* isolates (Bowen *et al.* 1996; Gomez *et al.* 1997; Grondona *et al.* 1997; Muthumeenakshi *et al,* 1994). In fact, it has been demonstrated that at least four distinct species are present within the biocontrol *T. harzianum* aggregate (–) : *T. harzianum s.str., T. atroviride, T. longibrachiatum* and *T. asperellum* (Hermosa *et al.* 2000). Also, it has been confirmed that there are biotypes within *T. harzianum s.str.* linked to biocontrol and mycoparasitic activity [*T. harzianum* and *T. inhamatum* (Gams and Meyer 1998)] whereas others [*T. aggressivum* (Samuels *et al.* 2002)] are pathogenic to cultivated mushrooms. The possibility to separate biocontrol and mushroom colonizer strains into different species using genetic techniques is important, for example, for registration of *Trichoderma* formulations as biofungicides. There are many other species and groups that have not been analyzed to the same degree as *T. harzianum*, but that are likely to show further important attributes and activities, as well as indicate the existence of further genetically diversity (Lieckfeldt *et al.*1998).

Phylogenetic studies based on the 18S rDNA sequence analysis suggest that *Trichoderma* is a monophyletic branch within the *Hypocreaceae* (Kullnig-Gradinger *et al.* 2002). The multicopy loci ITS1 5.8S rDNA, ITS2 5.8S rDNA and 28S rDNA, the small mitochondrial rDNA subunit, and fragments of the single-copy translation elongation factor 1 and endochitinase 42, have served to establish a phylogenetic tree (fig. 2) of 46 different species pertaining to the sections *Trichoderma, Pachybasium* and *Longibrachiatum* (Kullnig-Gradinger *et al.* 2002).

The information compiled from DNA analysis to date is of great interest in the search for new genes with biotechnological potential. The definition of accurate taxonomical groups is a key factor for the identification of *Trichoderma* strains, which will lead to the exploration and development of technologies based on this great genetic biodiversity.

4. FUNCTIONAL GENOMICS OF BIOCONTROL STRAINS

In the year 2001 a functional genomic project on antagonistic/mycoparasitic strains of *Trichoderma* was started to explore genetic biodiversity, with the aim to develop basic knowledge and commercial applications relatively to these fungi. The initiative is supported by the biotech company NewBioTechnic and two academic groups in Spain, and it has recently spread into a wider consortium with several EU groups. The objective is to obtain an extensive ESTs library including at least 7.000 unique ESTs, from conditions related to biocontrol. This has been attempted by using: i) strains from different molecular taxonomic groups, ecological origin and biocontrol abilities; ii) ESTs produced from a diverse set of conditions: different stresses, antagonistic interactions, plant-pathogen-biocontrol agent (BCA) interactions, etc; iii) high throughput expression studies from a wide spectrum of functional conditions related to biocontrol/antagonism/mycoparasitism; iv) functional analysis focused on commercial applications mainly in the agro-food and commercial enzyme sectors. The research strategy is summarized below (Fig. 3).

4.1 cDNA Library Construction

Three strains representing different molecular taxonomic groups (clades A, B and C), ecological origin and biocontrol abilities were selected initially. A set of more than 10 different conditions related to biocontrol activities (different stresses, antagonistic interactions, plant-pathogen-biocontrol agent interactions, etc.) have been used to build specific and mixed cDNA libraries. The work will be extended in the immediate future to ten more strains representing a wider biodiversity spectrum. For this purpose, total cellular RNA was extracted from different culture conditions by the guanidium isothiocyanate procedure, and Poly(A)+ RNA was purified using oligo(dT) chromatography. A unidirectional cDNA library was constructed in the Uni-ZAP XR vector. In vivo excision of pBluescript plasmids was performed in *E. coli* SOLR (Stratagene). To assess the quality of the library, the ratio of recombinants to non-recombinants and the average size of the cDNA inserts were determined by polymerase chain reaction (PCR) analysis of the DNA from 96 individual clones.

4.2 Clone Selection

Cell clones were isolated from the different cDNA libraries available and their 5' ends sequenced. A system (virtual substraction) (Katsuma *et al.* 2002) has been applied to reduce redundancy, by using high density membrane arrays to display isolated clones before sequencing and hybridization with an updated multiprobe including all the unique clones already sequenced. This method was able to reduce the redundancy from 60 to 25%, thus reaching over 1000 unique ESTs.

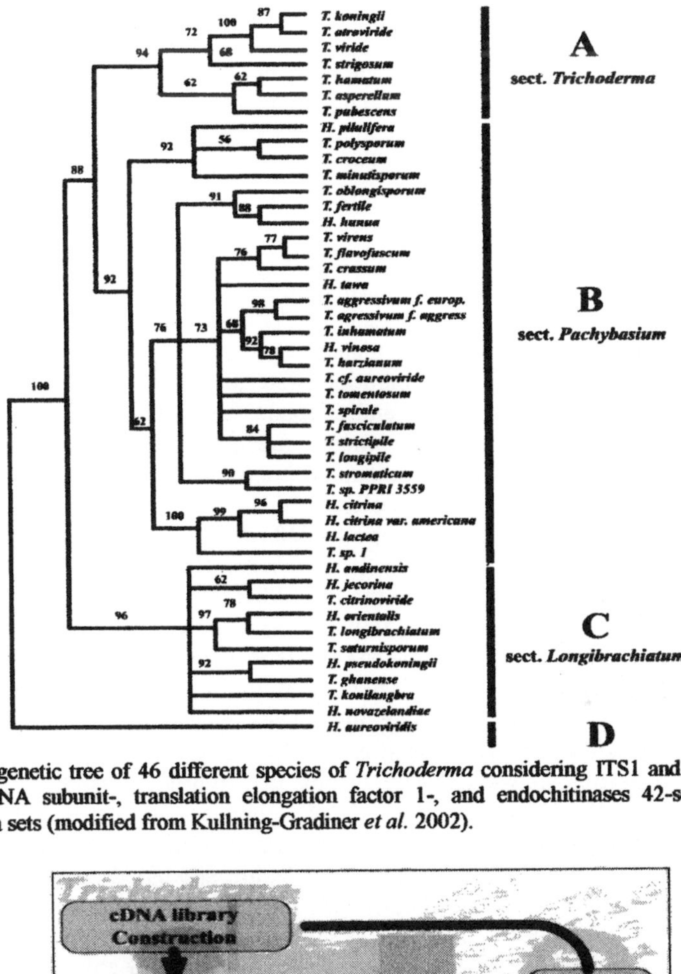

Fig. 2. Phylogenetic tree of 46 different species of *Trichoderma* considering ITS1 and ITS2-, 28S-, small mitochondrial DNA subunit-, translation elongation factor 1-, and endochitinases 42-sequences, and 500 bootstrapped data sets (modified from Kullning-Gradiner *et al.* 2002).

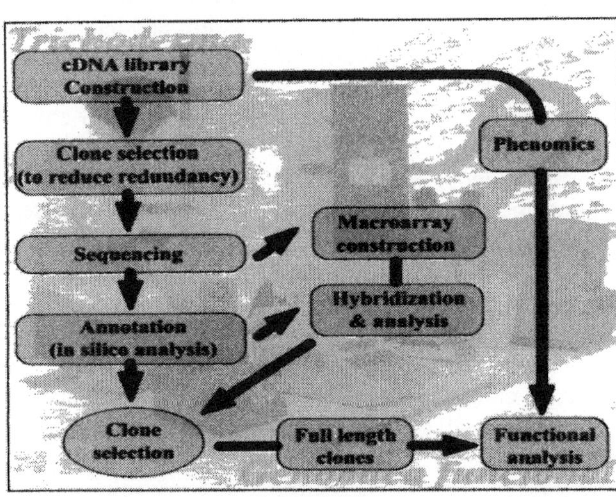

Fig. 3. Flow chart of a functional genomic project on biocontrol *Trichoderma* strains

4.3 Sequencing, Sequence Annotation and *in silico* Analysis

The data are managed and stored using software specifically developed for the project. A system for sample tracking and integration of sequencing and expression data and their annotation in a relational data base are the core of the bioinformatics analysis. The data are processed using a tailored software package that includes automatic routines for sequence curing and trimming, internal redundancy analysis, external sequence comparison and functional assignment (fig. 4). The package integrates commercial, public and inhouse developed bioinformatics tools. The *in silico* data management and tools are: i) SGSM: this is used for sequence sample tracking, enabling user registration of each clone and results annotation for each EST from the sequencing lab. SGSM runs on Solaris with MySQL as RDBMS and Java, PHP and Python as preprocessing languages. Web interfaces for all the user-required functions have been developed. SGSM interacts with ASA, sharing results to be supervised by the user through SGSM web interfaces; ii); DBEST TH Extractor: this Java tool has been developed to parse useful information from BLAST and FASTA results from the EST database (db) comparisons. Annotation of parsing works is made continuously by ASA into SGSM; iii); ASA: this is the automated sequence analysis heart of the project.

The analysis procedures are: i) ESTs curation : (steps: 1 to 3 in fig. 4) it is necessary to erase sequences corresponding to vector and adapters, annotate parsed results and special features. A part of this curation is made manually adding personal technician appreciations. *Phred* scripts make the automatic translation of traces into fasta format sequences, while masking of vector and adapters sequence is made by using *Cross_Match;* ii) Cluster analysis and redundancy report: (steps 4 to 20 in fig. 4) internal homology analysis is applied to each ESTs vs. the rest of the ESTs. *d2_cluster* algorithm and fragment assembly using *phrap* generates a cluster analysis of the ESTs that is refreshed as sequences are being added to the project. Parts of the *stackPACK* package classify the clusters by the highest homology (E-Value < 1e-130) of its sequences, determining redundancy reports, consensus sequences and singletons in the project. After the alignments of the clusters, a selection of the best clone is made to be used in further study. The consensus sequence of each cluster is used for *in silico* analysis. Best clone selection and singletons (alone not redundant sequences with any cluster) determine single sequence clones; iii) Homology analysis through searches in public and private DB: (steps: 22 to 25 in fig. 4) sequences are distributed in Workgroups that are spread in a grid responsible for the comparison of each sequence versus genomic databases using BLAST algorithms. Negative results are annotated and compared again by using FASTA. Both results are parsed by DBEST_TH Extractor and annotated on SGSM. iv) Phenomic analysis: (steps: 26 to 37 in fig. 4) several analysis are made in this part of the project, starting with signal peptide identification and initial ATG environment study. After this, frame translation is compared with gene prediction made by *GrailExp*, both for homologous and non-homologous sequences. The quality of the analysis is reported again adding hydropathy plots and codon usage analysis. If necessary, the user can make a manual ORF correction after evaluating this reports.

4.4 Expression Analysis

This preliminary expression study (steps: 38 to 42 in fig. 4) adds information about gene expression of each clone, under several conditions in which *Trichoderma* grows. Macroarrays of selected clones are used in Northern experiments with RNAs from the various conditions. Image analysis and clustering made with Array^2 package are also annotated on SGSM.

4.5 Cloning of Full Length Clones and Functional Analysis

The strategy considers the preliminary selection of specific ESTs based on their homologies to ORFs with known biological activities. For sequences of unknown function, expression profiles will be used as criteria for the initial selection. In all the selected cases, the biological activity of the gene products can be assessed by expressing the full length cDNA in different expression systems (bacterial and fungal) and/or using gene knockout strategies.

The results produced during the first two years can be summarized as follows. About 2500 sequences that meet the quality criteria parameters (a minimum length of 150 nucleotide and a Pher value of at least 20) were obtained. Of these sequences, 1011 EST's remained as singletons and 1483 sequences formed 315 clusters. Thus, 1326 partial sequences of expressed genes of *T. harzianum* and *T. atroviride* have been obtained to date. The clusters ranged in size from 2 to 68 sequences and 230 of them contain 2 to 4 sequences, 51 of them 5 to 8 sequences, 15 of them 9 to 12, and 19 of them had more than 12 sequences (table 4).

Table 4. *Trichoderma* largest clusters and their putative function

Cluster	Number of sequences	Database similarity
QA	96	Elongation factor 1-alpha
PA	61	Hydrophobin II precursor (HFBII)
WA	34	40S ribosomal protein S5
VA	32	Conserved hypothetical protein
RA	31	No similarity
TA	31	No similarity
SA	29	DDR48 stress protein
YA	25	No similarity
CB	21	Vacuolar-ATPase
GB	17	Histone H3
XX	17	40S ribosomal protein 5S18
IF	17	peptidilprolyl isomerase A, cytosolic - fungus
ZA	16	vacuolar-ATPase
BB	16	Histone H4
TE	15	60S acidic ribosomal protein P2-β (A4)
OE	14	40S ribosomal protein S9 (S7)
ZZ	13	ATP synthase protein 9
AB	13	Transcription factor
HB	12	No similarity

Using BLASTX (Altschul *et al.* 1990) and a stringency score ≥ 80, the total number of ESTs to which a cellular role could be assigned on the basis of sequence similarity to proteins with known function, was 714. The remaining ESTs are either unclassified or show similarity to sequences of unknown function, or have no significant similarity to any protein sequences in the databases (no matches, 921 sequences). The ESTs encoding putative protein sequences that show similarity to products in the NCBI non-redundant database were classified into different functional groups (fig. 5). Most of the known transcripts belong to groups related to housekeeping genes, such as those involved in protein and RNA synthesis.

Interestingly, a comparison between the available data from the *T. reesei* genomic initiative and the ones described here concerning *T. harzianum* shows a 71% of sequences without similarity (fig. 6). This can be explained by both differences in the genetic background of the strains and in expression conditions used.

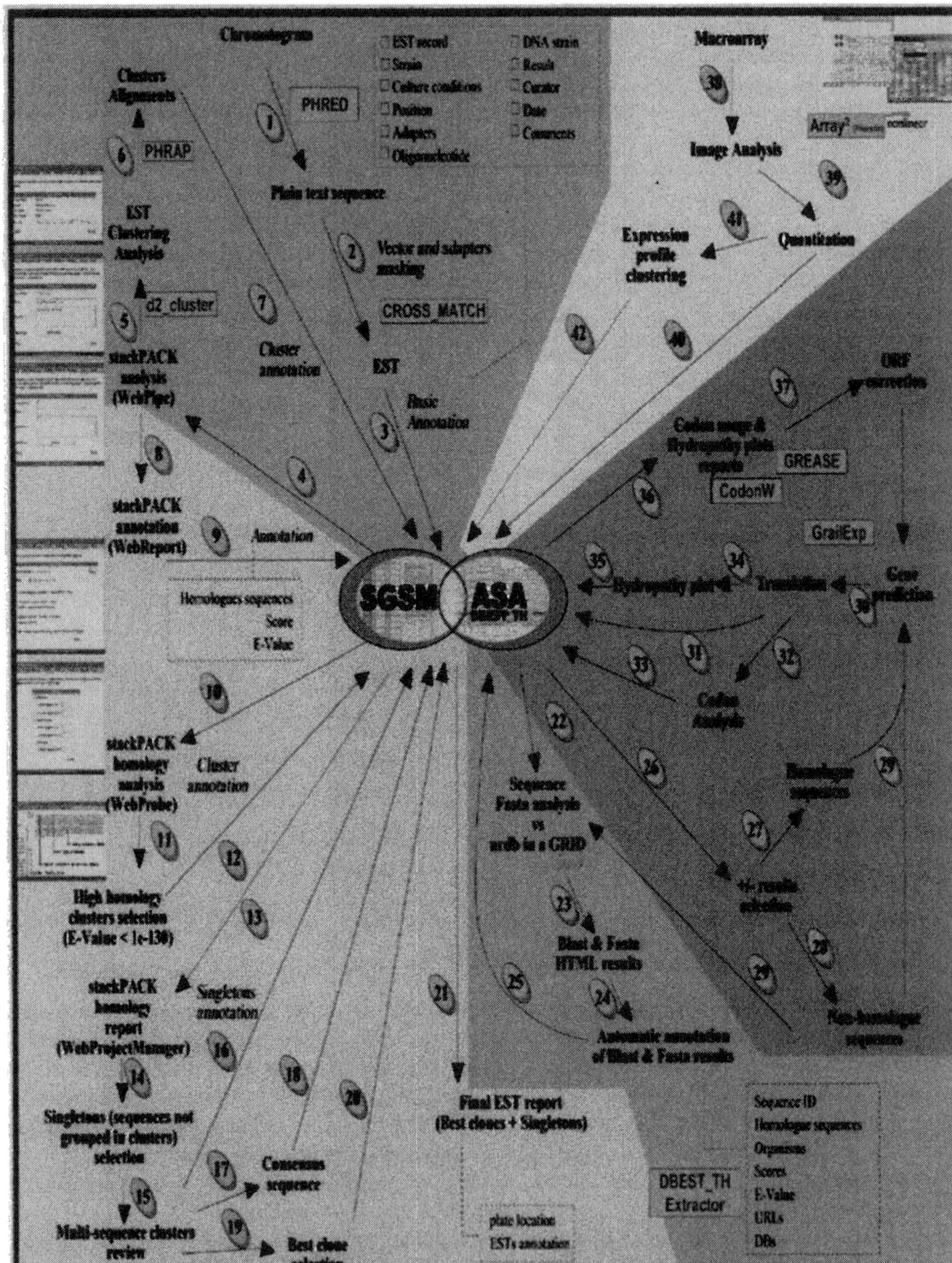

Fig. 4. Flow chart of the in silico analysis procedure. Steps 1, 2, 3: ESTs curing and trimming; steps 4 to 20: cluster analysis and redundancy; steps 22 to 25: homology analysis through searches in public and private DB; steps: 26 to 37: phenomic analysis; steps: 38 to 42: expression analysis; Final results are obtained by integration of data from other analysis (proteomics, biological activity screening, etc.).

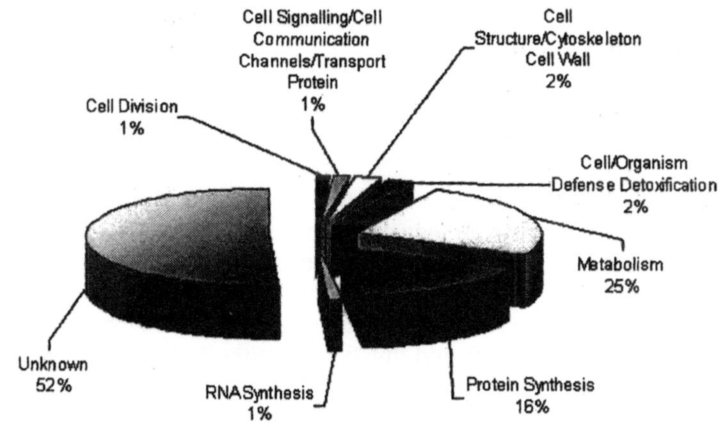

Fig. 5. Functional assignment of EST from biocontrol *Trichoderma* strains, according to the classification developed at the Institute for Genomic Research (TIGR, Rockville, USA) (White and Kerlavage 1996). Unknown refers to those sequences that show no significant similarity to protein or DNA sequences in the databases.

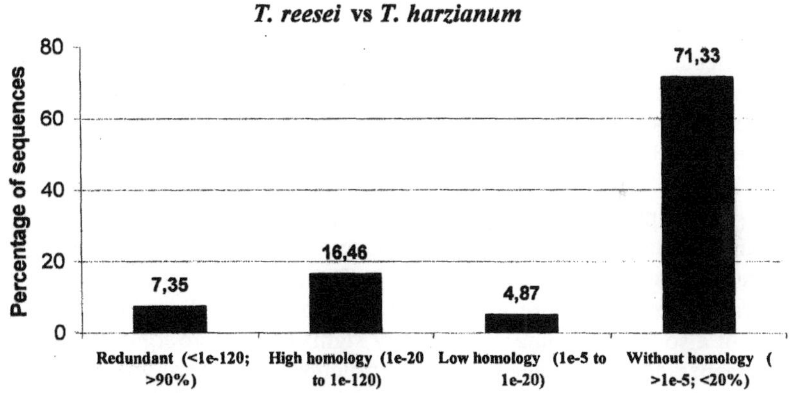

Fig. 6. A comparison between the available data from genomic projects on *T. reesei* and *T. harzianum*

A significant part of this effort is also devoted to the use of two dimension electrophoresis (2D-GE) and mass spectrometry analysis for identification of both extracellular and intracellular proteins produced by biocontrol strains of *T. harzianum* and *T. atroviride* during repressive and inducing conditions (i.e. when biocontrol, plant growthpromoting and SAR-inducing activities are turned on). Induction maps are being obtained and compared, stained spots are recovered robotically, subjected to hydrolysis and MALDI-TOF mass spectrometry, and characterized by comparison with data bases and EST data or direct amino acid sequencing (if no matches are found). By using this method it is expected to have a complete inventory of the proteins differentially expressed under each condition for each strain, and already several proteins expressed solely under specific conditions have been identified (unpublished data). This type of proteome study has also been applied to *T. reesei* (Pakula *et al.* 2002) by using 2D-GE associated to staining with specific fluorescent probes or labelling with 35S-methionine (to obtain more sensitivity). It also allowed to study the synthesis rate of different proteins, and identify changes in protein phosphorylation by using antibodies against phosphorylated serine or threonine. This research has been focused on the effect of secretion stresses, and has provided a wide view of the changes that affect the expression of a large number of proteins in this condition, as well as indicated major differences occurring in the phosphorylation pattern (Pakula *et al.* 2002).

5. CONCLUSIONS

Genomics, the generation of massive genetic information from living organisms by systematic approaches performed on an industrial scale, can be considered as a new scientific discipline due to the sheer scale of these projects and the technical innovations required to accomplish these tasks. Genomics is also changing the basis of biological investigation leading to the fusion of several scientific research areas into a new one. To date, genomics has provided insights into the structure and function of genes, polymorphisms within species, protein interaction, evolution, mRNA expression, protein expression and localization, and signaling pathways. As the genomic inventories approach closure, this mass of data will spur attempts to devise computational frameworks that integrate biological knowledge about cellular components as well as attempt to predict system behavior. This will have positive consequences not only from an academic point of view, but also for industrial applications in health and agriculture, thus speeding up the development of design-based biological engineering of cells and organisms to perform new functions.

All the potential of these new scientific approaches for understanding and exploiting complex biological systems is clearly evident when applied to fungi like *Trichoderma*. Their genome clearly represents an attractive and rich field to investigate which will lead to the discovery of mechanisms of general interest for biology (see section on genomics and function genomics of *Trichoderma*), especially considering the complexity and extent of biological interactions that these fungi are able to establish in nature with plants, animals and other microbes. Further, these fungi are virtually a still untapped source of genes and geneproducts of high and broad utility for the development of new biotechnology; with the potential to aid in pest control, to increase food production and quality, to clean both industrially and naturally polluted areas and materials, and to recover useful products from organic wastes. In fact, after a general search for released patent applications and issued patents, *Trichoderma* species are among the most represented, not only in comparison to other fungi, but also to most other industrially significant microorganisms. All the possible 'omics' approaches, including functional and structural genomics, will be very instrumental in this effort to obtain essential information and beneficial products. It is expected that the usefulness of the information enclosed inside the genome of *Trichoderma* (estimated to be about 40 Mb) will fully appear when the first sequencing projects will be completed and the differences among the fungi belonging to this multivariate group will be highlighted. Moreover, it is rare to find such a variety of microorganisms capable of performing so many different biological tasks in so many different environments (industrial enzyme production, biocontrol of plant diseases, plant growth promotion, bioremediation, mushroom pathogen etc.) which are members of the same genus. A large number of unique or species specific sequences are expected to be found when, for instance, the genome of powerful enzyme producers like some *T. reesei* strains will be compared to that of the biocontrol agents *T. harziamun* or mushroom colonizers *T. aggressivum*, and this will lead to at least a more effective use or control of these microbes.

Finally, *Trichoderma* also represent an excellent target for a strictly functional genomic project aimed at identifying and exploiting, in a holistic approach, the genes expressed, for instance, during the interaction with the plant or with plant pathogens, during growth on specific inducing/repressing substrates for industrial enzyme or biomass production; or during the utilization of polyphenol, hydrocarbons, pesticide compounds and other pollutants metabolized by these fungi as nutrients. Furthermore, from the simple isolation of differentially expressed proteins by using 2D-GE methods, it will be possible to set up a high throughput assay system to select from within the proteome that *Trichoderma* uses, for example to attack a plant pathogen, the genes or secondary metabolism products showing interesting and useful biological properties. In addition, new tools for transcriptome analysis in fungal species with unsequenced genomes based on simulations and restriction enzyme combination selection, can be used to obtain a good coverage of the genome. This can be applied to other *Trichoderma* spp. following methods which have been already tested on *T. reesei*. For all these reasons, it is inevitable that in the next years both genomic-based studies as well as the understanding and usefulness of *Trichoderma* spp. will be improved significantly, and receive strong support both from public and commercial institutions for further research and development.

Acknowledgments. The authors are grateful to Sheri Woo for the critical revision of the manuscript and to Santiago Gutiérrez, Rosa Hermosa, Rafael Jiménez, Fran González and Isabel Grondona for their help providing data. They also recognize the support of the EU-funded TRICHOEST project (QLRT-2001--02032) and grants FIRB 2002 and MIPAF "Risorse genetiche di organismi utili per il miglioramento di specie di interesse agrario e per un'agricoltura sostenibile" to ML, and Fundación Ramón Areces, Fundación El Monte and Fundación Andaluza de I+D for grants to ALL, MR and EM.

REFERENCES

Adams M, Kelly J, Gocayne J, Dubnick M, Polymeroppoulos M, Xiao H, Merril C, Wu A, Olde B, Moneno R, Kerlavage A, McCrombie R, and Venter JC (1991). Complementary DNA sequencing expressed sequence tags and Human Genome Project. Science 252:1651-1656.

Ait-Lahsen H, Soler A, Rey M, de la Cruz J, Monte E, and Llobell A (2001). An antifungal exo-α-1,3-glucanase (AGN13.1) from the biocontrol fungus Trichoderma harzianum. Appl Environ Microbiol 67:5833-5839.

Altomare C, Norvell WA, Björkman T, and Harman GE (1999). Solubilization of phosphates and micronutrients by the plant-growth promoting and biocontrol fungus Trichoderma harzianum Rifai strain 1295-22. Appl Environ Microbiol 65:2926-2933.

Altschul SF, Gish W, Miller W, Myers EW, and Lipman DJ (1990). Basic local alignment search tool. J Mol Biol 215:403-410

Arisan-Atac I, Heidenreich E, and Kubicek CP (1995). Randomly amplified polymorphic DNA fingerprinting identifies subgroups of Trichoderma viride and other Trichoderma sp. capable of chestnut blight biocontrol. FEMS Microbiol Lett 126:249-256

Audic S and Claverie JM (1997). The significance of digital gene expression profiles. Genome Res 10:986-995.

Benítez T, Limón C, Delgado-Jarana J, and Rey M (1998). Glucanolytic and other enzymes and their genes. In: CP Kubicek and GE Harman, ed. Trichoderma and Gliocladium Vol. 2. London: Taylor and Francis, pp 101-127.

Birren B, Fink G, and Lander E (2002). Fungal Genome Initiative. White Paper developed by the Fungal Genome Initiative. White Head Institute, Center For Genomic Research.

Bisby GR (1939). *Trichoderma viride* Pers. Ex Fries, and notes on Hypocrea. Trans Br Mycol Soc 23:149-168.

Bissett J (1991). A revision of the genus *Trichoderma* III. Sect. Pachybasium. Can J Bot 69:2373-2417.

Bonaldo MF, Lennon G, and Soares MB (1996). Normalization and subtraction: Two approaches to facilitate gene discovery. Genome Res 6:791-806.

Bowen JK, Franicevic SC, Crowhurst RN, Templeton MD, and Stewart A (1996). Differentiation of a specific Trichoderma biocontrol agent by restriction fragment length polymorphism (RFLP) analysis. N Z J Crop Hort Sci 24:207-217.

Bowtell D and Sambrook J. (2003). DNA microarrays. A molecular cloning manual. New York: Cold Spring Harbor Laboratories. pp 1-691.

Brent R (2000). Genomic Biology. Cell 100:169-183.

Bruns TD, White TJ, and Taylor JW (1991). Fungal molecular systematics Ann Rev Ecol Syst 22:525-564.

Butler BA (1998). Sequence analysis using GCG. Methods Biochem Anal 39:74-97.

Calvet C, Pera J, and Barea JM (1993). Growth response of marigold *(Tagetes erecta L.)* to inoculation with Glomus mosseae, Trichoderma aureoviride and Pythium ultimum in a peat-perlite mixture. Plant and Soil 148:1-6.

Celis JE, Kruhoffer M, Gromova I, Frederiksen C, Ostergaard M, Thykjaer T, Gromov P, Yu J, Palsdottir H, Magnusson N, and Orntoft TF (2000). Gene expression profiling: monitoring transcription and translation products using DNA microarrays and proteomics. FEBS Lett 480:2-16.

Chambergo FS, Bonaccorsi ED, Ferreira AJ, Ramos AS, Ferreira JR, Junior JR, Abrahão-Neto J, Farah JP, El-Dorry H (2002). Elucidation of the metabolic fate of glucose in the filamentous fungus Trichoderma reesei using EST analysis and cDNA microarrays. J Biol Chem 277(16): 13983-13984.

Chee M, Yang R, Hubbell E, Berno A, Huang XA, Stern D, Winkler J, Lockhart DJ, Morris MS, and Fodor SA (1996). Accessing genetic information with high-density DNA arrays. Science 274:610-614.

Chellappan M, Dunn-Coleman N, Hillian A, Houfek T, Mitchell T, Solingen PV, Teunnissen P, D. Wang D, Ward M, Yao J, and Dean RA (2001). Durable resources for discovery and development of new gene products from Trichoderma reesei. Phase I cDNA and BAC end sequencing. XXI Fungal Genetics Conference Asilomar, California.

Chen X, Romaine CP, Tan Q, Schlagnhaufer B, Ospina-Giraldo MD, Royse DJ, and Huff DR (1999). PCR-based genotyping of epidemic and pre-epidemic Trichoderma isolates associated with green mold of Agaricus bisporus. Appl Environ Microbiol 65:2674-2678

Chet I (1987). Trichoderma - application, mode of action, and potential as biocontrol agent of soilborne plant pathogenic fungi. In: I. Chet, ed. Innovative Approaches to Plant Disease Control. New York: John Wiley, pp 137-160.

de la Cruz J and Llobell A (1999). Purification and properties of a basic endo-β-1,6-glucanase (BGN16.1) from the antagonistic fungus Trichoderma harzianum. Eur J Biochem 265:145-151.

de la Cruz J, Hidalgo-Gallego A, Lora JM, Benítez T, Pintor-Toro JA, and Llobell A (1992). Isolation and characterization of three chitinases from Trichoderma harzianum. Eur J Biochem 206:859-867.

de la Cruz J, Pintor-Toro JA, Benítez T, and Llobell A (1995a). Purification and characterization of an endo-β-1,6-Glucanase from Trichoderma harzianum that is related to its mycoparasitism. J Bacteriol 177:1864-1871.

de la Cruz J, Pintor-Toro JA, Benítez T, Llobell A, and Romero L (1995b). A novel endo-β-1,3-glucanase, BGN13.1, involved in the mycoparasitism of Trichoderma harzianum. J Bacteriol 177:6937-6945.

Dennis C and Webster J (1971). Antagonistic properties of species groups of Trichoderma. III:Hyphal interactions. Trans Br Mycol Soc 57:363-369.

Ekins R and Chu FW (1999). Microarrays: their origins and applications. Trends in Biotechnology 17(6):217-218

Elad Y (1996). Mechanisms involved in the biological control of Botrytis cinerea incited diseases. Eur J Plant Pathol 102:719-732.

Enkerli J, Felix G, and Boller T (1999). Elicitor activity of fungal xylanase does not depend on enzymatic activity. Plant Physiol 121:391-398.

Evans WE and Relling MV (1999). Pharmacogenomics: translating functional genomics into rational therapeutics. Science 286:487-491.

Gams W and Meyer W (1998). What exactly is Trichoderma harzianum Rifai? Mycologia 90:904-915.

Garcia I, Lora JM, de la Cruz J, Benítez T, Llobell A, and Pintor-Toro JA (1994). Cloning and characterization of a chitinase (CHIT42) cDNA from the mycoparasitic fungus Trichoderma harzianum. Curr Genet 27:83-89.

Geremia RA, Goldman GH, Jacobs D, Ardiles W, Vila SB, van Montagu M, and Herrera-Estrella A (1993). Molecular characterization of the proteinase-encoding gene, prb1, related to mycoparasitism by Trichoderma harzianum. Mol Microbiol 8:603-613.

Gómez I, Chet I, and Herrera-Estrella A (1997). Genetic diversity and vegetative compatibility among Trichoderma harzianum isolates. Mol Gen Genet 256:127-135.

Grisham J (2000). New rules for gene patents. Nature Biotech 18: 921.

Grishin NV, Wolf YI, and Koonin EV (2000). From complete genomes to measures of substitution rate variability within and between proteins. Genome Res 10:991-1000.

Grondona I, Hermosa MR, Tejada M, Gomis MD, Mateos PF, Bridge PD, Monte E, and García-Acha I (1997). Physiological and biochemical characterization of Trichoderma harzianum, a biological control agent against soilborne fungal plant pathogens. Appl Environ Microbiol 63:3189-3198.

Ham KS, Wu SC, Darvill AG, and Albersheim P (1997). Fungal pathogens secrete an inhibitor protein that distinguishes isoforms of plant pathogenesis-related endo-β-1,3-glucanases. Plant J 11:169-179.

Harman GE (2000). Myths and dogmas of biocontrol. Plant Dis 84:377-393.

Harman GE and Björkman T (1998). Potential and existing uses of Trichoderma and Gliocladium for plant disease control and plant growth enhancement. In: CP Kubicek and GE Harman, eds. Trichoderma and Gliocladium Vol.2. London: Taylor and Francis, pp 229-265.

Harman GE and Kubicek PK (1998) Trichoderma & Gliocladium Vol 2. Enzymes, biological control and commercial applications. London: Taylor and Francis. pp 1-393.

Hellman U (2000). Proteomics and functional genomics. In P Jolles and H Jörnvall eds. Protein structure analysis. Basel: Birkhauser Verlag. pp. 43-54.
Hermosa MR, Grondona I, Díaz-Mínguez JM, Iturriaga EA, and Monte E (2001). Development of a strain-specific SCAR marker for the detection of Trichoderma atroviride 11, a biological control agent against soilborne fungal plant pathogens. Curr Genet 38:343-350.
Hermosa MR, Grondona I, Iturriaga EA, Díaz-Mínguez JM, Castro C, Monte E, and García-Acha I (2000). Molecular characterization and identification of biocontrol isolates of Trichoderma spp. Appl Environ Microbiol 66:1890-1898.
Hoheisel J (2002). Chip Technology. Berlin: Springer-Verlag. pp. 1-156
Houseman BT, Huh JH, Kron SJ, Deng C, and Mrksich M (2002). Peptide chips for the quantitative evaluation of protein kinase activity. Nature Biotechnol 20:270-274.
Inbar J, Abramsky M, Coen D, and Chet I (1994). Plant growth enhancement and disease control by Trichoderma harzianum in vegetable seedlings grown under commercial conditions. Eur J Plant Pathol 100:337-346.
Iyer VR, Horak CE, Scafe CS, Botstein D, Snyder M, and Brown PO (2001). Genomic binding sites of the yeast cell-cycle transcription factors SBF and MBF. Nature 409:533-538.
Kapteyn JC, Montijn RC, Vink E, de la Cruz J, Llobell A, Douwes JE, Shimoi H, Lipke PN, and Klis FM (1996). Retention of Saccharomyces cerevisiae cell wall proteins through a phosphodiester-linked β-1,3-/β-1,6-glucan heteropolymer. Glycobiology 6:337-345.
Kapteyn JC, Ram AFJ, Groos EM, Kollar R, Montijn RC, Van den Ende H, Llobell A, Cabib E, and Klis FM (1997). Altered extent of coss-linking of β-1,6-glucosylated mannoproteins to chitin in Saccharomyces cerevisiae mutants with reduced cell wall β-1,3-glucan content. J Bacteriol 179:6279-6284.
Katsuma S, Shiojima S, Hirasawa A, Suzuki Y, Ikawa H, Takagaki K, Kaminishi Y, Murai M, Ohgi T, Yano J, and Tsujimoto G (2002). Functional genomic search of G-protein-coupled receptors using microarrays with normalized cDNA library. Methods in Enzymology 345:585-600.
Keszler A, Forgacs E, Kotai L, Vizcaíno JA, Monte E, and García-Acha I (2000). Separation and identification of volatile components in the fermentation broth of Trichoderma atroviride by solid-phase extraction and gas chromatography-mass spectroscopy. J Chromatograph Sci 38:421-424.
Kubicek CP, Mach RL, Peterbauer CK, and Lorito M (2001). Trichoderma: from genes to biocontrol. J Plant Path 83:11-23.
Kuhls K, Lieckfeldt E, Samuels GJ, Kovacs W, Meyer W, Petrini O, Gams W, Börner T, and Kubicek CP (1996). Molecular evidence that the asexual industrial fungus Trichoderma reesei is a clonal derivative of the ascomycete Hypocrea jecorina. Proc Natl Acad Sci USA 93:7755-7760.
Kuhls K, Lieckfeldt E, Samuels GJ, Meyer W, Kubicek CP, and Börner T (1997). Revision of Trichoderma section Longibrachiatum including related teleomorphs based on an analysis of ribosomal DNA internal transcribed spacer sequences. Mycologia 89:442-460.
Kullnig-Gradinger CM, Szakacs G, and Kubicek CP (2002). Phylogeny and evolution of the genus Trichoderma: multigene approach. Mycol Res. 106:757-767.
Leming S, (1999). http.//www.gen-chips.com.
Lenoir T (2000). Science and the Academy of the 21st Century: Does Their Past Have a Future in an Age Computer-Mediated Networks?. Conference Paper for Ideale Akademie: Vergangene Zukunft oder konkrete Utopie? Berlin Akademie der Wissenschaften.
Lieckfeldt E and Seifert KA (2000). An evaluation of the use of ITS sequences in the taxonomy of the Hypocreales. Stud Mycol 45:35-44.
Lieckfeldt E, Kuhls K, and Muthumeenakshi S (1998). Molecular taxonomy of Trichoderma and Gliocladium and their telemorphs. In CP Kubicek and GE Harman eds. Trichoderma and Gliocladium Vol 1. London: Taylor and Francis, pp 35-56.
Lieckfeldt E, Kullnig CM, Kubicek CP, Samuels GJ, and Börner T (2001). Trichoderma aureoviride: phylogenetic position and characterization. Mycol Res 105:313-322.
Lieckfeldt E, Samuels GJ, Nirenberg HI, and Petrini O (1999). A morphological and molecular perspective of Trichoderma viride: is it one or two species. Appl Environ Microbiol 65:2418–2428.
Limón MC, Lora JM, García I, de la Cruz J, Llobell A, Benítez T, and Pintor-Toro JA (1995). Primary structure and expression pattern of the 33-kDa-chitinase gene from the mycoparasitic fungus Trichoderma harzianum. Curr Genet 28:478-483.
Linko M, Haikara A, Ritala A, and Penttilla M (1998). Recent advances in the malting and brewing industry. J Biotech 65:85-98.
Lora JM, de la Cruz, J, Llobell A, Benitez T, and Pintor-Toro JA (1995). Molecular characterization and heterologous expression of an endo-β-1,6-glucanase gene from the mycoparasitic fungus Trichoderma harzianum. Mol Gen Genet 247:639-645.

Lorito M (1998). Chitinolytic enzymes and their genes. In CP Kubicek and GE Harman eds. Trichoderma and Gliocladium Vol 2. London: Taylor and Francis,. pp 73-99.
Lorito M, Farkas V, Rebuffat S, Bodo B, and Kubicek CP (1996a). Cell-wall synthesis is a major target of mycoparasitic antagonism by Trichoderma harzianum. J Bacteriol 178:6382-6385.
Lorito M, Hayes CK, Di Pietro A, Woo SL, and Harman GE (1994a). Purification, characterization and synergistic activity of a glucan 1,3-β-glucosidase and an N-acetyl-β-glucosaminidase from Trichoderma harzianum. Phytopathology 84:398-405.
Lorito M, Hayes CK, Peterbauer C, Tronsmo A, Klemsdal S, and Harman GE (1993). Antifungal chitinolytic enzymes from T. harzianum and G. virens: purification, characterization, biological activity and molecular cloning. In: RAA Muzzarelli ed. Chitin Enzymology. Ancona: European Chitin Society, pp. 383-392.
Lorito M, Peterbauer C, Hayes CK, and Harman GE (1994b). Synergistic interaction between fungal cell wall-degrading enzymes and different antifungal compounds enhances inhibition of spore germination. Microbiology 140:623-629.
Lorito M, Woo SL, D'Ambrosio M, Harman GE, Hayes CK, Kubicek CP, and Scala F (1996b). Synergistic interaction between cell wall degrading enzymes and membrane affecting compounds. Molec Plant Microbe Interact 9:206-213.
Lorito M, Woo SL, García Fernández I, Colucci G, Harman GE, Pintor-Toro J.A, Filippone E, Mucciflora S, Lawrence C, Zoina A, Tuzun S, and Scala F (1998). Genes from mycoparasitic fungi as a source for improving plant resistance to fungal pathogens. Proc Natl Acad Sci USA 95:7860-7865.
Lübeck M, Poulsen SK, Lübeck PS, Jensen DF, and Thrane U (2000). Identification of Trichoderma strains from building materials by ITS1 ribotyping, UP-PCR fingerprinting and UP-PCR cross hybridization. FEMS Microbiol Lett 185:129-134.
Maier E, Meier-Ewert S, Bancroft, and Lehrach H (1997). Automated array technologies for gene expression profiling. Drug Discovery Today 2: 315-324.
Maskos U and Southern EM (1992). Oligonucleotide hybridizations on glass supports: A novel linker for oligonucleotide synthesis and hybridization properties of oligonucleotides synthesized in situ. Nucleic Acid Res 20:1679-1684.
Meyer W, Morawetz R, Börner T, and Kubicek CP (1992). The use of DNA-fingerprint analysis in the classification of some species of the Trichoderma aggregate. Curr Genet 21:27-30.
Monte E (2001). Editorial Paper: Understanding Trichoderma: Between Agricultural Biotechnology and Microbial Ecology. Int Microbiol 4:1-4.
Montero M (2001). Caracterización y clonación de actividades β-1,6 glucanasa de Trichoderma harzianum. PhD dissertation, Salamanca University, Salamanca, Spain.
Muthumeenakshi S, Mills PR, Brown AE, and Seaby DA (1994). Intraspecific molecular variation among Trichoderma harzianum isolates colonizing mushroom compost in British Isles. Microbiology 140:769-777.
Nakari T, Onnela M, Ilmen J, Nevalainen K, and Penttila M (2000). Fungal promoters active in the presence of glucose. United States Patent 6011147.
Noronha EF and Ulhoa CJ (1996). Purification and characterization of an endo-β1,3-glucanase from Trichoderma harzianum. Can J Microbiol 42:1039-1044
Ospina-Giraldo MD, Royse DJ, Chen X, and Romaine CP (1998). Molecular phylogenetic analyses of biological control strains of Trichoderma harzianum and other biotypes of Trichoderma spp. associated with mushroom green mold. Phytopathology 89:308-313.
Ospina-Giraldo MD, Royse DJ, Thon MR, Chen X, and Romaine CP (1999). Phylogenetic relationships of Trichoderma harzianum causing mushroom green mould in Europe and North America to other species of Trichoderma of world-wide sources. Mycologia 90:76-81.
Ouzounis C, Casari G, Sander C, Tamames J, and Valencia A (1996). Computational comparisons of model genomes. Trends Biotechnol 14:280-285.
Pakula T, Saloheimo M, and Penttilä M (2002) Proteomics studies in Trichoderma reesei. Proceedings of the First International Trichoderma-Tricel Joint Meeting, Cancun, pp 67.
Pandey A and Mann M (2000). Proteomics to study genes and genomes. Nature 405:837-846.
Papavizas GC (1985). Trichoderma and Gliocladium: biology, ecology and potential for biocontrol. Annu Rev Phytopathol 23:23-54.
Pérez-González JA, González R, Querol A, Sendra J, and Ramón D (1993). Construction of a recombinant wine yeast strain expressing β-(1,4)-endoglucanase and its use in microvinification processes. Appl Environ Microbiol 59:2801-2806.
Persoon CH (1794). Dispositio methodica fungorum. Romer's neues Mag Bot 1:81-128.
Peterbauer C, Lorito M, Hayes CK, Harman GE, and Kubicek CP (1996). Molecular cloning and expression of nag1 (N-acetyl-ß-D-glucosaminidase-encoding) gene from Trichoderma harzianum P1. Curr Genet 29:812-820.

Raamsdonk LM, Teusink B, Broadhurst D, Zhang N, Hayes A, Walsh M, Berden JA, Brindle KM, Kell DB, Rowland JJ, Westerhoff HV, van Dam K, and Oliver SG (2001). A functional genomics strategy that uses metabolome data to reveal the phenotype of silent mutations. Nature Biotechnol 19:45-50

Rashidi H and Buehler L (2000). Bioinformatics Basics: Applications in biological science and medicine. Boca raton: CRC Press.

Rifai MA (1969). A revision of the genus Trichoderma. Mycol Pap 116:1-116.

Rossignol M (2001). Analysis of plant proteome. Curr Op Biotechnol 12:131-134.

Rossman AY, Samuels GJ, Rogerson C, and Lowen R (1999). Genera of the Bionectriaceae, Hypocreaceae and Nectriaceae (Hypocreales, Ascomycetes). Stud Mycol 42:1-248.

Samuels GJ (1996). Trichoderma: a review of biology and systematics of the genus. Mycol Res 100:923-935.

Samuels GJ, Dodd SL, Gams W, Castlebury LA, and Petrini O (2002). Trichoderma species associated with the green mold epidemic of commercially grown Agaricus bisporus. Mycologia 94:146-170.

Sanger F, Coulson AR, Friedmann T, Air GM, Barrell BG, Brown NL, Fiddes JC, Hutchinson CA III, Slocombe PM, and Smith M (1978). The nucleotide sequence of bacteriophage phiX174. J Mol Biol 125:225-246.

Schena M, Shalon D, Davis RW, and Brown PO (1995). Quantitative monitoring of gene expression patterns with a complementary DNA microarray. Science 270:467-470.

Schirmböck M, Lorito M, Wang Y-L, Hayes CK, Arisan-Atac I, Scala F, Harman GE, and Kubicek CP (1994). Parallel formation and synergism of hydrolytic enzymes and peptaibol antibiotics, molecular mechanisms involved in the antagonistic action of Trichoderma harzianum against phytopathogenic fungi. Appl Environ Microbiol 60:4364-4370.

Sharon E, Bar-Eyal M, Chet I, Herrera-Estrella A, Kleifeld O, and Spiegel Y (2001). Biocontrol of the Root-Knot Nematode Meloidogyne javanica by Trichoderma harzianum. Phytopathology 91:687-693

Sherman DH (2002). New enzymes for "warheads". Nature Biotechnol 20:984-985.

Simon A and Sivasithamparam K (1989). Pathogen suppression: a case study in biological suppression of Gaeumannomyces graminis var. tritici in soil. Soil Biol Biochem 21:331-337.

Sivasithamparam KY and Ghisalberti EL (1998). Secondary metabolism in Trichoderma and Gliocladium. In: CP Kubicek and GE Harman eds. Trichoderma and Gliocladium vol. 2. London: Taylor & Francis. pp. 139-191.

Soundy P, Wheeler C, and Latham H (2001). Preparing highly fluorescent, evenly labelled probes for microarray hybridisation using the amino allyl method with CyScribe post-labelling kit. Life Sci News 9:17-19

Stasyk T, Hellman U, and Souchelnytskyi S (2001). Optimizing sample preparation for 2-D electrophoresis. Life Sci News 9:8-11.

Suárez B (2001). Caracterización y detección molecular de cepas de Colletotrichum causantes de antracnosis en fresa. Búsqueda de proteasas de Trichoderma implicadas en su biocontrol. PhD dissertation, Salamanca University, Salamanca, Spain.

Viterbo A, Haran S, Friesem D, Ramot O, and Chet I (2001). Antifungal activity of a novel endochitinase gene (chit36) from Trichoderma harzianum Rifai TM. FEMS Microbiol Letters 200:169-74.

Viterbo A, Montero M, Ramot O, Friesem D, Monte E, Llobell A, and Chet I (2002). Expression regulation of the endochitinase chit36 from Trichoderma asperellum (T. harzianum T-203). Curr Genet 42:114-122.

Vrany J, Rasochova M, Fiker A, and Dobias K (1990). Inoculation of potatoes with microorganisms under field conditions. I. Effect on plant growth, yields and physiological properties on tubers in potato and sugar-beet regions. Folia Microbiologica 35:326-335.

Wang D, Liu S, Trummer BJ, Deng C, and Wang A (2002). Carbohydrate microarrays for the recognition of cross-reactive molecular markers of microbes and host cells. Nature Biotechnol 20:275-281.

Wardle DA, Parkinson D, and Waller JE (1993). Interspecific competitive interactions between pairs of fungal species in natural substrates. Oecologia 94:165-172.

Weinstein JN (1998). Fishing expeditions. Science 282:5389.

White O and Kerlavage AR (1996). Methods in Enzymology 266, 27-40.

Wiest A, Grzegorski D, Xu BW, Goulard C, Rebouffat S, Ebbole DJ, Bodo B, and Kenerley CM (2002). Identification of peptaibols from Trichoderma virens and cloning of a peptaibol synthetase. J Biol Chem 277:20862-20868.

Yedidia I, Benhamou N, and Chet I (1999). Induction of defense responses in cucumber plants (Cucumis sativus L.) by the biocontrol agent Trichoderma harzianum. Appl Environ Microbiol 65:1061-1070.

Yoder OC and Turgeon BG (2001). Fungal genomics and pathogenicity. Curr Opin Plant Biol 4:315-321.

Zhu H, Bilgin M, Bangham R, Hall D, Casamayor A, Bertone P, Lan N, Jansen R, Bidlingmaier S, Houfek T, Mitchell T, Miller P, Dean RA, Gerstein M, and Snyder M (2001). Global analysis of protein activities using proteome chips. Science 293:2101-2105.
Zimand G, Valinsky L, Elad Y, Chet I, and Manulis S (1994). Use of the RAPD procedure for the identification of Trichoderma strains. Mycol Res 98:531-534.

Applied Mycology & Biotechnology
An International Series. Volume 4. Fungal Genomics

10

Genomics of Economically Significant *Aspergillus* and *Fusarium* Species

Jiujiang Yu[a]*, Robert H. Proctor[b], Daren W. Brown[b], Keietsu Abe[c,e], Katsuya Gomi[c,e], Masayuki Machida[d], Fumihiko Hasegawa[e], William C. Nierman[f], Deepak Bhatnagar[a] and Thomas E. Cleveland[a]

[a]Food and Feed Safety Research Unit U.S. Department of Agriculture, Agricultural Research Service, Southern Regional Research Center, New Orleans, Louisiana 70124 USA (jiuyu@srrc.ars.usda.gov); [b]Mycotoxin Research Unit, U.S. Department of Agriculture, Agricultural Research Service, National Center for Agricultural Utilization Research, Peoria, Illinois 61604 USA; [c]Department of Molecular and Cell Biology, Graduate School of Agricultural Science, Tohoku University, Sendai 981-8555 Japan; [d]Research Center for Glycoscience, National Institute of Advanced Industrial Science and Technology (AIST), Tsukuba, Ibaraki 305-8566 Japan; [e]The New Industry Creation Hatchery Center (NICHe) Tohoku University , Sendai 980-8579 Japan; [f]Institute for Genomic Research, Rockville, Maryland 20850 USA.

Mycotoxins are fungal secondary metabolites that are harmful to the health of humans and/or animals. Aflatoxins, trichothecenes (T-2 toxin and DON toxin) and fumonisins are the major mycotoxins that contaminate crop plants and, as a result, are of great importance to agricultural economics and in food and feed safety. Aflatoxins are produced mainly by the two *Aspergillus* species in section *Flavi*, *A. flavus* and *A. parasiticus*. Sterigmatocystins (precursor of aflatoxins) are produced by some strains of *A. nidulans*. Trichothecenes and fumonisin are produced by *Fusarium* species. In the genus *Aspergillus*, the nonaflatoxin-producing species *A. fumigatus*, which is a human pathogen; *A. oryzae* and *A. sojae*, which are used in food fermentation, and *A. niger*, which is used in industrial fermentation, are close relatives of aflatoxin-producing species *A. flavus* and *A. parasiticus*. The genetics and biology of aflatoxin, trichothecene and fumonisin biosynthesis have been investigated in significant detail, and many of the genes and/or enzymes involved in toxin formation have been identified. Genomic efforts, such as Expressed Sequence Tag (EST), cosmid clone sequencing, chromosome sequencing, and large-scale whole genome sequencing, on toxigenic and non-toxigenic *Aspergillus* and *Fusarium* species have been made in

*Corresponding author: Jiujiang Yu

recent years. The technological breakthroughs in genomics research will almost certainly promote revolution in our understanding of the biology and genetics of these filamentous fungi for the control of mycotoxin contamination in food and feed and for the improvement in yield and quality of industrial fermentation products. In this chapter, we review advances in genomics research on those *Aspergillus* and *Fusarium* species.

1. INTRODUCTION

Aspergillus flavus, *A. parasiticus*, *F. sporotrichioides*, *F. graminearum* and *F. verticillioides* are plant pathogens that infect agricultural crops (corn, cotton, peanut and tree nuts) and contaminate them with mycotoxins. *A. flavus* and *A. parasiticus* produce aflatoxins. Some strains of *A. nidulans*, which is used for drug development and also as a model organism in developmental biology, produce sterigmatocystins (ST), precursors of aflatoxins. *F. verticillioides* (formerly *F. moniliforme*) produces fumonisins, while *F. sporotrichioides* and *F. graminearum* produce trichothecenes.

When consumed, aflatoxins, ST, fumonisins and trichothecenes can pose a severe health hazard to animals and humans. *A. oryzae*, *A. sojae* and *A. niger* are close relatives of *A. flavus*, but do not produce aflatoxins. *A. oryzae* and *A. niger* are widely used for industrial purposes for enzymes, peptides and other organic compound productions, whereas *A. sojae* is used for soy sauce fermentation, which is a billion dollar industry worldwide. These three species are food grade organisms having GRAS status (Generally Regarded As Safe) according to the U.S. Food and Drug Administration and are of great economic importance because of their industrial uses. *A. fumigatus* neither infects plants nor is of any industrial value, instead it is a human pathogen that infects human lungs causing both invasive and noninvasive human aspergillosis, especially to those who are immunocompromised patients and can be fatal (Rankin 1953; Denning 1998). Aspergillosis is a clinical infection caused by fungi of the genus *Aspergillus* which can lead to allergic, superficial, saprophytic or invasive diseases. It produces gliotoxin and is significant medically.

Mycotoxins are low molecular weight secondary metabolites produced by fungi. These metabolites are of great concern to agriculture because they can accumulate in edible crop plants infected with mycotoxin-producing (mycotoxigenic) fungi. Mycotoxins can pose a severe health hazard to animals and humans. Mycotoxins vary greatly in their potency and toxic effects. Some are mutagenic, teratogenic and carcinogenic (Squire 1989; Bennett 1987; Eaton and Groopman 1994; Hall and Wild 1994 Bhatnagar *et al.* 2002a). Modern research on mycotoxins gained significant momentum after the incidence of "Turkey-X" disease in 1960 when 10,000 turkeys died from consumption of peanut-meal feed contaminated with the group of mycotoxins known as aflatoxins (Lancaster *et al.* 1961). Due to the health hazards of aflatoxins to humans and livestock productivity, aflatoxin content in food and feed is regulated in many countries (Eaton and Groopman 1994; Brown *et al.* 1998). Twenty parts aflatoxin per billion parts of food or feed substrate (ppb) is the maximum allowable limit imposed by the U.S. Food and Drug Administration for interstate shipment of foods and feeds. European regulatory agencies are more stringent and restrict aflatoxin levels to as low as 5 ppb (Sharma and Salunkhe 1991; Bhatnagar *et al.* 2002a). Today in the U.S., aflatoxin contamination is a chronic problem in cotton grown in the Southwest and peanuts grown in the Southeast. Sporadic, but severe, outbreaks of aflatoxin contamination also occur in corn grown in the Midwest. The total costs associated with aflatoxin contamination in corn have been estimated to be over $200 million during years with severe outbreaks.

Within the last decade, significant advances have been made in mycotoxin detection methods and control strategies as well as in understanding the biochemistry, genetics and regulation of mycotoxin biosynthesis. This has been especially true for the economically important aflatoxins, produced by *Aspergillus* species, and fumonisins and trichothecenes, produced by *Fusarium* species. The biosynthetic pathways for these mycotoxins, the clustering of biosynthetic genes, and the functions of these genes have been elucidated in great detail (Payne and Brown 1998; Bhatnagar *et al.* 2002a; Brown *et al.* 2002a; Yu *et al.* 2002f; Yu 2003). Some of the advances in the understanding of the genetics and regulation of aflatoxin biosynthesis have been facilitated by work on *A. nidulans*, which has served as a model organism in genetics and developmental biology for decades and some strains of which produce the aflatoxin precursors sterigmatocystins. Despite the many advances described above, mycotoxin contamination problems are far from being solved. There remains a vast gap in our understanding of the coordinated global regulation of toxin formation, of the signal transduction pathways underlying primary and secondary metabolisms, of the biotic and abiotic factors that affect toxin formation, and of the interactions of mycotoxigenic fungi and their host plants during infection.

Tremendous advances have also been made in understanding the genetics of four non-aflatoxigenic *Aspergillus* species, *A. oryzae*, *A. sojae*, *A. niger* and *A. fumigatus*. The three former species are economically important because of their industrial applications. For example, *A. oryzae* and *A. niger* are used in the production of enzymes, peptides and other organic compounds, and *A. sojae* is used in the fermentation of soy sauce, which is a billion dollar industry worldwide. In contrast, *A. fumigatus* is a human and animal pathogen and is the most common cause worldwide of human aspergillosis, an often-fatal disease that affects primarily immunocompromised individuals (Denning 1998).

The rapid development of high throughput sequencing made it possible in genetic research to advance from single gene cloning to whole genome sequencing such as the sequencing of the yeast genome (Dujon 1996; Goffeau *et al.* 1996; 1997a, 1997b). The technological breakthrough allows scientists to study the genome of an organism possibly in a very short time frame compared with traditional genetic studies. Currently, the whole genome sequencing and/or Expressed Sequence Tag (EST) projects for some of the filamentous fungi such as *Magnaporthe grisea*, *A. fumigatus*, *A. nidulans*, *A. oryzae*, *A. niger*, *A. flavus*, *F. verticillioides*, are well underway (Bennett 1997a, 1997b; Bennett and Arnold 2001). In this chapter, we review the genetics and genomics of mycotoxigenic fungi *A. flavus*, *A. parasiticus*, *F. graminearum*, *F. sporotrichioides* and *F. verticillioides* and the atoxigenic *A. oryzae*, *A. sojae*, *A. niger*, as well as a medically important human pathogen, *A. fumigatus*.

2. GENOMICS OF MYCOTOXIGENIC *ASPERGILLUS* SPECIES

2.1 *Aspergillus* Species

Species in *Aspergillus* section *Flavi*, which is more commonly known as the *Aspergillus flavus* group, consists of a metabolically diverse group of fungi (Bennett and Klich 1992). Because of their industrial applications and mycotoxin production, they are of great economic importance and health concern. Several species, especially *A. flavus*, cause food spoilage but also produce carcinogenic mycotoxins such as aflatoxins (Fig. 1). *A. nidulans* is a member of

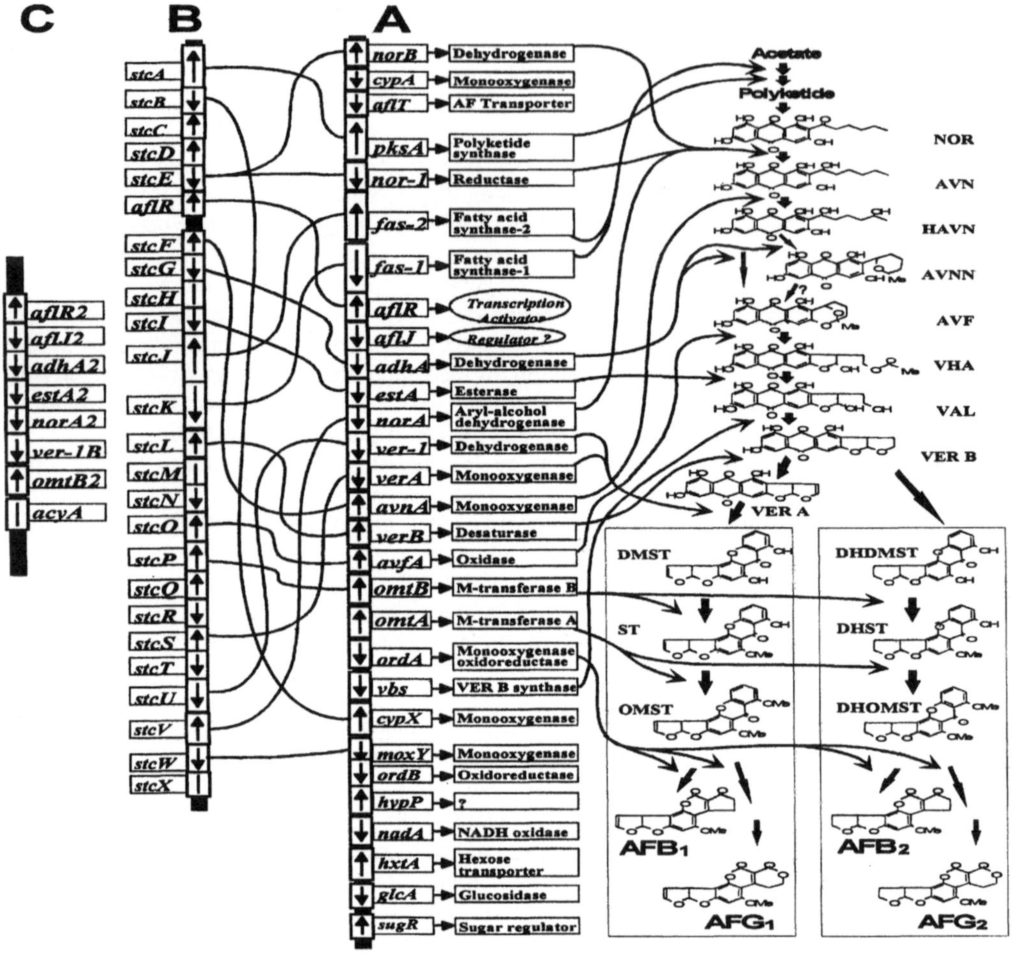

Fig. 1. Aflatoxin and sterigmatocystin biosynthetic pathway, pathway genes and gene cluster.
Proposed and generally accepted pathway for aflatoxin B_1, B_2, G_1 and G_2 biosynthesis and the corresponding genes and their enzymes are presented. The aflatoxin biosynthetic pathway gene cluster in *A. parasiticus* and *A. flavus* (Panel A) and the non-functional, partially duplicated aflatoxin gene cluster in *A. parasiticus* (Panel C) are shown. The gene names are labeled on the side of the cluster. Arrows indicate the direction of gene transcription. The homologous genes between the sterigmatocystin pathway gene cluster in *A. nidulans* (Panel B) and aflatoxin pathway gene cluster in *A. parasiticus* and *A. flavus* is connected by line. The four sugar utilization genes linked to the aflatoxin pathway gene cluster are on the bottom of panel A. Abbreviations for the intermediates are: norsolorinic acid (NOR), averantin (AVN), 5'-hydroxyaverantin (HAVN), averufanin (AVNN), averufin (AVF), versiconal hemiacetal acetate (VHA), versiconal (VAL), versicolorin B (VER B), versicolorin A (VER A), demethylsterigmatocystin (DMST), sterigmatocystin (ST), *O*-methylsterigmatocystin (OMST), aflatoxin B_1 (AFB_1), aflatoxin G_1 (AFG_1), dihydrodemethylsterigmatocystin (DHDMST), dihydrosterigmatocystin (DHST), dihydro-*O*-methylsterigmatocystin (DHOMST), aflatoxin B_2 (AFB_2), and aflatoxin G_2 (AFG_2), methyltransferase (M-transferase).

the closely related *Aspergillus* section *Nidulantes*. *A. nidulans* has a long history as a model system for the study of developmental and secondary metabolic processes. Sterigmatocystin (ST) and dihydrosterigmatocystin (DHST), the penultimate precursors of aflatoxins are produced by various *Aspergilli* including *A. nidulans* (Cole and Cox 1987; Betina 1989; Chu 1991).

2.2 Aflatoxin and Sterigmatocystin Biosynthesis

Aflatoxins B_1, B_2, G_1 and G_2 (AFB_1, AFB_2, AFG_1 and AFG_2) are the four major aflatoxin homologues among at least 16 structurally related metabolites (Goldblatt 1969) characterized to date. Aflatoxins are produced primarily by *A. flavus* and *A. parasiticus* as well as *A. nomius*. One isolate of each of *A. tamarii* (Goto *et al.* 1996) and *A. ochraceoroseus* (Frisvad and Samson 1999) have been reported to produce aflatoxins (Klich et. al. 2000). The two aflatoxigenic species *A. flavus* and *A. parasiticus* are closely related. *A flavus* can produce AFB_1 and AFB_2 and cyclopiazonic acid (CPA). *A. parasiticus* produces AFG_1 and AFG_2, in addition to AFB_1 and AFB_2. *A. flavus* is the most common cause of aflatoxin contamination in corn, cotton, peanuts and tree nuts. Both *A. flavus* and *A. parasiticus* are seed-inhabiting fungi and contaminate a wide variety of other crops with aflatoxins before harvest in the field and after harvest during handling and processing.

Because of the toxicity and extreme carcinogenicity of aflatoxins (Busby and Wogan 1981; Cullen *et al.* 1987; Van Egmond 1989; Eaton and Groopman 1994; Galvano *et al.* 1996) there are worldwide efforts to better understand the genetics, biochemistry and regulation of aflatoxin biosynthesis as well as the taxonomy, biology, toxicology, evolution of aflatoxigenic fungi. During the last decade, significant progress has been made in deciphering the aflatoxin biosynthetic pathway and the genetics of their regulation (Fig. 1; Minto and Townsend 1997; Bhatnagar *et al.* 2002a; Yu *et al.* 2002f; Yu 2003). The first aflatoxin biosynthetic gene to be characterized molecularly was *nor-1* (Chang *et al.* 1992). Three more biosynthetic genes, *ver-1* (Skory *et al.* 1992), *aflR* (Chang *et al.* 1993; Payne *et al.* 1993; Woloshuk *et al.* 1994), and *omtA* (Yu *et al.* 1993, 1995b) were characterized soon after. By mapping overlapping cosmid clones in *A. parasiticus* and *A. flavus* Yu *et al.* (1995a) established that aflatoxin biosynthetic pathway genes were clustered. The concept of aflatoxin pathway gene cluster greatly accelerated the speed of gene discovery (Chang *et al.* 1995a; Cleveland *et al.* 1997; Yabe *et al.* 1998; Yu *et al.* 1997, 1998, 2000a, 2000b; see review articles by Townsend 1997; Payne and Brown 1998; Bhatnagar *et al.* 2002a; Yu 2003). In *A. parasiticus* and *A. flavus*, the aflatoxin gene cluster consists of 24 genes or ORFs, within an approximately 70 kb region of DNA. In *A. parasiticus*, a partial duplicated aflatoxin gene cluster consisting of seven genes, *aflR2, aflJ2, adhA2, estA2, norA2, ver1B, omtB2*, has been confirmed and characterized by Chang and Yu (2002). In *A. nidulans*, the ST gene cluster, comprised of 25 co-regulated genes, have been characterized (Brown et. al. 1996). Homologs to almost all of the *A. parasiticus* aflatoxin ORFs have been found in the *A. nidulans* ST cluster. However, the order of these genes responsible for the biosynthesis of aflatoxins in *A. parasiticus* and *A. flavus* and those for the biosynthesis of sterigmatocystins in *A. nidulans* are quite different.

Genetic components are the major factors that regulate aflatoxin synthesis. In both the aflatoxin and sterigmatocystin gene clusters, there is a positive regulatory gene, *aflR*, located in the middle of the gene clusters for activating pathway gene transcription. The *aflR* gene, coding for a sequence specific zinc DNA-binding protein has been shown to be required for transcriptional activation of most, if not all, of the structural genes (Payne *et al.* 1993; Chang *et al.* 1993, 1995b, 1999a, 1999b; Woloshuk *et al.* 1994; Yu et. al. 1996; Flaherty and Payne 1997;

Fernandes et. al. 1998; Ehrlich et. al. 1998, 1999a, 1999b) in *A parasiticus, A. flavus* and *A. nidulans*. Adjacent to the *aflR* gene in the aflatoxin gene cluster, a divergently transcribed gene, *aflJ*, was also found to be involved in the regulation of transcription of other cluster genes (Flaherty and Payne 1997; Meyer *et al.* 1998; Chang and Yu 2002; Brown unpublished observation). However, the exact mechanism by which *aflJ* modulates transcription of these pathway genes in concert with *aflR* is to be studied. A gene, *aflT*, encoding a membrane bound protein with homology to antibiotic efflux genes presumed to be involved in some way in aflatoxin secretion, was discovered in *A. parasiticus* (Chang *et al.* unpublished observation). Other than the genetic factors, nutritional and environmental factors are also important in stimulating aflatoxin formation such as carbon and nitrogen sources, pH, temperature, drought, volatile compounds released from host plants etc.

Aspergillus flavus is the most common cause of aflatoxin contamination in pre-harvest field crops and post-harvest grains. Currently, there are no effective control strategies to prevent aflatoxin accumulation in the field. A better understanding of the genetics of aflatoxin biosynthesis in this fungus will contribute to development of new control strategies to eliminate pre-harvest aflatoxin contamination. These strategies will result in a safer, economically viable food and feed supply. A genomic analysis of this fungus will contribute to the understanding of the field ecology of the fungus, the diversity of its population, fungal-host plant interactions, genetic regulation and the effects of environmental factors on aflatoxin production.

2.3 *Aspergillus flavus* Genomics

Karyotyping studies by CHEF gel indicate that the *A. flavus* genome has 6-8 chromosomes that range in size from 3 to ≥ 7 Mb (Keller *et al.* 1992; Kitamoto *et al.* 1994; Foutz *et al.* 1995). By referencing the karyotyping data of *A. oryzae*, which contains 8 chromosomes (Kitamoto *et al.* 1994), a close relative of *A. flavus* or the domesticated strain of *A. flavus* as some scientists claim, it appears that there are most likely 8 chromosomes in *A. flavus*. The estimated genome size is about 33 to 36 Mb and contains approximately 10,000 to 12,000 functional genes. Based on our research experiences with *Aspergillus,* the *A. flavus* genome is compacted with less duplicated sequence or multiple copies of genes. The non-coding sequences between genes are much shorter than in higher plants and only a few small introns within each gene if any.

The goal of *A. flavus* genomics is to understand the genetic control and regulation of aflatoxin production and the evolution of *Aspergillus* section *Flavi*. The relationships between aflatoxin production and survival in *A. flavus*, between primary and secondary metabolism, and between aflatoxin formation and regulator proteins upstream of the *aflR* protein are important factors to understand in order to control aflatoxin contamination. More importantly, we need to understand the mechanism by which environmental conditions such as nutritional status of crops, temperature, water stress, pH, and volatiles, affect aflatoxin production.

The *A. flavus* genomics, either by whole genome sequencing or Expressed Sequence Tag (EST), can help to identify the whole set or majority of the genes, respectively, in the fungal genome. EST could identify as many as 7,000 - 9,000 genes. The fragments of these genes can be amplified for fabricating microarray to obtain more specific clues for targeting critical regulatory components and genes involved in aflatoxin formation.

A large-scale *A. flavus* EST/Microarray project is being carried out at the USDA/ARS, Southern Regional Research Center (SRRC), New Orleans, Louisiana, USA (Bhatnagar *et al.* 2002b; Yu *et al.* 2002c, 2002d, 2002e). The strain of *A. flavus* used in this project was wild-type aflatoxin-producing strain NRRL 3357. This strain was chosen not only because of its genetic representation of the species but also for the fact that this strain has been extensively used in field experiments and laboratory studies worldwide (Payne 1998).

It is estimated that at least 5 fold clone coverage is needed to obtain 80% of the expressed genes in a standard *A. flavus* cDNA library. However, this percentage can be achieved with less coverage (2-3 X) by reducing the copy numbers of highly expressed genes in the cDNA library via the normalization procedure. Normalization is based on the re-association kinetics theory to effectively reduce the high variation in abundance among the clones of a cDNA library that represent individual mRNA species (Soares *et al.* 1994). The highly abundant cDNA clones will be hybridized faster than the low abundant cDNA clones (rare copy genes). The re-associated (hybridized) double stranded clones are subsequently removed through column purification. After one or two rounds of normalization, the number of copies of highly abundant cDNA clones is reduced significantly whereas the number of copies of less abundant clones remains about the same. As a result, the percentage of the rare copy genes is improved drastically within a population of cDNA clones. Thus, normalization process alters the number of copies of individual clones within a cDNA library without changing the number of different cDNA species in the library. The normalization protocol used to construct the *A. flavus* cDNA library was a modified procedure of Soares (1994), which is called "Rare Cloned Biased Library." Such a normalized cDNA expression library was prepared to obtain 80% of expressed functional genes within it but with only 2-3X coverage to reduce sequencing costs and time. In order to make the library as representative as possible, the cDNA library was prepared from combined mycelia grown under several conditions (e.g. aflatoxin supportive and non-supportive media) and harvested at different growth phases. The normalized cDNA library was prepared in such a way that the poly-A sequence was trimmed off prior to cloning into a vector. This allowed cDNA clones to be sequenced from both 5' and 3' ends with equal efficiency and to obtain the same length of good quality sequence on both ends. Sequencing of the cDNA clones is being carried out at The Institute for Genomic Research (TIGR, Maryland, USA). The cDNA sequencing phase has been completed. The average of usable sequence length is about 550 bases. The discovery rate for unique genes is several folds higher than the rate from a standard library. Preliminary BLAST results indicated that more than 7,000 expressed unique genes were identified. Among the genes identified, many are rare copy genes potentially involved in secondary metabolism and gene regulation. All known aflatoxin biosynthetic genes have been identified among the sequenced clones in the library, an indication that the library is of good quality. As expected, with only 2-3X sequence coverage about 80% of the estimated 10,000 to 12,000 functional genes in the *A. flavus* genome have been obtained. The cost and time of sequencing have been reduced in at least half employing the normalized cDNA library as compared to a standard library.

TIGR's bioinformatics software, CAP3 (Huang and Madan 1999), was used to construct an *A. flavus* gene index. The *A. flavus* gene index consists of unique cDNA sequences. These unique sequences are made up of both tentative consensus sequences (TCs), which consist of overlapping cDNA sequences, and singletons, which consist of sequences identified only once from the sequence analysis. The ESTs within the gene index are classified based on their functions. Overlapping cDNA clones were combined into consensus sequences (TC). Both TC and singleton ESTs are considered unique gene sequences. The preliminary functions of TCs were classified into three categories (molecular function, cellular component and biological process) which were subsequently divided into sub-categories. Most genes in the TC group exhibit some identity/similarity to genes in the GenBank Database. There are only less than 3%

of the genes within the TC group that did not show any sequence homology to previously identified genes (function unknown). However, there is much higher percentage of the ESTs whose functions are unknown if all of the unique ESTs are considered since partial sequence is less likely to find a homolog in the database. Within the unique ESTs, we have identified many of the genes that may be involved directly or indirectly in aflatoxin formation (Yu *et al.* 2002e). The genes of interest can be summarized in the following three categories: 1) previously identified aflatoxin biosynthetic genes, e. g. *fas-1, fas-2, pksA, nor-1, aflR, aflJ, estA, avnA, ver-1, omtA, omtB, ordA* etc.; 2) regulatory genes that have the potential to regulate aflatoxin production or signal transduction, e.g. genes encoding DNA-binding proteins, RNA-binding proteins, zinc-finger proteins, transcription regulators, transducins, cAMP receptors, protein kinases etc.; 3) genes that have the potential to contribute to fungal virulence or pathogenicity. The latter category of genes could be involved in processes such as sporulation, conidiation, hyphal growth and hydrolytic activities. One unique gene identified in the *A. flavus* EST library exhibits a high degree of sequence homology to the pathogenicity protein in *Magnaporthe grisea*. Some unique genes in the EST library also show sequence homology to genes encoding hydrolytic enzymes, including amylase, cellulase, pectinases, proteases, chitinase, chitosanases, pectin methylesterases, endoglucanase C precursor, glucoamylase S1/S2 precursors, β–1,3-glucanase precursor, 1,4-β-D-glucan cellobiohydrolase A precursor, glycogen debranching enzyme and xyloglucan-specific endo-β-1,4-glucanase precursor. Such hydrolytic enzymes could be highly expressed virulence factors during invasion of *A. flavus* into crops and, if so, have the potential to be useful targets for inhibiting aflatoxin production or for antifungal growth through genetic engineering.

Microarrays (Gross *et al.* 2000) can be used to detect a whole set of genes transcribed under specific conditions and can be highly useful for studies, not only for biological functions of interested genes, for studies on gene expression and regulation, but for identifying factors involved in plant-microbe (crop-fungus) interaction. Using microarray technologies, we can screen and identify the critical gene or genes involved in aflatoxin production and fungal invasion to host plants, as well as better understand evolutionary biology of the aflatoxigenic and non-toxigenic strains and field isolates. Currently, a microarray containing 7,000 - 8,000 expressed *A. flavus* genes (an estimated ~80% of the genes in the *A. flavus* genome) is being prepared. This microarray will be used to screen genes that could be targeted in fungal systems for inhibiting aflatoxin formation or antifungal growth. Thus, microarray analysis could provide information that could lead to the identification of potent antifungal genes or genes that inhibit aflatoxin formation. It may then be possible to engineer these genes into crop plants in an attempt to eliminate or reduce aflatoxin contamination in the crop. The annotated sequence information will be available to the public once it is deposited into the GenBank Database.

Concurrent to USDA/ARS/SRRC effort, Gary Payne at North Carolina State University (NCSU) in collaboration with Ralph Dean, Fungal Genomic Laboratory at NCSU, has also carried out an *A. flavus* EST/Microarray project using the same *A. flavus* strain NRRL 3357 as used in the USDA *A. flavus* project (O'Brian and Payne 2001; Payne *et al.* 2002) (http://www.fungalgenomics.ncsu.edu/Projects/aspergillus.htm). Their objective is to identify those genes differentially expressed during aflatoxin biosynthesis and to profile their expression during aflatoxin biosynthesis. A total of approximately 100,000 *A. flavus* clones from a cDNA library constructed using mycelia grown under aflatoxin supportive medium, were gridded onto nylon filters. An initial screening of these cDNA clones on filters was carried out by hybridization with mRNA probes prepared from cultures of the fungus in the log phase of aflatoxin production. Over 10,000 positive clones that were expressed under aflatoxin-producing conditions were identified. Approximately 2,200 quality sequences, with Phred scores of >20 for at least 100 bases, have been obtained. BLAST searches against NCBI databases identified 753 unique ESTs. Many of the unique ESTs showed no sequence homology to genes in public databases. Among ESTs with homology to genes of known functions are those coding for signal transduction pathways, secondary metabolism, glucose regulation, cell wall biosynthesis, and cell cycle control. These unique ESTs were arrayed on glass slides (microarray) for further analysis. A tentative gene expression profile, e.g. up or down regulation of these genes under aflatoxin supportive and non-supportive conditions, has been established (Payne *et al.* 2002). This is the first effort spearheading functional genomics study on aflatoxin biosynthesis and its regulation using *A. flavus* EST/Microarray technologies even though it is relatively at small scale.

In addition, a small scale *A. flavus* EST project has been done by Nancy Keller (Department of Plant Pathology, University of Wisconsin-Madison) in collaboration with Doris Kupfer (University of Oklahoma's Advanced Center for Genome Technology (ACGT) http://www.genome.ou.edu/fungal.html). About 1,400 cDNA clones have been sequenced from a wild type, *A. flavus* vegetative mycelia cDNA library. The sequencing was performed from the 3' end of the directionally cloned inserts after excision into pBlueScript SK$^-$ vector and 1253 quality sequences were obtained. A unigene (unique gene) database has been generated and the results of a BLAST search of GenBank with this database are available at ACGT (http://www.genome.ou.edu/fungal.html).

Efforts have also been made to obtain funding to carry out whole genome sequencing of *A. flavus*. A proposal to sequence the entire *A. flavus* genome by a shotgun approach was submitted to USDA/NSF Microbial Genome Project by Gary Payne (North Carolina State University, Raleigh, North Carolina, USA) in cooperation with scientists from universities and USDA/ARS as steering committee members. In the meantime, USDA/ARS, Southern Regional Research Center is also attempting to obtain funds from its own agency to sequence the *A. flavus* genome.

Efforts directed at *A. parasiticus* genomics are limited at this time. Karyotyping studies have demonstrated that the *A. parasiticus* genome contains fewer chromosomes (5-7) than *A. flavus* (6-8) (Keller *et al.* 1992). However, the estimated genome size may be larger (40 Mb) than *A. flavus* and may reflect more functional genes. Researchers at ACGT have carried out a small scale *A. parasiticus* sequence project involving a small number of cosmids. No data have yet been released to the public.

2.4 *Aspergillus nidulans* Genomics

Aspergillus nidulans (*Emericella nidulans*) has a long and productive history as a source of industrial chemicals and enzymes and as a developmental model system to study genetic regulation, developmental biology, signal transduction and secondary metabolism (Pontecorvo *et al.* 1952; Timberlake 1990; Brown *et al.* 1996; Adams and Yu 1998). *A. nidulans* is well amenable to molecular manipulation and is the most extensively studied fungal species. Spontaneous and induced mutations have been generated in hundreds of genes and *A. nidulans* is amenable to Mendelian genetics. Mutation analysis is an important tool in the characterization of gene function and protein biological roles.

The *A. nidulans* genome consists of eight well-characterized chromosomes ranging in size from 2.8 Mb to 4.4 Mb. Estimates of genome size vary slightly from 28.4 Mb (U. Oklahoma's ACGT) to 31 Mb (Brody and Carbon 1989). Chromosome IV was targeted for sequencing at

Texas A & M University by an industrial consortium headed by Genecor Corp. (Dunn-Coleman and Prade 1998) at 4.5X of average genome coverage since it is the smallest in size (2.8 Mb). The chromosome IV contains many genes involved in primary and secondary metabolisms (e.g. sterigmatocystin gene cluster) (Brown *et al.* 1996).

An *A. nidulans* genomics EST project was carried on at U. of Oklahoma's ACGT (http://www.genome.ou.edu/fungal.html). Over 14,900 cDNA clones were isolated from a cDNA library constructed by Dr. Rodolfo Aramayo, at Texas A&M University, College Station, Texas. The library was made from a mixed vegetative and 24 hour asexual development culture of *A. nidulans* strain FGSC A26. The *A. nidulans* EST's was generated by Doris Kupfer at ACGT in collaboration with Rolf Prade and Rodolfo Aramayo (Oklahoma State University). They sequenced approximately 6000 templates (ca. 12,000 reactions) from both ends of the directionally cloned inserts which resulted in a total of 12,485 usable sequences. The EST collection is available to the public at ACGT for BLAST analysis.

An *A. nidulans* gene index (AnGI) has been constructed at TIGR (http://www.tigr.org) integrating data from international *A. nidulans* EST sequencing and gene research projects, which have been deposited in the databases. The most recent data release (June 1, 2002) includes a total of 5013 unique *A. nidulans* genes assembled from 12,857 cDNA sequences. Among these unique genes, 3263 are singletons and 1750 are tentative consensus sequences (TCs). The ultimate goal of the TIGR *A. nidulans* Gene Index project (AnGI) is to represent a non-redundant view of all *A. nidulans* genes and data on their expression patterns, cellular roles, functions and evolutionary relationships.

A 3X coverage of the *A. nidulans* genome by shotgun sequencing has been carried out by Cereon Corp. (Formerly Millenium and currently a subsidiary of Monsanto Corp.) (http://www.cereon.com). As reported, approximately 381,000 raw sequences have been completed with sequence annotation assembled into 11,186 contigs and 4,658 singletons. The longest contig is 18,933 bp with an average contig length of 2350 bp. A 10X coverage of the *A. nidulans* whole genome sequence has been completed by the Whitehead Institute/MIT. A total of 13X genome assembly of combined *A. nidulans* sequence from 10X Whitehead Institute, Center for Genome Research (WICGR) and 3X Monsanto was released on February 28, 2003 (http://www-genome.wi.mit.edu/annotation/fungi/aspergillus/). The total size of the genome is about 30 Mb based on the length of the combined 248 contigs (30,068,514 bp). The *A. nidulans* genome sequence data are available for download and for BLAST search.

3. GENOMICS OF *ASPERGILLUS FUMIGATUS*

3.1 *Aspergillus fumigatus*

Among the 182 recognized species of Aspergilli (Pitt *et al.* 2000), *Aspergillus fumigatus* is the most common human and animal pathogen (Denning 1998), and can be the most common thermophilic species found in the air in human dwellings (Latgé 1999). It is a ubiquitous organism in the air and its distribution is worldwide. The natural niche of the mold is decaying organic matter. It is remarkably thermotolerant, as are many of its enzymes including phytase (Pasamontes *et al.* 1997), which gives it the capacity to survive and thrive in composting materials (Beffa *et al.* 1998). It is capable of micro-aerophilic growth (Hall and Denning 1994) and its growth rate is substantially increased in the presence of hydrocortisone so that it is among the fastest growing fungi known, with a hyphal linear extension rate of 1-cm/hr (Ng *et al.* 1994). Aspergilli possess numerous secondary metabolic pathways that produce compounds with biotechnological and pharmacological properties. *A. fumigatus* produces the angiogenesis inhibitor fumagillin as well as other secondary metabolites including gliotoxin, mannitol, phthioic acid, fumitremorgin, verruculogen, fumigaclavine, helvolic acid, and sphingofungin (Moss 1989).

3.2 *Aspergillus fumigatus* and Aspergillosis

Aspergillus fumigatus causes rapidly fatal invasive infection in immunocompromised patients, occasional infections in non-immunocompromised patients and is the most common microbial allergen in atopic individuals (Casadevall & Pirofski 1999). Invasive aspergillosis is the leading cause of infectious death in leukaemia and transplant patients. Patients at risk of invasive aspergillosis, based on disease frequency, include those with chronic granulomatous disease (25-40% lifetime risk), transplant recipients including lung (17-26%), allogeneic bone marrow and stem cell (4-30%), heart (2-13%), pancreas (1-4%) and kidney (~1% in Europe and the USA; ~10% in India), neutropenic patients with leukaemia (5-25%) and patients with AIDS, multiple myeloma and severe combined immunodeficiency (~4%) (Denning 1998). Patients treated with corticosteroids are also at risk. The mortality from invasive aspergillosis exceeds 50% even with treatment (Denning 1996). In the U.S. in 1996, the cost of treating diagnosed cases of invasive aspergillosis was estimated at $633M, with each case costing approximately $64,500 (Dasbach 2000).

In addition to invasive disease, *Aspergillus* causes a number of other diseases in man. These include aspergilloma ("colonisation" of existing pulmonary cavities), sinusitis in normal people, allergic bronchopulmonary and sinus infections, keratitis (which usually leads to blindness in the affected eye and is common in the developing world) and post-operative infections in immunocompetent patients. Recent data suggest that airborne fungi are responsible for most cases (93%) of chronic sinusitis (Ponikau *et al.* 1999), which has very substantial implications as a recently completed study in allergic bronchopulmonary aspergillosis indicated a major role for antifungal therapy in this condition (Stevens *et al.* 2000).

The currently available antifungals (amphotericin B and its lipid formulations, and azoles) have mediocre efficacy, especially in heavily immunosuppressed hosts. Furthermore, the toxicities associated with amphotericin B and the increasing incidence of azole resistance (Moore *et al.* 2000) contribute to the need for development of less toxic and more efficacious antifungals. Caspofungin (CAS), a promising novel echinocandin lipopeptide, selectively inhibits 1,3-β-D-glucan synthetase, a fungus-specific enzyme that is critical for fungal cell wall biosynthesis. Preliminary experience with CAS as salvage therapy in patients with refractory aspergillosis has shown that the drug is effective in 40% of patients (Maertens *et al.* 2000). In view of the apparent lack of toxicity of CAS, this novel antifungal holds promise for the treatment of candidiasis and aspergillosis.

Understanding pathogenicity in *A. fumigatus* has been problematic, probably because it is multifactorial, and the organism has many extraordinary features. Traditional research tools are out paced with the urgent need for solving the problem of aspergillosis. *A. fumigatus* genomics opens a new frontier at the molecular level for dissection of its biology and pathogenicity and offers hope for development of effective antifungal drugs. At the same time, it provides an opportunity to study the evolutionary process within *Aspergilli* by integrating the genomic data obtained from related Aspergillus species such as *A. flavus*, *A. parasiticus*, *A. nidulans*, *A. oryzae*, *A. niger* and *A. sojae*.

3.3 Sequencing of *Aspergillus fumigatus*

An international project to sequence the genome of *A. fumigatus* AF 293 was initiated in 2001. It is being accomplished by an international consortium composed of The Institute for Genomic Research (TIGR, USA), the Wellcome Trust Sanger Institute (UK), the University of Salamanca (Spain), Complutense University (Spain), the Centro de Investigaciones Biológicas, (Spain) and the Pasteur Institute (France). The project is coordinated by Dr. David Denning of the University of Manchester (UK). Details of progress and links to the sequence data are posted on the *Aspergillus* website (http://www.aspergillus.man.ac.uk/index.htm). A pilot project to sequence BACs covering a small region of the genome was started at the Sanger Centre and the University of Manchester in July 1999 and is nearing completion. TIGR and the Sanger Centre are providing the major contribution to sequencing (See http://www.tigr.org/tdb/e2k1/afu1/and http://www.sanger.ac.uk). The project is based on a whole genome shotgun approach with random sequencing now completed to 10X sequence coverage, approximately 510,000 sequences. The shotgun sequencing at TIGR was performed to 6X sequence coverage with an additional 4X sequence coverage from the Sanger Centre and 25,000 sequences from the Spanish participants. The TIGR sequencing was accomplished from libraries with insert sizes of 3-4 kb, 10-12 kb, and 50 kb. The Sanger sequences were from a 3-4 kb pUC library and from BAC libraries with inserts averaging 75 kb.

The 10X assemblies provide an estimated genome size of 29.4 Mb with a 49.6% G+C. The largest contig was 1.3 Mb and 97% of the genome was in contigs greater than 10 kb. By analysis of paired-end sequences with their sequences in different contigs, 122 scaffolds were identified. The largest scaffold was 2.9 Mb and 4 scaffolds were larger than 1 Mb. An electronic annotation of the assemblies identified the presence of 11,000 candidate open reading frames.

The mitochondrial genome closed when the nuclear genome was at 4X sequence coverage. The mitochondrial DNA is a 32,237 bp circle with a 25.3% G+C content. Based on the redundancy of the mitochondrial sequences, the copy number is estimated to be 5. It contains a complete set of 26 tRNA genes including a UGA-tyr tRNA. It appears to be a typical fungal mitochondrial DNA with genes for 1 ATPase subunit, an apocytochrome b, 6 NADH dehydrogenase complex subunits, 1 cytochrome oxidase subunit and 1 ribosomal protein subunit, *rps5*.

Completion of the genome is in progress at TIGR and the Sanger Centre. Telomeres as well as a region that appears to completely span a centromere have been identified in this process, suggesting that it will be possible to determine the complete sequence. Following the process of closing all the gaps and quality checking the sequence, full manual annotation will be accomplished at TIGR and Sanger with input from community scientists. The project is scheduled to be completed in 2003.

3. GENOMICS OF MYCOTOXIGENIC *FUSARIUM* SPECIES

4.1 Fumonisin and Trichothecene-Producing Fungi

Fusarium is a large and diverse genus of filamentous fungi that includes a number of economically important plant pathogens, including *F. graminearum*, *F. oxysporum*, and *F. solani*. Although several species can be opportunistic pathogens in humans with severely compromised immune systems, *Fusarium* has more often been associated with human and animal health problems as a result of mycotoxicoses. *Fusarium* produces a remarkable diversity of mycotoxins and other biologically active secondary metabolites. These include enniatins, fumonisins, fusaric acid, fusarins, gibberellins, moniliformin, trichothecenes, and zearalenone

(De Nijs *et al.* 1996; Desjardins and Proctor 1999). Two of these groups of metabolites are used commercially: the zearalenone derivative α-zearalanol is used to enhance growth of cattle (Pusateri and Kenison 1993) and gibberellins, which are plant growth regulators, have numerous applications in horticulture (Tudzynski 2002). The toxicity and widespread occurrence of the mycotoxins trichothecenes and fumonisins has lead to significant efforts over the past decade and a half to understand the genetics and biochemistry of their biosynthesis.

4.2 Trichothecenes: Structure, Toxicity and Role in Plant Disease

Trichothecenes are a large and structurally diverse group of mycotoxins that are synthesized by a variety of fungal genera, including *Fusarium*, *Myrothecium*, *Stachybotrys*, *Cephalosporium*, *Trichoderma* and *Trichothecium* (Sharma and Kim 1991). The structural diversity of trichothecenes results from the type and number of functional groups attached to the tricyclic trichothecene skeleton, 12, 13-epoxy-trichothec-9-ene. Typically, a given *Fusarium* strain or species produces only a limited number of trichothecenes. For example, *F. sporotrichioides* produces primarily T-2 toxin and scirpenol derivatives, while *F. graminearum* (teleomorph *Gibberella zeae*) produces primarily deoxynivalenol (DON) or nivalenol (NIV) and their acetylated derivatives. The generally accepted pathway leading from farnesyl pyrophosphate to T-2 in *F. sporotrichioides* (Fig. 2) was determined through biochemical and genetic studies (Desjardins *et al.* 1993).

Trichothecenes have gained notoriety due to their toxicity to both plants and animals that results from their ability to inhibit protein synthesis. For some *Fusarium* species, trichothecenes are a critical component of virulence on certain crop plants. For example, DON production contributes to the ability of *F. graminearum* to cause wheat head blight and maize ear rot (Desjardins *et al.* 1996; Harris *et al.* 1999). The mycotoxicity of trichothecenes, coupled with their frequent occurrence on a variety of agricultural commodities, has precipitated significant effort to understand their biosynthesis.

4.3 Genetics of Trichothecene Biosynthesis

Trichothecene biosynthesis requires gene products from at least three loci in the *Fusarium* genome: the core trichothecene biosynthetic gene cluster (Brown *et al.* 2002a), the *TRI101* locus (McCormick *et al.* 1999) and the *TRI1* locus (Brown *et al.* 2003). The core trichothecene gene cluster consists of 12 open reading frames (ORFs) that encode both structural and regulatory genes (Fig. 3B). Gene deletion and metabolite feeding studies have resulted in the assignment of seven genes (Fig. 3C; *TRI5*, *TRI4*, *TRI11*, *TRI3*, *TRI13*, *TRI7*, *TRI8*) to specific biochemical steps in the biosynthetic pathway (Desjardins *et al.* 1993; Brown *et al.* 2002a; McCormick and Alexander 2002 and references therein), the *TRI6* and *TRI10* genes to regulatory roles (Proctor *et al.* 1995; Tag *et al.* 2001), and *TRI12* to a self-protective role (Alexander *et al.* 1999). The role of *TRI9* and *TRI14* is unclear. Although *TRI14* expression is similar to other cluster genes, its predicted protein and disruption did not indicate a direct role for it in T-2 toxin biosynthesis (Brown *et al.* 2002a). The second trichothecene locus in the *Fusarium* genome contains a single structural gene, *TRI101*, required for the acetylation of the C-3 oxygen (Kimura *et al.* 1998; McCormick *et al.* 1999). *TRI101* is flanked by genes that are likely involved in other cellular functions and thus are not part of a second trichothecene biosynthetic gene cluster (Kimura *et al.* 1998). The third trichothecene locus in the *Fusarium* genome includes a structural gene, *TRI1*,

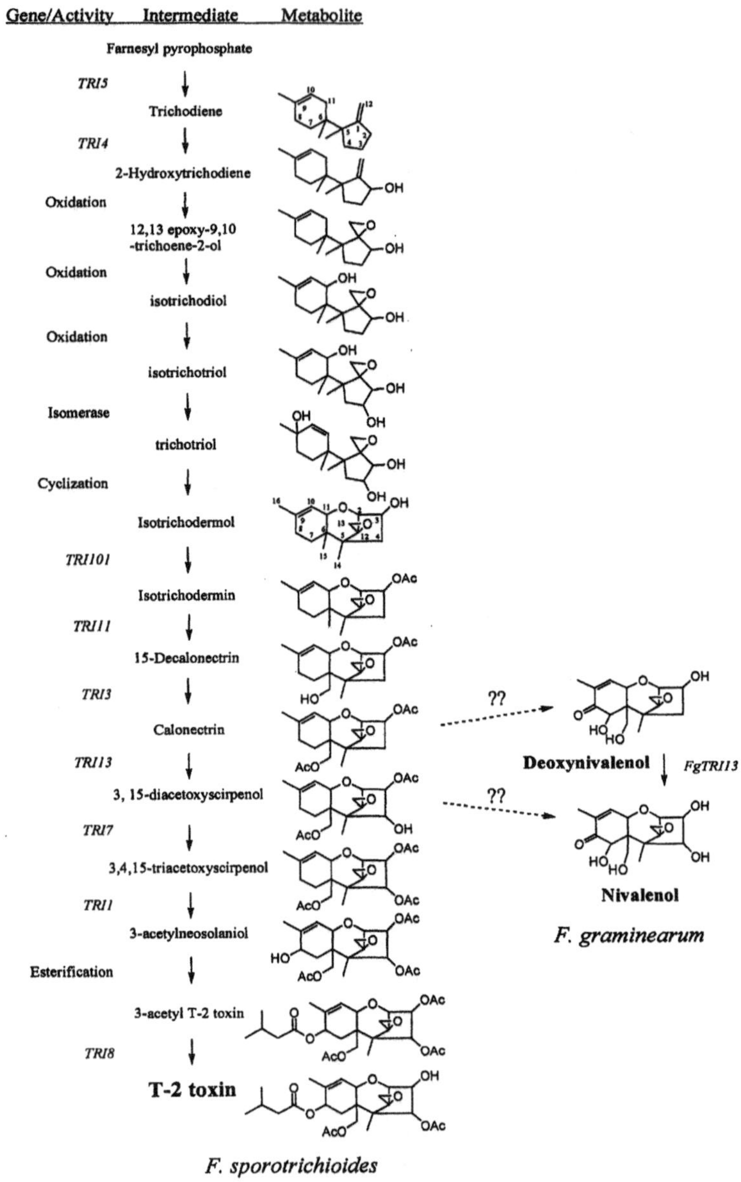

Fig. 2. Trichothecene biosynthetic pathway. Biochemical and genetic pathway for trichothecene biosynthesis in *Fusarium sporotrichioides* leading to T-2 toxin and in *F. graminearum* leading to deoxynivalenol (DON) and nivalenol (NIV).

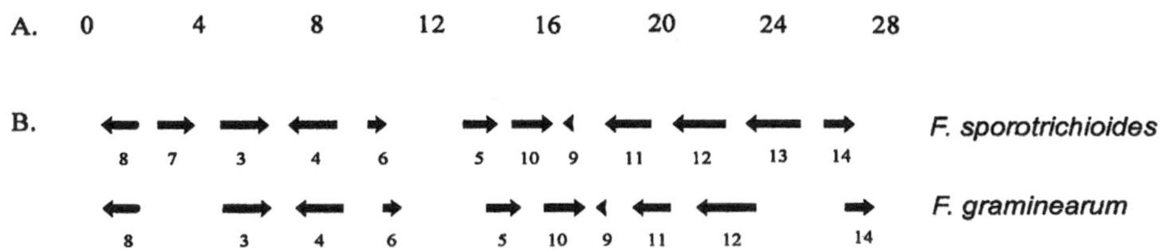

C.

Gene	Function	Group
TRI8	deacetylase	C-3 acetate
TRI7	acetylesterase	C-4 oxygen
TRI3	acetyltransferase	C-15 oxygen
TRI4	P450 hydroxylase	C-2
TRI6	regulation	
TRI5	sequiterpene cyclase	
TRI10	regulation	
TRI9	unknown	
TRI11	P450 hydroxylase	C-15
TRI12	trichothecene efflux pump	
TRI13	P450 hydroxylase	C-4
TRI14	unknown	

Fig. 3. Genomic organization of the *Fusarium* trichothecene gene cluster. A. Relative scale in kilobases. B. Organization of the predicted ORFs for *F. sporotrichioides* and *F. graminearum*. The arrows heads indicate direction of transcription. Numbers underneath each large arrow refer to the specific genes (e.g. 5 indicates *TRI5*). Genes from different *Fusarium* species with the same number are homologues. *F. graminearum* lack a functional *TRI7* and *TRI13* genes. C. Table listing all of the trichothecene genes, their predicted enzyme function and the carbon molecule affected on the trichothecene skeleton.

required for C-8 oxygenation during T-2 biosynthesis (Brown *et al.* 2003). Sequence data indicate that *TRI1* is not located within 20-40 kb of the core cluster or *TRI101* and that it is flanked by a putative regulatory gene on one side and a putative acetyltransferase on the other. Experiments are underway to determine if these additional genes are required for trichothecene biosynthesis. The requirement of two different gene clusters for trichothecene biosynthesis is not unique in fungal secondary metabolism. Cephalosporin, produced by *Acremonium chrysogenum* requires at least two gene clusters for its synthesis (Keller and Hohn 1997).

The characterization of trichothecene biosynthetic genes has provided evidence for two mechanisms by which the structural diversity of secondary metabolites may arise: gene inactivation and gene recruitment. The first mechanism is evident from intra- and interspecies comparison of genes responsible for hydroxylation (*TRI13*) and acetylation (*TRI7*) of the trichothecene skeleton at C-4. In *F. graminearum*, strains that only produce trichothecenes without a C-4 oxygen (e.g. DON), *TRI13* and *TRI7* are inactivated as a result of multiple nucleotide insertions and deletions in their coding regions (Brown *et al.* 2002a; Lee *et al.* 2002). However, in *F. graminearum* and *F. sporotrichioides*, strains that produce trichothecenes with C-4 hydroxyl and/or acetyl functions (e.g. T-2 toxin, NIV and 4-acetyl NIV) both *TRI13* and *TRI7* are functional. Evidence for the second mechanism of structure evolution comes from characterization of *TRI1*, the first *TRI* gene characterized that is required for a stable modification of the trichothecene skeleton and that is not located within the core trichothecene cluster. The location of *TRI1* outside the core cluster suggests that it may have been recruited, perhaps through co-regulation via the acquisition of TRI6-binding sites within its promoter region, to modify the trichothecene skeleton.

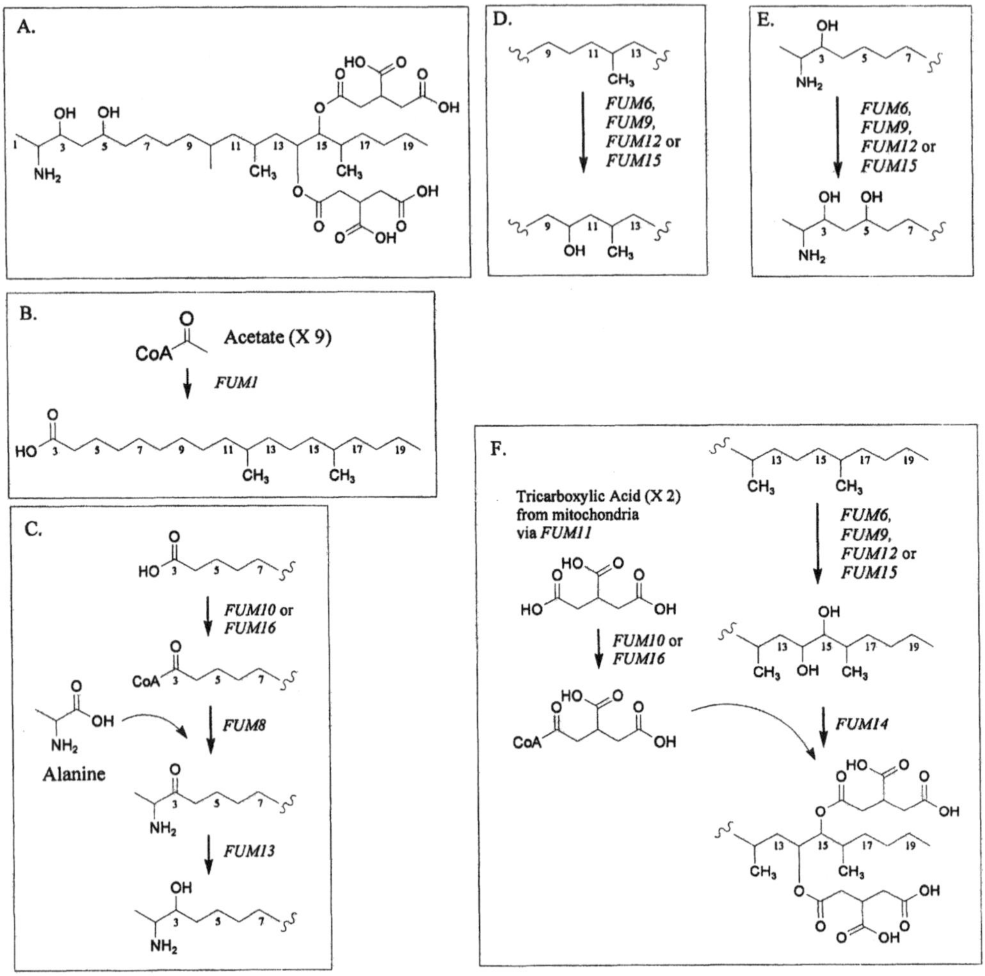

Fig. 4. Chemical structure of Fumonisin B_1. Chemical structure of Fumonisin B_1 (A), and predicted biochemical steps and predicted functions of *FUM* proteins in fumonisin biosynthesis (B-F). (B), formation of the linear polyketide; (C) CoA activation, condensation with alanine, and carbonyl reduction of polyketide; (D) hydroxylation of backbone at C-10; (E) hydroxylation of backbone at C-5; (F) CoA activation of tricarboxylic acids (TCAs), and hydroxylation and TCA esterification of the fumonisin backbone at C-14 and C-15. The order of most of the reactions following the formation of the linear polyketide is not known. Therefore the order of reactions indicated by the lettering of schemes B-F may not reflect the order of reactions during fumonisin biosynthesis.

4.4 Fumonisins: Occurrence, Toxicity and Structure

Fumonisins are produced by *F. verticillioides* (teleomorph *Gibberella moniliformis*) and several related species. *F. verticillioides* is one of the most common ear and stalk rot pathogens of maize worldwide, and as a result, fumonisins are frequent contaminants of maize (Munkvold and Desjardins 1997). Fumonisin B_1 (FB_1) is generally present at the highest levels (70%) in naturally contaminated maize with the other B-series fumonisins (FB_2, FB_3 and FB_4) present at lower levels (5-20%) (Nelson *et al.* 1993). Consumption of fumonisin-contaminated feed can cause several fatal animal diseases, including cancer in rodents (Howard *et al.* 2001), and there are epidemiological correlations between the consumption of *F. verticillioides*-rotted maize and human esophageal cancer in some regions of the world where maize is a dietary staple (Marasas 2001). Fumonisins are structurally similar to the sphingolipid intermediates sphinganine and sphingosine and disrupt sphingolipid metabolism by inhibiting the enzyme ceramide synthase (Wang *et al.* 1991). A causal relationship between the disruption of sphingolipid metabolism and fumonisin-associated diseases has not been established but seems likely given the multiple functions of sphingolipids in cells.

B-series fumonisins consist of a linear 20-carbon backbone with an amine group at carbon 2 (C-2), a hydroxyl at C-3, methyl groups at C-12 and C-16, and 6-carbon tricarballylic esters at C-14 and C-15 (Fig. 4A) (Nelson *et al.* 1993). B-series fumonisins differ from one another by the presence or absence of hydroxyl groups at C-5 and C-10.

4.5. Genetics of Fumonisin Biosynthesis

Fumonisin biosynthetic genes (*FUM*) are clustered in a manner similar to genes required for aflatoxin, sterigmatocystin and trichothecene biosynthesis. The fumonisin biosynthetic gene cluster consists of 15 co-regulated genes (*FUM1* and *FUM6* through *FUM19*) and occupies a ~45-kb region of DNA on linkage group 1 of *F. verticillioides* (Proctor *et al.* 2003) (Fig. 5). Although the exact functions of the *FUM* genes have not been demonstrated experimentally, it has been possible to assign putative functions to most of them based on their sequence similarities to genes with known functions, knowledge of the chemical structures of fumonisins, and results of precursor feeding studies, and by analogy to early steps in *de novo* sphingolipid biosynthesis.

The first step in the synthesis of FB_1 is most likely the formation of a linear polyketide corresponding to C-3 to C-20 of the fumonisin backbone. Synthesis of the polyketide is likely catalyzed by the *FUM1* (formerly *FUM5*) protein, which is predicted to be a polyketide synthase (PKS) (Proctor *et al.* 1999). The presence of a putative methyltransferase domain in the PKS suggests that the methyl groups at C-12 and C-16 of fumonisins are incorporated into the polyketide during its synthesis. Thus, the putative product of the PKS is a fatty acid-like 18-carbon chain with two methyl groups and a terminal carboxylic acid function (Proctor *et al.* 2003) (Fig. 4B). The order of most subsequent steps in fumonisin biosynthesis is not known. However, precursor feeding studies indicate that the polyketide undergoes the following modifications once it is formed: 1) condensation with alanine to form C-1, C-2 of the backbone and the amine function at C-2; 2) reduction of the polyketide carbonyl to form the C-3 hydroxyl; 3) hydroxylation of C-5, C-10, C-14 and C-15 of the fumonisin backbone; and 4) esterification of tricarballylic acids to the C-14 and C-15 hydroxyls (Branham and Plattner 1993; Blackwell *et al.* 1994; Caldas *et al.* 1998).

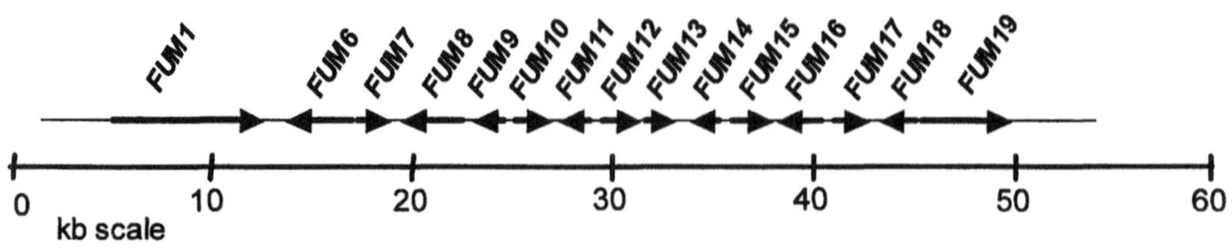

Gene	Predicted Function
FUM1	Polyketide Synthase
FUM6	Cytochrome P450 Monooxygenase/Reductase
FUM7	Dehydrogenase
FUM8	Aminotransferase
FUM9	Dioxygenase
FUM10	Fatty Acyl-CoA Synthetase
FUM11	Tricarboxylate Transporter
FUM12	Cytochrome P450 Monooxygenase
FUM13	Carbonyl Reductase
FUM14	Condensation Domain Peptide Synthetase
FUM15	Cytochrome P450 Monooxygenase
FUM16	Fatty Acyl-CoA Synthetase
FUM17	Longevity Assurance Factor
FUM18	Longevity Assurance Factor
FUM19	ABC Transporter

Fig. 5. The fumonisin biosynthetic gene cluster. The fumonisin biosynthetic gene cluster and predicted function of *FUM* genes based on BLAST sequence comparisons. The functions or proposed functions of these genes are listed above.

Condensation of the polyketide and alanine is most likely catalyzed by the *FUM8* protein, which shares significant similarity to the enzyme that catalyzes the condensation of serine and palmitoyl-CoA during *de novo* sphingolipid biosynthesis (Seo *et al.* 2001) (Fig. 4C). Reduction of the carbonyl group to a hydroxyl is likely catalyzed by the *FUM13* protein, which is most similar to short-chain dehydrogenases/reductases, enzymes that often catalyze reduction of carbonyl functions (Proctor *et al.* 2003) (Fig. 4C). Hydroxylations of the fumonisin backbone at C-5, C-10, C-14 and C-15 is likely to be catalyzed by the putative cytochrome P450 monooxygenases encoded by *FUM6*, *FUM12* and *FUM15* and/or the dioxygenase encoded by *FUM9* (Fig. 4D, E, F). Both types of oxygenases commonly catalyze hydroxylation reactions.

Which gene catalyzes esterification of the tricarballylic acids to the fumonisin backbone is less easily predicted from sequence similarities because none of the *FUM* genes show similarity to well characterized esterase genes. However, the predicted *FUM14* protein, which is moderately similar to condensation domains of nonribosomal peptide synthetases, could be an esterase (Proctor *et al.* 2003) (Fig. 4F). Condensation domains typically catalyze peptide bond formation, but sometimes, as in the case of beavericin and enniatin synthesis, they appear to be able to catalyze ester bond formation (Marahiel *et al.* 1997). The putative tricarboxylate transporter encoded by *FUM11* may contribute to the formation of the esters by transporting tricarboxylic acids precursors of the esters out of the mitochondria and thereby making them available for fumonisin biosynthesis (Fig. 4F). The putative fatty acyl-CoA synthetase encoded by *FUM10* and/or *FUM16* may also contribute to formation of the tricarballylic esters by activating the tricarboxylic acid precursors with CoA prior to their esterification of the fumonisin backbone (Fig. 4F).

Gene disruption experiments in fungi often provide conclusive evidence for the function of the disrupted gene. To date, there are published reports on disruptions of six *FUM* genes. However, the phenotypes of disruption mutants did not clearly demonstrate gene function. As expected, disruption of the polyketide synthase gene, *FUM1*, blocked fumonisin production and did not lead to the accumulation of detectible intermediates. Disruption of both *FUM6*, which encodes a putative cytochrome P450 monooxygenase, and *FUM8*, which encodes a putative aminotransferase, also blocked fumonisin production, but unexpectedly did not lead to the accumulation of detectible intermediates. Thus, while disruption experiments demonstrated that *FUM6* and *FUM8* are required for fumonisin biosynthesis, their exact functions in this process remain to be determined.

Disruption of *FUM17* and *FUM18*, which are predicted to encode longevity assurance factors, did not affect fumonisin production (Proctor *et al.* 2003). These genes are predicted to be involved in self-protection because of their similarity to *Asc-1*, a tomato longevity assurance factor gene that confers resistance to FB_1 and the structurally similar AAL toxin produced by the tomato pathogen *Alternaria alternata* (Brandwagt *et al.* 2000). Disruption of the ABC transporter gene *FUM19*, which may also have a self-protection function, resulted in production of slightly altered ratios of fumonisins (Proctor *et al.* 2003). Studies on whether *FUM17*, *FUM18* and *FUM19* provide protection from fumonisins are in progress.

In contrast to the trichothecene, aflatoxin and sterigmatocystin biosynthetic gene clusters, the fumonisin cluster does not include any obvious regulatory genes. As a result, efforts are underway to identify genes outside the cluster that are involved in the regulation of fumonisin biosynthesis. The only published reports on such genes is that of the *F. verticillioides FCC1* gene, which is predicted to encode a cyclin-like protein involved in signal transduction (Shim and Woloshuk 2001).

4.6. *Fusarium* Genomics

The goal of *Fusarium* genomic programs is to identify critical genes involved in host-parasite interactions and mycotoxin production. At present, there are two *Fusarium* EST programs available over the Internet. The *F. sporotrichioides* EST project is a collaboration between researchers at the University of Oklahoma's Advanced Center for Genome Technology (ACGT) and the Department of Plant Pathology and Microbiology at Texas A&M University (http://www.genome.ou.edu/fsporo.html). This effort is directed, at least in part, at identifying genes involved in trichothecene biosynthesis. The cDNA library used to generate the ESTs was prepared from a strain of *F. sporotrichioides* that overexpresses *TRI10*, which causes increased expression of other trichothecene pathway genes and increased T-2 toxin production (Tag *et al.* 2001). The EST library consists of a total of 7495 ESTs representing 3238 unique sequences. Enrichment for trichothecene genes appears to have been successful as almost 5% of the ESTs recovered represent genes located in the core trichothecene gene cluster. The *F. sporotrichioides* EST library has already been useful in identifying additional genes involved in trichothecene biosynthesis. *TRI1*, which, as noted above, is not located in the core cluster and encodes a trichothecene-8-hydroxylase, was originally identified from the EST collection (Brown *et al.* 2002b).

A second EST collection has taken a different tact to identify *Fusarium* genes. Researchers at Michigan State University, Purdue University and the USDA Cereal Disease Laboratory have collaborated to generate over 10,000 *F. graminearum* ESTs from three cDNA libraries prepared from different culture conditions and developmental stages of the fungus. The ESTs represent over 2500 unique sequences and are available at http://www.genomics.purdue.edu/~jxu/Fgr/. A major goal of this work is to better understand the biology of *F. graminearum* in order to develop more effective wheat head blight control strategies.

The USDA's National Center for Agricultural Utilization Research and The Institute for Genomic Research (TIGR) are currently collaborating to generate an extensive *F. verticillioides* EST collection. This project is focused on generating sequences expressed during several different growth conditions (e.g. mycotoxin non-production and production conditions and developing maize kernels) and growth stages (e.g. germinating conidia). To date (December, 2002), the project has generated 17,000 ESTs by sequencing the 5' ends of cDNA clones. These ESTs represent over 4,800 different genes, including all of the genes in the fumonisin biosynthetic gene cluster.

The three *Fusarium* EST programs described above are at relatively early stages and have not yet been used extensively to identify genes involved in mycotoxin production or host-parasite interactions. However, the presence of sequences corresponding to trichothecene and fumonisin biosynthetic genes in the *F. sporotrichioides* and *F. verticillioides* EST libraries, respectively, demonstrates their potential utility.

5 GENOMICS OF NON-TOXIGENIC INDUSTRIAL *ASPERGILLI*

5.1 *Aspergillus oryzae*

Aspergillus Section Flavi complex consists of mycotoxin producers such as *A. flavus* and *A. parasiticus* described earlier that contaminate post-harvest crops (corn, peanut, cotton etc.) in low latitudes, and thus agribusiness is often overwhelmed by the fungal contamination. On the other hand, other fungi such as *A. oryzae* and *A. sojae*, so-called *koji* molds, have been extensively used for indigenous Japanese fermentation products such as *sake* (rice wine) and *shochu* (spirits), *shoyu* (soy sauce), and *miso* (soybean paste) for over 1,000 years. The long history of extensive use in the food industries placed *A. oryzae* on the list of Generally Regarded as Safe (GRAS) organisms by the Food and Drug Administration (FDA) in the United States (Tailor and Richardson 1979). The safety of *A. oryzae* as a food-grade organism is also supported by the World Health Organization (FAO/WHO 1987). Today, the *koji*-molds are also used as host cells for enzyme production by DNA recombination technology. The beneficial involvement of *koji*-molds in Japanese society has been the driving force for research and development activities in fields including academia, industry, medicine and agriculture.

Host cells for recombinant protein production have been extended from prokaryotes to eukaryotes since the mid 1980's. Because the potential of hyper protein secretion found in filamentous fungi including *A. oryzae*, *A. sojae*, and *A. niger* has attracted the fermentation industry, several production systems for recombinant proteins have been developed and industrialized (Machida 2002). At an early stage of recombinant protein production, the target proteins were restricted to homologous proteins found in host cells, subsequently the expression system has become heterologous for host cells and the commercial targets are extending to proteins for medical use.

For the above reasons, genomics on industrial *Aspergilli* has been studied since late 1990's. Although *A. sojae* has been well industrialized as well as *A. oryzae*, research into *A. sojae* genomics are very limited. Therefore, this section mainly covers genomics of *A. oryzae*.

5.1.1 Gene manipulation

Gene manipulation in *A. oryzae* can be achieved by incubating *A. oryzae* protoplasts with DNA in the presence of polyethylene glycol (Gomi *et al.* 1987; Iimura *et al.* 1987). The introduced DNA is integrated into chromosomes through homologous/non-homologous recombination or double crossover displacement. There are nearly 10 host-marker gene systems available, among which *niaD* (Unkles *et al.* 1989), *argB* (Berka *et al.* 1990), *pyrG* (Ruiter-Jacobs *et al.* 1989; Wu and Linz 1993) and *amdS* (Gomi *et al.* 1992) based systems are preferred. Recently, a pyrithiamine resistant gene (*ptrA*) was successfully used as a dominant selectable marker for the transformation of wild-type *A. oryzae* strains (Kubodera *et al.* 2000). The typical transformation efficiency of *A. oryzae* is 10-100 transformants per microgram of plasmid DNA. The low transformation frequency is apparently due to the integration of DNA into an indispensable gene on the chromosome and/or the removal of the marker DNA during integration. The autonomously replicating fragment, *AMA1*, from *A. nidulans* functions in *A. oryzae* is able to maintain the external DNA without integration into the chromosome and allow a 30-fold increase in the transformation efficiency (Gems *et al.* 1991).

5.1.2 EST sequences of *A. oryzae*

A large-scale EST sequencing project was initiated by the collaboration of M. Machida at the National Institute of Advanced Industrial Science and Technology (AIST) (Tsukuba, Japan), O. Akita at the National Research Institute of Brewing (NRIB) (Higashi-Hiroshima, Japan), Y. Kashiwagi at the National Food Research Institute (NFRI) (Tsukuba, Japan), T. Kobayashi at the Nagoya University (Nagoya, Japan), N. Kitamoto at the Food Research Institute of Aichi Prefectural Government (Nagoya, Japan), K. Kitamoto and H. Horiuchi at The University of Tokyo (Tokyo, Japan), M. Takeuchi at Tokyo University of Agricultural Technology (Tokyo, Japan) and K. Gomi and K. Abe at Tohoku University (Sendai, Japan). The project was partly supported by private companies, Amano Enzyme (Nagoya, Japan), Ozeki (Nishinomiya, Japan), Gekkeikan Sake (Kyoto, Japan), Higashimaru (Tatsuno, Japan), Higeta (Chyoshi, Japan), Kikkoman (Noda, Japan), Yamasa (Chyoshi, Japan), Miso-Kyokai (the society of soybean paste producing companies) and Tanekoji-Kumiai (the association of *koji* seed companies which produce *A. oryzae* conidiophores). Most of these companies have been involved in traditional Japanese fermentation industries.

For the sequencing of ESTs, the *A. oryzae* strain RIB40 (ATCC-42149) was selected. In general, soy sauce companies have their own strains, selected after extensive breeding. *Sake* brewers also have their own strains different from the strains used for soy sauce fermentation. *A. oryzae* RIB40 is a wild type strain, similar to those used for *sake* brewing, but still have the ability to produce proteinases required for soy sauce fermentation. mRNA was prepared from *A. oryzae* mycelia grown in several different culture conditions including in complete medium, at high temperature and without any carbon source (see Table 1). It was expected that the chance of finding new genes from a limited number of ESTs would be increased using these different culture conditions. In addition, two libraries were prepared from the mycelia grown in solid-state culture, using rice bran or wheat bran. These two solid-state culture conditions have been extensively and traditionally used in Japanese fungal fermentation industry. More amylases and proteases are produced in solid-state culture than in liquid culture, thus, the information from the libraries is expected to be important to improve the productivity of *A. oryzae*.

Table 1. *A. oryzae* ESTs and their libraries.

Culture condition	Analyzed ESTs
Liquid complete medium (+glucose)	2,693
Liquid complete medium (+glucose, 37°C)	2,072
Liquid synthetic medium (-glucose)	1,953
Liquid complete medium (+maltose)	932
Liquid complete medium (pH 10)	751
Solid-state cultivation (wheat bran)	6,309
Solid-state cultivation (Shoyu *koji* 25°C) [a]	1,049
Liquid complete medium (germination)	1,049
Total	16,808

[a]solid-state cultivation is the same as that using wheat bran except that a mashed and steamed complex of soybean and wheat was used as the medium instead of mashed and steamed wheat.

cDNAs synthesized from the mRNA using oligo(dT) as a primer were unidirectionally inserted onto plasmid vectors such as pBlueScript and were sequenced specifically from their 5' termini to cover protein-coding regions effectively. To decrease an overestimation of identifying new genes, the average insert size of the cDNA library was maintained for as long as possible, typically exceeding 1.5 kb. The total number of ESTs sequenced reached 16,808 and the total length analyzed was 9.83 Mb with the average length of a single pass sequence being 585 bp. After clustering, the total number of the non-redundant sequences was approximately 6,000 with the total length of the contigs (non-redundant sequence) being 4.5 Mb (Fig. 6) (Machida 2002). The length of the non-redundant sequence was equivalent to 13-15% of the *A. oryzae* genome (estimated size = 30-35 Mb). Thus, the nucleotide sequence of approximately 40% of the coding region has been sequenced, assuming that the average length of the coding region is 1.5 kb as in *Saccharomyces cerevisiae*. The total number of genes already sequenced has not been precisely determined due to possible incomplete clustering as mentioned above. Considering the number of contigs after clustering, and the ratio of already sequenced protein coding regions, approximately 50% of the total genes is estimated to have been sequenced. The polypeptides encoded by 47% of the contigs had significant homology to those found in the public database by a BLAST search. The contigs of the *A. oryzae* ESTs are made available from the "Database of genomes and transcriptional regulations for filamentous fungi" on the web site of the Research Information Database (RIO-DB) at the Advanced Institute of Industrial Science and Technology (AIST) (http://www.aist.go.jp/RIODB/ffdb/index.html).

The contigs were sorted according to the number of ESTs involved in the contig and were plotted versus the redundancy of the ESTs (Fig. 6). The number of contigs containing highly redundant ESTs (i.e. highly expressed genes) is less than 500, which is roughly 5% of the total number of estimated *A. oryzae* genes. The same result was obtained from the ESTs from any particular condition, but also from all the ESTs sequenced using the 8 different libraries. The number of the highly expressed genes, the ESTs of which were found to occupy greater than 10% of the total number of ESTs, was only 167, which was 2.5% of the total number of contigs. The number of singletons still occupied 66% of the total contigs, even after the collection of 16,808 ESTs from various culture conditions (Fig. 6).

5.1.3 Genome Sequencing of *A. oryzae*

5.1.3.1 Pilot project

For comprehensive analysis of the nucleotide sequence of *A. oryzae* genes and their organization on the chromosomes, an *A. oryzae* genome sequencing project was promoted since

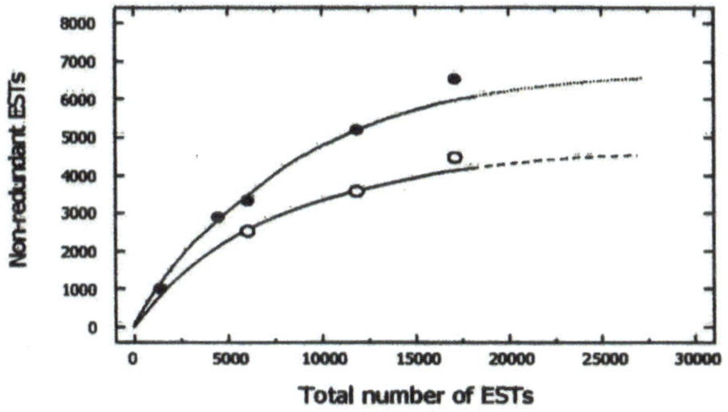

Fig. 6. Growth of non-redundant ESTs versus total number of analyzed ESTs. The *A. oryzae* ESTs were clustered according to the overlap of their nucleotide sequences. Closed and open circles indicate the numbers of resulting contigs and singletons, respectively.

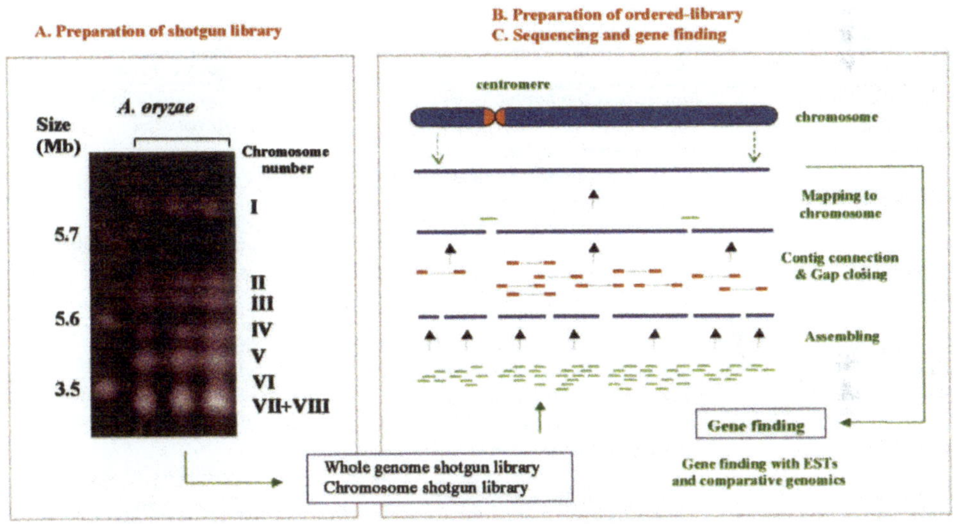

Fig. 7. *A. oryzae* genome project. A) Pulse field gel electrophoresis of the *A. oryzae* chromosomes. Protoplasts and agarose plugs containing *A. oryzae* chromosomes were prepared according to the method as described by Kitamoto *et al.* (1994). The *A. oryzae* chromosomes were electrophoresed on a 0.8% agarose gel in 0.5X TEA buffer by CHEFF Mapper (Bio-Rad, Hercules, CA) at 1.5 V/cm with 50 and 45 min intervals of angle switch (60°/-60°) for an initial 36 h and the successive 300 h, respectively. The temperature was kept at 12°C. After electrophoresis for 14 days, the migration positions of the chromosomes were visualized by ethidium bromide staining under UV-irradiation. The chromosomes from *Schizosaccharomyces pombe* were used as the size standard. The chromosome numbers are indicated on the right side of the gel image. A whole genome shotgun library and shotgun libraries for each chromosome were prepared from total genomic DNA and chromosome DNAs extracted from the electrophoretically separated chromosomes respectively. B) Ordered libraries for each chromosome were also prepared and sequenced. C) Both the shotgun and the ordered libraries were sequenced and assembled to form contigs. ESTs were available for contig formation, mapping, and gene finding.

1998. At the same time, the large-scale EST sequencing project was started as described above. The *A. oryzae* genome consists of 8 chromosomes ranging from 2.8 Mb to 7 Mb in length, and is estimated to have a total genome size of 35 Mb (Kitamoto *et al.* 1994). The shortest band at 2.8 Mb indicates approximately twofold stronger intensity than the other bands, indicating that the band derives from two chromosomes, VII and VIII (Fig. 7). The second shortest band (VI) is obviously weaker than the others and is smeared. The band was found to hybridize with the DNA fragment that has a ribosomal DNA (rDNA) sequence from *A. oryzae* (Chang *et al.* 1991), suggesting that chromosome VI possesses rDNA which repeats in variable numbers (Machida 2002).

Machida *et al.* have constructed an *A. oryzae* genomic DNA library containing *Hin*dIII or *Eco*RI complete digests specific to chromosome V. The clones were randomly selected and sequenced from either or both ends of the inserts, yielding 525 single pass sequences of chromosome V (Machida unpublished data). The total analyzed sequence was 423 kb with an average length of 807 b. Approximately 35% of the sequences had sequence similarity with nucleotide sequences in the public databases. The overall GC-content was 45.6%. These results indicate that the *A. oryzae* genome is an appropriate target for a large-scale sequencing project.

Table 2. Genomic resources of interested fungal species.

Institution	Website	Organism, Project
TIGR, USDA/ARS/SRRC	http://www.tigr.org/tdb/tgi/afgi/	*A. flavus* gene index
USDA/ARS/MSA	http://199.133.85.14/unique/blast.html	*A. flavus* EST
North Carolina St. Univ.	http://www.cifr.ncsu.edu/ aspergillusflavus/Genomics.html	*A. flavus* genomics
Whitehead Institute/MIT	http://www-genome.wi.mit.edu/ annotation/fungi/aspergillus/	*A. nidulans* genome
TIGR	http://www.tigr.org/tdb/tgi/angi/	*A. nidulans* gene index
Univ. of Oklahoma, TAMU	http://www.genome.ou.edu/fungal.html	*A. nidulans* EST
AIST, Japan	http://www.nrib.go.jp/ken/EST/ db/blast.html	*A. oryzae* EST
Integrated Genomics	http://www.integratedgenomics.com/	*A. niger* genome
Gene Alliance	http://www.gene-alliance.com/	*A. niger* genome
Univ. of Manchester, UK	http://www.aspergillus.man.ac.uk/	*A. fumigatus* genome
TIGR	http://www.tigr.org/tdb/e2k1/afu1/	*A. fumigatus* genome
The Sanger Institute	http://www.sanger.ac.uk/Projects/	*A_fumigatus/A.fumigatus*
Oklahoma St. Univ., TAMU	http://aspergillus-genomics.org/	*Aspergillus* genome
TIGR, USDA/ARS/NCAUR	http://www.tigr.org/	*F. vercitillioides* EST
Fungal Genetics Stock Center	http://www.fgsc.net	Fungal genomes

TIGR: The Institute for Genomic Research; TAMU: Texas A & M University; AIST: The Advanced Institute of Industrial Science and Technology, Japan.

5.1.3.2 Whole genome sequencing project

Since 2001, a whole genome sequencing project for *A. oryzae* has been underway at the National Institute of Technology and Evaluation (NITE) (Tokyo, Japan) by the cooperation of The Consortium for *A. oryzae* Genomics consisting of the National Institute of the Advanced Institute of Industrial Science and Technology (AIST), the National Research Institute of Brewing (NRIB), the National Food Research Institute (NFRI), The University of Tokyo, Tokyo University of Agricultural Technology, Tohoku University, Nagoya University, Axiohelix (Tokyo, Japan), Amano Enzyme, Gekkeikan Sake, Higeta, Intec Web and Genome Informatics (Tokyo, Japan), Kikkoman, Kyowa-Hakko Kogyo (Tokyo, Japan), Ozeki and the Brewing

Society of Japan (Tokyo, Japan). The DNA libraries were prepared and supplied by AIST, NRIB and NFRI. The large-scale sequencing is being done by NITE. The members of the consortium will focus on the analysis of gene function and the utilization of data derived from the *A. oryzae* genome project.

Because there are no genetic maps of the *A. oryzae*, the sequencing will be done mainly by a whole genome shotgun sequencing approach in combination with some chromosome specific shotgun sequencing (Fig. 7). A rough draft of *A. oryzae* genome was completed in January 2002 by accumulating sequences of approximately 6X depth of coverage after 600,000 sequence reactions. Tentatively, the total genome size of *A. oryzae* is estimated to be 37 Mb with 954 contigs and 36 supercontigs (max. 3.3 Mb). Asai *et al.* at Computational Biology Research Center in AIST (http://www.cbrc.jp/) has conducted computational gene finding using GeneDecoder algorithm that is a gene finding technology for eukaryotes, based on hidden Markov models (http://www.genedecoder.org/). They predict presence of over 13,000 genes including ca. 1,700 for membrane proteins, ca. 380 for transcription factors, ca. 160 for translation factors, and ca. 70 for protein secretory pathways besides abundant genes for hydrolytic enzymes. Approximately 40% of genes are speculated to have introns. The completely fixed sequences will be published for each chromosome without significant delay.

5.1.4. Functional Genomics

5.1.4.1 Gene disruption

Gene disruption is the key technology for functional genomics. However, gene disruption in *A. oryzae* is significantly difficult for the following reasons. (i) Efficiency at both transformation and homologous recombination is low. (ii) *A. oryzae* possesses multiple nuclei not only in vegetative filamentous cells but also in reproductive cells, conidiospores (Ushijima and Nakadai, 1987; Ushijima *et al.* 1990). Even after several rounds of single spore isolation, transformed cells may still possess a nucleus in which the target gene remains undisrupted. Recently, researchers in Kikkoman developed a novel method for efficient isolation of the single gene knockout in *A. oryzae* and *A. sojae* (Hatamoto *et al.* 1999a). The vector consists of a positive selection marker, a negative selection marker and a marker for homokaryon selection based on a chromogenic marker, which are *pyrG*, *oliC31*, and the polyphenol oxidase (laccase) gene (Hatamoto *et al.* 1999b). The two markers, *pyrG* and the laccase gene are encompassed with the two DNA fragments derived from upstream or downstream of the target gene, yielding a cassette to be integrated in the genome. The 5' and 3' arms of the target gene on the vector have to be partially deleted for keeping the gene disrupted at the target locus in the chromosome, because the markers (*pyrG* and the laccase gene) are selectively removed and reused for the subsequent second round knockout as described bellow. The negative selection marker, *oliC31*, is located outside the integration cassette and is removed during homologous recombination. The transformants that possess the integration cassette are positively selected as prototrophs using the *pyrG*-deleted *A. oryzae* strain as a host, which prevents homologous recombination at the authentic *pyrG* locus in *A. oryzae*. *oliC31* encodes a mutant Subunit 9 in FoF1-ATPase of *A. nidulans*, and the protein OliC31 is resistant to oligomycin but hypersensitive to triethyltin (Ward *et al.* 1986). Homologous integration at the target site excises the *oliC31* fragment from the cassette, however non-homologous recombination leaves *oliC31* in chromosomes. Consequently, the transformnts with the cassette by homologous integration can survive, and are selectable on the medium containing triethyltin. Laccase can be expressed on gallic acid plates and color formation of the transformants is proportional to the number of the integrated laccase gene. The color maker enables tracking of homokaryon generation through repetitive isolation of progenies (Hatamoto *et al.* unpublished data). The *pyrG* and the laccase gene markers integrated in the target site can be eliminated by selection as 5-fluoroorotic acid (5FO) resistant progenies. Then, the first round knockout strains, in which *pyrG* and the laccase gene markers have already been eliminated after the 5FO selection, are ready for the second round of gene targeting. The knockout system enables (i) multiple knockouts and (ii) easy elimination of markers after gene targeting. Therefore, the targeting system is completely food-grade, and can be applicable to other fungi.

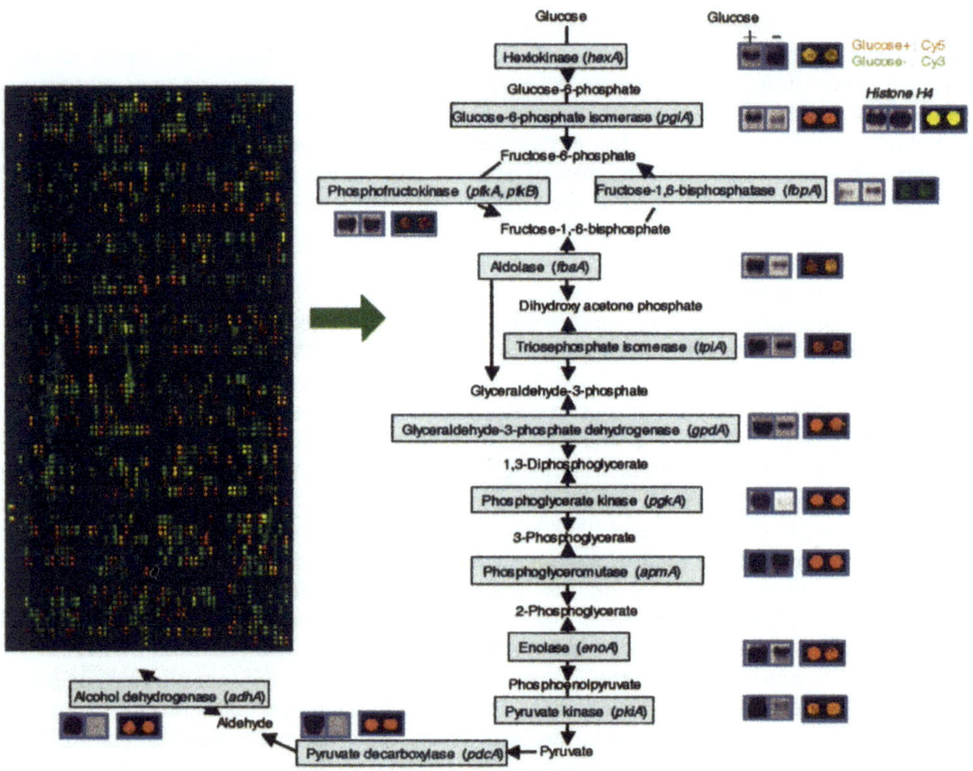

Fig. 8. Expression profiling of *A. oryzae* glycolytic genes by Northern hybridization and a cDNA microarray. *A. oryzae* cDNAs were amplified by PCR using the *A. oryzae* EST clones as templates and were spotted onto a glass slide. mRNAs from mycelia grown in the modified CD medium (Czapek-Dox medium with 1% polypepton) (Hata *et al.* 1992) containing 3% glucose and CD medium without any carbon source were labeled by Cy3 and Cy5 fluorescent dyes, respectively, combined, and then, were hybridized to the cDNAs immobilized on the slide. The results from duplicate spots of the DNA microarray for each glycolytic gene (Maeda, H., Maruyama, Y., Abe, K., Gomi, K., Hasegawa, F., Sano, M., Machida, M., Akao, T., Akita, O., Nakajima, T. and Iguchi, Y., unpublished data) were compared with those from Northern hybridization (Nakajima *et al.* 2000).

5.1.4.2 Expression analysis

A first generation *A. oryzae* DNA microarray was manufactured by Maeda *et al.* at Tohoku University (Sendai, Japan) consisting of approximately 2,000 cDNAs amplified from EST clones. The 2,000 cDNAs were highly expressed clones in the 6,000 non-redundant EST clones prepared in the *A oryzae* EST project described above. Several genes in the aflatoxin biosynthesis gene cluster of *A. flavus* were supplied from Yu *et al.* at U.S. Department of Agriculture, Agricultural Research Service, Southern Regional Research Center (New Orleans, LA) and from Payne *et al.* at North Carolina State University (Raleigh, NC), and also were put on the array. Fig. 8 shows the comparison of expression of some of the glycolytic genes between the cDNA microarray and Northern hybridization. Most of the glycolytic genes except *fbpA*, which encodes fructosebisphosphatase, were confirmed to be induced by glucose. The expression profile derived from the DNA microarray analysis was in good agreement with that from Northern analyses. Chigira *et al.* used the DNA microarrays for expression analysis of *A. oryzae* chitin synthase genes, and speculated that family I-III *chs* genes are highly expressed in solid culture conditions and that family V and VI *chs* genes are highly expressed in liquid culture conditions (Chigira *et al.* 2002). Very recently, Yamada *et al.* at NRIB manufactured second-generation cDNA microarays (so called NRIB 3000 DNA microarray) that cover additional 3,000 EST clones from the *A. oryzae* 6,000 non-redundant contigs. At this moment, approximately 5,000 independent EST clones have been covered by the two sets of DNA microarrays.

5.1.4.3 Comprehensive analyses of *cis*-elements and transcription factors

Analyses of the regulatory elements and the transcription factor of a particular gene are important for utilizing its function by predicting expression in the production condition. However, the difficulty in the *in vivo* analysis, including the reporter gene assays significantly delays the analysis of the promoters of *A. oryzae* genes. One way to circumvent the problem is to use *A. nidulans* for *in vivo* analysis. Alternatively, *in vitro* analysis using an electrophoretic mobility shift assay (EMSA) generates useful information about transcription regulatory elements (*cis*-elements). Scanning of the element(s) based on sequence-specific binding of cellular factors(s) by EMSA with highly sensitive fluorescence detection has remarkable potential for rapid determination of these elements (Sano *et al.* 2001). Identification of cellular factor(s) associated with the elements can also be accelerated by the use of an *in vitro* technique such as phage display (Zhang *et al.* 2000; Hagiwara *et al.* 2002). High throughput identification of *A. oryzae* factors from a cDNA phage display library is underway. The combination of the DNA microarray and analyses based on DNA-protein interaction may be a useful way to generate information about transcription of industrially important organisms for which *in vivo* analysis techniques have not yet been well established.

5.2 *Aspergiilus niger*

Aspergillus niger is a species of great industrial importance besides *koji*-molds. Classified as GRAS by the U.S. Food and Drug Administration, it is widely used in the industrial production of citric acid and a variety of industrial enzymes such as amylases, pectinases, and proteases (Godfrey and West 1996). Eight linkage groups have been defined and rudimentary genetic maps are available (Debets *et al.* 1993) as well as an electrophoretic karyotype (Verdoes *et al.* 1994).

5.2.1 Gene manipulations

Transformation techniques used in *A. niger* are almost the same as those available for *A. oryzae* as described earlier. Recently, de Groot *et al.* developed a novel transformation tool for filamentous fungi including *A. niger*, using *Agrobacterium tumefaciens* Ti plasmid (de Groot *et al.* 1998). *A. tumefaciens* transfers part of its Ti plasmid, the T-DNA, to plant cells during tumorigenesis, and it is routinely used for genetic modification of a wide range of plant species. *A. tumefaciens* can also transfer its T-DNA efficiently to the filamentous fungus *Aspergillus awamori*, demonstrating DNA transfer between a prokaryote and a filamentous fungus. de Groot *et al.* transformed both protoplasts and conidia with frequencies that were improved up to 600-fold as compared with conventional techniques for transformation of *A. awamori* protoplasts. The majority of the transformants contained a single T-DNA copy randomly integrated at a chromosomal locus. The T-DNA is integrated into the *A. awamori* genome in a manner similar to that described for plants. The T-DNA dependent tranformation system can be applicable to a variety of filamentous fungi, including *A. niger*, *Fusarium venenatum*, *Trichoderma reesei*, *Colletotrichum gloeosporioides*, *Neurospora crassa*, and the mushroom *Agaricus bisporus* (de Groot *et al.* 1998).

5.2.2 Genome sequencing

Wild-type strains of *A. niger* have the capacity of secreting large amounts of various enzymes and are potentially suitable host strains for homologous and heterologous gene expression. Since *A. niger* also industrially supplies citric acid, the species is believed to be a potent producer of various metabolites available for commodity and specialty chemicals including pharmaceuticals using metabolic engineering technology. Although high production levels of homologous proteins originating from host strains can be readily achieved, production levels of heterologous proteins are several orders of magnitude lower. Moreover, since *Aspergilli* fungi including *A. niger* show different morphogenesis in the fermentation process ranging from distinct spherules (pellets) to long hyphae, it causes high viscosity and limitation of oxygen transfer in fermenters, resulting in reduced enzyme productivity. In order to rationalize and facilitate improvement of stains and processes as well as to discover potential new products, *A. niger* genome has been sequenced.

DSM Research, a division of DSM N.V. (Amstèrdam: DSM.ASX), has commissioned the Dutch and German genomics consortium Gene Alliance (Geleen, The Netherlands and Hilden, Germany) to determine the DNA sequence of *A. niger* (http://www.gene-alliance.com/). The Gene Alliance, which consists of AGOWA (Berlin) Biomax Informatics (Munich), GATC (Konstanz), MediGenomix (Munich), and QIAGEN GmbH (Hilden) and coordinated by QIAGEN, has completed the DNA sequencing in 2001 employing a BAC by BAC sequencing approach. The 7.5X coverage random sequencing of selected large "insert BACS" allowed assembly of the 8 linkage groups into 19 large supercontigs that contained only small sequence gaps (van Peij *et al.* 2002). After computational gene finding, presence of over 14,000 genes in 37.5 Mb was predicted. The genome was annotated by Biomax Informatix (www.biomax.de), using the "GeneReliancea" bioinformatics system. The GeneReliance system is based on the "Pedant-Proa" sequence analysis software developed at Biomax. The system identifies all genetic elements within DNA sequences and assigns functional classification of all open reading frames (ORFs) which gives DSM full use of the data. DSM is offering to make the results available to commercial partners within its existing licensing arrangements. In addition, a low-barrier access program has been made available to academic organizations from 1 January 2002 (van Peij *et al.* 2002).

Genencor has also announced gaining access to the *A. niger* genome sequence data of Integrated Genomics (http://www.genencor.com/webpage_templates/sec.php3?

page_name=pr_1001434970). The details have not been made publicly available. Genencor is also sponsoring a cDNA sequencing project for *A. niger* at the North Carolina State University.

6. CONCLUSION

Studies on economically important fungi at the genomic scale are an innovative strategy to unravel the mystery of mycotoxin biosynthesis and also to help better understand the biology, evolution, biochemical function and genetic regulation of the genes in fungal systems. Expressed Sequence Tag (EST) technology allows rapid identification of the majority, if not all, of the genes expressed in fungal genomes and leads to better understanding of their functions, regulation, coordination of gene expression in response to internal and external factors, the relationship between primary and secondary metabolism, plant-fungal interaction and fungal pathogenicity as well as evolutionary biology. A microarray, made from the EST sequences, can be used to detect a whole set of genes expressed under specific environmental conditions. This technology allows us to study, simultaneously, a complete set of fungal genes that are responsible for or related to mycotoxin production. The genomics of mycotoxigenic, industrial and pathogenic fungi are expected to provide valuable information on turning on and off toxin production in fungal system (Table 2). It will provide vital clues for identifying antifungal gene(s), eliminating mycotoxin contamination of pre-harvest crops, improving high efficiency production of industrial enzymes and accelerating drug development in the near future.

REFERENCES

Adams TH and Yu JH (1998). Coordinate control of secondary metabolite production and asexual sporulation in *Aspergillus nidulans*. Curr Opin Microbiol 1:674-677.

Alexander NJ, McCormick SP, and Hohn TM (1999). *TRI12*, a trichothecene efflux pump from Fusarium sporotrichioides: gene isolation and expression in yeast. Mol Gen Genet 261:977-984.

Beffa T, Staib F, Lott Fischer J, Lyon PF, Gumowski P, Marfenina OE, Dunoyer-Geindre S, Georgen F, Roch-Susuki R, Gallaz L, and Latge JP (1988). Mycological control and surveillance of biological waste and compost. Med Mycol 36 Suppl 1:137-145. Review.

Bennett JW (1987). Mycotoxigenic fungi in southern crops: interaction of fungal genetics and environmental factors that influence selection of toxin-producing populations. Final Report Cooperative Agreement 58-7B30-3-556.

Bennett JW (1997a). Open letter to fungal researchers. Fungal Genet Biol 21:22.

Bennett JW (1997b). White paper: genomics for filamentous fungi. Fungal Genet Biol 21:3-7.

Bennett JW and Arnold J (2001). Genomics for fungi. In: Howard/Gow, eds. The Mycota VIII, Biology of the Fungal Cell. Berlin: Springer-Verlag, Heidelberg pp 267-297.

Bennett JW and Klich MA (1992). *Aspergillus* - Biology and Industrial Applications. Stoneham, Massachussetts: Butterworth-Heinemann.

Berka RM, Ward M, Wilson LJ, Hayenga KJ, Kodama KH, Carlomagno LP, and Thompson SA (1990). Molecular cloning and deletion of the gene encoding aspergillopepsin A from *Aspergillus awamori*. Gene 86:153-162.

Betina V (1989). Mycotoxins: Chemical, Biological and Environmental Aspects. Amsterdam-Oxford-New York, Tokyo: Elsevier Publisher.

Bhatnagar D, Yu J, and Ehrlich KC (2002a). Toxins of filamentous fungi. In: M Breitenbach, R Crameri, and S Lehrer, eds. Fungal Allergy and Pathogenicity. Chem Immunol. Vol 81, Basel, Karger, pp167-206.

Bhatnagar D, Yu J, Cleveland TE (2002b). Applying the genomic wrench - New tool for an old problem. Mycopathologia 155:159.

Blackwell BA, Miller JD, and Savard ME (1994). Production of carbon 14-labeled fumonisin in liquid culture. J AOAC Int 77:506-511.

Brandwagt BF, Mesbah LA, Takken FL, Laurent PL, Kneppers TJ, Hille J, and Nijkamp HJ (2000). A longevity assurance gene homolog of tomato mediates resistance to *Alternaria alternata* f. sp. lycopersici toxins and fumonisin B1. Proc Natl Acad Sci USA 97:4961-4966.

Branham BE and Plattner RD (1993). Alanine is a precursor in the biosynthesis of fumonisin B_1 by *Fusarium moniliforme*. Mycopathologia 124:99-104.

Brody H and Carbon J (1989). Electrophoretic karyotype of *Aspergillus nidulans*. Proc Natl Acad Sci USA 86:6260-6263.

Brown DW, Yu J-H, Kelkar HS, Fernandes M, Nesbitt TC, Keller NP, Adams TH, and Leonard TJ (1996). Twenty-five coregulated transcripts define a sterigmatocystin gene cluster in *Aspergillus nidulans*. Proc Natl Acad Sci USA 93:1418-1422.

Brown RL, Bhatnagar D, Cleveland TE, and Cary JW (1998). Recent advances in preharvest prevention of mycotoxin contamination. In: KK Sinha and D Bhatnagar, eds. Mycotoxins in Agriculture and Food Safety. New York: Marcel Dekker, pp 351-379.

Brown DW, McCormick SP, Alexander NJ, Proctor RH, and Desjardins AE (2002). Inactivation of a cytochrome P-450 is a determinant of trichothecene diversity in Fusarium species. Fungal Genet Biol 36:224-233.

Brown DW, Proctor RH, Dyer RB, and Plattner RD (2003). Unpublished observations.

Busby WF Jr and Wogan GN (1981). Aflatoxins. In: RC Shank, ed. Mycotoxins and N-nitrosocompounds: Environmental Risks. Vol 2, Boca Raton: CRC Press, pp 3-45.

Caldas ED, Sadilkova K, Ward BL, Jones AD, Winter CK, and Gilchrist DG (1998). Biosynthetic studies of fumonisin B_1 and AAL toxins. J Agric Food Chem 46:4734-4743.

Casadevall A and Pirofski L-A (1999). Host-pathogen interactions: Redefining the basic concepts of virulence and pathogenicity. Infect Immun 67:3703-3713.

Chang JM, Oyaizu H, and Sugiyama J (1991). Phylogenetic relationship among eleven selected species of *Aspergillus* and associated teleomorphic genera estimated from 18S ribosomal RNA partial sequences. J Gen Appl Microbiol 37:289-308.

Chang PK, Skory CD, and Linz JE (1992). Cloning of a gene associated with aflatoxin B1 biosynthesis in *Aspergillus parasiticus*. Curr Genet 21:231-233.

Chang P-K, Cary JW, Bhatnagar D, Cleveland TE, Bennett JW, Linz JE, Woloshuk CP, and Payne GA (1993). Cloning of the *Aspergillus parasiticus apa-2* gene associated with the regulation of aflatoxin biosynthesis. Appl Environ Microbiol 59:3273-3279.

Chang P-K, Cary JW, Yu J, Bhatnagar D, and Cleveland TE (1995a). *Aspergillus parasiticus* polyketide synthase gene, *pksA,* a homolog of *Aspergillus nidulans wA,* is required for aflatoxin B_1. Mol Gen Genet 248:270-277.

Chang P-K, Ehrlich KC, Yu J, Bhatnagar D, and Cleveland TE (1995b). Increased expression of *Aspergillus parasiticus aflR*, encoding a sequence-specific DNA binding protein, relieves nitrate inhibition of aflatoxin biosynthesis. Appl Environ Microbiol 61:2372-2377.

Chang P-K, Yu J, Bhatnagar D, and Cleveland TE (1999a). Repressor-AFLR interaction modulates aflatoxin biosynthesis in *Aspergillus parasiticus*. Mycopathologia 147:105-112.

Chang P-K, Yu J, Bhatnagar D, and Cleveland TE (1999b). The carboxy-terminal portion of the aflatoxin pathway regulatory protein *AFLR* of *Aspergillus parasiticus* activates *GAL1::lacZ* gene expression in *Saccharomyces cerevisiae*. Appl Environ Microbiol 65:2058-2512.

Chang P-K and Yu J (2002). Characterization of a partial duplication of the aflatoxin gene cluster in *Aspergillus parasiticus* ATCC 56775. Appl Microbiol Biotechnol 58:632-636.

Chigira Y, Abe K, Gomi K, and Nakajima T (2002). *chsZ*, a gene for a novel class of chitin synthase from *Aspergillus oryzae*. Curr Genet 41:261-267.

Chu FS (1991). Mycotoxins: food contamination, mechanism, carcinogenic potential and preventive measures. Mutat Res 259:291-306.

Cleveland TE, Cary JW, Brown RL, Bhatnagar D, Yu J, Chang PK, Chlan CA, and Rajasekaran K (1997). Use of biotechnology to eliminate aflatoxin in preharvest crops. Bull Inst Compr Agr Sci Kinki Univ 5:75-90.

Cole RJ and Cox EH (1987). Handbook of Toxic Fungal Metabolites. New York: Academic Press.

Cullen JM, Ruebrier BH, Hsieh LS, Hyde DM, and Hsieh DP (1987). Carcinogenicity of dietary aflatoxin M I in male Fischer rats compared to aflatoxin B_1. Cancer Res 47:1913-1917.

Dasbach EJ, Davies GM, and Teutsch SM (2000). Burden of aspergillosis-related hospitalizations in the United States. Clin Infect Dis 31:1524-1528.

de Groot MJ, Bundock P, Hooykaas PJ, and Beijersbergen AG (1998). *Agrobacterium tumefaciens*-mediated transformation of filamentous fungi. Nat Biotechnol 16:839-842.

De Nijs M, Rombouts F, and Notermans S (1996). *Fusarium* molds and their mycotoxins. J Food Saf 16:15-58.

Debets F, Swart K, Hockstra RF, and Bos CJ (1993). Genetic maps of eight linkage groups of *Aspergillus niger* based on mitotic mapping. Curr Genet 23:47-53.

Denning DW (1996). Therapeutic outcome of invasive aspergillosis. Clin Infect Dis 23:608-615.

Denning DW (1998). Invasive aspergillosis. Clin Infect Dis 26:781-805.

Desjardins AE and Proctor RH (1999). Biochemistry and genetics of *Fusarium* toxins. In: B Summerell, JF Leslie, D Backhouse, WL Bryden, and LW Burgess, eds. Fusarium: Paul E. Nelson Symposium. St Paul: APS Press,

Desjardins AE, Hohn TM, and McCormick SP (1993). Trichothecene biosynthesis in *Fusarium* species: chemistry, genetics, and significance. Microbiol Rev 57:595-604.

Desjardins AE, Proctor RH, Bai G, McCormick SP, Shaner G, Buechley G, and Hohn TM (1996). Reduced virulence of trichothecene-nonproducing mutants of *Gibberella zeae* in wheat field tests. Mol Plant-Microbe Interact 9:775-781.

Dujon B (1996). The yeast genome project: what did we learn? Trends Genet 12:263-270.

Dunn-Coleman N and Prade R (1998). Toward a global filamentous fungus genome sequencing effort. Nature Biotech 16:5.

Eaton DL and Groopman JD (1994). The Toxicology of Aflatoxins-Human Health. Veterinary and Agricultural Significance. New York: Academic Press.

Ehrlich KC, Montalbano BG, Bhatnagar D, and Cleveland TE (1998). Alteration of different domains in AFLR affects aflatoxin pathway metabolism in *Aspergillus parasiticus* transformants. Fungal Genet Biol 23:279-287.

Ehrlich KC, Cary JW, and Montalbano BG (1999a). Characterization of the promoter for the gene encoding the aflatoxin biosynthetic pathway regulatory protein AFLR. Biochim Biophys Acta 1444:412-417.

Ehrlich KC, Montalbano BG, and Cary JW (1999b). Binding of the C6-zinc cluster protein, AFLR, to the promoters of aflatoxin pathway biosynthesis genes in *Aspergillus parasiticus*. Gene 230:249-257.

Fernandes M, Keller NP, and Adams TH (1998). Sequence-specific binding by *Aspergillus nidulans AflR*, a C6 zinc cluster protein regulating mycotoxin biosynthesis. Mol Microbiol 28:1355-1365.

Flaherty JE and Payne GA (1997). Overexpression of *aflR* leads to upregulation of pathway gene expression and increased aflatoxin production in *Aspergillus flavus*. Appl Environ Microbiol 63:3995-4000.

Foutz KR, Woloshuk CP, and Payne GA (1995). Cloning and assignment of linkage group loci to a karyotypic map of the filamentous fungus *Aspergillus flavus*. Mycologia 87:787-794.

Frisvad JC and Samson RA (1999). New producers of aflatoxin. In: RA Samson and JI Pitt, eds. Interaction of molecular and morphological approaches to *Aspergillus* and *Penicillium* taxonomy. Harwood, Reading.

Galvano F, Galofaro V, and Galvano G (1996). Occurrence and stability of aflatoxin MI in milk and milk products: A worldwide review. J Food Protect 59:1079-1090.

Gems D, JohnstoneIL, and Clutterbuck AJ (1991). An autonomously replicating plasmid transforms *Aspergillus nidulans* at high frequency. Gene 98:61-67.

Godfrey T and West S (1996). Industrial Enzymology. 2nd ed. New York: Stockton Press.

Goffeau A, Barrell BG, Bussey H, Davis RW, Dujon B, Feldmann H, Galibert F, Hoheisel JD, Jacq C, Johnston M, Louis EJ, Mewes HW, Murakami Y, Philippsen P, Tettelin H, and Oliver SG (1996). Life with 6000 genes. Science. 274(5287):546, 563-567. Review.

Goffeau A, Aert R, Agostini-Carbone ML, Ahmed A, Aigle M, Alberghina L, Albermann K, Albers M, *et al.* (1997a). The yeast genome directory. Nature 387(6632 Suppl):5.

Goffeau A, Park J, Paulsen IT, Jonniaux JL, Dinh T, Mordant P, and Saier MH Jr (1997b). Multidrug-resistant transport proteins in yeast: complete inventory and phylogenetic characterization of yeast open reading frames with the major facilitator superfamily. Yeast 13:43-54.

Goldblatt LA (1969). Aflatoxin-Scientific Background, Control and Implications. New York: Acaderitic Press.

Gomi K, Iimura Y, and Hara S (1987). Integrative transformation of *Aspergillus oryzae* with a plasmid containing the *Aspergillus nidulans argB* gene. Agric Biol Chem 51:2549-2555.

Gomi K, Kitamoto K, and Kumagai C (1992). Transformation of the industrial strain of *Aspergillus oryzae* with the homologous *amdS* gene as a dominant selectable marker. J Ferment Technol 74:389-391.

Goto T, Wicklow DT, and Ito Y (1996). Aflatoxin and cyclopiazonic acid production by sclerotium-producing *Aspergillus tamarii* strain. Appl Environ Microbiol 62:4036-4038.

Gross C, Kelleher M, Vyer VR, Brown PO, and Winge DR (2000). Identification of copper regulation of *Saccharomyces cerevisiae* by DNA microarrays. J Biol Chem 275:32310-32316.

Hagiwara H, Kunihiro S, Nakajima K, Sano M, Masaki H, Yamamoto M, Pak JW, Zhang Y, Takase K, Kuwabara I, Maruyama IN, and Machida M (2002). Affinity selection of DNA-binding proteins from yeast genomic DNA libraries by improved lambda phage display vector. J Biochem (Tokyo) 132:975-982.

Hall AJ and Wild CP (1994). Epidemiology of aflatoxin-related disease. In: DL Eaton and JD Groopman, eds. The Toxicology of Aflatoxins: Human Health, Veterinary, and Agricultural Significance. San Diego: Academic Press, pp 233-258.

Hall LA and Denning DW (1994). Oxygen requirements of *Aspergillus* species. J Med Microbiol 41:311-315.

Harris, LJ, Desjardins AE, Plattner RD, Nicholson P, Butler G, Young JC, Weston G, Proctor RH, and Hohn TM (1999). Possible role of trichothecene mycotoxins in virulence of *Fusarium graminearum* on maize Plant Dis 83:954-960.

Hata Y, Kitamoto K, Gomi K, Kumagai C, and Tamura G (1992). Functional elements of the promoter region of the *Aspergillus oryzae glaA* gene encoding glucoamylase. Curr Genet 22:85-91.

Hatamoto O, Matsushima K, Takahashi O, Umitsuki G, and Abe K (1999a). Modification of genome target site of multinucleated cell, targeting vector used thereof and modified multinucleated cell. In patent #: JP19990228915.

Hatamoto O, Sekine H, Nakano E, and Abe K (1999b). Cloning and expression of a cDNA encoding the laccase from *Schizophyllum commune*. Biosci Biotechnol Biochem 63:58-64.

Howard PC, Eppley RM, Stack ME, Warbritton A, Voss KA, Lorentzen RJ, Kovach RM, and Bucci TJ (2001). Fumonisin B1carcinogenicity in a two-year feeding study using F344 rats and B6C3F1 mice. Environ Health Perspect 109 S2:277-282.

Huang X and Madan A (1999). CAP3: A DNA sequence assembly program. Genome Res 9:868-877.

Iimura Y, Gomi K, Uzu H, and Hara S (1987). Transformation of *Aspergillus oryzae* through plasmid-mediated complementation of the methionine-auxotrophic mutation. Agric Biol Chem 51:323-328.

Keller NP, Cleveland TE, and BhatnagarD (1992). Variable electrophoretic karyotypes of members of *Aspergillus* section Flavi. Curr Genet 21:371-375.

Keller NP and Hohn TM (1997). Metabolic pathway gene clusters in filamentous fungi. Fungal Genet Biol 21:17-29.

Kimura M, Matsumoto G, Shingu Y, Yoneyama K, and Yamaguchi I (1998). The mystery of the trichothecene 3-O-acetyltransferase gene. Analysis of the region around *Tri101* and characterization of its homologue from *Fusarium sporotrichioides*. FEBS Lett 435:163-168.

Kitamoto K, Kimura K, Gomi K, and Kumagai C (1994). Electrophoretic karyotype and gene assignment to chromosomes of *Aspergillus oryzae*. Biosci Biotechnol Biochem 58:1467-1470.

Klich MA, Mullaney EJ, Daly CB, and Cary JW (2000). Molecular and physiological aspects of aflatoxin and sterigmatocystin biosynthesis by *Aspergillus tamarii* and *A. ochraceoroseus*. Appl Microbiol Biotechnol 53:605-609.

Kubodera T, Yamashita N, and Nishimura A (2000). Pyrithiamine resistance gene (*ptrA*) of *Aspergillus oryzae*: cloning, characterization and application as a dominant selectable marker for transformation. Biosci Biotechnol Biochem 64:1416-1421.

Lancaster MD, Jenkins FP, and Philip JM (1961). Toxicity associated with certain samples of groundnuts. Nature 192:1095-1096.

Latgé JP (1999). *Aspergillus fumigatus* and aspergillosis. Clin Microbiol Rev 12:310-350.

Lee T, Han Y-K, Kim K-H, Yun S-H, and Lee Y-W (2002). *Tri13* and *Tri7* determine deoxynivalenol- and nivalenol-producing chemotypes of *Gibberella zeae*. Appl Environ Microbiol 68:2148-2154.

Machida M (2002). Progress of *Aspergillus* genomics. Advances in Applied Microbiology. In: AI Laskin, JW Bennett, and GM Gadd, eds. New York: Academic Press, 51:81-106.

Maertens J, Raad I, Sable CA, Ngai A, Berman R, Patterson T F, Denning DW, and Walsh T (2000). Multicenter, non-comparative study to evaluate safety and efficacy of caspofungin (CAS) in adults with invasive aspergillosis refractory or intolerant to amphotericin B, AMB lipid formulations or azoles. Interscience Conference on Antimicrobial Agents and Chemotherapy.

Marahiel MA, Stachelhaus T, and Mootz HD (1997). Modular peptide synthetases involved in nonribosomal peptide synthesis. Chem Rev 97:2651-2673.

Marasas WFO (2001). Discovery and occurrence of the fumonisins:a historical perspective. Environ Health Perspect 109 S2:239-243.

McCormick SP and Alexander NJ (2002). *Fusarium Tri8* encodes a trichothecene C-3 esterase. Appl Environ Microbiol 68:2959-2964.

McCormick SP, Alexander NJ, Trapp SE, and Hohn TM (1999). Disruption of *TRI101*, the gene encoding trichothecene 3-O- acetyltransferase, from *Fusarium sporotrichioides* Appl Environ Microbiol 65:5252-5256.

Meyers DM, Obrian G, Du WL, Bhatnagar D, and Payne GA (1998). Characterization of *aflJ*, a gene required for conversion of pathway intermediates to aflatoxin. Appl Environ Microbiol 64:3713-3717.

Minto RE and Townsend CA (1997). Enzymology and molecular biology of aflatoxin biosynthesis. Chem Rev 97:2537-2555.

Moore CB, Sayers N, Mosquero J, Slaven J, and Denning DW (2000). Antifungal drug resistance in *Aspergillus*. J Infection 41:203-220.

Moss MO (1989). Mycotoxins of Aspergillus and other filamentous fungi. Soc. Appl Bacteriol Symp Ser 18:69S-87S.

Munkvold GP and Desjardins AE (1997). Fumonisins in maize: can we reduce their occurrence? Plant Dis 81:556-565.

Nakajima K, Kunihiro S, Sano M, Zhang Y, Eto S, Chang YC, Suzuki T, Jigami Y, and Machida M (2000). Comprehensive cloning and expression analysis of glycolytic genes from the filamentous fungus, *Aspergillus oryzae*. Curr Genet 37:322-327.

Nelson PE, Desjardins AE, and Plattner RD (1993). Fumonisins, mycotoxins produced by *Fusarium* species: biology, chemistry, and significance. Annu Rev Phytopathol 31:233-252.

Ng TTC, Robson GD, Denning DW (1994). Hydrocortisone-enhanced growth of *Aspergillus* spp: implications for pathogenesis. Microbiol 140:2475-2480.

O'Brian GR and Payne GA (2001). Identification of differentially expressed genes during aflatoxin biosynthesis in *Aspergillus flavus*. Proceedings of the 21st Fungal Genetics Conference, Asilomar, California (abs).

Pasamontes L, Haiker M, Wyss M, Tessier M, and van Loon AP (1997). Gene cloning, purification, and characterization of a heat-stable phytase from the fungus *Aspergillus fumigatus*. Appl Environ Microbiol 63:1696-1700.

Payne GA, Nystrom GJ, Bhatnagar D, Cleveland TE, and Woloshuk CP (1993). Cloning of the *afl-2* gene involved in aflatoxin biosynthesis from *Aspergillus flavus*. Appl Environ Microbiol 59:156-162.

Payne GA (1998). Process of contamination by aflatoxin-producing fungi and their impacts on crops. In: KK Sinha and D Bhatnagar, eds. Mycotoxins in Agriculture and Food Safety, New York: Marcel Dekker, pp 279-306.

Payne GA and Brown MP (1998). Genetics and physiology of aflatoxin biosynthesis. Annu Rev Phytopathol 36:329-362.

Payne GA, Fakhoury A, and O'Brian GR (2002). Gene expression analysis of *Aspergillus flavus* and *A. parasiticus*. Proceedings of the 2nd Fungal Genomics, 3rd Fumonisin Elimination and 15th Aflatoxin Elimination Workshops, San Antonio, Texas (abs).

Pitt JI, Samson RA, and Frisvad JC (2000). List of accepted species and their synonyms in the family Trichocomaceae. In: Integration of modern taxonomic methods for *Penicillium* and *Aspergillus* classification. In: RA Samson and JI Pitt, eds. Amsterdam: Harwood Publishers, pp 9-49.

Ponikau JU, Sherris DA, Kern EB, Homburger HA, Frigas E, Gaffey TA, and Roberts GD (1999). The diagnosis and incidence of allergic fungal sinusitis. Mayo Clin Proc 74:877-884.

Pontecorvo G, Roper JA, Hemmons LM, Macdonald KD, and Bufton AWJ (1952). The genetics of *Aspergillus nidulans*. Adv Genet 5:141-238.

Proctor RH, Hohn TM, McCormick SP, and Desjardins AE (1995). *TRI6* encodes an unusual zinc finger protein involved in regulation of trichothecene biosynthesis in *Fusarium sporotrichioides*. Appl Environ Microbiol 61:1923-1930.

Proctor RH, Desjardins AE, Plattner RD, and Hohn TM (1999). A polyketide synthase gene required for biosynthesis of fumonisin mycotoxins in *Gibberella fukiuroi* mating population A. Fungal Genet Biol 27:100-112.

Proctor RH, Brown DW, Plattner RD, and Desjardins AE (2003). Co-expression of 15 contiguous genes delineates a fumonisin biosynthetic gene cluster in *Gibberella moniliformis*. Fungal Genet Biol 38:237-249.

Pusateri AE and Kenison DC (1993). Measurement of zeranol in plasma from three blood vessels in steers implanted with zeranol. J Anim Sci 71:415-419.

Rankin NE (1953). Disseminated aspergillosis and moniliasis associated with agranulocytosis and antibiotic therapy. Br Med J 25:918-919.

Ruiter-Jacobs YM, Broekhuijsen M, Unkles SE, Campbell EI, Kinghorn JR, Contreras R, Pouwels PH, and van den Hondel CA (1989). A gene transfer system based on the homologous *pyrG* gene and efficient expression of bacterial genes in *Aspergillus oryzae*. Curr Genet 16:159-163.

Sano M, Ohyama A, Takase K, Yamamoto M, and Machida M (2001). Electrophoretic mobility shift scanning using an automated infrared DNA sequencer. Biotechniques 31:1056-1058, 1060, 1062.

Seo J-A, Proctor RH, and Plattner RD (2001). Characterization of four clustered and coregulated genes associated with fumonisin biosynthesis in *Fusarium verticillioides*. Fungal Genet Biol 34:155-165.

Sharma RP and Kim Y-W (1991). In: RP Sharma and DK Salunkhe, eds. Mycotoxins and Phytoalexins. Boca Raton: CRC Press, pp 339-359.

Sharma RP and Salunkhe DK (1991). Mycotoxin and Phytoalexins in Human and Animal Health. Boca Raton: CRC Press.

Shim W-B and Woloshuk CP (2001). Regulation of fumonisin B1 biosynthesis and conidiation in *Fusarium verticillioides* by a cyclin-like (C-type) gene, *FCC1*. Appl Environ Microbiol 67:1607-1612.
Skory CD, Chang P-K, Cary J, and Linz JE (1992). Isolation and characterization of a gene from *Aspergillus parasiticus* associated with the conversion of versicolorin A to sterigmatocystin in aflatoxin biosynthesis. Appl Environ Microbiol 58:3527-3537.
Soares MB, Bonaldo MF, Jelene P, Su L, Lawton L, and Efstratiadis A (1994). Construction and characterization of a normalized cDNA library. Proc Natl Acad Sci USA 91:9228-9232.
Squire RA (1989). Ranking animal carcinogens: a proposed regulatory approach. Science 214:887-891.
Stevens DA, HJ Schwartz, JY Lee, BL Moskovitz, DC Jerome, A Catanzaro, DM Bamberger, AJ Weinmann, CU Tuazon, MA Judson, TA Platts-Mills, and AC DeGraff Jr (2000). A randomized trial of itraconazole in allergic bronchopulmonary aspergillosis. N Engl J Med 342:756-762.
Tag AG, Garifullina GF, Peplow AW, Ake C Jr, Phillips TD, Hohn TM, and Beremand MN (2001). A novel regulatory gene, *Tri10*, controls trichothecene toxin production and gene expression. Appl Environ Microbiol 67:5294-302.
Tailor MJ and Richardson T (1979). Applications of microbial enzymes in food systems and in biotechnology. Adv Appl Microbiol 25:7-35.
Timberlake WE (1990). Molecular genetics of Aspergillus development. Annu Rev Genet 24:5-36.
Townsend CA (1997). Progress towards a biosynthetic rationale of the aflatoxin pathway. Pure Appl Chem 58:227-238.
Tudzynski B (2002). Biosynthesis of gibberellins in *Gibberella fujikuroi*: biomolecular aspects. Appl Microbiol Biotechnol 52:298-310.
Unkles SE, Campbell EI, Ruiter-Jacobs YM, Broekhuijsen M, Macro JA, Carrez D, Contreras R, van den Hondel CA, and Kinghorn JR (1989). The development of a homologous transformation system for *Aspergillus oryzae* based on the nitrate assimilation pathway: A convenient and general selection system for filamentous fungal transformation. Mol Gen Genet 218:99-104.
Ushijima S and Nakadai T (1987). Breeding by protoplast fusion of *koji* mold *Aspergillus sojae*. Agric Biol Chem 51:1051-1057.
Ushijima S, Nakadai T and Uchida K (1990). Breeding of new *koji*-molds through interspecific hybridization between *Aspergillus oryzae* and *Aspergillus sojae* by protoplast fusion. Agric Biol Chem 54:1667-1676.
Van Egmond HP (1989). Current situation on regulations for mycotoxins: Overview of tolerances and status of standard methods of sampling and analysis. Food Addit Contam 6:134-188.
van Peij N, Pel H, van den Hombergh H, van den Berg M, van Rooijen R, van Ooyen A, and van Dijk A (2002). The DSM *Aspergillus niger* genome project. In 6th European Conference on Fungal Genetics. Pisa, Italy.
Verdoes JC, Calil MR, Punt PJ, Debets F, Swart K, Stouthamer AH, and van den Hondel CA (1994). The complete karyotype of Aspergillus niger: the use of introduced electrophoretic mobility variation of chromosomes for gene assignment studies. Mol Gen Genet 244:75-80.
Wang E, Norred WP, Bacon CW, Riley RT, and Merrill AH Jr (1991). Inhibition of sphingolipid biosynthesis by fumonisins. Implications for diseases associated with *Fusarium moniliforme*. J Biol Chem 266:14486-14490.
Woloshuk CP, Foutz KR, Brewer JF, Bhatnagar D, Cleveland TE, and Payne GA (1994). Molecular characterization of *aflR*, a regulatory locus for aflatoxin biosynthesis. Appl Environ Microbiol 60:2408-2414.
Wu TS and Linz JE (1993). Recombinational inactivation of the gene encoding nitrate reductase in *Aspergillus parasiticus*. Appl Environ Microbiol 59:2998-3002.
Yabe K, Matsushima K, Koyama T, and Hamasaki T (1998). Purification and characterization of *O*-methyltransferase I involved in conversion of demethylsterigmatocystin to sterigmatocystin and of dihydrodemethylsterigmatocystin to dihydrosterigmatocystin during aflatoxin biosynthesis. Appl Environ Microbiol 64:166-171.
Yu J, Cary JW, Bhatnagar D, Cleveland TE, Keller NP, and Chu FS (1993). Cloning and characterization of a cDNA from *Aspergillus parasiticus* encoding an *O*-methyltransferase involved in aflatoxin biosynthesis. Appl Environ Microbiol 59:3564-3571.
Yu J, Chang P-K, Cary JW, Wright M, Bhatnagar D, Cleveland TE, Payne GA, and Linz JE (1995a). Comparative mapping of aflatoxin pathway gene clusters in *Aspergillus parasiticus* and *Aspergillus flavus*. Appl Environ Microbiol 61(6):2365-2371.
Yu J, Chang P-K, Payne GA, Cary JW, Bhatnagar D, and Cleveland TE (1995b). Comparison of the *omtA* genes encoding *O*-methyltransferases involved in aflatoxin biosynthesis from *Aspergillus parasiticus* and *A. flavus*. Gene 163:121-125.

Yu J, Chang P-K, Cary JW, Bhatnagar D, and Cleveland TE (1997). *avnA*, a gene encoding a cytochrome P-450 monooxygenase is involved in the conversion of averantin to averufin in aflatoxin biosynthesis in *Aspergillus parasiticus*. Appl Environ Microbiol 63:1349-1356.

Yu J, Chang P-K, Cary JW, Ehrlich KC, Montalbano B, Dyer JM, Bhatnagar D, and Cleveland TE (1998). Characterization of the critical amino acids of an *Aspergillus parasiticus* cytochrome P450 monooxygenase encoded by *ordA* involved in aflatoxin B_1, G_1, B_2, and G_2 biosynthesis. Appl Environ Microbiol 64:4834-4841.

Yu J (2003). Genetics and Biochemistry of Mycotoxin Synthesis. In: DK Arora, ed. Handbook of Fungal Biotechnology, 2nd ed. New York: Marcel Dekker. In press.

Yu J, Chang P-K, Bhatnagar D, and Cleveland TE (2000a). Genes encoding cytochrome P450 and monooxigenase enzymes define one end of the aflatoxin pathway gene cluster in *Aspergillus parasiticus*. Appl Microbiol Biotechnol 53:583-590.

Yu J, Woloshuk CP, Bhatnagar D, and Cleveland TE (2000b). Cloning and characterization of *avfA* and *omtB* genes involved in aflatoxin biosynthesis in three *Aspergillus* species. Gene 248:157-167.

Yu J, Bhatnagar D, and Cleveland TE (2002a). *Aspergillus flavus* genomics for elimination of aflatoxin contamination. Mycopathologia (Special Issue) 155:10.

Yu J, Bhatnagar D, Cleveland TE, and Nierman WC (2002b). *Aspergillus flavus* EST technology and its applications for eliminating aflatoxin contamination. Mycopathologia (Special Issue) 155:6.

Yu J, Whitelaw CA, Bhatnagar D, Cleveland TE, and Nierman WC (2002c). *Aspergillus flavus* EST project – A tool for eliminating aflatoxin contamination. Proceedings of the 2nd Fungal Genomics, 3rd Fumonisin Elimination and 15th Aflatoxin Elimination Workshops, San Antonio, Texas. (Abs).

Yu J, Bhatnagar D, and Ehrlich KC(2002d). Aflatoxin Biosynthesis. Rev Iberoam Microl (RIAM), Spain 19:191-200.

Yu J-H, Butchko RA, Fernandes M, Keller NP, Leonard TJ, and Adams TH (1996). Conservation of structure and function of the aflatoxin regulatory gene *aflR* from *Aspergillus nidulans* and *A. flavus*. Curr Genet 29:549-555.

Zhang Y, Pak JW, Maruyama IN, and Machida M (2000). Affinity selection of DNA-binding proteins displayed on bacteriophage lambda. J Biochem 127:1057-1063.

Applied Mycology & Biotechnology
An International Series. Volume 4. Fungal Genomics

Penicillium Genomics

John C. Royer, Kevin T. Madden, Thea C. Norman, Katherine F. LoBuglio[1]
Microbia, Inc., 320 Bent St., Cambridge MA 02142 U.S.A. (jroyer@microbia.com); [1]Harvard University Herbaria, 22 Divinity Ave., Cambridge, MA 02138 U.S.A.

Penicillium species exhibit wide variability in metabolite production, morphology, pathogenicity and lifestyle. Despite their economic importance, relatively little publicly available genomic analysis has been performed on *Penicillium* species. In this review, we summarize karyotype analysis of *Penicillium* species, and compare penicillin and statin gene clusters of *Penicillium* and closely related *Aspergillus* species. In addition, we present a preliminary genomic comparison between wild type and commercial penicillin producing strains of *P. chrysogenum* generated using genomic fragment microarrays. Finally, we summarize initial results of an EST sequencing project for the human pathogen *P. marneffei*.

1. INTRODUCTION

The genus *Penicillium* comprises a ubiquitous group of fungi commonly found in soil, and as decomposers of various types of organic matter (Pitt 1985). The genus is characterized by the production of asexual spores (conidia) from verticels of phialides supported on a conidiophore of varying complexity termed the penicillus (Raper and Thom 1949; Pitt 1979). Pitt and Samson (1993) list 162 *Penicillium* species for which only asexual spores are produced, and 61 species that have a sexual cycle (either a *Talaromyces* or *Eupenicillium* teleomorph) as well as the asexual *Penicillium* state.

Penicillium species are phylogenetically associated with the order of sexual ascomycete fungi known as the Eurotiales (Geiser and LoBuglio 2001). Phylogenetic analyses of DNA sequence data has demonstrated that anamorphic (asexual) genera such as *Penicillium* and its close relatives in the genus *Aspergillus* have not evolved independently as asexual lineages but rather have evolved multiple times from meiotic genera (Geiser *et al.* 1996; LoBuglio *et al.* 1993; Peterson *et al.* 1993). It has been estimated that *Aspergillus* and *Penicillium* diverged at least 60 million years ago based on 18S ribosomal DNA analysis under the constraint of a molecular clock (Berbee and Taylor 1993).

Penicillium strains are important producers of secondary metabolites; many of which have been developed into antibiotics and other pharmaceuticals. *Penicillium chrysogenum* has been utilized for commercial production of penicillin for over 50 years. Intensive random mutagenesis and selection, coupled with improvements in fermentation processes have resulted in an increase of approximately 3 orders of magnitude in the titre of penicillin produced by *P. chrysogenum* (Backus and Stauffer 1955; Lein 1986; Demain and Elander 1999). *Penicillium citrinum* is currently utilized for production of the statin compactin, which is a precursor for the cholesterol lowering drug pravastatin (Endo *et al.* 1976; Brown *et al.* 1976). In addition, *Penicillium* strains are important in the food industry, both in food production and food contamination (Pitt 1985).

One member of the genus, *Penicillium marneffei*, has become a well-recognized human pathogen in Southeast Asia (Wong *et al.* 1999). It typically infects immunocompromised individuals afflicted with diseases such as Hodgkin's disease, tuberculosis, or AIDS as well as those undergoing treatment with cortical steroids (Deng and Connor 1985, Hilmardottir *et al.* 1993, Viviani *et al.* 1993). *Penicillium marneffei* appears to be unique among *Penicillium* species in its capability to undergo a temperature-dependent dimorphic growth switch (Andrianopoulis 2002; Garrison and Boyd 1973). A symbiotic interaction occurs with another member of the genus, *Penicillium nodositatum.* This species has been shown to induce root nodule formation on roots of alder trees (Valla *et al.* 1989) and to exist as a neutral microsymbiont within the host tissue (Sequerra *et al.* 1995). Interestingly, RFLP analysis of the nuclear ribosomal DNA region containing the two internal transcribed spacers and 5.8S rRNA gene suggest that *P. nodositatum* is phylogenetically related to the *Penicillium* subgenus *Biverticillium* (Pitt 1979), which includes *P. marneffei* (LoBuglio and Taylor 1995; Sequerra *et al.* 1997). Thus, considerable variability with regard to metabolite production, pathogenicity, morphology, and lifestyle exists within the genus *Penicillium.* Genomic tools offer powerful methods for elucidating the genetic basis for these variations. Despite its importance, genomic analysis of *Penicillium* appears to have lagged behind that of other model, and economically important fungi. In this review, we summarize results on karyotype analysis of various *Penicillium* strains and compare structure of penicillin and statin gene clusters in *Penicillium* and *Aspergillus* strains. We also present preliminary genomic profiling analysis of penicillin production strains of *P. chrysogenum*, and briefly summarize initial results of an EST (expressed sequence tag) sequencing project underway for *P. marneffei.*

2. ELECTROPHORETIC KARYOTYPE AND GENOME SIZE

Electrophoretic karyotype analysis has been performed on a number of *Penicillium* species (Table 1). Most strains examined, including *P. chrysogenum* possess 4 or 5 resolvable bands (Farber and Geisen 2000; Fierro *et al.* 1993; Chavez *et al.* 2001). *Penicillium paxilli* contains at least 6 chromosomes (Itoh *et al.* 1994) while the *P. janthinellum* genome has been resolved into 8-10 bands (Kayser and Shulz 1991). Genome sizes vary from 17.8-26.2 MB (*P. marneffei*) and 22.1 MB (*P. purpurogenum)* to 39-46 MB for *P. janthinellum.*

Table 1. Electrophoretic karyotypes of various *Penicillium* species.

Penicillium Species	Chromosome Number	Chromosome Size Range (Mbp)	Genome Size (Mbp)	Reference
P. chrysogenum	4	6.8-10.4	34.1	Fierro *et al.* 1993
P. notatum	4	5.4-10.8	32.1	Fierro *et al.* 1993
P. janthinellum	6-8	2.0-8.0	39.0-49.0	Kayser and Shulz 1991
P. paxilli	8	2.5-6.0	23.4	Young *et al.* 1998
P. nalgiovense	4	4.1-9.1	26.5	Farber and Geisen 2000
P. purpurogenum	5	2.3-7.1	21.2	Chavez *et al.* 2001
P. marneffei	3-6	2.2-5.0	17.8-26.2	Wong and Yuen, pers. comm., 2003

3. SECONDARY METABOLITE GENE CLUSTERS

Genes involved in the synthesis of complex secondary metabolites are often grouped together as clusters in filamentous fungi (Keller and Hohn 1997). This clustering has been proposed to be a remnant of horizontal gene transfer from prokaryotes, where genes involved in a particular function are often physically linked and transcribed together as operons (Landan *et al.* 1990; Penalva *et al.* 1990; Weigel *et al.* 1988). Clustering of secondary metabolite genes has also been proposed to facilitate coordinate regulation, and to allow for horizontal gene transfer between fungi, which may be particularly important in species that lack a sexual cycle (Keller and Hohn 1997). Walton (2000) proposed a "selfish cluster" hypothesis to explain the evolution and maintenance of secondary metabolite gene clusters in fungi.

3.1 Penicillin Clusters in *Penicillium* and *Aspergillus*

Penicillins and cephalosporins are B-lactam containing antibiotics produced by filamentous fungi from a number of genera. The penicillin gene cluster is composed of 3 genes; ACV synthase (pcbAB), IPN synthase (pcbC) and acyltransferase synthase (penDE) (Diez *et al.* 1990; Barredo *et al.* 1989a; Barredo *et al.* 1989b; Samson *et al.* 1985; Carr *et al.* 1986). These genes are clustered together in an approximately 20 kb region in both *P. chrysogenum* and *A. nidulans* (Fig.1.; Diez *et al.* 1990; MacCabe *et al.* 1990). Gene order and direction of transcription are maintained in these fungi, and protein identity is quite high; approximately 80% for the IPN synthetases. The degree of protein identity between the fungal genes and bacterial IPN synthetases is approximately 60%, and the fungal genes lack introns. These results have been cited as evidence for a horizontal gene transfer of this cluster from bacteria to fungi (Penalva *et al.* 1990; Weigel *et al.* 1988). Recently, Laich *et al.* (2002) demonstrated that the presence of the penicillin gene cluster is variable among *Penicillium* strains used in food production. *Penicillium griseofulvum* contains a complete gene cluster, *P. verrucosom* contains a truncated cluster, while many strains (including *P. roquefortii* and *P. camembertii*) lack any of the cluster genes. Interestingly, the cephalosporin production strain *Acremonium chrysogenum* contains two of the penicillin biosynthetic genes as a cluster. This partial cluster along with a second, separate cluster of genes is involved in synthesis of cephalosporin C.

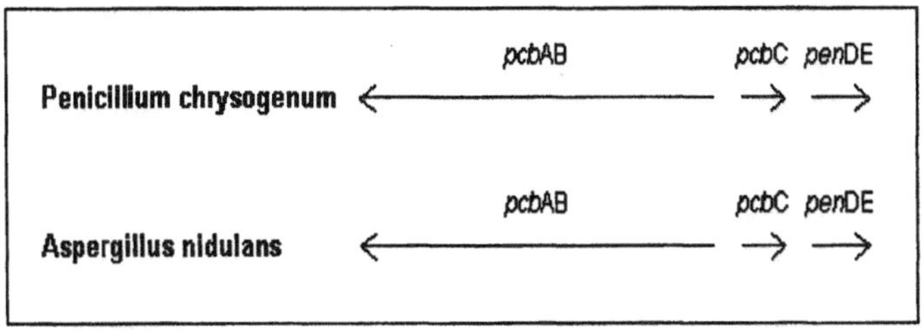

Fig. 1. Genes of the penicillin cluster of *P. chrysogenum* and *A. nidulans*.

Industrial penicillin production strains generated by random mutagenesis have been shown to contain amplifications of a region of either 106.5 or 57.6 kb units, which contain the 3 penicillin biosynthetic genes, and are flanked by conserved hexanucleotide repeats (Fierro *et al.* 1995). Thus, the increased penicillin production capability of these strains is linked with increased copy number of the biosynthetic genes. The variation in presence and copy number of the penicillin cluster in wild type and production strains suggests great plasticity of this cluster (Gutierrez *et al.* 1999; Laich *et al.* 2002).

3.2 Statin Gene Clusters in *Penicillium* and *Aspergillus*

Members of the statin family of secondary metabolites are potent inhibitors of HMGCoA reductase, the key enzyme in cholesterol biosynthesis in humans. These polyketides have found utility for reduction of cholesterol, and natural and modified statins represent multibillion-dollar drugs. *Penicillium citrinum* produces the statin compactin, which is used as a substrate for microbial conversion to the commercial product pravastatin sodium. *Aspergillus terreus* naturally produces the statin lovastatin, which is identical to compactin except that a methyl group is present at the C-6 position (Moore *et al.* 1985). As with many fungal secondary metabolites, the genes for these polyketides are clustered together; and the gene clusters for both compactin and lovastatin have been cloned (Kennedy *et al.* 1999; Abe *et al.* 2002). As expected, many homologous gene products (with 57-75% identity) are shared between the two clusters. Identity between MlcR and LovE, transcription factors that regulate compactin and lovastatin production, respectively, is significantly lower at 34%. The presence of introns, which are lacking in the genes of the penicillin cluster, argues against a "recent" horizontal gene transfer from a prokaryotic source. The low level of homology between the proteins of the two clusters (similar to the 60% identity between the homologous nitrogen regulatory genes Nre of *P. chrysogenum* and AreA of *A. nidulans,* (Haas *et al.* 1995)) suggests that the cluster may have been present in the shared ancestor of these two fungi and has evolved considerably since the divergence of the genera *Penicllium* and *Aspergillus*.

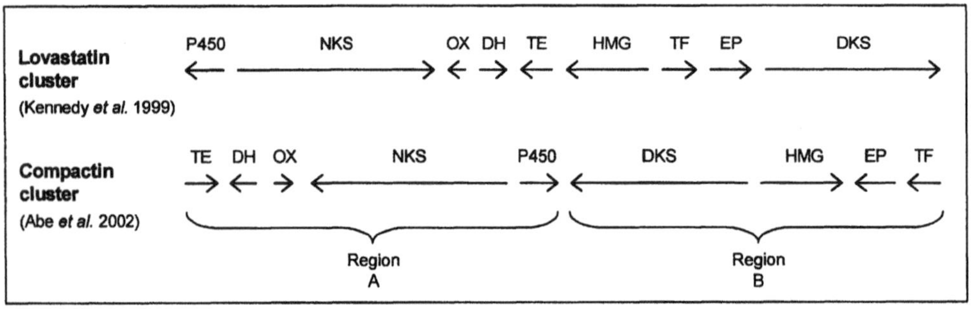

Fig. 2. Comparison of the compactin cluster of *P. citrinum* with the lovastatin cluster of *A. terreus.* Region A is flipped between the two clusters, while the relationship between region B is more complex. P450, P450 monooxygenase; NKS, nonaketide synthase; OX, oxidoreductase; DH, dehydrogenase; TE, transesterase; HMG, HMG-CoA reductase; TF, transcription factor; EP, efflux pump; DKS, diketide synthase.

While gene functions are conserved and the genes are maintained as clusters in these two fungi of differing genera, gene order and direction of transcription have been altered (Fig 2). Interestingly, the 5' portion of the two clusters (comprising 5 genes) maintains synteny if the orientation is flipped. The 3' end of the cluster can be restored to synteny by a more complex flipping and insertion. These results contrast with comparisons of penicillin clusters, where the degree of protein identity, as well as synteny within the cluster is maintained to a higher degree between *Penicillium* and *Aspergillus* species.

4. RANDOM FRAGMENT GENOMIC ARRAYS TO CHARACTERIZE GENOMES OF PENICILLIN PRODUCTION STRAINS

The lineage of strains with varying penicillin production potentials generated during the penicillin strain improvement process represents a unique source of genetic material to examine directed evolution, and to identify genes whose modifications affect metabolite production. DNA microarray technology is a particularly well-suited tool for uncovering variations between closely related strains.

Microarrays are often generated using DNA representing known open reading frames (where genomic sequence is available) or ESTs. Since neither sequence information nor ESTs were available for *P. chrysogenum*, we utilized a random fragment genomic microarray approach to assess the changes that have occurred within the *Penicillium* production strains (Fig. 3, see Askenazi *et al.* 2003 for experimental details). DNA from the progenitor strain (ATCC 9480, NRRL 1951) was isolated, digested to produce fragments of approximately 2 KB, and cloned. Approximately 13,000 PCR products were generated from the resulting fragment library using a common primer pair. In addition, PCR fragments were generated for all known *P. chrysogenum* genes. Resulting PCR products were purified and transferred to 384-well plates to generate the microarray.

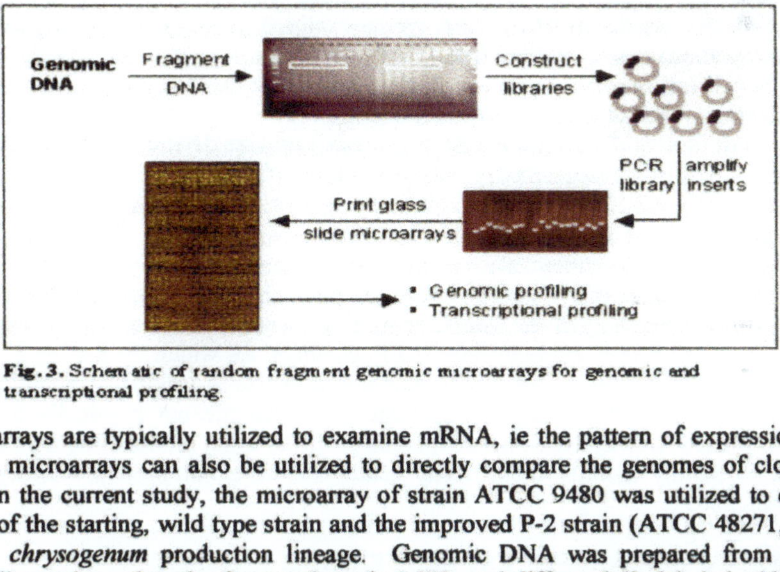

Fig. 3. Schematic of random fragment genomic microarrays for genomic and transcriptional profiling.

Microarrays are typically utilized to examine mRNA, ie the pattern of expression of genes. However, microarrays can also be utilized to directly compare the genomes of closely related strains. In the current study, the microarray of strain ATCC 9480 was utilized to compare the genomes of the starting, wild type strain and the improved P-2 strain (ATCC 48271, Lein 1986) of the *P. chrysogenum* production lineage. Genomic DNA was prepared from each strain, partially digested to a length of approximately 1 KB, and differentially labeled with fluorescent nucleotides (Cy3-dCTP or Cy5-dCTP) using a random priming reaction. Competitive microarray hybridizations were performed using DNA from the starting strain and differentially labeled DNA from the P2 strain. Wild type DNA was also competitively hybridized to itself to determine the experimental reproducibility. Genomic fragments that displayed signal intensities with absolute value log2 ratios >0.8 were considered significantly amplified or deleted, relative to the wild type strain, and were selected for sequence analysis (See Askenazi *et al.* 2003 for experimental details).

A number of genomic fragments were identified which were associated with stronger hybridization signals in the improved P-2 strain. Not surprisingly, sequence analysis revealed that many of these genomic fragments contained genes of the penicillin biosynthetic cluster. These results were consistent with results of previous studies (Fierro *et al.* 1995) in which Southern blot hybridizations and sequencing techniques were used to show that the penicillin biosynthetic cluster has been amplified from 1 copy in the progenitor strain to 5 to 7 copies in the P-2 strain. The genomic fragment microarray approach has, in addition identified at least 2 more genes that are either contained within or closely linked to the 55kb amplified region and could be associated with penicillin production. These include a putative isoamyl alcohol oxidase, which could regulate formation of valine, a precursor of penicillin; and a zinc binuclear cluster transcription factor that could regulate transcription of genes involved in penicillin production.

The microarray approach also identified a number of amplified genes that do not reside within the penicillin amplification unit, and might be important for penicillin production. Specific ABC transporter efflux pumps have been either amplified or deleted. This seemingly contradictory finding is plausible, since certain pumps could positively impact penicillin production by increasing penicillin efflux into the medium, while other pumps could negatively impact penicillin production by transporting precursors into the medium. Multiple amplified fragments were identified that encode a glucose transporter/sensor which may regulate the flux of carbon into central metabolism or the transition from primary to secondary metabolism. A fragment that contains a putative HMG CoA synthase; which can regulate flux into branch chain amino acid biosynthesis was also identified. Finally, two clones were found to contain sequence that suggests amplification of a transposon. It is possible that this transposon could be involved in the amplification of the penicillin biosynthetic cluster.

Genomic profiling is well suited for detecting significant size deletions and insertions in wild type, and particularly in mutagenized production strains. Clearly, such gross changes can be detected in enhanced penicillin strains, and similarly, genomic rearrangements are expected to be associated with the increased production of other secondary metabolites. Furthermore, genomic profiling using genomic fragment microarrays can be employed in combination with gene expression studies, more sensitive genomic profiling methods, and sequencing efforts in order to gain a detailed understanding of the genetic control of secondary metabolite production. For example, in strains such as *P. chrysogenum* and *P. citrinum*, for which the secondary metabolite gene clusters have been cloned, but whole genomic sequence is lacking, genomic profiling is particularly well suited for identifying important genes that are not physically linked to the cluster. In uncharacterized systems, transcriptional profiling using random fragment microarrays can be used to rapidly identify biosynthetic genes that have not been previously identified (Askenazi *et al.* 2003).

5. *PENICILLIUM MARNEFFEI* EST SEQUENCING PROJECT

An EST sequencing project has recently been initiated for *P. marneffei* (S. S. Y.Wong and K. Y. Yuen, personal communication, 2003). Initial analysis of 2303 sequence tags has revealed an overall G+C content of 48.8%. More than 3.4% of the genes encode secondary metabolism genes for non-ribosomal peptide synthesis and polyketide synthesis; among these is a homologue of the lovastatin nonaketide synthase gene of *A. terreus* (Kennedy *et al.* 1999). This finding is particularly interesting, given the observation that plant pathogens also appear to contain more genes dedicated to secondary metabolism than do saprophytes (Yoder and Turgeon 2001). Significantly, the sequencing project has revealed the presence of homologues to mating type and pheromone genes in this asexual species. In addition, nuclear small (18S) and large (28S) ribosomal gene sequences were highly similar to ribosomal genes of *Talaromyces*. These results concur with a previous phylogeny study and suggest that if a cryptic sexual stage exists for *P. marneffei* it would most likely be a species of *Talaromyces* (LoBuglio and Taylor 1995).

6. CONCLUSIONS

Examination of the penicillin gene cluster in *P. chrysogenum* and *A. nidulans* reveals a conservation of synteny, high degree of sequence similarity and lack of introns. A comparison of the two statin clusters; compactin in *P. citrinum*, and lovastatin in *A. terreus*, reveals a reduced level of protein identity, lack of synteny, and the presence of introns. These results are consistent with a relatively recent introduction of the penicillin cluster from a bacterial source, and a more ancient evolution of the statin clusters. Preliminary genomic profiling of *P. chrysogenum* in the current study has identified a number of genes that are amplified or deleted in the mutated, high producing strains. These include, but are not limited to the biosynthetic genes directly involved in penicillin production. Despites it's importance, relatively little publicly available genomic analysis has been performed with *Penicillium* species. Surprisingly, there is no publicly available sequence, or sequencing project underway for the penicillin production strain *P. chrysogenum*. Detailed results from the *P. marneffei* sequencing project are forthcoming, and will highlight potential benefits of genomic analyses of other *Penicillium* species.

REFERENCES

Abe Y, Suzuki T, Ono C, Iwamoto K, Hosobuchi M, Yoshikawa H (2002). Molecular cloning and characterization of an ML-236B (compactin) biosynthetic gene cluster in *Penicillium citrinum*. Mol. Genet. Genomics 267: 636-646.

Andrianopoulos A (2002). Control of morphogenesis in the human fungal pathogen *Penicillium marneffei*. Int J Med Microbiol 292:331-347.

Askenazi M, Driggers EM, Holtzman DA, Norman TC, Iverson S, Zimmer DP, Boers M, Blomquist PR, Martinez EJ, Monreal AW, Feibelman TP, Mayorga ME, Maxon ME, Sykes K, Tobin JV, Cordero E, Salama SR, Trueheart J, Royer JC, and Madden K (2003) Integrating transcriptional and metabolite profiles to direct the engineering of lovastatin-producing strains. Nature Biotechnol. 21:1-7.

Backus MP, Stauffer JF (1955). The production and selection of a family of strains in *Penicillium chrysogenum*. Mycologia 47: 429-463.

Barredo, JL, Cantoral JM, Alvarez E, Diez B, and Martin JF (1989a). Cloning, sequence analysis and transcriptional study of the isopenicillin N synthase of *Penicillium chrysogenum* AS-P-78. Mol Gen Genet 216:91-98

Barredo JL, van Solingen P, Diez B, Alvarez E, Cantoral JM, Katevilder A, Smaal EB, Groenen MAM, Veenstra AE and Martin JF (1989b) Cloning and characterization of acyl-CoA:6-APA acyltransferase gene of *Penicillium chrysogenum*. Gene 83:291-300.

Brown AG, Smale TC, King TJ, Hasenkamp R, Thompson RH (1976). Crystal and molecular structure of compactin, a new antifungal metabolite from *Penicillium brevicompactin*. J Chem Soc Perkin Trans 1: 1165-1170.

Berbee ML, and Taylor JW (1993). Dating the evolutionary radiations of the true fungi. Can. J Bot 71(8):1114-1127.

Carr LG, Skatrud PL, Scheetz ME II, Queener SW and Ingolia TD. 1986. Cloning and expression of the isopenicillin N synthetase gene from *Penicillium chrysogenum*. Gene 48:257-266.

Chavez R, Fierro F, Gordillo F, Martin JF, and Eyzaguirre J (2001). Electrophoretic karyotype of the filamentous fungus *Penicillium purpurogenum* and chromosomal location of several xylanolytic genes. FEMS Microbiol. Letters 205:379-383.

Demain AL, and Elander RP (1999). The *B*-lactam antibiotics: past, present, and future. Antonie van Leeuwenhoek 75: 5-19.

Deng Z, and Connor DH (1985). Progressive disseminated penicilliosis caused by *Penicillium marneffei*. Am. J. Clin. Pathol. 84:323-327.

Diez B, Gutierrez S, Barredo J, van Solingen P, van der Vort LHM, and Martin JF (1990). The cluster of penicillin biosynthetic genes. Identification and characterization of the *pcb* AB gene encoding the alpha-aminoadipyl-cysteinyl-valine synthetase and linkage to the *pcb*C and *pen*DE genes. J Biol Chem 265:16358-16365

Endo A, Kuroda M, Tsujita M (1976). ML-236A, ML236B, and ML236C, new inhibitors of cholesterogenesis produced by *Penicillium citrinum*. J Antibiot 29: 1346-1348.

Farber P, and Geisen R (2000). Karyotype of *Penicillium nalgiovense* and assignment of the penicillin biosynthetic genes to chromosome IV. Int J of Food Microbiol. 58:59-63.

Fierro F, Gutierrez S, Diez B, and Martin JF (1993). Resolution of four large chromosomes in penicillin-producing filamentous fungi: the penicillin gene cluster is located on chromosome II (9.6 Mb) in *Penicillium notatum* and chromosome I (10.4 Mb) in *Penicillium chrysogenum*. Mol Gen Genet 241:573-578.

Fierro F, Barredo JL, Diez B, Gutierrez S, Fernandez F, and Martin JF (1995). The penicillin gene cluster is amplified in tandem repeats linked by conserved hexanucleotide sequences. Proc Natl Acad Sci 92: 6200-6204.

Garrison RG, and Boyd KS (1973). Dimorphism of *Penicillium marneffei* as observed by electron microscopy. Can J Microbiol 19:1305-1309.

Geiser DM, Timberlake WE, and Arnold ML (1996). Loss of Meiosis in *Aspergillus*. Mol. Biol. Evol. 13:809-817.

Geiser DM, and LoBuglio KF (2001). The monophyletic Plectomycetes: Ascosphaerales, Onygenales, Eurotiales. In: McLaughlin, McLaughlin, Lemke, eds. The Mycota VII Part A, Systematics and Evolution. Berlin Heidelberg: Springer-Verlag, pp.201-219.

Gutierrez S, Fierro F, Casqueiro J, and JF Martin (1999). Gene organization and plasticity of the *B*-lactam genes in different filamentous fungi. Antonie van Leeuwenhoek 75: 81-94.

Haas H, Bauer B, Redl B, Stoffler G, and Marzluf GA (1995). Molecular cloning and analysis of nre, the major nitrogen regulatory gene of *Penicillium chrysogenum*. Curr Genet 27: 150-158.

Hilmarsdottit I, Meynard JL, Rogeaux O, Guermonprez G, Datry A, Katlama C, Brucker G, Coutellier A, Danis M, and Gentilini M (1993). Disseminated *Penicillium marneffei* infection associated with human immunodeficiency virus: a report of two cases and a review of 35 published cases. J Acquired Immune Defic Syndr 6:466-471.

Itoh Y, Johnson R, and Scott B (1994). Integrative transformation of the mycotoxin-producing fungus, *Penicillium paxilli*. Curr Genet 25:508-513.

Kayser T, and Schulz G (1991). Electrophoretic karyotype of cellulolytic *Penicillium janthinellum* strains. Curr Genet 20:289-291.

Keller NP, and Hohn TM (1997). Metabolic pathway gene clusters in filamentous fungi. Fungal Genetics and Biol 21:17-29.

Kennedy J, Auclair K, Kendrew SG, Park C, Vederas JC, and Hutchinson CR (1999). Modulation of polyketide synthase activity by accessory proteins during lovastatin biosynthesis. Science 284: 1368-1372.

Laich F, Fierro F, and Martin JF (2002). Production of Penicillin by fungi growing on food products: Identification of a complete Penicillin gene cluster in *Penicillium griseofulvum* and a truncated cluster in *Penicillium verrucosum*. Applied and Environ Microbiol 68: 1211-1219.

Landan G, Cohen F, Aharonowitz Y, Shuali Y, Graur D, and Shiffman D (1990). Evolution of isopenicillin N synthase genes may have involved horizontal gene transfer. Mol Biol Evol 7: 399-406.

Lein, J. 1986. The Panlabs penicillin strain improvement program, p. 105-139. *In* Z. Vanek and A. Hostakek (ed.), Overproduction of microbial metabolites. Butterworths, Boston, Mass.

LoBuglio KF, Pitt JI, and Taylor JW (1993). Phylogentic analysis of two ribosomal DNA regions indicates multiple independent losses of a sexual *Talaromyces* state among asexual *Penicillium* species in subgenus *Biverticillium*. Mycologia 85:592-604.

LoBuglio KF, and Taylor JW (1995). Phylogeny and PCR identification of the human pathogenic fungus *Penicillium marneffei*. J Clinical Microbiol 33:85-89.

MacCabe AP, Riach MBR, Unkles SE and Kinghorn JR (1990) The *Aspergillus nidulans npe*A locus consists of three contiguous genes required for penicillin biosynthesis. EMBO J 9:279-287

Moore RN, Gigam G, Chan JK, Hogg AM, Nakashima TT, and Vederas JC (1985). Biosynthesis of the hypercholesterolemic agent mevinolin by *Aspergillus terreus*: determination of the origin of carbon, hydrogen and oxygen atoms by carbon-13 NMR and mass spectrometry. J Am Chem Soc 107:3694-3701.

Penalva MA, Moya A, Dopazo J, and Ramon D (1990). Sequences of isopenicillin N synthetase genes suggest horizontal gene transfer from prokaryotes to eukaryotes. Proc R Soc London 241:164-169.

Peterson SW (1993). Molecular genetic assessment of relatedness of Penicillium subgenus Penicillium. In: DR Reynolds, and JW Taylor eds. The Fungal Holomorph: Mitotic, Meiotic and Pleomorphic Speciation in Fungal Systematics. Wallingford: CAB International, pp 121-128.

Pitt JI (1979). The genus *Penicillium* and its Teleomorphic States *Eupenicillium* and *Talaromyces*. London: Academic Press.

Pitt JI (1985). A Laboratory Guide to Common *Penicillium* Species. North Ryde: CSIRO Division of Food Processing.

Pitt JI and Samson RA (1993). Trichocomaceae. In: W Greuter ed. Names in Current Use in the Families Trichocomaceae, Cladonaceae, Pinaceae, and Lemnaceae. Konigstein: Koeltz Scientififc Books, pp. 13-57.

Raper KB, and Thom C (1949). A manual of *Penicillia*. Baltimore:Williams and Wilkins.

Samson SM, Belagaje R, Blankenship DT, Chapman JL, Perry D, Skatrud PL, Frank RM, Abraham EP, Baldwin JE, Queener SE and Ingolia TD (1985). Isolation, sequence determination and expression in *E. coli* of the isopenicillin N synthetase gene from *Cephalosporium acremonium*. Nature 318: 191-194.

Sequerra J, Capellano A, Gianinazzi, Pearson V, and Moiroud A (1995). Ultrastructure of cortical root cells of *Alnus incana* infected by *Penicillium nodositatum*. New Phytol 130:545-555.

Sequerra J, Marmeisse R, Valla G, Normand P, Capellano A, and Moiroud A (1997). Taxonomic position and intraspecific variability of the nodule forming *Penicillium nodositatum* inferred from RFLP analysis of the ribosomal intergenic spacer and Random Amplified Polymorphic DNA. Mycol Res 101:465-472.

Valla G, Capellano A, Hugueny R, and Moiroud A (1989). *Penicillium nodositatum* Valla, a new species inducing myconodules on *Alnus* roots. Plant Soil 114:142-146.

Viviani MA, Hill JO, and Dixon DM (1993). *Penicillium marneffei*: dimorphisim and treatment. HV Bossche, FC Odds, and D Kerridge, eds. Dimorphic fungi in biology and medicine. New York: Plenum Press.

Walton JD (2000). Horizontal gene transfer and the evolution of secondary metabolite gene clusters in fungi: An hypothesis. Fungal Genetics and Biol. 30:167-171.

Wong SSY, Slau H, and Yuen KY (1999). Penicilliosis marneffei - West meets East J Med Microbiol 48: 973-975.

Weigel BJ, Burgett SG, Chen VJ, Skatrud PL, Frolik CA, Queener SW, and Ingolia TD (1988). Cloning and expression in *Escherichia coli* of isopenicillin N synthetase genes from *Streptomyces lipmanii* and *Aspergillus nidulans*. J Bacteriol 170:3817-3826.

Yoder OC, and Turgeon BG (2001). Fungal genomics and pathogenicity. Current Opinions in Plant Biol 4: 315-321.

Applied Mycology & Biotechnology
An International Series. Volume 4. Fungal Genomics

12

Genomics in *Neurospora crassa*: From One-Gene-One-Enzyme to 10,000 Genes

Edward L. Braun[a], Donald O. Natvig[b], Margaret Werner-Washburne[b] and Mary Anne Nelson[b]
[a]Department of Zoology, University of Florida, Gainesville, Florida 32611, USA; [b]Department of Biology, University of New Mexico, Albuquerque, New Mexico 87131, USA (manelson@unm.edu).

Neurospora crassa was the central organism in the development of biochemical genetics, providing a model system that established the relationship between genes and enzymes; it remains the best-studied filamentous fungus. This review focuses on the impact that the recent publication of a high-quality draft sequence of the *N. crassa* genome will have upon efforts to understand the biology of the filamentous fungi. Although several fungal genomes have been sequenced, annotated, and published, the organisms that have been examined are yeasts with relatively small genomes. In sharp contrast, *N. crassa* contains about 10,000 protein-coding genes, approximately twice as many genes as the yeasts and only slightly fewer than the invertebrate animals. Analysis of this gene set suggests that several different processes have led to the differences in gene content between *N. crassa* and the yeasts. Evidence for the loss of genes in the yeasts and the acquisition of novel genes in *Neurospora* lineage is described, as well as details regarding the biological processes that have led to these changes. Analyses of the *N. crassa* genome sequence revealed the widest array of genome defense mechanisms known for any organism, and one of these defense mechanisms (RIP) appears to have blocked the productive duplication of genes. Since gene duplication is the most common pathway for the origin of novel genes, it seems likely that *N. crassa* will provide an excellent model system for understanding alternative ways in which novel genes arise. A number of unexpected genes were identified when the complete genome sequence was analyzed, indicating that *N. crassa* produces secondary metabolites, shares apparent "pathogenicity" genes with plant pathogens, and responds to environmental cues such as light in novel ways. The genome sequence for *N. crassa* is the first exciting step toward a detailed understanding of the biology of filamentous fungi, and it will allow fungal biologists to establish which features of the filamentous fungi are shared with non-fungal organisms and which features are unique.

Corresponding author: Mary Anne Nelson

1. INTRODUCTION

N. crassa has a long and distinguished history in classical and biochemical genetics. C. L. Shear and B. O. Dodge discovered the sexual cycle of *Neurospora* and named the genus nearly eighty years ago (Shear and Dodge 1927). Less than ten years later, C. C. Lindegren identified by mutation and mapped six loci on linkage group (LG) I, which he referred to as the "sex chromosome" of *Neurospora* (Lindegren *et al.* 1939). The first biochemical mutants in any organism were identified in *Neurospora* by G. W. Beadle and E. L. Tatum; their work initiated the science of biochemical genetics (Beadle and Tatum 1941). In 1945, Beadle proposed the one-gene-one-enzyme hypothesis (Beadle 1945). Beadle and Tatum were awarded the Nobel Prize for this body of work. Also in 1945, B. McClintock studied the meiotic cytology of *Neurospora* and showed that there are seven chromosomes, the same as the number of linkage groups (McClintock 1945). A few years later, the first temperature-sensitive mutants were isolated in *Neurospora* (Houlahan *et al.* 1949). In 1955, M. B. Mitchell was the first to demonstrate the process of gene conversion (Mitchell 1955).

Since these early studies, a detailed genetic map of the seven linkage groups has been developed, including over 1,000 mapped genes; a fairly detailed restriction fragment length polymorphism (RFLP) map is also available (Perkins *et al.* 2001). Efficient DNA-mediated transformation is possible in *Neurospora* (Case *et al.* 1979), so that cloned genes can be introduced and their properties analyzed. The *Neurospora* genome sequence was completed and analyzed recently; this represents the first available sequence of a filamentous fungus (Galagan *et al.* 2003). Coordination of the physical (sequence) and genetic maps is greatly speeding gene discovery and characterization.

2. FUNGAL GENOMICS

In the era of genomics, one of the major questions is how many and which complete genome sequences will be necessary (Miller 2000). This is an especially important question for fungal biologists because of the diversity of fungi and the potential for these eukaryotic organisms to contribute to understanding fundamental, questions in biology. Some of the debate regarding the appropriate organisms to sequence has focused on how many genomes to obtain from representatives of the ascomycete fungi. Complete or nearly complete genome sequences now exist for three ascomycetes that exhibit predominantly yeast-like growth, *Saccharomyces cerevisiae* (Goffeau *et al.* 1996), *Candida albicans* (Berman and Sudbery 2002), and *Schizosaccharomyces pombe* (Wood *et al.* 2002). All three organisms have served as important models in cell biology and genetics.

The community of scientists who work on filamentous fungi have argued that additional complete genomes, especially genomes of filamentous fungi, are necessary because yeasts such as *S. cerevisiae* and *S. pombe* lack the complex life cycles typical of many filamentous fungi (Bennett 1997). The genome of *Ashbya gossypii*, a close relative of *S. cerevisiae* that possesses a type of filamentous growth, has also been sequenced (Brachat *et al.* 2003). However, *A. gossypii* is a plant parasite that has fewer genes and a smaller genome than does *S. cerevisiae* (Fig. 1). Thus, the genome sizes and total gene numbers of these model ascomycetes are much smaller than those typical of most filamentous fungi (Fig. 1), suggesting that they lack certain gene functions important to more developmentally complex filamentous fungi and, therefore, may not be "models" for many critical processes.

It is clear that the absence of filamentous growth in the yeasts results in major differences, in terms of life history and ecology, between yeasts and filamentous fungi. The filamentous ascomycetes exhibit spectacular diversity in terms of morphology and ecology and include strict saprobes, facultative parasites, obligate parasites of plants and animals, mutualists of plants, and the fungal components of lichens (Alexopoulos *et al.* 1996). The yeasts also lack some specific responses to the environment, such as responses to light (Roenneberg and Merrow 2001), that are characteristic of many other fungi. Thus, there are important physiological questions to be answered by studying a variety of genomes within the ascomycetes.

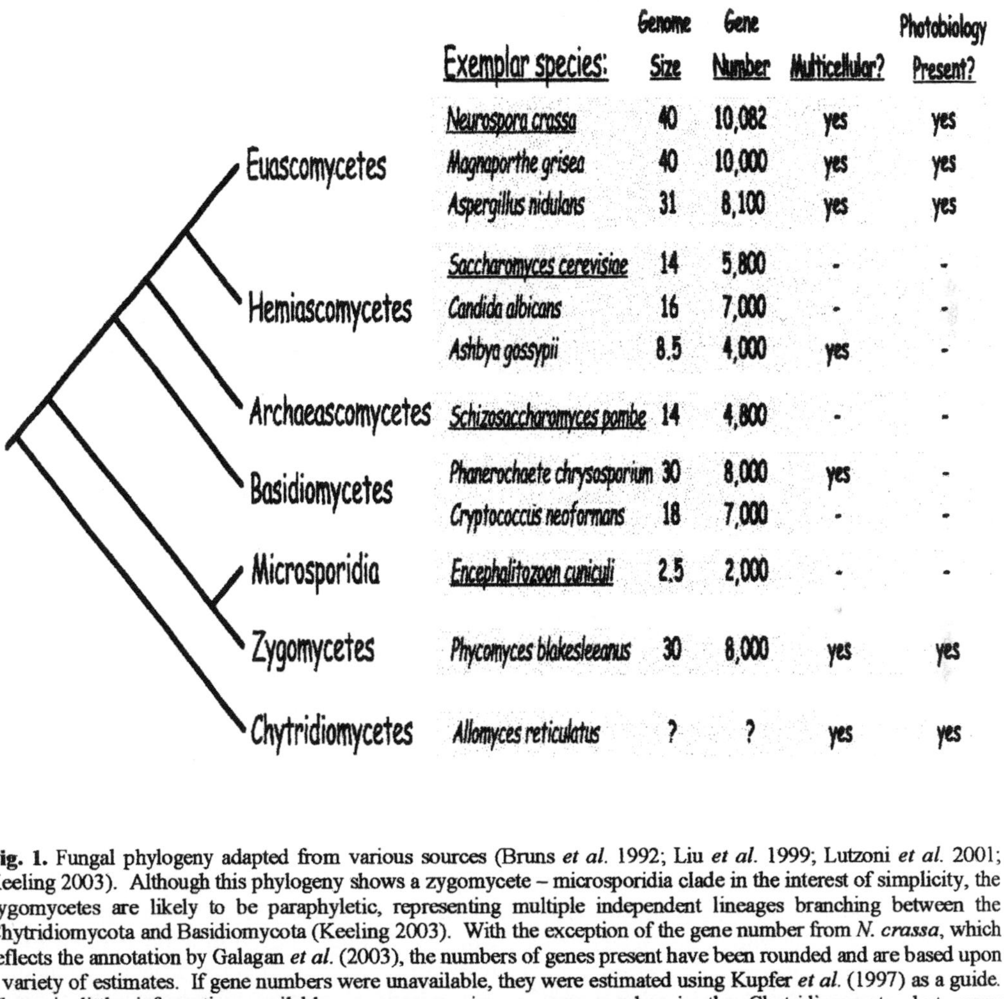

Fig. 1. Fungal phylogeny adapted from various sources (Bruns *et al.* 1992; Liu *et al.* 1999; Lutzoni *et al.* 2001; Keeling 2003). Although this phylogeny shows a zygomycete – microsporidia clade in the interest of simplicity, the zygomycetes are likely to be paraphyletic, representing multiple independent lineages branching between the Chytridiomycota and Basidiomycota (Keeling 2003). With the exception of the gene number from *N. crassa*, which reflects the annotation by Galagan *et al.* (2003), the numbers of genes present have been rounded and are based upon a variety of estimates. If gene numbers were unavailable, they were estimated using Kupfer *et al.* (1997) as a guide. There is little information available on genome size or gene number in the Chytridiomycota, but many chytridiomycetes exhibit substantial developmental complexity and exhibit features (e.g., flagella) absent in other fungi. For these reasons, at least some chytridiomycetes are likely to have large genomes and gene numbers similar to other developmentally complex fungi. The presence or absence of photobiology (responses to light and circadian rhythms) is shown as an example of a complex trait that has been lost in a variety of lineages. Although neither of the basidiomycetes shown here has a described photobiology, many basidiomycetes are known to exhibit responses to light.

A number of phylogenetic analyses (Bruns *et al.* 1992; Berbee and Taylor 1993; Liu *et al.* 1999) have suggested a model in which the yeasts arose from more complex filamentous ancestors, and some extremely simplified unicellular organisms are clearly derived from more complex fungal ancestors (Keeling 2003). The hypothesis that yeasts arose from multicellular ancestors by simplification suggests that developmentally complex filamentous fungi are likely to share genes that have been lost in the yeasts. In fact, analyses of *N. crassa* expressed sequence tags (ESTs) indicate that *S. cerevisiae* has undergone substantial gene loss (Braun *et al.* 2000), consistent with this hypothesis. However, rigorous tests of the hypothesis of genome simplification will require comparison of genomes from yeasts and diverse filamentous fungi.

The release of the *N. crassa* genome sequence (Galagan *et al.* 2003), representing the first complete genome sequence for a developmentally complex filamentous fungus, has revealed the presence of a number of surprising genes and opened the door to genomics of the filamentous fungi. In this chapter we review the impact of adding *N. crassa*, a model filamentous ascomycete, to the list of completely sequenced genomes and discuss the prospects for using this sequence information to investigate fungi and other groups of organisms.

3. *NEUROSPORA CRASSA* – A MODEL FILAMENTOUS FUNGUS

One of the major morphological differences between filamentous fungi like *N. crassa* and typical yeasts is the constitutive nature of hyphal growth. The fundamental cellular structure of filamentous fungi is the hypha (pl. hyphae), a multinucleate tube that grows by elongation from the apex (Heath and Steinberg 1999). This type of tip growth has only been found in the fungi and specific plant cells (root hairs and pollen tubes). Fungal hyphae typically have crosswalls called septa that divide the filament into a linear array of cells. Hyphal networks can remain relatively undifferentiated morphologically, as is typical of vegetative stages, or they can form very complex structures with highly specialized tissues, most notably in fungal fruiting bodies. Although the fruiting bodies (perithecia) of *N. crassa* are much smaller than the fruiting bodies of typical mushrooms or puffballs, they are complex multicellular structures that exhibit a characteristic shape (Nelson 1996).

Fungal hyphae allow a number of specialized functions that are not available to yeasts by providing a robust and intricate structure that supports continuous, directed growth. Hyphae form a network called the mycelium that gives filamentous fungi their structure. The mycelium of an individual fungus is a complex system in which the hyphae can transport nutrients and organelles, and these networks can exhibit trophic responses to a variety of signals such as nutrients and light. Hyphae also have a great facility for penetrating substrates, reflecting the combination of physical forces and the release of digestive enzymes at the growing hyphal tip. In the case of pathogenic fungi, penetration through healthy tissues such as leaf surfaces often plays an important role in the initiation of infection (Gow *et al.* 2002). For saprobes like *N. crassa*, hyphal penetration is important in spreading a mycelial network through a substrate.

The mycelial network that results from the growth of hyphae is a true individual that can occupy a large space and live for extremely long periods of time. The most impressive examples of such individuals are found among basidiomycetes growing below ground, sometimes observed as fairy rings representing growth over hundreds of years starting from a single point. It has been suggested that the largest known living organism, covering 2,200 acres and with an estimated age of 2,400 years, is an individual of the basidiomycete *Armillaria bulbosa* (Smith *et al.* 1992). Mycelia also allow fungi to spread in and on living hosts; examples of this type of growth include plant pathogens, such as *Magnaporthe grisea* (rice blast fungus), and animal pathogens that can infect humans, such as *Trichophyton* spp. (ringworm).

3.1 *Neurospora* in the Environment

The life histories of *Neurospora* species are an interesting mix of specialization and generalization. A specialized aspect of *Neurospora* species is their adaptation to colonize dead or dying plants after fires, and most *Neurospora* isolates have been obtained from burned vegetation. Consistent with this observation, ascospores are routinely treated with either heat or furfuryl alcohol, a product of burned vegetation, in order to elicit germination in the laboratory. However, the generalist nature of *Neurospora* species is reflected in their lack of host specificity and their broad geographic ranges, which extend from the tropics to northern conifer forests (Perkins and Turner 1988; Jacobson *et al.* 2003).

Although *Neurospora* species are not pathogenic, they are likely to be excellent models for filamentous fungal pathogens, since closely related groups of perithecial ascomycetes include some of the most important plant pathogens (e.g., *Magnaporthe grisea*, the rice blast fungus). It is clear that *Neurospora* species have evolved in close association with plants and have life histories dependent on plants. The availability of the *N. crassa* genome presents an opportunity to assess whether plant pathogenicity and specialization for a narrow host range depend primarily on the acquisition of new genes or the modification of gene function and expression during the course of evolution.

3.2 The *Neurospora* Genome and its Impact on Fungal Genomics

In the genomic era, there are two approaches to the study of model organisms. One is to examine aspects of biology that are conserved in large portions of the "tree of life". Many experimental studies with members of the genus *Neurospora* have focused on fundamental aspects of cell biology and genetics likely to have such relevance (Davis and Perkins 2002), as have studies using yeast (*S. cerevisiae* and *S. pombe*) model systems. A second approach is to examine processes that are unique to specific groups of organisms, toward the goal of understanding processes that are unique to fungi or specific groups of fungi. Such studies will be extremely important to the understanding and control of fungal pathogens, for example. In this context, studies of hyphal growth and sexual development in *Neurospora* species have been responsible for a number of important contributions with special relevance to fungal biology. The complete sequence of the *N. crassa* genome will accelerate both types of research, but the unique features of *N. crassa* as a model filamentous fungus are expected to have an especially significant impact on the second research approach.

As a model filamentous fungus, *N. crassa* also brings with it a considerable body of information regarding the population biology and evolution of the genus, and progress has been made in understanding the ecology of certain species (Jacobson *et al.* 2003). When the information on the natural history of *Neurospora* species is combined with past laboratory studies (reviewed by Davis and Perkins 2002) and the new genome sequence (Galagan *et al.* 2003), it should be possible to build a solid foundation for understanding how filamentous fungi respond to the environment at a molecular level. The *N. crassa* genome will provide a reference for comparison with other filamentous fungi exhibiting different lifestyles, including pathogens. Overall, the availability of the genome sequence from *N. crassa* should greatly accelerate the expansion of experimental biology into realms with specific relevance to fungal biology.

4. COMPARATIVE GENOMICS USING *NEUROSPORA CRASSA*

Analyses of the complete genome sequence for *N. crassa* revealed 10,082 protein-coding genes (Galagan *et al.* 2003), in agreement with previous predictions of gene number based upon more limited samples of genomic DNA that ranged from 9,200 to 13,000 (Kupfer *et al.* 1997; Nelson *et al.* 1997; Bean *et al.* 2001; Kelkar *et al.* 2001; Schulte *et al.* 2002; Mannhaupt *et al.* 2003). Thus, it was clear before the completion of the *N. crassa* genome that this filamentous fungus would have a substantially larger number of genes than the 4500 to 6000 genes that are typical of the yeasts (Goffeau *et al.* 1996; Wood *et al.* 2001; Wood *et al.* 2002; Kellis *et al.* 2003; also see Fig. 1).

There are three distinct processes that can account for the differences in gene numbers between the yeasts and the filamentous fungi (Fig. 2). One process, gene loss in the yeast lineages, would reduce the sizes of the yeast genomes. In contrast, two alternative processes, the creation of novel genes and lateral gene transfer (LGT) into the *Neurospora* lineage, would lead to an increase in the number of genes present in the *Neurospora* lineage. Prior to the completion of the *N. crassa* genome sequence, large-scale sequence comparisons had provided evidence that all of these processes have contributed to the differences in gene content between the *N. crassa* and *S. cerevisiae* genomes (detailed by Braun *et al.* 2000). The availability of the complete *N. crassa* genome sequence has allowed these basic conclusions to be expanded in several different ways.

4.1 Genetic Innovation in the *Neurospora* Lineage

Genetic innovation can be broadly defined as the origin of novel genes in specific lineages. Gene duplication is thought to represent the primary explanation for the process of genetic innovation in most lineages (Ohno 1970), and a small number of gene families are evident in the complete *N. crassa* genome sequence (Galagan *et al.* 2003). However, the proportion of *N.*

Table 1. Percentage of genes in various organisms that are members of gene families

Organism	Number of Genes[a]	% Genes in Families[b]	Expected % Genes in Families[c]
Mycoplasma genitalium	500	16%	—
Haemophlius influenzae	1,700	17%	18%
Pyrococcus abyssi	1,800	28%	20%
Encephalitozoon cuniculi	2,000	21%	22%
Escherichia coli	4,400	33%	35%
Schizosaccharomyces pombe	4,80	26%	36%
Pseudomonas aeruginosa	5,600	44%	38%
Saccharomyces cerevisiae	5,800	33%	38%
Neurospora crassa	10,000	17%	41%
Drosophila melanogaster	14,500	39%	43%
Caenorhabditis elegans	18,000	46%	43%
Arabidopsis thaliana	25,000	71%	44%

[a] Due to the problems associated with gene prediction (Wood *et al.* 2001; Braun 2003; Kellis *et al.* 2003), these values have been rounded to the nearest hundred. [b] These data are from analyses conducted for Galagan *et al.* (2003) and presented in Fig. 1 and S6.1 from that paper.[c] The expected percentage of genes in families was calculated using expected number of paralogs (n_P) given a specific number of genes (n_G) using the equation $n_P = 0.46n_G - 471$ (from Hooper *et al.* 2003) and converting this value to a percentage. This equation does not yield an appropriate prediction for organisms with fewer than 1,000 genes, so the value for *M. genitalium* is not presented.

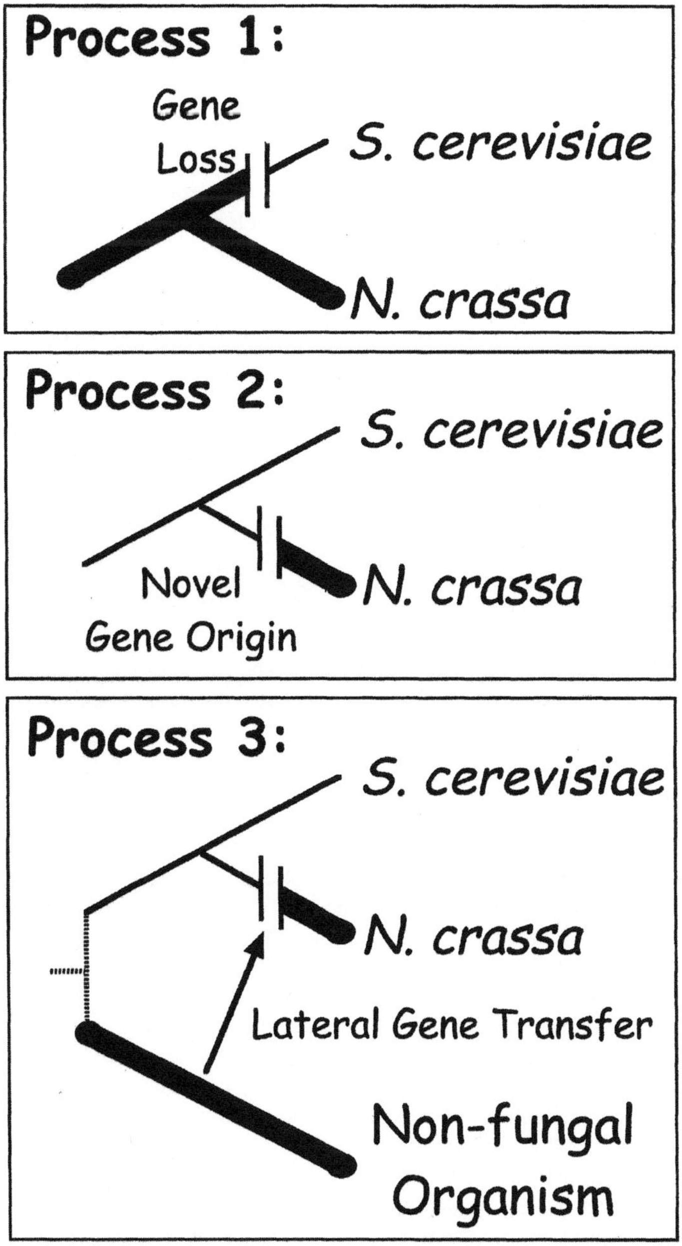

Fig. 2. Schematic of three processes that could give rise to the higher gene numbers observed in *N. crassa* compared to *S. cerevisiae.* The presence or absence of a specific gene is indicated using the thickness of the line in the evolutionary tree. Similar processes can also be invoked to explain the differences in gene numbers between *N. crassa* and fungi like *S. pombe* or *E. cuniculi.* Evidence for the action of all three processes has been obtained (see text).

crassa genes that are members of gene families is substantially lower than expected for an organism with approximately 10,000 genes (Table 1), and most of the gene duplications observed in *N. crassa* are relatively ancient. Although there was existing evidence that *N. crassa* had few gene families (Nelson *et al.* 1997), the striking near absence of gene families with closely-related members in the *N. crassa* genome only became clear upon analysis of the complete genome sequence (Galagan *et al.* 2003).

The absence of genes that arose by recent duplications probably reflects the activity of a process in *Neurospora* called RIP (repeat-induced point mutation), which causes multiple C to T transition mutations on both strands of duplicated sequences (Cambareri *et al.* 1989). RIP is one of the set of processes that operate as genome defense mechanisms in *Neurospora* species; additional defense mechanisms include the vegetative quelling of repeated sequences (Cogoni 2001) and the meiotic silencing of unpaired DNA (MSUD; Shiu *et al.* 2001). These genomic defense mechanisms have left a clear signature on the complete *N. crassa* genome sequence, with the extreme paucity of gene families. Analysis of the genome sequence has revealed that RIP has affected all transposable elements and the majority of other duplicated sequences present in the genome (Galagan *et al.* 2003). The notable exceptions are the ribosomal DNA repeat units present at the nucleolus organizer, which for reasons that remain unknown have not been affected by RIP. However, the virtual absence of recent protein-coding gene duplications in the *N. crassa* lineage is undoubtedly another consequence of RIP.

The complete *Neurospora* genome sequence also allowed examination of alternative hypotheses regarding processes involved in genetic innovation. Although it seems clear that the fixation of duplicated genes in the *Neurospora* lineage stopped after the process of RIP arose, detailed understanding of the RIP process may provide clues about types of duplication that can occur despite the existence of RIP. For example, duplications of relatively short genes or gene segments might occur, since RIP does not act on duplicated sequences unless they exceed a certain minimum size (~400 bp; Watters *et al.* 1999). Likewise, duplication coupled with loss of a segment of the gene (e.g., due to domain loss; Braun and Grotewold 2001) might allow survival of both duplicates if the region exhibiting a high degree of nucleotide sequence identity is shorter than 400 bp and thus not susceptible to the process of RIP. The fixation of duplicates by any of these alternative pathways should still result in the duplication of individual protein domains. Examination of the complete proteome inferred using the *N. crassa* genome sequence revealed few closely related domains (Galagan *et al.* 2003), suggesting that these pathways to gene duplication are uncommon and that RIP has stopped the productive duplication of genes rather than altering the pathways that result in duplicated genes.

This leaves unanswered the major question of how *Neurospora* accumulated so many genes (prior to the establishment of the RIP process). The majority of *N. crassa* sequences that lack homologs in the yeast genomes (57% of predicted *N. crassa* proteins) correspond to "orphan" genes, defined as genes that lack identifiable homologs in available databases (Nelson *et al.* 1997; Braun *et al.* 2000; Schulte *et al.* 2002; Galagan *et al.* 2003; Mannhaupt *et al.* 2003). The number of true orphan genes will likely be reduced substantially when full comparisons are made between the genomes of *N. crassa* and closely-related ascomycetes. In fact, comparisons of *N. crassa* genes on linkage groups 2 and 5 to an assembly of the *Aspergillus fumigatus* genome (available at http://www.tigr.org/tdb/e2k1/afu1) greatly reduced the percentage of genes classified as orphans (Mannhaupt *et al.* 2003). However, it is quite clear that *N. crassa* possesses many genes with no identifiable homologs in other groups of organisms.

What processes could generate such a large proportion of orphan genes? In principle, orphan genes can be generated by duplication followed by rapid divergence of one member of the duplicated pair. However, the active genomic defense mechanisms of *Neurospora* species have blocked the productive duplication of genes. This raises the question of when the orphan genes present in the *N. crassa* genome arose. If many orphans arose prior to the origin of RIP, then they might reflect duplication followed by rapid divergence. Alternatively, if the orphan genes arose after the origin of RIP and they reflect rapid evolutionary divergences, they would have to be divergent orthologs of genes present in other organisms. In this model, the divergence would not be coupled with duplication, so *N. crassa* would lack less divergent copies of the genes. Since the extreme sequence divergence necessary to obscure evolutionary relationships among protein-coding genes is likely to be correlated with functional changes (Aravind *et al.* 2000; Braun 2003), this pathway seem implausible as a major pathway of innovation.

Assuming a limited number of genes have undergone divergence in the absence of duplication, only two pathways remain for genetic innovation after the evolution of RIP: 1) lateral (horizontal) gene transfer (LGT); and 2) "overprinting," a process defined as the generation of novel genes from noncoding sequences (Fig. 2; for details see Ohno 1984; Keese and Gibbs 1992). LGT would not tend to produce orphans unless there were no sequences available from the donor organism, and LGT into the *Neurospora* lineage appears relatively limited (Braun *et al.* 2000; also see Section 4.2 below), suggesting that it does not represent a major source of orphan genes. Overprinting, therefore, appears to be a possible explanation for the high proportion of orphan genes if many of the orphan genes arose after the evolution of RIP. Although the source of the requisite unexpressed ORFs that represent the raw material for overprinting remains obscure, the existence of specific processes that block the fixation of duplicated genes (e.g., RIP) should make *Neurospora* species excellent model systems to examine this process. Alternatively, if the orphans largely reflect duplication prior to the origin of RIP followed by divergence, then other ascomycetes that diverged after the origin of RIP should have the a set of genes very similar to that found in *N. crassa*. Since the rice blast fungus *M. grisea* is known to RIP duplicated sequences (Ikeda *et al.* 2002) and a *M. grisea* sequence assembly is virtually complete (available from http://www-genome.wi.mit.edu/annotation/fungi/magnaporthe/), comparison of *N. crassa* sequences to this organism will be especially informative. Regardless of the evolutionary origin of the orphan genes in *Neurospora*, the observation that ESTs from specific developmental stages in *N. crassa* differ in the proportion of orphan genes (Nelson *et al.* 1997) is likely to provide a means to connect these unknown sequences with specific aspects of fungal biology.

4.2 Lateral Transfer into the *Neurospora* Genome

The explosion of genome sequence data has revealed surprising numbers of genes that are shared among genomes by the process of LGT (Doolittle 1999). As described above, the absence of productive gene duplication in the *Neurospora* lineage suggests that the origin of novel genes in this lineage after the origin of RIP has been limited largely to the poorly understood process of overprinting and to LGT. Although the impact of LGT on the evolution of prokaryotic genomes is fairly clear (Doolittle 1999), the number of genes that originate by LGT in eukaryotic lineages has been more controversial (Doolittle 1998; Salzberg *et al.* 2001; Braun 2003; Gogarten 2003). There have been suggestions that large numbers of eukaryotic genes originated by lateral transfers from prokaryotes (Doolittle 1998) and that the clustering of functionally related genes evident in some fungi reflects LGT (Prade *et al.* 1997). However, there are also reasons to suspect that the probability of prokaryote-to-eukaryote LGT is substantially lower than the probability of LGT among prokaryotes (Braun 2003). Likewise, the observed clustering of functionally related genes in the fungi could reflect processes other than LGT.

The existence of barriers to prokaryote-to-eukaryote LGT have been supported by phylogenetic analyses of many candidates for prokaryote-to-vertebrate LGT (Roelofs and Van Haastert 2001; Salzberg *et al.* 2001; Stanhope *et al.* 2001). These candidates for LGT from various prokaryotes to vertebrates were identified using the initial annotation of the draft sequence of the human genome (The International Human Genome Sequencing Consortium 2001) as those genes with possible orthologs in prokaryotic genomes (Salzberg *et al.* 2001). However, genes with this phylogenetic distribution could also reflect loss in multiple non-human eukaryotic lineages. In fact, the observed number of genes with this distribution is similar to that expected under a simple model of gene loss (Salzberg *et al.* 2001), so gene loss in several eukaryotic lineages represents a simpler explanation for these genes than does prokaryote-to-vertebrate LGT. Despite these results, the barriers to prokaryote-to-eukaryote LGT that exist are probably not absolute. Instead, they probably act to reduce the probability of prokaryote-to-eukaryote LGT. Thus, establishing an upper limit for the impact of prokaryote-to-eukaryote LGT upon the *N. crassa* lineage represents a useful exercise. Using a criterion to identify candidates for prokaryote-to-eukaryote LGT comparable to that used by the International Human Genome Sequencing Consortium (presence of a homolog only in the focal eukaryote and prokaryotes), we found 95 candidates for LGT from a prokaryotic organism to *N. crassa* (0.9% of the *N. crassa* proteome). If we relax this criterion slightly to include genes with substantially better BLASTP hits to prokaryotes than to eukaryotes (using a 10 log difference in *E*-values as a cut-off), then an additional 297 genes would be included (2.9% of the *N. crassa* proteome).

Estimates of the numbers of *N. crassa* genes that arose by LGT from prokaryotes using the sets of protein-coding genes that are present in prokaryotes and absent in other eukaryotes should exclude instances of fungus-to-prokaryote LGT, which would result in the same distribution of homologs. Unfortunately, accurate estimates of the proportions of genes reflecting each direction of LGT are difficult to obtain without conducting detailed phylogenetic analyses. It may be necessary to include sequences from additional filamentous fungi in these analyses, and they would have to involve rooted phylogenetic trees despite the difficulties associated with establishing the position of the root in phylogenetic trees (Bieszke *et al.* 1999). Given these limitations, the values presented in this chapter should be viewed as upper limits for the number of *N. crassa* genes originating by LGT from prokaryotes.

The current lack of genome sequences from additional filamentous fungi makes it impossible to determine whether LGT among fungi has made a substantial contribution to the *N. crassa* genome. However, some aspects of the potential contribution of LGT within the fungi to the *N. crassa* genome can be examined by searching for clusters of functionally related genes, since it has been suggested that LGT may drive the clustering of certain genes in the fungi (Prade *et al.* 1997). By analogy with the "selfish operon" model (Lawrence and Roth 1996), sets of genes that carry out similar biological functions could be transferred together in tightly-defined clusters, while LGT involving individual genes would not result in transfer of the complete pathway. This phenomenon has been suggested to largely involve nonessential genes, such as those involved in secondary metabolism (Prade *et al.* 1997). Some clusters of functionally-related genes are evident in *N. crassa*, including the well-studied *qa* cluster (Geever *et al.* 1989) and a cluster of putative laccase and melanin biosynthetic genes on linkage group 5 (Mannhaupt *et al.*

2003). However, various analyses of the complete *N. crassa* genome did not reveal large numbers of clusters characterized by the presence of functionally-related genes (Galagan *et al.* 2003).

There are several additional reasons for the clustering of functionally-related genes in the fungi and other eukaryotes. These include the presence of shared regulatory elements (Rosewich and Kistler 2000), the existence of chromosomal domains containing multiple genes with similar patterns of expression (Cohen *et al.* 2000), and the possibility that the clustering may be an ancestral feature of the genes in question. For example, the *snz* and *sno* genes responsible for pyridoxine synthesis in some organisms are present in a cluster in *N. crassa*, *S. cerevisiae*, and the marine sponge *Suberites domuncula* (Padilla *et al.* 1998; Bean *et al.* 2001; Seack *et al.* 2001), and the divergence observed among the *snz* and *sno* homologs in these organisms is consistent with divergence due to speciation. Although it is likely that the selective pressure maintaining the *snz* – *sno* cluster in these diverse organisms is the presence of regulatory elements that drive the expression of these genes under the same conditions (Padilla *et al.* 1998), the evolutionary origin of this clustering appears to be ancient and unrelated to LGT. Thus, these analyses only serve to place an upper limit on LGT within the fungi. Nonetheless, they serve to emphasize the relatively limited contribution of LGT to the *N. crassa* genome, especially when they are combined with the limited number of genes potentially reflecting prokaryote-to-eukaryote LGT in *N. crassa*.

4.3 Gene Loss in the Yeasts

The initial analyses of the complete *N. crassa* genome (Galagan *et al.* 2003) revealed a total of 584 *N. crassa* genes that have matches in other groups of eukaryotes but lack a clear homolog in either of the completely sequenced yeasts (*S. cerevisiae* and *S. pombe*). These genes are likely to have been lost in both of the yeasts, suggesting that about 5% of the ancestral proteome was lost in both yeasts, assuming the ancestral proteome for the ascomycetes was similar in size to the *N. crassa* proteome. Although the small number of complete genome sequences currently available for fungi limits our ability to reconstruct the presence or absence of specific gene products in the ancestral fungal proteome, the hypothesis that the yeasts arose from more complex filamentous ancestors (Bruns *et al.* 1992; Berbee and Taylor 1993; Liu *et al.* 1999) and the existence of developmentally complex fungi in all groups of fungi (Fig. 1) are consistent with the common ancestor of the ascomycetes having a relatively large proteome.

To extend the initial comparative analyses of the *N. crassa* genome, we examined the numbers of genes with distributions suggesting loss in the yeasts. Using strict criteria outlined in previous studies of gene loss (a significant hit in the non-fungal database [*E*-value $\leq 10^{-5}$] and no potential homolog in the yeast databases [*E*-value > 0.1]; Braun *et al.* 2000; Braun 2003), we found a slightly smaller number of candidates for gene loss (301 *N. crassa* queries). This smaller number of genes identified may reflect the loss of some sequences by restricting our consideration to eukaryotes for which complete and extensively annotated genome sequences are available, but it is more likely to reflect the restrictive criterion used to score loss (sequences must have no potential homolog in BLASTP searches of either yeast). If we consider *N. crassa* genes with homologs in the yeasts but non-fungal homologs that are substantially more closely related (the non-fungal homolog is at least 10-logs better than the yeast homolog; Braun *et al.* 2000; Braun 2003), then 438 additional candidates for loss can be identified. These sequences are expected to include genes that have undergone accelerated evolution in the yeasts, so the set of protein-coding genes present in *N. crassa* that were lost in both of the completely-sequenced yeasts falls somewhere between 301 and 739 genes (approximately 3% – 8% of the proteome).

Although detailed analyses of the *N. crassa* genes that have been lost in yeasts will be presented elsewhere (manuscript in preparation), several patterns emerge from a consideration of the most conservative set of 301 protein-coding genes. Using the same criteria to score loss revealed the potential loss of 535 genes in *S. cerevisiae* and 499 genes in *S. pombe*. Thus, more than half of the genes that have been lost in a single yeast species have been lost in both yeast species. The loss of these genes would have occurred independently in each yeast lineage if the fungal phylogeny presented in Fig. 1 is correct, as suggested by a variety of analyses (Bruns *et al.* 1992; Berbee and Taylor 1993; Liu *et al.* 1999; Lutzoni *et al.* 2001; Vivarès *et al.* 2002). However, since some phylogenetic analyses support a *S. pombe* – *S. cerevisiae* clade (Bullerwell *et al.* 2003), it remains possible that these genes were lost in the common ancestor of *S. cerevisiae* and *S. pombe*. If the loss in each yeast lineage did occur independently, the large size for the set of genes lost in both lineages relative to the sets of genes lost in each lineage suggests that certain genes may be more likely to undergo loss in the yeasts. As complete genome sequences become available from additional yeasts that are distantly related to *S. cerevisiae* and *S. pombe*, such as *Pneumocystis carinii* (a basal ascomycete) and *Cryptococcus neoformans* (a basidiomycete yeast), this question may be resolved.

In addition to the ascomycete yeasts, the complete genome sequence of one other fungus is currently available. The microsporidian *Encephelitozoon cuniculi* is an obligate intracellular parasite with a very small genome (2.9 x 10^6 bp) originally thought to be an ancient eukaryote but now recognized as a highly reduced fungus (Keeling and Fast 2002; Vivarès *et al.* 2002). More recent phylogenetic analyses suggest a specific relationship between the microsporidia and certain zygomycetes (Keeling 2003), indicating that the extreme reduction of the microsporidia occurred independently of the more modest reduction that occurred in the ascomycete yeasts. Not surprisingly, the set of *N. crassa* genes that have non-fungal homologs but lack *E. cuniculi* homologs is very large (2334 protein-coding genes). However, the set of *N. crassa* genes with non-fungal homologs that are clearly absent in the genomes of *S. cerevisiae*, *S. pombe*, and *E. cuniculi* is only 284 genes. Thus, there are 17 genes that were likely to have been lost in both of the ascomycete yeasts that were retained in the genomes of the highly reduced microsporidian *E. cuniculi*. The degree to which these genes share similar functions is unclear, although it is surprising that genes retained in a highly-reduced genome would be dispensable in the yeasts. It is possible that the functions of these genes are mediated by non-orthologous genes in the yeasts (non-orthologous displacement; Koonin *et al.* 1996). Alternatively, these genes may provide information about the differences in the selective pressures that have acted to reduce the gene number and genome size in the free-living yeasts and parasitic microsporidia.

5. UNEXPECTED GENES IN THE *NEUROSPORA* GENOME SEQUENCE

One of the most exciting aspects of genomic sequencing is the potential to find genes that were completely unanticipated in the organism. The *Neurospora* genes that were lost in *S. cerevisiae* (Braun *et al.* 1998; Braun *et al.* 2000) might fall into this category, since it is possible to find many papers in molecular biology journals that assert specific features are universal to the eukaryotes because they are "conserved from yeast to man." However, the availability of multiple eukaryotic genome sequences has made the contribution of gene loss to eukaryotic genome evolution clear (Salzberg *et al.* 2001; Braun 2003), so these genes should not be considered surprising. Nonetheless, the completion of the *N. crassa* genome sequence did reveal a number of genes that were completely unanticipated despite the long history of experimental biology using *Neurospora* species.

5.1 Genes Associated with Secondary Metabolism

Despite the absence of characterized secondary metabolites synthesized by *Neurospora* species, the earliest large-scale sequence data sets revealed the existence of specific genes involved in the synthesis of secondary compounds (Nelson *et al.* 1997). Subsequent analyses revealed a number of genes involved in the synthesis and transport of compounds ranging from trichothecene, lovastatin, aflatoxin, and penicillin in other fungi (Mannhaupt *et al.* 2003). Thus, the diversity of genes involved in secondary metabolism identified using the complete genome sequence (Galagan *et al.* 2003) was consistent with previous large-scale sequence data sets, but they had been unanticipated based upon the biology of *Neurospora* species. A possible resolution of the apparent paradox of finding secondary metabolism genes in an organism may be the production of relatively small amounts of these compounds, possibly as signalling molecules.

One particularly interesting class of secondary metabolism genes found in the complete *N. crassa* genome includes those associated with the biosynthesis of diterpenes and, in particular, gibberellins in other organisms. Gibberellins are best known as regulators of stem elongation and other developmental processes in plants (Graebe 1987; Yamaguchi and Kamiya 2000). However, gibberellic acid (GA) was first identified as a metabolic product of the plant pathogen *Gibberella* (*Fusarium*) *fujikuroi*, a pyrenomycete relative of *N. crassa* that causes "foolish seedling" disease in rice. GA was shown to be responsible for the over elongation of shoots (preceding seedling death), a phenomenon from which the disease received its name (Kurosawa 1926; Yabuta and Sumiki 1938). GA was only later shown to be a normal growth regulator in plants.

Plants and *G. fujikuroi* share certain steps in GA synthesis but differ in others (Hedden *et al.* 2001). The proteins predicted for *N. crassa* include at least one member of each of three enzyme classes required for gibberellin biosynthesis in plants (Yamaguchi and Kamiya 2000), as well as homologs of all enzymes required for GA biosynthesis in *G. fujikuroi*. The enzymes encoded by the *N. crassa* genes include an apparent terpene synthase (aka terpene cyclase; NCU09272.1), several members of the cytochrome P450 monooxygenase family closely related to enzymes involved in gibberellin biosynthesis in both plants (NCU02852.1) and fungi (NCU05376.1, NCU09274.1, NCU05967.1), three apparent homologs of GA4 desaturase from *G. fujikuroi* (NCU00751.1, NCU01598.1, NCU00847.1), and others. (Numbers in parentheses are annotation numbers associated with the *N. crassa* genome sequence reported by Galagan *et al.* 2003.)

Although these proteins have unknown functions in *N. crassa*, their presence has important implications for the evolution of pathogenicity in *G. fujikuroi* and other fungi, because it indicates that the components necessary for GA production were present in the fungal ancestors of pathogens and non-pathogens alike. Quite clearly, *G. fujikuroi* has evolved to use these genes in a special manner in the context of pathogenicity. This is evident in that in *G. fujikuroi* the genes encoding enzymes for GA synthesis reside in a cluster (Tudzynski and Holter 1998), whereas the *N. crassa* homologs do not. It is possible that *N. crassa* and its non-pathogen relatives synthesize gibberellins or related compounds for reasons unrelated to plant pathogenesis, possibly as signals during development. In fact, there is a report of GA3 from *N. crassa* (Kawanabe *et al.* 1983), as well as evidence that GA has effects on hyphal growth in this organism (Tomita *et al.* 1984). In this context, the growing availability of sequence data from the fungi may allow us to determine whether the clustering of GA biosynthetic enzymes in *G. fujikuroi* reflects LGT within the fungi, selection for a novel pattern of expression for a set of genes common to many pyrenomycetes, or a distinct selective pressure.

Alternatively, it is entirely possible that in *N. crassa* these genes have functions unrelated to GA biosynthesis. Throughout the fungal and plant kingdoms, the enzyme families in question participate in the synthesis of large numbers of secondary compounds, including certain antibiotics. In conifers, for example, the terpene synthase family includes a diverse group of enzymes, some with roles in GA synthesis and others that participate in the synthesis of terpene components of resins (Phillips and Croteau 1999; Trapp and Croteau 2001). In addition, specific enzymes can be both multifunctional (catalyzing multiple steps in a pathway) and promiscuous in terms of substrate (Yamaguchi and Kamiya 2000). Therefore, it is not possible to conclude precisely what roles the *N. crassa* enzymes have in diterpene-related secondary metabolism. Regardless of the specific functions of these enzymes, the discovery of genes related to terpene metabolism in *N. crassa* provides strong support for additional studies in this organism, focused on the possible roles of secondary compounds in defense and signaling.

5.2 Genes Associated with Pathogenicity in other Fungi

Neurospora species are not plant or animal pathogens, and they are not known to produce mycotoxins (Perkins and Davis 2000). However, analysis of the complete genome sequence revealed 12 genes associated with pathogenicity in other fungi (Galagan *et al.* 2003). Surprisingly, these potential pathogenicity-related genes include one of the few gene families present in *N. crassa*, with a total of five genes showing homology to the ECP2 gene of the tomato pathogen *Cladosporium fulvum*. ECP2 is a secreted protein that elicits the hypersensitive response and necrosis in some plants (Wubben *et al.* 1994; Takken *et al.* 2000), but its function in *Neurospora* is currently unclear. Some of the genes involved in secondary metabolism (see above, section 5.1) also play a role in the pathogenic phenotype of certain fungi, and their presence in *N. crassa* was unanticipated. Establishing the functions of these pathogenicity-related genes in a non-pathogenic group like *Neurospora* presents both challenges and exciting opportunities to learn about the ways in which pathogenic and non-pathogenic fungi differ.

5.3 Genes Associated with Light Sensing

Although the existence of circadian rhythms and blue-light responses in members of the genus Neurospora has been known for some time (Loros and Dunlap 2001), large-scale sequencing in *N. crassa* has revealed several surprising genes involved in light sensing. All the known light-induced phenotypes in *Neurospora* are regulated by blue light, and all of these phenotypes can be blocked in mutants of either *white collar-1* (*wc-1*) or *white collar-2* (*wc-2*; (Lee *et al.* 2003). The products of the *wc* genes form a flavin mononucleotide-containing complex and act as a photoreceptor (Froehlich *et al.* 2002; He *et al.* 2002). Since the product of *wc-1* is a photoreceptor, and all known responses to light in *Neurospora* are abolished by the *wc* mutations, the potential roles of additional photoreceptors is unclear and their existence was unexpected.

Additional photoreceptors encoded in the *N. crassa* genome include homologs of the archaeal rhodopsins, cryptochromes, and phytochromes (Bieszke *et al.* 1999; Galagan *et al.* 2003). The archaeal rhodopsin homolog (*nop-1*) was identified using EST data before completion of the *N. crassa* genome sequence (Bieszke *et al.* 1999), but it also represents an unexpected gene revealed by large-scale sequencing. Exposure of *nop-1* mutants to the mitochondrial ATPase inhibitor oligomycin produces a light-dependent change in growth pattern (Bieszke *et al.* 1999), although it is unclear whether the change in growth and conidiation reflects an indirect effect of oligomycin or redundancy between the product of *nop-1* and the mitochondrial H+-ATPase for a light-regulated conidiation function. Regardless, these data suggest the existence of light-regulated processes in *Neurospora* that were not anticipated based upon the results of genetic screens.

The presence of phytochromes in *Neurospora*, revealed by analyses of the complete genome sequence (Galagan *et al.* 2003), is also unexpected because all known responses to light in *Neurospora* involve blue-light. However, the bacteriophytochromes related to the *N. crassa* phytochrome homologs and the *Aspergillus nidulans velvet* gene are involved in responses to red light. In fact, *N. crassa* also has a clear ortholog of the *A. nidulans fluG* gene (NCU04264.1), which is known to functionally interact with the *velvet* gene (Yagera *et al.* 1998). Establishing whether the genes that are involved in responses to red light in other organisms that are present in *Neurospora* regulate additional light regulated processes (that were not detected by genetic screens), or whether they interact functionally with the products of the *wc* genes should prove interesting.

6. CONCLUSIONS

The completion of the *N. crassa* genome sequence represents a landmark in fungal genetics and genomics. This event constitutes a fundamental step in the remarkable journey that has taken the biology of *Neurospora* from the experiments of Beadle and Tatum that established a direct relationship between genes and proteins to a preliminary description of *Neurospora*'s 10,000 genes. A number of challenges remain for the future, most notably that of establishing the ways in which these 10,000 genes interact to product the developmental complexity evident in *Neurospora*. However, the availability of the complete genome sequence of *N. crassa* has allowed us to understand more precisely the ways in which this organism differs from the yeasts.

The first fungi for which complete genome sequences were available (the yeasts *S. cerevisiae* and *S. pombe* and the microsporidian *E. cuniculi*) are organisms that have undergone substantial reduction by gene loss during evolution. Thus, it is not surprising that gene loss in the yeast lineages has made a substantial contribution to the differences between the genomes of these organisms and the genome of *N. crassa*. However, loss in the yeast lineages cannot explain all of the differences between the genomes of *Neurospora* species and the yeasts. Instead, some type of genetic innovation must be invoked to explain the large number of genes present in *N. crassa*. Gene duplication, the most common pathway of genetic innovation, has been blocked in the *Neurospora* lineage, and other pathways of genetic innovation such as LGT and partial duplications do not appear to be especially active in the *Neurospora* lineage. Thus, either much of the innovation occurred prior to the origin of the processes that block gene duplication (e.g., RIP), or pathways of innovation that are poorly understood at present are responsible for many of the differences between *N. crassa* and the yeasts.

The difficulties associated with the analysis of the *N. crassa* genome raise the question of to what extent comparative genomics will ultimately help unravel the nature of fungal diversity. Our analysis of the *N. crassa* genome suggests that much, if not most, of the genetic framework required for this diversity is reflected in the genome of this one organism. The genome sequence of *N. crassa* has revealed genes thought to have functions that *Neurospora* was not known to possess, particularly in the realms of secondary metabolism, interactions with plants, and the transduction of signals from the environment. The *N. crassa* genome may well have a nearly complete complement of the gene families present among the filamentous fungi. This presents both opportunities and challenges for experimental biologists, because it implies that major differences in biology may result from small changes in gene structure and expression.

In the eighty years since Shear and Dodge named the genus *Neurospora*, experiments using this organism established the fundamental relationship between genetics and biochemistry. With the publication of the *N. crassa* genome we now have the tools to elucidate the nature of the subtle changes in genetic function that have led to the diversity of phenotypes evident in the filamentous fungi. We believe this will ultimately lead to another exciting eighty years of *Neurospora* research.

Acknowledgements: This research was supported by the NSF Grant MCB-9874488 to M.A.N.

REFERENCES

Alexopoulos CJ, Mims CW, and Blackwell M (1996). Introductory Mycology. New York: John Wiley & Sons.

Aravind L, Watanabe H, Lipman DJ, and Koonin E (2000). Lineage-specific loss and divergence of functionally linked genes in eukaryotes. Proc Natl Acad Sci USA 97:11319-11324.

Beadle GW (1945). Biochemical genetics. Chem Revs 37:15-96.

Beadle GW and Tatum EL (1941). Genetic control of biochemical reactions in *Neurospora*. Proc Natl Acad Sci USA 27:499-506.

Bean LE, Dvorachek WH, Braun EL, Errett A, Saenz GS, Giles MD, Werner-Washburne M, Nelson MA, and Natvig DO (2001). Analysis of the *pdx-1* (*snz-1/sno-1*) region of the *Neurospora crassa* genome: Correlation of pyridoxine-requiring phenotypes with mutations in two structural genes. Genetics 157:1067-1075.

Bennett JW (1997). White paper: Genomics for filamentous fungi. Fungal Genet Biol 21:3-7.

Berbee ML and Taylor JW (1993). Dating the evolutionary radiations of the true fungi. Can J Bot 71:1114-1127.

Berman J and Sudbery PE (2002). *Candida albicans*: A molecular revolution built on lessons from budding yeast. Nature Rev Genet 3:918-930.

Bieszke JA, Braun EL, Bean LE, Kang S, Natvig DO, and Borkovich KA (1999). The *nop-1* gene of *Neurospora crassa* encodes a seven transmembrane helix retinal-binding protein homologous to archaeal rhodopsins. Proc Natl Acad Sci USA 96:8034-8039.

Brachat S, Dietrich FS, Voegeli S, Lerch A, Gaffney T, and Philippsen P (2003). The *Ashbya gossypii* genome: lessons learned by comparison to the *Saccharomyces cerevisiae* genome. Fungal Genet Newsl 50 (Suppl.):92.

Braun EL (2003). Innovation from reduction: Gene loss, domain loss, and sequence divergence in genome evolution. Appl Bioinform, in press.

Braun EL and Grotewold E (2001). Fungal Zuotin proteins evolved from MIDA1-like factors by lineage-specific loss of MYB domains. Mol Biol Evol 18:1401-1412.

Braun EL, Halpern AL, Nelson MA, and Natvig DO (2000). Large-scale comparison of fungal sequence information: Mechanisms of innovation in *Neurospora crassa* and gene loss in *Saccharomyces cerevisiae*. Genome Res 10:416-430.

Braun EL, Kang S, Nelson MA, and Natvig DO (1998). Identification of the first fungal annexin: Analysis of annexin gene duplications and implications for eukaryotic evolution. J Mol Evol 47:531-543.

Bruns TD, Vilgalys R, Barns SM, Gonzalez D, Hibbett DS, Lane DJ, Simon L, Stickel S, Szaro TM, Weisburg WG, and Sogin ML (1992). Evolutionary relationships within the fungi: Analyses of nuclear small subunit rRNA sequences. Mol Phylogenet Evol 1:231-241.

Bullerwell CE, Leigh J, Forget L, and Lang BF (2003). A comparison of three fission yeast mitochondrial genomes. Nucl Acids Res 31:759-768.

Cambareri EB, Jensen BC, Schabtach E, and Selker EU (1989). Repeat-induced G-C to A-T mutations in *Neurospora*. Science 244:1571-1575.

Case ME, Schweizer M, Kushner SR, and Giles NH (1979). Efficient transformation of *Neurospora crassa* by utilizing hybrid plasmid DNA. Proc Natl Acad Sci USA 76:5259-5263.

Cogoni C (2001). Homology-dependent gene silencing mechanisms in fungi. Annu Rev Microbiol 55:381-406.

Cohen BA, Mitra RD, Hughes JD, and Church GM (2000). A computational analysis of whole genome expression data reveals chromosomal domains of gene expression. Nature Genet 26:183 186.

Davis RH and Perkins DD (2002). *Neurospora*: A model of model microbes. Nature Rev Genet 3:397-403.

Doolittle WF (1998). You are what you eat: A gene transfer ratchet could account for bacterial genes in eukaryotic nuclear genomes. Trends Genet 14:307-311.

Doolittle WF (1999). Lateral genomics. Trends Genet 15:5-8.

Froehlich AC, Liu Y, Loros JJ, and Dunlap JC (2002). White Collar-1, a circadian blue light photoreceptor, binding to the frequency promoter. Science 297:815-819.

Galagan JE, Calvo SE, Borkovich KA, Selker EU, Read ND, Jaffe D, FitzHugh W, Ma L-J, Smirnov S, Purcell S, Rehman B, Elkins T, Engels R, Wang S, Nielsen C, Butler J, Endrizzi M, Qui D, Ianakiev P, Bell-Pedersen D, Nelson MA, Werner-Washburne M, Selitrennikoff CP, Kinsey JA, Braun EL, Zelter A, Schulte U, Kothe GO, Jedd G, Mewes W, Staben C, Marcotte E, Greenberg D, Roy A, Foley K, Naylor J, Stange-Thomann N, Barrett R, Gnerre S, Kamal M, Kamvysselis M, Mauceli E, Bielke C, Rudd S, Frishman D, Krystofova S, Rasmussen C, Metzenberg RL, Perkins DD, Kroken S, Cogoni C, Macino G, Catcheside D, Li W, Pratt RJ, Osmani SA, DeSouza CP, Glass L, Orbach MJ, Berglund JA, Voelker R, Yarden O, Plamann M, Seiler S, Dunlap J, Radford A, Aramayo R, Natvig DO, Alex LA, Mannhaupt G, Ebbole DJ, Freitag M, Paulsen I, Sachs MS, Lander ES, Nusbaum C, and Birren B (2003). The genome sequence of the filamentous fungus *Neurospora crassa*. Nature 422:859-868.

Geever RF, Huiet L, Baum JA, Tyler BM, Patel VB, Rutledge BJ, Case ME, and Giles NH (1989). DNA sequence, organization and regulation of the *qa* gene cluster of *Neurospora crassa*. J Mol Biol 207:15-34.

Goffeau A, Barrell BG, Bussey H, Davis RW, Dujon B, Feldmann H, Galibert F, Hoheisel JD, Jacq C, Johnston M, Louis EJ, Mewes HW, Murakami Y, Philippsen P, Tettelin H, and Oliver SG (1996). Life with 6000 genes. Science 274:546-567.

Gogarten JP (2003). Gene transfer: gene swapping craze reaches eukaryotes. Curr Biol 13:R53 R54.

Gow NAR, Brown AJP, and Odds FC (2002). Fungal morphogenesis and host invasion. Curr Opin Microbiol 5:366-371.

Graebe JE (1987). Gibberellin biosynthesis and control. Annu Rev Plant Physiol Plant Mol Biol 38:419-465.

He Q, Cheng P, Yang Y, Wang L, Gardner KH, and Liu Y (2002). White Collar-1, a DNA binding transcription factor and a light sensor. Science 297:840-843.

Heath IB and Steinberg G (1999). Mechanisms of hyphal tip growth: tube dwelling amebae revisited. Fungal Genet Biol 28:79-93.

Hedden P, Phillips AL, Rojas MC, Carrera E, and Tudzynski B (2001). Gibberellin biosynthesis in plants and fungi: A case of convergent evolution? J Plant Growth Regul 20:319-331.

Hooper SD and Berg OG (2003). On the nature of gene innovation: Duplication patterns in microbial genomes. Mol Biol Evol 20:945-954.

Houlahan MB, Beadle GW, and Calhoun HG (1949). Linkage studies with biochemical mutants of *Neurospora crassa*. Genetics 34:493-507.

Ikeda K, Nakayashiki H, Kataoka T, Tamba H, Hashimoto Y, Tosa Y, and Mayama S (2002). Repeat-induced point mutation (RIP) in *Magnaporthe grisea*: implications for its sexual cycle in the natural field context. Mol Microbiol 45:1355-1364.

Jacobson DJ, Powell AJ, Dettman JR, Saenz GS, Barton MM, Hiltz MD, Dvorachek WH, Glass NL, Taylor JW, and Natvig DO (2003). *Neurospora* in temperate forests of western North America. Mycologia, in press.

Kawanabe Y, Yamane H, Murayama T, Takahashi N, and Nakamura T (1983). Identification of gibberellin A3 in mycelia of *Neurospora crassa*. Agric. Biol. Chem. 47:1693-1694.

Keeling PJ (2003). Congruent evidence from alpha-tubulin and beta-tubulin gene phylogenies for a zygomycete origin of microsporidia. Fungal Genet Biol 38:298-309.

Keeling PJ and Fast NM (2002). Microsporidia: Biology and evolution of highly reduced intracellular parasites. Annu Rev Microbiol 56:93-116.

Keese PK and Gibbs A (1992). Origins of genes: "Big bang" or continuous creation? Proc Natl Acad Sci USA 89:9489-9493.

Kelkar HS, Griffith J, Case ME, Covert SF, Hall RD, Keith CH, Oliver JS, Orbach MJ, Sachs MS, Wagner JR, Weise MJ, Wunderlich JK, and Arnold J (2001). The *Neurospora crassa* genome: Cosmid libraries sorted by chromosome. Genetics 157:979-990.

Kellis M, Patterson N, Endrizzi M, Birren B, and Lander ES (2003). Sequencing and comparison of yeast species to identify genes and regulatory elements. Nature 423:241-254.

Koonin EV, Mushegian AR, and Bork P (1996). Non-orthologous gene displacement. Trends Genet 12:334-336.

Kupfer DM, Reece CA, Clifton SW, Roe BA, and Prade RA (1997). Multicellular ascomycetous fungal genomes contain more than 8000 genes. Fungal Genet Biol 21:364-372.

Kurosawa E (1926). Experimental studies on the nature of the substance excreted by the "bakanae" fungus (in Japanese). Trans Nat His Soc Formosa 16:213-227.

Lawrence JG and Roth JR (1996). Selfish operons: Horizontal transfer may drive the evolution of gene clusters. Genetics 143:1843-1860.

Lee K, Dunlap JC, and Loros JJ (2003). Roles for WHITE COLLAR-1 in circadian and general photoperception in *Neurospora crassa*. Genetics 163:103-114.

Lindegren CC, Beanfield V, and Barber R (1939). Increasing the fertility of *Neurospora* by selective inbreeding. Bot Gaz 100:592-599.

Liu YJ, Whelen S, and Hall BD (1999). Phylogenetic relationships among ascomycetes: Evidence from an RNA polymerase II subunit. Mol Biol Evol 16:1799-1808.

Loros JJ and Dunlap JC (2001). Genetic and molecular analysis of circadian rhythms in *Neurospora*. Annu Rev Physiol 63:757-794.

Lutzoni F, Pagel M, and Reeb V (2001). Major fungal lineages are derived from lichen symbiotic ancestors. Nature 411:937–940.

Mannhaupt G, Montrone C, Haase D, Mewes HW, Aign V, Hoheisel JD, Fartmann B, Nyakatura G, Kempken F, Maier J, and Schulte U (2003). What's in the genome of a filamentous fungus? Analysis of the *Neurospora* genome sequence. Nucl Acids Res 31:1944-1954.

McClintock B (1945). *Neurospora*. I. Preliminary observations of the chromosomes of *Neurospora crassa*. Am J Bot 32:671-678.

Miller W (2000). So many genomes, so little time. Nature Biotechnol 18:148-149.

Mitchell MB (1955). Aberrant recombination of pyridoxine mutants of *Neurospora*. Proc Natl Acad Sci USA 41:215-220.

Nelson MA (1996). Mating systems in ascomycetes: A romp in the sac. Trends Genet 12:69-74.

Nelson MA, Kang S, Braun EL, Crawford ME, Dolan PL, Leonard PM, Mitchell J, Armijo AM, Bean L, Blueyes E, Cushing T, Errett A, Fleharty M, Gorman M, Judson K, Miller R, Ortega J, Pavlova I, Perea J, Todisco S, Trujillo R, Valentine J, Wells A, Werner-Washburne M, and Natvig DO (1997). Expressed sequences from conidial, mycelial, and sexual stages of *Neurospora crassa*. Fungal Genet Biol 21:348-363.

Ohno S (1970). Evolution by Gene Duplication. New York: Springer.

Ohno S (1984). Birth of a unique enzyme from an alternative reading frame of the preexisted, internally repetitious coding sequence. Proc Natl Acad Sci USA 81:2421-2425.

Padilla PA, Fuge EK, Crawford ME, Errett A, and Werner-Washburne M (1998). The highly conserved, coregulated *SNO* and *SNZ* gene families in *Saccharomyces cerevisiae* respond to nutrient limitation. J Bacteriol 180:5718-5726.

Perkins DD and Davis RH (2000). Evidence for safety of *Neurospora* species for academic and commercial uses. Appl Environ Microbiol 66:5107-5109.

Perkins DD, Radford A, and Sachs MS (2001). The *Neurospora* Compendium: Chromosomal Loci. San Diego: Academic Press.

Perkins DD and Turner BC (1988). *Neurospora* from natural populations: toward population biology of a haploid eukaryote. Exp Mycol 12:91-131.

Phillips MA and Croteau RB (1999). Resin-based defenses in conifers. Trends Plant Sci 4:184 190.

Prade RA, Griffith J, Kochut K, Arnold J, and Timberlake WE (1997). *In vitro* reconstruction of the *Aspergillus* (= *Emericella*) *nidulans* genome. Proc Natl Acad Sci USA 94:4564-14569.

Roelofs J and Van Haastert PJ (2001). Genes lost during evolution. Nature 411:1013-1014.

Roenneberg T and Merrow M (2001). Seasonality and photoperiodism in fungi. J Biol Rhythms 16:403-414.

Rosewich UL and Kistler HC (2000). Role of horizontal gene transfer in the evolution of fungi. Annu Rev Phytopathol 38:325-363.

Salzberg SL, White O, Peterson J, and Eisen JA (2001). Microbial genes in the human genome: Lateral transfer or gene loss? Science 292:1903-1906.

Schulte U, Becker I, Mewes HW, and Mannhaupt G (2002). Large scale analysis of sequences from *Neurospora crassa*. J Biotechnol 94:3-13.

Seack J, Perovic S, Gamulin V, Schröder HC, Beutelman P, Müller IM, and Müller WEG (2001). Identification of highly conserved genes: SNZ and SNO in the marine sponge *Suberites domuncula*: their gene structure and promoter activity in mammalian cells. Biochim Biophys Acta 1520:21-34.

Shear CL and Dodge BO (1927). Life histories and heterothallism of the red bread-mold fungi of the *Monilia sitophila* group. J Agric Res 34:1019-1042.

Shiu PKT, Raju NB, Zickler D, and Metzenberg RL (2001). Meiotic silencing by unpaired DNA. Cell 107:905-916.

Smith ML, Bruhn JN, and Anderson JB (1992). The fungus *Armillaria bulbosa* is among the largest and oldest living organisms. Nature 356:428-431.

Stanhope MJ, Lupas A, Italia MJ, Koretke KK, Volker C, and Brown JR (2001). Phylogenetic analyses do not support horizontal gene transfers from bacteria to vertebrates. Nature 411:940 944.

Takken FL, Luderer R, Gabriels SH, Westerink N, Lu R, deWit PJ, and Joosten MH (2000). A functional cloning strategy, based on a binary PVX-expression vector, to isolate HR-inducing cDNAs of plant pathogens. Plant J 24:275-283.

The International Human Genome Sequencing Consortium (2001). Initial sequencing and analysis of the human genome. Nature 409:860-921.

Tomita K, Murayama T, and Nakamura T (1984). Effects of auxin and gibberellin on elongation of young hyphae in *Neurospora crassa*. Plant Cell Physiol 25:355-358.

Trapp S and Croteau R (2001). Defensive resin biosynthesis in conifers. Annu Rev Plant Physiol Plant Mol Biol 52:689-724.

Tudzynski B and Holter K (1998). Gibberellin biosynthetic pathway in *Gibberella fujikuroi*: Evidence for a gene cluster. Fungal Genet Biol 25:157-170.

Vivarès CP, Gouy M, Thomarat F, and Méténier G (2002). Functional and evolutionary analysis of a eukaryotic parasitic genome. Curr Opin Microbiol 5:499-505.

Watters MK, Randall TA, Margolin BS, Selker EU, and Stadler DR (1999). Action of repeat induced point mutation on both strands of a duplex and on tandem duplications of various sizes in *Neurospora*. Genetics 153:705-714.

Wood V, Gwilliam R, Rajandream MA, Lyne M, Lyne R, Stewart A, Sgouros J, Peat N, Hayles J, Baker S, Basham D, Bowman S, Brooks K, Brown D, Brown S, Chillingworth T, Churcher C, Collins M, Connor R, Cronin A, Davis P, Feltwell T, Fraser A, Gentles S, Goble A, Hamlin N, Harris D, Hidalgo J, Hodgson G, Holroyd S, Hornsby T, Howarth S, Huckle EJ, Hunt S, Jagels K, James K, Jones L, Jones M, Leather S, McDonald S, McLean J, Mooney P, Moule S, Mungall K, Murphy L, Niblett D, Odell C, Oliver K, O'Neil S, Pearson D, Quail MA, Rabbinowitsch E, Rutherford K, Rutter S, Saunders D, Seeger K, Sharp S, Skelton J, Simmonds M, Squares R, Squares S, Stevens K, Taylor K, Taylor R, Tivey A, Walsh S, Warren T, Whitehead S, Woodward J, Volckaert G, Aert R, Robben J, Grymonprez B, Weltjens I, Vanstreels E, Rieger M, Schafer M, Muller-Auer S, Gabel C, Fuchs M, Fritzc C, Holzer E, Moestl D, Hilbert H, Borzym K, Langer I, Beck A, Lehrach H, Reinhardt R, Pohl TM, Eger P, Zimmermann W, Wedler H, Wambutt R, Purnelle B, Goffeau A, Cadieu E, Dreano S, Gloux S, Lelaure V, Mottier S, Galibert F, Aves SJ, Xiang Z, Hunt C, Moore K, Hurst SM, Lucas M, Rochet M, Gaillardin C, Tallada VA, Garzon A, Thode G, Daga RR, Cruzado L, Jimenez J, Sanchez M, delRey F, Benito J, Dominguez A, Revuelta JL, Moreno S, Armstrong J, Forsburg SL, Cerrutti L, Lowe T, McCombie WR, Paulsen I, Potashkin J, Shpakovski GV, Ussery D, Barrell BG and Nurse P (2002). The genome sequence of *Schizosaccharomyces pombe*. Nature 415:871-880.

Wood V, Rutherford KM, Ivens A, Rajandream M-A, and Barrell BG (2001). A re-annotation of the *Saccharomyces cerevisiae* genome. Comp Funct Genom 2:143-154.

Wubben JP, Joosten MH, and deWit PJ (1994). Expression and localization of two *in planta* induced extracellular proteins of the fungal tomato pathogen *Cladosporium fulvum*. Mol Plant Microbe Interact 7:516-524.

Yabuta T and Sumiki Y (1938). On the crystal of gibberellin, a substance to promote plant growth. J Agric Chem Soc Japan 14:1526.

Yagera LN, Lee H-O, Nagle DL, and Zimmerman JE (1998). Analysis of *fluG* mutations that affect light-dependent conidiation in *Aspergillus nidulans*. Genetics 149:1777-1786.

Yamaguchi S and Kamiya Y (2000). Gibberellin biosynthesis: its regulation by endogenous and environmental signals. Plant Cell Physiol 41:251-257.

Applied Mycology & Biotechnology
An International Series. Volume 4. Fungal Genomics

Genetics and Genomics of *Mycosphaerella graminicola*: A Model for the Dothideales

Stephen B. Goodwin[1], Cees Waalwijk[2] and Gert H. J. Kema[2]
[1] U. S. Department of Agriculture, Agricultural Research Service, Department of Botany and Plant Pathology, 915 West State Street, Purdue University, West Lafayette, IN 47907-2054, USA (sgoodwin@purdue.edu); [2] Plant Research International B.V., P.O. Box 16, 6700 AA Wageningen, The Netherlands.

The genus *Mycosphaerella* and its associated anamorphs comprise one of the largest groups of plant pathogenic fungi that also includes a few opportunistic pathogens of humans and other mammals. Despite the high economic importance of this group, it has no representatives among the current model host-pathogen systems. Because the current and developing model systems are too distantly related, a model within the genus *Mycosphaerella* is greatly needed. Recent discoveries in genetics and the availability of extensive molecular tools in both *M. graminicola* and its wheat host make this the best candidate for a new model system for fungi in the genus *Mycosphaerella* and the order Dothideales.

1. INTRODUCTION

Mycosphaerella graminicola is developing rapidly as a model for fungi in the order Dothideales. Most fungi in this order are pleomorphic, with two names, one for the teleomorph (sexual stage) and another for the anamorph (asexual stage). The Dothideales and associated anamorph genera form one of the largest orders of plant pathogenic fungi, containing several thousand species at least. One or more pathogens from this order are found on hosts spanning the range of plant diversity, including dicots and monocots, gymnosperms, ferns, horsetails and lycopods in both temperate and tropical climates (Farr *et al.* 1995). Most cause leaf-spot symptoms that damage but do not kill their hosts. However, many species cause economically significant damage on important crop species. In addition to plant hosts, a few fungi in this order such as black yeasts in several genera (de Hoog *et al.* 1999) are opportunistic pathogens of humans and other animal species.

Mycosphaerella is one of the largest genera in the Dothideales containing at least 500 species (Corlett 1991) and probably more than 1,000. So far, more than 40 anamorph genera have been associated with *Mycosphaerella*, some of which also contain large numbers of species. For example, named species of *Septoria* number more than 1,000 (Hawksworth *et al.* 1995) while those of *Cercospora* are more than 3,000 (Pollack 1987). Even if many of these names are synonyms, the number of species of *Mycosphaerella* and its associated anamorphs easily could be between three and five thousand.

Despite this large number of species, no good model system exists currently for fungi in the genus *Mycosphaerella* or the order Dothideales. This is unfortunate, because the economic damage caused by these loculoascomycete fungi can be huge. Septoria tritici leaf blotch, caused by *M. graminicola* (anamorph *S. tritici*), is one of the most important diseases of wheat worldwide, occurring in most wheat-growing regions every year and causing yield losses of up to 50% (Eyal *et al.* 1987). Black sigatoka, caused by *M. fijiensis*, is one of the most important diseases of banana requiring costly and environmentally damaging fungicide sprays to minimize crop loss. Many if not most cultivated plants have at least one disease caused by a member of this economically important order of pathogens. A genetic and genomic model organism to represent this order is desperately needed to focus research and help design improved disease-management strategies for the future.

The purpose of this chapter is to review recent advances in our understanding of the genetics and genomics of *M. graminicola* and its wheat host. These advances position the *M. graminicola* -wheat pathosystem as an excellent model for fungi in general and for the Dothideales in particular.

2. NEED FOR A NEW MODEL SYSTEM

Existing and developing model systems for fungal host-pathogen interactions are too few, too distantly related and cover too small a portion of the phylogenetic spectrum to model biological traits that may be unique to the Dothideales. The primary model systems for fungal plant pathogens are the rice blast pathogen *Magnaporthe oryzae* and the aflatoxin-producing fungus *Emericella* (anamorph *Aspergillus*) *nidulans*. Other plant pathogenic fungi that have been studied intensively and are likely to serve as models in the future include the powdery mildew fungus of wheat, *Blumeria graminis* f. sp. *tritici*, the wheat stem rust pathogen *Puccinia graminis*, the head scab pathogen *Fusarium graminearum* and the discomycetes *Sclerotinia sclerotiorum* and *Botryotinia fuckeliana* (anamorph *Botrytis cinerea*). The only other loculoascomycete being developed as a model system is the southern corn leaf blight pathogen *Cochliobolus heterostrophus* (anamorph *Bipolaris maydis*). However, all of these are distantly related phylogenetically and have many biological differences from fungi in the Dothideales.

2.1. Phylogenetic Relationships of *Mycosphaerella*

Taxonomic placement of the genus *Mycosphaerella* has been confusing and contradictory. Currently, the phylogenetic relationships of this genus are considered uncertain (Eriksson and Winka 1998). Most taxonomists place *Mycosphaerella* in the Dothideales on the basis of shared morphological features of the teleomorphs (Barr 1987; Luttrell 1973), and this is supported by recent phylogenetic analyses of *Mycosphaerella* and related taxa (Crous *et al.* 2001; Goodwin *et al.* 2001b). In a comprehensive analysis of sequences of the internal transcribed spacer (ITS) region of the ribosomal DNA, all of the *Mycosphaerella* species tested had a common evolutionary origin and were contained within a large cluster derived from a single branch with high bootstrap support (Goodwin *et al.* 2001b). These species included a large number of anamorphs, many of which had not been associated with *Mycosphaerella* previously. Species that were closely related to *M. graminicola* included the tomato pathogen *Cladosporium fulvum* (often used as a model to study host-pathogen interactions, but with no known sexual stage), the strawberry leaf spot pathogen *Ramularia brunnea* (teleomorph *M. fragariae*), *Cercospora* species that cause diseases on numerous important crops including beets (*C. beticola*), maize (*C. zeae-maydis*) and soybean (*C. kikuchii*), other species of *Cladosporium* including the iris leaf spot pathogen *C. iridis*, and the human pathogens *Hortaea werneckii*, *Lacazia loboi* and *Trimmatostroma abietina* (Goodwin *et al.* 2001b). Similar results were reported for different species of *Mycosphaerella* by Crous *et al.* (2001). Analyses of ITS data also confirmed that *Mycosphaerella* is much more closely related to *Dothidea* and belongs in the Dothideales rather than the Pleosporales (Goodwin *et al.* 2001b: Goodwin and Zismann 2001).

In contrast, molecular taxonomists sometimes have included *Mycosphaerella* in the Pleosporales. In these cases only one *Mycosphaerella* species represented the genus: *M. citrullina* (Silva-Hanlin and Hanlin 1999); *M. mycopappi* (Berbee 2001); or *M. zeae-maydis*

Fig. 1. Phylogenetic analysis of 18S ribosomal RNA gene sequences from *Mycosphaerella graminicola* and other fungi. All bootstrap values of 70 or greater (percent of 1000 replications) are indicated at the appropriate nodes. Branch lengths are proportional to genetic distance, which is indicated by the bar at the upper left. Fungi being developed as models are indicated in bold. Brackets to the right of the dendrogram indicate major groups: M, *Mycosphaerella*; D, Dothideales; and P, Pleosorales.

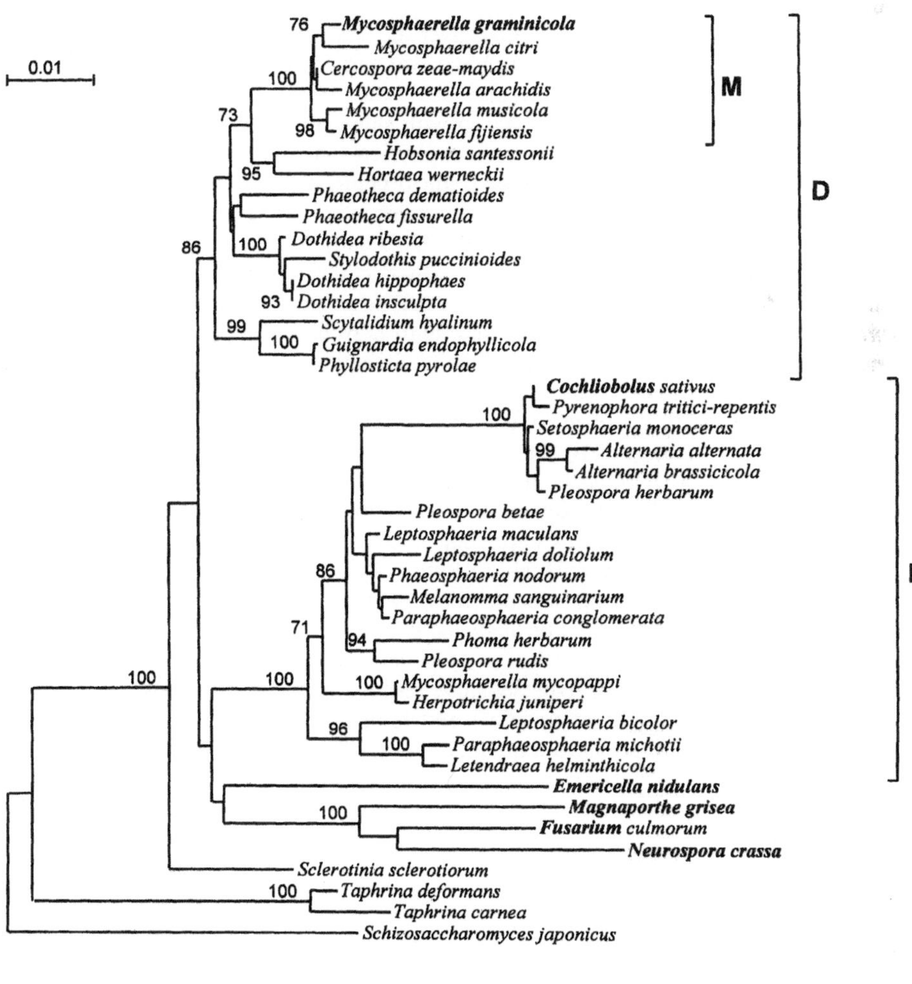

(Arie *et al.* 1997). In all of these cases the single *Mycosphaerella* species analyzed was more similar to fungi in the Pleosporales than to those in the Dothideales. Most of these contradictions can be explained easily by misclassification. The species *M. citrullina* (Silva-Hanlin and Hanlin 1999) and *M. zeae-maydis* (Arie *et al.* 1997) actually belong in *Didymella*, not *Mycosphaerella* (Corlett 1991), and are expected to cluster in the Pleosporales, not the Dothideales. The seemingly anomalous result with *M. mycopappi* (Berbee 2001) so far has not been explained, but it also may be a simple case of misidentification or misclassification. Placement of *Mycosphaerella* in the Dothideales was further tested by analysis of 18S ribosomal RNA sequences (Goodwin and Cavaletto 2002). This region evolves more slowly than the ITS so is more appropriate for analyzing ancient evolutionary events. The 18S analysis confirmed that *Mycosphaerella* is related to *Dothidea* and belongs in the Dothideales (Fig. 1). Furthermore, the Dothideales was phylogenetically distinct from the Pleosporales and other orders containing model organisms. Thus, there is a great need for a representative of the Dothideales to augment the phylogenetic coverage of current model systems.

2.2. Unusual Biology of *Mycosphaerella*

Histopathological studies have revealed the major characteristics of the wheat-*M. graminicola* pathosystem (Cohen and Eyal 1993; Hilu and Bever 1957; Kema *et al.* 1996d; Weber 1922), some of which are unusual. Both pycnidiospores and ascospores have a very high germination rate and their germ tubes grow over the leaf surface in an undirected way until they encounter a stoma. Recently, some emphasis was put on the capability of *M. graminicola* to penetrate the host directly through the anticlinal walls of epidermal cells (Rohel *et al.* 2001). However, detailed studies of early hyphal growth showed conclusively that most infections are established through the stomatal pores without physical damage to nearby host cells. Germ tubes may produce appressorium-like structures prior to penetration of the stomatal opening (Duncan and Howard 2000). The function of these structures is not yet understood, although they are not required for penetration (Kema *et al.* 1996d). This contrasts to the appressoria of other model fungi such as *Magnaporthe grisea* which generate sufficient turgor pressure to physically penetrate host tissues directly (Dixon *et al.* 1999). After penetration, further colonization by *M. graminicola* is symptomless and intercellular until 8-10 days post inoculation (dpi), when rapid collapse of mesophyll tissue occurs in compatible interactions (Duncan and Howard 2000; Kema *et al.* 1996d). The rapid tissue collapse suggests an active role of toxic compounds (Kema *et al.* 1996d), which are produced *in vitro* and could act as virulence factors (Perrone *et al.* 2000). Lack of symptoms during early pathogenesis suggests that, if toxins are involved in the pathogenicity of *M. graminicola*, their production may be initiated by morphological or developmental triggers. Toxin production in some other fungi is regulated by infection-related morphogenesis, but at the stage of appressorium formation (Weiergang *et al.* 1996) or earlier (Dunkle *et al.* 1991; Jones and Dunkle 1995). Possible developmental regulation of toxin production late in the infection process may be an unusual characteristic of the wheat-*M. graminicola* interaction that cannot be analyzed in other model systems.

The resistance mechanism of wheat to *M. graminicola* is largely unknown but appears to be different from those of classical hypersensitive responses. Host responses to the pathogen range from near immunity through intermediate and high susceptibility. Because the responses of many host-isolate interactions have a quantitative mode of expression, it is likely that resistance in this pathosystem, at least partly, results from processes that retard rather than eliminate fungal colonization (Kema *et al.* 1996d). This may explain why the compartmentalization described in typical hypersensitive responses of other cereal-fungus interactions are absent in this pathosystem. These aspects of the biology of the *M. graminicola*-wheat interaction are different from and expand those found in other model systems.

3. DEVELOPMENT OF *M. GRAMINCIOLA* AS A MODEL ORGANISM

A useful model system for fungal host-pathogen interactions should possess several characteristics. First, both host and pathogen should be amenable to thorough genetic analyses, both by classical approaches and molecular technologies. Second, it is desirable if the interactions between the host and pathogen span the range from highly susceptible, through moderately resistant, to immune including both single-gene and quantitative resistances to allow analysis of interactions with applicability to the broadest range of biological questions. Third, both host and pathogen should be easy to grow and manipulate in the laboratory or greenhouse. A pathogen causing an economically important disease of a major crop is attractive as new insights can be applied to real-world problems directly without translation; information derived from non-crop models such as various pathogens on *Arabidopsis* may or may not apply to wheat. Availability of tools such as mutant strains, linkage maps, molecular markers, or other genetic stocks is essential. Technologies for assessing and altering gene expression in host and pathogen are useful for testing the function of genes potentially involved in pathogenicity or resistance. A small genome size and short generation time will increase the speed and efficiency of research. For fungi, models representing each of the major phylogenetic lineages and major variants in life history or other interesting biological traits are desirable. Availability of a gene catalog, EST or other publicly available sequence databases will greatly foster research. All of these requirements are met with the *M. graminicola*-wheat pathosystem except for the genome size of the host which is extremely large and complicates functional analyses.

3.1. Basic Genetics of *M. graminicola*

Genetic analyses of this pathogen were not possible until Sanderson (1972; 1976) identified *M. graminicola* as the teleomorph of *S. tritici*. Since then, *M. graminicola* has been found in virtually all wheat-producing areas of the world (Garcia and Marshall 1992; Madariaga 1986; Scott *et al.* 1988). The first attempt at genetic analysis occurred when Sanderson *et al.* (1986) tested eight ascospore isolates for virulence to a number of wheat cultivars. Although the parents of the ascospore isolates were unknown, the amount of genetic variation was consistent with segregation during meiosis.

In the 30 years since Sanderson's initial discovery of its teleomorph, *M. graminicola* has been developed rapidly into an excellent genetic model. Development of a laboratory crossing protocol resulted in the determination of a bipolar heterothallic mating system in *M. graminicola* (Kema *et al.* 1996c), consistent with the overwhelming evidence for recombination among molecular markers in natural populations (Chen and McDonald 1996; McDonald *et al.* 1999). Mendelian inheritance of both biological and molecular markers occurred despite the large karyotypic differentiation observed among isolates. The *MAT* alleles, *MAT1-1* and *MAT1-2,* were recently cloned and characterized (Waalwijk *et al.* 2002b). Specific PCR primers allow the mating type of any isolate to be determined accurately and rapidly (Waalwijk *et al.* 2002b). Interestingly, both *MAT* alleles occur in approximately equal frequencies in most populations tested worldwide including a small sample of isolates from durum wheat (Waalwijk *et al.* 2002b; Zhan *et al.* 2002).

3.1.1 Mating System

With its heterothallic, bipolar mating system (Kema *et al.* 1996c) crosses can be made by pairing isolates of opposite mating type on a compatible wheat cultivar. The first step is a mating-type characterization of potential parent isolates with diagnostic PCR primers (Waalwijk *et al.* 2002b). Isolates of opposite mating type are co-inoculated onto seedlings of susceptible wheat cultivars, grown in a greenhouse until symptoms are visible and subsequently transferred outdoors. After four to six weeks, the inoculated leaves are sampled at random and tested for discharge of ascospores. Discharge is facilitated by drying, so the leaf tissue is first soaked in water, then transferred to filter paper in the lid of a water-agar filled Petri plate for discharge (Kema *et al.* 1996c). Usually, ascospores start to discharge after 15-20 minutes and can be collected individually using a sterile needle. All isolates tested so far have been fertile and produced viable progeny. Hence, the generation of large segregating populations, including backcross and F_2 progeny for mapping purposes have become standard laboratory practice.

3.1.2 Avirulence

Genetic analyses of avirulence have mainly involved crosses between the Dutch field isolates IPO323 and IPO94269 (Brading *et al.* 2002; Kema *et al.* 2000). These isolates show characteristic phenotypic differences towards numerous wheat cultivars, including the differential cultivars Shafir, Kavkaz, and Veranopolis. Inoculations of F_1, BC_1 and F_2 progenies in whole-plant and detached-leaf assays induced regular and typical symptoms. The differential cultivars Veranopolis, Kavkaz and Shafir were either susceptible or resistant with similar symptoms as the respective checks. Therefore, qualitative scores (virulent or avirulent) could be taken for each isolate. The segregation for avirulence on each of the differential cultivars in the F_1 population was according to a 1:1 ratio, indicating that avirulence in IPO323 is controlled by a single locus. Surprisingly, all F_1 isolates were either virulent or avirulent for the three differential cultivars Veranopolis, Kavkaz and Shafir, even though these cultivars have different pedigrees and are unlikely to share resistance genes. Amplified fragment length polymorphism (AFLP) markers and mating-type data confirmed that the F_1 was a segregating population. Analysis of BC_1 and F_2 populations confirmed the 1:1 segregation ratio for avirulence in 221 isolates from diverse segregating progenies. Crosses between durum- and bread-wheat adapted strains, which comprise the strongest interactions observed in the *M. graminicola*-wheat system, also produced viable progenies. Abundant recombination was observed in this population resulting in isolates virulent or avirulent to both hosts.

3.1.3 Genetic map

The aforementioned cross between *M. graminicola* isolates IPO323 and IPO94269 also was used to create a genetic linkage map (Kema *et al.* 2002). The total F_1 progeny population consisted of 202 isolates, among which a sub-set of 68 isolates was chosen for the mapping population. The framework of the map consisted of AFLP markers generated with the restriction enzymes *Eco*RI and *Msp*I, chosen because of the relatively high CG content of *M. graminicola*. The enzyme *Msp*I should cleave more sites than commonly used alternatives which usually are specific for AT-rich regions. After screening the parental isolates for optimal polymorphisms with 64 primer combinations, 11 were chosen for the mapping analysis along with 104 random amplified polymorphic DNA (RAPD) markers. After removing loci with distorted segregation ratios, linkage analysis was performed on the remaining markers. In total, data for 255 AFLP and 84 RAPD markers were used to construct the map, which ultimately contained 23 linkage groups.

Utility of the map was demonstrated by analysis of avirulence and mating type, two traits of biological interest. Both mating-type idiomorphs (Waalwijk *et al.* 2002b) appeared in nearly equal frequencies among the progeny, confirming the heterothallic, bipolar nature of the mating system (Kema *et al.* 1996c). The *MAT* locus mapped to linkage group 16, a small cluster which also contained four AFLP and two RAPD markers. The avirulence locus mapped to the second-smallest linkage group, number 22, along with four AFLP markers. A subsequent bulked segregant analysis was conducted to identify additional markers closely linked to the avirulence locus. The two bulks contained equal amounts of DNA from 34 virulent and 33 avirulent isolates, respectively. Two additional AFLP markers were generated that both cosegregated with the avirulence locus. One of these was used to screen a bacterial artificial chromosome (BAC) library of the avirulent isolate IPO323 and lead to the cloning of a putative avirulence locus.

Pulsed-field gel electrophoresis was used to compare the electrophoretic karyotype with the number of identified linkage groups. An 850-bp fragment from the *MAT1-2* gene of *Tapesia yallundae* was used as a probe to map the *MAT* locus (Kema *et al.* 2002), and a 145-bp AFLP marker, designated AVIR2, was used to map the *AVR* locus on the electrophoretic karyotypes. Interestingly, both *MAT* and *AVR* mapped on small linkage groups but hybridized to the larger chromosomal bands, indicating a discrepancy between the genetic and physical maps. Although much work clearly remains to be done, this first genetic linkage map for any species of the Dothideales can provide a starting point for analyses of biologically interesting traits in related species infecting a wide range of economically important crops.

3.2. Molecular Genetics

During the last decade most molecular genetic studies of *M. graminicola* concentrated on using anonymous markers, primarily arbitrarily cloned DNA fragments. Much of this work was reviewed elsewhere (McDonald *et al.* 1995; 1999). Estimates of the chromosome number in different field isolates using pulsed-field gel electrophoresis (PFGE) ranged from 17-18 in one study (McDonald and Martinez 1991) to 13-15 discernable chromosomes in another (Kema *et al.* 2002). Discrepancies between these estimates may be due to comigration of chromosomes, which is substantiated by bands with different intensities on PFGE gels. New technologies, such as the germ-tube burst method, which is based on the cytological analysis of metaphase chromosomes in somatic fungal cells (Tsuchiya and Taga 2001), will be helpful in determining the exact chromosome number; preliminary analyses with this technique indicate that *M. graminicola* may contain as many as 22 chromosomes.

Confirming the function of predicted genes will be facilitated by efficient transformation techniques. Payne *et al.* (1998) reported the first transformation of *M. graminicola*. Their strategy was based on transformation of protoplasts with naked DNA. Unfortunately, single integrations and gene disruption through homologous recombination were not easily established and the transformation protocol seemed to depend on the *M. graminicola* isolate used. Nevertheless, useful transformants were generated such as those harboring Green Flourescent Protein, which enabled sophisticated histological studies into the penetration and colonization process of *M. graminicola* in wheat (Duncan and Howard 2000; Rohel *et al.* 2001). Recently, Zwiers and De Waard (2001) used a protocol based on *Agrobacterium tumefaciens* (De Groot *et al.* 1998), which resulted in a high percentage of single integration events (40-75% of all transformants analyzed) and therefore targeted disruption became feasible. This opens up the way for high-throughput functional analyses of identified genes or of reading frames with predicted functions.

3.2.1 Cloning and analysis of mating-type genes

Mating-type genes have been the subject of study in many fungi, including ascomycetes and basidiomycetes. Most plant pathogenic true fungi belong to the ascomycetes and mating-type genes from loculoascomycetes, pyrenomycetes and discomycetes have been cloned and characterized (Turgeon and Yoder 2000). Mating-type genes appear to evolve rapidly; probes from *Magnaporthe grisea* or *Cochliobolus heterostrophus* did not generate a signal when hybridized to DNA from *M. graminicola*. However, a fragment of the *Mat1-2* gene from *Tapesia yallundiae* hybridized with DNA from isolate IPO 94269 of *M. graminicola* but not with DNA from the other parent of the mapping population (Kema *et al.* 2002). Subsequent screening of a phage library identified two clones encompassing a 15-kb region. Sequencing this stretch of DNA identified the central portion as the putative *Mat1-2* locus of *M. graminicola*.

Primers derived from the regions flanking this idiomorph were used to clone the opposite mating type and sequencing identified the *Mat1-1* idiomorph (Waalwijk *et al.* 2002b). The *MAT* idiomorphs of *M. graminicola* are similar in structure to those of other ascomycetes. Each is about 3 kb in size and contains a single reading frame interrupted by one (*Mat1-2*) or two (*Mat1-1*) introns. Cloning of the mating-type idiomorphs will facilitate future research and lays the groundwork for cloning these regions from related species in the Dothideales.

3.2.2. Cloning and analysis of virulence genes

Previous progress in genetic mapping (Kema *et al.* 2002) facilitated cloning the first avirulence locus from *M. graminicola*. DNA from the avirulent parent of the mapping population, IPO323, was used to construct a BAC library in vector pBELOBAC. For fast screening of this library, DNA from 6,144 clones with an insert size of 80 - 100 kb was pooled in three dimensions. A total of 56 pools (16 plates, 16 rows and 24 columns) was analyzed with primers derived from the cosegregating AFLP marker, *AVIR2*. In this way four overlapping clones were identified spanning a total length of 96 kb. Sequencing and annotation of the entire region led to the identification of many open reading frames (ORF) longer than 100 bp. Several putative candidate ORFs for the avirulence gene were identified. Comparison with the sequence of the corresponding region from the virulent parent IPO94269 reduced this number to a major candidate gene that was deleted in the virulent isolate. Disruption of this gene in the avirulent background should lead to a change in phenotype, e.g., virulence on cultivar Shafir. Heterologous expression in the virulent parent, on the other hand, should lead to loss of virulence on Shafir. These experiments, currently in progress, will prove the nature of this gene.

3.2.3 Transposons in *M. graminicola* and related species

Transposons are DNA elements that can move from one region of a genome to another, either with or without replication. Many transposable elements have been identified in fungi (Kempken and Kück, 1998), but only a handful has been shown to be active. Two major classes of transposons are known: class I elements transpose by an RNA intermediate, usually with an increase in copy number (often called retrotransposons or retroelements); while class II transposons move by a cut-and-paste mechanism mediated by DNA-DNA interactions. Both types are known in fungi. Transposons are of particular interest because their activities can alter or eliminate gene function. Thus, active transposons can be developed into tools for functional analysis of genes.

So far, only three transposons have been identified in *M. graminicola* or related species. The first of these was a retroelement in the genome of the tomato pathogen *Cladosporium fulvum* (McHale *et al.* 1992). The second transposon was found from an analysis of EST sequences of *M. graminicola* that identified 11 copies of a sequence similar to COPIA, a class II retrotransposon from *Drosophila* (Keon *et al.* 2000). However, this transposon has not been analyzed further and neither of these transposons appears to be active.

More recently, analysis of a DNA fingerprint probe from *M. graminicola* (McDonald *et al.* 1995) revealed that it contained direct repeats and a partial coding region for a reverse-transcriptase gene (Goodwin *et al.* 2001a) which are characteristics of known transposons. Analysis of the progeny of the *M. graminicola* mapping population with the DNA fingerprint probe identified numerous additional bands that were not present in the parents, indicating that this transposable element is stimulated to move during meiosis (Goodwin *et al.* 2001a). More surprisingly, analysis of single-spore isolates representing 100 asexual transfers from each of two "parental" isolates indicated that 40% of the descendants from one of the parents had gained or lost one or more bands (Goodwin *et al.* 2001a). Therefore, the transposable element identified by the DNA fingerprinting probe appears to be active during both sexual and asexual reproduction. Other evidence indicates that it probably was acquired recently in *M. graminicola*, possibly since the dispersal of the pathogen from a center of origin in the Middle East (Goodwin *et al.* 2001a; McDonald *et al.* 1995). This is the first apparently active transposon identified in the genus *Mycosphaerella* or in the class loculoascomycetes. Recently, evidence for repeat-induced point mutation was found in *M. graminicola* (Tian and Goodwin 2002), which could be an important mechanism of inactivating transposons in this fungus.

4. GENOMICS OF *M. GRAMINICOLA*

Large-scale genomic and expressed sequence tag (EST) sequencing projects have been started to identify as many genes as possible that are expressed during growth in culture and in the plant (Keon *et al.* 2000). Collaboration between scientists from Syngenta and Plant Research International B.V. generated more than 30,000 EST sequences from both *in vitro* and *in vivo* libraries (Kema *et al. unpublished*). Such projects will greatly enhance our understanding of *M. graminicola*, both *in vitro* and in the plant. Keon *et al.* (2000) found that among 704 ESTs with a predicted function about 27% was involved in primary metabolism, 18% was associated with protein and RNA metabolism, 5% was involved in signal transduction and 6% was associated with transport and secretion.

The generation of a large *M. graminicola* mutant library adds another tool for functional analysis (Hamer *et al.* 2001). This project used a strategy based on transposon technology to knock out virtually every *M. graminicola* gene *in vitro*. Transformation of the resulting cosmid constructs into *M. graminicola* can be used to assess the function of the disrupted gene in a high-throughput manner. The active transposon identified recently in asexual progenies from isolate IPO94269 (Goodwin *et al.* 2001a) possibly also could be used to generate large mutant libraries for functional analysis of crucial genes in the fungus.

An interesting category of genes is represented by the so-called ABC transporters. They encode for ATP Binding Cassette proteins that are responsible for the transport of compounds through the cell walls. In *Magnaporthe grisea*, disruption of the ABC transporter, ABC1, led to a severe reduction in pathogenicity (Urban *et al.* 1999). However, ABC transporters are also involved in the transport of fungicides and/or antifungal compounds produced by plants. Until now five ABC transporters have been isolated and their function has been determined in *M. graminicola* (Stergiopoulos *et al.* 2002). Several other genes that are crucial for fungicide resistance also have been cloned. Rohel *et al.* (1998) and Payne *et al.* (1998) cloned the α-tubulin and β-tubulin genes, respectively, and Skinner *et al.* (1998) identified a single amino-acid change in the iron-sulphur protein subunit of succinate dehydrogenase that determined resistance to the fungicide carboxin.

It is evident that functional genomics of plant pathogenic fungi will lead to large-scale genome-wide comparisons to eventually solve major disease problems in crops that feed the world. An example of the power of comparative genomics is provided by the mating-type locus of *M. graminicola*. In the flanking regions of both idiomorphs, several coding regions were identified that are in close vicinity to the *MAT* idiomorphs of the unrelated fungi *Fusarium proliferatum* and *Leptosphaeria maculans* (Waalwijk *et al.* 2002a). Using the complete genome sequences of *Neuropora crassa* and *Magnaporthe grisea* this (micro)synteny could well be expanded to other ascomycetes. This example of synteny prompted Waalwijk *et al* (2002b) to predict that the *MAT* sequences from *M. graminicola* could be used to clone the homologous loci from closely related species. This prediction was confirmed recently with the cloning of both *MAT* idiomorphs from *Septoria passerinii*, a closely related barley pathogen with no known teleomorph (Goodwin *et al.* 2003b).

5. HOST-PATHOGEN INTERACTIONS

The nature of pathogenic variability within populations of *M. graminicola* and of resistance in wheat has long been debated. Absence of strong cultivar × isolate interactions led some authors to conclude that host specificity is lacking, and that isolates differed only in aggressiveness to quantitatively inherited resistance (Van Ginkel and Scharen 1987; 1988a; 1988b). Other studies identified large numbers of putative virulence factors by analysis of many individual cultivar-isolate combinations (Eyal *et al.* 1985; Kema *et al.* 1996a). All studies identified a large difference between isolates from bread *versus* durum wheat. Isolates from bread wheat generally were not able to overcome the resistance of durum wheat and *vice versa.* Resistance in different cultivars can vary from that due to single dominant genes (reviewed in Nelson and Marshall 1990), to genes that are partially dominant with additive effects, while in other cultivars the resistance may be recessive (Rosielle and Brown 1979) or quantitative (Van Ginkel and Scharen 1988b). Most studies on inheritance of resistance to septoria tritici leaf blotch did not use carefully characterized pathogen isolates and this might explain the variable results. These complex relationships provide a rich source of biologically interesting research problems and help make the *M. graminicola*-wheat interaction an excellent model system.

5.1 Types of Resistance and Gene-for-Gene Interactions

Evidence for specific interactions between *M. graminicola* and wheat was first provided by Eyal *et al.* (1973). Extensive tests with five isolates of *M. graminicola* on wheat cultivars purported to possess resistance genes revealed a strong differential interaction between durum and bread wheat and on the bread wheat Bulgaria 88, confirming a differential interaction initially noted by Rillo and Caldwell (1966).

Additional evidence for specific interactions came when large numbers of isolates were inoculated onto expanded sets of cultivars. Eyal *et al.* (1985) inoculated 97 isolates of *M. graminicola* from 22 countries onto seedlings of 35 wheat and triticale cultivars. Strong cultivar

× isolate interactions were identified, indicating differential responses. Assuming gene-for-gene interactions, 28 putative complementary genes were identified among the interactions studied. Further support for cultivar specificity came from an extensive study of 80 isolates and 47 host cultivars (Kema *et al.* 1996b), which identified a large number of putative interacting gene pairs.

Conclusive evidence for gene-for-gene interactions came from recent genetic analyses of both host and pathogen. A 1:1 segregation for avirulence, coupled with single-gene segregations for resistance in the hosts indicates gene-for-gene interactions. These were shown conclusively in a recent analysis of both host and pathogen, in which pathogen isolates segregated for virulence to the same resistance that segregated in the host (Brading *et al.* 2001; Kema *et al.* 1996c; 2000; 2002). These experiments removed any doubt that gene-for-gene interactions exist in the *M. graminicola*-wheat pathosystem, at least for some of the known resistance genes. However, this does not preclude non-specific, quantitative variation in aggressiveness in the pathogen or polygenic, additive resistance in the host. Certainly, both types of interaction are present and important in the *M. graminicola* - wheat pathosystem.

5.2 Genetics and Mapping of Genes for Specific Resistance

Good sources of resistance are rare but have been identified. For example, testing of more than 13,600 wheat accessions (Rosielle 1972; Krupinsky *et al.* 1977) identified 52 cultivars of bread wheat and 266 of durum wheat which were consistently resistant. From these and other studies eight genes for resistance to septoria tritici blotch have been identified and named (Table 1). Two of these have been successfully used in breeding programs. The *Stb1* gene from Bulgaria 88 was transferred into the soft red winter wheat cultivar Oasis (Patterson *et al.* 1975), which was grown widely in Indiana, USA and adjacent states. This gene has remained effective for more than 25 years. The *Stb4* gene was transferred from the winter wheat cultivar Tadorna into the spring wheat Tadinia and was grown throughout California for more than 15 years with no loss of resistance (Somasco *et al.* 1996). However, susceptibility of *Stb4*-containing lines was noted in the field in central California during 2000 (Jackson *et al.* 2000). Therefore, each resistance gene may be effective only in a particular geographical area and for a short period of time before it may be overcome, generating a constant need for new and more durable sources of resistance.

Table 1. Named major genes for resistance against the septoria tritici leaf blotch pathogen *Mycosphaerella graminicola* in wheat.

Gene	Original source	Derived cultivars	Chromosomal location	Molecular markers[a]	Reference(s)[b]
Stb1	Bulgaria 88	Oasis, Sullivan	5BL	RAPD	1, 2, 3
Stb2	Veranopolis	—[c]	—	—	4
Stb3	Israel 493	—	—	—	4
Stb4	Tadorna	Tadinia	7D	AFLP, SSR	5, 6
Stb5	Synthetic 6x	—	7DS		7
Stb6	Flame, Hereward		3AS		8
Stb7	Estanzuela Federal	—	4A	—	9
Stb8	Synthetic W7984	—	7BL	AFLP, SSR	10

[a] The most closely linked molecular markers are indicated. RAPD = random amplified polymorphic DNA; AFLP = amplified fragment length polymorphism; SSR = simple-sequence repeat or microsatellite. [b]1 = Rillo and Caldwell 1966; 2 = Patterson *et al.* 1975; 3 = Yang 2000; 4 = Wilson 1985; 5 = Somasco *et al.* 1996; 6 = Adhikari *et al.* 2003b; 7 = Arraiano *et al.* 2001; 8 = Brading *et al.* 2001; 9 = RA McIntosh, *personal communication*; and 10 = Adhikari *et al.* 2003a. [c]Not known or not determined.

Neither the *Stb2* and *Stb3* genes, nor the four genes named most recently, have been used extensively in breeding programs, possibly due to the difficulty in transferring resistance without the aid of linked molecular markers. None of these genes had been mapped until recently. During the past two years, six of the eight named genes were mapped to wheat chromosomes (Table 1) and molecular markers are becoming available. These will provide the tools for marker-assisted selection in plant breeding programs. More importantly, the availability of mapped resistance genes provides the biological material for analysis of resistance gene expression and function. These genes map to several wheat chromosomes and may function by different mechanisms.

5.3 Genomics of Host Resistance Responses

Genomic resources for wheat have expanded dramatically during the past few years. During 1998, the entire wheat EST database in GenBank consisted of only eight sequences. Currently (December 2002), more than 300,000 EST sequences from wheat representing more than 22,000 genes are available in GenBank, approximately 10% of which are derived from pathogen-treated tissues. The database contains collectively more than 32,000 sequences from tissue treated with *Puccinia recondita* (the cause of leaf rust: 5,600 sequences), *Blumeria graminis* (powdery mildew: 13,800 sequences), *Fusarium graminearum* (head scab: 6,200 sequences), and *M. graminicola* (6,400 sequences).

To further augment the number of wheat EST sequences from pathogen-stressed tissue, a collaborative project was initiated between scientists at the USDA-Agricultural Research Service, Purdue University, and CuraGen Corporation. For this project, wheat tissue treated with various pests and pathogens was analyzed by GeneCalling®, an open-architecture transcript-profiling technology. This approach is complementary to the subtractive approaches used in other laboratories with a higher representation of rare transcripts (www.curagen.com). The following pests or pathogens were inoculated onto wheat head or leaf tissue: *M. graminicola*; *Fusarium graminearum*; *Mayetiola destructor* (Hessian fly); and Barley and Cereal Yellow Dwarf Viruses. The GeneCalling experiments profiled more than 12,000 cDNA fragments per sample, many of which were differentially modulated (Goodwin *et al.* 2003a). Among all four treatments, 11,719 fragments were differentially modulated by at least 1.5 times at one or more time points, divided nearly equally between those that were up or down regulated. Responses began by 3 hours after inoculation and continued through each time sampled up to the maximum of four days for the *M. graminicola* treatments. The first 685 cDNA fragments sequenced revealed many genes known to be activated during defense responses including phenylalanine ammonia lyase, glucanases, chitinases, peroxidases, and thaumatin-like proteins. Additional genes that probably were involved in cell signaling included kinases and phosphatases. Approximately 10% of the sequences had no matches in any of six databases searched and an additional 8% had only weak matches with expected values greater than e^{-08}. These may represent previously unknown mechanisms of resistance that may be specific to one or more treatments. Only two cDNA fragments were modulated in response to all four treatments. Sequencing of an additional 5,000 cDNA fragments is continuing. These sequences, when added to those already in GenBank, will provide an extremely valuable resource that can be used for microarray-based analyses of host-pathogen interactions in the future.

6. CONCLUSION

The *M. graminicola*-wheat pathosystem is developing rapidly as a new model for host-pathogen interactions, particularly for fungi in the Dothideales. Recent advances in genetics, molecular biology and host-pathogen relations have provided the tools needed for functional analysis of gene expression in both host and pathogen. As the community of scientists working on this pathosystem continues to grow, prospects seem great for understanding some of the unusual interactions between *M. graminicola* and its host which should lead to better disease-management strategies for the future.

Acknowledgements: We thank the Wageningen Mycosphaerella Group for discussion and support and Dr. Tika Adhikari for access to manuscripts in press. This work was supported in part by USDA-ARS CRIS project 3602-22000-013-00D.

REFERENCES

Adhikari TB, Anderson JM, and Goodwin SB (2003a). Identification and molecular mapping of a gene in wheat conferring resistance to *Mycosphaerella graminicola*. Phytopathology 93 (in press).

Adhikari TB, Cavaletto JR, Dubcovsky J, Gieco J, and Goodwin SB (2003b). Molecular mapping of the *Stb4* gene for resistance to septoria tritici leaf blotch in wheat. Theor Appl Genet (submitted May 2003).

Arie T, Christiansen SK, Yoder OC, and Turgeon BG (1997). Efficient cloning of ascomycete mating type genes by PCR amplification of the conserved MAT HMG box. Fungal Genet Biol 21:118-130.

Arraiano LS, Worland AJ, Ellerbrook C, and Brown JKM (2001). Chromosomal location of a gene for resistance to *Septoria tritici* blotch (*Mycosphaerella graminicola*) in the hexaploid wheat 'Synthetic 6x'. Theor Appl Genet 103:758-764.

Barr ME (1987). Prodromus to Class Loculoascomycetes. Amherst, MA. 168 pp.

Berbee ML (2001). The phylogeny of plant and animal pathogens in the Ascomycota. Physiol Mol Plant Pathol 59:165–187.

Brading PA, Verstappen ECP, Kema GHJ, and Brown JKM (2002). A gene-for-gene relationship between wheat and *Mycosphaerella graminicola*, the Septoria tritici blotch pathogen. Phytopathology 92:439-445.

Chen R-S, and McDonald BA (1996). Sexual reproduction plays a major role in the genetic structure of populations of the fungus *Mycosphaerella graminicola*. Genetics 142:1119-1127.

Cohen L, and Eyal Z (1993). The histology of processes associated with the infection of resistant and susceptible wheat cultivars with *Septoria tritici*. Plant Pathol 42:737-743.

Corlett M (1991). An annotated list of the published names in *Mycosphaerella* and *Sphaerella*. Mycol Mem 18:1-328.

Crous PW, Kang J-C, and Braun U (2001). A phylogenetic redefinition of anamorph genera in *Mycosphaerella* based on ITS rDNA sequence and morphology. Mycologia 93:1081-1101.

De Groot MJA, Bundock P, Hooykaas PJJ, and Beijersbergen AGM (1998). *Agrobacterium tumefaciens*-mediated transformation of filamentous fungi. Nature Biotech 16:839-842.

De Hoog GS, Zalar P, Urzì C, de Leo F, Yurlova NA, and Sterflinger K (1999). Relationships of dothideaceous black yeasts and meristematic fungi based on 5.8S and ITS2 rDNA sequence comparison. Stud Mycol 43:31-37.

Dixon KP, Xu JR, Smirnoff N, and Talbot NJ (1999). Independent signaling pathways regulate cellular turgor during hyperosmotic stress and appressorium-mediated plant infection by *Magnaporthe grisea*. Plant Cell 11:2045-2058.

Duncan KE, and Howard RJ (2000). Cytological analysis of wheat infection by the leaf blotch pathogen, *Mycosphaerella graminicola*. Mycol Res 104:1074-1082.

Dunkle LD, Cantone FA, and Ciuffetti LM (1991). Accumulation of host-specific toxin produced by *Cochliobolus carbonum* during pathogenesis of maize. Physiol Mol Plant Pathol 38:265-273.

Eriksson OE, and Winka K (1998). Families and higher taxa of Ascomycota. Myconet 1:17-24.

Eyal Z, Amiri Z, and Wahl I (1973). Physiologic specialization of *Septoria tritici*. Phytopathology 63:1087-1091.

Eyal Z, Scharen AL, Huffman MD, and Prescott JM (1985). Global insights into virulence frequencies of *Mycosphaerella graminicola*. Phytopathology 75:1456-1462.

Eyal Z, Scharen AL, Prescott JM, and Van Ginkel M (1987). The Septoria Diseases of Wheat: Concepts and Methods of Disease Management. Mexico, D. F: CIMMYT.

Farr DF, Bills GF, Chamuris GP, and Rossman AY (1995). Fungi on Plants and Plant Products in the United States. St. Paul, MN, USA: APS Press.

Garcia C, and Marshall D (1992). Observations on the ascogenous stage of *Septoria tritici* in Texas. Mycol Res 96:65-70.

Goodwin SB, Anderson JM, Ohm HW, Lohret TA, Crasta OR, and Williams CE (2003a). Use of high-throughput transcript analysis in wheat to characterize genes responding to diverse pests and pathogens. Abstracts of the Plant and Animal Genome XI meetings, San Diego, CA.

Goodwin SB, and Cavaletto JR (2002). Analysis of 18S ribosomal RNA gene sequences reveals the phylogenetic relationships of the genus *Mycosphaerella*. Abstracts of the 7th International Mycological Congress, Oslo, Norway, p 202.

Goodwin SB, Cavaletto JR, Waalwijk C, and Kema GHJ (2001a). DNA fingerprint probe from *Mycosphaerella graminicola* identifies an active transposable element. Phytopathology 91:1181-1188.

Goodwin SB, Dunkle LD, and Zismann VL (2001b). Phylogenetic analysis of *Cercospora* and *Mycosphaerella* based on the internal transcribed spacer region of ribosomal DNA. Phytopathology 91:648-658.

Goodwin SB, Waalwijk C, Kema GHJ, Cavaletto JR, and Zhang G (2003b). Cloning and analysis of the mating-type idiomorphs from the barley pathogen *Septoria passerinii*. Mol Genet Genomics 269:1-12.

Goodwin SB, and Zismann VL (2001). Phylogenetic analyses of the ITS region of ribosomal DNA reveal that *Septoria passerinii* from barley is closely related to the wheat pathogen *Mycosphaerella graminicola*. Mycologia 93:934-946.

Hamer L, Adachi K, Montenegro-Chamorro MV, Tanzer MM, Mahanty SK, Lo C, Tarpey RW, Skalchunes AR, Heiniger RW, Frank SA, Darveaux BA, Lampe DJ, Slater TM, Ramamurthy L, DeZwaan TM, Nelson GH, Shuster JR, Woessner J, and Hamer J (2001). Gene discovery and gene function assignment in filamentous fungi. Proc Natl Acad Sci USA 98:5110-5115.

Hawksworth D, Kirk PM, Sutton BC, and Pegler DN (1995). Ainsworth & Bisby's Dictionary of the Fungi. 8th edn. Wallingford, UK: CAB International.

Hilu HM, and Bever WM (1957). Inoculation, oversummering, and suscept-pathogen relationship of *Septoria tritici* on Triticum species. Phytopathology 47:474-480.

Jackson LF, Dubcovsky J, Gallagher LW, Wennig RL, Heaton J, Vogt H, Gibbs LK, Kirby D, Canevari M, Carlson H, Kearney T, Marsh B, Munier D, Mutters C, Orloff S, Schmierer J, Vargas R, Williams J, and Wright S (2000). 2000 Regional barley and common and durum wheat performance tests in California. Agronomy Progress Report 272:1-56.

Jones MJ, and Dunke LD (1995). Virulence gene expression during conidial germination in *Cochliobolus carbonum*. Mol Plant-Microbe Interact 8:476-479.

Kema GHJ, Annone JG, Sayoud R, Van Silfhout CH, Van Ginkel M, and De Bree J (1996a). Genetic variation for virulence and resistance in the wheat-*Mycosphaerella graminicola* pathosystem. I. Interactions between pathogen isolates and host cultivars. Phytopathology 86:200-212.

Kema GHJ, Goodwin SB, Hamza S, Verstappen ECP, Cavaletto JR, van der Lee TAJ, Hagenaar-de Weerdt M, Bonants PJM, and Waalwijk C (2002). A combined AFLP and RAPD genetic linkage map of *Mycosphaerella graminicola*, the septoria tritici leaf blotch pathogen of wheat. Genetics 161:1497-1505.

Kema GHJ, Sayoud R, Annone JG, and Van Silfhout CH (1996b). Genetic variation for virulence and resistance in the wheat-*Mycosphaerella graminicola* pathosystem. II. Analysis of interactions between pathogen isolates and host cultivars. Phytopathology 86: 213-220.

Kema GHJ, Verstappen ECP, Todorova M, and Waalwijk C (1996c). Successful crosses and molecular tetrad and progeny analyses demonstrate heterothallism in *Mycosphaerella graminicola*. Curr Genet 30:251-258.

Kema GHJ, Verstappen ECP, and Waalwijk C (2000). Avirulence in the wheat septoria tritici leaf blotch fungus *Mycosphaerella graminicola* is controlled by a single locus. Mol Plant-Microbe Interact 13:1375-1379.

Kema GHJ, Yu DZ, Rijkenberg FHJ, Shaw MW, and Baayen RP (1996d). Histology of the pathogenesis of *Mycosphaerella graminicola* in wheat. Phytopathology 86:777-786.

Kempken F, and Kück U (1998). Transposons in filamentous fungi—facts and perspectives. BioEssays 20:652-659.

Keon J, Bailey A, and Hargreaves J (2000). A group of expressed cDNA sequences from the wheat fungal leaf blotch pathogen, *Mycosphaerella graminicola (Septoria tritici)*. Fungal Genet Biol 29:118-133.

Krupinsky JM, Craddock JC, and Scharen AL (1977). Septoria resistance in wheat. Plant Dis Reporter 61:632-636.

Luttrell ES (1973). Loculoascomycetes. In: GC Ainsworth, FK Sparrow, and AS Sussman, eds. The Fungi, an Advanced Treatise. Vol. IVA. New York: Academic Press, pp 135-219.

Madariaga BR (1986). Presencia en Chile de *Mycosphaerella graminicola* (Fuckel) Schroeter, estado sexuado de *Septoria tritici* Rob. ex. Desm. Agricultura Tecnica 46:209-211.

McDonald BA, and Martinez JP (1991). Chromosome length polymorphisms in a *Septoria tritici* population. Curr Genet 19:265-271.

McDonald BA, Pettway RE, Chen RS, Boeger JM, and Martinez JP (1995). The population genetics of *Septoria tritici* (teleomorph *Mycosphaerella graminicola*). Can J Bot 73(Suppl. 1):S292-S301.

McDonald BA, Zhan J, Yarden O, Hogan K, Garton J, and Pettway RE (1999). The population genetics of *Mycosphaerella graminicola* and *Stagonospora nodorum*. In: Lucas JA, Bowyer P and Anderson HM, eds. Septoria on Cereals: a Study of Pathosystems. Wallingford, UK and New York: CABI Publishing, pp 44-69.

McHale MT, Roberts IN, Noble SM, Beaumont C, Whitehead MP, Seth D, and Oliver RP (1992). CfT-I: an LTR-retrotransposon in *Cladosporium fulvum*, a fungal pathogen of tomato. Mol Gen Genet 233:337-347.

Nelson LR, and Marshall D (1990). Breeding for resistance to *Septoria nodorum* and *Septoria tritici*. Adv Agron 44:257-277.

Patterson FL, Gallun RL, Roberts JJ, Finney RE, and Shaner GE (1975). Registration of Arthur 71 and Abe wheat. Crop Sci 15:736-737.

Payne AC, Grosjean-Cournoyer MC, and Hollomon DW (1998). Transformation of the phytopathogen *Mycosphaerella graminicola* to carbendazim and hygromycin B resistance. Curr Genet 34:100-104.

Perrone G, Logrieco A, Kema GHJ, Ritieni A, and Bottalico A (2000). Phytotoxic activity of *Mycosphaerella graminicola* culture filtrates. In: De Vries GE and Metzlaff K, eds. Phytosphere '99-Highlights in European Plant Biotechnology Research and Technology Transfer. Rome, pp 39-45.

Pollack FG (1987). An annotated compilation of *Cercospora* names. Mycolog Mem 12:1-212.

Rillo AO, and Caldwell RM (1966). Inheritance of resistance to *Septoria tritici* in *Triticum aestivum* subsp. *vulgare* 'Bulgaria 88'. Phytopathology 56:897.

Rohel EA, Payne AC, Fraaije BA, and Hollomon DW (2001). Exploring infection of wheat and carbohydrate metabolism in *Mycosphaerella graminicola* transformants with differentially regulated green fluorescent protein expression. Mol Plant-Microbe Interact 14:156-163.

Rohel EA, Payne AC, Hall L, Barker H, Butters J, and Hollomon DN (1998). Isolation and characterization of alpha-tubulin genes from *Septoria tritici* and *Rhynchosporium secalis*, and comparative analysis of fungal alpha-tubulin sequences. Cell Motil Cytoskeleton 41:247-253.

Rosielle AA (1972). Sources of resistance in wheat to speckled leaf blotch caused by *Septoria tritici*. Euphytica 21:152-161.

Rosielle AA, and Brown AGP (1979). Inheritance, heritability and breeding behaviour of three sources of resistance to *Septoria tritici* in wheat. Euphytica 28:385-392.

Sanderson FR (1972). A *Mycosphaerella* species as the ascogenous state of *Septoria tritici* Rob. and Desm. NZ J Bot 10:707-709.

Sanderson FR (1976). *Mycosphaerella graminicola* (Fuckel) Sanderson comb. nov., the ascogenous state of *Septoria tritici* Rob. and Desm. NZ J Bot 14:359-360.

Sanderson FR, Scharen AL, Eyal Z, and King AC (1986). A study of genetic segregation using single ascospore isolates of *Mycosphaerella graminicola*. Special Publication, Agronomy Society of New Zealand 5:195-201.

Scott PR, Sanderson FR, and Benedikz PW (1988). Occurrence of *Mycosphaerella graminicola*, teleomorph of *Septoria tritici*, on wheat debris in the UK. Plant Pathol 37:285-290.

Silva-Hanlin DMW, and Hanlin RT (1999). Small subunit ribosomal RNA gene phylogeny of several loculoascomycetes and its taxonomic implications. Mycol Res 103:153–160.

Skinner W, Bailey A, Renwick A, Keon J, Gurr S, and Hargreaves J (1998). A single amino-acid substitution in the iron-sulphur protein subunit of succinate dehydrogenase determines resistance to carboxin in *Mycosphaerella graminicola*. Curr Genet 34:393-398.

Somasco OA, Qualset CO, and Gilchrist DG (1996). Single-gene resistance to *Septoria tritici* blotch in the spring wheat cultivar Tadinia. Plant Breeding 115:261-267.

Stergiopoulos I, Zwiers LH, and de Waard MA (2002) Secretion of natural and synthetic toxic compounds from filamentous fungi by membrane transporters of the ATP-binding cassette and major facilitator superfamily. Eur J Plant Pathol 108:719-734.

Tian Y, and Goodwin SB (2002). Possible repeat induced point mutation (RIP) in coding and flanking regions of a transposable element from the wheat pathogen *Mycosphaerella graminicola*. Phytopathology 92:S81.

Tsuchiya D, and Taga M (2001). Cytological karyotyping of three *Cochliobolus* spp. by the germ tube burst method. Phytopathology 91:354-360.

Turgeon BG, and Yoder OC (2000). Proposed nomenclature for mating type genes of filamentous ascomycetes. Fungal Genet Biol 31:1-5.

Urban M, Bhargava T, and Hamer JE (1999). An ATP-driven efflux pump is a novel pathogenicity factor in rice blast disease. EMBO J 18:512-521.

Van Ginkel M, and Scharen AL (1987). Generation mean analysis and heritabilities of resistance to *Septoria tritici* in durum wheat. Phytopathology 77:1629-1633.

Van Ginkel M, and Scharen AL (1988a). Diallel analysis of resistance to *Septoria tritici* isolates in durum wheat. Euphytica 38:31-37.

Van Ginkel M, and Scharen AL (1988b). Host-pathogen relationships of wheat and *Septoria tritici*. Phytopathology 78:762-766.

Waalwijk C, Arts J, Howlett B, Goodwin S, van der Lee T, de Vries I, Mendes O, Hesselink T, Verstappen E, and Kema G (2002a). The mating type locus, an example of synteny among ascomycetes. Abstracts of the 6th European Conference on Fungal Genetics, Pisa, Italy, p 99.

Waalwijk C, Mendes O, Verstappen ECP, de Waard MA, and Kema GHJ (2002b). Isolation and characterization of the mating-type idiomorphs from the wheat septoria leaf blotch fungus *Mycosphaerella graminicola*. Fungal Genet Biol 35:277-286.

Weber GF (1922). II. Septoria diseases of wheat. Phytopathology 12:537-588.

Weiergang I, Dunkle LD, Wood KV, and Nicholson RL (1996). Morphogenetic regulation of pathotoxin synthesis in *Cochliobolus carbonum*. Fungal Genet Biol 20:74-78.

Wilson RE (1985). Inheritance of resistance to *Septoria tritici* in wheat. In: AL Scharen, ed. Proceedings of Septoria of Cereals Workshop, USDA-ARS Publ 12. Washington, DC: U. S. Gov Print Office, pp 33-35.

Yang X (2000). Genetic analysis and development of molecular markers for a resistance locus to *Mycosphaerella graminicola* in wheat. Ph.D. dissertation, Purdue University, West Lafayette, IN.

Zhan J, Kema GHJ, Waalwijk C, and McDonald BA (2002). Distribution of mating type alleles in the wheat pathogen *Mycosphaerella graminicola* over spatial scales from lesions to continents. Fungal Genet Biol 36:128-136.

Zwiers LH, and De Waard MA (2001). Efficient *Agrobacterium tumefaciens*-mediated gene disruption in the phytopathogen *Mycosphaerella graminicola*. Curr Genet 39:388-393.

Applied Mycology & Biotechnology
An International Series. Volume 4. Fungal Genomics

14

Functional Genomic Analysis of the Rice Blast Fungus *Magnaporthe grisea*

Martin J. Gilbert, Darren M. Soanes and Nicholas J Talbot
School of Biological Sciences, University of Exeter, Washington Singer Laboratories, Perry Road, Exeter, EX4 4QG, UK. (N.J.Talbot@exeter.ac.uk).

Magnaporthe grisea, a filamentous, heterothallic ascomycete fungus, which infects a wide variety of grasses, has emerged as a model system to study genes involved in fungal pathogenesis. This is largely due to the successful application of numerous molecular genetic techniques for the manipulation of this fungus as well as the use of classical genetic techniques to examine all aspects of its development. The following chapter describes the current position of fungal genomics in *M. grisea* focusing on examples of how the use of fungal genomics has furthered our understanding of pathogenesis in the fungus, particularly the identification of signalling genes involved in appressorium formation. The genomic resources available to researchers studying phytopathogenic fungi and the bioinformatic strategies used for the investigation of *M.grisea* will also be reviewed. This includes the recently released *Magnaporthe* Genome Project that signals a new era in *Magnaporthe* research. The chapter continues by focusing on some of the technical advances made in an attempt to increase the throughput of large single gene mutant collections in *M.grisea.* Restriction-enzyme mediated integration (REMI), *Agrobacterium tumefaciens* Mediated Transformation (ATMT), and transposon based mutagenesis techniques will all be reviewed. Finally some of the techniques being developed which will make it possible to investigate a wide variety of parasitic and symbiotic fungi using the wealth of genomic sequence information now available will be described.

1 INTRODUCTION

We are entering a new and exciting era in fungal biology. For the first time publicly accessible full genome sequences of many fungi will be available, strategies to generate gene disruption mutants at a high throughput will be developed, and conventional methods for determining gene function will be refined. In this chapter we will review the current state of fungal genomics in a model phytopathogenic fungus *Magnaporthe grisea*. We will describe some of the molecular genetic approaches that have led to identification of genes involved in plant infection by *M. grisea*, particularly focusing on those genes that have been shown to encode products with a role in signal transduction pathways that lead to appressorium formation. In addition, we will look at the new genomic resources available to researchers studying phytopathogenic fungi and new genomic techniques being developed which may enable more rapid identification of virulence determinants.

Corresponding author: N. J. Talbot

1.1 Introduction to Genomics of Phytopathogenic Fungi

The genome sizes of fungi are relatively small compared to those of other eukaryotes such as animals and plants. The budding yeast *Saccharomyces cerevisiae* and the fission yeast *Schizosaccharomyces pombe*, for example, have genome sizes of 13.7Mb and 13.8Mb, respectively (Goffeau *et al.* 1996; Wood *et al.* 2002). With a few exceptions, filamentous ascomycetes and basidiomycetes have genome sizes that range between 13Mb and 42Mb (Kupfer *et al.* 1997; Yoder and Turgeon, 2001). These fungal genome sizes are therefore approximately a third of those, for example, the model plant species *Arabidopsis thaliana* and the metazoan *Caenorhabditis elegans*.

The *S. cerevisiae* genome project was completed in 1996 and was the first eukaryote genome to be fully sequenced (Goffeau *et al.* 1996). Recently two fungal genome sequences have been generated, those of the fission yeast *Saccharomyces pombe* (http://www.sanger.ac.uk/Projects/S_pombe/) and the filamentous fungus *Neurospora crassa* (http://www-genome.wi.mit.edu/annotation/fungi/neurospora). Optimising the utility of the enormous amount of data generated in these projects has required the establishment of specifically designed databases, such as the yeast protein database (http://www.incyte.com/sequence/proteome/index.shtml) and the Stanford yeast genome database (http://genome-www.stanford.edu/Saccharomyces/). These databases enable researchers to identify and retrieve specific gene sequences and associated functional information from the 6,215 open reading frames that have been identified so far in the *S. cerevisiae* genome sequence. The information available includes transcriptional profile analysis, which monitors the expression of the genome under different environmental or developmental conditions, and information regarding mutant phenotypes. The availability of a full genome sequence for budding yeast, coupled with the genetic tractability of the species and the development of genome-wide analytical processes has resulted in very rapid advances in this field of research. At the same time a dramatic gap in research tools and resources has emerged between the model yeasts *S. cerevisiae* and *S. pombe* and the less tractable filamentous fungi. However, as we shall see in this review, much will change as the genomic revolution begins to encompass pathogenic and saprophytic filamentous fungi.

Fungal pathogens have a number of characteristics that make them difficult to study. This is perhaps best highlighted by the obligate parasitic fungi such as the rusts and powdery mildews where the fungus cannot be grown in culture. In addition many filamentous fungal species lack classical sexual genetics and homologous recombination, used to generate gene replacements, occurs at very low frequency (1-20%). To obtain targeted insertion events in filamentous fungi, large gene fragments must be used spanning at least 1kb on each side of a given gene (Asch and Kinsey 1990). In contrast, using *S. cerevisiae* to create gene-specific mutations by targeted integration is very efficient and requires only 50-bp fragments of target gene homology on either side of a selectable marker (Baudin *et al.* 1993).

Despite these intrinsic problems of working with phytopathogenic fungi, *Magnaporthe grisea* has emerged as a model system to study genes involved in fungal pathogenesis (Talbot, 1995; Tucker and Talbot, 2001). Research has been facilitated by extensive use of classical genetic techniques and development, during the 1980s and 1990s of molecular genetic methods for manipulation of the fungus (Valent *et al.* 1991; Valent and Chumley 1991). *M. grisea* can be cultured on simple, well-defined media under a variety of laboratory conditions. The disease can be studied on host plants with a relatively quick regeneration time with legions visible on infected rice seedlings within five days of infection (Talbot 1995). Conidia can be germinated and form appressoria *in vitro* on hydrophobic surfaces. Fertile, hermaphroditic laboratory strains such as Guy 11, a strain isolated from French Guiana by Jean-Loup Notteghem and colleagues (1992), are available and genetic crossing is possible under laboratory conditions, which allows genetic analysis to be performed (Hamer *et al.* 1989). A simple transformation system has also been developed either by complementation of an auxotrophic marker or by bestowing antibiotic resistance (Valent and Chumley, 1991), and targeted gene disruption experiments are routinely performed (Sweigard *et al.* 1992; Talbot *et al.* 1993; Xu and Hamer 1996). In parallel, a variety of mutants affected in fungal pathogenicity have been isolated by insertional mutagenesis using restriction enzyme-mediated integration (REMI) (Sweigard *et al.* 1998; Balhadère *et al.* 1999). As a consequence of these forward and reverse genetic approaches, *Magnaporthe grisea* has become better understood and more widely studied than most ascomycete foliar pathogens.

2. *MAGNAPORTHE GRISEA*: THE DISEASE

M. grisea (anamorph: *Pyricularia grisea*) is a filamentous, heterothallic ascomycete fungus which infects over 50 species of grasses including barley, wheat and rice (Ou, 1985). *M. grisea* is best known, however, as the causal agent of the rice blast disease, a leaf spot disease that is characterised by large ellipsoid lesions on the surface of rice leaves. In mature rice plants, the fungus spreads into the panicle (the inflorescence that holds the rice grain), causing neck blast symptoms that can result in complete loss of the rice crop (Talbot 1995).

The infection process of *M. grisea* begins when three-celled spores, or conidia, are produced from lesions during periods of high humidity and carried to the surface of a rice leaf by wind or rain dispersal. Upon contact with the waxy rice cuticle, an adhesive is released from the spore apex that enables the conidium to attach to the leaf surface. It has been shown that the release of this mucilage is triggered by wetting of the conidium (Hamer *et al.* 1988). Once attached, the spore germinates within one hour, forming a short germ tube. The germ tube is sensitive to the leaf surface and a number of modifications have been described that may be associated with leaf surface recognition (see Bourett and Howard, 1990). Within 4 hours, an appressorium differentiates; the germ tube stops growing at the tip and the formation of a swollen hook appears at the tip of the hypha. This marks the beginning of cellular differentiation required for the infection of the plant.

Appressorium development is associated with a series of morphogenetic events. Within the developing germ tube, a linked mitotic division occurs, producing two daughter nuclei. One of these nuclei migrates into the developing appressorium, while the other returns to the cell of the conidium from where the mother nucleus originated (Bourett and Howard, 1990). A specialised septum then forms which separates the appressorium from the germ tube and conidium, and as appressorium development continues, both the germ tube and conidium eventually collapse (Howard, 1997).

Multiple signals operate to induce appressoria development. Examples of inductive triggers for appressorium development include exposure to cutin monomers (Gilbert *et al.* 1996), exposure to plant lipid and other wax compounds (Uchiyama *et al.* 1979; Gilbert *et al.* 1996; deZwaan *et al.* 1999) and starvation stress (Talbot *et al.* 1993; Howard 1994; Jelitto *et al.* 1994). Studies of appressorium formation have also revealed two fundamentally important details about the developmental process. These are the importance of hydrophobicity and the contribution of cyclic AMP (cAMP) signalling to appressorium morphogenesis (Lee and Dean 1993; 1994). The importance of hydrophobicity is perhaps best illustrated by the ease with which appressoria form on hydrophobic surfaces (Hamer *et al.* 1988; Lee and Dean 1994; Jelitto *et al.* 1994), and cAMP is able to induce appressorium formation on normally non-inductive surfaces when exogenously applied to germinating conidia (Lee and Dean 1993).

The appressorium produces the physical force required to penetrate the host leaf surface (Howard *et al.* 1991). It generates enormous turgor pressure during the infection process, in fact it is estimated to be the highest pressure recorded in a living cell (Talbot 1999). When appressoria were incubated in increasing hyperosmotic solutions of polyethylene glycol an estimate of intracellular hydrostatic pressure could be made based on the point that cytorrhysis (cell collapse) was observed. Using this approach, it was suggested that appressoria were capable of generating up to 8MPa of turgor pressure (Howard *et al.* 1991). Additionally, artificially lowering turgor pressure prevents penetration, indicating that appressorium turgor is an absolute requirement for infection (Howard *et al.* 1991).

The huge internal turgor pressure is thought to be generated by an influx of water caused by an osmotic gradient produced in the appressorium. The appressorium is freely permeable to water, but the efflux of any larger molecules is prevented by the presence of a melanin layer in the cell wall, which significantly reduces the porosity of the appressoria (Bourett and Howard 1990; Howard *et al.* 1991). Melanin is a polyketide that is synthesised via the polymerisation of 1,9-dihydroxynaphthalene. Genetic analysis has revealed the importance of the melanin layer during *M. grisea* infection. Three mutants have been identified that are deficient in melanin biosynthesis and those strains carrying mutations at any one of these loci make completely non-functional appressoria (Woloshuk *et al.* 1980; Howard and Ferrari, 1989) and are consequently non-pathogenic due to their inability to maintain turgor (Chumley and Valent 1990).

Biochemical analysis of the contents of conidia during germination and subsequent appressorium differentiation has shown the mechanism by which this huge turgor pressure is generated (de Jong *et al.* 1997). Glycerol accumulates rapidly during spore germination and decreases during germ tube elongation, before accumulating to extremely high concentrations during appressorium maturation. The concentration of glycerol has been estimated, based on an enzymatic assay, to be as high as 3.2M, sufficient in theory to generate turgor of 8.7MPa (de Jong *et al.* 1997).

Appressorium maturation and turgor pressure lead to the production of a penetration peg, which has been shown to break the surface of the rice leaf (Howard *et al.* 1991; Sweigard *et al.* 1992). Appressoria can puncture inert plastic surfaces (Howard *et al.* 1991) and cutinase, an enzyme that degrades the major component of plant cuticles, appears to be dispensable for pathogenicity (Sweigard *et al.* 1992). It is possible, however that as the penetration peg is forced through the surface leaf layers, its passage is aided by secretion of enzymes that soften the wall layers enhancing the rate of infection. A major obstacle for addressing the molecular basis for the precise function of cell-wall degrading enzymes in pathogenesis has been that pathogens are known to have multiple genes encoding a number of these enzymes. Therefore, fungal strains mutated in cell wall-degrading enzyme genes still retain some residual enzyme activity (for example Cooper, 1987; Scott-Craig *et al.* 1990). Recently, however an alternative approach has been to identify the genetic regulatory elements which, when disrupted, result in the simultaneous loss of multiple enzymes. Tonukari and co-workers (2000) used this approach to investigate the role of cell-wall degrading enzymes in the maize pathogen *Cochliobolus carbonum*. The expression of most cell-wall degrading enzymes by most fungi, including phytopathogenic fungi is inhibited by glucose in a well-studied metabolic process known as catabolite repression (Ruijter and Visser 1997). In *S. cerevisiae* a protein kinase called Snf1p is required to stop catabolite repression (Celenza and Carlson 1984; Hardie *et al.* 1998; Treitel *et al.* 1998) and has been shown to control the response to glucose through phosphorylation of transcription factors (Ronne 1995; DeVit *et al.* 1997; Lesage *et al.* 1996; Vincent and Carlson 1998). A targeted gene replacement mutant of *ccSNF1* from *C. carbonum*, an ortholog of *SNF1*, showed reduced growth on complex and simple sugar media and activities of β-1,3-glucanase, pectinase and xylanase were all reduced in *Δccsnf1* mutants (Tonukari *et al.* 2000). As a result, *Δccsnf1* mutants are much less virulent on plants (Tonukari *et al.* 2000). This suggests that the regulation of cell wall-degrading enzymes is important for the pathogenicity of *C. carbonum*. It will be interesting in the future to investigate the interaction between the glucose repression pathway and cAMP signalling in phytopathogens given the clear linkages between these processes in yeast (Banuett 1998; Lengeler *et al.* 2000).

Cytological examination shows that penetration pegs rupture the plant cuticle after 28-31 hours, breaking through the epidermal cell wall and forming an infection hypha within the first cell encountered (Talbot 1995). The penetration peg widens into a primary infection hypha and the fungus takes on a beaded, bulbous appearance as it colonises the host tissues. Infection continues with the appearance of large lesions visible three to four days after inoculation. Conidiation of these large coalescing lesions occurs only when the relative humidity exceeds 93% (Talbot 1995) and spores are released to infect new hosts.

2.1 The PMK1 and MAP Kinase Pathway

A good example of how fungal genomics has furthered our understanding *M.grisea* pathogenesis is the identification of signalling genes involved in appressorium formation. We have seen that appressoria develop in response to precise environmental stimuli, which enables the infectious growth stage of the fungus to be initiated. The perception of environmental signals has received considerable attention mainly as a consequence of being able to clone and characterise genes encoding signalling components based on homology to genes from *Saccharomyces cerevisiae* and other model fungi (Dean, 1997; Lengeler, 2000). It has been relatively straightforward to test the involvement of a given signal transduction pathway in infection-related development because of this conservation of signalling molecules among eukaryotes. By engineering null mutants using targeted gene deletion, very rapid and significant advances have been made in dissecting molecules directly implicated in regulating infection-related development in *M. grisea*. It is now clear that cyclic AMP and mitogen-activated protein kinases regulate key steps in fungal pathogenesis (for reviews see Dean, 1997; Kronstad 1997; Hamer and Talbot 1998).

Cyclic AMP is a well-characterised secondary messenger in both prokaryotes and eukaryotes (Kronstad 1997; Adachi and Hamer 1998). In pathogenic fungi, cAMP has been shown to be important in growth and morphogenesis as well as the regulation of appressorium formation (Kronstad 1997). The role of cAMP during *M.grisea* appressorium development was confirmed following the identification of *MAC1* a gene encoding adenylate cyclase (Choi and Dean, 1997). Targeted gene deletion resulted in *Δmac1* mutants that are non-pathogenic and do not form appressoria (Choi and Dean 1997). The activity of adenylate cyclase results in changes in the levels of cAMP and therefore the control of downstream cAMP-dependent proteins such as protein kinase A (PKA). The *Δmac1* phenotype can be overcome by exogenous application of cAMP, confirming that the catalytic activity of *MAC1* is required for pathogenesis (Choi and Dean 1997; Kronstad 1997; Adachi and Hamer 1998).

In some strains of *M. grisea*, the *Δmac1* phenotype is unstable and can be overcome by an extragenic suppressor mutation. One identified suppressor mutation, found to be caused by a single base change in the first cAMP-binding domain of the regulatory subunit of PKA (Adachi and Hamer 1998), named *mac1 sum1-99*, showed wild-type growth, morphology and appressorium formation. However the suppression of the *Δmac1* phenotype was not complete because the *sum1-99* strain was not fully pathogenic (Adachi and Hamer, 1998). This is consistent with PKA activity being required for appressorium elaboration.

The role of PKA has been investigated further with the deletion of the *M. grisea CPKA* gene, which encodes a catalytic subunit of PKA. Resulting *Δcpka* mutants are non-pathogenic. However they do produce appressoria but development is delayed and appressoria are variable in size (Mitchell and Dean 1995; Xu *et al.* 1997). It has been suggested that *Δcpka* mutants display a turgor or penetration defect because they have the ability to colonise plant tissues if injected through a wound (Kronstad 1997; Adachi and Hamer 1998; Knogge 1998; Kang *et al.* 1999; Balhadère and Talbot 2000). The findings with

Δmac1 and *Δcpka* mutant strains suggest that cAMP signalling is involved in the early stages of appressorial morphogenesis. It is also becoming clear that there may be many components of cAMP signalling pathways yet to be identified. In *S.cerevisiae* for example, there are three PKA catalytic sub-units, each shown to have a distinct role in morphogenesis (for example Kronstad 1997; Hamer and Talbot 1998; Lengeler *et al.* 2000), and it is likely that the genomic sequence of *M. grisea* will reveal further genetic components of these pathways.

Protein phosphorylation is an important process for intracellular signal transduction in eukaryotic and prokaryotic cells and is catalysed by protein kinases. A particularly important component of signal transduction pathways in a variety of developmental processes in all eukaryotes are the mitogen-activated protein kinases (MAPKs). MAPKs are serine/threonine kinases which are responsible for transducing signals from the cell surface to the nucleus resulting, via transcription factors, in the increased transcription of a defined set of target genes (Banuett, 1998; Lev *et al.* 1999).

In *Magnaporthe grisea* the *PMK1* (Pathogenicity MAP Kinase 1) MAPK gene was initially identified based on its homology to the *FUS3* and *KSS1* MAPK genes from *S. cerevisiae* (Xu and Hamer, 1996). *FUS3* and *KSS1* play a role in the yeast pheromone signalling pathway and are involved with pseudo-hyphal growth regulation (Herskowitz, 1995; Gustin *et al.* 1998). The *PMK1* gene was shown to be a functional homologue of the *FUS3/KSS1* genes and targeted gene deletion of *PMK1* generated mutants unable to form appressoria and an inability to grow invasively in rice plants (Xu and Hamer 1996). Interestingly, microscopic techniques showed that *Δpmk1* mutants germinated normally and appeared to initiate appressorium formation, however they failed to form a melanised cell wall and underwent an unusual pattern of nuclear division and hyphal elongation (Xu and Hamer 1996; Hamer and Talbot 1998). When *Δpmk1* mutants were injected into rice leaves no spreading lesions were formed and the fungus could not be detected on the plant. These, and other studies, indicate that *PMK1* is necessary for invasive growth and fungal viability *in planta* (Xu and Hamer 1996). It should be noted that the mating ability, vegetative growth rate and conidiation of the *Δpmk1* mutant strain was not altered by the mutation. This indicates that the *PMK1* gene is only involved in specific virulence-associated processes and is important in transducing external signals resulting in infection-related development (Xu and Hamer 1996; Deising *et al.* 2000).

The effects on appressorium morphogenesis of *Δpmk1, Δcpka, Δmac1, and Δmac1 Δsum1-99 M. grisea* mutants have led to biochemical experiments to determine the roles of each of these genes in regulating metabolic pathways associated with appressorium function. Appressorium turgor in *M. grisea* results from the generation of large concentrations of glycerol (de Jong *et al.* 1997) and to investigate the biochemical mechanisms for turgor generation glycerol biosynthetic enzymes have been assayed during appressorium formation (Thines *et al.* 2000). Glycogen localises to the appressorium during its formation but rapidly disappears at the onset of turgor generation (Thines *et al.* 2000). Lipid droplets were observed to move into the developing appressorium and coalesced into a central vacuole before degrading at the onset of turgor generation (Thines *et al.* 2000). Rapid lipolysis then occurs and is associated with abundant triacylglycerol lipase activity. In *Δpmk1* mutants, glycogen and lipid mobilisation to the developing appressorium did not occur and in *Δcpka* mutants glycogen and lipid breakdown were both retarded. In *Δmac1 Δsum1-99* mutants glycogen and lipid degradation were very rapid (Thines *et al.* 2000). These findings indicate that the transfer of storage carbohydrate and lipid reserves to the appressorium occurs under the control of *PMK1* MAPK pathway and that their breakdown is dependent on the cAMP-dependent PKA encoded by *CPKA* (Weber *et al.* 2001). The signalling pathways regulating virulence in *M. grisea* can be seen below in Figure 1.

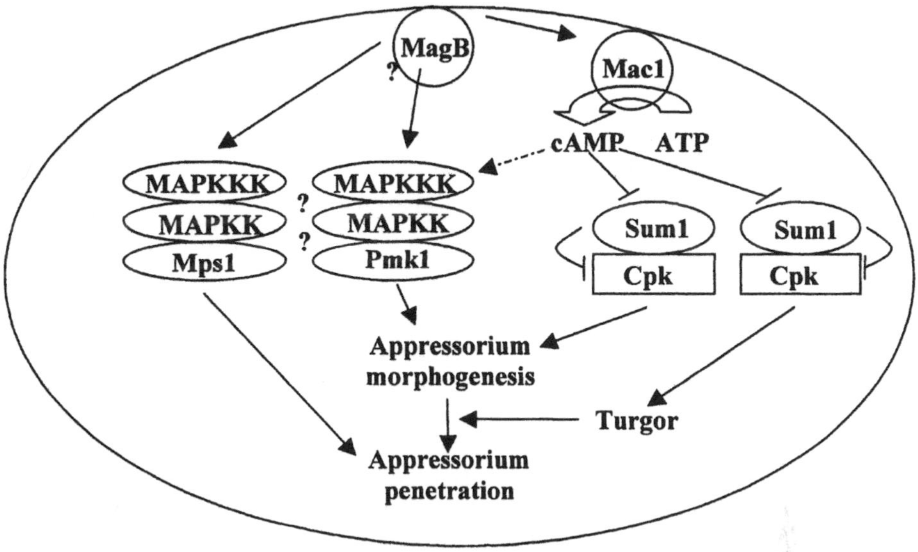

Fig. 1. The signalling pathways regulating virulence of the rice blast fungus *M. grisea*. Appressorium formation and virulence of *M.grisea* involves both conserved MAP kinase signalling cascades and the cAMP-PKA signalling pathway (Adapted from Lengeler *et al.* 2000).

The corn smut fungus *Ustilago maydis* has emerged as a widespread model for studying the interaction between the cAMP response pathway and MAP kinase signalling during pathogenesis and has been the subject of recent reviews (Kronstad 1997; Kahmann *et al.* 1999). The formation of the infectious, filamentous dikaryon in *U. maydis* follows the meeting of haploid sporidia of opposite mating types on the surface of a corn leaf and is triggered by pheromone signalling. The cellular response to pheromone exposure is transmitted via a MAP kinase signalling cascade involving the PMK1-related MAPK *Kpp2* (Mayorga and Gold, 1999; Muller *et al.* 1999). *Kpp2* is a MAPK that is closely related to the *M. grisea PMK1* gene. Loss-of-function *Δkpp2* mutants are still able to cause disease but are deficient in pheromone production and mating, and also show defects in filamentation, tumour formation and virulence (Mayorga and Gold, 1999; Muller *et al.* 1999). This indicates that *Kpp2* contributes to the regulation of pathogenesis but is not absolutely required. The MAPK-kinase-encoding gene *Fuz7* may act upstream of *Kpp2*, or in a parallel pathway and shows a similar mutant phenotype (Banuett and Herskowitz 1994). Interestingly the *Δkpp2* mutation is less dramatic than the *Δfuz7* mutation and it is possible that there may be another partially redundant MAPK homologue that compensates for the *Δkpp2* mutation (Lengeler *et al.* 2000; Tucker and Talbot 2001).

The role of cAMP in the virulence and morphogenesis of *U.maydis* was first suggested following the isolation of mutants that formed filamentous hypha without the requirement for mating. These haploid mutants displayed filamentous growth similar to the aerial hyphae that resulted from mating (Barrett *et al.* 1993). Subsequent analysis revealed that a defect in adenylyl cyclase, encoded by *UAC1*, was responsible for this phenotype, and that exogenous cAMP addition could restore budding growth (Gold 1994). The *Δuac1* mutants were unable to cause disease on corn seedlings, suggesting that a functional cAMP signalling pathway is required for infection, either at the mating stage or at a subsequent stage in the infection process (Kronstad 1997).

Further research using the *Δuac1* mutant led to the identification of *UBC1*, a gene encoding the regulatory subunit of PKA (Gold, 1994). Inactivation of *UBC1* restored budding growth to *Δuac1* mutants indicating that the elevated PKA activity promoted budding (Gold, 1994). More recently, two genes encoding catalytic subunits of PKA have been characterised in *U. maydis*. Targeted gene disruption of one, *ADR1* resulted in non-pathogenic mutants that showed a filamentous phenotype, however inactivation of the other gene, *UKA1* had little effect on growth, morphology or pathogenicity (Kronstad 1997). These results suggest that a cAMP signalling pathway is required for *U. maydis* pathogenesis, and that a major role of the pathway involves the morphological switch from budding to filamentous growth (Kronstad 1997).

Although MAPK-mediated signalling pathways have been directly implicated in regulating infection-related development in phytopathogenic fungi, it is not clear what kind of genes are regulated in this process. In *S. cerevisiae*, the pheromone response pathway results in the activation of *Fus3*, which phosphorylates several substrates including *STE12*, *DIG1* and *DIG2* (Sprague and Thorner 1992; Elion *et al.* 1993; Tedford *et al.* 1997; Bardwell *et al.* 1998a). *STE12* is the major transcription factor that binds to the pheromone response element found in the upstream region of many pheromone-responsive genes (Dolan *et al.* 1989) and *KSS1* regulates the repression and transcriptional activation of *STE12* for filamentous growth in yeast (Bardwell *et al.* 1998b). Recently *STE12* homologues have been characterised in several phytopathogenic fungi including *M. grisea* (Park *et al.* 2002). Targeted gene deletion of *MST12*, the *STE12* homologue in *M. grisea*, results in no obvious defects in vegetative growth or conidia germination, however *Δmst12* mutants are non-pathogenic. *Δmst12* mutants do produced typical dome-shaped melanised appressoria unlike *Δpmk1* mutants and, when inoculated through wound sites, *Δmst12* mutants are defective in invasive growth. These observations suggest that *MST12* may function downstream of *PMK1* to regulate genes involved in infectious hyphal growth (Park *et al.* 2002).

Two further *M. grisea* genes, *GAS1* and *GAS2* have been isolated using a subtraction library enriched for genes regulated by *PMK1* (Xue *et al.* 2002). Both *GAS1* and *GAS2* encode small proteins that are homologous to powdery mildew fungus *gEgh16* gene (Justesen *et al.* 1996), which is expressed during the early stages of *Blumeria graminis* infection. *GAS1* and *GAS2* are expressed specifically during appressorium formation, but neither is expressed in the *Δpmk1* mutant (Xue *et al.* 2002). Targeted gene deletions of GAS1 and GAS2 showed no defect in vegetative growth or appressorium formation but they were reduced in appressorial penetration and lesion development (Xue *et al.* 2002). Using GFP-tagging, GAS1 and GAS2 proteins were observed only in the cytoplasm of appressoria (Xue *et al.* 2002). It is suggested that *GAS1* and *GAS2* encode a class of proteins specific for filamentous fungi that may function as novel virulence factors in fungal phytopathogens. It will be interesting in the future to investigate the transcription factors that may regulate their expression during appressorium formation.

3 GENOMIC STRATEGIES FOR THE INVESTIGATION OF *M. GRISEA*

3.1 Expressed Sequence Tags (ESTs)

Prior to the availability of fully sequenced genomes, the genome diversity of several fungal pathogens has been explored by generation of collections of expressed sequence tags (ESTs). ESTs are partial length DNA sequences that have been sequences from either the 5' or 3' ends of cDNA clones. As the sequences have been derived from genes expressed at a specific developmental stage, or from a particular tissue, EST collections have the potential to provide information on the expression levels of particular genes in specific tissue at a specific stage of an organism's development. It should be noted, however, that a large amount of EST data is required before any differences in expression levels can be shown statistically (Skinner *et al.* 2001).

3.2 The *Magnaporthe* Genome Project

The *Magnaporthe* Genome Project has now been publicly released as a result of a partnership between The International Rice Blast Genome Consortium and The Whitehead Institute for Biomedical research (Cambridge, MA,USA) (www-genome.wi.mit.edu/annotation/fungi/magnaporthe/index.html). This obviously heralds a new era in *Magnaporthe* research.

The rice-infecting *M. grisea* strain 70-15 was chosen as the isolate for genome sequencing which was developed through numerous back-crosses to the wild rice pathogenic isolate Guy11 (Leung *et al.* 1988; Chao and Ellingboe 1991). The current assembly available on the database contains over 2000 sequences or contigs and is estimated to represent 97% of the *M.grisea* genome. Sequence information can be accessed in several ways including BLAST searches with an option for contig sub-sequence retrieval or the ability to retrieve defined areas of contiguous assembled DNA sequence and it is possible to download the entire genome.

3.3 Bioinformatics

The application of molecular genetic analysis to the study of phytopathogenic fungi has led to the identification and characterisation of a number of genes involved in fungal pathogenicity (Knogge 1998; Sweigard *et al.* 1998; Idnurm and Howlett 2001). These include genes involved in the detoxification of antifungal compounds produced by plants (Bowyer *et al.* 1995; Straney and VanEtten 1994), biosynthesis of phytotoxic compounds (Panaccione *et al.* 1992), breakdown of the host plant cuticle (Tonukari *et al.* 2000; Walton, 1994), conidiogenesis (Hamer and Givan 1990), appressorium formation and function (Talbot *et al.* 1993; Silué *et al.* 1998; Balhadère and Talbot 2001; Clergeot *et al.* 2001), amino acid metabolism (Balhadère *et al.* 1999), as well as those involved in conserved signalling pathways (Xu and Hamer 1996; Kronstad 1997). The advent of genome-wide analysis promises to provide a new and powerful means of investigating fungal pathogens. The application of genomic approaches to study fungal pathogens has so far lagged behind that of many other organisms, mainly due to the lack of publicly available sequence data. In a recent review, for example, out of 379 genome project websites surveyed, only 16 (4%) involved fungi, of which only seven involved pathogenic species (Yoder and Turgeon 2001).

To fully sequence a fungal genome is currently beyond the resources of most fungal research groups. The sequencing of the *M. grisea* genome is therefore a significant event. Single pass, partial sequencing of either 3′ or 5′ ends of complementary DNA (cDNA) clones to generate a set of expressed sequence tags (ESTs), however, has allowed a low-cost strategy to identify gene inventories in phytopathogenic fungi. EST collections from pathogenic fungi infecting both animals and plants are becoming more widespread (Keon *et al.* 2000, Skinner *et al.* 2001). Table 1 shows a list of websites containing genome sequence information and EST collections for pathogenic fungi.

Generally EST inventories are in a flat file format with only limited annotation. As part of the COGEME (Consortium for the Genomics of Microbial Eukaryotes) project publicly funded in the United Kingdom, all available EST sequence information from ten phytopathogenic fungal and oomycete species has been deposited in a new relational database (Soanes *et al.* 2002). To reduce redundancy and to improve sequence quality, the EST

Table 1: Publicly accessible pathogenic fungal and oomycete genomic resources.

Fungus	URL(s)	Status December 2002
Aspergillus fumigatus	http://www.sanger.ac.uk/Projects/A_fumigatus/	5548 BAC end sequences
Blumeria graminis	http://cogeme.ex.ac.uk	2701 unisequences
Botryotinia fuckeliana	http://www.genoscope.cns.fr/externe/English/Projets/Projet_W/W.html	6558 ESTs
	http://cogeme.ex.ac.uk	2859 unisequences
Candida albicans	http://sequence-www.stanford.edu/group/candida/	1.5-fold genome coverage
	http://www.sanger.ac.uk/Projects/C_albicans/	10 cosmids
Colletotrichum trifolii	http://cogeme.ex.ac.uk	550 unisequences
Cryptococcus neoformans	http://www.genome.ou.edu/cneo.html	4000 ESTs
	http://baggage.stanford.edu/group/C.neoformans/menu.html	156 genomic DNA contigs, 18.1 Mb
	http://www.tigr.org/tdb/e2k1/cna1/index.shtml http:cgt.genetics.duke.edu/data/index.html	nominal 6-fold genome coverage
	http://rcweb.bcgsc.bc.ca/cgi-bin/cryptococcus/cn.pl	4867 BAC end sequences
Fusarium sporotrichioides	http://www.genome.ou.edu/fsporo.html	5000 ESTs
	http://cogeme.ex.ac.uk	3238 unisequences
Gibberella zeae	http://cogeme.ex.ac.uk	2403 unisequences
Magnaporthe grisea	http://www-genome.wi.mit.edu/annotation/fungi/magnaporthe/	39,000 kb genomic sequence
	http://cogeme.ex.ac.uk	7245 unisequences
Mycosphaerella graminicola	http://cogeme.ex.ac.uk	2926 unisequences
Phytophthora infestans	https://xgi.ncgr.org/pgc/	13,234 ESTs
	http://cogeme.ex.ac.uk	2798 unisequences
Phytophthora sojae		
Pneumocystis carinii	http://biology.uky.edu/Pc/	3896 ESTs
	http://www.sanger.ac.uk/Projects/P_carinii/	1 telomeric cosmid
Verticillium dahliae	http://cogeme.ex.ac.uk	1455 unisequences

sequences were assembled into a set of unique sequences (unisequences) for each organism through cluster assembly (Huang and Madan 1999). A putative function was then assigned to each unisequence based on homology to known genes. Based on these assignments each unisequence was classified by function according to a hierarchical scheme used by the Munich Information Centre for Protein Sequences (MIPS) (Mewes *et al.* 1997). The web-based front end of the database (http://cogeme.ex.ac.uk/) allows the user to search for unisequences based on their functional classification group or annotation and the retrieval of those sets of sequences (Fig. 2). BLAST algorithms (Altschul *et al.* 1997) can also be used to compare unisequences in the database with sequences entered by the user.

The primary use of the database was envisaged to be as a tool for gene discovery based on homology of unisequences to known genes. One striking observation, though, is that a significant proportion (40-60%) of the sequences exhibit no similarity to protein or DNA sequences present in publicly available databases. Such genes have often been called orphans (Oliver 1996) and are commonly found in the genomes of eukaryotic model organisms. For example, it has been estimated that about one-third of all predicted protein-coding regions in *S. cerevisiae* are orphans (Oliver 1996). Orphan genes may represent genes whose phylogenetic distribution is restricted to certain evolutionary lineages or genes that rapidly diverge between closely related species (Schmid and Aquadro 2001). Therefore more sophisticated analysis is needed to make full use of the data present in the EST collections.

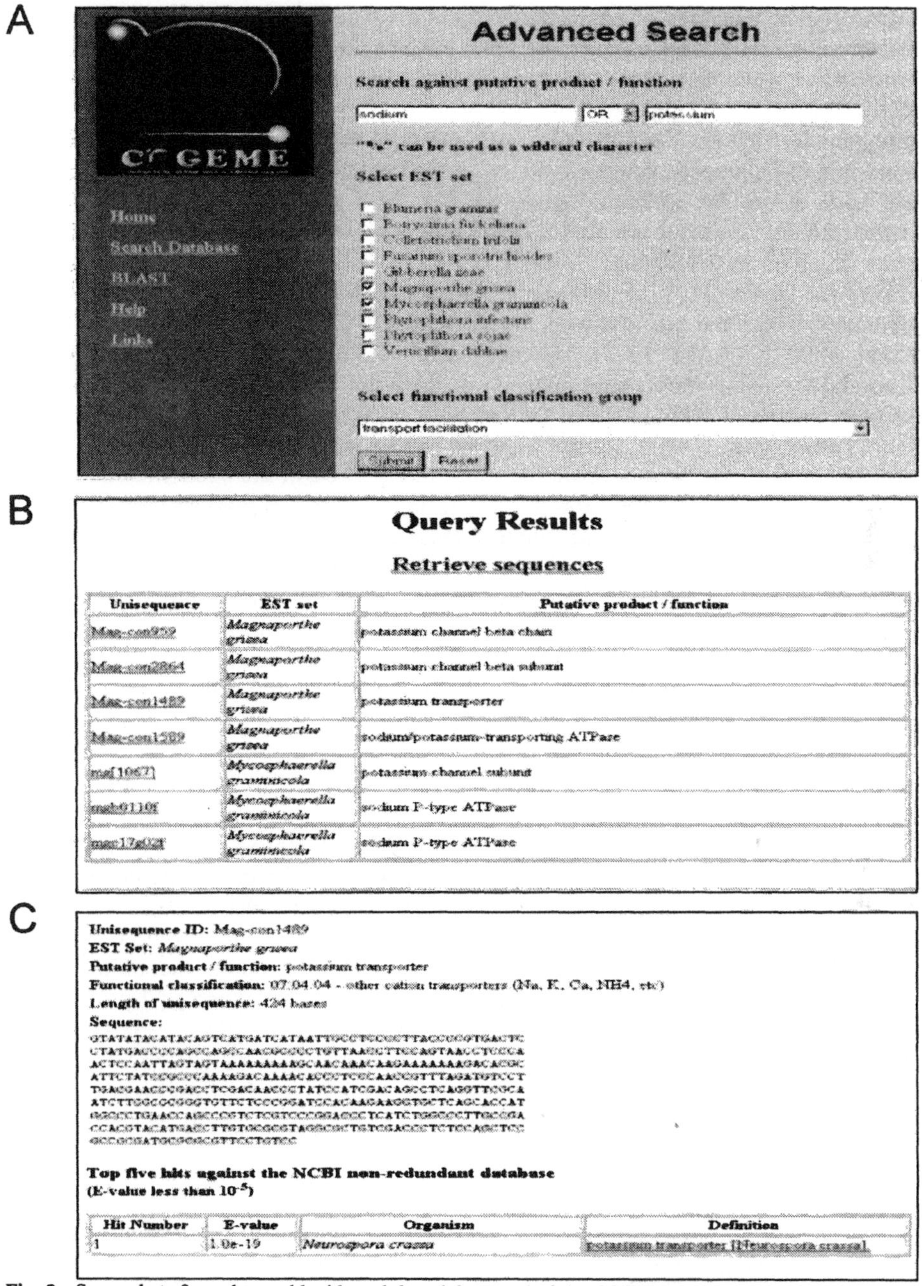

Fig. 2. Screenshots from the world-wide web based front end of the COGEME EST database, illustrating an advanced query. A) The advanced search web page. This enables users to search for unisequences from specified organisms based on putative product / function and functional classification group. In this screenshot the user is searching for unisequences from *Magnaporthe grisea* or *Mycosphaerella graminicola*, in the functional classification group "transport facilitation", whose putative product / function annotation contains the words "sodium" or "potassium". B) Results of running the query, showing a list of unisequences, the EST set they are from and a description of their putative product / function. Clicking on the name of a unisequence brings up a page showing details about that unisequence (C).

One goal in the study of plant pathogens is to identify pathogenicity factors, a set of genes that are essential for invasion of the host but not for saprotrophic growth (Knogge, 1998). Comparative genomics between non-pathogenic fungi such as *Neurospora crassa* or*Saccharomyces cerevisiae* and pathogenic fungi may help to identify potential pathogenicity factors. For example, pathogenicity factors may be novel genes that are not found in non-pathogenic fungi but are conserved in pathogenic species. Some of these factors may have arisen by horizontal gene transfer from prokaryotic organisms (Rosewich and Kistler 2000). Evidence for this may include abnormal GC-content, clustering together of genes involved in a common pathway (Han *et al.* 2001), or sequence similarity to known prokaryotic genes. Most pathogenicity factors so far identified have homologues in non-pathogenic fungi and are involved, for example, in conserved signalling pathways (Xu and Hamer 1996; Kronstad 1997). The study of the expression profiles of genes under a variety of conditions using microarray analysis could point to potential pathogenicity factors. For example, potential pathogenicity factors may be up-regulated during invasion of the host plant (Talbot *et al.* 1993). Cluster analysis can be used to look for groups of genes that are regulated in the same way and are therefore likely to be in the same metabolic or regulatory pathway. Clustering of gene expression data in yeast and in humans, for example, has been used to classify genes of similar function (Eisen *et al.* 1998). In the absence of microarray data, the frequency of ESTs representing particular genes in sequence data obtained from different libraries can be used to produce digital 'northerns blots'. Unfortunately, a very large number of EST sequences from a number of different libraries are needed to produce statistically significant results (Audic and Claverie 1997). Comparative genomics could also be used to identify families of genes that are more highly represented in the genomes of pathogenic fungi as compared to non-pathogenic fungi. This has been shown in preliminary comparisons between the genomes of *M. grisea* and *N. crassa.* The results for example suggest that the genome of *M. grisea* encodes a greater abundance of proteins that may be involved in breakdown of the host cell wall such as cutinases and xylanases.

Although EST collections represent an incomplete sample of the transcriptome, Bayesian probability analysis can be used to infer whether a particular metabolic or signalling pathway is present in an organism. Giles *et al.* (in press) looked for the presence or absence of EST sequences representing enzymes from ten amino acid biosynthesis pathways in six phytopathogens. A Bayesian metric was used to calculate whether the number of enzymes in a particular pathway was higher or lower than would be expected by chance. The most significant result in this study was the likely presence of all amino acid biosynthetic pathways in the obligate pathogen *Blumeria graminis*, which causes barley powdery mildew. The results from this study have been used to make a database that can be accessed via a web-front end. (http://cogeme.ex.ac.uk/biosynthesis.html) An example of the interface of this database is given in Figure 3.

4 FUNCTIONAL ANALYSIS

The large scale gene disruption strategies developed in *S. cerevisiae* have not yet been transferred to most filamentous fungi because of their larger genomes and significantly lower rates of targeted integration during transformation. Here we focus on some of the techniques used to modify the fungal genome through the integration of non-homologous DNA and describe some of the new methods that are being developed in an attempt to increase the throughput of large annotated single gene mutant collections.

4.1 Restriction-Enzyme-Mediated Integration (REMI)

The efficiency of insertional mutagenesis studies has been significantly improved with the introduction of a procedure known as restriction-enzyme-mediated integration (REMI). In

A

Cysteine biosynthesis

The table below shows for which combinations of enzyme and organism significant homologies were found. See here for details of the comparisons.
Numbers in the table indicate the degree of significance of the blast hits. See here for details.
Details of the homologies are accessed by clicking on the link
Combinations for which no significant homologies were found are indicated by '-'. The blast report may be accessed by clicking on the link.
Key:
Blu= *Blumeria graminis*
Bot= *Botrytis cinerea*
Fus =*Fusarium graminearum*
Mag= *Magnaporthe grisea*
Myc= *Mycosphaerella graminicola*
Neu= *Neurospora crassa*
Phy= *Phytophthora infestans*
Sch= *Schizosaccharomyces pombe*

Analysis of pathway results
Analysis of pathway results for enzymes that occur in this pathway only

Enzyme	Blu	Bot	Fus	Mag	Myc	Neu	Phy	Sch
2.6.1.1	39	47	52	49	49	50	-	49
4.2.1.13	-	-	-	-	-	28	-	21
4.2.1.13	-	-	-	-	-	28	-	26
4.2.99.10	-	-	56	31	50	35	-	36
4.2.99.2	-	66	48	-	-	54	-	58
4.2.99.8	-	59	31	-	37	42	35	44
4.2.99.8	-	-	56	31	50	35	-	36
4.2.99.9	-	-	-	-	-	34	-	35
4.2.99.9	-	-	-	-	-	33	-	37
4.4.1.8	-	-	31	34	-	28	-	28

B

Analysis of Pathway Results for All

The table below shows the probabilities of the observed numbers of enzymes from the amino acid biosynthesis pathways for each species having occurred by chance. These values are strictly only valid if the sequence samples are random. This is unlikely to be the case with EST data. However, the values offer an indication of the likely significance of the observed results.

Show results for one organism Blumeria graminis go

Show results for all pathways for enzymes that occur in only one pathway
Show results for all pathways for all enzymes

Pathway	Blu	Bot	Fus	Mag	Myc	Neu	Phy	Sch
Alanine and aspartate biosynthesis	0.2385	0.0536	0.2258	0.2848	0.0222	1.0000	0.1937	0.0000
Arginine and proline biosynthesis	0.2696	0.1093	0.1093	0.2925	0.2416	1.0000	0.3116	1.0000
Cysteine biosynthesis	0.1996	0.2735	0.1640	0.0279	0.2109	1.0000	0.3115	1.0000
Glutamate biosynthesis	0.2585	0.1171	0.2462	0.0053	0.0290	1.0000	0.0583	0.0000
Glycine, serine and threonine biosynthesis	0.1508	0.1834	0.1222	0.1700	0.0494	1.0000	0.2204	1.0000
Histidine biosynthesis	0.2749	0.0312	0.0039	0.1704	0.0689	1.0000	0.1000	1.0000
Lysine biosynthesis	0.2732	0.2462	0.2462	0.3015	0.0003	1.0000	0.3005	1.0000
Methionine biosynthesis	0.2088	0.2462	0.0702	0.1741	0.0162	1.0000	0.1168	1.0000
Phenylalanine, tyrosine and tryptophan biosynthesis	0.1901	0.1171	0.1171	0.2712	0.2506	1.0000	0.0562	1.0000
Valine, leucine and isoleucine biosynthesis	0.2749	0.2188	0.2188	0.3377	0.0009	1.0000	0.2670	1.0000

Fig. 3: Screenshots from the world-wide web based front end of the Amino Acid Biosynthesis Gene Discovery Database. (A) Screenshot showing the occurrence of EST sequences encoding enzymes in the cysteine biosynthetic pathway for each organism; (B) Screenshot showing the probability that the number of EST sequences encoding biosynthetic enzymes in each pathway, for each organism could have occurred by chance. A probability of less than 0.05 suggests that the number of enzyme encoding sequences is significantly different than would be expected by chance sampling of the transcriptome.

this method, linearized plasmid DNA is transformed into the organism of choice in the presence of a restriction enzyme, which generates compatible or incompatible ends. By treating the host genome with the same restriction enzyme used to linearise the introduced plasmid, a high number of exposed ends are generated in the genome, which are compatible with the already cleaved ends of the transforming vector. This can result in a significantly higher percentage of single-copy integrations compared to conventional insertional mutagenesis approaches as well as an increase transformation rate as shown in *A. nidulans* (Tilburn *et al.* 1990) and *Colletotrichum graminicola* (Thon *et al.* 2000). This technique has also been used successfully in *M.grisea.* Sweigard and co-workers (1998) identified 27 *M.grisea* mutants with reduced host pathogenicity representing possible fungicide targets. In addition Balhadère *et al.* (2001), identified five novel pathogenicity mutants using REMI mutagenesis, one of which showed reduced appressorium-mediated penetration on susceptible plant hosts.

Mutagenesis approaches using REMI offers distinct advantages over conventional mutagenesis techniques in plant pathogenic fungi as it produces an easy identification of genes through the introduced tag (Kahnmann and Basse 1999). However in some systems a low percentage of genes appear to be tagged and the identification of insertion events can be very labour intensive. For example, Balhadère and coworkers (1999) generated a REMI library of 3527 M. grisea transformants and 1150 of these were screened for defects in pathogenicity. From these only five mutants were identified and characterised of which only two were tagged (Balhadère *et al.* 1999).

It is also still uncertain whether REMI instigates random integration throughout the host genome. It has been suggested that highly transcribed areas of the genome are likely to be more receptive to restriction digest than those corresponding to transcriptionally inactive areas (Maier and Schafer 1999). Sweigard and co-workers (1998) reported that they tagged the same locus twice out of a population of 5538 *M. grisea* REMI transformants and similar results were obtained with *Cochliobolus heterostrophus* (Lu *et al.* 1994). In addition REMI experiments rely on the incorporation of the introduced vector mediated by cellular DNA repair mechanisms. If this process is not efficient or accurate enough it can lead to undesired rearrangements during transformation. For example mutants generated by REMI may found to be untagged with the selectable marker and can cause problems when attempting to clone the gene causing the mutant phenotype.

4.2 *Agrobacterium tumefaciens* Mediated Transformation (ATMT)

Transformation mediated through the plant pathogenic bacteria *A.tumefaciens* (ATMT) is a widely used and established procedure to transfer foreign genes into plants species, particularly *Arabidopsis thaliana* (for example Feldmann 1991; Koncz *et al.* 1992; Zupan and Zambryski 1995). More recently ATMT approaches have also been transferred to several fungal species (Bundock *et al.* 1995; de Groot *et al.* 1998; Gouka *et al.* 1999; Abuodeh *et al.* 2000; Mikosch *et al.* 2001; Mullins *et al.* 2001). ATMT involves the insertion of a plasmid carrying a selectable marker situated between a left and right border of the transforming DNA sequence (T-DNA). The bacterium can initiate a process that results in the transfer and incorporation of a piece of single-stranded bacterial DNA into the host genome. One of the main advantages of ATMT is that various tissues can be chosen as the starting material for transformation. For example conidia of *Fusarium oxysporum* have been transformed using ATMT avoiding the problems associated with protoplast isolation (Mullins *et al.* 2001). Recently the successful transformation of *M.grisea* has been reported mediated by *A. tumefaciens* (Rho *et al.* 2001). Using the *A. nidulans* trpC promoter linked to the hygromycin B phosphotransferase gene it was estimated that 60% of the transformants contained a single T-DNA insert on their genome (Rho *et al.* 2001).

4.3 Transposon Based *in vitro* Mutagenesis

An elegant method for genome-wide mutagenesis studies in filamentous fungi has been demonstrated by Hamer and co-workers. (2001). They have developed a transposon-arrayed gene knockout (TAG-KO) strategy to discover genes and simultaneously create gene disruption cassettes for subsequent transformation and mutant analysis (Hamer *et al.* 2001) (Fig. 4).

The method uses in vitro transposition (IVT) into cosmid libraries to create gene-sequencing templates. The transposon carries the hygromycin resistance selectable marker for the expression in the fungus and the insertion site can be subsequently determined by DNA sequencing. This creates an annotated collection of insertional gene disruption vectors with targeted integration shown to occur at much higher frequencies than with conventional gene disruption vectors (Hamer *et al.* 2001).

Villalba and co-workers (2001) have developed another novel method for gene tagging in *M. grisea*. The *Fusarium oxysporum* transposable element *impala* (Langin *et al.* 1995) was introduced into *M.grisea* to develop transposon-based insertional mutagenesis. An autonomous copy of *impala* was inserted in the promoter of *niaD*, a nitrate reductase gene from *Aspergillus nidulans* and transformed into a *M.grisea* nitrate reductase-deficient mutant. Analysis of the resulting mutants revealed that *impala* was able to excise and reinsert at new loci in *M.grisea*. In addition *impala* transposition was shown to be autonomous in *M. grisea* because a defective *impala* element was active only when complemented in *trans* by a functional *impala* transposase. This method could prove to be very useful for insertional mutagenesis in fungi because it appears not to have severe host restriction and is easily modified without losing its transposition properties (Villalba *et al.* 2001).

Cosmids have also been used to improve homologous recombination in *Aspergillus nidulans* (Chaveroche *et al.* 2000). This method involves the use of an improved homologous recombination system of an *Escherichia coli* strain. This particular *E. coli* strain contains an inducible form of RedγBα from λ phage which enables precise allelic exchange between a cassette containing a fungal and bacterial selectable marker and known *A. nidulans* sequences on a cosmid (Chaveroche *et al.* 2000). The circular or linearized cosmid is then used to transform *A. nidulans* yielding transformants with the appropriate gene replacement at frequencies up to 60% (Chaveroche *et al.* 2000). This strategy could be applied to most filamentous fungi and to introduce highly specific gene modifications in fungal genomes.

5 GENOMIC TOOLS

5.1 Microarrays

Microarrays provide a powerful and rapid technique to analyse large numbers of genes expression profiles (for reviews see Epstein and Butow, 2000; Lee and Lee 2000). Microarrays consist of ordered sets of DNA fixed to a solid surface and subsequent hybridisation analysis allows the expression of sets of genes to be investigated simultaneously. In addition, microarray readers can automatically collect this expression data in a digital image format which can then be converted to numeric expression data by appropriate computer software. This technique is ideal for comparing the differential expression of genes under varying environmental or genetic conditions. Microarray analysis has been used extensively in a number of studies (e.g. Desprez *et al.* 1998; Eisen *et al.* 1998; Reymond *et al.* 2000), and has provided the grounding for large-scale functional analysis of thousands of genes. Due to the wide spectrum of genes and signals involved, these microarray approaches are well-suited for analysing the development and functioning of complex interactions such as phytopathogenic fungi and their hosts. Indeed, as the complete

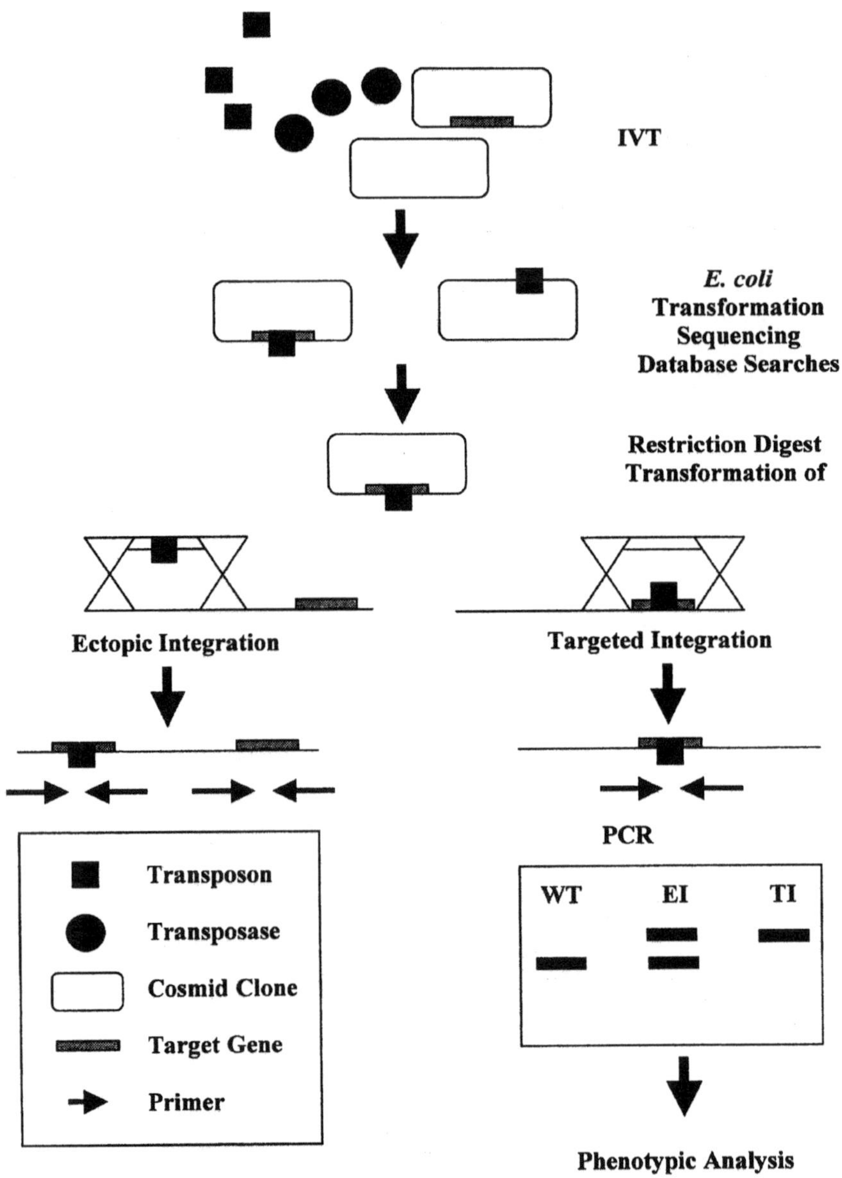

Fig. 4. Transposon-arrayed gene knockouts. The IVT reaction includes a transposon, the corresponding transposase, and recipient DNA. The transposon carries the hygromycin phosphotransferase gene that confers resistance to hygromycin in *Escherichia coli*, *M. grisea* and other filamentous fungi. The recipient DNA consists of cosmid clones. The cosmid vector contains homing endonuclease sites flanking the cloning site and the β-lactamase gene. The IVT products are transformed into *E. coli*, and individual transposon insertion sites are determined by sequencing. Cosmids containing an appropriate transposon insertion are digested with the homing endonuclease to release full-length inserts for transformation. Ectopic and TI events are distinguished by PCR analysis, and mutants are subjected to phenotype analysis. EI, ectopic integration.

genomic sequence data of both host and pathogen become available, it provides an opportunity for extensive research in fungal pathogenicity using microarray analysis.

The feasibility of following gene expression changes in this way has been shown in a number of studies. For example Voiblet and co-workers (2001) investigated the symbiosis-regulated genes in the ectomycorrhizal interaction *Eucalyptus globulus/Pisolithus tinitorius*. Microarrays were screened to identify symbiosis-related genes using differential hybridisation. It was shown that 17% of the genes analysed were differentially regulated in the mycorrhizal root tissue by comparing the signals from free-living partners with those from symbiotic tissues (Voiblet *et al.* 2001). Results like these will allow the basis for a more precise molecular dissection of the complex genetic networks that control symbiosis development and function.

Several other recent papers describe the use of microarrays to investigate fungal pathogens. An array based on the *C.albicans* genome sequence data (Tzung *et al.* 2001) was used to identify new cellular targets of several transcriptional regulators that play a key role in the control of metabolism and yeast to hypha morphogenesis in *C.albicans* (Murad *et al.* 2001). *C.albicans* gene expression microarrays have also been used to identify possible targets in the fungus to itraconazole treatment (De Backer *et al.* 2001). The wealth of genome sequence information now available will make it possible to use microarray analysis to investigate a wider range of parasitic and symbiotic fungi in the future.

6. CONCLUSION

In summary, genomic analysis of the rice blast fungus is already beginning to further our understanding of the biology of pathogenesis. Clearly, however, deployment of the full complement of genomic tools to address fungal pathogenesis, will provide a very rapid increase in our knowledge of the biology of this fascinating organism and of the fundamental basis of plant disease.

REFERENCES

Abuodeh RO, Orbach MJ, Mandel MA, Das A and Galgiani JN (2000). Genetic transformation of *Coccidioides immitis* facilitated by *Agrobacterium tumefaciens*. J Infect. Dis 181: 2106-2110.

Adachi K and Hamer JE (1998). Divergent cAMP signalling pathways regulate growth and pathogenesis in the rice blast fungus *Magnaporthe grisea*. Plant Cell 10: 1361-1373.

Altschul SF, Madden TL, Schaffer AA, Zhang J, Zhang Z, Miller W and Lipman DJ (1997). Gapped BLAST and PSI-BLAST: a new generation of protein database search programs. Nucleic Acids Res 25: 3389-402.

Asch DK and Kinsey JA (1990) Relationship of vector insert size to homologous integration during transformation of *Neurospora crassa* with the cloned *AM* (*GDH*) gene. Molecular and General Genetics 221: 37-43.

Audic S and Claverie JM (1997). The significance of digital gene expression profiles. Genome Res. 7: 986-95.

Balhadère PV, Foster AJ and Talbot NJ (1999) Identification of pathogenicity mutants of the rice blast fungus *Magnaporthe grisea* by insertional mutagenesis. Molecular Plant-Microbe Interactions 12: 129-142.

Balhadère PV and Talbot NJ (2000). *PDE1* encodes a P-type ATPase involved in appressorium-mediated plant infection by the rice blast fungus *Magnaporthe grisea*. The Plant Cell 13: 1987-2004.

Banuett F and Herskowitz I (1994). Identification of *Fuz7*, a *Ustilago maydis* MEK/MAPKK homologue required for a locus dependent and a locus independent steps in the fungal life-cycle. Genes Dev 8: 1367-1378.

Banuett F (1998). Signalling in the yeasts: An informational cascade with links to the filamentous fungi. Microbiology and Molecular Biology Reviews 62: 249-255.

Bardwell L, Cook JG, Voora D, Baggott DM, Martinez AR and Thorner J (1998a). Repression of yeast Ste12 transcription factor by direct binding of unphosphorylated *Kss1* MAPK and its regulation by the Ste7 MEK. Genes Dev. 12: 2887-2898.

Bardwell L, Cook JG, Zhu-Shimoni JX, Voora D and Thorner J (1998b). Differential regulation of transcription: Repression by unactivated mitogen-activated protein kinase Kss1 requires the Dig1 and Dig2 proteins. Proc Natl Acad Sci USA 95: 15400-15405.

Barrett K, Gold S and Kronstad JW (1993). Identification and complementation of a mutation to constitutive filamentous growth in *Ustilago maydis*. Molecular Plant-Microbe Interactions 6: 274-283.

Baudin A, Ozierkalogeropoulos O, Denouel A, Lacroute F and Cullin C. (1993). A simple and efficient method for direct gene deletion in *Saccharomyces cerevisiae* Nucleic Acids Res 21: 3329-3330.
Bourett TM and Howard RJ (1990) *In vitro* development of penetration structures in the rice blast fungus *Magnaporthe grisea*. Can J Bot. 68: 329-342.
Bowyer P, Clarke BR, Lunness P, Daniels MJ and Osbourn AE (1995).. Host range of a plant pathogenic fungus determined by a saponin detoxifying enzyme. Science 267: 371-374
Bundock P, den Dulk-Ras A, Beijersbergen A and Hooykaas PJJ (1995). Trans-kingdom T-DNA transfer from *Agrobacterium tumefaciens* to *Saccharomyces cerevisiae*. EMBO J. 14: 3206-3214.
Celenza JL and Carlson M (1984) Cloning and genetic mapping of *SNF1*, a gene required for expression of glucose-repressible genes in *Saccharomyces cerevisiae*. Mol Cell Biol 4: 49-53.
Chao CCT and Ellingboe AH (1991) Selection for mating competence in *Magnaporthe grisea* pathogenic to rice. Can J Bot 69: 2130-2134.
Chaveroche M-K, Ghigo J-M and d'Enfert C. (2000). A rapid method for efficient gene replacement in the filamentous fungus *Aspergillus nidulans*. Nucleaic Acids Res 28: E97.
Choi W and Dean RA (1997). The adenylate cyclase gene *MAC1 of Magnaporthe grisea* controls appressorium formation and other aspects of growth and development. Plant Cell 9: 1973-1983.
Chumley FG and Valent B (1990). Genetic analysis of melanin deficient non-pathogenic mutants of *Magnaporthe grisea*. Molecular Plant-Microbe Interactions 3: 135-143.
Clergeot PH, Gourgues M, Cots J, Laurans F, Latorse MP, Pepin R, Tharreau D, Notteghem JL and Lebrun MH (2001). *PLS1*, a gene encoding a tetraspanin-like protein, is required for penetration of rice leaf by the fungal pathogen *Magnaporthe grisea*. Proc Natl Acad Sci USA 98: 6963-6968.
Cooper RM (1987). The use of mutants in exploring depolymerises as determinants of pathogenicity. In: PR Day and GJ Jellis, ed. Genetics and Plant Pathogenesis. Oxford, UK: Blackwell, pp261-281.
De Backer MD, Ilyiana T, Ma X-J, Vandoninck S, Luyten WHML and Vanden Bossche H (2001) Genomic profiling of the response of *Candida albicans* to itraconazole treatment using a DNA microarray. Antimicrob. Agents Chemother 45: 1660-1670.
De Groot MJA, Bundock P, Hooykaas PJJ and Beijersbergen AGM (1998). *Agrobacterium tumefaciens*-mediated transformation of filamentous fungi. Nat Biotech 16: 839-842.
de Jong JC, McCormack BJ, Smirnoff N and Talbot NJ (1997). Glycerol generates turgor in rice blast. Nature 389: 244-245.
DeVit MJ, Waddle JA and Johnston M (1997). Regulated nuclear translocation of the Mig1 glucose repressor. Mol. Biol. Cell 8: 1603-1618.
DeZwaan TM, Carroll AM, Valent B and Sweigard JA (1999). *Magnaporthe grisea* Pth11p is a novel plasma membrane protein that mediates appressorium differentiation in response to inductive substrate cues. The Plant Cell 11: 2013-2030.
Dean RA (1997). Signal pathways and appressorium morphogenesis. Annual Review of Phytopathology 35: 211-234.
Deising HB, Werner S and Wernitz M (2000). The role of fungal appressoria in plant infection. Microbes and Infection 2: 1631-1641.
Desprez T, Amselem J, Caboche M and Hofte H (1998). Differential gene expression in *Arabidopsis* monitored using cDNA arrays. Plant Journal 14: 643-652.
Dolan JW, Kirkman C and Fields S (1989). The yeast STE12 protein binds to the DNA sequence mediating pheromone induction. Proc. Natl Acad Sci USA 86: 5703-5707.
Eisen MB, Spellman PT, Brown PO and Botstein D (1998). Cluster analysis and display of genome-wide expression patterns. Proc Natl Acad Sci USA 9525: 14863-14868.
Elion EA, Satterberg B and Kranz JE (1993). *FUS3* phosphorylates multiple components of the mating signal transduction cascade: Evidence for *STE12* and *FAR1*. Mol. Biol. Cell 4: 495-510.
Epstein CB and Butow RA (2000). Microarray technology - enhanced versatility, persistent challenge. Current Opinion in Biotechnology 11: 36-41.
Feldmann KA (1991). T-DNA insertion mutagenesis in *Arabidopsis*: mutational spectrum. Plant J. 1: 71-82.
Gilbert RD, Johnson AM and Dean RA (1996) Chemical signals responsible for appressorium formation in the rice blast fungus. Physiological and Molecular Plant Pathology 48: 335-346.
Gold SE (1994). cAMP regulates morphogenesis in the fungal pathogen *Ustilago maydis*. Genes Dev. 8: 2805-2816.
Goffeau A, Barrell BG, Bussey H, Davis RW, Dujon B, Feldmann H, Galibert F, Hoheisel JD, Jacq C, Johnston M, Louis EJ, Mewes HW, Murakami Y, Philippsen P, Tettelin H and Oliver SG (1996). Life with 6000 genes. Science 274: 546-563.
Gouka RJ, Gerk C, Hooykaas PJJ, Bundock P, Musters W, Verrips CT *et al.* (1999). Transformation of *Aspergillus awamori* by *Agrobacterium tumefaciens*-mediated homologous recombination. Nat. Biotech. 17: 598-601.

Gustin MC, Albertyn J, Alexander M and Davenport K (1998). MAP kinase pathways in the yeast *Saccharomyces cerevisiae*. Microbiology and Molecular Biology Reviews 62: 1264-1300.

Hamer JE, Howard RJ, Chumley FG and Valent B (1988). A mechanism for surface attachment of spores of a plant pathogenic fungus. Science 239: 288-290.

Hamer JE, Valent B and Chumley FG (1989). Mutations at the *SMO* locus affect the shapes of diverse cell types in the rice blast fungus. Genetics 122: 351-361.

Hamer JE and Givan S (1990). Genetic mapping with dispersed repeated sequences in the rice blast fungus: mapping the SMO locus. Molecular and General Genetics 223: 487-95.

Hamer JE and Talbot NJ (1998). Infection-related development in the rice blast fungus *Magnaporthe grisea*. Curr Opin Microbiol 1: 693-697

Hamer L, Adachi K, Montenegro-Chamorro MV, Tanzer MM, Mahanty SK, Lo C, *et al.* (2001). Gene discovery and gene function assignment in filamentous fungi. Proc Natl Acad Sci USA 98: 5110-5115.

Han Y, Liu X, Benny U, Kistler HC and VanEtten, HD (2001). Genes determining pathogenicity to pea are clustered on a supernumerary chromosome in the fungal plant pathogen *Nectria haematococca*. Plant J. 25: 305-14.

Hardie DG, Carling D and Carlson M (1998). The AMP-activated/Snf1 protein kinase subfamily: Metabolic sensors of the eukaryotic cell? Annual Review of Biochemistry 67: 821-855.

Herskowitz I (1995) MAP kinases in yeast: For mating and more. Cell 80: 187-197.

Howard RJ and Ferrari MA (1989) Role of melanin in appressorium formation. Experimental Mycology 13: 403-418.

Howard RJ, Ferrari MA, Roach DH and Money NP (1991). Penetration of hard substrates by a fungus employing enormous turgor pressures. Proc Natl Acad Sci USA 88: 11281-11284.

Howard RJ (1994) Cell biology of pathogenesis. In: RS Zeigler, SA Leong, PS Teng, ed. The Rice Blast Disease. CAB International, Wallingford, Oxon, UK. pp 3-22.

Howard RJ (1997). Breaching the outer barriers – Cuticle and cell wall penetration. In: G.C. Carroll and P. Tudzynski, eds. The Mycota V Part B Plant Relationships. Springer-Verlag, Berlin, Heidelberg, pp 43-60.

Huang X and Madan A (1999). CAP3: A DNA sequence assembly program. Genome Res 9: 868-877.

Idnurm A and Howlett BJ (2001). Pathogenicity genes of phytopathogenic fungi. Mol. Plant Path. 2: 241-255.

Jelitto TC, Page HA and Read ND (1994) Role of external signals in regulating the pre-penetration phase of infection by the rice blast fungus, *Magnaporthe grisea*. Planta 194: 471-477.

Justesen A, Somerville S and Giese H (1996). Isolation and characterization of two novel genes expressed in germinating conidia of the obligate biotroph *Erysiphe graminis f sp hordei*. Gene 170: 131-135.

Kahmann R and Basse C (1999). REMI (restriction enzyme-mediated integration) and its impact on the isolation of pathogenicity genes in fungi attacking plants. European Journal of Plant Pathology 105: 221-229.

Kahmann R, Basse C and Feldbrugge M (1999) Fungal-plant signalling in the *Ustilago maydis*-maize athosystem. Current Opinion in Microbiology 2: 647-650.

Kang SH, Khang CH and Lee YH (1999). Regulation of cAMP-dependent protein kinase during appressorium formation in *Magnaporthe grisea*. FEMS Microbiology Letters 170: 419-423.

Keon J, Bailey A and Hargreaves J (2000). A group of expressed cDNA sequences from the wheat fungal leaf blotch pathogen, *Mycosphaerella graminicola* (*Septoria tritici*). Fungal Genetics and Biology 29: 118 33.

Knogge W (1998) Fungal pathogenicity. Current Opinion in Plant Biology 1: 324-328.

Koncz C, Németh K, Rédei GP and Schell J (1992). T-DNA insertional mutagenesis in *Arabidopsis*. Plant Mol. Biol. 20: 963-976.

Kronstad JW (1997). Virulence and cAMP in smuts, blasts and blights. Trends in Plant Science 2: 193-199.

Kupfer DM, Reece A, Clifton SW, Roe BA and Prade RA (1997) Multicellular ascomycetous fungal genomes contain more than 8000 genes. Fungal Genetics and Biology 21: 364-372.

Langin T, Capy P and Daboussi MJ (1995) The transposable element *impala*, a fungal member of the *tc1*-mariner superfamily. Molecular and General Genetics 246: 19-28.

Lee Y-H and Dean RA (1993). cAMP regulates infection structure formation in the plant pathogenic fungus *Magnaporthe grisea*. Plant Cell 5: 693-700.

Lee Y-H and Dean RA (1994). Hydrophobicity of contact surface induces appressorium formation in *Magnaporthe grisea*. FEMS Microbiology Letters 115: 71-75.

Lee PS and Lee KH (2000). Genomic analysis. Current Opinion in Biotechnology 11: 171-175.

Lengeler KB, Davidson RC, D'souza C, Harashima T, Shen WC, Wa Pan X, Waugh M and Heitman J (2000). Signal transduction cascades regulating fungal development and virulence. Microbiology and Molecular Biology Reviews 64: 746-785.

Lesage P, Yang X and Carlson M (1996). Yeast SNF1 protein kinase interacts with SIP4, a C_6 zinc cluster transcriptional activator: A new role for SNF1 in the glucose response. Mol Cell Biol 16: 1921-1928.

Leung H, Borromeo S, Bernardo MA and Notteghem JL (1998). Genetic analysis of virulence in the rice blast fungus *Magnaporthe grisea* Phytopathology 78: 1227-1233.

Lev S, Sharon A, Hadar R, Ma H and Horwitz BA (1999) A mitogen-activated protein kinase of the corn leaf pathogen *Cochliobolus heterostrophus* is involved in conidiation, appressorium formation, and pathogenicity: Diverse roles for mitogen-activated protein kinase homologs in foliar pathogens. Proc Natl Acad Sci USA 96: 13542-13547.

Lu S, Lyngholm L, Yang G, Bronson C, Yoder OC and Turgeon BG (1994) Tagged mutations at the *Tox1* locus of *Cochliobolus heterostrophus* by restriction enzyme-mediated integration. Proc Natl Acad Sci USA 91: 12649-12653.

Maier FJ and Schafer W (1999) Mutagenesis via insertional restriction enzyme-mediated-integration (REMI) as a tool to tag pathogenicity related genes in plant pathogenic fungi. Biol Chem 380: 855-864.

Mayorga ME and Gold SEA (1999) MAP kinase encoded by the *ubc*3 gene of *Ustilago maydis* is required for filamentous growth and full virulence. Molecular Microbiology 34: 485-497.

Mewes HW, Albermann K, Heumann K, Liebl S and Pfeiffer F (1997). MIPS: a database for protein sequences, homology data and yeast genome information. Nucleic Acids Res 25: 28-30.

Mikosch TSP, Lavrijssen B, Sonnenberg ASM and van Griensven LJLD (2001) Transformation of the cultivated mushroom, *Agaricus bisporus* (large) using T-DNA from *Agrobacterium tumefaciens*. Curr Genet 39: 35-39.

Mitchell TK and Dean RA (1995) The cAMP-dependent protein kinase catalytic sub-unit is required for appressorium formation and pathogenesis by the rice blast fungus *Magnaporthe grisea*. Plant Cell 7: 1869-1878.

Muller P, Aichinger C, Feldbrugge M and Kahmann R (1999). The MAP kinase *kpp2* regulates mating and pathogenic development in *Ustilago maydis*. Molecular Microbiology. 34: 1007-1017.

Mullins ED, Chen X, Romaine P, Raina R, Geiser DM and Kang S (2001) *Agrobacterium*-mediated transformation of *Fusarium oxysporum*: an efficient tool for insertional mutagenesis and gene transfer. Phytopathology 91: 173-180.

Murad AMA, d'Enfert C, Gaillardin C, Tournu H, Tekaia F, Talibi D, Marechal D, Marchais V, Cottin J and Brown AJP (2001) Transcript profiling in *Candida albicans* reveals new cellular functions for the transcriptional repressors CaTup1, CaMig1 and CaNrg1. Mol Microbiol 42: 981-993.

Notteghem JL and Silué D (1992) Distribution of the mating type alleles in *Magnaporthe grisea* populations pathogenic on rice. Phytopathology 82: 421-424.

Oliver SG (1996). From DNA sequence to biological function. Nature 379: 597-600.

Ou SH (1985) Rice Diseases 2nd Edition. Kew, Surrey: Commonwealth Mycological Institute, CABI. pp 109-201.

Panaccione DG, Scott-Craig JS, Pocard JA and Walton JD (1992). A cyclic peptide synthetase gene required for pathogenicity of the fungus *Cochliobolus carbonum* on maize. Proc Natl Acad Sci USA 89: 6590-6594.

Park G, Xue C, Zheng L, Lam S and Xu J-R (2002) *MST12* regulates infectious growth but not appressorium formation in the rice blast fungus *Magnaporthe grisea*. Molecular Plant-Microbe Interactions 15: 183-192.

Reymond P, Weber H, Damond M and Farmer EE (2000) Differential gene expression in response to mechanical wounding and insect feeding in *Arabidopsis*. Plant Cell 12: 707-719.

Rho HS, Kang S and Lee YH (2001) *Agrobacterium tumefaciens*-mediated transformation of the plant pathogenic fungus, *Magnaporthe grisea*. Molecules and Cells 12:407-411.

Ronne H (1995) Glucose repression in fungi. Trends in Genetics 11: 12-17.

Rosewich UL and Kistler HC (2000). Role of horizontal gene transfer in fungi. Annual Review of Phytopathology: 325-363.

Ruijter GJG and Visser J (1997) Carbon repression in *Aspergilli*. FEMS Microbiol. Lett. 151: 103-114.

Schmid KJ and Aquadro CF (2001). The evolutionary analysis of "orphans" from the *Drosophila* genome identifies rapidly diverging and incorrectly annotated genes. Genetics 159: 589-598.

Scott-Craig JS, Panaccione DG, Cervone F and Walton JD (1990) Endopolygalacturonase is not required for pathogenicity of *Cochliobolus carbonum* on maize. Plant Cell 2: 1191-1200.

Silué D, Tharreau D, Talbot NJ, Clergeot PH, Notteghem JL, and Lebrun MH (1998). Identification and characterization of *apf1*⁻ in a non-pathogenic mutant of the rice blast fungus *Magnaporthe grisea* which is unable to differentiate appressoria. Physiol. Mol. Plant Pathol. 53: 239-251.

Skinner W, Keon J and Hargreaves J (2001) Gene information for fungal pathogens from expressed sequences. Curr Opin Microbiol 4: 381-386.

Soanes DM, Skinner W, Keon J, Hargreaves J and Talbot NJ (2002). Genomics of phytopathogenic fungi and the development of bioinformatic resources. Molecular and Plant-Microbe Interactions 15: 421-427.

Sprague GF and Thorner JW (1992) Pheromone response and signal transduction during the mating process of Saccharomyces cerevisiae. In: EW Jones, JR Pringle, JR Broach, eds. The Molecular Cellular Biology of the Yeast *Saccharomyces cerevisiae*. Cold Spring Harbor Laboratory Press, Cold Spring Harbor, NY. USA. pp 657-744.

Straney DC and VanEtten HD (1994). Characterization of the *PDA1* promoter of *Nectria haematococca* and identification of a region that binds a pisatin-responsive DNA binding factor. Molecular and Plant-Microbe Interactions 7: 256-66.

Sweigard JA, Chumley FG and Valent B (1992) Disruption of a *Magnaporthe grisea* cutinase gene. Molecular and General Genetics 232: 183-190.

Sweigard JA, Carroll AM, Farrall L, Chumley FG and Valent B (1998) *Magnaporthe grisea* pathogenicity genes obtained through insertional mutagenesis. Molecular Plant-Microbe Interactions 11: 404-412.

Talbot NJ, Ebbole DJ and Hamer JE (1993) Identification and characterization of *MPG1*, a gene involved in pathogenicity from the rice blast fungus *Magnaporthe grisea*. Plant Cell 5: 1575-1590.

Talbot NJ (1995) Having a blast: exploring the pathogenicity of *Magnaporthe grisea*. Trends in Microbiology 3: 9-16.

Talbot NJ (1999) Plant biophysics - Forcible entry. Science 285: 1860-1861.

Tedford K, Kim S, Sa D, Stevens K and Tyers M (1997) Regulation of the mating pheromone and invasive growth responses in yeast by two MAP kinase substrates. Curr Biol 7: 228-238.

Thines E, Weber RWS and Talbot NJ (2000) MAP kinase and protein kinase A-dependent mobilization of triacylglycerol and glycogen during appressorium turgor generation by *Magnaporthe grisea*. The Plant Cell 12: 1703-1718.

Thon MR, Nuckles EM and Vaillancourt LJ (2000) Restiction enzyme-mediated integration used to produce pathogenicity mutants of *Collototrichum graminicola*. Molecular Plant-Microbe Interactions 13: 1356-1365.

Tilburn J, Roussel F and Scazzocchio C (1990) Insertional in-activation and cloning of the wA gene of *Aspergillus nidulans*. Genetics 126: 81-90.

Tonukari NJ, Scott-Craig JS and Walton JD (2000) The *Cochliobolus carbonum SNF1* gene is required for cell wall-degrading enzyme expression and virulence on maize. The Plant Cell 12: 237-247.

Treitel MA, Kuchin S and Carlson M (1998) Snfl protein kinase regulates phosphorylation of the Migl repressor in *Saccharomyces cerevisiae*. Mol Cell Biol 18: 6273-6280.

Tucker SL and Talbot NJ (2001). Surface attachment and pre-penetration stage development by plant pathogenic fungi. Annual Review of Phytopathology. 39: 385-417.

Tzung KW, Williams RM, Scherer S, Federspiel N, Jones T, Hansen N, Bivolarevic V, Huizar L, Komp C, Surzycki R, *et al.* (2001) Genomic evidence for a complete sexual cycle in Candida albicans. Proc Natl Acad Sci USA 98: 3249-3253.

Uchiyama T, Ogasawara N, Nanba Y and Ito H (1979) Conidial germination and appressorial formation on the coverglass or cellophane coated with various lipid components of plant leaf waxes. Agricultural Biological Chemistry 43: 383-384.

Valent B and Chumley FG (1991) Molecular genetic analysis of the rice blast fungus *Magnaporthe grisea*. Annual Review of Phytopathology 29: 443-367.

Valent B, Farrall L and Chumley FG (1991) *Magnaporthe grisea* genes for pathogenicity and virulence identified through a series of backcrosses. Genetics 127: 87-101.

Villalba F, Lebrun MH, Hua-Van A, Daboussi MJ and Grosjean-Cournoyer MC (2001) Transposon *impala*, a novel tool for gene tagging in the rice blast fungus *Magnaporthe grisea*. Molecular Plant-Microbe Interactions 14: 308-315

Vincent O and Carlson M (1998) Sip4, a Snfl kinase-dependent transcriptional activator, binds to the carbon source-responsive element of gluconeogenic genes. EMBO J. 17: 7002-7008.

Voiblet C, Duplessis S, Encelot N and Martin F (2001) Identification of symbiosis-regulated genes in *Eucalyptus globules-Pisolithus tinctorius* ectomycorriha by differential hybridisation of arrayed cDNAs. Plant J. 25: 181-191.

Walton JD (1994). Deconstructing the cell wall. Plant Physiol. 104: 1113-1118.

Weber RWS, Wakley GE, Thines E and Talbot NJ (2001) The vacuole as central element of the lytic system and sink for lipid droplets in maturing appressoria of *Magnaporthe grisea*. Protoplasma 216: 101-112.

Woloshuk CP, Sisler HD, Tokousbalides MC and Dutky SR (1980) Melanin biosynthesis in *Pyricularia oryzae* – site of tricyclazole inhibition and pathogenicity of melanin-deficient mutants. Pesticide Biochemistry and Physiology 14: 256-264.

Wood V, Gwilliam R, Rajandream MA, Lyne M, Lyne R, Stewart A, Sgouros J, Peat N, Hayles J, Baker S *et al.* (2002) The genome sequence of *Schizosaccharomyces pombe*. Nature 415: 871-880.

Xu J-R and Hamer JE (1996) MAP Kinase and cAMP signalling regulate infection structure formation and pathogenic growth in the rice blast fungus *Magnaporthe grisea*. Genes and Development 10: 2696-2706.

Xu J-R, Urban M, Sweigard JA and Hamer JE (1997) The *CPKA* gene of *Magnaporthe grisea* is essential for appressorial penetration. Molecular Plant Microbe Interactions 10: 187-194.

Xue CY, Park G, Choi WB, Zheng L, Dean RA and Xu JR (2002) Two novel fungal virulence genes specifically expressed in appressoria of the rice blast fungus. Plant Cell 14: 2107-2119.

Yoder OC and Turgeon BG (2001) Fungal genomics and pathogenicity. Current Opinion in Plant Biology. 4: 315-321.

Zupan JR and Zambryski P. (1995) Transfer of T-DNA from *Agrobacterium* to the plant cell. Plant Physiol. 107: 1041-1047.

Applied Mycology & Biotechnology
An International Series. Volume 4. Fungal Genomics

15

Genomics of Entomopathogenic Fungi

George G. Khachatourians[a] and Daniel Uribe[a,b]
[a]BioInsecticide Research Laboratory, Department of Applied Microbiology and Food Science, University of Saskatchewan, Saskatoon, S7N 5A8, SK. Canada (khachatouria@usask.ca) and [b]Biotechnology Institute, National University of Colombia, Bogotá, Colombia.

Entomopathogenic fungi (EPF) are important in mycology and entomology because of the dyad of interactions between their environment and their insect host(s). The applied uses of EPF are in insect pest management, industrial and pharmaceutical bio-product formation. The general and molecular genetics, genomics and population level analysis of only three genera *Beauveria, Metarhizium* and *Verticillium* and then only one species in each case (*B. bassiana, M. anisopliae*, and *V. lecanii*) is well established. In this chapter we review general and molecular genetics of EPF. The most complete molecular and biophysical characterization of chromosomal DNA is that for *B. bassiana.* Molecular karyotyping, RFLP analysis and chromosomal genes cloning, sequencing, heterologous expression PCR fingerprinting, probes and DNA/DNA hybridization and comparative genomics, of a few protein-coding genes are known for half a dozen EPF. Genomic efforts, such as EST, chromosome sequencing, and whole genome sequencing are emerging. The mitochondrial (mt) genome studies have determined genome sizes for *B. bassiana,* 28.5 kb, *and M. anisopliae* with aprox. 32Kb and *V. lecanii*, 24.5 kb, placing these EPF at the small end of the fungal mitochondrial genomes, 19 kb (*Torulopsis glabrata*) and 176Kb *(Agaricus bitorquis)*. RFLP mapping of mtDNA and sequencing of the Mtc encoded rRNA genes have been used for taxonomic purposes and characterization of populations' subgroups. The future research in the use of genomics in population, taxonomic and phylogenetic studies will be important in the production and use of EPF and their contribution to alternatives in pest management and the history of agriculture and forestry sciences.

1. INTRODUCTION

Fungal disease of insects represents an important opportunity for understanding the molecular basis of the interaction between entomopathogenic fungi (EPF) and their insect hosts and ultimately their applied uses in mycology, biotechnology and entomology, which is insect biocontrol. The genetic, molecular genetics, genomics and population level analysis of only 5-6 of the entomophthorales and hyphomycetes have been studied (Khachatourians 1991, 1996; Bidochka *et al.* 2000; Bidochka 2001; St-Leger and Screen 2001; Khachatourians *et al.* 2002). In

Corresponding author: George G. Khachatourians

spite of some early start by the authors group in late 80's there has been a rather unusual and long delay in the recognition of the long range potential contemporary biotechnological and related sciences of genomic and proteomics that EPF have held for insect pest management. Since the mid-late 90's a small but growing number of investigators have begun active research and generated the much needed information and data in this regard. With such knowledge, physiological manipulations, enhancement of virulence, construction of environmentally safe strains with limited persistence and issues of intellectual property aspects of the EPF should be possible. This chapter reviews the primary literature on molecular genetics and genomics of EPF.

2. CHROMOSOMES

As with other fungi, there are many approaches in describing the genetic apparatus of EPF. At the cytogenetic level the chromosomes can be visualized with the light and electron microscope. Karyotyping generates information on the chromosome(s) number, size, shape, origin, evolution, and systematics. There can be physical genetic maps constructed by the localizing genes for which DNA hybridization probes are available.

Cytological studies of entomopathogenic fungal nuclei have been performed on *Metarhizium anisopliae* and *Beauveria bassiana* showing the nuclei divide immediately before or during spore germination. One can assume the nucleus in these fungi are haploid even though some hyphal cells may contain two or more nuclei or heterokaryons from two nuclei of different types presumably due to anastomosis and nuclear migration. The earlier literature (see Khachatourians 1991, 1996) concluded that the nucleus in a conidium divided just before germination and that one of the two daughter nuclei subsequently emerged into the germ tube while the other stayed in the spore. During the germination process each conidium therefore contain two nuclei and during conidial development, both daughter nuclei occasionally moved into the young conidium.

As early as mid 1980's this was attempted through the formation, fusion and regeneration of protoplasts from several EPFs (Khachatourians 1991). Cytogenetic evidence was used to indicate relationship between ploidy and nuclear size of EPF (Khachatourians 1991). Drummond and Heale (1988) demonstrated that complementary parental diauxotrophic strains of *Verticillium lecanii* could be paired on minimal medium using hyphal anastomosis and protoplast fusion techniques to produce diploid strains. The pathogenicity of *V. lecanii* diploids was demonstrated against the white fly (*Trialeurodas vaporariorum*). From cadavers of white flies infected with heterozygous diploid conidia, prototrophic and auxotrophic progeny haploid strains were re-isolated. These results indicated recombination for genes involved in pathogenicity, sporulation and germination.

Classical genetic studies in EPF that would be contextual to chromosomal structure and recombinational uses have been few and have not been extended to rigorous genetic analysis. For example, the production of mutants and forced heterokaryon formation tried in the 1970- 80's have not been continued rigorously or been extended to a degree to create generalized models (Khachatourians 1991). Kotlyarevskii and Levites (1992) demonstrated a parasexual process in *V. lecanii* verified through the patterns of isozymes. Bello and Paccola-Meirelles (1998) found *B. bassiana* parasexual crosses among strains with complementary genetic markers yielded heterokaryons but no diploids implying high instability of the diploid nucleus. Analysis of genetic markers was localized in four linkage groups by the parasexual crosses. In the first group markers *nic*4, *nic*3, *thi*2, *bio*3, *ade*2, *ths*2, and *ben*1R were localized; in the second, the marker *met*1 in the third, *pab*1; and in the fourth, *bio*1. The parental strain 196/A11/3 is a carrier of translocation among the linkage groups III and I. Viaud *et al*. (1998) study of *B. bassiana* and *B. sulfurescens* protoplasts showed that hybrids appeared to be diploid or aneuploid with portions of the genome being heterozygous while mitochondrial molecular marker indicated homoplasmy of the hybrids and inheritance of mitochondria. More recently, Leal *et al.* (2000) demonstrated genetic exchange in *M. anisopliae* strains co-infecting *Phaedon cochleariae* as revealed by molecular markers. Together these studies have extended the notion that genetic exchange and recombination of chromosomal markers occurs. Confirmation of these results by conventional and molecular markers remains open.

Study of chromosomal genetics through protoplast fusion has demonstrated both recombination-like events and karyogamy. Since early 1990s a number of in vitro transformation systems have been established for various EPF (Khachatourians 1996). Barreto *et al.* (1997) compared high frequency gene transfer by microprojectile bombardment of intact conidia and PEG methods to transform protoplasts *Paecilomyces fumosoroseus*. Transformation system used hygromycin resistance as the selectable marker. They found frequencies varied from 1.9 to 2.5 transformants μg of DNA by the PEG method, and from 33 to 153 transformants μg of DNA by the biolistic procedure.

Use of genetically multiple marked isolates is important for study of both pathogenicity and improvements in virulence as much that of dispersal and carryover of released fungi. Berretta *et al.* (1998) designed fluorescent deoxynucleotides used for high-resolution RAPD DNA fingerprint analysis should be especially useful to the study of genetic diversity and to screening many isolates of EPF in population studies. There are several investigators who have used a number of marker genes- green fluorescent protein, beta-glucuronidase, antibiotic resistance e.g. benomyl resistant, and nitrate resistance- to perform taxonomic, population genetics or transformation studies (Sandhu *et al.* 2001a, b; Thorvilson 2002; Bello and Paccola-Meirelles 1998, Inglis *et al.* 2000; Cantone and Vandenberg 1999). Inglis *et al.* (2000) demonstrated the transformation of *M. anisopliae* var. *acridum* (syn. *M. flavoviride*) by using multiple plasmid borne markers, i.e., the bar gene fused to *trp*C promoter, resistance to glufosinate ammonium, the telomeric repeat (TTAGGG) 18 and the gene for *Aequorea victoria* green fluorescent protein fused to *gpd* promoter and *trp*C terminator. A recent field release and study using an *A. victoria* green fluorescent protein gene alone or with additional protease genes in a recombinant *M. anisopliae* revealed rhizosphere competency of the construct (Hu and St-Leger 2002). This study confirmed the utility of gfp for monitoring pathogen strains in field populations over time and the little dissemination of transgenic strains and its stability under field conditions.

Viaud *et al.* (1998) demonstrated molecular analysis of hypervirulent somatic hybrids of the entomopathogenic fungi *B. bassiana* and *B. sulfurescens* by protoplast fusion of diauxotrophic mutants. Some of the hybrids were significantly hypervirulent and killed insects more quickly. By using six nuclear genes and a telomeric fingerprint probe, these authors demonstrated the occurrence of parasexual events. Hybrids appeared to be diploid or aneuploid with portions of the genome being heterozygous. Variations in nuclear small subunit rRNA group I intron among *B. bassiana* and *B. brongniartii* and evidence of horizontal transfer was shown by Coates-*et al.*, (2002a). These authors found insertion points were conserved among nuSSU rRNA genes from 35 *Beauveria* isolates. PCR-RFLP and DNA sequencing identified 12 group I intron variants and were applied to the identification of strains isolated from insect hosts. Alignment of 383-404-nt subgroup IB3 group I intron indicated that four insertion/deletion mutations were the main basis of fragment length variation. Phylogeny reconstruction using parsimony and neighbor-joining methods suggested six lineages may be present among nuSSU rRNA group I intron sequences from *Beauveria* and related ascomycete fungi. Terminal node placement of *Beauveria* introns conflicted with previously published phylogenies constructed from gene sequences, suggesting horizontal transfer of group I introns. PCR-RFLP among introns provided a means for the differentiation of *Beauveria* isolates.

In *B. bassiana* cross incubation of 'compatible' vegetative strains were shown to result in a heterokaryon, which undergoes successive losses of chromosomes or its fragments and haploidization (Wang *et al.* 2000). However a rigorous back cross analysis or multiple marker analysis has not occurred. Against the backdrop of the age of molecular genetics the question of natural anastomosis or heterokaryon formation in EPF is still valid. This is because in their natural environments, genetic exchange events are important determinants for genetic -equilibrium and -drift.

Wang *et al.* (2000, 2001) produced heterokaryons of *B. bassiana* by cross incubation of two compatible vegetative strains with genetic markers of actinione resistance and 34°C tolerance and subsequently performed molecular identification of heterokaryosis in population using the presence or absence of an intron at domain D11 of 28S rRNA gene by the PCR product sizes of this region. Population investigations indicated that the heterokaryotic frequencies varied greatly associated with different geographical and ecological habitats. Random amplified polymorphic DNA analysis of representative segregants suggested that genetic exchange and recombination occurred during heterokaryon parasexual cycling. However, high level of genetic instability was also found during subculturing of heterokaryotic strains through saprophytic media.

2.1. Genome and Chromosomes

The first study of the genome of any EPF was that of *Entomophaga aulicae* examined by Murrin *et al.* (1986). This genome was shown to contain 8 x 10^6 kb DNA per nucleus. The base composition of *E. aulicae* chromosomal DNA was 38% G+C. There were 15 chromosomes and 11 pairs of kinetocores. The ratio of nuclei to mitochondria in 12 protoplasts was estimated to be 1: 25 ± 13. In comparing the nuclei of this fungus with those of the others, Murrin *et al.* (1986) suggest that the genome size of this fungus is two orders of magnitude greater than that traditionally attributed to fungi.

The most comprehensive studied description of biophysical and a biochemical characteristic of any EPF chromosomal DNA is that of *B. bassiana.* Pfeifer and Khachatourians (1989) determined the G+C content of this fungus by CsCl buoyant density centrifugation and thermal denaturation and found to be 56.9 ± 1.9%, a value in line with 53.0% G+C ratio of *B. tenella* (Storck and Alexopoulos 1970). No unusual modified bases were found.

Descriptions of chromosomes of EPF can also be based on electrophoretic karyotyping techniques. By using alternating electric field gel electrophoresis, chromosomes of EPF can now be separated and their numbers and sizes determined. There are several apparatuses for separation, but as their names indicate pulsed field gel electrophoresis (PFGE), clamped or contoured homogeneous electric field (CHEF), crossed field gel electrophoresis and pulsed homogeneous orthogonal field gel electrophoresis (PHOGE), they differ in electrogeometry.

Electrophoretic karyotyping and chromosomal sizes of *B. bassiana* isolates F12, F13 and F8807 obtained from *B. mori* differ in all chromosomal bands except the 2.4 Mb DNA (Shimizu *et al.* 1993a). Using CHEF Pfeifer and Khachatourians (1993b) showed conditions for *B. bassiana* preparation of chromosome samples, their separation and karyotyping. The electrophoretic karyotype showed the presence of seven chromosomal bands of which one was a doublet resulting in a total of eight chromosomes. The total genome size for *B. bassiana* strain GK 2016 was estimated to be 40.6 ± 1.1 Mbp which is within the range for other filamentous fungi, but larger than the 26.1 to 30.4 Mbp for other *B. bassiana* isolates and the estimated 27.6 to 30.1 Mbp genome of *P. fumosoroseus* (Shimizu *et al.* 1991a,b). Additional differences in the electrophoretic karyotyping of *M. anisopliae* but not in *P. fumosoroseus* and *P. farinosus* were found by Shimizu *et al.* (1991). Valadares-Inglis and Peberdy (1998) used PFGE to separate chromosome-sized DNA molecules of four Brazilian strains of *M. anisopliae*. These authors identified eight chromosomal DNA bands, varying in size between 23.39 to 31.88 Mb, with two chromosomes migrating as doublets. Viaud *et al.* (1996) described multiplicity of approaches and techniques, electrophoretic karyotype, nitrate reductase gene for gene mapping, RFLP analysis, and telomeric fingerprinting to study genome organization in *B. bassiana*. The total genome size was between 27.6-28.1 Mb. The study by Shi (2000) used PFGE to estimate 6 chromosome bands with *B. bassiana* isolate F1 showing karyotypes of 2.5-6.6 and 26.5 Mbp while those of F2 and F3 karyotype sizes of 2.8, and 6.7, and 2.9-7.2 and 29 Mbp respectively.

2.2. Chromosomal DNA: Restriction Fragment Length Polymorphism

Demonstrations of differences in the genetic make up of various EPF isolates and strains would be another approach to begin in the dissection of the genetic basis for fungal disease of insects. Restriction fragment length polymorphisms especially between those EPF showing differences in their virulence or host specificity could be a direct way of elucidation of the genetics and the molecular biology of insect pathogenesis. Further, study of RFLP would contribute to the taxonomic status of various EPF isolates and strains.

The initial examples of the use of the restriction enzymes to analyze RFLP for EPF were demonstrated for *Entomophaga maimaiga* (Hajek *et al.* 1990). RFLP has been demonstrated to show sympatric occurrence of two *E. aulicae* isolates (Hajek *et al.* 1991). Differences in RFLP between virulent and less virulent mutant isogenic strains of *B. bassiana* were reported (Kosir *et al.* 1991). In the latter study the RFLP of genomic DNA of two strains of *B. bassiana*, representing strain GK2016, a 'wild type' (virulent) and strain GK2051, a less virulent mutant was demonstrated. The data showed the loss of some DNA sequences from the mutant strain, which may be responsible for loss of virulence.

Interstrain and interspecies RFLP comparison of the genus *Beauveria* by Kosir *et al.* (1991) where the genomic DNA of *B. bassiana*, *B. brongniartii* and *B. cylindrospora* were tested for RFLP banding patterns and the discrimination between *Hirsutella longicolla* var. *longicolla* and *Hirsutella longicolla* var. *cornuta* were performed by Strongman and MacKay (1993). Fegan *et al.* (1993) and Cobb and Clarkson (1993) used RAPD DNA markers to show high degree of diversity between *M. anisopliae* varieties. Leal and co-workers (1994b, 1997) studied isolates of *M. anisopliae* from England, France, Brazil, Australia, Japan, Finland, Benin, Philippines, Wales by nested RAPD-PCR of the Pr1 gene sequence to show four major clustering.

RFLP differences could be used in taxonomy and host specificity determinations. Phenotypic traits and their variabilities, which make the task of pathogen identification challenging, can be much better handled by RFLP. Chew *et al.* (1997, 1998), Walsh *et al.* (1990), and Rakotonirainy *et al.* (1991, 1994) respectively used rDNA and rRNA sequence comparisons to define RFLP patterns for several species *P. farinosus, Entomophaga* and *Metarhizium* spp. and the relatedness of *M. anisopliae, M. flavoviride, B. bassiana* to *Tolypocladium cylindrosporum* and *T. extinguens*. Interisolate RFLP differences of > 20 bands have been shown in restriction endonuclease *Hae*III digestion of 11 isolates of *M. flavoviride* and *M. anisopliae* (Bridge *et al.* 1993).

Thomsen and Beauvais (1995) cloned two chitin synthase gene fragments from hyphal bodies of *Entomophaga aulicae*. Two chitin synthase gene fragments EaCHS1 and EaCHS2 of 600 bp were obtained using PCR amplification of genomic DNA. Compared with other fungal chitin synthases, they belong to class II. EaCHS1 and EaCHS2 were used to probe total RNA from *E. aulicae* hyphal bodies and protoplasts. A single transcript of 2.4 kb hybridized only with EaCHS1 in protoplasts and hyphal bodies. Recently, Thomsen and Bruun-Jenssen (2002) demonstrated nested primers for PCR amplification to resting spores of *E. muscae* a pathogen of adult Diptera and other insects. These authors demonstrated the combined use of RFLP and PCR for analysis of host-pathogen population interactions. Hajek *et al.* (2003) used the criteria of the shape and nucleation of primary conidia within two genera, *Entomophaga* and *Eryniopsis* of the order Entomophthorales as they correlated with PCR-RFLP data. Results of the molecular DNA and morphological data indicated that at least one species, *Ery. ptychopterae* belonged to *Entomophaga.*

2.3. Cloned and Sequenced Chromosomal Genes

In the last decade there have been several publications that describe the isolation, characterization and expression of chromosomal genes of *B. bassiana,* and *M. anisopliae.* These data are summarized in Table 1. Only a few reports have tested the manipulation of such genes in homologous systems for pathogenicity attributes (Screen *et al.* 2001). The results of such experiments have provided mixed results from early induction of insect pathogenesis related proteins without altered in pathogenicity to a net increase in killing.

Table 1. Entomopathogenic fungal protein coding genes isolated and sequenced.

Fungus	Gene	Enzyme	Reference
Metarhizium	sod	Superoxide dismutase	Shrank *et al.* 1993
anisopliae	Pr1B	Subtilisin-like protease	Joshi *et al.* 1997
	CRR1	DNA-binding protein	Screen *et al.* 1997
	nrr1	Nitrogen response regulator	Screen *et al.* 1998
	chit1	Chitinase	Bogo et al, 1998
	---	Chitinase	Kang *et al.*, 1998
	---	Chitin synthase	Nam *et al.*, 1998
	MeCPA	A zinc carboxypeptidase	Joshi and St-Leger 1999
	ssgA	Hydrophobin	Bidochka *et al.*2001
	---	Trehalase	Xia *et al.* 2002
Beauveria	prt1	Protease	Joshi *et al.* 1995
bassiana	prt1	Protease	Joshi *et al.* 1995
	---	Bassiasin I	Kim *et al.* 1999
	prt1-like	Serine endoprotease	Fang *et al.* 2002
		Endonuclease	Yokoyama *et al.* 2002

Ribosomal DNA sequences have been used in several studies of phylogenetic relationship between EPF. Shih *et al.* (1995) performed complete nucleotide sequence of 5.8s rRNA coding gene and flanking transcribed spacer in 2 clones of *B. bassiana.* The overall sequence similarity of these 2 clones was 96%. The identities of the internal transcribed spacer (ITS) regions were 91% (ITSI) and 100% (ITSII), respectively. Both 5.8s ribosomal RNA sequences had 98% homology. Obornik *et al.* (2001) analyzed sequences of the divergent domain at the 5' end of the large subunit rRNA gene from entomopathogenic fungi *Aschersonia* sp., *Aschersonia placenta, B. bassiana, P. fumosoroseus, P. farinosus, P. lilacinus, V. lecanii, V. psalliotae,* and ascomycetous *Cordyceps* sp., and *C. militaris.* Despite of the fact that *M. anisopliae* and *C. paradoxa* are EPF they were clustered with some mycoparasites in the same clad. The relationship between the two fungi is also reported by Liu *et al.* (2001) indicating that a different species, *Cordyceps brittlebankisoides* is an anamorph, *M. anisopliae* var. *majus.* Liu *et al.* (2002) present molecular evidence for teleomorph-anamorph connections in *Cordyceps* and several EPF based on ITS-5.8S rDNA sequences. The morphological and sequence data confirm that *Paecilomyces hawkesii* is the anamorph of

Cordyceps gunnii, and several presumed connections are also confirmed: *B. brongniartii* is the anamorph of *C. brongniartii*, *M. anisopliae* var. *majus* is the anamorph of *C. brittlebankisoides*, *B. sobolifera* is the anamorph of *C. sobolifera*, *Mariannaea pruinosa* is the anamorph of *C. pruinosa*, *P. militaris* is the anamorph of *C. militaris*, and *Hirsutella sinensis* is the anamorph of *C. sinensis*. Their phylogenetic analysis showed *P. fumosoroseus* as the best characterized out of the analyzed species with the *B. bassiana* clad as its sister group. Two of the *P. farinosus* isolates were invariably placed within the *Verticillium* cluster, which also contained *C. militaris*. Boucias *et al.* (2000) analyzed the ITS-5.8s and 28s regions of *N. rileyi* isolates by RAPD, AFLP, and telomeric fingerprinting methods. Analysis of suggested that this species was more closely related to *M. anisopliae* and *M. flavoviride* than to other *Nomuraea* species. In all cases, the non-*rileyi* isolates (*N. anemonoides* and *N. atypicola*) appeared unrelated to the *N. rileyi*.

3. MITOCHONDRIAL GENOME OF EPF

Only two laboratories, those of the authors' and M. Typas have been lead in the mitochondrial (mt) genome, including sequencing studies of EPF. Mt genome size determination was first described for *B. bassiana,* 28.5Kb (Pfeifer *et al*., 1993; Pfeifer and Khachatourians 1989), and *M. anisopliae* with aprox. 32Kb (Mavridou and Typas 1998), and *V. lecanii*, 24.5Kb (Kouvelis and Typas 2003). These mitochondrial genome sizes place them at the small end of the fungal mitochondrial genomes, where we can find ranges between 19Kb (*Torulopsis glabrata*) and 176Kb (*Agaricus bitorquis*) (Leblanc *et al.* 1997; Hintz *et al.* 1985; Clark-Walker *et al.* 1981). Taken in account the mt genome of organisms from other kingdoms, the fungal mt genome has an intermediate complexity which ranges between the small and compacted animal mt genome which present sizes of 14 kb (*Caenorhabditis elegans*) and 42 kb (*Placopecten magellanicus*) and the larger and more complex plant mt genomes with sizes between 184Kb (*Marchantia polymorpha*) and 2,400 kb (*Cucumis melo*) (Leblanc *et al.* 1997).

In spite of a tenfold variation in genome size, similar numbers of functional genes are found in fungal mtDNA. These include the translation-apparatus genes (a single set of large- and small-ribosomal subunits RNA (lrRNA and srRNA), 23-28 transfer RNA (tRNA) genes -with the exception of some members of the cytridiomycete lineages which encode only 8 tRNA- and in some cases a ribosomal protein); respiratory chain proteins (cytochrome c oxidase, NADH dehydrogenase subunits, and apocytochrome b); ATP synthase complex and often several unique open reading frames (ORFs) and unidentified reading frames (URFs) (Lang *et al.* 1999; Paquin *et al.* 1997; Leblanc *et al.* 1997). The URFs have been found in most of the fungal mtDNA examined, in *Aspergillus niger* for instance a total of 15 URFs have been found (Grossman and Hudspeth 1985), clearly suggesting that we still do not know fully about the information contained in the mtDNA. The mt-genes of EPF that are cloned and sequenced are shown in Table 2.

All fungal genomes so far examined contain a single set of ribosomal RNA genes, a minimal number of tRNA species, usually 23-26, and several protein encoding genes. The only comprehensively studied mtDNA of EPF is that of *B. bassiana* by Pfeifer and Khachatourians (1989) who showed that; 1) the mitochondrion was quasi-spherical (d= 0.21 to 0.68 μm), 2) mtDNA had a buoyant density of 1.7118 (52.1% G + C), 3) mtDNA was a covalently closed circular molecule with an Mw=17.6 ± 0.6 MDa and an average contour length of 8.84 ± 0.51 μm, and 4) mtDNA was biphasic with Tm values of 76.2 ± 0.5 and 93.1 ± 0.7°C. Studies by Hegedus *et al.* (1991) on *B. bassiana* mt genes for valine-, isoleucine-, serine-, tryptophan- and proline-accepting tRNAs were found clustered in the region 5' to the large ribosomal (Lr) RNA gene. These genes were 64-77% homologous to the equivalent genes from other filamentous fungi, 49-58% to yeasts, with the exception of the valine-accepting tRNA gene, which was 76%, and only slight homology to *Escherichia coli.* The *B. bassiana* mt genetic code was similar to that of other fungal mitochondria in that the UGA codon was used as a signal for tryptophan rather than as a

Table 2. Isolated and sequenced protein coding mitochondrial genes from EPF.

Fungus	Gene	Enzymatic activity /Function	Reference
Beauveria bassiana	NAD1	NADH dehydrogenase subunit 1	Pfeifer, *et al.* 1993
	NAD6	NADH dehydrogenase subunit 6	Pfeifer, *et al.* 1993
	ATP6	ATP synthase subunit 6	Pfeifer, *et al.* 1993
	CO3	Cytochrome oxidase subunit 3	Pfeifer, *et al.* 1993
Verticillium lecanii	NAD1	NADH dehydrogenase subunit 1	Kouvelis and Typas*
	NAD2	NADH dehydrogenase subunit 2	Kouvelis and Typas*
	NAD3	NADH dehydrogenase subunit 3	Kouvelis and Typas*
	NAD4	NADH dehydrogenase subunit	Kouvelis and Typas*
	NAD4L	NADH dehydrogenase subunit 4L	Kouvelis and Typas*
	NAD5	NADH dehydrogenase subunit	Kouvelis and Typas*
	NAD6	NADH dehydrogenase subunit 6	Kouvelis and Typas*
	COB	Apocytochrome b	Kouvelis and Typas*
	CO1	Cytochrome oxidase subunit 1	Kouvelis and Typas*
	CO2	Cytochrome oxidase subunit 2	Kouvelis and Typas*
	CO3	Cytochrome oxidase subunit 3	Kouvelis and Typas*
	ATP6	ATP synthase subunit 6	Kouvelis and Typas*
	ATP8	ATP synthase subunit 8	Kouvelis and Typas*
	ATP9	ATP synthase subunit 9	Kouvelis and Typas*

* Kouvelis and Typas, 2003- see: locus NC_004514 at www.ncbi.nlm.nih.gov: 80

termination codon. Transcript analysis revealed that the genes present in the tRNA cluster were transcribed and processed into tRNA-size products. Secondary structure models proposed for the gene products showed that conservation of tRNA secondary structure also existed. The presence of a GGC sequence rather than a GGU sequence in the D-loop of the $tRNA^{Trp}$ gene was a feature unique to the *B. bassiana* mitochondrion. An unconventional G-A base pair present in the D-stem of the $tRNA^{Ser}$ gene was a feature conserved in the mitochondria of other filamentous fungi. Comparison of the *B. bassiana* tRNA genes with those of *A. nidulans* and *P. anserina* and the yeasts *S. cerevisieae* and *T. glabrata* revealed that the differences between closely related tRNA clusters were mostly due to transition-type mutations. Further, by performing extensive physical analysis via cloning, restriction enzyme mapping Pfeifer *et al.* (1993) and sequencing we showed 11 structural and transfer RNA as well as protein-encoding genes (Hegedus *et al.* 1991; Pfeifer *et al.* 1993). Hegedus and Khachatourians (1993b) identified mtDNA variations within isolates of *B. bassiana*, and between three other *Beauveria* species as well as *T. nivea*, *T. cylindrosporum*, *M. anisopliae*, *V. lecanii* and *P. farinosus*. Hybridizations carried out with multiple probes present simultaneously produced unique patterns, which characterized the *B. bassiana* group from all other fungi tested. Further, Hegedus and Khachatourians (1993b) categorized the mtDNA of *B. bassiana* into two types designated A and B (see below) and provided a means for testing mt compatibility in EPF. These results indicate how mtDNA polymorphisms in *B. bassiana* may relate to natural population structures, mt transmission in Deuteromycetes and their use for structural analysis of mtDNA.

Due to the small size of the EPF mt genomes, one would expect a reduced number of introns and AT-rich intergenic regions. This is the case for the EPF *B. bassiana* and *V. lecanii* where previous studies suggested the presence of very few introns in their mt genome (Pfeifer *et al.* 1993; Kouvelis and Typas, 2003). For instance in the total sequence of *V. lecanii* mt genome, only one intron located at the lrRNA gene has been described (Kouvelis and Typas 2003). This intron seems to be similar to the non-optional intron of *Neurospora* and *Aspergillus* lrRNA gene, which code for a putative ribosomal protein (Grossman and Hudspeth 1985; Kouvelis and Typas 2003). As in most fungi, non- introns have been described in mt tRNA genes of *B. bassiana* and *V. lecanii* (Hegedus *et al.* 1991; Kouvelis and Typas 2003).

In filamentous fungi, introns are inserted in many different mt genes, with a strong preference for protein coding genes, most frequently cytochrome c oxidase and apocytochrome b. No introns have been found in mt tRNA genes and relatively few are located in rRNA genes (Paquin *et al.* 1997). The ascomycete mitochondria often contain large numbers of introns. The most extreme example is probably *Podospora anserina* where 36 introns occupy about 60% of the mt genome (Cummings *et al.* 1990). The introns together with the intergenic spaces and non-coding regions make the differences in terms of the mt size. For instance even accounting for all the introns, protein coding, rRNA and tRNA genes, of the races of *P. anserina* genome of 94,192 bp about 23% of the mtDNA is not known to have a purpose (Cummings *et al.* 1990).

Most filamentous fungi contain clusters of tRNA genes flanking the mt lrRNA and srRNA genes with some specific protein coding genes around this area. Protein-coding genes in mtDNA of *P. farinosus*, *T. cylindrosporum*, *V. lecanii* and *B. bassiana* are similarly clustered with respect to the two rRNA regions (Pfeifer *et al.* 1993). However, the availability of more complete sequences data of this region in the EPF *V. lecanii* and *B. bassiana* have shown few divergences in the mt gene arrangement of these two EPF. The most important difference is the presence of two more genes (NAD4 and ATP8 subunits) in the 5' flanking region of the ATP6 gene and the mt location of the ATP9 subunit gene in *V. lecanii* (Kouvelis and Typas 2003), in contrast of the nuclear location of the same gene in *B. bassiana* (Pfeifer *et al.*, 1993). The functional gene of the ATP9 is usually located in the nuclear genome for filamentous fungi, despite of the presence of the gene in both the mt and nuclear genome for *Neurospora* (Van den Boogaart *et al.* 1982) and *Aspergillus* (Brown *et al.* 1984). Therefore, further studies should be carried out to determine if such structural difference is also functional.

The rRNA-tRNA gene region in A. *nidulans*, which is 10 Kb, is apparently the condensed version of the N. *crassa*, 20Kb region. Both have the same number of tRNA genes but contain different length introns (larger in N. *crassa*), intergenic and rRNA sequences (Kochel *et al.* 1981; Heckman *et al.* 1979). *Aspergillus nidulans*, and more strongly, *P. anserina* and *N. crassa* have a different gene arrangement in this region in relation to *B. bassiana* (Pfeifer *et al.*, 1993).

The 5' flanking region of the *B. bassiana, V. lecanii* and the mycoparasite *Hypocrea jecorina* lrRNA gene contains the Val-, Ile-, Ser-, Trp-, and Pro- accepting tRNA genes (Chambergo *et al.* 2002; Hegedus *et al.* 1991; Kouvelis and Typas 2003). The same region have been compared between *B. bassiana* and other filamentous fungi (Hegedus *et al.*, 1991) showing clustering of these genes to be more like *P. anserina* (Ile, Ser, and Pro- accepting tRNA) and *A. nidulans* (Ile, Ser, Trp and Pro- accepting tRNA) tRNA clusters, but different to *N. crassa.* The information presented here suggests that there is a lack of selection pressure in nature to keep any specific conservation in terms of the mt genes arrangements, in opposition to the highly conservative set of mt genes even at high taxonomic levels.

In relation to the sequence of the tRNA genes it has been shown that five tRNA genes of *B. bassiana* share between 64-77% of homology to that of *A. nidulans* and *P. anserina*, whereas four of these five genes of *Saccharomyces cerevisieae* and *T. glabrata* shows a sequence similarity between 49 and 58%. This result make this yeast taxonomically more distantly related to *B.*

bassiana than the filamentous fungi analyzed (Hegedus *et al.* 1991). As it has been shown in other studies, around 80% of the mutational changes in the tRNA genes of *B. bassiana,* occurs in the stem regions, areas that are vital to both the structure and function of the tRNA molecule (Hegedus *et al.*, 1991, Cedergren *et al.* 1981).

Despite the variability in the sequence, all *B. bassiana* tRNAs studied form a typical secondary stem-and-loop structure. However, other unique features were found for *B. bassiana* such as the presence of a GGC sequence in the D-loop of the Trp- accepting tRNA, instead of the highly conserved GGU sequence. Additionally, it was found the unconventional G-A base pair in the D-stem of the Ser-accepting tRNA, which are characteristics of Ser- tRNA from both *A. nidulans* and *P. anserina* (Hegedus *et al.* 1991), supporting again some level of similarity between those filamentous fungi.

Hegedus *et al.* (1998) performed characterization and structure of the mitochondrial small rRNA gene of *B. bassiana.* The entire mitochondrial (mt) small ribosomal RNA (srRNA) gene from the entomopathogenic fungus *B. bassiana* was sequenced. Alignment of the sequence to those of other filamentous fungi revealed gross length differences in their respective products. Construction of a secondary structural model showed that these differences were restricted to known variable srRNA sub domains. Several features were identified that were common only to the hyphomycetous fungi examined. Phylogenetic analysis indicated that the anamorph *B. bassiana* was more closely related to the pyrenomycete than to the plectomycete ascomycetous fungi.

3.1. A Note on DNA Isolation

When the introduction of molecular biology studies in EPF started, there was a substantial lack in suitable protocols for the study of the molecular genetics of EPF. Some protocols were available for a wide variety of fungi including *Cochliobolus heterostrophus* (Garber and Yoder 1983), *P. anserina* (Stahl *et al.* 1982) *S. cerevisiae* (Hudspeth *et al.* 1980) and *Schizophyllum commune* (Ullrich *et al.*, 1980). However, these protocols were time consuming when they were applied to filamentous fungi, due to the high levels of contamination caused by fungal nucleases, complex polysaccharides and several pigments. Starting from there, our research group at the University of Saskatchewan was first to develop a set of protocols with rapid methods, for isolation of DNA genomic or mt and molecular genetic studies of SNPs and pathogen tracking through host infection (Pfeifer *et al* 1993; Pfeifer and Khachatourians 1993, Pfeifer and Khachatourians 1989). The yields of 0.12- 0.53% DNA (μg DNA/mg dry weight x100%) was obtained from 10ml of culture. The quality of the DNAs isolated by this method was sufficiently high for restriction and electroblotting for Southern hybridization for EPF identification (Pfeifer *et al.* 1993; Pfeifer and Khachatourians 1993 a-c; Hegedus and Khachatourians 1993a,b).

3.2. Diversity of mt Genome in EPF

The use of EPF as biological control agents necessitates development of reliable genomic typing systems of these microorganisms at the species or strain level, to be able to monitor their impact in the environment and also to protect intellectual property rights claims of strains of commercial interest (Khachatourians 1996, Bidochka 2001).

Burns *et al.* (1991) pointed out the attributes of the mtDNA such as the small but informative size of its genome; the lack of methylation of bases, which avoid confusion with nuclear DNA; the high copy number and the A+T biased composition which makes isolation, purification and analysis relatively easy; and the diversity of the mtDNA at the intraspecific level, mostly due to the presence of length mutations. All together these features makes the mtDNA the molecule of choice for use in typing and evaluative studies. Moreover, comparative mtDNA maps determining location of protein or t- or rRNA encoding genes have been reported for members of the genera *Beauveria*, *Verticillium*, *Paecilomyces*, *Tolypocladium* and *Metarhizium* (Hegedus and Khachatourians, 1993). Additionally, the mt DNA, is the best studied genomic element in EPF, where the complete sequence of *V. lecanii* mtDNA (Kouvelis and Typas, 2003), partial (60%) mtDNA sequence of *B. bassiana* (Hegedus *et al.* 1991; Pfeifer *et al.* 1993) and being completed in authors' laboratory and several small partial sequences such as the PCR fragments of the mt srRNA of *Metarhizium, Paecilomyces* and some *Ascomycetes* of the genus *Cordyceps* have been sequenced (Nikoh and Fakutso 2000).

3.3. Mapping of mtDNA in EPF

The construction of the molecular map of a mtDNA, consist in the analysis in terms of physical location of different patterns produced by a set of restriction enzymes, which allow us to deduce a map of the enzyme sites. Those kinds of maps are the only way to determine the physical relationship of restriction fragments to each other, and as a result it is the tool of choice for investigating structural variation in mtDNA (Burns *et al.*, 1991). When such analyses are accompanied by the utilization of heterologous probes containing mt-coded genes, physical location of those genes into the restriction enzyme map can be deduced.

The above approach was used for the development of *B. bassiana* mtDNA map and to deduce the general arrangement of 11 specific mt genes (Hegedus and Khachatourians 1993; Pfeifer *et al.* 1993; Hegedus *et al.* 1991). When the *B. bassiana* mt map was compared with other filamentous fungi, high diversity in terms of gene arrangements was found. These data suggest not only an absence of selection pressure to follow a specific gene distribution within the mt genome, but also the presence of intra molecular recombination events associated with the mt genome. The latter is probably due to the presence of insertional elements or transposons (Paquin *et al.* 1997; Grossman and Hudspeth 1985). Another interesting feature found in *B. bassiana,* was the location of the ATP9 subunit gene in the nuclear genome, as is the case for *P. anserina* (Riddler *et al.* 1991) but contrary to *V. lecanii* where it is located in the mt genome like in *S. cerevisiae* (Pfeifer *et al.* 1993; Kouvelis and Typas 2003).

Due to the high degree of divergence in the mt genome, the heterologous probes use for mapping should be selected from species closely related, if we want to obtain reliable hybridization patterns. For instance, when probes from protein coding genes from mtDNA of other fungi, such as *P. anserina*, *T. glabrata* or *A. niger* were used to identify mt genes in *B. bassiana,* no hybridization was obtained, despite of the use of low stringency conditions (Pfeifer *et al.* 1993). These results stress the need for additional characterization of the mtDNA of EPF.

3.3.1. mtDNA restriction fragment length polymorphism analysis

The restriction enzyme analysis of several strains, coupled with the utilization of known or unknown specific probes in a southern blot system is very useful to identify genomic polymorphisms. This technique, known as RFLP, has been widely used for strain identification and to establish phylogenetic relationships between populations, specie and even genera in filamentous fungi (Burns *et al.* 1991). The reason for its popularity resides in its reliability, speed, and multiple sampling processing in an electrophoretic system. However, there are some constrains with the utilization of RFLP in fungal mt genomes, due to the presence of length mutations on it. It is because a single nucleotide variation (SNV) may be counted several times, especially when multiple restriction enzymes are used for the analysis (Burns *et al.* 1991; Taylor 1986).

In EPF the situation described above is a minor problem because no length mutations have been identified so far in the mapped genomes (Hegedus and Khachatourians 1993a; Pfeifer and Khachatourians 1989). Additionally, the availability of complete and partial maps of the mt genome of different EPF, allows the design of more reliable strategies by screening regions less prone to the presence of those kind of mutations.

Mitochondrial probes derived from *B. bassiana* were used to evaluate the similarity of mtDNAs from 15 isolates of *B. bassiana, B. caledonica* and *B. densa*, and five genera of EPF (Hegedus and Khachatourians 1993b). The probes produced identical hybridization patterns in *Eco*RI digested DNA from nearly all isolates. The isolates of *B. caledonica* and *B. densa* DNAs could be differentiated from each other and from *B. bassiana* on the basis of a *Hind*III digestion and probing with pBbmtE3. Probe pBbmtE2 produced either a 5.0 kb or 4.1 kb band in all of the *B. bassiana* isolates, aiding us to categorize the mtDNA of *B. bassiana* into two types designated A and B. Hybridization of the probes produced distinct banding patterns with *B. brongniartii, T. cylindrosporum, T. nivea, M. anisopliae, V. lecanii* and *P. farinosus* and those carried out with multiple probes presented simultaneously, produced unique patterns to characterize the group from all other fungi tested. Further, Pfeifer *et al.* (2003) evaluated the utility of *B. bassiana* mtDNA derived probes to provide mtDNA RFLPs that identify mtDNA variations within several isolates of EPF.

The mtDNA is a suitable tool for the taxonomic characterization of EPF (Typas *et al.*, 1992; Hegedus and Khachatourians 1993; Kouvelis *et al.* 1999). Typas *et al.* (1992) for example presented a work with the genus *Verticillium* that contains a heterogeneous group of species that are relevant in agriculture, because they can be both, plant pathogens or insect pathogen. In that work they were able to differentiate seven species of the genus *Verticillum* and also clearly separated alfalfa pathogenic strains of *V. albo-atrum* from not pathogenic strains by using a mtDNA RFLPs analysis of 29 *Verticillum* isolates. A wider study presented later by the same research group, showed 20 different band patterns after the analysis of 54 isolates mainly of *V. lecanii* (51 out of the 54), using a mtDNA RFLPs (Kouvelis *et al.*, 1999). Some interesting suggestions such as the subtropical origin of the specie, was supported by the higher level of divergences found in mt genotypes, from strains belonging to this region of the world, in comparison to the ones found in temperate countries.

By using a similar approach Mavridou and Typas (1998), found a high degree of polymorphism between 25 isolates of the EPF *M. anisopliae*, obtained from at least 16 different hosts and 15 different countries from five continents. In that study, they used a double digestion of total DNA with endonucleases (*Hae*III, *Cfo*I, *Hpa*II, *Kpn*I and *Sac*I), which recognize GC-rich nucleotide regions. Then, the obtained restriction patterns were hybridized with a total mtDNA probe obtained from a specific isolate. Twenty different groups were suggested for those 25 isolates, 16 of which presented a unique mt RFLP pattern, showing the huge power of the technique to identify DNA polymorphism. Despite of the high level of resolution found in the system, it was not possible to identify any correlation between the isolates in terms of neither host nor geographical origin. Contrary to this study, Bidochka *et al.* (2001), showed no variation in 83 soil isolates of *M. anisopliae* from two different habitats (agricultural and forest) in Ontario Canada. This analysis employed RFLP of total mtDNA and even the sequence analysis of the mt lrRNA from 15 randomly chosen isolates. This result suggests a clonal distribution of the mt genome of the specie in this particular area. However, the overall genetic variability analysis of this study including allozymes, random amplified polymorphic DNAs and RFLP of a subtilisin like protease-encoding gene, divided the 83 isolates in two major groups (Bidochka *et al.* 2001). The studies mentioned above clearly support on one hand the clonal distribution of the mt genome, expected for fungi lacking of sexual reproduction. On the other hand it shows that despite of clonality, the mt genome may be exposed to individual mutagenic events that can be perpetuated due to the presence of mt incompatibility and even geographic isolation (Hegedus and Khachatourians 1993a).

In terms of the taxonomic characterization of the *Beauveria* genus Hegedus and Khachatourians (1993a) differentiated mtDNA RFLP patterns of four different species of *Beauveria* (*B. bassiana, B. brongniartii, B. nivea* and *B. cylindrospora*) and those species from *M. anisopliae*, *V. lecanii* and *P. farinosus*. Furthermore the mtDNA patterns of those species support previous studies based in morphological, biochemical and genetic data (Hegedus and Khachatourians 1993b; St. Leger *et al.* 1992; Mugnai *et al.* 1989), suggesting the assertion that *B. densa* and *B. caledonica* are subspecies of *B. bassiana* or at least sister species sharing ancestral and/or transmitted mtDNA (Hegedus and Khachatourians 1993a). In another work four different populations in terms of the mtDNA were identified for *B. brongniartii* (Hegedus *et al.* 1998).

The polymorphic analysis of *B. bassiana* mtDNA using 15 different isolates mostly from North America (two from Russia) showed only two populations of mtDNA (Hegedus and Khachatourians 1993a), suggesting a very conservative mt genome in comparison to other EPFs (Mavridou and Typas 1998, Kouvelis *et al.*, 1999). In a recent study Uribe and Khachatourians (2003) used whole genomic DNA from 18 *B. bassiana,* one each of *B. amorpha, B. cylindrospora* and *B. nivea,* which came from 3 continents and 4 target insect species. Single and double restriction enzyme digestion of total genomic or mtDNA with *Eco*RI, *Eco*RI-*Hind*III or *Eco*RI-*Bgl*II and probed with BbmtE2 showed the predominance of mito-types A-B reported earlier through the use of BbmtE2 DNA probe (Hegedus and Khachatourians 1993a). However, by using whole *B. bassiana* mtDNA digested with *Hpa*II as probe, we further demonstrate up to additional 9 types (C-K) with *B. bassiana.* Members of the genus *Beauveria* were clearly distinguishable between them and from *P. farinosus* and *M. anisopliae*. Phylogenic analysis could not be resolved *B. amorpha* and *B. nivea* isolates from *B. bassiana*, suggesting a close genetic relation between the three species of the genus.

What lies ahead in the uses of mtDNA probes can be many. Such probes can be used to study; 1) the mtDNA during infectious processes, 2) polymorphisms of mtDNA in natural population structures, 3) mode of inheritance and transmission of mtDNA in Deuteromycetes and 4) for tracking after release of EPF in the environment.

3.3.2. Sequencing the mitochondrial genome

The ultimate goal for the characterization of the mt genome is the production of complete sequence of the mtDNA. Further sequencing data allows to confirm gene arrangements in a specific region of the genome, presence or absence of introns and URF, distinction between conformational mutations and those due to length variation, identification of particular kind of mutations (transitions vs. transversions) or the extent of nucleotide bias. Additionally, studies designed to determine the segregation of the mitochondria; parasexual reproduction in EPF and the production of protoplast hybrids can be carried out with the availability of mt markers (Viaud *et al.* 1998). The fact however is that only *Verticillium* mtDNA has been fully and 60% of *B. bassiana* mt genome has been sequenced.

The mtDNA is ideal at the moment for doing phylogenetic studies due to its intrinsic features such as, it is a fast evolving molecule; it is an haploid molecule, therefore any allele is present as a single copy and finally most of those alleles have the same function in different taxonomic levels (Bruns *et al.* 1991). Additionally, due to the presence of universally conserved regions in some genes such as rRNA, it can be easily amplified in a polymerase chain reaction system and used in comparative studies, suitable to include numerous individuals.

Wang *et al.* (2002) have cloned and performed sequence analysis of a mitochondrial gene cluster encoding cytochrome c oxidase subunit III from *Trichoderma pseudokoningii.* Phylogenetic analysis showed that cytochrome c oxidase subunits III exhibited high degree of similarity to sequences from *H. jecorina, V. lecanii, P. anserina, N. crassa* and *Magnaporthe grisea* (99, 90, 84, 82 and 79% identity, respectively). Figure 1 shows our compilation of the phylogenetic analysis of 14 filamentous fungal mtDNA for srRNA sequence, seven of which are EPF. As shown in Figure 1 both deuteromycetes (*B. bassiana, B. brongniartii, V. lecanii, P. tenuipes* and *M. anisopliae*), *Ascomycetes* (*C. militaris* and *C. paradoxa*), and some other Ascomycetes considered mycoparasite fungi (*Hypocrea lutea, H. jecorina* and *Hyphomyces chrysospermus*) were also included; additionally the sequences of *N. crassa, P. anserina* and *A. nidulans* were included as a point of reference. There are four clusters separating these fungi. The relationship between EPF, *B. bassiana* and *B. brongniartii* is much closer to *V. lecanii, C. militaris* and *Paecilomyces* than *Metarhizium* and *C. paradoxa*. Mitochondrial DNA differences between *Metarhizium* and other EPF were previously pointed out in the study of Hegedus and Khachatourians (1993), where mtDNA probes from *B. bassiana* easily hybridized with isolates from *Beauveria, Verticillium* and *Paecilomyces* but not with *Metarhizium.* One may think that species sharing the strategy of including host insects as part of their life cycle may be more closely related than with mycoparasites. As this is not the case, this result is probably suggesting that the strategy of entomopathogenicity came from different ancestral roots in filamentous fungi. However, analysis including a broader number of taxa should be performed in order to bring such hypothesis to a robust conclusion. *Cordyceps* are members of Ascomycetes, which unlike deuteromycetes have known sexual cycles. These results point out to the association of an ancestral mtDNA within the fungal kingdom, where in spite of evolutionary development, the same srRNA is shared between sexual and asexual fungal genera. All EPF except *C. paradoxa* and *M. anisopliae* were clustered in the same clad. Interestingly enough the two *Cordyceps* species were aligned separately and were even more closely related with other anamorphic fungi, than with each other. This result suggests not only a large degree of genetic variability in the *Cordyceps* genus, but also some similarity between some species of the genus and Deuteromycete EPF, supporting previous studies that claim several species of these fungi (*M. taii, P. tenuipes, B. bassiana*) as the anamorph of *Cordyceps* spp. (Bo *et al.* 2002; Kukatsu *et al.* 1997; Zong-Qui *et al.* 1991).

4. TRANSPOSONS AND DOUBLE STRANDED RNA VIRUSES

As yet no plasmids other than a few transposons have been found in EPF. Maurer *et al.* (1997) first showed the isolation of the transposable element hupfer from the entomopathogenic fungus *B. bassiana* by insertion mutagenesis of the nitrate reductase structural gene. Kempken *et al.* (1998) examined the distribution of the fungal transposon Restless, a member of the hAT family of mobile DNA elements, full-length and truncated copies in 13 closely related strains of *Tolypocladium inflatum,* which are taxonomically related. Three strains, which show identical banding patterns in a comparative RAPD analysis with strain ATCC34921, similarly carry multiple copies of Restless. In addition, one *T. inflatum* strain and two *B. nivea* strains contain only a few or even single copies of the transposon. The presence of a single transposon copy of

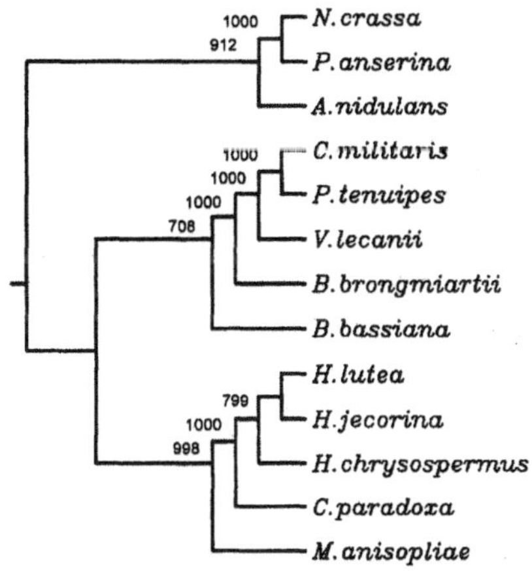

Fig. 1. Phylogenetic tree based on mt srRNA gene sequences from filamentous fungi. A total of 743 unambiguously aligned nucleotide sites were subjected to neighborhood-joining analysis using CLUSTAL W. Bootstrap values for 1000 re-sampling are shown at the nodes. The source of the sequences was those from our laboratory for *B. bassiana* GK 2016 and the Gene Bank (Nikoh and Fukatsu 2000) for *B. brongniartii, P. tenuipes, V. lecanii, M. anisopliae, Hypocrea sp.* and *Cordyceps* with the following accession numbers: *A. nidulans* (V00653.1), *P. anserina* (X55026.1), *N. crassa* (Z34001.1), *B. bassiana* (S55623.1; S55619S5), *B. brongniartii* (AB027359), *P. tenuipes* (AB027358.1), *V. lecanii* (AF487277.1), *C. militaris* (AB027357.1), *C. paradoxa* (AB027345.1), *M. anisopliae* (AB027361), *H. lutea* (AB027362.1), *H. jecorina* (AF447590), *H. chrysospermus* (AB027363.1).

Restless in a defined *Beauveria* strain indicates recent acquisition of this transposon, since class II transposons usually occur in several copies per haploid genome. The authors conclude that a transposon copy in strain ATCC 34921 does not occupy the corresponding genomic location. Jacobsen (1999) searched for the presence of the restless transposon in *T. inflatum* to find the first fungal hAT transposon to be identified. The distribution of the restless transposon in *T. inflatum, B. nivea, T. varium, B. rosariensis, T. geodes, T. cylindrosporum, V. chlamydosporium, V. suchlasporium* and *Sordaria macrospora*. Their results showed a close relationship between *Tolypocladium* spp. and *Beauveria* spp., the existence of Restless copies without the 5' transposon element in 2 strains and transposition activity of a cloned restless copy in strains of *T. inflatum*. It should be feasible to generate hAT transposon induced fungal mutants, which could serve as sources for isolating functional genes using the Restless element as a molecular marker. Indeed auxotrophic mutants were derived from *M. anisopliae* strain RJ exhibited the same DsRNA band pattern and similar virus-like particles (Bogo *et al.* 1996).

The presence of dsRNA viruses was first shown by Leal *et al.* (1994). Inglis and Valdares-Inglis (1997) described rapid extraction of double-stranded RNAs from *Paecilomyces*. Lyophilized and ground mycelium was incubated with 6 M guanidine thiocyanate, centrifuged, and the cleared lysates applied to a QIAGEN silica-based mini-spin column. Following washing with 70% isopropanol, bound nucleic acids were eluted under low salt conditions and treated with DNAse I prior to analysis by non-denaturing agarose gel electrophoresis. The presence of dsRNA viruses was subsequently reported by Melzer and Bidochka (1998) and Bidochka *et al.* (2000) *for M. anisopliae, M. flavoviride* and *B. bassiana* and discuss the implications of diversity of dsRNA viruses within populations of EPF and for their growth and virulence. Gimenez *et al.* (2002)

described *M. anisopliae* strains containing dsRNA mycoviruses MaV-A1, MaV-M5, MaV-RJ. Two isolates, MaV-A1 and MaV-M5 were identified as members in the family Totiviridae and MaV-RJ in the family Partitiviridae. Interesting features associated with one spontaneous mutant, which lost some of the high molecular weight dsRNA components, showed significant alterations in colony morphology and spore production. A comparison between dsRNA mycoviruses-free and infected *M. anisopliae* isolates showed that virus-free isolates have increased endochitinase secretion. These data indicate that viral genes interfere with fungal phenotype yet, Azevedo and coworkers (2000) showed the opposite. Gimenez *et al.* (2002) identified bands of double-stranded RNA (dsRNA) in three out of twelve isolates of *P. fumosoroseus*. Isogenic strains, with or without (cured) dsRNA, when tested for virulence against the whitefly *Bemisia tabaci* strain B showed these dsRNA fragments did not cause hypo virulence in *P. fumosoroseus*. The applied uses of dsRNA remain unexplored.

5. GENOMICS: PROBES AND PATHOGENESIS DETERMINANATS

As the genomics of EPF are beginning to emerge, DNA sequences, probes for DNA analysis of genome or the mitochondria become important tools for, 1) the phylogenetic identification, and 2) detection and manipulation of the activity of particular genes involved in pathogenesis.

Barrientos *et al.* (2002) show the use of a comparative analysis on RAPD DNA patterns. They found that the two Mexican isolates of *Metarhizium*, MaPL40 and MaPL32, and an Australian isolate of *M. anisopliae* var. *acridum* (FI-985), have similar DNA fingerprints, suggesting they may belong to the same variety. Devi *et al.* (2001) characterization by RAPD-PCR in *B. bassiana* isolates showed correlation between the RAPD grouping and the phenotypic classification of the lepidopteran isolates grouped into one major cluster. Their results showed most sub clusters were constituted by isolates from the same climatic zone. Bidochka *et al.* (2002) study of *B. bassiana* genotypic variations are associated with habitat and thermal growth preferences. The wide host range of the fungus suggests that it is a facultative insect pathogen. However, analysis of the fungal isolates from forested, agricultural and Arctic habitats from Canada for their abilities to grow at 8, 15, 25 and 37°C and for their tolerances to UV exposure showed distinct genetic groups associated with the three different habitats. Bidochka *et al.* (2002) study showed genetic groups from the Arctic and from the forested habitats grew at lower temperatures, while the genetic group from the agricultural habitat grew at 37°C and was tolerant to UV exposure. They did not find clear associations between the genetic group and the ability to infect coleopteran or lepidopteran insect larvae. Therefore concluding that habitat selection, not insect host selection drives the population structure of deuteromycetous insect-pathogenic fungi. An earlier study of Bidochka *et al.* (1999) concerning nuclear rDNA phylogeny, virulence, extracellular proteases and carbohydrases in the fungal genus *Verticillium*, a pathogen of insects and plants, shows polyphyletic in origin and is therefore a form genus while, the phylogenetic tree supported three species, *V. dahliae, V. albo-atrum* and *V. nigrescens* which are plant pathogens are that of a clad. Strains of *V. lecanii* and *V. indicum* were able to infect insects and are present in divergent groups in the consensus tree, suggesting that the ability to infect insects may have evolved independently many times but the nematophagous species to have evolved independently along several different routes. In the plant pathogens, *V. albo-atrum, V. nigrescens* and *V. dahliae* all produced high levels of pectinase while insect pathogens and mushroom pathogen (*V. fungicola*) were characterized by production of high levels of subtilisin-like proteases but not pectinases. In a subsequent study, Bidochka and Meizer (2000) show genetic polymorphisms in three subtilisin-like protease Pr1A, Pr1B, and Pr1C from *M. anisopliae* and *M. flavoviride* strains are isoforms of an ancestral genes otherwise the data would suggest that a combination of selection mechanisms to be operative. Chew *et al.* (1998) reached the conclusion that RAPD banding patterns of *P. farinosus* isolates did not correlate with ecological backgrounds or morphological phenotypes of isolates from eastern Canada. Therefore Canadian *P. farinosus* isolates are not composed of strains, which can be separated on the basis of the ecological or morphological criteria selected. What is clear is that it appears that a more complex mechanism not well understood to date, where even the habitat may play a role in determining the final outcomes of species variation and population genetic structure.

5.1. Taxonomic and Phylogenetic Studies

Classical taxonomic criteria are simple but have their shortcomings when it comes to the classification of EPF. For example, chemotaxonomic approaches have suggested that *B. densa* and *B. caledonica* are distinct classes within a highly variant *B. bassiana* population and thus constitute sub-species. Similarly, chemotaxonomic studies generated the proposal for a classification of the members of the subdivisions of the genus *Entomophaga* and the genus *Erynia* (Keller 1991). Without a molecular genetic analysis, while the generic and group level reclassification may clarify the situation, at species level, the questions will be outstanding. On the other hand, rRNA sequence comparisons are clear in that they showed evidence for distinct differences between the two genera, *Beauveria* and *Tolypocladium* (Rakotonirainy *et al.* 1991; Pfeifer *et al.* 2003) and within the genus of *Metarhizium* or in comparison with *B. bassiana* (Rakotonirainy *et al.* 1994). Thus, rDNA analysis has given credibility to the chemotaxonomic (Mugnai *et al.* 1989) but not the morphological analysis in resolving members of the EPF. Therefore various tools of genomics can be used for studies on genetic relatedness and possibly those of phylogeny and population diversity.

There have been several recent reports on DNA sequence alignment of the nuclear 5.8S rRNA gene and internal transcribed spacers (ITS) for genomic studies in many EPF. Yazawa and Shimizu (2002) reported the sequence comparison of the 5.8S rDNA region with the flanking among *M. anisopliae* and the related species. Fargues *et al.* (2002) study of variability in the rDNA-ITS was studied in 48 isolates of *P. fumosoroseus* from various geographical and host insect origins e.g., *Bemisia tabaci-argentifolii*. Of the three distinct groups only one representing 25 isolates came only from the host *B. tabaci-argentifolii* whereas those in the group 3 were more diffuse and came from various insect host and geographical origins. Huang *et al.* (2002) used the genes of the 5.8S rRNA and the complete ITS regions from *Cordyceps bassiana* and *B. bassiana* for molecular identification of the teleomorph of *B. bassiana.* Their results showed that *C. bassiana* and *B. bassiana* have the same ITS1-5.8S-ITS2 nucleotide sequences, which strongly supports the fact that *C. bassiana* is the teleomorph of *B. bassiana.* Coates *et al.* (2002b) were able to determine *B. bassiana* haplotypes based on nuclear rDNA internal transcribed spacer PCR-RFLP. They demonstrated that 6.62% sequence variation existed between nine isolates. A higher level of mutation was observed within the ITS regions, where 8.39% divergence occurred. The allelic frequency of each genetic marker was determined from 96 isolates. PCR-RFLP defined 24 *B. bassiana* genotypes within the sample set, from which eight phylogenetic clusters were predicted to exist. Analysis of the data found no significant correlation existed between *B. bassiana* haplotypes and insect host range as defined by insect order from which each isolate was derived. In a subsequent study Coates *et al.* (2002b) found a minisatellite locus, BbMin1, from a partial *B. bassiana* genomic library that consisted of poly (GA) flanked inserts. PCR amplification of the BbMin1 repeats demonstrated allele size variation among 95 isolates and a single isolates of *B. amorpha, B. brongniartii* and *B. caledonica.* AMOVA and theta (Fst) indicated that fixation of repeat number has not occurred within pathogenic ecotypes or geographically isolated samples of

B. bassiana. However, they suggest presence of alleles with such a large number of repeat units may be indicative of rare occurrence of somatic recombination or DNA replication error.

Fungaro *et al.* (1996) tested diversity among soil and insect isolates by RAPD analysis of 13 isolates of *M. anisopliae* var. *anisopliae* from different regions of Brazil was analyzed using Dice similarity index. The results showed that their isolates were extremely diverse (47% similarity) and suggested that *M. anisopliae* var. *anisopliae* have developed host specificity. Driver *et al.* (2001) suggest taxonomic revision of the species of *Metarhizium* based on a phylogenetic analysis of rDNA sequence data. The taxonomy of *Metarhizium* species was reassessed using the ITS and 28S rDNA D3 regions' sequence data and RAPD patterns from 123 isolates recognized as *M. anisopliae, M. flavoviride* or *M. album*. A high level of genetic diversity was found which was best resolved at the species/variety level by sequence data. Ten distinct clads were revealed by the cladogram based on the combined sequence data set. The data support the monopoly of the *M. anisopliae* group, and recognize four clads within it. Two correspond with *M. anisopliae* var. *anisopliae* and *M. anisopliae* var. *majus* and the other two are described as new varieties based on their distinctive ITS sequence data: *M. anisopliae* var. *lepidiotum* and *M. anisopliae* var. *acridum* vars. *nov.* Three clads represent two new varieties, *M. flavoviride* var. *novazealandicum* and *M. flavoviride* var. *pemphigum* vars. *nov,* based on their distinct ITS sequence data.

Obornik *et al.* (2000) reported on genetic variability and phylogeny from RAPD DNA data of EPF, *P. fumosoroseus. P. farinosus, and V. lecanii.* Zare *et al.* (1999) reported a 20 bp insertion/deletion in the ITS1 region of *V. lecanii*. Zare *et al.* (2000) have proposed a revision of *Verticillium* based on ITS sequences because the anamorph genus is extremely heterogeneous with biological differences, plant pathogens vs. parasites of fungi, insects, nematodes and rotifers. A subdivision into more natural entities was examined morphologically, and sequences of the ITS regions of nrDNA. Four clads (A-D) were sharply delimited, where clad B, which contains species with white fluffy colonies lacking any resting structures; the most important members are the aggregates of *V. lecanii* with cylindrical conidia and *V. psalliotae* with falcate conidia. Sugimoto *et al.* (2001, 2003) estimated the genetic diversity in 30 isolates of *V. lecanii* from aphids, whiteflies, mite and black pine in Japan, including two commercialized strains (Mycotal and Vertalec) by DNA polymorphisms in ribosomal DNA by PCR. The ITS and intergenic spacer (IGS) regions of the nuclear ribosomal RNA gene of each isolate were analyzed by PCR-RFLP. Ten distinct IGS haplotypes were detected in the IGS region, some of which were associated with aphid and whitefly origins. The size of the PCR product from the ITS region was -580 bp in 27 of the isolates. A 600 bp ITS product was detected in EPF from Mycotal and Vertalec. One Japanese isolate produced both the 580- and 600-bp products. Enzymatic digestion of the ITS region with *Sau*3A I, *Msp* I, *Hae* III and *Rsa* I revealed RFLPs that consisted of eight haplotypes. Species of EPF in Mycotal and Vertalec were specific haplotypes that differed from other isolates. The Japanese isolates had a complex relationship with the original host, but we identified several specific haplotypes common to an aphid origin.

5.2. Probes and Pathogenesis Related Genes

Chromosomal DNA probes from EPF could have utility in either taxonomy, tracking of released biocontrol fungi or dissecting the progression of the infection (Khachatourians 1996 Bidochka 2001; Khachatourians *et al.* 2002). The DNA based technology systems include genomic DNA probes or differentiate between isolates rDNA and rRNA or mtDNA sequence comparisons, PCR amplification RAPD, RFLPs, single stranded conformational polymorphism (SSCP), which can identify base change per 100 bases, (Hegedus and Khachatourians 1996b; Urtz and Rice 1997; Sugimoto *et al.* 2003). A series of four DNA probes derived from moderately repetitive DNA and ranging in size from 1,180 to 2,700 bp was described by Hegedus and Khachatourians (1993a). These probes specifically hybridized to *B. bassiana* but not to DNA from other *Beauveria* spp. (*brongniartii*, *caledonica* and *densa*), *T. nivea*, *T. cylindrospora*, *M. anisopliae* strains, *V. lecanii*, *P. farinosus*, or other non-entomopathogenic fungi (*A. niger* and *N. crassa*). The probes produced distinct hybridization patterns with 15 strains of *B. bassiana* worldwide. The probes pBb22 without PCR amplification detected a minimum of 152 ng and with PCR amplification, 5 pg of total fungal DNA within several species of field collected grasshoppers infected with *B. bassiana*. These types of probes can be useful for field studies of released EPF as most recently demonstrated by Castrillo *et al.* (2003).

There are many catabolic genes responsible for the interaction with and determination of the mechanism and or outcome of pathogenesis. In the last decade several insect cuticle-degrading genes have been identified (Table 1). Cloned and characterized extracellular subtilisin-like serine endoprotease has been identified in *M. anisopliae* Pr1 and in *B. bassiana* Pr1 and bassiasin I and can be used as probe DNA sequences. Although the conventions of genetics require that confirmation of gene functions to demonstrate through null mutations their indispensability, mutant strain showing the contrary have also been reported. Often the presence of alternate gene copy or an isozyme is the most obvious explanation. Bidochka and Khachatourians (1990) and Wang *et al.* (2002) showed *B. bassiana* and *M. anisopliae* isolates respectively lacking the most important pathogenesis gene product, serine protease, were still able to infect hosts though with much reduced lethal activity. With respect to extracellular chitinase, which is also implicated as a pathogenesis determinant, Screen *et al.* (2001) showed that *M. anisopliae* sf. *acridum* transformants over expressing enzyme levels were not altered in pathogenicity to the caterpillars of *Manduca sexta*.

The use of contemporary genomics, gene data mining sources has made it easier for the search and identification through expressed sequence tag (EST) the genomes of many organisms. It is not a surprise that with the EST analysis of two subspecies of *M. anisopliae* and *M. anisopliae* sf. *acridum* as reported by Freimoser *et al.* (2003). Approximately 1,700 5' end sequences from each subspecies were generated from cDNA libraries representing fungi grown under conditions that maximize secretion of cuticle-degrading enzymes. Both subspecies had ESTs for virtually all pathogenicity-related genes cloned to date from *M. anisopliae*, but many novel genes encoding potential virulence factors were also tagged. Enzymes with potential targets in the insect host included proteases, chitinases, phospholipases, lipases, esterases, phosphatases and enzymes producing toxic secondary metabolites. New breakthroughs of the post-Human Genome Project era should facilitate the discovery of other fungal pathogenesis genes for wider genomic and applied entomological work.

6. CONCLUSIONS

A great deal of work in the area of the genomics of EPF lies ahead for mycologists and biotechnologists alike. With advances in molecular biology and the science of genomics better understanding of the biotechnological potential of EPF in insect pest control programs will be realized. As stated before, this becomes more evident when we examine the contribution of such control agents in agriculture, forestry and the world economy in the decade ahead. Thus the challenge of applied use of EPF can be met with enhanced research activities in genomics. With better understanding of genetics of host-pathogen interaction a more specific and better use in applied mycology and entomology could be realized. It is not to say that certain degree of success even in applied genomics may not be derived from incremental as much as serendipitous discoveries. Genomics will be an important and strategic technology in the production and use of

EPF and their contribution to alternatives in pest management and the history of agriculture and forestry sciences.

Acknowledgements: This work was made possible through Natural Sciences and Engineering Research Council of Canada and Agricultural Biotechnology Initiative Funds to GGK.

REFERENCES

Azevedo ACS, Sosa-Gomez DR, Faria MR, and Fungaro MHP (2000). Effects of double-stranded RNA on virulence of *Paecilomyces fumosoroseus* (Deuteromycotina: Hyphomycetes) against the silverleaf whitefly, *Bemisia tabaci* strain B (Homoptera: Aleyrodidae) Genet Molec Biol 23: 61-63.

Barreto CC, Alves LC, Aragao FJL, Rech E, Schrank A, and Vainstein MH (1997). High frequency gene transfer by microprojectile bombardment of intact conidia from the entomopathogenic fungus *Paecilomyces fumosoroseus*. FEMS-Microbiol Lett 156:95-99.

Berretta MF, Lecuona RE, Zandomeni RO and Grau O (1998). Genotyping isolates of the entomopathogenic fungus *Beauveria bassiana* by RAPD with fluorescent labels. J. Invertebr. Pathol 71: 145-150.

Barrientos LL, Hernandez-Velazquez VM, Milner-RJ and Hunter DM (2002). Advances in biological control of locusts and grasshoppers in Mexico. J Orthopt Res 11: 77-82.

Bello VA and Paccola-Meirelles LD (1998). Localization of auxotrophic and benomyl resistance markers through the parasexual cycle in the *Beauveria bassiana* (Bals.) Vuill entomopathogen. J Invertebr Pathol 72:119-125.

Bidochka MJ (2001) Monitoring the fate of biocontrol fungi In: Fungi as biocontrol agents: Progress, problems and potential. (Eds) Butt TM, Jackson C and Magan N. CABI Publishing, Wallingford, UK. Pp.193-218.

Bidochka MJ and Khachatourians GG (1990). Identification of *Beauveria bassiana* extracellular protease as a virulence factoring pathogenesis toward the migratory grasshopper, *Melanoplus sanguinipes*. J Invertebr Pathol 56:362-370.

Bidochka MJ, Kamp AM and de Croos JNA (2000). Insect pathogenic fungi: from genes to populations. In Fungal pathology. (Ed) Kronstad JW. Kluwer Academic Publishers Group, Dordrecht, Netherlands. Pp. 171-193.

Bidochka MJ, Kamp AM, Lavender TM, Dekoning J and Amritha De Croos JN (2001). Habitat association in two genetic groups of the insect pathogenic fungus *Metarhizium anisopliae*: Uncovering cryptic species. Appl Environ Microbiol 67:1335-1342.

Bidochka MJ, Melzer MJ, Lavender TM and Kamp AM (2000). Genetically related isolates of the entomopathogenic fungus *Metarhizium anisopliae* harbour homologous dsRNA viruses. Appl Environ Microbiol 104: 1094-1097.

Bidochka MJ, Menzies FV and Kamp AM (2002.). Genetic groups of the insect-pathogenic fungus *Beauveria bassiana* are associated with habitat and thermal growth preferences. Arch Microbiol 178: 531-537.

Bidochka MJ, St-Leger RJ, Stuart A and Gowanlock K (1999). Nuclear rDNA phylogeny in the fungal genus *Verticillium* and its relationships to insect and plant virulence, extracellular proteases and carbohydrases. Microbiology 145: 955-963.

Bidochka, MJ, McDonald MA, St Leger RJ and Roberts DW (1994). Differentiation of species and strains of entomopathogenic fungi by random amplification of polymorphic DNA (RAPD) Curr Genet 25:107-113

Bo HL, Chun RL, Zheng G, Mei ZF and Li ZZ (2002). Molecular identification of the teleomorph of *Beauveria bassiana*. Mycotaxon. 81:229-236.

Bogo MR, Queiroz MV, Silva DM, Gimenez MP, Azevedo JL and Schrank A (1996). Double-stranded RNA and isometric virus-like particles in the entomopathogenic fungus *Metarhizium anisopliae*. Mycol Res 100:1468-1472.

Bogo MR, Rota CA, Pinto H Jr., Ocampos M, Correa CT, Vainstein MH and Schrank A (1998). A chitinase-encoding gene (chit1 gene) from the entomopathogen *Metarhizium anisopliae:* isolation and characterization of genomic and full-length cDNA. Curr Microbiol 37: 221-225.

Bridge PD, Williams MAJ, Prior C and Paterson RRM (1993). Morphological, biochemical and molecular characteristics of *Metarhizium anisopliae* and *M. flavoviride*. J Gen Microbiol 139:1163-1169.

Brown TA, JA Ray, RB Waring, C Scazzocchio and RW Davies (1984). A mitochondrial reading frame which may code for a second form of ATPase subunit 9 in *Aspergillus nidulans*. Curr Genet 8:488-492.

Bruns TD, TJ White and JW Taylor (1991) Fungal molecular systematics. Annu Rev Ecol Sys. 22: 525-564.

Boucias DG, Tigano MS, Sosa-Gomez DR, Glare TR and Inglis PW (2000). Genotypic properties of the entomopathogenic fungus *Nomuraea rileyi*. Biol Control 19: 124-138.

Cantone FA and Vandenberg JD (1999). Use of the green fluorescent protein for investigations of *Paecilomyces fumosoroseus* in insect hosts. J Invertebr Pathol 74: 193-197.

Castrillo LA, Vandenberg JD and Wraight SP (2003). Strain-specific detection of introduced *Beauveria bassiana* in agricultural fields by use of sequence-characterized amplified region markers. J Invertebr Pathol 82:75-83.

Cedergren R, Sankoff D, La Rue B and Grosjean H (1981). The evolving tRNA molecule. CRC Crit Rev Biochem 11:25-104.

Chambergo F, Bonaccorsi ED, Ferreira AJ, Ramos AS, Ferreira JR, Habrahao-Neto J, Farah JP and El-Dorry H (2002). Elucidation of the metabolic fate of glucose. In the filamentous fungus *Trichoderma reesei* using EST analysis and cDNA microarrays. J Biol Chem 277: 13983-13988.

Chew JSK, Strongman DB and MacKay RM (1997). RFLP analysis of rRNA intergenic spacer regions of 23 isolates of the entomopathogen *Paecilomyces farinosus*. Can J Bot 75:2038-2044.

Chew JSK, Strongman DB and MacKay RM (1998). Comparisons of twenty isolates of the entomopathogen *Paecilomyces farinosus* by analysis of RAPD markers. Mycol Res 102:1254-1258.

Clark-Walker GB, McArthur CR and Sriprakash KS (1985). Location of transcriptional control signals and transfer RNA sequences in *Torulopsis glabrata* mitochondrial DNA. EMBO J 4:465-473.

Coates BS, Hellmich RL, Lewis LC (2002a). Nuclear small subunit rRNA group I intron variation among Beauveria spp provide tools for strain identification and evidence of horizontal transfer. Curr Genet 41: 414-424.

Coates BS, Hellmich RL and Lewis LC (2002b). *Beauveria bassiana* haplotype determination based on nuclear rDNA internal transcribed spacer PCR-RFLP. Mycol Res 106: 40-50.

Coates BS, Hellmich RL, Lewis LC (2002c). Allelic variation of a *Beauveria bassiana* (Ascomycota: Hypocreales) minisatellite is independent of host range and geographic origin. Genome. 45: 125-132.

Cobb BD and Clarkson JM (1993). Detection of molecular variation in the insect pathogenic *Metarhizium anisopliae* using RAPD-PCR. FEMS Microbiol Lett 112:319-324.

Cummings DJ, McNally KL, Domenico JM and Matsuura ET (1990). The complete DNA sequence of the mitochondrial genome of *Podospora anserina*. Curr Genet 17:375-402.

Devi KU, Padmavathi J, Sharma HC and Seetharama N (2001). Laboratory evaluation of the virulence of *Beauveria bassiana* isolates to the sorghum shoot borer *Chilo partellus* Swinhoe (Lepidoptera: Pyralidae) and their characterization by RAPD-PCR. World J Microbiol Biotechnol 17: 131-137.

Driver F, Milner RJ and Trueman JWH (2001). A taxonomic revision of *Metarhizium* based on a phylogenetic analysis of rDNA sequence data. Mycol Res 104: 134-150.

Drummond J and Heale JB (1988). Genetic studies on the inheritance of pathogenicity in *Verticillium lecanii* against *Trialeurodes vaporariorum*. J Invertebr Pathol 52:57-65.

Fang WG, Zhang YJ, Yang Y, Wang K and Pei Y (2002). Cloning and characterization of several cuticle-degrading enzymes CDEP-1 from *Beauveria bassiana*. Acta Genet Sinica 29: 278-282.

Fargues J, Bon-Marie C, Manguin S and Couteaudier Y (2002). Genetic variability among *Paecilomyces fumosoroseus* isolates from various geographical and host insect origins based on the rDNA-ITS regions. Mycol Res 106: 1066-1074.

Fegan M, Manners JM, MacLean DJ, Irwin JAG, Samuels KDZ, Holdom DE and Li DP (1993). Random amplified polymorphic DNA markers reveal a high degree of genetic diversity in the entomopathogenic fungus *Metarhizium anisopliae* var. *anisopliae*. J Gen Microbiol 139:2075-2081.

Feng M-G, Poprawski TJ and Khachatourians GG (1994). Production, formulation and application of the entomopathogenic fungus *Beauveria bassiana* for insect control. Biocontrol Sci Technol 4:3-34.

Freimoser FM, Screen S, Bagga S, Hu G and St-Leger RJ (2003). Expressed sequence tag (EST) analysis of two subspecies of *M. anisopliae* reveals a plethora of secreted proteins with potential activity in insect hosts. Microbiology. 149: 239-247.

Fungaro MHP, Vieira MLC, Pizzirani-Kleiner AA and de Azevedo JL (1996). Diversity among soil and insect isolates of *Metarhizium anisopliae* var. *anisopliae* detected by RAPD. Lett Appl Microbiol 22: 389-392.

Garber RC and Yoder OC (1983). Isolation of DNA from filamentous fungi and separation into nuclear, mitochondrial, ribosomal and plasmid components. Anal. Biochem 135:416-422.

Gimenez PMP, Bogo MR, Santi L, de-Moraes CK, Correa CT, Vainstein MH and Schrank A. (2002). Characterization of mycoviruses and analyses of chitinase secretion in the biocontrol fungus *Metarhizium anisopliae*. Curr Microbiol 45: 334-339.

Grossman LI, and Hudspeth MES (1985). Fungal mitochondrial plasmids In: Gene Manipulations in Fungi (Eds. Bennett JW and Lasure LL) Academic Press, New York. Pp. 65-103.

Hajek AE, Bruun-Jenssen A, Thomsen L, Hodge KT and Eilenberg J (2003). PCR-RFLP is used to investigate relations among species in the entomopathogenic genera *Eryniopsis* and *Entomophaga*. Mycologia 95:262-268.

Hajek AE, Humber RA, Elkinton JS, May B, Walsh SRA and Silver JC (1990). Allozyme and RFLP analyses confirm *Entomophaga maimaiga* responsible for 1989 epizootics in North American Gypsy moth populations. Proc Natl Acad Sci USA 87:6979-6982.

Hajek AE, Humber RA, Walsh SRA and Silver JC (1991). Sympatric occurrence of two *Entomophaga aulicae* (Zygomycetes: Entomophthorales) complex species attacking forest Lepidoptera. J Invertbr Pathol 58:373-380.

Hajek AE and St. Leger RJ (1994) Interactions between fungal pathogens and insect hosts. Ann Rev Entomol 39:293-322.

Heckman JES, Yin B, Alzner DW and RajBhadary UL (1979). Mapping and cloning of *Neurospora crassa* mitochondrial transfer RNA genes. J Biol Chem 254:12694-12700.

Hegedus DD and Khachatourians GG (1993a). Construction of cloned DNA probes for the specific detection of the entomopathogenic fungus *Beauveria bassiana* in grasshoppers. J Invertebr Pathol 62: 233-240.

Hegedus DD and Khachatourians GG (1993b). Identification of molecular variants in mitochondrial DNAs of members of the genera *Beauveria, Verticillium, Paecilomyces, Tolypocladium*, and *Metarhizium*. Appl Environ Microbiol 59: 4283-4288.

Hegedus DD and Khachatourians GG (1994). Production and characterization of conditional lethal mutants of *Beauveria bassiana*. Can J Microbiol 40:766-776.

Hegedus DD and Khachatourians GG (1996a). Detection of the entomopathogenic fungus *Beauveria bassiana* within infected migratory grasshoppers (*Melanoplus sanguinipes*) using polymerase chain reaction and DNA probe. J Invertebr Pathol 67:21-27.

Hegedus DD and Khachatourians GG (1996b). Identification and differentiation of the entomopathogenic fungus *Beauveria bassiana* using polymerase chain reaction and single strand conformation polymorphism analysis. J Invertebr Pathol 67:289 299.

Hegedus DD and Khachatourians GG (1996c). The effect of temperature on the pathogenicity of heat sensitive mutants of the entomopathogenic fungus *Beauveria bassiana* toward the migratory grasshopper, *Melanoplus sanguinipes*. J Invertebr Pathol 68:160-165.

Hegedus DD, Pfeifer TA, MacPherson JM and Khachatourians GG (1991). Cloning and analysis of five mitochondrial tRNA-encoding genes from the fungus *Beauveria bassiana*. Gene, 109:149-154.

Hegedus DD, Pfeifer TA, Mulyk DS and Khachatourians GG (1998) Characterization and structure of the mitochondrial srRNA gene of the entomopathogenic fungus *Beauveria bassiana*. Genome 41: 471-476.

Hintz, WEA, Mohan M, Anderson JB and Horgan, PA (1985). The mitochondrial DNAs of Agaricus: heterogeneity in A. bitorquis and homogeneity in *A. brunnescens*. Cur Genet 9:127-132.

Hu G and St-Leger RJ (2002) Field studies using a recombinant mycoinsecticide (*Metarhizium anisopliae*) reveal that it is rhizosphere competent. Appl Environ Microbiol. 68: 6383-6387.

Huang B, Ru LC, Gang LZ, Zhen FM and Zhi LZ (2002). Molecular identification of the teleomorph of *Beauveria bassiana*. Mycotaxon. 81: 229-236.

Hudspeth MES, Shmard DS, Tatti KM, and Grossman LI (1980). Rapid purification of yeast mitochondrial DNA in high yield. Biochim Biophys Acta 610: 221-228.

Iijima R, Kurata S and Natori S (1993). Purification, characterization and cDNA cloning of an antifungal protein from the hemolymph *of Sarcophaga peregrina* (flesh fly) larvae. J Biol Chem 268:12055-12061.

Inglis PW and Valadares-Inglis MC (1997). Rapid isolation of double-stranded RNAs from entomopathogenic species of the fungus *Paecilomyces* using a commercial minicolumn system. J Virol Methods 67:113-116.

Inglis PW, Aragao FJL, Frazao H, Magalhaes BP and Valadares-Inglis MC (2000). Biolistic co-transformation of *Metarhizium anisopliae* var. *acridum* strain CG423 with green fluorescent protein and resistance to glufosinate ammonium. FEMS-Microbiol Lett 191:249-254.

Jacobsen S (1999). Distribution and activity of the transposon restless in biotechnically important Hyphomycetes. Mycologica. 182: 1-126.

Joshi L and St.Leger RJ (1999). Cloning, expression, and substrate specificity of MeCPA, that is secreted into infected tissues by the fungal entomopathogen *Metarhizium anisopliae*. J Biol Chem 274: 9803-9811.

Joshi L, St.Leger RJ and Roberts DW (1997). Isolation of a cDNA encoding a novel subtilisin-like protease (Pr1B) from the entomopathogenic fungus, *Metarhizium anisopliae* using differential display-RT-PCR. Gene197: 1-8.

Joshi L, St-Leger RJ and Bidochka MJ (1995). Cloning of a cuticle-degrading protease from the entomopathogenic fungus, *Beauveria bassiana*. FEMS Microbiol Lett 125: 211-218.

Kang SC, Park S and Lee DG (1998). Isolation and characterization of a chitinase cDNA from the entomopathogenic fungus, *Metarhizium anisopliae*. FEMS Microbiol Lett. 165:267-271.

Keller S (1991). Arthropod-pathogenic Entomophthorales of Switzerland. II. *Erynia, Eryniopsis, Neozygites, Zoopteora* and *Tarichium*. Sydowia 43:39-122.

Kempken F, Jacobsen S and Kueck U (1998). Distribution of the fungal transposon restless: Full-length and truncated copies in closely related strains. Fungal Genet Biol 25:110-118.

Khachatourians GG (1986). Production and use of biological pest control agents. Trends Biotechnol 4:120-124.

Khachatourians GG (1991). Physiology and genetics of entomopathogenic fungi. In: Arora DK, Ajello L, Mukerji KG (eds) Handbook of applied mycology, vol 2: Humans, animals, and insects. Marcel Dekker Inc, New York, Pp613-661.

Khachatourians GG (1996). The relationship between biochemistry and molecular biology of entomopathogenic fungi and insect diseases. *In* The Mycota, Vol. VI: Animal and Human Relationships (D.H. Howard, and J.D. Miller, Eds.) Springer-Verlag, Berlin. Pp. 331-363.

Khachatourians GG (2002). Agriculture and food crops: Development, science and society. Chapter 1. Pp. 1-29. *In:* Transgenic Plants and Crops. (Eds.) Khachatourians GG, McHughen A, Nip WK, Scorza R, and Hui YH. Marcel Dekker Inc., New York. Pp.876+xxii.

Khachatourians GG, E Valencia and Miranpuri GS (2002). *Beauveria bassiana* and other entomopathogenic fungi in the management of insect pests. In Advances In Biopesticide Research Ed by O Koul. Taylor & Francis, London, UK. 239-275.

Kim HK, Hoe HS, Suh DS, Kang SC, Hwang C and Kwon ST (1999). Gene structure and expression of the gene from *Beauveria bassiana* encoding bassiasin I, an insect cuticle-degrading serine protease. Biotechnol Lett 21:777-783.

Kochel HG, Lazarus CM, Basak N and Kuntzel H (1981). Mitochondrial tRNA gene clusters in *Aspergillus nidulans*: organization and nucleotide sequence. Cell 23: 3705-3713.

Kosir JM, MacPherson JM and Khachatourians GG (1991). Genomic analysis of virulent and less virulent strain of the entomopathogenic fungus *Beauveria bassiana* using RFLP. Can J Microbiol 37:534-541.

Kotlyarevskii DI and Levites EV (1992). Use of isoenzymes to detect parasexual process in the fungus *Verticillium lecanii* (Zim) Viegas. Doklady: Biol Sci 319:427-428.

Kouvelis VN, Zare R, Bridge PD and Typas MA (1999). Differentiation of mitochondrial subgroups in the *Verticillium lecanii* species complex. Lett Appl Microbiol. 28: 263-268.

Kouvelis VN and Typas MA (2003) see: locus NC_004514 at www.ncbi.nlm.nih.gov:8 NCBI Submission date 8 January 2003.

Kukatsu T, Sato H and Kuriyama H (1997). Isolation inoculation to insect host and molecular phylogeny of an entomopathogenous fungus *Paecilomyces tenuipes*. J Invertebr Pathol 70:203-208.

Lang BF, Gray MW and Gurger G (1999). Mitochondrial genome evolution and the origin of eukaryotes. Ann Rev Genet 33:351-397.

Leal SCM, Bertioli DJ, Ball BV and Butt TM (1994). Presence of double-stranded RNAs and virus-like particles in the entomopathogenic fungus *Metarhizium anisopliae*. Biocontrol Sci Technol 4:89-94.

Leal SCM, Bertioli DJ, Butt TM and Peberdy JF (1994). Characterization of isolates of the entomopathogenic fungus *Metarhizium anisopliae* by RARD-PCR. Mycol Res 98: 1077-1081.

Leal SCM, Bertioli DJ, Butt TM, Carder JH, Burrows PR and Peberdy JF (1997). Amplification and restriction endonuclease digestion of the Pr1 gene for the detection and characterization of *Metarhizium* strains. Mycol Res 101: 257-265.

Leal SCM, Butt TM, Peberdy JF and Bertioli DJ (2000). Genetic exchange in *Metarhizium anisopliae* strains co-infecting *Phaedon cochleariae* is revealed by molecular markers. Mycol Res 104: 409-414.

Leblanc C, Richard O, Kloareg B, Viehmann S, Zetsche K and Boyen C (1997). Origin and evolution of mitochondria: what have we learnt from red algae. Curr Genet 31:193-207.

Liu ZY, Liang ZQ, Whalley AJS, Yao YJ and Liu AY (2001). *Cordyceps brittlebankisoides*, a new pathogen of grubs and its anamorph, *Metarhizium anisopliae* var. *majus*. J Invertebr Pathol 78:178-182.

Liu ZY, Liang Z-qi, Liu AY, Yao YJ, Hyde KD and Yu ZN (2002). Molecular evidence for teleomorph-anamorph connections in *Cordyceps* based on ITS-5.8S rDNA sequences. Mycol Res106: 1100-1108.

Maurer P, Rejasse A, Capy P, Langin T and Riba G (1997). Isolation of the transposable element hupfer from the entomopathogenic fungus *Beauveria bassiana* by insertion mutagenesis of the nitrate reductase structural gene. Molec Gen Genet 256:195-202.

Mavridou A and Typas MA (1998). Intraspecific polymorphism in *Metarhizium anisopliae* var *anisopliae* revealed by analysis of rRNA gene complex and mtDNA RFLPs. Mycol Res 102: 1233-1241.

Melzer MJ and Bidochka MJ (1998). Diversity of double stranded RNA viruses within populations of entomopathogenic fungi and potential implications for fungal growth and virulence. Mycologia 90:586-594.

Mugnai L, Bridge PD and Evans HC (1989). A chemotaxonomic evaluation of the genus *Beauveria*. Mycol Res 92:199-209.

Murrin F, Holtby J, Nolan RA, and Davidson WS (1986). The genome of *Entomophaga aulicae* (Entomophthorales, Zygomycetes): Base composition and size. Exp Mycol 10:67-75.
Nam JS, Lee DH, Lee KH, Park HM and Bae KS (1998) Cloning and phylogenic analysis of chitin synthase genes from the insect pathogenic fungus, *Metarhizium anisopliae* var. *anisopliae*. FEMS Microbiol Lett. 159: 77-84.
Neuveglise C and Brygoo Y (1994). Identification of group-I introns in the 28S rDNA of the entomopathogenic fungus *Beauveria brongniartii*. Curr Genet 27:38-45.
Nikoh N and Fakutso T (2000). Interkingdom host jumping underground: Phylogenetic analysis of the entomoparasitic fungi of the genus *Cordyceps*. Molec Biol Evol 17:629-638.
Obornik M, Klic M. and Zizka L (2000). Genetic variability and phylogeny inferred from random amplified polymorphic DNA data reflect life strategy of entomopathogenic fungi. Can J Bot 78: 1150-1155.
Obornik M, Jirku M and Dolezel D (2001). Phylogeny of mitosporic entomopathogenic fungi: is the genus *Paecilomyces* polyphyletic? Can J Microbiol 47: 813-819.
Paquin B, Laforest MJ, Forget L, Roewer I, Wang Z, Longcore J and Lang BF (1997). The fungal mitochondrial genome project: evolution of fungal mitochondrial genomes and their gene expression. Curr Genet 31:380-395.
Pfeifer TA and Khachatourians GG (1989) Isolation and characterization of DNA from the entomopathogen *Beauveria bassiana*. Exp Mycol 13: 392-402.
Pfeifer TA and Khachatourians GG (1993a). Electrophoretic karyotype of the entomopathogenic Deuteromycetes, *Beauveria bassiana* J Invertebr Pathol 62: 231-235.
Pfeifer TA and Khachatourians GG (1993b) Isolation of DNA from entomopathogenic fungi grown in liquid cultures. J Invertebr Pathol 61:113-116.
Pfeifer TA, Hegedus DD and Khachatourians GG (1993). The mitochondrial genome of the entomopathogen fungus *Beauveria bassiana*: analysis of the ribosomal RNA region. Can J Microbiol 39: 25-31.
Pfeifer TA, Hegedus DD and Khachatourians GG (2003). Assessment of morphological, enzymatic and molecular characteristics for the differentiation of the genera *Beauveria and Tolypocladium*. *In prep*.
Rakotonirainy M, Cariou ML, Brygoo Y and Riba G (1994). Phylogenetic relationships within the genus *Metarhizium* based on 28S rRNA sequences and isozyme comparison. Mycol Res 98:225-230.
Rakotonirainy M, Dutertre M, Brygoo Y and Riba G (1991). rRNA sequence comparison of *Beauveria bassiana*, *Tolypocladium cylindrospora* and *Tolypocladium extinguens*. J Invertebr Pathol 57:17-22.
Riddler R, Kunkele KP and Osiewacz HD (1991). Sequence of the nuclear ATP synthase subunit 9 gene from *Podospora anserina*: lack of similarity to the mitochondrial genome. Curr Genet 20:349-351.
Sandhu SS, Uncles SE, Rajak RC and Kinghorn JR (2001a). Generation of benomyl resistant *Beauveria bassiana* strains and their infectivity against *Helicoverpa armigera*. Biocont Sci Technol 11: 245-250.
Sandhu SS, Kinghorn JR, Rajak RC and Unkles SE (2001b). Transformation system of *Beauveria bassiana* and *Metarhizium anisopliae* using nitrate reductase gene of *Aspergillus nidulans*. Indian J Exp Biol 39: 650-653.
Schrank A, Bassanesi MC Pinto Jr. H, Costa SV, Bogo MR and Silva MSN (1993). Superoxide dismutases in the entomopathogenic fungus *Metarhizium anisopliae*. Ciencia e-Cultura-Sao-Paulo. 45: 200-205.
Screen S, Bailey A, Charnley K, Cooper R and Clarkson J (1997) .Carbon regulation of the cuticle-degrading enzyme PR1 from *Metarhizium anisopliae* may involve a trans-acting DNA-binding protein CRR1, a functional equivalent of the Aspergillus nidulans CREA protein. Curr Genet 31: 511-518.
Screen S, Bailey A, Charnley K, Cooper R and Clarkson J (1998). Isolation of a nitrogen response regulator gene (nrr1) from *Metarhizium anisopliae*. Gene 221:17-24.
Screen SE, Hu G and St-Leger RJ (2001). Transformants of *Metarhizium anisopliae* sf. *anisopliae* over expressing chitinase from *Metarhizium anisopliae* sf. *acridum* show early induction of native chitinase but are not altered in pathogenicity to *Manduca sexta*. J Invertebr Pathol 78: 260-266.
Shi L-G (2000) Study on the karyotype of pathogenic *Beauveria bassiana* to *Bombyx mori*. Acta Seric Sin 26:224-227.
Shih HL, Lin CP, Liou RF and Tzean SS (1995). Complete nucleotide sequence of *Beauveria bassiana* 5.8s rRNA coding gene and flanking transcribed spacers. DNA-Sequence. 5: 381-383.
Shimizu S, Nishida Y, Yoshioka H and Matsumoto T (1991) Separation of chromosomal DNA molecules from *Paecilomyces fumosoroseus* by pulse field electrophoresis. J Invertebr Pathol 58:461-463.
Shimizu S, Higashiyama R and Matsumoto T (1993a). Chromosome length polymorphism in *Beauveria bassiana*. J Seric Sci Jpn 62:45-49.
Shimizu S, Yoshioka H and Matsumoto T (1993b). Electrophoretic karyotyping of the entomogenous fungus *Paecilomyces fumosoroseus*. Lett Appl Microbiol 16:183-186.
St. Leger RJ, Allee LL, May B, Staples RC and Roberts DW (1992) Worldwide distribution of genetic variation among isolates of *Beauveria* spp. Mycol Res 96: 1007-1015.

St-Leger RJ and Screen SE (2001). Prospects for strain improvement of fungal pathogens of insects and weeds. In: Fungi as biocontrol agents: Progress, problems and potential. (Eds) Butt TM, Jackson C and Magan N. CABI Publishing, Wallingford, UK. Pp219-238.

Sthal U, Tudzynski P, Kuck U and Esser K (1982). Replication and expression of a bacterial-mitochondrial hybrid plasmid in the fungus *Podospora anserina*. Proc Natl Acad Sci USA 79: 3641-3645.

Storck R and Alexopoulos CJ (1970). Deoxyribonucleic acid of fungi. Bacteriol Rev 34:126-154.

Strongman DB and MacKay RM (1993) Discrimination between *Hirsutella longicolla* var. *longicolla* and *Hirsutella longicolla* var. *cornuta* using random amplified polymorphic DNA fingerprinting. Mycologia 85:65-70.

Sugimoto M, Koike M and Nagao H (2001). Ribotyping of entomopathogenic *Verticillium lecanii* in Japan. Phytoparasitica 29: 413-420.

Sugimoto M, Koike M, Hiyama N and Nagao H (2003). Genetic, morphological, and virulence characterization of entomopathogenic *Verticillium lecanii*. J Invertebr Pathol 82:176-187.

Taylor JW (1986) Fungal evolutionary biology and mitochondrial DNA. Exp. Mycol. 10:259-269.

Thomsen L and Beauvais A (1995). Cloning of two-chitin synthase gene fragments from a protoplastic entomophthorale. FEMS-Microbiology-Letters. 129: 115-120.

Thomsen L and Bruun-Jenssen A (2002). Application of nested-PCR technique to resting spores from *Entomophthora muscae* species complex: implications for analyses of host-pathogen population interactions. Mycologia 94:794-802.

Thorvilson H, Wheeler D, Bextine B and Francisco MS (2002). Development of *Beauveria bassiana* formulations and genetically marked strains as a potential biopesticide for imported fire ant control. Southwest Entomol 25(Supplement): 19-29.

Tigano-Milani MS, Gomes ACMM and Sobral BWS (1995b). Genetic variability among Brazilian isolates of the entomopathogenic fungus *Metarhizium anisopliae*. J Invertebr Pathol 65:206-210.

Typas MA, Griffin AM, Bainbridge BW and Heale JB (1992). Restriction fragment length polymorphism in mitochondrial DNA and ribosomal RNA gene complexes as an aid to the characterization of species and sub-species populations in the genus *Verticillium*. FEMS Microbiol Lett 95:157-162.

Ullrich RC, Dtoms KA, Doyon JD and Specht CA (1980). Characterization of DNA from the basidiomycete *Schizophyllum commune*. Exp Mycol 4:123-134.

Uribe D and Khachatourians GG (2003). Novel *Beauveria bassiana* mitochondrial genotypes. *Submitted.*

Urtz BE and Rice WC (1997) RAPD-PCR characterization of *Beauveria bassiana* isolates from the rice water weevil *Lissorhoptrus oryzophilus*. Lett Appl Microbiol 25:405-409.

Valadares-Inglis MC and Peberdy JF (1998). Variation in the electrophoretic karyotype of Brazilian strains of *Metarhizium anisopliae*. Genet Molec Biol 21: 11-14.

Van den Boogaart P, Sanallo J and Agsteribbe E (1982). Similar genes for a mitochondrial ATPase subunit in the nuclear and mitochondrial genome of *Neurospora crassa*. Nature 298:187-189.

Viaud M, Couteaudier Y and Riba G (1998). Molecular analysis of hypervirulent somatic hybrids of the entomopathogenic fungi *Beauveria bassiana* and *Beauveria sulfurensis*. Appl Environ Microbiol 64:88-93.

Viaud M, Couteaudier Y, Levis C and Riba G (1996). Genome organization in *Beauveria bassiana*: electrophoretic karyotype, gene mapping and telomeric fingerprint. Fungal Gen Biol 20:175-183.

von Arx JA (1986) *Tolypocladium*, a synonym of *Beauveria*. Mycotaxon 25:153-158.

Walsh SRA, Tyrell D, Humber RA and Silver JC (1990). DNA restriction fragment length polymorphisms in the rDNA repeat unit of *Entomophaga*. Exp Mycol 14:381-392.

Wang C-S, Ding DG, Wang SB and Li ZZ (2000). Genetic analysis on cross incubation of *Beauveria bassiana*. Acta Microbiol Sinica 40:545-550.

Wang CS, Typas MA, Butt TM and Wang CS (2002.) Detection and characterization of pr1 virulent gene deficiencies in the insect pathogenic fungus *Metarhizium anisopliae*. FEMS Microbiol Lett 2002, 213: 251-255.

Wang CS, Wang SB, Li ZZ, Butt TM, Wang CS, Wang SB and Li ZZ (2001). Molecular identification of heterokaryosis in population of wild type isolates of *Beauveria bassiana*. Mycosystema 20: 475-481.

Wang TH, Wu ZH, Liu SL, Huang W, Wang TH, Wu ZH, Liu SL and Huang W (2002). Cloning and sequence analysis of a mitochondrial gene cluster encoding cytochrome c oxidase subunit III from *Trichoderma pseudokoningii*. DNA-Sequence. 13: 271-275.

Xia Y, Gao M, Clarkson J, and Charnley KA (2002). Molecular cloning, characterization, and expression of a neutral trehalase from the insect pathogenic fungus *Metarhizium anisopliae*. J Invertebr Pathol 80: 127-37.

Yazawa C and Shimizu S (2002). Sequence comparison of the 5.8S rDNA region with the flanking among *Metarhizium anisopliae* and the related species. J Insect Biotechnol Sericol. 71: 55-59.

Yokoyama E, Yamagishi K and Hara A (2002). Group-I intron containing a putative homing endonuclease gene in the small subunit ribosomal DNA of *Beauveria bassiana* IFO 31676. Molec Biol Evol 19: 2022-2025.

Zare R, Kouvelis VN, Typas MA and Bridge PD (1999) Presence of a 20 bp insertion/ deletion in the ITS1 region of *Verticillium lecanii*. Lett Appl Microbiol 28:258-262.
Zare R, Gams W and Culham A (2000). A revision of *Verticillium* sect Prostrata. I. Phylogenetic studies using ITS sequences. Nova-Hedwigia 71: 465-480.
Zong-Qui L, Ai-Ying L and Jie-Ling L. (1991). A new species of the genus *Cordyceps* and its *Metarhizium* anamorph. Acta Mycol Sin 10:257-262.

Applied Mycology & Biotechnology
An International Series. Volume 4. Fungal Genomics

16

Genomics of Arbuscular Mycorrhizal Fungi

Nuria Ferrol*[1], Concepción Azcón-Aguilar[1], Bert Bago[2], Philipp Franken[3], Armelle Gollotte[4], Manuel González-Guerrero[1], Lucy Alexandra Harrier[5], Luisa Lanfranco[6], Diederik van Tuinen[4] and Vivienne Gianinazzi-Pearson[4]
[1]Estación Experimental del Zaidín, CSIC, Granada, Spain, [2]Centro de Investigaciones sobre Desertificación, CSIC, Valencia, Spain; [3]Institute for Vegetable and Ornamental Plants, Grossbeeren, Germany; [4]INRA, Dijon, France; [5]The Scottish Agricultural College, Edinburgh, United Kingdom; [6]Università degli Studi di Torino, Torino, Italy.

Arbuscular mycorrhizal fungi are soilborne microorganisms that form a mutualistic symbiotic association with most land plants. As obligate biotrophs these fungi are unable to complete their life cycle in the absence of the host plant. This symbiosis is increasingly being recognised as an integral and important part of natural ecosystems throughout the word. Because of the incalcitrance of arbuscular mycorrhizal fungi to grow in pure culture and consequently the difficulties in obtaining sufficiently large quantities of fungal material, the analysis of gene products has remained an extremely challenging but unexplored area. Until recently, little was known about the genomics of these fungi and it is only with the advent of powerful molecular techniques that it has been possible to venture research into their genetic makeup. This review surveys the most recent molecular genetics of arbuscular mycorrhizal fungi and their contributions to basic knowledge of the biology of this group of organisms.

1. INTRODUCTION

Arbuscular mycorrhizal (AM) fungi are soilborne microorganisms that form a symbiotic association named arbuscular mycorrhiza with most land plants. Fossil and molecular data indicate that these fungi originated in the Ordovician about 460 million years ago and significantly contributed to the colonization of land by plants (Redecker *et al.* 2000). Since their appearance, they have persisted through periods of important environmental change and spread to newly evolving plant species to become abundant symbionts in terrestrial ecosystems across the globe (Smith and Read 1997). More than 150 species of AM fungi have been described and they have been traditionally included in the order Glomales (Zygomycota) which encompass six genera: *Glomus*, *Sclerocystis*, *Gigaspora*, *Scutellospora*, *Acaulospora* and *Entrophospora* (Morton and Benny 1990). Nevertheless, recently the taxonomic concept of AM fungi has been reviewed and they have been placed into a new monophyletic phylum, the Glomeromycota comprising four new orders: Glomerales, Diversisporales, Paraglomerales and Archaeasporales (Schüßler *et al.* 2001).

*Corresponding author: Nuria Ferrol
Departamento de Microbiología del Suelo y Sistemas Simbióticos, Estación Experimental del Zaidín, Consejo Superior de Investigaciones Científicas, C. Profesor Albareda 1, 18008, Granada, Spain (nferrol@eez.csic.es).

AM fungi are obligate biotrophs whose completion of their life cycle depends on their ability to colonize the root of a host plant. The AM fungal-plant association is a mutually beneficial event: the plant supplies the fungus with carbon (from its fixed phostosynthates, reviewed by Bago *et al.* 2000; Douds *et al.* 2000) while the fungus assists the plant in its uptake of phosphate and other mineral nutrients from the soil (Smith and Gianinazzi-Pearson 1988; Smith and Read 1997; Ferrol *et al.* 2002a). In addition to an improved nutrition, mycorrhizal plants also show an increased resistance to root pathogens (Azcón-Aguilar and Barea 1996; Cordier *et al.* 1998; Pozo *et al.* 2002) and a higher tolerance to abiotic stresses (Al-kariki and Hammard 2001; Cantrel and Linderman 2001; Leyval et *al.* 2002). This symbiosis is increasingly being recognized as an integral and important part of natural ecosystems throughout the word (Jeffries and Barea 2001; Gianinazzi *et al.* 2002). Despite the large amount of information obtained from physiological studies, knowledge of the mechanisms which lead to the beneficial effects exerted by fungal symbionts is still limited. This is partly due to the complexity of the biological system and to the obligate symbiotic character of AM fungi.

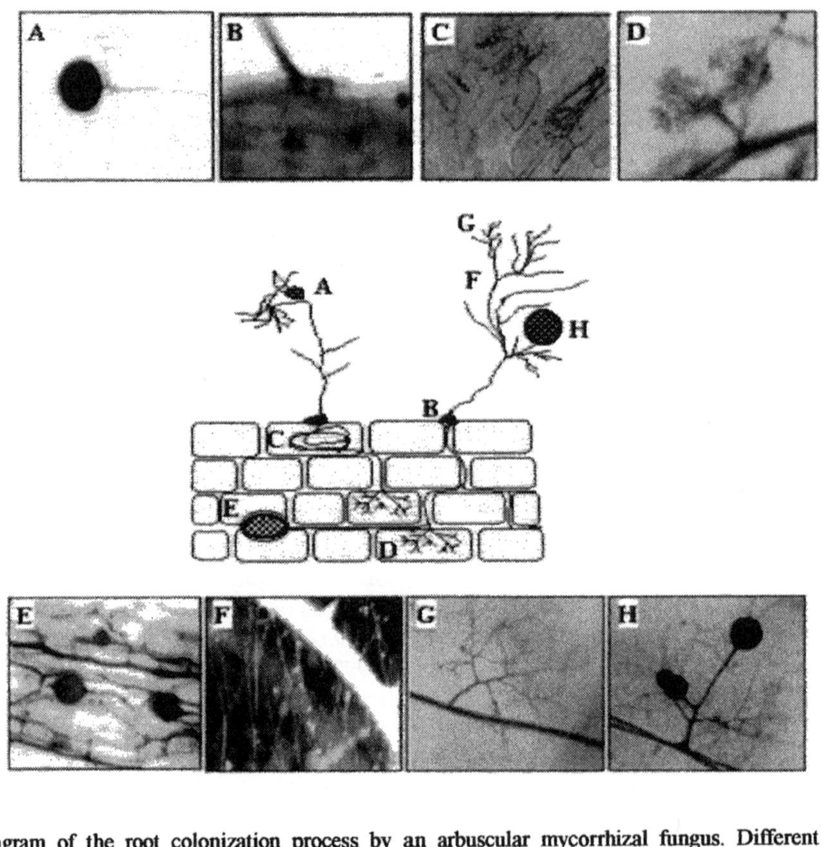

Fig. 1. Diagram of the root colonization process by an arbuscular mycorrhizal fungus. Different fungal structures are illustrated in the micrographs. A: Germinated resting spore, B: Appressorium, C: Intracellular coils, D: Arbuscule, E: Vesicles, F: External mycelium, G: Branching absorbing structures, H: Developing spore.

AM fungi germinate from asexual spores and form a limited, asymbiotic mycelium at the expenses of their internal spore reserves in the absence of a host plant (Bago *et al.* 1999a) (Fig. 1A). The process of root colonization by AM fungi has been reviewed elsewhere in detail (Bonfante and Perotto 1995; Harrison 1999) and it is schematically shown in Fig. 1. Briefly, the interaction begins when the fungal hyphae, arising from spores or from adjacent colonized roots are intensely stimulated to grow and ramify in the rhizosphere of a host plant and finally contact the root surface. Here they differentiate to form appressoria via which they penetrate the root (Fig. 1B). Once inside the root, the fungus may produce intracellular coils in the subepidermal cell layers (Fig. 1C), followed by intercellular growth into the inner cortex of the root (although sometimes only intracellular hyphae are observed). On reaching the inner cortex, hyphal branches penetrate the cortical cell walls and differentiate within the cells to form highly branched structures, known as arbuscules (Fig. 1D). These fungal structures, which establish a large surface of contact with the plant protoplast, play a key role in reciprocal nutrient exchange between the plant cells and the AM fungal symbionts. Later on in the colonisation process, certain AM fungi produce storage structures within the roots (vesicles) (Fig. 1E). Simultaneously to intraradical colonisation, the fungus develops an extensive network of hyphae in the soil surrounding the root (Fig. 1F). This extraradical mycelium, able to start new colonization events, explores and exploits soil microhabitats for nutrient acquisition. Different structures are formed on this extraradical mycelium: branching absorbing structures (BAS) (Fig. 1G), probably involved in nutrient uptake, and large multinucleate resting spores able to survive under unfavourable conditions (Fig. 1H).

Given the effects of AM fungi on plant growth and health, it is generally accepted that appropriate management of this symbiosis should permit a reduction of agrochemical inputs, and thus provide for sustainable plant productivity (Gianinazzi *et al.* 2002). Maximum benefits will only be obtained from inoculation with efficient AM fungi and a careful selection of compatible host/fungus/soil combination. Although AM biotechnology is feasible for many crop production systems, further research into the mechanisms involved in AM development and function is essential to acquire the scientific background for successful exploitation of this symbiosis in agriculture. Numerous plant mycorrhiza-regulated genes have been identified through molecular analysis of the symbiosis (for a review readers are referred to Harrison 1999; Gollotte *et al.* 2002a). However, because of the incalcitrance of AM fungi to grow in pure culture and consequently the difficulties in obtaining sufficiently large quantities of fungal material, the analysis of gene products has remained an extremely challenging but unexplored area. Until recently, little was known about AM fungal genomics and it is only with the advent of powerful molecular techniques that it has been possible to venture research into their genetic makeup. This review surveys the most recent molecular genetics of the AM fungi and their contributions to basic knowledge of the biology of this group of organisms.

2. GENERAL CHARACTERISTICS OF THE AM FUNGAL GENOME

The close association of AM fungi with land plants for more than 460 Myrs (Redecker *et al.* 2000) must have greatly influenced both plant and fungal biology. AM fungi do not have any known sexual stage of reproduction and, as indicated previously, their developmental cycle starts by the germination of multinucleate asexual spores. Depending on the species or isolate analysed, the number of nuclei per spore can vary from about 720 for *Scutellospora castanea* (Hosny *et al.* 1998) to 2600 for *Gigaspora* (Bécard and Pfeffer 1993; Cooke *et al.* 1987). For this latter genus, one study (Viera and Glenn 1990) even reported the number of nuclei to be as high as 20000 per spore. These asexual spores are the sole source from which fungal DNA can be extracted in sufficient amounts for genome size and structure analysis, and for which the origin can be controlled. The spores of AM fungi can however harbour a large number of endosymbiotic bacteria (Bianciotto *et al.* 1996). This requires that the fungal DNA is separated from any prokaryotic DNA prior to any genome analyses (Hosny *et al.* 1999b).

Several experiments have been undertaken to estimate the DNA content of the nuclei of AM fungi. The DNA content per nuclei of 11 species, measured by flow cytometry using chicken red blood cells as a calibrating standard, ranged from 0.13 pg DNA/genome (1.3 10^8 bp) for *S. pellucida* (INVAM 337) to 1.08 pg DNA/genome (1.06 10^9 bp) for *S. gregaria* (W1727) (Hosny *et al.* 1998). Previous studies by Bianciotto and Bonfante (1992) had determined the DNA content of *Glomus versiforme* and *Gigaspora margarita* to be between these two extremes (0.27 and 0.65 respectively). Although this range of genome size may seem broad, it is low when compared to the 167 fold range observed overall in the fungal kingdom. Nevertheless, the genome size of AM fungi is high when compared to most other true fungi and it approaches the genome size of Zygomycetes in the genus *Entomophaga* with 8.1 10^9 bp, the largest reported in fungi (Murrin *et al.* 1986). All the data concerning the genome size of AM fungi have been obtained from species with relatively large-sized nuclei. To date no information concerning the DNA content of species with smaller nuclei, such as *G. mosseae* or *G. intraradices* is available, leaving the question of the range of the genome size open. Still, it has to be borne in mind that following the C value paradox (Thomas 1971), there is no correlation between genome size and the morphological complexity of an organism.

An idea of genome complexity can be obtained from rehybridization (Cot) curves performed on DNA. Genomic DNA from fungal spores is sheared ultrasonically and the rehybridization curve, after thermal denaturation, compared with that of *E. coli*. The renaturation curve for the genomic DNA of *S. castanea* presents several slopes in contrast to the renaturation curve observed with DNA of *E. coli,* which gives a simple sigmoid curve, indicating a more complex genome in the AM fungus (Hosny 1997). The rehybridization curve of *S. castanea* suggests the existence of four sequence families: one (about 7% of the genome) generally believed to be made of palindromic sequences and three others representing three levels of repeated sequences. Although complete rehybridization could not be achieved with the DNA of *S. castanea,* due to the high amount of DNA required for this type of experiment, it can be assumed from the partial renaturation curve that the genome of *S. castanea* is made up of at least of 50% of repetitive DNA. This high number of repetitive DNA could account for the large genome size of this fungus (0.88 pg, 8.7 10^8 bp).

A number of repeated sequences have been isolated by screening of genomic libraries obtained from several AM fungi (Hosny *et al.* 1999b). One of these sequences (SC1), from *S. castanea*, is tandemly repeated with 2600 copies per genome, contains a number of short repeated sequences and accounts for about 0.24% of the genome of *S. castanea.* This sequence did not hybridize with DNA of other AM fungi, and it was used to generate specific primers allowing identification of this fungus specifically in plant roots (Zézé *et al.* 1996). Other repetitive DNA sequences that are presently being characterized in *Gi. rosea, Gi. margarita* and *G. mosseae* appear mainly to be moderately repeated DNA elements. Most of them have no homology to known sequences and no open reading frame. Southern blot analyses and fluorescence *in situ* hybridization indicate that these sequences are mostly dispersed in the genome (Gollotte, van Tuinen and Gianinazzi-Pearson, unpublished results). Another repeated sequence *Mycdire* (170 copies per genome), which is dispersed in the genome of *S. castanea*, contains two perfect CDEIs (Centromeric DNA Element) found in *Saccharomyces cerevisiae* (Hieter *et al.* 1985) and one with a single C to T transition which may be the result from methylation of cytosine (Zézé *et al.* 1999). This *Mycdire* sequence contains also 3 sequences with 10/11 identities with an autonomously replicating sequence (*ars*) of *S. cerevisiae* (Broach *et al.* 1983). These molecular features suggest that the *Mycdire* sequence could have had some transposable element characteristics. Furthermore, *Gi. rosea* and *G. caledonium* harbour a sequence with a similarity of 97% to *Mycdire* while *Acaulospora laevis* has an element with a sequence similarity of only 65%. The identity of the *Mycdire* sequence between *S. castanea* and *G. caledonium* is higher than the identity observed at the ribosomal level (91% identity over 1546 nucleotides of the small ribosomal subunit), suggesting a separate evolution of this element. Although no data are available on the possible biological activity of these sequences, the occurrence of autonomously replicating elements in the genome of AM fungi strongly suggests the possible existence of transposable elements in these fungi. No active transposable element has been isolated from these fungi so far, but recent work has confirmed the presence of sequences showing homologies to gypsy and Non-Long Terminal Repeat retrotransposons in the genome of *Gi. margarita* (Gollotte *et al.*, unpublished results). Whether such elements have played a role in the evolution of AM fungal genomes remains to be elucidated.

The genome of AM fungi is also characterized by an exceptionally low GC content. Values determined for different species of four families range from 29.9 to 35.25% (Hosny *et al.* 1997), which are in the lower limit of those reported for Zygomycetes (27.5-59%) (Storck and Alexopoulos 1970). However, an even more interesting feature of the genome of AM fungi is the high content of methylated cytosine. For nine species that have been studied, values range from 2.44% to 4.20% which indicates that up to 24.85% of the cytosine can be methylated in the genome (Hosny *et al.* 1997). The level of methylated cytosine is generally much lower in fungi. For example, only 1-2% of the cytosine are methylated in *Neurospora crassa* (Russell *et al.* 1987), where methylated cytosine has been reported to be involved in the RIP (Repeated Induced Point mutation) mechanism associated with inactivation of duplicated genes (Singer *et al.* 1995). Methylated cytosine could favour the mutational rate of cytosine to thymidine and could account for a rapid divergence among repeated sequences. Furthermore, methylated cytosine plays an important role in the silencing of transposable elements and in gene expression regulation (for review see Bird 2002).

3. GENETIC DIVERSITY OF AM FUNGI

Ribosomal gene sequences were the first to be targeted for in detail studies of genetic diversity in AM fungi (Simon *et al.* 1992). Small rDNA subunit sequences have been useful to clarify the taxonomic position of AM fungi allowing the creation of the new phylum, Glomeromycota. Sequence data from ribosomal sequences has represented an additional tool for the correct identification of AM fungal taxa showing low morphological divergence (Lanfranco *et al.* 2001; Redecker 2002). At the same time, analyses of these sequences have revealed the genetic structure of AM fungi to be complex, with an unexpected high level of genetic diversity despite their low morphological diversity. Several characteristics of the biology of AM fungi make them a particularly interesting but also an extremely difficult subject for genetic studies. Not only AM fungi lack a known sexual stage, their life cycle is also devoid of a uninucleate stage, an important bottleneck that in other fungi operates to homogenize genomes within an individual. The coenocytic cellular organization of AM fungi allows nuclei to freely move along hyphae (Bago *et al.* 1998, 1999b) and into newly produced spores to form multinucleate structures. Nothing is known about nuclear segregation inside the forming spores. Moreover, anastomosis events, which have been reported to occur under *in vitro* conditions (Giovannetti *et al.* 1999) and in mycorrhizal roots (Casana and Bonfante 1982), may increase exchange of genetic material between hyphae.

Since the first molecular analyses which targeted anonymous repetitive DNA as well as ribosomal sequences a considerable and unexpected genetic variability has been detected in AM fungal spore DNA (Wyss and Bonfante 1993; Rosendahl and Taylor 1997; Zézé *et al.* 1997). Sequence diversity within a single spore was first reported in ITS regions of *Glomus* spp. (Sanders *et al.* 1995 ; Lloyd-MacGilp *et al.* 1996), and later in the ITS regions of *Gi. margarita* (Lanfranco *et al.* 1999a) and *S. castanea* (Hosny *et al.* 1999a). The diversity is not restricted to the ITS region, but has also been demonstrated in the small (18S) and large (28S) rDNA subunit sequences (Hijri *et al.* 1998; Clapp *et al.* 1999; Hosny *et al.* 1999a; Schüβler 1999). These studies have shown that intrasporal diversity can be extensive: four distinct sequences were found in a single *Scutellospora* spore (Clapp *et al.* 1999). The ITS sequence variations inside a single spore may be higher than that observed between spores of the same isolate (Lanfranco *et al.* 1999a). Although intra-isolate ITS variation could be as great as that of geographically disjunct isolates, sequences obtained from the same species still cluster together (Lloyd-MacGilp *et al.* 1996; Lanfranco *et al.* 2001).

Recently, an extensive investigation of *G. coronatum* isolates has been performed by Clapp *et al.* (2001). Five hundred clones from seven *G. coronatum* isolates, and one isolate each of *G. geosporum, G. constrictum* and *G. mosseae*, were screened for variation across 460 bp of the large subunit D2 region by PCR-single-strand conformation polymorphism and sequencing. Distinct sequence clusters were present within each isolate and although a *G. coronatum* core cluster could be identified, several sequences from other species occurred in the *G. coronatum* clusters. At least considering this portion of the 28S rDNA, sequence variability among isolates therefore obscured species level resolution.

All these studies indicate that, contrary to most organisms, different ribosomal sequences are present in the same multinucleated spore in AM fungi. This finding is of great interest as it suggests a possible polymorphism at the nuclear level and, consequently, raises the question of the reproduction strategy of these fungi. As the mechanisms underlying spore formation by AM fungi are still largely unknown, this research area is very challenging. Various studies have been undertaken to address the questions whether nuclei of a same spore are genetically different or whether gene sequences are different in a single nucleus. FISH (Fluorescence *In Situ* Hybridization) experiments showed that the number of rDNA loci within a spore nuclei differ (Trouvelot *et al.* 1999) and that ribosomal ITS variants are dispersed among nuclei of *S. castanea* with different frequencies (Kuhn *et al.* 2001). These data suggest that AM spores contain a population of genetically different nuclei, although additional experimental evidences need to be obtained to confirm this hypothesis. In contrast, other experiments based on the analyses of variant sequence segregation from single spores obtained in *in vitro* cultures suggest a homokaryotic model of nuclear organization where variants are distributed within single nuclei (Pawlowska and Taylor 2002). Whatever the localization of variants, this genetic variability has led to a new view of AM fungi where the spore, and not the isolate, is considered an individual.

Likewise, there are contradictory data concerning the clonal (Rosendahl and Taylor 1997) or non clonal (Vandenkoornhuyse *et al.* 2001) reproduction mode of AM fungi. An artificial recombination of AFLP (Amplified Fragment Length Polymorphism) data obtained from single spores suggested that AM fungi reproduce clonally (Rosendahl and Taylor 1997). Recently, Kuhn and colleagues (2001) used the phylogenetic technique of character in compatibility analysis to detect whether genetically divergent sequences have arisen by the accumulation of mutations in clonal nuclear lineage or by recombination events. It was concluded, based on ITS and 28S rDNA analyses, that variant sequences are the result of accumulation of mutations in a clonal genome and that recombination events might be very rare. However, other studies point out the possible existence of recombination events (Vandenkoornhuyse *et al.* 2001). These data illustrate the difficulty of defining the reproduction strategy of AM fungi due to the biological characteristics of these symbiotic organisms.

4. TOWARDS SEQUENCING THE FUNGAL GENOME

4.1. Genomic Sequencing

It is clear from the above sections that much more information is needed concerning the structure and composition of the AM fungal genome. To fully sequence an AM fungal genome is beyond the scope of most research groups because of the high requirements in human and environmental inputs. A low-cost strategy to catalogue functional genes of a given organism is to randomly and partially sequence, in a single pass, many cDNA clones and build a collection of expressed sequence tags (ESTs). Despite the interesting information which has been generated through such an approach (see next section), ESTs do not give any information about gene structure and regulation or genome organization. In this context, mycorrhizal research would greatly benefit from the complementary approach of genome sequencing to identify fungal genes essential for symbiosis establishment and functioning. Moreover, the comparison of AM fungal genome with those of other fungi, and particularly pathogenic fungi, may give a better insight in the molecular bases of compatibility and incompatibility in plant-microbe interactions, which will be important for practical applications in sustainable agriculture.

For a genome sequencing project, it will be essential to chose an appropriate fungal model which has a small genome, does not harbour endosymbiotic bacteria, is easy to grow in sterile conditions, widespread in nature and known for its beneficial effect on plant growth and health. The AM fungus *G. intraradices* is an efficient symbiont, it can easily be grown monoxenically in root organ cultures, and its cell cycle has been well studied under these controlled conditions (Bago *et al.* 1998, 1999b; Fortin *et al.* 2002). Endosymbiotic bacteria have not been reported in *Glomus* species so far (Hosny *et al.* 1999b), and the nuclei of *G. intraradices* are small (average diameter 1.7 μm) as compared to other AM fungi like *Gi. rosea* (2.7 μm) and *S. castanea* (3.5 μm). If the genome size of *G. intraradices* is small enough, this fungus could be a good candidate for a genome sequencing project.

The two alternative approaches for genome sequencing are clone-by-clone shotgun and whole-genome shotgun sequencing (Zhang and Wu 2001). In the whole genome shotgun approach, genomic DNA is sheared and cloned into plasmid vectors which can integrate inserts up to 20 kb long. The sequences generated are then assembled into the whole genome sequence. This may be difficult for the genome of AM fungi as it contains a high proportion of repetitive DNA. A more accurate assembly of sequences is generally obtained by using the clone-by-clone shotgun approach. This would require the construction of a bacterial artificial chromosome (BAC) library containing inserts 100-200 kb long. The sequencing of BAC clone ends would enable the construction of a physical map of the genome prior to complete sequencing of selected BAC clones by a shotgun approach.

Genomic libraries have already been obtained for *S. castanea*, *G. mosseae* (Franken and Gianinazzi-Pearson 1996), *G. versiforme* and *Gi. margarita* (van Buuren *et al.* 1999), in lambda DASH and lambda ZAP vectors which can take inserts up to 20 kb. As it was observed that libraries contained in some cases endosymbiotic bacterial genes or genes from contaminant surface associated bacteria (van Buuren *et al.* 1999), a new approach was developed to separate fungal DNA from bacterial DNA by cesium chloride centrifugation prior to library construction (Hosny *et al.* 1999b). The different libraries have been used to isolate repetitive DNA sequences, including ribosomal DNA sequences (Franken and Gianinazzi-Pearson 1996; Hosny *et al.* 1999a; 1999b), as well as genes with interesting functions and/or their regulatory elements such as a chitin synthase gene from *G. versiforme* (Lanfranco *et al.* 1999b; van Buuren *et al.* 1999), a phosphoglycerate kinase gene from *G. mosseae* (Harrier 2001) and a *Gi. margarita* metallothionein (MT1) gene (Lanfranco *et al.* 2002). This technology will have to be developed further to construct BAC libraries which require high molecular weight DNA that has to be restricted with an appropriate enzyme and size-fractionated by pulse field gel electrophoresis (PFGE).

To study the synteny of the sequenced genome with that of another AM fungus, it may be interesting to isolate gene-rich regions of the genome while excluding repetitive DNA sequences (Mayer and Mewes 2002). For this, a methylation-sensitive restriction enzyme may be used to eliminate repetitive DNA sequences during library construction. This can be followed by the elimination of library clones giving strong hybridization signals with genomic DNA and therefore rich in repetitive elements (Rabinowicz *et al.* 1999). Alternatively, DNA enriched with low copy genes could be obtained through Cot DNA analysis and elimination of DNA fractions corresponding to highly and moderately repetitive DNA sequences (Peterson *et al.* 2002).

4.2. Expressed Sequence Tags

ESTs are usually partially sequenced inserts from clones of cDNA libraries; however, they can be derived by alternative techniques. The first ESTs from AM fungi were obtained by differential RNA display analyses which were carried out in order to identify plant genes induced during symbiotic development. Some fragments which did not appear in control roots turned out to be of fungal origin. Using this technique, an EST from *G. mosseae* with similarities to phosphoglycerate kinase genes was identified in tomato mycorrhiza and further investigations indicated higher accumulation of the protein during symbiotic interaction with the plant, compared to asymbiotic spore germination (Harrier *et al.* 1998). A fragment from *G. mosseae* with no homologies with known genes, but with an interesting expression pattern was identified in pea mycorrhiza (Lapopin *et al.* 1999). The corresponding gene seemed to be induced during interaction with the plant and this induction was more pronounced in the wild-type pea compared to an arbuscule-less mutant (Lapopin, Gianinazzi-Pearson and Franken, unpublished results). Three differential display fragments of *G. intraradices* showing interesting similarities to proteins involved in gene regulation were found in barley mycorrhiza (Delp *et al.* 2000). Recently, a fungal gene harbouring an open reading frame encoding a peptide with weak similarities to glutamine synthetases was isolated from a mycorrhiza by a different non-targeted approach, differential screening of cDNA libraries (Ruiz-Lozano *et al.* 2002).

Improvement of molecular techniques made it possible to extract RNA and synthesize cDNA from minute amounts of fungal material. The influence of a *Bacillus* isolate, which was able to promote hyphal development of *G. mosseae* after spore germination, was investigated by differential display. Two genes could be identified: one encoded a putative fatty acid oxidase (Requena *et al.* 1999), and the second one might be involved in the control of the cell cycle arrest in AM fungi in the absence of the plant (Requena *et al.* 2000).

In order to access more expressed sequence tags, libraries of cDNA fragments were established from activated spores of *Gi. rosea* (Stommel *et al.* 2001), germinated spores of *Gi. margarita* (Lanfranco *et al.* 2000) and *G. intraradices* (Lammers *et al.* 2001), as well as from extraradical hyphae of *G. intraradices* (Sawaki and Saito 2001). These libraries mostly contained ESTs from constitutively expressed genes. In contrast, cDNA libraries enriched for regulated genes can be achieved by the method of suppressive subtractive hybridization (Diatchenko *et al.* 1996). Combination of this technique with subsequent screening of the clones with complex cDNA probes results in a relatively fast identification of differentially expressed genes. This technique was used to compare RNA accumulation patterns from germinating spores and extraradical hyphae of *G. mosseae* (Requena *et al.* 2002). Numerous genes were identified which are preferentially expressed during the asymbiotic development of the fungus in the absence of the plant. A combination of differential display, subtractive hybridization and expression profiling to analyse the developmental switch in *Gi. rosea* from asymbiotic hyphal elongation to presymbiotic branching induced by plant root exudates revealed numerous changes in transcript patterns already one hour after addition of root exudates, although the morphological response of the fungus to these was not detectable before five hours (Tamasloukht *et al.* 2002).

The flood of information coming from these different technologies has started to be exploited. One approach has been to choose single genes for further analysis. For example, the EST library from germinated spores of *G. intraradices* has been exploited for genes involved in carbohydrate metabolism, and gene identifications were consistent with the postulated carbon fluxes (Lammers *et al.* 2001 ; Bago *et al.* 2002; 2003) (see section 5.2.3.). A metallothionein-encoding cDNA was isolated among the ESTs from *Gi. margarita* spores (Lanfranco *et al.* 2002). RNA accumulation analyses showed the corresponding gene to be induced by copper and to provide tolerance against cadmium and copper in a heavy metal-sensitive yeast mutant. The gene might therefore be involved in heavy metal resistance of the AM fungus. In addition to work on single genes, clustering of ESTs into functional groups can be used to formulate working hypotheses for further investigations. Among the *Gi. rosea* genes identified to be induced by root exudates, one group encoded proteins putatively involved in mitochondrial functioning. Subsequent physiological and cell biological experiments showed that the branching factor in root exudates indeed induced respiration (Tamasloukht *et al.* 2003).

Although the use of ESTs to analyse AM fungal development and functioning is recent, it has already provided new insights as well as confirmed results obtained before by other technologies. Analyses of high number of ESTs from total cDNA libraries will give access to many more genes, whilst subtractive cDNA libraries will give information about the behaviour of AM fungi at specific developmental steps and under particular environmental conditions.

5. LINKING AM FUNGAL FUNCTION TO GENE EXPRESSION: TOWARDS AN OVERALL PICTURE

As outlined above, EST sequencing has allowed identification of numerous fungal genes that are expressed in germinated spores and extraradical mycelium produced in monoxenic cultures. However, the identification of fungal genes expressed in intraradical hyphae and arbuscules has been hampered by the low amount of fungal RNA extracted from colonized root tissues (Maldonado-Mendoza *et al.* 2002). The overflow of sequence data generated by analyses of ESTs, will require compiling an accurate set of genes that are up- or down-regulated during the different phases of the life cycle of the fungus and under different environmental conditions. Such information should itself lead to high throughput, coordinated analyses of gene expression temporally across ranges of conditions by techniques such as microarrays (DeRisi et *al.* 1997) or SAGE (serial analyses of gene expression) (Velculescu *et al.* 1995).

In order to fully understand the functionality of AM fungi, it is necessary to first have a clear idea of the physiological/cellular behaviour of these organisms, so that the expression/repression of fungal genes can be interpreted in a global manner. In this section we divide studies of AM fungal gene expression into two major groups: developmental processes and metabolic processes. Table 1 gives a list of AM fungal genes which have been identified together with the specific structures and AM fungal species that have been targeted.

5.1. Expression of Genes Involved in AM Fungal Development

The establishment of symbiosis is crucial for an AM fungus to fulfil its life cycle and to become 'differentiated' both morphologically (i.e. formation of structures adapted to the intra-/extraradical environment) and biochemically (i.e. formation of a 'metabolic bipole').

Therefore, such a morphological/biochemical differentiation may be at the basis of AM fungal obligate biotrophy (Bago and Bécard 2002). The bases of such 'differentiation' remain completely unknown. The life cycle of AM fungi comprises several well defined developmental stages for which the genetic determinants are still unknown. Early developmental stages of spore germination and the presymbiotic hyphal growth must be controlled by genes involved in plant recognition and signal transduction, while others must be active in regulation of the growth arrest and cytoplasm retraction that takes place in the

Table 1. Developmental / metabolic AM fungal genes.

	Intraradical Hyphae	Extraradical Hyphae	Germ Tubes (Spores)
METABOLIC GENES			
PHOSPHORUS			
P transporter	-(8)	+(8, 13)	--
NITROGEN			
Nitrate reductase (NADH) (E.C. 1.6.6.1)	+(9)	--	--
Nitrate reductase (NADPH) (E.C. 1.6.6.3)	+(9)	--	+(9)
Glutamine synthetase (E.C. 6.3.1.2)	+(15*)	--	--
Asp amino transferase (E.C. 2.6.1.1)	--	--	+(10)
CARBON			
Glycolysis			
Glyceraldehyde-3-P-dehydrogenase (E.C. 1.2.1.12)	--	--	+(5)
3-phosphoglycerate kinase (E.C. 2.7.2.3)	+(7)	--	+(7)
Glycogen metabolism			
Glycogen synthase (E.C. 2.4.1.11)	--	+(2)	+(2)
1-4 α-glucan branching enzyme (E.C. 2.4.1.18)	--	+(2)	+(2)
Lipid metabolism			
Fatty acid coenzyme A ligase	--	--	+(1)
Acyl-coenzyme A dehydrogenase	--	+(1)	+(1)
Glyoxilate cycle			
Isocitrate lyase (E.C. 4.1.3.1)	--	+(10)	+(10)
Malate synthase (E.C. 4.1.3.2)	--	+(10)	+(10)
METABOLIC PUMPS			
H^+-ATPase (E.C. 3.6.1.x)	--	+(3, 4)	+(3, 5)
STRESS RESPONSE GENES			
Metallothionein	--	+(6, 12)	+(12)
STRUCTURAL GENES			
Cell wall metabolism			
Chitin synthase (E.C. 2.4.1.16)	+(11)	+(16)	-(11)
Gln-Fru 6P-transaminase (E.C. 2.6.1.16)	--	+(2)	+(2)
Cytoskeletal genes			
Beta-tubulin	--	+(14)	+(14)

+, expressed gene; -, gene repressed/absent; --, gene expression not determined in that tissue; *This gene presents weak homology with *Helicobacter pylori* GS, and it is induced by N fertilization; [1]Bago *et al.*, 2002: *G. intraradices;* [2]Bago *et al.* 2003: *G. intraradices*; [3]Ferrol *et al.*, 2002a: *G. mosseae*; [4]Ferrol *et al.* (unpublished): *G. intraradices*; [5]Franken *et al.* 1997: *Gi. rosea*; [6]Gonzalez-Guerrero *et al.*, 2002: *G. intraradices;* [7]Harrier *et al.*, 1998: *G. mosseae*; [8]Harrison and van Buuren 1995: *G. versiforme*; [9]Kaldorf *et al.*, 1998: *G. intraradices.*; [10]Lammers *et al.*, 2001: *G. intraradices;* [11]Lanfranco *et al.*, 1999b, c: *Gi. margarita*; [12]Lanfranco *et al.*, 2002: *Gi.* margarita; [13]Maldonado-Mendoza *et al.*, 2001: *G. intraradices*; [14]Rhody *et al.*, 2003: *Different AMF*; [15]Ruiz-Lozano *et al.*, 2002: *G. intraradices*; [16]Ubalijoro *et al.*, 2001: *G. mosseae, G. intraradices, Gi. margarita, S. calospora, A. scrobiculata, E. colombiana*; Different ESTs libraries have been prepared and analysed, giving a full range of DNA sequences presenting homology with structural/functional genes (Sawaki *et al.*, 2001; www.chemistry.nmsu.edu/glomus/).

absence of a host root. One may also hypothesize that the root apoplast, where the intraradical mycelium develops, provides an environment where the specific physical or chemical factors induce or repress fungal genes so regulating the programme of morphological, biochemical and physiological differentiation of the AM fungus within the root tissues. Recently, a novel gene (*GmGIN1*) was identified in *G. mosseae* which was only expressed in the early developmental stages and which might participate in the signalling events prior to symbiosis formation (Requena *et al.* 2002). In addition, a gene encoding a protein homolog involved in cell cycle and actin cytoskeleton (*GmTOR2*), which might have a role in the control of the cell cycle arrest in AM fungi in the absence of the plant, has also been identified in *G. mosseae* (Requena *et al.* 2000). The anti-inflamatory drug Rapamycin, which specifically inhibits the cell cycle controlling activity of TOR2, affected germ tube development but not spore germination. These results concord with the findings that nuclear replication is not a prerequisite for germination, but it is necessary for presymbiotic hyphal growth (Bécard and Pfeffer 1993; Bianciotto and Bonfante 1993).

Microtubules, as an important component of the cytoskeleton, play an important role in AM fungal development and in the extensive morphogenesis they undergo during their asymbiotic and symbiotic development (Timonen *et al.* 2001). To obtain insight into the mechanisms controlling AM fungal development, genes coding for cytoskeleton components have been investigated. Two different types of β-tubulin genes have been identified in a number of AM fungal isolates (Franken *et al.* 1997; Butehorn *et al.* 1999; Rhody *et al.* 2003). The finding that the two sequence types display different expression patterns in asymbiotic germinating spores compared to symbiotic extraradical hyphae indicates that different isoforms of β-tubulin may be involved in specific cellular needs (Rhody *et al.* 2003).

Attention has also been focused on the analysis of genes encoding chitin synthases, enzymes that play key roles in fungal morphogenetic processes. Complex spatial and temporal changes in chitin structure of the AM fungal cell wall have been observed in different developmental stages of the symbiotic interaction (Bonfante 2001). The differential gene expression pattern of chitin synthases during the life cycle of two AM fungi suggests that regulation of chitin synthesis may contribute to the control of fungal morphogenesis during different steps of mycorrhiza formation (Lanfranco *et al.* 1999b; 1999c; Ubalijoro *et al.* 2001).

5.2. Expression of Genes Involved in Nutrient Acquisition and Metabolism

Bidirectional nutrient transfer between the plant and the fungus is the key physiological feature of the AM symbiosis. Until recently our knowledge about metabolic processes occurring in AM fungi during both asymbiosis and symbiosis was hindered due to the lack of appropriate methodology to obtain sufficient amount of fungal material in which experiments could be carried out. This has been overcome in recent years with the development and extensive use of AM monoxenic cultures (Bécard and Fortin 1988; St Arnaud *et al.* 1996), where large amounts of contaminat-free extra and intraradical AM fungal mycelium have been obtained. Here we focus on the recent studies which have provided new insights into gene expression/regulation of metabolic processes taking place in AM fungi throughout their life cycle.

5.2.1. Phosphorus metabolism

One of the major benefits associated with AM symbiosis is the enhanced P status of mycorrhizal plants (Smith and Read 1997; Jakobsen *et al.* 2001). The AM fungal-mediated translocation of phosphate from soil to plant requires specific transporters that function in at least two fungal structures. The external hyphae which are expected to mediate uptake from the soil, and arbuscules located within the root cortical cells and expected to have an efflux transporter or channel (Ferrol *et al.* 2002b). A gene encoding a phosphate transporter for phosphorus uptake from the soil solution has been identified in *G. versiforme* (*GvPT*) and *G. intraradices* (*GiPT*) (Harrison and van Buuren 1995; Maldonado-Mendoza *et al.* 2001). *GiPT* expression in the extraradical mycelium of *G. intraradices* is induced in response to phosphate concentrations of 1 to 5 μM, typical of those found in the soil solution and is modulated by the overall phosphate status of the mycorrhiza (Maldonado-Mendoza *et al.* 2001). Moreover, functional analysis of the *GvPT* in yeast showed that the *GvPT* gene product is a high-affinity phosphate transporter that operates by proton-coupled symport (Harrison and van Buuren 1995), indicating that *GvPT* phosphate transport activity requires the activity of a plasma membrane H^+-ATPase. In this context, Ferrol *et al.* (2002a) showed that one of the five genes encoding the plasma membrane H^+-ATPase of *G. mosseae* (Ferrol *et al.* 2000) and two of *G. intraradices* (Ferrol, unpublished results) are only expressed in the extraradical mycelium. Although transfer of phosphate to the host plant has been shown to be metabolically dependent (Smith and Read 1997), the mechanisms have yet to be identified.

5.2.2. Nitrogen metabolism

Although the contribution of AM fungi in the acquisition of nitrogen by plants is not so clear as in the case of phosphorus, measurements of the $^{15}NO_3^-$ or $^{15}NH_4^+$ uptake indicate that AM fungi significantly contribute to the N-budget of the plant (Johansen *et al.* 1992; Frey and Schüepp 1993; Tobar *et al.* 1994). The genetic determinants of this contribution are currently unknown, but there is some evidence that AM fungi possess the enzymatic machinery involved in nitrogen metabolism. Kaldorf *et al.* (1998) showed by *in situ* hybridization that the gene encoding a nitrate reductase of *G. intraradices* is preferentially expressed in the arbuscules. The demonstration that nitrate reduction occurs in the arbuscules indicates that AM fungi must have other enzymes involved in nitrogen metabolism. As previously indicated, Ruiz-Lozano *et al.* (2002) have identified a *G. intraradices* gene harbouring an open reading frame encoding a peptide with weak similarities to glutamine synthetases, which is only expressed in the symbiotic stage of the fungus and that is up-regulated by nitrogen-fertilization.

5.2.3. Carbon metabolism

The need for AM fungi to differentiate their carbon metabolism in order to become fully functional has been known for a long time. Recently crucial metabolic and cellular processes have been studied in depth (Saito 1995; Shachar-Hill *et al.* 1995; Bago *et al.* 1999a; Pfeffer *et al.* 1999; Bago *et al.* 2002; 2003). These studies have shown that the symbiotic (both intraradical and extraradical) and the asymbiotic AM fungus present quite distinctive characteristics in respect to C metabolism.

Carbon utilization by the symbiotic AM fungus starts when photosynthetically-fixed plant C is actively taken up (as hexose) by intraradical structures of the fungus. Unfortunately none of the numerous efforts carried out up to now to identify and characterize the hexose transporters involved in these crucial processes have proved successful. More research is needed in this sense, since a more profound knowledge of the mechanism for C acquisition would no doubt open new possibilities for manipulating and selecting AM fungi for applied purposes.

Several of the major carbohydrate metabolic pathways such as glycolysis, gluconeogenesis and the tricarboxylic acid cycle are active in the symbiotic AM fungus, and some of the key genes encoding for enzymes involved in these pathways have been identified and studied [glyceraldehyde-3-phosphate dehydrogenase (Franken *et al.* 1997); 3-phosphoglycerate kinase (Harrier *et al.* 1998)]. Data are in agreement that when the AM fungus is in symbiosis behaves as a metabolic bipole: intraradical structures are mainly glycolytic, whereas the extraradical mycelium presents little or no glycolysis, but large fluxes of C are utilised via gluconeogenesis and lipogenesis (reviewed by Bago *et al.* 2000; Jun *et al.* 2002).

Since glycogen and trehalose are the most plausible candidates to act as buffers of the cytoplasmic hexose levels within fungal hyphae (Bago *et al.* 1999a), genes involved in the metabolic routes undertaken by these compounds should be active in the symbiotic fungus. This has been confirmed recently by the identification and expression studies of two genes encoding two key enzymes for glycogen metabolism, glycogen synthase and 1-4 alpha-glucan branching enzyme (Bago *et al.* 2003). However, although glycogen appears to play an important role in C translocation during early stages of symbiosis establishment, storage lipids (triacylglycerides, TAGs) are the dominant form for AM fungi to store carbon (Beilby and Kidby 1980; Jabaji-Hare 1988; Gaspar *et al.* 1997). Recent studies which have combined AM monoxenic cultures, NMR spectroscopy and multiphoton microscopy have made it clear that TAG is synthesised by the fungus within the root and then exported to the extraradical mycelium (Pfeffer *et al.* 1999; Lammers *et al.* 2001; Bago *et al.* 2002). Some key genes coding for enzymes involved in these processes (fatty acid coenzyme A ligase, acyl-coenzyme A dehydrogenase) are active in oleolytic fungal structures such as extraradical hyphae and germinating spores (Bago *et al.* 2002). Nothing is known, however, about the mechanisms governing such a massive export of lipids, and much less of the genes regulating these complicated but necessary processes.

Once TAGs arrive in the extraradical hyphae they undergo one of two possible fates: i) become C storage deposits within the spores, or ii) are utilized, via the glyoxylate cycle, to obtain carbohydrates. The latter is crucial to extraradical mycelium metabolism, since the AM fungus is not able to acquire exogenous hexose from the external medium (Pfeffer *et al.* 1999). Thus, the glyoxylate cycle should be one of the most important metabolic pathways at work in AM fungi. This has been corroborated by NMR spectroscopy analysis and by the fact that genes coding the two major enzymes of the glyoxylate cycle, isocitrate lyase and malate synthase, are actively expressed in extraradical hyphae of *G. intraradices* (Lammers *et al.* 2001).

As has been stated above, the C metabolism in the asymbiotic stage of AM fungi presents some characteristics which could give clues to better understand the inherent obligate biotrophism of these organisms. Carbon metabolism in germinating spores presents mixed characteristics of both intraradical and extraradical symbiotic hyphae. Thus, germ-tubes are able to take up exogenous hexose, but at a very discrete level which could in no way support germ-tube development. Germinating spores therefore depend for growth on their lipid (TAG) stores, so that they present a substantial gluconeogenetic flux by mobilizing these via glyoxylate cycle (Bago *et al.* 1999a). This has been corroborated by measuring expression of isocitrate lyase and malate synthase genes in germinating hyphae (Lammers *et al.* 2001). On the other hand, biochemical analysis revealed that glycolysis, TAC cycle and the pentose phosphate pathway are active in germ-tubes (MacDonald and Lewis 1978; Saito 1995). An important translocation of lipid globules has also been found, together with high levels of transcripts of fatty acid coenzyme A ligase and acyl-coenzyme A dehydrogenase, both implicated in fatty acid metabolism (Bago *et al.* 2002). Other metabolic pathways active in germinating spores are dark fixation of CO_2 and non-photosynthetic one-carbon metabolism (Bago *et al.* 1999a), although no genes involved in these pathways have been characterized up to date. Interestingly, ^{13}C experiments on germinating spores strongly suggest that the synthesis of fatty acids (FA) (crucial component of TAGs) does not represent a significant C flux in this stage of the fungus (Bago *et al.* 1999a). This has led to the speculation that it is the absence of FA synthesis that prevents the asymbiotic fungus from forming new propagules, making it an obligate symbiont (Bago and Bécard 2002). Therefore genes involved in FA metabolism should be certainly sought and their regulation studied in order to test such a hypothesis, perhaps one of the most challenging ones in AM fungal biology.

6. FUNCTIONAL ANALYSIS OF EXPRESSED GENES

Functional analysis of expressed genes identified within AM fungi is necessary to confirm their biological role. This is usually accomplished in other organisms by either functional complementation of mutants or transformation of the organism of interest. Functional complementation, the restoration of the normal phenotype of a mutant by the introduction of the wild-type allele is one of the most common techniques in genetic analysis to prove the function of a gene. Since a sexual cycle and mutant phenotypes are absent from AM fungi, heterologous complementation assays have to be used and, to date, all the functional complementation studies on AM fungal genes have been carried out with *S. cerevisiae*. No mutants from *E. coli* and/or other fungal species have been complemented by AM fungal genes. Table 2 is a compilation of AM fungal genes for which the function has been validated by functional complementation of yeast mutants.

Table 2. AM fungal genes demonstrated to complement *Saccharomyces cerevisiae* mutants.

Gene	AM fungal species	Mutant complemented	Reference
Phosphate transporter	*G. versiforme*	*Pho84*	Harrison and van Buuren 1995
Metallothionein like	*Gi. margarita*	*Δyap-1*, *Δcup-2*	Lanfranco *et al.* 2002
	G. intraradices	*Δcup-2*	González-Guerrero *et al.* 2002
3-phosphoglycerate kinase	*G. mosseae*	*pgk⁻*	Harrier and Paterson 2002

In addition to proving the function of AM fungal genes, studies can be undertaken within the complemented *S. cerevisiae* to establish functional attributes of the expressed gene. For example, Harrison and van Buuren (1995) demonstrated that phosphate uptake by *S. cerevisiae* cells expressing the *G. versiforme* phosphate transporter accumulated more ^{33}Pi than control cells. Furthermore, they demonstrated that phosphate uptake by these transformed cells followed Michaelis-Menten kinetics with an apparent K_m of 18 μM, that were similar to values of *S. cerevisiae* high affinity transporters.

Although AM fungal genes can complement *S. cerevisiae* mutants, it is not known whether regulation and control elements are functional and recognized in a similar manner to the homologous situation within the AM fungi. For example, the promoter P_{GmPGK} contains motifs that may be responsible for specific C source inductions, but it is not known whether the response of the *GmPGK*-encoding gene is mediated through the action of transcription factors like GCR1, RAP1 and GAL11 which are involved in modulating *S. cerevisiae PGK* gene activity (Henry *et al.* 1994; Stanway *et al.* 1994). Future work should aim to identify the regions of promoters that are responsible for the specific inductions or repressions in AM fungi and investigate whether these responses are mediated in a similar way (Harrier 2001; Harrier and Paterson 2002).These types of studies would help to elucidate the evolutionary differences of transcriptional regulation between different AM fungal isolates and *S. cerevisiae* and determine whether control elements are functional and recognized in the same way.

The development of AM fungal transformation strategies will be a core research tool in AM fungal biology, and a practical tool for AM fungal isolate improvement. An essential pre-requisite for successful transformation is the successful delivery of foreign DNA into the organism to be transformed. Traditionally transformation of fungi has involved the production of protoplasts, electroporation and chemical based transformation strategies. However, AM fungi being aseptate, protoplast fusion based techniques cannot be used as a mean of introducing DNA into these fungi, and other procedures require to be established. There are three potential processes for transformation of AM fungi based on biolistic, *Agrobacterium* or endosymbiotic bacteria mediated transformation.

Biolistic transformation otherwise known as particle bombardment involves the explosive acceleration of microscopic particles coated with DNA into tissue of the organism to be transformed. This is an effective way to introduce foreign DNA into the spores of AM fungi and/or mycorrhiza from *in vitro* culture systems in order to study the functional attributes of given genes. Biolistic transformation has been used successfully to introduce genes into *Gi. rosea* (Forbes *et al.* 1998; Harrier and Millam, 2001; Harrier *et al.* 2002). In these cases, the plasmid vector used to transform the AM fungus contained a heterologous promoter and terminator. Results showed that expression was relatively weak, and the authors attributed this to the use of the heterologous promoter.

Optimization of the transformation vectors is required to achieve optimal transgene expression and maximal stability, although the later is not required in studies that only require transient gene expression (Bergero *et al.* 2003). Optimal transgene expression requires several pre-requisites including strong homologous promoter and terminator regions. The expression of the transgene can be enhanced by the presence of genetic elements within the vector which enhance stability of the transgene. A successful strategy used within fungi is the incorporation of repetitive sequences into plasmid vectors. Repetitive sequences such as ribosomal DNA genes and genetic elements like *seg1*, a single copy region that leads to high mitotic stability, or *rag1*, a highly repetitive interspersed DNA sequence that promotes plasmid integration, have been used to enhance the stability of transformants in other fungi (Ruiz-Díez and Martínez-Suárez 1999; Mackenzie *et al.* 2000; Schilde *et al.* 2001).

Genetic elements such as transposable sequences can be used to enhance stability of transformants. A subclass of Class II transposons that are short inverted repeat-type elements have proven to be useful in constructing gene vectors for *Drosophila*, fungi and are becoming increasingly important for plant genome manipulation. Short inverted repeat type elements are particularly amenable to be introduced into transformation vectors because the two components of the element can function *in trans* (Walbot 1992; Hehl 1994; O'Brochta and Atkinson, 1996). These characteristics were first exploited in eukaryotes for the purpose of gene vector development using the *P*-element from *Drosophila melanogaster* (Rubin and Spradling 1982; Spradling and Rubin 1982) and is referred to as the *P*-element paradigm. The terminal sequences are attached to the gene or DNA sequence to be integrated forming a chimeric transposable element. Upon entry into a nucleus, transposase promotes the cutting and joining of the vector to the chromosome of the host resulting in chromosomal integration. As movement of the vector does not require an RNA intermediate, the types of sequences that can be included in these types of vectors is less restricted and may include introns. These types of vectors may be particularly important for transformation of AM fungi because recombination within these fungi is thought to be a rare event. Transposable element-like sequences have recently been identified in the genome of AM fungi in particular gypsy and Non-Long Terminal Repeat retrotransposons (Gollotte *et al.* 2002b) and these may provide an alternative source of DNA to be included into transformation vectors in order to improve transgene integration.

Different type of vectors could be used within the biolistic process that would facilitate studying gene expression patterns and/or the function of the gene through gene silencing. Post transcriptional gene silencing (PTGS) involves the silencing of an endogenous gene by the introduction of homologous double stranded RNA, transgenes or viruses (Hannon 2002). These tools may facilitate the evaluation of AM fungal genetics, and may be particularly useful to prove the function of the AM fungal gene sequences which do not show significant homology to other sequences present in the database.

A potential strategy to generate stable transformants of AM fungi may be to utilize the plant pathogenic soil bacterium *Agrobacterium tumefaciens*. This type of transformation strategy has been used successfully to transform a range of fungi (Bundock *et al.* 1995; de-Groot *et al.* 1998; Gouka *et al.* 1999; Abuodeh *et al.* 2000; Covert *et al.* 2001; Malonek and Meinhardt 2001; Mikosch *et al.* 2001; Mullins *et al.* 2001; Rho *et al.* 2001; Bundock *et al.* 2002; Pardo *et al.* 2002). The advantage of this type of transformation strategy is that it has been shown that integration of the transgene only occurs once within the fungal genome.

A completely novel strategy for transforming AM fungi may be possible through the genetic engineering of the endosymbiotic bacteria *Glomeribacter* present within some AM fungi (Bianciotto *et al.* 1996). This may be possible by either re-introducing genetically modified endosymbiotic bacterial species into AM fungal isolates that lack such bacteria such as *Gi. rosea* and/or by transforming the bacteria within the AM fungi. These approaches could both utilize biolistic technology. This type of approach has enabled genetic transformation of bacterial symbionts from insects (Beard *et al.* 1992, 1993). The genetically modified bacterial symbionts were maintained stably in their hosts, expressing the antibiotic marker gene throughout the entire developmental cycle of the host, even in the absence of a selectable marker. Such an approach may be possible with *Glomeribacter* species as a means of tagging AM fungi.

7. AM FUNGAL PROTEOMICS

Whilst studies of the AM fungal transcriptome can provide substantial information about genes that are expressed in different developmental stages, they do not give any insight into whether transcripts are translated to proteins, or if constitutively expressed genes are differentially post-translationally modified. Thus, high throughput analyses of proteins synthesised during the different phases of the life cycle of the fungus is also a requirement for full characterization of the biochemical and physiological events that are occurring.

The first analyses of AM fungal proteins were performed for taxonomic purposes. Preliminary investigations using native PAGE indicated species differences in spore protein profiles from *A. laevis*, *G. fasciculatum* and *G. mosseae* (Schellenbaum *et al.* 1992). Detection and profile resolution was much improved by SDS-PAGE and distinct protein profiles have subsequently been established at the genus, species and isolate level for different AM fungi (Dodd *et al.* 1996; Avio and Giovanetti 1998). The feasibility of using taxon-discriminating fungal protein profiles as a support for taxonomic studies of these fungi has been illustrated in the detailed study on different isolates of *G. mosseae* and *G. coronatum* by Dodd *et al.* (1996). The existence of taxon-specific fungal proteins has prompted the use of protein fractions from spores as antigens to produce antibodies against various AM fungal species (Sanders *et al.* 1992; Gobel *et al.* 1995; Hahn *et al.* 2001). However, serological identification of AM fungi has met with problems of antibody specificity. Substantial progress was made in the resolution of fungal polypeptides by applying high resolution 2D-PAGE analysis (Samra *et al.* 1996). Important qualitative differences were found between polypeptide profiles of spore extracts from four fungal species belonging to different genera of AM fungi (*Gi. rosea*, *S. castanea*, *A. laevis*, *G. mosseae*). Although some polypetides were common to the four fungi, some others were specific for some of them.

First attempts to globally identify the AM fungal proteome targeted qualitative protein modifications during spore germination and/or hyphal growth prior to plant colonisation. When polypeptide profiles of ungerminated spores of *G. mosseae* were compared with those of spores germinated in water, a strong increase in polypeptide number was observed following germination (Samra *et al.* 1996). This activation of protein synthesis during spore germination and hyphal growth of *G. mosseae* corroborated previous observations by Beilby and Kidby (1982) that protein synthesis is essential to these processes. However, in the study by Samra *et al.* (1996), no significant modifications were elicited in polypeptide profiles of germinating spores by root exudates from pea genotypes differing in their ability to form mycorrhizas.

The symbiotic proteome of arbuscular mycorrhizas in different plant species has also been identified by 2D-PAGE analyses (Dumas-Gaudot *et al.* 1994; Simoneau *et al.* 1994; Samra *et al.* 1997; Benabdellah *et al.* 1998; 2000). However, a major problem associated with studying proteins of AM fungi during the symbiotic stage is differentiating them from those of plant origin. Attempts have been made to study fungal polypeptides in the symbiotic stage by using enzymatic digestion of host roots to liberate intraradical mycelium (Simoneau *et al.* 1994), but this strategy could lead to artefactual alterations in polypeptides of AM fungi. An alternative approach to discriminate fungal from plant polypeptides has been based on comparing protein profiles from AM roots with those of asymbiotic roots, of extraradical hyphae, or a mix of germinating spores and hyphae (Benabdellah *et al.* 1998; Dassi *et al.* 1999). Nevertheless, this only gives an approximative picture since it is likely that protein profiles of spores and extraradical hyphae will differ from that of fungal structures growing inside roots. Moreover, these studies suggested that the additional polypeptides detected in AM roots are most likely of plant origin, and those that have since been assigned a biologcical function are all plant polypeptides (Benabdellah *et al.* 2000; Bestel-Corre *et al.* 2002).

The failure to detect fungal proteins in the symbiotic phase could be due to the low abundance of the fungal proteins in the extract of a mycorrhizal root. When large proteomes consisting of thousands of proteins are analyzed, the dynamic resolution is limited and only the most abundant proteins can be detected. There is a general consensus that analysis of the total proteome of an organism at significant depth can only be obtained after sub-fractionation into smaller sub-proteomes according to cell type and subcellular compartments (van Wijk 2001). Differential fractionation of cell organelles has already been shown to increase the number of proteins resolved by 2D-PAGE in mycorrhizal tomato roots, where only nine polypeptides were differentially displayed in crude protein extracts of the symbiosis, whilst 44 were identified when proteins were fractionated into soluble and microsomal proteins (Benabdellah *et al.* 1998). Moreover, isolation of plasma membrane fractions by two-phase partitioning of microsomal membranes isolated from mycorrhizal tomato roots led to detection of 21 newly synthesized polypeptides versus two new polypeptides in crude extracts (Benabdellah *et al.* 2000). Such subcellular approaches together with mass spectrometry for protein characterization may enable identification of the fungal proteome in the symbiotic phase. Recently, Dumas-Gaudot *et al.* (2002) analyzed by mass spectrometry some polypeptides from the extraradical mycelium of *G. intraradices* developing in monoxenic root cultures. However, few were attributed a function due to the lack of protein databases for AM fungi.

With the recent advances in proteomic techniques and increasing genomic and EST sequence data for AM fungi and other organisms, it should be possible to envisage a more detailed characterization of the fungal proteome in both presymbiotic (spores and germinating spores) and symbiotic phases (intraradical and extraradical mycelium).

8. CONCLUSIONS

Genome sequence information is currently being generated for AM fungi through the development of appropriate genome technologies and model experimental systems. Genome-wide comparisons coupled with RNA and protein profiling is certain to provide unique insights into the biology of these fungi and into the processes controlling development and/or functioning of AM symbiosis. Equally attractive are the prospects of finding genes unique to AM fungi that determine their obligate symbiotic character. It should be emphasized that although genome technologies are powerful, their value is substantially reduced without a genetic system that allows gene validation. Therefore, development of AM fungal transformation strategies should be a core for the future research, not only for gene validation but also as a practical tool for AM fungal isolate improvement.

From an ecological point of view it is clear that complete monitoring of entire AM fungal communities in field samples requires probes which have to be both general enough to recognize all AM fungi and sufficiently specific to exclude other fungi or plants. The design of such probes, which will greatly benefit from new and larger sequence data, is not an easy task because of the existence of multiple sequence types, at least for ribosomal genes, within single spores. The extent of intraspecific genetic variation must be clearly defined before sequence data can be used to imply AM species diversity. This will require exploring target genes with different rates of evolution from rDNA and considering higher numbers of isolates. Further sequence data, especially for protein-coding genes, are needed to evaluate the number of alleles eventually present within a spore/individual and to draw conclusions about relations between genetic diversity and functional traits. Understanding the relationships between genetic diversity and functional characteristics will also help us to know whether the genome plasticity of AM fungi may explain the phenotypic plasticity in an ecological context i.e. their adaptation to a wide range of hosts and environments.

From a biotechnological point of view, the identification of fungal genes responsible for symbiotic functioning and efficiency will enable the development of molecular markers to accurately monitor mycorrhizal benefits and to assess the contribution of host genotypes to the induction of key genes. Knowledge of the functional genome of AM fungi is crucial for the identification and exploitation of genes that could be central to optimize sustainable plant production systems in the future.

Acknowledgments: We are grateful to Drs. José Miguel Barea and Silvio Gianinazzi for valuable comments on the manuscript. Financial support was partially provided by the EU project Genomyca (QLK5-CT-2000-01319).

REFERENCES

Abuodeh RO, Orbach MJ, Mandel MA, Das A, and Galgiani JN (2000). Genetic transformation of *Coccodioides immitis* facilitated by *Agrobacterium tumefaciens*. J Infect Dis 181:2106-2110.

Al-kariki GN and Hammard R (2001). Mycorrhizal influence on fruit yield and mineral content of tomato grown under salt stress. J Plant Nutrit 24:1311-1323.

Avio L and Giovannetti M (1998). The protein pattern of spores of arbuscular mycorrhizal fungi: a comparison of species, isolates and physiological stages. Mycol Res 102:985-990.

Azcón-Aguilar C and Barea JM (1996). Arbuscular mycorrhizas and biological control of soil-borne plant pathogens - An overview of the mechanisms involved. Mycorrhiza 6:457-464.

Bago B and Bécard G (2002). Bases of the obligate biotrophy of arbuscular mycorrhizal fungi. In : S Gianinazzi, H Schüepp, JM Barea and K Haselwandter, eds. Mycorrhizal Technology in Agriculture: From Genes to Bioproducts. Basel, Boston, Berlin: Birkhäuser, pp 33-48.

Bago B, Pfeffer PE, Abubaker J, Jun J, Allen JW, Brouillette J, Douds DD, Lammers PJ, and Shachar-Hill Y (2003). Carbon export from arbuscular mycorrhizal roots involves the translocation of carbohydrate as well as lipid. Plant Physiol 131:1496-1507.

Bago B, Pfeffer PE, Douds DD, Bécard G, and Shachar-Hill Y (1999a). Carbon metabolism in spores of the arbuscular mycorrhizal fungus *Glomus intraradices* as revealed by nuclear magnetic resonance spectroscopy. Plant Physiol 121:263-271.

Bago B, Pfeffer PE, and Shachar-Hill Y (2000). Carbon metabolism and transport in arbuscular mycorrhizas. Plant Physiol 124:949-957.

Bago B, Zipfel W, Williams RC, Chamberland H, Lafontaine J-G, Webb WW, and Piché Y (1998). *In vivo* studies on the nuclear behavior of the arbuscular mycorrhizal fungus *Gigaspora rosea* grown under axenic conditions. Protoplasma 203:1-15.

Bago B, Zipfel W, Williams RM, Jun J, Arreola R, Lammers PJ, Pfeffer PE, and Shachar-Hill Y (2002). Translocation and utilization of fungal storage lipid in the arbuscular mycorrhizal symbiosis. Plant Physiol 128:108-124.

Bago B, Zipfel W, Williams RC, and Piché Y (1999b). Nuclei of symbiotic arbuscular mycorrhizal fungi, as revealed by *in vivo* two-photon microscopy. Protoplasma 203:1-15.

Beard CB, Mason P, Askoy S, Tesh RB, and Richards FF (1992). Transformation of an insect symbiont and expression of a foreign gene in the Chagas disease vector *Rhodnius prolixus*. Am J Trop Med Hyg 46:195-200.

Beard CB, O'Neill SL, Mason P, Mandelco L, Woese CR, Tesh RB, Richards FF, and Askoy S (1993). Genetic transformation and phylogeny of bacterial symbionts from tsetse. Insect Molec Biol 1:123-131.

Bécard G, Doner LW, Rolin DB, Douds DD, and Pfeffer PE (1991). Identification and quantification of trehalose in vesicular-arbuscular mycorrhizal fungi by *in vivo* ^{13}C NMR and HPLC analyses. New Phytol 118:547-552.

Bécard G and Fortin A (1988). Early events of vesicular-arbuscular mycorrhiza formation on Ri T-DNA transformed roots. New Phytol 108:211-218.

Bécard G and Pfeffer PE (1993). Status of nuclear division in arbuscular mycorrhizal fungi during *in vitro* development. Protoplasma 194:62-68.

Beilby JP and Kidby DK (1980). Biochemistry of ungerminated spores of the vesicular-arbuscular mycorrhizal fungus *Glomus caledonium*: changes in neutral and polar lipids. J Lipid Res 21:739-750.

Beilby JP and Kidby DK (1982). The early synthesis of RNA protein and some associated metabolic events in germinating vesicular-arbuscular fungal spores of *Glomus caledonium*. Can J Bot 28:623-628.

Benabdellah K, Azcón-Aguilar C, and Ferrol N (1998). Soluble and membrane symbiosis-related polypeptides associated with the development of arbuscular mycorrhizas in tomato (*Lycopersicon esculentum*). New Phytol 140:135-143.

Benabdellah K, Azcón-Aguilar C, and Ferrol N (2000). Alterations in the plasma membrane polypeptide pattern of tomato roots (*Lycopersicon esculentum*) during the development of arbuscular mycorrhiza. J Exp Bot 51:747-754.

Bergero R, Harrier LA, and Franken P (2003). Reporter genes: applications to the study of arbuscular mycorrhizal (AM) fungi and their symbiotic interactions with plant roots. Plant Soil, in press.

Bestel-Corre G, Dumas-Gaudot E, Poinsot V, Dieu M, Dirick J-F, van Tuinen D, Remacle J, Gianinazzi-Pearson V, and Gianinazzi S (2002). Proteome analysis and identification of symbiosis-related proteins from *Medicago truncatula* Gaertn. by two-dimensional electrophoresis and mass spectrometry. Electrophoresis 23:122-137.

Bianciotto V and Bonfante P (1992). Quantification of the nuclear DNA content of two arbuscular mycorrhizal fungi. Mycol Res 96:1071-1076.

Bianciotto V and Bonfante P (1993). Evidence of DNA replication in an arbuscular mycorrhizal fungus in the absence of the host plant. Protoplasma 176:100-105.

Bianciotto V, Bandi C, Minerdi D, Sironi M, Tichy HV, and Bonfante P (1996). An obligately endosymbiotic mycorrhizal fungus itself harbors obligately intracellular bacteria. Appl Environ Microbiol 62:3005-3010.

Bird A (2002). DNA methylation patterns and epigenetic memory. Genes and Development 16:6-21.

Bonfante P (2001). At the interface between mycorrhizal fungi and plants: the structural organization of cell wall, plasma membrane and cytoskeleton. Fungal Assoc. 9:45-61.

Bonfante P and Perotto S (1995). Strategies of arbuscular mycorrhizal fungi when infecting host plants. New Phytol 139:3-21.

Broach JR, Li YY, Feldman J, Jayaram M, Abraham J, Nasmyth KA, and Hicks JB (1983). Localization and sequence analysis of yeast origins of DNA replication. Proceedings of Cold Spring Harbor Symposia on Quantitative Biology 47 Pt 2, pp1165-1173.

Bundock P, den Dulk-Ras A, Beijersbergen A, and Hooykaas PJJ (1995). Trans-kingdom T-DNA transfer from *Agrobacterium tumefaciens* to *Saccharomyces cerevisiae*. EMBO J 14:3206-3214.

Bundock P, van Attikum H, den Dulk-Ras A, and Hooykaas PJJ (2002). Insertional mutagenesis in yeasts using T-DNA from *Agrobacterium tumefaciens*. Yeast 19:529-536.

Butehorn B, Gianinazzi-Pearson V, and Franken P (1999). Quantification of beta-tubulin RNA expression during asymbiotic and symbiotic development of the arbuscular mycorrhizal fungus *Glomus mosseae*. Mycol Res 103:360-364.

Cantrel IC and Linderman RG (2001). Preinoculation of lettuce and onion with VA mycorrhizal fungi reduces deleterious effects of soil salinity. Plant Soil 233:269-281.

Casana M and Bonfante P (1982). Ife intracellulari ed arbuscoli di *Glomus fasciculatum* (Thaxter) Gerd. et Trappe isolato con digestione enzimatica. Allionia 25:17-25.

Clapp JP, Rodriguez A, and Dodd JC (2001). Inter- and intra-isolate rRNA large sub-unit variation in spores of *Glomus coronatum*. New Phytol 149:539-554.

Clapp JP, Young JPW, and Fitter AH (1999). Ribosomal small subunit sequence variation within spores of an arbuscular mycorrhizal fungus, *Scutellospora* sp. Mol Ecol 8:915-921.

Cooke JC, Gemma JN, and Koske RE (1987). Observations of nuclei in vesicular-arbuscular mycorrhizal fungi. Mycologia 79:331-333.

Cordier C, Pozo MJ, Barea JM, Gianinazzi S, and Gianinazzi-Pearson V (1998). Cell defense responses associated with localized and systemic resistance to *Phytophthora parasitica* induced in tomato by an arbuscular mycorrhizal fungus. Mol Plant-Microbe Interact 11:1017-1028.

Covert SF, Kapoor P, Lee M, Briley A, and Nairn CJ (2001). *Agrobacterium tumefaciens*-mediated transformation of *Fusarium circinatum*. Mycol Res 105:259-265.

Dassi B, Samra A, Dumas-Gaudot E, Gianinazzi-Pearson V, and Gianinazzi S (1999). Different polypeptide profiles from tomato roots following interactions with arbuscular mycorrhizal (*Glomus mosseae*) or pathogenic (*Phytophthora parasitica*) fungi. Symbiosis 26:65-77.

De-Groot MJ, Bundock P, Hooykaas PJ, and Beijersbergen AG (1998). *Agrobacterium tumefaciens*-mediated transformation of filamentous fungi. Nat Biotechnol 16:839-42.

Delp G, Smith SE, and Barker SJ (2000). Isolation by differential display of three partial cDNAs potentially coding for proteins from the VA mycorrhizal *Glomus intraradices*. Mycol Res 104:293-300.

DeRisi JL, Iyver VR, and Brown PO (1997). Exploring the metabolic and genetic control of gene expression on a genomic scale. Science 278:680-686.

Diatchenko L, Lau YF, Campbell AP, Chenchik A, Moqadam F, Huang B, Lukyanov S, Lukyanov K, Guskaya N, Sverdlov ED, and Siebert PD (1996). Suppression subtractive hybridization: a method for generating diferentially regulated or tissue specific cDNA probes and libraries. Proc Natl Acad Sci USA 93:6025-6030.

Dodd JC, Rosendahl S, Giovannetti M, Broome A, Lanfranco L, and Walker C (1996). Inter- and intraspecific variation within the morphologically similar arbuscular mycorrhizal fungi *Glomus mosseae* and *Glomus coronatum*. New Phytol 133:113-122.

Douds DD, Pfeffer PE, and Shachar-Hill Y (2000). Application of *in vitro* methods to study carbon uptake and transport by AM fungi. Plant Soil 226:255-261.

Dumas-Gaudot E, Bestel-Corre G, Valot B, St-Arnaud M, Dieu M, and Gianinazzi S (2002). Proteomics to study plant-AM fungi interactions. In: 10th New Phytologist Symposium on Functional Genomics of Plant-Microbe Interactions, Nancy, France.

Dumas-Gaudot E, Guillaume P, Tahiri-Alaoui A, Gianinazzi-Pearson V, and Gianinazzi S (1994). Changes in polypeptide patterns in tobacco roots colonized by two *Glomus* species. Mycorrhiza 4:215-221.

Ferrol N, Barea JM, and Azcón-Aguilar C (2000). Plasma membrane H^+-ATPase gene family in the arbuscular mycorrhizal fungus *Glomus mosseae*. Curr Genet 37:112-118.

Ferrol N, Barea JM, and Azcón-Aguilar C (2002a). Mechanisms of nutrient transport across interfaces in arbuscular mycorrhizas. Plant Soil 244:231-237.

Ferrol N, Gianinazzi S, and Gianinazzi-Pearson V (2002b). Arbuscular mycorrhiza induced ATPases and membrane nutrient transport mechanims. In : S Gianinazzi, H Schüepp, JM Barea and K Haselwandter, eds. Mycorrhizal Technology in Agriculture: From Genes to Bioproducts. Basel, Boston, Berlin: Birkhäuser, pp 113-122.

Forbes PJ, Millam S, Hooker JE, and Harrier LA (1998). Transformation of the arbuscular mycorrhizal fungus *Gigaspora rosea* Nicolson and Schenck using particle bombardment. Mycol Res 102:497-501.

Fortin JA, Becard G, Declerck S, Dalpe Y, St-Arnaud M, Coughlan AP, and Piche Y (2002). Arbuscular mycorrhiza on root-organ cultures. Can J Bot 80:1-20.

Franken P and Gianinazzi-Pearson V (1996). Construction of genomic phage libraries of the arbuscular mycorrhizal fungi *Glomus mosseae* and *Scutellospora castanea* and isolation of ribosomal RNA genes. Mycorrhiza 6:167-173.

Franken P, Lapopin L, Meyer-Gauen G, and Gianinazzi-Pearson V (1997). RNA accumulation and genes expressed in spores of the arbuscular mycorrhizal fungus *Gigaspora rosea*. Mycologia 89:295-299.

Frey B and Schüepp H (1993). Acquisition of nitrogen by external hyphae of arbuscular mycorrhizal fungi associated with *Zea mays* L. New Phytol 124:221-230.

Gaspar L, Pollero R, and Cabello M (1997). Partial purification and characterisation of a lipolytic enzyme from spores of the arbuscular mycorrhizal fungus *Glomus versiforme*. Mycologia 89:610-614.

Gianinazzi S, Schüepp H, Barea JM, and Haselwandter K (2002). Mycorrhizal Technology in Agriculture: from Genes to Bioproducts. Basel, Boston, Berlin: Birkhäuser.

Giovannetti M, Azzolini D, and Citernesi AS (1999). Anastomosis formation and nuclear and protoplasmic exchange in arbuscular mycorrhizal fungi. Appl Environ Microbiol 65:5571-5575.

Gobel C, Hahn A, and Hock B (1995). Production of polyclonal and monoclonal antibodies against hyphae from arbuscular mycorrhizal fungi. Crit Rev Biotech 15:3-4.

Gollotte A, Brechenmacher L, Weidmann S, Franken P, and Gianinazzi-Pearson V (2002a). Plant genes involved in arbuscular mycorrhiza formation and functioning. In : S Gianinazzi, H Schüepp, JM Barea and K Haselwandter, eds. Mycorrhizal Technology in Agriculture: From Genes to Bioproducts. Basel, Boston, Berlin: Birkhäuser, pp 87-102.

Gollotte A, Chatagnier O, Arnould D, van Tuinen D, Gianinazzi S, and Gianinazzi-Pearson V (2002b). Identification of transposon-like sequences in the genome of fungi belonging to the Glomales. Proceedings of the 7th International Mycological Congress, Oslo, Norway, p335.

González-Guerrero M, Rhody D, Azcón-Aguilar C, Franken P, and Ferrol N (2002). Isolation and characterizatiom of two metallothionein genes from Glomalean fungi. Proceedings of the 10th New Phytologist Symposium on Functional Genomics of Plant-Microbe Interactions, Nancy, France.

Gouka RJ, Gerk C, Hooykaas PJJ, Bundock P, Musters W, and Verrips CT (1999). Transformation of *Aspergillus awamori* by *Agrobacterium tumefaciens* -mediated homologous recombination. Nat Biotechnol 17:98-601.

Hahn A, Wright S, and Hock B (2001). Immunochemical characterization of mycorrhizal fungi. Fungal Assoc 9:29-43.

Hannon GJ (2002). RNA interference. Nature 418:244-251.

Harrier LA (2001). Isolation and sequence analysis of the arbuscular mycorrhizal fungus *Glomus mosseae* (Nicol & Gerd.) Gerdemann & Trappe 3-phosphoglycerate kinase (*PGK*) gene promoter region. DNA Seq 11:463-473.

Harrier LA and Millam S (2001). Biolistic Transformation of Arbuscular Mycorrhizal Fungi: Progress and Perspectives. Mol Biotech 18:25-33.

Harrier LA, Millam S, and Franken P (2002). Biolistic transformation of AM fungi: advances and applications. In : S Gianinazzi, H Schüepp, JM Barea and K Haselwandter, eds. Mycorrhizal Technology in Agriculture: From Genes to Bioproducts. Basel, Boston, Berlin: Birkhäuser, pp 59-71.

Harrier LA and Paterson L (2002). Inducibility studies of the arbuscular mycorrhizal fungus *Glomus mosseae* 3-phosphoglycerate kinase (*PGK*) gene promoter. Curr Genet 42:169-178.

Harrier LA, Wright F, and Hooker JE (1998). Isolation of the 3-phosphoglycerate kinase gene from the arbuscular mycorrhizal fungus *Glomus mosseae* (Nicol. & Gerd.) Gerdemann & Trappe. Curr Genet 34:386-392.

Harrison MJ (1999). Molecular and cellular aspects of the arbuscular mycorrhizal symbiosis. Ann Rev Plant Physiol Plant Mol Biol 50:361-389.

Harrison MJ and van Buuren ML (1995). A phosphate transporter from the mycorrhizal fungus *Glomus versiforme*. Nature 378:626-629.

Hehl R (1994). Transposon tagging in heterologous host plants. Trends Genet 10:385-386.

Henry YAL, Lopez MC, Gibbs JM, Chambers A, Kingsman SM, Baker HV, and Stanway CA (1994). The yeast protein Gcr1p binds to the PGK UAS and contributes to the activation of transcription of the *PGK* gene. Mol Gen Genet 245:506-511.

Hieter P, Pridmore D, Hegemann JH, Thomas M, Davis RW, and Philippsen P (1985). Functional selection and analysis of yeast centromeric DNA. Cell 42:913-21.

Hijri M, Hosny M, van Tuinen D, and Dulieu H (1998). Intraspecific ITS polymorphism in *Scutellospora castanea* (Glomales, Zygomycetes) is structured within multinucleate spores. Fungal Genet Biol 26:141-151.

Hosny M (1997). Tailles et contenus en (G+C) des Glomales. Complexité du génome et polymorphisme des ADN ribosomiques chez une espèce-modèle. Doctoral Dissertation. University of Burgundy, France.

Hosny M, de Barros JPP, Gianinazzi-Pearson V, and Dulieu H (1997). Base composition of DNA from glomalean fungi: high amounts of methylated cytosine. Fung Genet Biol 22:103-111.

Hosny M, Gianinazzi-Pearson V, and Dulieu H (1998). Nuclear DNA content of 11 fungal species in Glomales. Genome 41:422-428.

Hosny M, Hijri M, Passerieux E, and Dulieu H (1999a). rDNA units are highly polymorphic in *Scutellospora castanea* (Glomales, Zygomycetes). Gene 226:61-71.

Hosny M, van Tuinen D, Jacquin F, Fuller P, Zhao B, Gianinazzi-Pearson V, and Franken P (1999b). Arbuscular mycorrhizal fungi and bacteria: how to construct prokaryotic DNA-free genomic libraries from the Glomales. FEMS Lett 170:425-430.

Jabaji-Hare S (1988). Lipid and fatty acid profiles of some vesicular-arbuscular mycorrhizal fungi: contribution to taxonomy. Mycologia 80:622-629.

Jakobsen I, Gazey C, and Abbott IK (2001). Phosphate transport by communities of arbuscular mycorrhizal fungi in intact soil cores. New Phytol 149:95-103.

Jeffries P and Barea JM (2001). Arbuscular Mycorrhiza - a key component of sustainable plant-soil ecosystems. In: B Hock, ed. The Mycota. Vol. IX. Fungal Associations. Springer-Verlag, Berlin, Heidelberg, pp 95-113.

Johansen A, Jakobsen I, and Jensen ES (1992). Hyphal transport of ^{15}N-labelled nitrogen by a vesicular-arbuscular mycorrhizal fungus and its effect on depletion of inorganic soil N. New Phytol 122:281-288.

Jun J, Abubaker J, Rehrer C, Pfeffer PE, Shachar-Hill Y, and Lammers PJ (2002). Expression in an arbuscular mycorrhizal fungus of genes putatively involved in metabolism, transport, the cytoskeleton and the cell cycle. Plant Soil 244:141-148.

Kaldorf M, Schmelzer E, and Bothe H (1998). Expression of maize and fungal nitrate reductase genes in arbuscular mycorrhiza. Mol Plant-Microbe Interact 11:439-448.

Kuhn G, Hijri M, and Sanders IR (2001). Evidence for the evolution of multiple genomes in arbuscular mycorrhizal fungi. Nature 414:745-8.

Lammers PJ, Jun J, Abubaker J, Arreola R, Gopalan A, Bago B, Hernandez-Sebastia C, Allen JW, Douds DD, Pfeffer PE, and Shachar-Hill Y (2001). The glyoxylate cycle in an arbuscular mycorrhizal fungus. Carbon flux and gene expression. Plant Physiol 127:1287-1298.

Lanfranco L, Bianciotto V, Lumini E, Souza M, Morton JB, and Bonfante P (2001). A combined morphological and molecular approach to characterize isolates of arbuscular mycorrhizal fungi in *Gigaspora* (Glomales). New Phytol 152:169-179.

Lanfranco L, Bolchi A, Ros EC, Ottonello S, and Bonfante P (2002). Differential expression of a metallothioneir gene during the presymbiotic versus the symbiotic phase of an arbuscular mycorrhizal fungus. Plant Physiol 130:58-67.

Lanfranco L, Delpero M, and Bonfante P (1999a). Intrasporal variability of ribosomal sequences in the endomycorrhizal fungus *Gigaspora margarita.* Mol Ecol 8:37-45.

Lanfranco L, Gabella S, and Bonfante P (2000). Expressed sequence tags from germinating spores of *Gigaspora margarita.* Proceedings of COST Action 8.38 Meeting on Managing Arbuscular Mycorrhizal Fungi for Improving Soil Quality and Plant Health in Agriculture, Santiago de Compostela, Galicia, Spain.

Lanfranco L, Garnero L, and Bonfante P (1999b). Chitin synthase genes in the arbuscular mycorrhizal fungus *Glomus versiforme*: full sequence of a gene encoding a class IV chitin synthase. FEMS Microbiol Lett 170:59-67.

Lanfranco L, Vallino M, and Bonfante P (1999c). Expression of chitin synthase genes in the arbuscular mycorrhizal fungus *Gigaspora margarita.* New Phytol 142:347-354.

Lapopin L, Gianinazzi-Pearson V, and Franken P (1999). Comparative differential RNA display analysis of arbuscular mycorrhiza in *Pisum sativum* wild type and a mutant defective in late stage development. Plant Mol Biol 41:669-677.

Leyval C, Joner EJ, del Val C, and Haselwandter K (2002). Potential or arbuscular mycorrhizal fungi for bioremediation. In : S Gianinazzi, H Schüepp, JM Barea and K Haselwandter, eds. Mycorrhizal Technology in Agriculture: From Genes to Bioproducts. Basel, Boston, Berlin: Birkhäuser, pp 175-186.

Lloyd-MacGilp SA, Chambers SM, Dodd JC, Fitter AH, Walker C, and Young JPW (1996). Diversity of the ribosomal internal transcribed spacers within and among isolates of *Glomus mosseae* and related mycorrhizal fungi. New Phytol 133:103-111.

MacDonald RM and Lewis M (1978). The occurrence of some acid phosphatases and dehydrogenases in the vesicular-arbuscular mycorrhizal fungus *Glomus mosseae.* New Phytol 80:135-141.

Mackenzie DA, Wongwathanarat P, Carter AT, and Archer DB (2000). Isolation and use of a homologous histone H4 promoter and a ribosomal DNA region in a transformation vector for the oil producing fungus *Mortierella alpina.* Appl Environ Microbiol 66:4655-4661.

Maldonado-Mendoza IE, Dewbre GR, van Buuren ML, Versaw WK, and Harrison MJ (2002). Methods to estimate the proportion of plant and fungal RNA in an arbuscular mycorrhiza. Mycorrhiza 12:67-74.

Maldonado-Mendoza IE, Dewbre GR, and Harrison MJ (2001). A phosphate transporter gene from the extra-radical mycelium of an arbuscular mycorrhizal fungus *Glomus intraradices* is regulated in response to phosphate in the environment. Mol Plant-Microbe Interact 14:1140-1148.

Malonek S and Meinhardt F (2001). *Agrobacterium tumefaciens*-mediated genetic transformation of the phytopathogenic ascomycete *Calonectria morganii.* Curr Genet 40:152-155.

Mayer K and Mewes HW (2002). How can we deliver the large plant genomes? Strategies and perspectives. Current Opin Plant Biol 5:173-177.

Mikosch TSP, Lavrijssen B, Sonnenberg ASM, and van Griensven LJLD (2001). Transformation of the cultivated mushroom *Agaricus bisporus* (Lange) using T-DNA from *Agrobacterium tumefaciens.* Mycol Res 39:35-39.

Morton JB and Benny GL (1990). Revised classification of arbuscular mycorrhizal fungi (zygomycetes): a new order, Glomales, two new suborders, Glomineae and Gigasporineae, and two new families, Acaulosporaceae and Gigasporaceae, with an emendation of glomaceae. Mycotaxon 37:471-491.

Mullins ED, Chen X, Romaine P, Raina R, Geizer DM, and Kang S (2001). *Agrobacterium*-mediated transformation of *Fusarium oxysporum*: an efficient tool for insertional mutagenesis and gene transfer. Phytopathol 91:173-180.

Murrin F, Holtby J, Nolan RA, and Davidson WS (1986). The genome of the *Entomophaga aulicae* (Entomorphtorales, Zygomycetes): base composition and size. Experim Mycol 67-75.

O'Brochta DA and Atkinson PW (1996) Transposable elements and gene transformation in non-drosophilid insects. Insect Biochem Molec Biol 26:739-753.

Pardo AG, Hanif M, Raudaskoski M, and Gorfer M (2002). Genetic transformation of ectomycorrhizal fungi mediated by *Agrobacterium tumefaciens*. Mycol Res 106:132-137.

Pawlowska TE and Taylor JW (2002). Organization of genetic variation within glomalean individuals. Proceedings of the 7th International Mycological Congress, Oslo, pp 73.

Peterson DG, Wessler SR, and Paterson AH (2002). Efficient capture of unique sequences from eukaryotic genomes. Trends Genet 18: 547-550.

Pfeffer PE, Douds DD, Bécard G, and Shachar-Hill Y (1999). Carbon uptake and the metabolism and transport of lipids in an arbuscular mycorrhiza. Plant Physiol 120:587-598.

Pozo MJ, Slezack-Deschaumes S, Dumas-Gaudot E, Gianinazzi S, and Azcón-Aguilar C (2002). Plant defense responses induced by arbuscular mycorrhizal fungi. In : S Gianinazzi, H Schüepp, JM Barea and K Haselwandter, eds. Mycorrhizal Technology in Agriculture: From Genes to Bioproducts. Basel, Boston, Berlin: Birkhäuser, pp 103-111.

Rabinowicz PD, Schultz K, Dedhia N, Yordan LD, Parnell L, Stein WR, McCombie WR, and Martiensen RA (1999). Differential methylation of genes and retrotransposons facilitates shotgun sequencing of the maize genome. Nature Genet 23:305-308.

Redecker D (2002). Molecular identification and phylogeny of arbuscular mycorrhizal fungi. Plant Soil 244:67-73.

Redecker D, Kodner R, and Graham LE (2000). Glomalean fungi from the Ordovician. Science 289:1920-1921.

Requena N, Fuller P, and Franken P (1999). Molecular characterization of *GmFOX2*, an evolutionarily highly conserved gene from the mycorrhizal fungus *Glomus mosseae*, down-regulated during interaction with rhizobacteria. Mol Plant-Microbe Interact 12:934-942.

Requena N, Mann P, and Franken P (2000). A homologue of the cell cycle check point TOR2 from *Saccharomyces cerevisiae* exists in the arbuscular mycorrrhizal fungus *Glomus mosseae*. Protoplasma 212:89-98.

Requena N, Mann P, Hampp R, and Franken P (2002). Early developmentally regulated genes in the arbuscular mycorrhizal fungus *Glomus mosseae*: Identification of GmGIN1, a novel gene with homology to the C-terminus of metazoan hedgehog proteins. Plant Soil 244:129-139.

Rho HS, Kang S, and Lee YH (2001). *Agrobacterium tumefaciens*-mediated transformation of the plant pathogenic fungus *Magnaporthe grizea*. Molec Cells 12:407-411.

Rhody D, Stommel M, Roeder C, Mann P, and Franken P (2003). Differential RNA accumulation of two β-tubulin genes in arbuscular mycorrhizal fungi. Mycorrhiza 13 :137-142.

Rosendahl S and Taylor JW (1997). Development of multiple genetic markers for studies of genetic variation in arbuscular mycorrhizal fungi using AFLP™. Mol Ecol 6:821-829.

Rubin GM and Spradling AC (1982). Genetic transformation of *Drosophila* with transposable element vectors. Science 218:348-353.

Ruiz-Díez B and Martínez-Suárez JV (1999). Electrotransformation of the human pathogenic fungus *Scedosporium prolificans* mediated by repetitive rDNA sequences. FEMS Immunol Medic Microbiol 25:275-282.

Ruiz-Lozano JM, Collados C, Porcel R, Azcon R, and Barea JM (2002). Identification of a cDNA from the arbuscular mycorrhizal fungus *Glomus intraradices* that is expressed during mycorrhizal symbiosis and up-regulated by N fertilization. Mol Plant-Microbe Interact 15: 360-367.

Russell PJ, Rodland KD, Rachlin EM, and McCloskey JA (1987). Differential DNA methylation during the vegetative life cycle of *Neurospora crassa*. J Bacteriol 169:2902-2905.

Saito M (1995). Enzyme activities of the internal hyphae and germinated spores of an arbuscular mycorrhizal fungus, *Gigaspora margarita* Becker & Hall. New Phytol 129: 425-431.

Samra A, Dumas-Gaudot E, and Gianinazzi S (1997). Detection of symbiosis-related polypeptides during the early stages of the establishment of arbuscular mycorrhiza between *Glomus mosseae* and *Pisum sativum* roots. New Phytol 135:711-722.

Samra A, Dumas-Gaudot E, Gianinazzi-Pearson V, and Gianinazzi S (1996). Soluble proteins and polypeptide profiles of spores of arbuscular mycorrhizal fungi. Interspecific variability and effects of host (myc^+) and non-host (myc^-) *Pisum sativum* root exudates. Agronomie 16:709-719.

Sanders IR, Alt M, Groppe K, Boller T, and Wiemken A (1995). Identification of ribosomal DNA polymorphisms among and within spores of the Glomales: application to studies on the genetic diversity of arbuscular mycorrhizal fungal communities. New Phytol 130: 419-427.

Sanders IR, Ravolanirina F, Gianinazzi-Pearson V, Gianinazzi S, and Lemoine MC (1992). Detection of specific antigens in the vesicular-arbuscular mycorrhizal fungi *Gigaspora margarita* and *Acaulospora laevis* using polyclonal antibodies to soluble spore fractions. Mycol Res 96:477-480.

Sawaki H and Saito M (2001). Expressed genes in the extraradical hyphae of an arbuscular mycorrhizal fungus, *Glomus intraradices*, in the symbiotic phase. FEMS Microbiol Letts 195:109-113.

Schellenbaum L, Gianinazzi S, and Gianinazzi-Pearson V (1992). Comparison of acid soluble protein synthesis in roots of endomycorrhizal wild type *Pisum sativum* and corresponding isogenic mutants. J Plant Physiol 141:2-6.

Schilde C, Wöstemeyer J, and Burmester A (2001). Green fluorescent protein as a reporter for gene expression in the mucoralean fungus *Absidia glauca*. Arch Microbiol 175:1-7.

Schüßler A (1999). Glomales SSU rRNA gene diversity. New Phytol 144:205-207.

Schüßler A, Schwarzott D, and Walker C (2001). A new fungal phylum, the *Glomeromycota*: phylogeny and evolution. Mycol Res 105:1413-1421.

Shachar-Hill Y, Pfeffer PE, Douds D, Osman SF, Doner LW, and Ratcliffe RG (1995). Partitioning of intermediate carbon metabolism in VAM colonized leek. Plant Physiol 108:7-15.

Simon L, Lalonde M, and Bruns TD (1992). Specific amplification of 18S fungal ribosomal genes from vesicular-arbuscular endomycorrhizal fungi colonizing roots. Appl Environ Microbiol 58:291-295.

Simoneau P, Louisy-Louis N, Plenchette C, and Strullu DG (1994). Accumulation of new polypeptides in Ri T-DNA-transformed roots of tomato (*Lycopersicon esculentum*) during the development of vesicular-arbuscular mycorhiza. Appl Environ Microbiol 60:1810-1813.

Singer MJ, Marcotte BA, and Selker EU (1995). DNA methylation associated with repeat-induced point mutation in *Neurospora crassa*. Mol Cell Biol 15:5586-97.

Smith SE and Gianinazzi-Pearson V (1988). Physiological interactions between symbiosis in vesicular arbuscular mycorrhizal plants. Ann Rev Plant Physiol Plant Mol Biol 39:221-244.

Smith SE and Read DJ (1997). Mycorrhizal Symbiosis. Academic Press, San Diego.

Spradling AC and Rubin GM (1982). Transposition of cloned P elements into *Drosophila* germ line chromosomes. Science 218:341-347.

Stanway CA, Gibbs JM, Kearsey SE, Lopez MC, and Baker HV (1994). The yeast co-activator GAL11 positively influences transcription of the phosphoglycerate kinase gene but only when RAP1 is bound to its upstream activation sequence. Mol Gen Genet 243:207-214.

St-Arnaud M, Hamel C, Vimard B, Caron M, and Fortin JA (1996). Enhanced hyphal growth and spore production of the arbuscular mycorrhizal fungus *Glomus intraradices* in an *in vitro* system in the absence of host roots. Mycol Res 100:328-332.

Stommel M, Mann P, and Franken P (2001). Construction and analysis of an EST library using RNA from activated spores of the arbuscular mycorrhizal fungus *Gigaspora rosea*. Mycorrhiza 10:281-285.

Storck R and Alexopoulos CJ (1970). Deoxyribonucleic acid of fungi. Bacteriol Rev 34:126-154.

Tamasloukht M, Séjalon-Delmas N, Kluever A, Jauneau A, Roux C, Bécard G, and Franken P (2003). Root factor induce mitochondrial-related gene expression and fungal respiration during the developmental switch from asymbiosis to presymbiosis in the arbuscular mycorrhizal fungus *Gigaspora rosea*. Plant Physiol 131:1468-1478.

Thomas CA (1971). The genetic organization of chromosomes. Ann Rev Genet 5:237-256.

Timonen S, Smith FA, and Smith SE (2001). Microtubules of the mycorrhizal fungus *Glomus intraradices* in symbiosis with tomato roots. Can J Bot 79:307-313.

Tobar R, Azcón R, and Barea JM (1994). Improved nitrogen uptake and transport from ^{15}N-labelled nitrate by external hyphae of arbuscular mycorrhiza under water-stressed conditions. New Phytol 126:119-122.

Trouvelot S, van Tuinen D, Hijri M, and Gianinazzi-Pearson V (1999). Visualization of DNA loci in interphasic nuclei of glomalean fungi by fluorescence in situ hybridization. Mycorrhiza 8:203-208.

Ubalijoro E, Hamel C, McClung CR, and Smith DL (2001). Detection of chitin synthase class I and II type sequences in six different arbuscular mycorrhizal fungi and gene expression in *Glomus intraradices*. Mycol Res 105:470-476.

van Buuren ML, Lanfranco L, Longato S, Minerdi D, Harrison MJ, and Bonfante P (1999). Construction and characterization of genomic libraries of two endomycorrhizal fungi: *Glomus versiforme* and *Gigaspora margarita*. Mycol Res 103:955-960.

van Wijk KJ (2001) Challenges and prospects of plant proteomics. Plant Physiol 126:501-508.

Vandenkoornhuyse P, Leyval C, and Bonnin I (2001). High genetic diversity in arbuscular mycorrhizal fungi: evidence for recombination events. Heredity 87:243-253.

Velculescu VE, Zhang L, Volgelstein B, and Kinzler KW (1995). Serial analysis of gene expression. Science 270:484-487.

Viera A and Glenn MG (1990). DNA content of vesicular-arbuscular mycorrhizal fungal spores. Mycologia 82: 263-267.

Walbot V (1992). Strategies for mutagenesis and gene cloning using transposon tagging and T-DNA insertional mutagenesis. Annu Rev Plant Physiol Plant Mol Biol 43:49-82.

Wyss P and Bonfante P (1993). Amplification of genomic DNA of arbuscular mycorrhizal (AM) fungi by PCR using short arbitrary primers. Mycol Res 97:1351-1357.

Zézé A, Hosny M, Gianinazzi-Pearson V, and Dulieu H (1996). Characterization of a highly repeated DNA sequence (SC1) from the arbuscular mycorrhizal fungus *Scutellospora castanea* and its detection in planta. Appl Environ Microbiol 62:2443-2448.
Zézé A, Hosny M, van Tuinen D, Gianinazzi-Pearson V, and Dulieu H (1999). MYCDIRE, a dispersed repetitive DNA element in arbuscular mycorrhizal fungi. Mycol Res 103:572-576.
Zézé A, Sulistyowati E, Ophel-Keller K, Barker S, and Smith S (1997). Intrasporal genetic variation of *Gigaspora margarita*, a vesicular arbuscular mycorrhizal fungus, revealed by M13 minisatellite-primed PCR. Appl Environ Microbiol 63:676-678.
Zhang HB and Wu C (2001). BAC as tools for genome sequencing. Plant Physiol Biochem 39:195-209.

Keyword Index

Addendum: **Volume 3**. Chapter 14. Fungal Germplasm and Databases by Kevin McCluskey.

Table 2. Online taxonomy, genome and DNA sequence databases.

Subject	Web location	Features
National Center for Biotechnology Information (NCBI)	http://www.ncbi.nlm.nih.gov/Taxonomy/tax.html	Entry for every species for which there is DNA or protein sequence data
DNA Databank of Japan TXsearch (DDBJ)	http://sakura.ddbj.nig.ac.jp/uniTax.html	Developed by DDBJ and includes nuclear and mitochondrial genetic codes
The CABI Bioscience and CBS Database of Fungal Names	http://www.indexfungorum.org/Names/Names.asp	Database of fungal names containing over 345,000 taxa. Includes citations to original literature
USDA Systematic Botany and Mycology Laboratory (SBML)	http://nt.ars-grin.gov/index.htm	Databases of fungal names, herbarium specimens, literature citations and biogeography
Integrated Taxonomic Information System	http://www.itis.usda.gov/index.html	A partnership between US, Canadian and Mexican agencies to provide taxonomic information
Checklists of Lichens and Lichenicolous Fungi of the World	http://141.84.65.132/ChecklistsDe/Lichens/index.html	Limited to Africa, offers species names and authorities
Index Nominum Genericorum	http://rathbun.si.edu/botany/ing/ingform.cfm	U.S. National Herbarium, Dept. of Systematic Biology , taxonomy of all plants listed in International Code of Botanical Nomenclature
USDA Germplasm Resources Information Network (GRIN)	http://www.ars-grin.gov/npgs/tax/	Plant taxonomy for plants in the GRIN repository system
Genbank	http://www.ncbi.nlm.nih.gov/entrez/query.fcgi	Nucleotide and protein sequence databases, as well as publications, taxonomy, protein structure and more. Over 20 billion nucleotides
DNA Data Bank of Japan	http://www.ddbj.nig.ac.jp/	18 million entries
European Molecular Biology Laboratory	http://www.ebi.ac.uk/embl/	DNA sequence, SWIS-PROT protein sequence database
Whitehead Institute Center for Genome Resources, Fungal Genome Initiative	http://www-genome.wi.mit.edu/annotation/fungi/fgi/	Broad and comprehensive Fungal Genome resource

Saccharomyces cerevisiae	http://genome-www.stanford.edu/Saccharomyces	Blast search, gene search, primer search and more
	http://cgsigma.cshl.org/jian/	Cold Spring Harbor Promoter Database
	http://www.cse.ucsc.edu/research/compbio/ yeast introns.html	Ares lab Yeast Intron Database
	http://mips.gsf.de/proj/yeast/CYGD/db/ index.html	MIPS Comprehensive Yeast Genome Database
	http://depts.washington.edu/~yeastrc/index.html	Yeast Resource Center
various Ascomycota	http://mips.gsf.de/proj/yeast/CYGD/hemi/	4 Saccharomyces sp, 3 Kluyveromyces sp, and others
Schizosaccharomyces pombe	http://www.ncbi.nlm.nih.gov/cgi-bin/Entrez/ map_search.cgi?chr=spombe.inf	Uses ENTREZ browser to link between genome and Genbank
*Neurospora crassa**	http://www-genome.wi.mit.edu/annotation/fungi/ neurospora/	Whitehead institute, genome browser, blast search, genetic map, links to strains and clones at FGSC
	http://mips.gsf.de/proj/neurospora/	MIPS Neurospora crassa database, annotated proteome
	http://www.unm.edu/~ngp/	Neurospora Genome Project, ESTs and proteome
	http://gene.genetics.uga.edu/	Physical map,
*Aspergillus nidulans**	http://www-genome.wi.mit.edu/annotation/fungi/aspergillus/	Genome database, blast search
	http://gene.genetics.uga.edu/	Physical map
A. fumigatus	http://www.sanger.ac.uk/Projects/A_fumigatus/	Blast search,
	http://www.aspergillus.man.ac.uk/	Free registration required, gene annotations,
	http://www.tigr.org/tdb/e2k1/afu1/	
A. parasiticus	http://www.genome.ou.edu/fungal.html	
A. flavus	http://www.genome.ou.edu/fungal.html	ESTs
Candida albicans	http://alces.med.umn.edu/candida.html	Physical map,
	http://www-sequence.stanford.edu/group/candida/index.html	Blast search, contig index,
*Cryptococcus neoformans**	http://www.tigr.org/tdb/e2k1/cna1/	Blast search, 10X coverage
	http://www-genome.wi.mit.edu/annotation/fungi/cryptococcus_neoformans/index.html	Blast, graphical genome viewer

Pneumocystis carinii	http://biology.uky.edu/Pc/	The Pneumocystis Genome Project, ESTs, link to physical map at UGA
Fusarium sporotrichioides	http://www.genome.ou.edu/fsporo.html	ESTs
*Fusarium graminearum**	http://www-genome.wi.mit.edu/annotation/fungi/fusarium/index.html	Blast, graphical genome viewer
*Magnaporthe grisea**	http://www-genome.wi.mit.edu/annotation/fungi/magnaporthe/	Blast, graphical genome viewer
Phanerochaete chrysosporium	http://www.jgi.doe.gov/programs/whiterot.htm	Blast search
*Ustialgo maydis**	http://www-genome.wi.mit.edu/annotation/fungi/ustilago_maydis/index.html	Blast, graphical genome viewer
*Coprinus cinereus**	http://www-genome.wi.mit.edu/annotation/fungi/coprinus_cinereus/	Blast, graphical genome viewer

*Biological materials available from FGSC (www.fgsc.net)

Kolibri 360 GmbH
Ferdinand-Jühlke-Straße 7
99095 Erfurt, Deutschland
produktsicherheit@kolibri360.de